METHODS IN MOLECULAR BIOLOGY™

Series Editor
John M. Walker
School of Life Sciences
University of Hertfordshire
Hatfield, Hertfordshire, AL10 9AB, UK

For further volumes:
http://www.springer.com/series/7651

Fluorescence Spectroscopy and Microscopy

Methods and Protocols

Edited by

Yves Engelborghs

Department of Chemistry, University of Leuven, Leuven-Heverlee, Belgium

Antonie J.W.G. Visser

Laboratory of Biochemistry, Microspectroscopy Centre, Wageningen University, Wageningen, The Netherlands

Humana Press

Editors
Yves Engelborghs
Department of Chemistry
University of Leuven
Leuven-Heverlee, Belgium

Antonie J.W.G. Visser
Laboratory of Biochemistry
Microspectroscopy Centre
Wageningen University
Wageningen, The Netherlands

ISSN 1064-3745 ISSN 1940-6029 (electronic)
ISBN 978-1-4939-6315-7 ISBN 978-1-62703-649-8 (eBook)
DOI 10.1007/978-1-62703-649-8
Springer New York Heidelberg Dordrecht London

Softcover reprint of the hardcover 1st edition 2014

Printed on acid-free paper

Humana Press is a brand of Springer
Springer is part of Springer Science+Business Media (www.springer.com)

Preface

At a first glance, it looks strange to have a book on protocols in Fluorescence Spectroscopy and Microscopy in a series on methods in molecular biology. However, with the discovery of the fluorescent proteins, molecular biology has become an integral part of the standard tools of modern fluorescence technologies, especially when applied to the living cell. Of course, molecular biology has already given the technique of site-directed mutagenesis, allowing to remove or introduce fluorescent amino acids in different positions of the peptide chain and, in this way, producing better insight into the fluorescent properties of proteins, their folding, and their dynamics.

Why Another Fluorescence Book At All?

After receiving the invitation from John and Jan Walker to be editors of a coming volume on fluorescence protocols, we were initially a little reluctant to accept that invitation. The reason is that fluorescence has already been established as a sensitive technique at the interface of chemistry, physics, and biology and has emerged as one of the most frequently used spectroscopic techniques in many research areas. Numerous excellent textbooks and edited books on fluorescence principles and fluorescence microscopy and its applications have already been published thus far (see, for instance, refs. [1–7], which form only a small fraction of the total amount of literature available on this subject). What will then be the impact of a fluorescence protocol book considering the large amount of time needed for chapter writing? The "raison d'être" for a protocol book is that the fluorescence field is still expanding and, in the last decade, methods and applications of fluorescence have made an important, synergistic jump forward extending the fluorescence toolbox. The increasing popularity of fluorescence techniques, however, creates the need for correct protocols; that is where this edition aims to contribute. Therefore, teaching both "novice" and established scientists in advanced fluorescence spectroscopy and microscopy with the help of this fluorescence protocol volume convinced us to accept the invitation. For the success of this volume, we have relied on the generous input of contributing authors, each specialized in a particular topic.

In the remaining part, we will give a brief overview of historic aspects of fluorescence. Although fluorescence spectroscopy is widely used in cuvette-based studies, exemplified in several chapters of this volume, it is mostly the combination of fluorescence spectroscopy and microscopy that is important in more advanced studies of more complex systems like cells. Technical advances are highlighted eventually leading to fluorescence lifetime imaging microscopy (FLIM), fluorescence correlation spectroscopy (FCS), and single-molecule fluorescence spectroscopy (smFS), which are methods discussed in various chapters of this volume. We will give a classification of the different chapters. Finally, we devote a paragraph to the memory of Prof. Dr. Robert (Bob) Clegg. During the process of editing, we were struck by the sad news of the death of Bob Clegg.

History

Fluorescence spectroscopy has a long history; a description of the phenomenon dates from previous centuries. The quantity known as the Stokes shift, which is the shift to lower energy of the fluorescence spectrum relative to the absorption spectrum of a compound was already observed and published in 1852 by Stokes, who also coined the term "fluorescence" [8]. In the first half of the previous century, fluorescence turned out to be already a settled method as illustrated by famous phenomena such as, among others, quenching of fluorescence [9], polarization of fluorescence [10, 11], the construction of a phase fluorometer for measuring fluorescence lifetimes of dye molecules [12], the Perrin-Jablonski diagram [11, 13], radiationless energy transfer [14], and so forth. Applications to biochemical and biophysical systems have been pioneered by Weber [15, 16]. The very first FLIM system has been based on a frequency-domain microscope setup [17]. The history of "Förster" resonance energy transfer (FRET, see below) has been excellently reviewed by Clegg [18].

Instrumentation, Probe, and GFP Technology

The popularity of fluorescence applications is greatly enhanced by newly developed techniques such as femtosecond and picosecond mode-locked lasers, laser diodes, and semiconductor light-emitting diodes (LEDs) as excitation sources, rapid proximity-focused microchannel plate (MCP) photomultipliers (PMTs) and single-photon avalanche photodiodes (SPADs) as photon detectors, electron-multiplied charge-coupled device (EMCCD) cameras as imaging detectors, new microscope setups, powerful computers, and data analysis software, rapid multidimensional data acquisition schemes, and so forth, as put in perspective by Birch [19]. The popularity also stems from the fact that the method is extremely sensitive, enabling the investigation of even single fluorescent molecules. In this respect many novel fluorescent probes have been developed with both high light absorption strength and high quantum yield of fluorescence. Another advantage is that the fluorescence technique is minimally invasive. For that reason, fluorescence is now an established technique in cell biology, because fluorescence can be influenced by timescales and distances that are directly relevant to monitor the intracellular biochemical status. To visualize and quantify cellular proteins of interest, advanced spectroscopic techniques are combined with microscopy and biosensors so that specific molecular information of cells is obtained. The development and application of the green fluorescent protein (GFP) technology has been of crucial importance for studying genetically encoded proteins and their interactions in their natural cellular habitat [20].

Towards Nanometer Resolution

The introduction of high spatial resolution confocal microscopy gave the opportunity to investigate the co-expression of different proteins in living cells. The spatial resolution of a microscope allows detection at a subcellular level, but physical molecular interactions between proteins or receptors on a nanometer scale cannot be visualized. The spatial resolution of a typical confocal or wide-field image is diffraction limited, meaning that the use of excitation light of 488 nm will result in an optical resolution of around 220 nm by applying the empirical relation of 0.61 λ/NA with λ the excitation wavelength (488 nm) and NA (1.4) the numerical aperture of the microscope objective. One possibility to go beyond the optical limitations is to apply super-resolution microscopy methods such as stimulated emission depletion (STED) microscopy, saturated structural illumination microscopy (SSIM), photoactivatable localization microscopy (PALM), or stochastic optical reconstruction microscopy (STORM) as reviewed by Hell [21, 22]. Another possibility is to combine Förster Resonance Energy Transfer (FRET) methodology with fluorescence microscopy. FRET is a photophysical process where the excited-state energy from a donor molecule is transferred nonradiatively to an acceptor molecule (in the ground state) at close distance via weak dipole–dipole coupling. Successful application of FRET makes this method very sensitive to intermolecular distances in the range between 1 and 10 nm. Many fluorescent biosensors are based on FRET offering a nanometer spectroscopic ruler. Monitoring protein–protein interactions via FRET forms the basis of studying intracellular signaling networks [23–25].

FLIM/FRET

The integration of fluorescence spectroscopy in light microscopy adds a new dimension to microscopy since spatial information about the molecular behavior of fluorescent molecules can be obtained. The microscopic contrast is provided by fluorescence parameters such as quantum yield (intensity), spectrum, lifetime, anisotropy (polarization), or combinations. Let us take the measurements of fluorescence lifetimes in the microscope as an example. The acronym "FLIM" can cover nearly identical meanings, namely Fluorescence Lifetime Imaging Microscopy, Fluorescence Lifetime IMaging, or Fluorescence LIfetime Microscopy. Fluorescence lifetimes are absolute quantities. The most important advantage of FLIM over fluorescence intensity imaging is that fluorescence lifetimes are independent of fluorophore concentration and excitation intensity. Since the fluorescence lifetime of a fluorophore is sensitive to the local environment (pH, charge, presence of quenchers, refractive index, temperature, and so forth), their measurements in a microscope offer the important advantage of contrast by spatial variations of lifetimes. Methods to visualize molecular interactions can be based on FRET, for example, measured by FLIM. Detection of FRET in a microscopic image using the quenching of donor fluorescence quantum yield is not straightforward, since one should have knowledge on the fraction of donor molecules having an acceptor in the vicinity as compared to those having no acceptor. Detection of FRET in a microscope using the shortening of the donor fluorescence lifetime is a much more direct method than that of fluorescence intensities. Using FLIM, the donor fluorescence lifetime as a function of space can be measured. FRET imaging by donor fluorescence lifetime reduction in the microscope allows quantitative detection of interacting proteins in live cells. Various review papers and edited books on FLIM and FRET combinations have appeared in the recent literature with many back references [26–32]. The two main methods for obtaining fluorescence lifetimes are time-correlated single photon counting (TCSPC) and multi-frequency phase-modulation fluorometry. Both techniques yield essentially the same information and differ mainly in how the time-resolved fluorescence data are obtained, for instance, in the time domain or in the frequency domain. For more detailed information on the two methods, we refer to several textbooks, edited books, and reviews [1, 2, 33–35]. For widefield FLIM detection, gated or modulated image intensifiers are used in combination with CCD readout. FLIM can be performed in either time- or frequency domain. For time-domain measurements, short pulses of excitation light are used, and the fluorescence decay is directly measured with a gated detection system or TCSPC. Frequency-domain FLIM measurements are performed by periodic intensity modulation of the excitation source and (homodyne or heterodyne) phase-sensitive detection. Both time- and frequency-domain FLIM systems require high repetition rate pulsed lasers or rapidly modulated continuous wave (CW) lasers. For FLIM with beam scanning, time-correlated single-photon counting (TCSPC) is widely employed [36].

Fluorescence Correlation Spectroscopy

In the last decades, fluorescence correlation spectroscopy (FCS) has become a widely used technique for studying various dynamic molecular processes. It has found applications in measuring local concentrations, diffusion coefficients, reaction rates, and detection of intermolecular interactions in vitro and in vivo, reviewed in [37–39]. The sensitivity and noninvasive nature of the technique has made it an important technique for studying molecular processes in living cells.

Fluorescence fluctuation methods are based on the detection of tiny, spontaneous fluctuations in fluorescence intensity caused by deviations from thermal equilibrium in an open system. These fluctuations can arise due to diffusion of fluorescent molecules (at nanomolar concentration) in and out of a subfemtoliter observation volume generated by a focused laser beam. Another cause of fluctuations arises from an intrinsic property of aromatic molecules in aqueous solution: the photocycle involving singlet ground state, singlet excited state, and the metastable triplet state. The intensity fluctuations are monitored and autocorrelated over time as in fluorescence correlation spectroscopy (FCS). Alternatively, the intensity fluctuations can be analyzed in a distribution of fluorescence intensity amplitudes as in photon counting histogram (PCH) analysis [40] or in fluorescence intensity distribution analysis (FIDA) [41]. Over time, several developments in fluorescence fluctuation spectroscopy have taken place. Fluorescence cross-correlation spectroscopy (FCCS), reviewed in [39], is an established technique to monitor intracellular protein–protein interaction involving different measuring strategies (one-photon and two-photon excitation with dual-color detection, dual-color one-photon excitation and dual-color detection). Dual-color PCH analysis has been developed [42] as well as 2D-FIDA [43]. Many technical improvements have been implemented in FCS. Dual-focus FCS has been employed to remove experimental artifacts [44]. Novel excitation strategies have been designed such as alternating laser excitation [45] or pulsed interleaved excitation [46] to remove cross-talk in dual channel measurements and modulated excitation to suppress triplet state buildup while keeping complete time range information in FCS [47]. Fluorescence lifetime correlation spectroscopy has been developed to observe single dye diffusion in two distinct environments [48, 49]. Scanning FCS has been implemented for precise measurements of diffusion coefficients [50]. Total internal reflection FCS has been utilized to measure diffusion processes in the small confines of an evanescent field [51].

Raster image correlation spectroscopy (RICS) is a novel image analysis technique that uncovers temporal information hidden in raster-scanned confocal images [52–56]. RICS combines high time resolution (>microseconds) with sufficient spatial resolution and can be considered as a bridge between FCS and image correlation spectroscopy (ICS). A major benefit is that the method can be applied to any commercial laser scanning microscope. RICS can be used in cellular research to follow dynamics of cytosolic proteins, membrane proteins, and multiprotein complexes. RICS provides an intrinsic method to quantify the immobile fraction of proteins as well as to distinguish diffusion from binding. The method has even been extended to cross-correlation RICS to detect protein complexes [57].

Single-Molecule Fluorescence Spectroscopy

Although FCS is a single-molecule spectroscopic method that is performed on low concentrations of molecules diffusing freely in solution in and out of a confocal laser point source, many single-molecule events are averaged to yield ensemble-like quantities. When single molecules are immobilized on a surface, one can employ either confocal beam scanning techniques or wide-field excitation using total internal reflection fluorescence microscopy (TIRF) to detect single molecules for extended periods of time (seconds to minutes), limited only by the photostability of the fluorescent biomolecules. Over the past decade, single-molecule fluorescence spectroscopy (smFS), often in combination with FRET, has enabled to obtain new dynamical and structural information about biomolecules such as proteins and nucleic acids at unprecedented level of detail. Conformational dynamics of these individual biomolecules can be determined for a variety of time scales revealing sometimes characteristic transient reaction states that will be lost in ensemble-type of experiments. Initially applied to surface-attached molecular systems, single-molecule methods have also found applications to observe single fluorescent molecules in more complex samples as in living cells illustrating that smFS has come of age. After seminal reviews on smFS by Weiss [58, 59] many innovative single-molecule fluorescence tools are reported as surveyed in a book [60] and in a comprehensive review [61].

Division of Chapters in Categories

We can broadly divide the submitted chapters into four main categories. Cuvette type of experiments can be conducted with steady-state fluorescence spectroscopy (SSFS, 1) or with time-resolved fluorescence spectroscopy (TRFS, 2). Fluorescent probe development (FPD, 3) is an indispensable part of successful fluorescence practice. The most extensive category is formed by fluorescence microscopy (FM, 4), which embodies various subcategories as presented in this volume. Fluorescence microscopy can be split up in different sections as fluorescence recovery after photobleaching (FRAP, 4-a), live cell FRET imaging (FRETim, 4-b), fluorescence lifetime imaging (FLIM, 4-c), fluorescence fluctuation spectroscopy (FFS, 4-d), and single-molecule fluorescence spectroscopy (smFS, 4-e). For convenience, we have classified time-resolved fluorescence anisotropy imaging (TRFAI) and multimode fluorescence imaging (mmFLIM) under FLIM. Fluorescence correlation spectroscopy (FCS) and photon counting histogram (PCH) analysis both belong to FFS, since both techniques use the same experimental fluorescence intensity fluctuations, but each analytical method focuses on a different property of the signal. RICS is also classified under FFS. Similarly, smFS encompasses single-molecule FRET spectroscopy (smFRET) and photoactivated localization microscopy (PALM). The order of appearance of the chapters follows the division of categories proposed above: 1, 2, 3, 4-a, 4-b, 4-c, 4-d, and 4-e.

Occasional chapters are not presented in the standard format. These chapters include review articles and are indicated as such. The nonstandard format also pertains to articles in which data analysis plays an important role or where experimental data are compared with results of molecular dynamics simulations or quantum chemical approaches. These chapters have obtained a tentative bioinformatics signature.

Altogether, 32 research groups have submitted 36 chapters, which are classified according to the four abovementioned implementations.

Robert M. Clegg

One of the chapters that we have received in the first half of September 2012 is about polar plot analysis of time-resolved fluorescence with principal author, Robert M. Clegg. We were shocked to hear that Bob Clegg passed away on October 15, 2012, in Urbana, Illinois, from complications arising from cancer. He was Professor of Physics and Bioengineering at the University of Illinois at Urbana-Champaign. Bob was a leading expert in applying fluorescence spectroscopy to biological research, in particular photosynthesis and structural dynamics of nucleic acids. He was a pioneer in developing wide-field fluorescence lifetime-resolved microscopy (FLIM). Bob was also an expert in the combined use of FLIM and Förster Resonance Energy Transfer (FRET) in cell biology and biophysics. He developed an enthusiastic interest in the history of science, and colleagues mentioned him as a "living library." In ref. [18], we have quoted his article "The history of FRET," which is a wonderful, historic account of a technique that is used by so many colleagues. Bob also communicated to us the article mentioning the first FLIM apparatus (ref. [17]). We also remember Bob as a lively speaker and a great teacher explaining complex physical problems in a simple way (very often by making analogies with events in daily life of the macroscopic world) to make them understandable for everybody. The last time we met Bob in person was at the meeting organized in Göttingen, Germany, 27–30 March 2011, dedicated to the memory of Theodor Förster. He explained with great enthusiasm how he had found, in the town of Göttingen, the house of Förster and how he had been sitting in the garden house, where Förster had been writing his famous book. It was always a pleasure to have Bob around. Bob will be sorely missed.

Leuven-Heverlee, Belgium *Yves Engelborghs*
Wageningen, The Netherlands *Antonie J.W.G. Visser*

References

1. Valeur B (2002) Molecular fluorescence. Principles and applications. Wiley-VCH, Weinheim
2. Lakowicz JR (2006). Principles of fluorescence spectroscopy, 3rd edn. Springer Science and Business Media, LLC, New York
3. Goldys EM (ed) (2009) Fluorescence applications in biotechnology and life sciences. Wiley-Blackwell, Hoboken
4. Wang XF, Herman B (eds) (1996) Fluorescence imaging spectroscopy and microscopy. Wiley, New York
5. Herman B (1998) Fluorescence microscopy. Springer, New York
6. Pawley PB (ed) (2006) Handbook of biological confocal microscopy. Springer, New York
7. Sauer M, Hofkens J, Enderlein J (eds) (2011) Handbook of fluorescence spectroscopy and imaging: from single molecules to ensembles. Wiley-VCH, Weinheim
8. Stokes GG (1852) On the change of refrangibility of light. Philos Trans R Soc Lond 30:463–562
9. Stern O, Volmer M (1919) Über die Abklingungszeit der Fluoreszenz. Phys Z 20:183–188
10. Wawilow SJ, Lewschin WL (1923) Beiträge zur Frage uber polarisiertes Fluoreszenzlicht von Farbstofflösungen. Z Phys 16:135–154
11. Perrin F (1929) La fluorescence des solutions. Ann Phys (Paris) 12:169–275
12. Gaviola E (1926) Die Abklingzeiten der Fluoreszenz von Farbsofflösungen. Z Phys 35:748–756
13. Jablonski A (1933) Efficiency of anti-Stokes fluorescence in dyes. Nature 131:839–840
14. Förster T (1948) Zwischenmolekulare Energiewanderung und Fluoreszenz. Ann Phys 2:55–75
15. Weber G (1953) Rotational Brownian motion and polarization of the fluorescence of solutions. Adv Protein Chem 8:415–459
16. Weber G (1997) Fluorescence in biophysics: accomplishments and deficiencies. Methods Enzymol 278:1–15
17. Venetta BD (1959) Microscope phase fluorometer for determining the fluorescence lifetimes of fluorochromes. Rev Sci Instrum 30:450–457
18. Clegg RM (2006) The history of FRET. In: Geddes CD, Lakowicz JR (eds) Reviews in fluorescence. Springer, New York. 1–45
19. Birch DJS (2011) Fluorescence detections and directions. Meas Sci Technol 22: 052002
20. Tsien RY (1998) The green fluorescent protein. Annu Rev Biochem 67:509–544
21. Hell SW (2007) Far-field optical nanoscopy. Science 316:1153–1158
22. Hell SW (2009) Microscopy and its focal switch. Nat Methods 6:24–32
23. Bastiaens PI, Pepperkok R (2000) Observing proteins in their natural habitat: the living cell. Trends Biochem Sci 25:631–637
24. Wouters FS, Verveer PJ, Bastiaens PI (2001) Imaging biochemistry inside cells. Trends Cell Biol 11:203–211
25. Verveer PJ, Bastiaens PI (2008) Quantitative microscopy and systems biology: seeing the whole picture. Histochem Cell Biol 130:833–843
26. Levitt JA, Matthews DR, Ameer-Beg SM, Suhling K (2009) Fluorescence lifetime and polarization-resolved imaging in cell biology. Curr Opin Biotechnol 20:28–36
27. Suhling K, FrenchPM, Phillips D (2005) Time-resolved fluorescence microscopy. Photochem Photobiol Sci 4:13–22
28. Festy F, Ameer-Beg SM, Ng T, Suhling K (2007) Imaging proteins in vivo using fluorescence lifetime microscopy. Mol Biosyst 3:381–391
29. Wallrabe H, Periasamy A (2005). Imaging protein molecules using FRET and FLIM microscopy. Curr Opin Biotechnol 16:19–27
30. Periasamy A, Clegg RM (eds) (2009) FLIM Microscopy and medicine. Chapman & Hall/CRC, Boca Raton
31. Gadella TWJ (ed) (2009) FRET and FLIM techniques. Elsevier, Amsterdam
32. Borst JW, Visser AJ (2010) Fluorescence lifetime imaging microscopy in life sciences. Meas Sci Technol 21:102002 (21 pp)
33. Gratton E, Jameson DM, Hall RD (1984). Multifrequency phase and modulation fluorometry. Annu Rev Biophys Bioeng 13:105–124
34. O'Connor DV, Phillips D (1984) Time-correlated single photon counting. Academic Press, London
35. Birch DJS, Imhof RE (1991) Time-domain fluorescence spectroscopy using time-correlated single-photon counting. In: Lakowicz JR (ed) Topics in fluorescence spectroscopy vol 1 techniques. Plenum, New York, p 1–95
36. Becker W (2005) Advanced time-correlated single photon counting techniques. Springer, Berlin
37. Hess ST, Huang S, Heikal AA, Webb WW (2002) Biological and chemical applications

of fluorescence correlation spectroscopy: a review. Biochemistry 41:697–705
38. Krichevsky O, Bonnet G (2002) Fluorescence correlation spectroscopy: The technique and its applications. Rep Prog Phys 65:251–297
39. Haustein E, Schwille P (2007) Fluorescence correlation spectroscopy: novel variations of an established technique. Annu Rev Biophys Biomol Struct 36:151–169
40. Chen Y, Müeller JD, So PT, Gratton E (1999) The photon counting histogram in fluorescence fluctuation spectroscopy. Biophys J 77:553–567
41. Kask P, Palo K, Ullmann D, Gall K (1999). Fluorescence-intensity distribution analysis and its application in biomolecular detection technology. Proc Natl Acad Sci U S A 96:13756–13761
42. Chen Y, Müeller JD (2007) Determining the stoichiometry of protein heterocomplexes in living cells with fluorescence fluctuation spectroscopy. Proc Natl Acad Sci U S A 104:3147–3152
43. Kask P, Palo K, Fay N, Brand L, Mets U, Ullmann D, Jungmann J, Pschorr J, Gall K (2000) Two-dimensional fluorescence intensity distribution analysis: theory and applications. Biophys J 78:1703–1713
44. Dertinger T, Pacheco V, von der Hocht I, Hartmann R, Gregor I, Enderlein J (2007) Two-focus fluorescence correlation spectroscopy: a new tool for accurate and absolute diffusion measurements. Chemphyschem 8:433–443
45. Kapanidis AN, Lee NK, Laurence TA, Doose S, Margeat E, Weiss S (2004) Fluorescence-aided molecule sorting: analysis of structure and interactions by alternating-laser excitation of single molecules. Proc Natl Acad Sci U S A 101:8936–8941
46. Müller BK, Zaychikov E, Brauchle C, Lamb DC (2005). Pulsed interleaved excitation. Biophys J 89:3508–3522
47. Persson G, Thyberg P, Widengren J (2008) Modulated fluorescence correlation spectroscopy with complete time range information. Biophys J 94:977–985
48. Benda A, Fagul'ova V, Deyneka A, Enderlein J, Hof M (2006) Fluorescence lifetime correlation spectroscopy combined with lifetime tuning: new perspectives in supported phospholipid bilayer research. Langmuir 22:9580–9585
49. Kapusta P, Wahl M, Benda A, Hof M, Enderlein J (2007) Fluorescence lifetime correlation spectroscopy. J Fluoresc 17:43–48
50. Petrasek Z, Schwille P (2008) Precise measurement of diffusion coefficients using scanning fluorescence correlation spectroscopy. Biophys J 94:1437–1448
51. Thompson NL, Steele BL (2007) Total internal reflection with fluorescence correlation spectroscopy. Nat Protoc 2:878–890
52. Digman MA, Brown CM, Sengupta P, Wiseman PW, Horwitz AR, Gratton E (2005) Measuring fast dynamics in solutions and cells with a laser scanning microscope. Biophys J 89:1317–1327
53. Digman MA, Sengupta P, Wiseman PW, Brown CM, Horwitz AR, Gratton E (2005) Fluctuation correlation spectroscopy with a laser-scanning microscope: exploiting the hidden time structure. Biophys J 88: L33–36
54. Brown CM, Dalal RB, Hebert B, Digman MA, Horwitz AR, Gratton E (2008). Raster image correlation spectroscopy (RICS) for measuring fast protein dynamics and concentrations with a commercial laser scanning confocal microscope. J Microsc 229:78–91
55. Digman MA, Brown CM, Horwitz AR, Mantulin WW, Gratton E (2008) Paxillin dynamics measured during adhesion assembly and disassembly by correlation spectroscopy. Biophys J 94:2819–2831
56. Digman MA, Wiseman PW, Choi C, Horwitz AR, Gratton E (2009) Stoichiometry of molecular complexes at adhesions in living cells. Proc Natl Acad Sci U S A 106:2170–2175
57. Digman MA, Wiseman PW, Horwitz AR, Gratton E (2009) Detecting protein complexes in living cells from laser scanning confocal image sequences by the cross correlation raster image spectroscopy method. Biophys J 96:707–716
58. Weiss S (1999) Fluorescence spectroscopy of single biomolecules. Science 283:1676–1683
59. Weiss S (2000) Measuring conformational dynamics of biomolecules by single molecule fluorescence spectroscopy. Nat Struct Biol 7:724–729
60. Selvin PR, Ha T (eds) (2008) Single-molecule techniques: a laboratory manual. Cold Spring Harbor Laboratory Press, Cold Spring Harbor
61. Hohlbein J, Gryte K, Heilemann M, Kapanidis AN (2010) Surfing on a new wave of single-molecule fluorescence methods. Phys Biol 7:031001 (22 pp)

Contents

Part VII Fluorescence Microscopy: Fluorescence Fluctuation Spectroscopy (FFS)

Part VIII Fluorescence Microscopy: Single-Molecule Fluorescence Spectroscopy (smFS)

Contributors

DAN AMIR • *The Goodman Faculty of Life Sciences, Bar Ilan University, Ramat Gan, Israel*

VLADIMIR V. APANASOVICH • *Department of Systems Analysis and Computer Simulation, Belarusian State University, Minsk, Belarus*

NICOLAS AUDUGÉ • *Institut Jacques Monod, UMR 7592, CNRS, Université Paris-Diderot, Paris, France*

ARIANNA BIESSO • *NIDDK, National Institutes of Health, Bethesda, MD, USA*

DAPHNE S. BINDELS • *Swammerdam Institute for Life Sciences, van Leeuwenhoek Centre for Advanced Microscopy, Section of Molecular Cytology, University of Amsterdam, Amsterdam, The Netherlands*

DAVID J.S. BIRCH • *The Photophysics Research Group, Department of Physics, Centre for Molecular Nanometrology, SUPA, University of Strathclyde, Glasgow, Scotland, UK*

CHRISTIAN BLUM • *Nanobiophysics Group, University of Twente, Enschede, The Netherlands*

JAN WILLEM BORST • *Laboratory of Biochemistry and Microspectroscopy Centre, Wageningen University, Wageningen, The Netherlands; Center for BioSystems Genomics, Wageningen, The Netherlands*

KEVIN BRAECKMANS • *Laboratory of General Chemistry and Physical Pharmacy, Ghent University and Centre of Nano- and Biophotonics, Ghent, Belgium*

JAAP BROOS • *Laboratory of Biophysical Chemistry and Groningen Biomolecular Science and Biotechnology Institute, University of Groningen, Groningen, The Netherlands*

CHRISTOPH A. BÜCHERL • *Laboratory of Biochemistry, Wageningen University, Wageningen, The Netherlands*

PATRIK R. CALLIS • *Department of Chemistry and Biochemistry, Montana State University, Bozeman, MT, USA*

YAN CHEN • *School of Physics and Astronomy, University of Minnesota, Minneapolis, MN, USA*

PEI-HUA CHUNG • *Department of Physics, King's College London, London, UK; Faculty of Life Sciences, University of Manchester, Manchester, UK*

ROBERT M. CLEGG • *Center of Biophysics and Computational Biology, University of Illinois at Urbana-Champaign, Urbana, IL, USA; Loomis Laboratory of Physics, Department of Physics, University of Illinois at Urbana-Champaign, Urbana, IL, USA*

MAÏTÉ COPPEY-MOISAN • *Institut Jacques Monod, UMR 7592, CNRS, Université Paris-Diderot, Paris, France*

ANA COUTINHO • *Centro de Química-Física Molecular and Institute of Nanoscience and Nanotechnology, Instituto Superior Técnico, Universidade Técnica de Lisboa, Lisbon, Portugal; Departamento Química e Bioquímica, Faculdade de Ciências da Universidade de Lisboa, Lisbon, Portugal*

ZEINAB DARWICH • *Faculté de Parmacie, Laboratoire de Biophotonique et Pharmacologie, UMR 7213 CNRS, Université de Strasbourg, Illkirch, France*

HERLINDE DE KEERSMAECKER • *Department of Chemistry, KU Leuven, Leuven, Belgium*
STEFAAN C. DE SMEDT • *Laboratory of General Chemistry and Physical Pharmacy, Ghent University, Ghent, Belgium*
MARIANO DELLAROLE • *INSERM U554, Centre de Biochimie Structurale, Montpellier, France*
JO DEMEESTER • *Laboratory of General Chemistry and Physical Pharmacy, Ghent University, Ghent, Belgium*
HENDRIK DESCHOUT • *Laboratory of General Chemistry and Physical Pharmacy, Ghent University and Centre of Nano- and Biophotonics, Ghent, Belgium*
ANATOLI V. DIGRIS • *Department of Systems Analysis and Computer Simulation, Belarusian State University, Minsk, Belarus*
JOHN PAUL EICHORST • *Center of Biophysics and Computational Biology, University of Illinois at Urbana-Champaign, Urbana, IL, USA*
YONG HWEE FOO • *Department of Chemistry, Faculty of Science, National University of Singapore, Singapore; Mechanobiology Institute, Singapore*
THEODORUS W. J. GADELLA, JR. • *Swammerdam Institute for Life Sciences, van Leeuwenhoek Centre for Advanced Microscopy, Section of Molecular Cytology, University of Amsterdam, Amsterdam, The Netherlands*
ADAM MATTHEW GILMORE • *Fluorescence Division, HORIBA Instruments Inc., Edison, NJ, USA*
JOACHIM GOEDHART • *Swammerdam Institute for Life Sciences, van Leeuwenhoek Centre for Advanced Microscopy, Section of Molecular Cytology, University of Amsterdam, Amsterdam, The Netherlands*
ELISHA HAAS • *The Goodman Faculty of Life Sciences, Bar Ilan University, Ramat Gan, Israel*
ROLAND HARTIG • *Multidimensional Microscopy and Cellular Diagnostics, Medical Faculty, Otto-von-Guericke-University, Magdeburg, Germany*
JELLE HENDRIX • *Physical Chemistry, Department of Chemistry, Munich Center for Integrated Protein Science (CiPSM) and Center for Nanoscience (CeNS), Ludwig-Maximilian-Universität München, Munich, Germany*
MARK A. HINK • *Swammerdam Institute for Life Sciences, van Leeuwenhoek Centre for Advanced Microscopy, Section of Molecular Cytology, University of Amsterdam, Amsterdam, The Netherlands*
MARTIN HOF • *J. Heyrovský Institute of Physical Chemistry v.v.i., Academy of Sciences of the Czech Republic, Prague, Czech Republic*
JOHAN HOFKENS • *Department of Chemistry, KU Leuven, Leuven, Belgium*
NICHOLAS G. JAMES • *Department Cell and Molecular Biology, John A. Burns School of Medicine, University of Hawaii, Honolulu, HI, USA*
DAVID M. JAMESON • *Department Cell and Molecular Biology, John A. Burns School of Medicine, University of Hawaii, Honolulu, HI, USA*
JOLENE JOHNSON • *School of Physics and Astronomy, University of Minnesota, Minneapolis, MN, USA*
LINDA JOOSEN • *Swammerdam Institute for Life Sciences, van Leeuwenhoek Centre for Advanced Microscopy, Section of Molecular Cytology, University of Amsterdam, Amsterdam, The Netherlands*

Clemens F. Kaminski • *Department of Chemical Engineering and Biotechnology, University of Cambridge, Cambridge, UK*

Andrey S. Klymchenko • *Faculté de Parmacie, Laboratoire de Biophotonique et Pharmacologie, UMR 7213 CNRS, Université de Strasbourg, Illkirch, France*

Jay R. Knutson • *NHLBI, National Institutes of Health, Bethesda, MD, USA*

Sirirat Kokpol • *Faculty of Science, Department of Chemistry, Center of Excellence for Petroleum, Petrochemicals, and Advanced Materials, Chulalongkorn University, Bangkok, Thailand*

Kai R. Konrad • *Molecular Plant Physiology and Biophysics, Julius-Maximilians-University Würzburg, Würzburg, Germany*

Daniel A. Lafontaine • *Faculté des Sciences, RNA Group, Département de Biologie, Université de Sherbrooke, Sherbrooke, QC, Canada*

Don C. Lamb • *Physical Chemistry, Department of Chemistry, Munich Center for Integrated Protein Science (CiPSM) and Center for Nanoscience (CeNS), Ludwig-Maximilian-Universität München, Munich, Germany; Loomis Laboratory of Physics, Department of Physics, University of Illinois at Urbana-Champaign, Urbana, IL, USA*

Sergey P. Laptenok • *Laboratoire d'Optique et Biosciences, INSERM U696-CNRS UMR7645, Ecole Polytechnique, Palaiseau, France*

Eitan Lerner • *The Goodman Faculty of Life Sciences, Bar Ilan University, Ramat Gan, Israel*

James Levitt • *Department of Physics, King's College London, London, UK*

Kiattisak Lugsanangarm • *Faculty of Science, Department of Chemistry, Chulalongkorn University, Bangkok, Thailand*

Andras Lukacs • *Laboratoire d'Optique et Biosciences, INSERM U696-CNRS UMR7645, Ecole Polytechnique, Palaiseau, France; Department of Biophysics, Medical School, University of Pécs, Pécs, Hungary*

Xiaoxiao Ma • *Faculty of Science, Department of Chemistry, National University of Singapore, Singapore*

Patrick J. Macdonald • *Department of Biomedical Engineering, University of Minnesota, Minneapolis, MN, USA*

Radek Macháň • *Faculty of Biomedical Engineering, Czech Technical University in Prague, Kladno, Czech Republic; J. Heyrovský Institute of Physical Chemistry v.v.i., Academy of Sciences of the Czech Republic, Prague, Czech Republic*

Kaley McCluskey • *SUPA School of Physics and Astronomy, University of St. Andrews, St. Andrews, Scotland, UK*

Ana M. Melo • *Centro de Química-Física Molecular and Institute of Nanoscience and Nanotechnology, Instituto Superior Técnico, Universidade Técnica de Lisboa, Lisbon, Portugal*

Yves Mély • *Faculté de Parmacie, Laboratoire de Biophotonique et Pharmacologie, UMR 7213 CNRS, Université de Strasbourg, Illkirch, France*

Hideaki Mizuno • *Department of Chemistry, KU Leuven, Leuven, Belgium*

Joachim D. Mueller • *Department of Biomedical Engineering, School of Physics and Astronomy, University of Minnesota, Minneapolis, MN, USA*

Paul Müller • *BIOTEC, Biophysics, Technische Universität Dresden, Dresden, Germany*

HANNES NEUWEILER • *Department of Biotechnology & Biophysics, Biozentrum, Julius-Maximilians-University Würzburg, Würzburg, Germany*
EUGENE G. NOVIKOV • *Department of Systems Analysis and Computer Simulation, Belarusian State University, Minsk, Belarus; Carl Zeiss Microscopy GmbH, Jena, Germany*
ARTHIT NUEANGAUDOM • *Faculty of Science, Department of Chemistry, Chulalongkorn University, Bangkok, Thailand*
NADTANET NUNTHABOOT • *Faculty of Science, Department of Chemistry, Mahasarakham University, Mahasarakham, Thailand*
PATRICK NUERNBERGER • *Laboratoire d'Optique et Biosciences, INSERM U696-CNRS UMR7645, Ecole Polytechnique, Palaiseau, France; Institut für Physikalische und Theoretische Chemie, Julius-Maximilians-Universität Würzburg, Würzburg, Germany*
TOMER OREVI • *The Goodman Faculty of Life Sciences, Bar Ilan University, Ramat Gan, Israel*
SERGI PADILLA-PARRA • *Institut de Génétique et Développement de Rennes, UMR 6290, CNRS, Université de Rennes 1, Rennes, France*
J. CARLOS PENEDO • *SUPA School of Physics and Astronomy, Biomedical Sciences Research Complex, University of St. Andrews, St. Andrews, Scotland, UK*
SOMSAK PIANWANIT • *Faculty of Science, Department of Chemistry, Chulalongkorn University, Bangkok, Thailand*
MANUEL PRIETO • *Centro de Química-Física Molecular and Institute of Nanoscience and Nanotechnology, Instituto Superior Técnico, Universidade Técnica de Lisboa, Lisbon, Portugal*
YURY PROKAZOV • *Special Laboratory Electron- and Laserscanning Microscopy, Leibniz Institute for Neurobiology (LIN), Magdeburg, Germany*
GIL RAHAMIM • *The Goodman Faculty of Life Sciences, Bar Ilan University, Ramat Gan, Israel*
ERIC J. REES • *Department of Chemical Engineering and Biotechnology, University of Cambridge, Cambridge, UK*
SUSANA ROCHA • *Department of Chemistry, KU Leuven, Leuven, Belgium*
CATHERINE A. ROYER • *Centre de Biochimie Structurale, INSERM U554, Montpellier, France*
MARKUS SAUER • *Department of Biotechnology & Biophysics, Biozentrum, Julius-Maximilians-University Würzburg, Würzburg, Germany*
GABRIELE S. KAMINSKI SCHIERLE • *Department of Chemical Engineering and Biotechnology, University of Cambridge, Cambridge, UK*
PETRA SCHWILLE • *BIOTEC, Biophysics, Technische Universität Dresden, Dresden, Germany; Cellular and Molecular Biophysics, Max Planck Institute of Biochemistry, Martinsried, Germany*
EUAN SHAW • *SUPA School of Physics and Astronomy, University of St. Andrews, St. Andrews, Scotland, UK*
IGOR P. SHINGARYOV • *Department of Systems Analysis and Computer Simulation, Belarusian State University, Minsk, Belarus*
VICTOR V. SKAKUN • *Department of Systems Analysis and Computer Simulation, Belarusian State University, Minsk, Belarus*

JORIS J. SNELLENBURG • *Faculty of Sciences, Department of Physics and Astronomy, VU University Amsterdam, Amsterdam, The Netherlands*
TOMÁŠ STEINBERGER • *Faculty of Biomedical Engineering, Czech Technical University in Prague, Kladno, Czech Republic; J. Heyrovský Institute of Physical Chemistry v.v.i., Academy of Sciences of the Czech Republic, Prague, Czech Republic*
MARTIJN H.W. STOPEL • *Nanobiophysics Group, University of Twente, Enschede, The Netherlands*
VINOD SUBRAMANIAM • *Nanobiophysics Group, University of Twente, Enschede, The Netherlands*
KLAUS SUHLING • *Department of Physics, King's College London, London, UK*
FUMIO TANAKA • *Faculty of Science, Department of Chemistry, Chulalongkorn University, Bangkok, Thailand; Division of Laser BioScience, Institute for Laser Technology, Osaka, Japan*
DMITRI TOPTYGIN • *Department of Biology, Johns Hopkins University, Baltimore, MD, USA*
MARC TRAMIER • *Institut de Génétique et Développement de Rennes, UMR 6290, CNRS, Université de Rennes 1, Rennes, France*
EVGENY TURBIN • *Special Laboratory Electron- and Laserscanning Microscopy, Leibniz Institute for Neurobiology (LIN), Magdeburg, Germany*
JOSE R. TUSELL • *Department of Chemistry and Biochemistry, Montana State University, Bozeman, MT, USA*
HIROSHI UJI-I • *Department of Chemistry, KU Leuven, Leuven, Belgium*
IVO H.M. VAN STOKKUM • *Faculty of Sciences, Department of Physics and Astronomy, VU University Amsterdam, Amsterdam, The Netherlands*
LAURA VAN WEEREN • *Swammerdam Institute for Life Sciences, van Leeuwenhoek Centre for Advanced Microscopy, Section of Molecular Cytology, University of Amsterdam, Amsterdam, The Netherlands*
CARISSA M. VETROMILE • *Department Cell and Molecular Biology, John A. Burns School of Medicine, University of Hawaii, Honolulu, HI, USA*
MARTEN H. VOS • *Laboratoire d'Optique et Biosciences, INSERM U696-CNRS UMR7645, Ecole Polytechnique, Palaiseau, France*
THOMAS WEIDEMANN • *BIOTEC, Biophysics, Technische Universität Dresden, Dresden, Germany; Cellular and Molecular Biophysics, Max Planck Institute of Biochemistry, Martinsried, Germany*
KAI WEN TENG • *Center of Biophysics and Computational Biology, University of Illinois at Urbana-Champaign, Urbana, IL, USA*
THORSTEN WOHLAND • *Faculty of Science, Department of Chemistry and Department of Biological Sciences, National University of Singapore, Singapore*
RANHUA XIONG • *Laboratory of General Chemistry and Physical Pharmacy, Ghent University and Centre of Nano- and Biophotonics, Ghent, Belgium*
JIANHUA XU • *NHLBI, National Institutes of Health, Bethesda, MD, USA*
PHILIP YIP • *The Photophysics Research Group, Department of Physics, Centre for Molecular Nanometrology, SUPA, University of Strathclyde, Glasgow, Scotland, UK*
WERNER ZUSCHRATTER • *Special Laboratory Electron- and Laserscanning Microscopy, Leibniz Institute for Neurobiology (LIN), Magdeburg, Germany*

Part I

Steady-State Fluorescence Spectroscopy (SSFS)

Chapter 1

How to Collect National Institute of Standards and Technology (NIST) Traceable Fluorescence Excitation and Emission Spectra

Adam Matthew Gilmore

Abstract

Contemporary spectrofluorimeters comprise exciting light sources, excitation and emission monochromators, and detectors that without correction yield data not conforming to an ideal spectral response. The correction of the spectral properties of the exciting and emission light paths first requires calibration of the wavelength and spectral accuracy. The exciting beam path can be corrected up to the sample position using a spectrally corrected reference detection system. The corrected reference response accounts for both the spectral intensity and drift of the exciting light source relative to emission and/or transmission detector responses. The emission detection path must also be corrected for the combined spectral bias of the sample compartment optics, emission monochromator, and detector. There are several crucial issues associated with both excitation and emission correction including the requirement to account for spectral band-pass and resolution, optical band-pass or neutral density filters, and the position and direction of polarizing elements in the light paths. In addition, secondary correction factors are described including (1) subtraction of the solvent's fluorescence background, (2) removal of Rayleigh and Raman scattering lines, as well as (3) correcting for sample concentration-dependent inner-filter effects. The importance of the National Institute of Standards and Technology (NIST) traceable calibration and correction protocols is explained in light of valid intra- and interlaboratory studies and effective spectral qualitative and quantitative analyses including multivariate spectral modeling.

Key words Absorbance, Band-pass, Grating orders, Inner-filter effects, Photobleaching, Quenching, Raman scatter, Rayleigh scatter, Spectral resolution, Wood's anomalies

1 Introduction

Fluorescence excitation and emission spectroscopy, compared to absorbance spectrophotometry, is a popular analytical instrument technique because it exhibits higher sensitivity and spectral selectivity. However, in contrast to absorbance spectrophotometry, where the spectral qualities of the light sources, dispersive elements, and detectors divide out commutatively in the final signal formulae, fluorescence data are subject to being highly convolved

Yves Engelborghs and Antonie J.W.G. Visser (eds.), *Fluorescence Spectroscopy and Microscopy: Methods and Protocols*, Methods in Molecular Biology, vol. 1076, DOI 10.1007/978-1-62703-649-8_1,

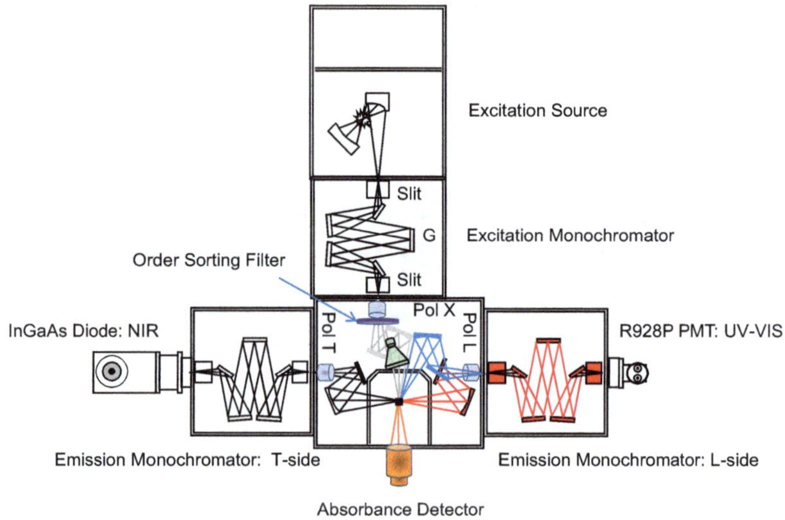

Fig. 1 Schematic of a typical modular Fluorolog® spectrofluorimeter manufactured by HORIBA Instruments Inc. illustrating the major components including the excitation source, the excitation monochromator, the sample compartment, and the L-side and T-side emission monochromators. The slit locations and dispersive grating (G) are labeled inside the excitation monochromator. Inside the sample compartment, the following components are labeled including the respective excitation, the L-side and T-side polarizers (Pol X, Pol L, and Pol T), the ideal location for order-sorting and optical filters in the excitation path, the beam-splitter and reference detector and beam path components sb influenced by the excitation correction factor (in *green*), the front-face 22.5° beam path and mirror components (in *light blue*), and the location and beam path for an absorbance detector (*orange*). The *red right-angle* sample compartment beam path and *black* L-side emission monochromator components illustrate those components influenced by their respective emission correction factors

with and potentially distorted by the instrument's spectral response function and sample conditions. Hence in order to facilitate valid intra- and interlaboratory comparisons of data collected on a fluorimeter, all components of the light path including the sample properties must be corrected in sequence to approach the "ideal" National Institute of Standards and Technology (NIST) traceable instrument response function [1–3] as is the subject of this chapter.

A conventional modular scanning spectrofluorimeter as shown in Fig. 1 comprises several components acting in concert to excite the sample with a chosen color of light followed by measuring the spectrum of light emitted by the sample. The initial essential element is the excitation light source, in this case a vertically mounted xenon arc lamp, which emits a broadband solar-like spectrum. The light source is optically coupled to an excitation monochromator, the design and internal components of which determine the spectral resolution, stray-light rejection, and peak wavelength efficiency of the light emitted into the sample compartment. The excitation monochromator contains the dispersive grating(s) and slits which regulate the color and band-pass of light. Ideally, the excitation monochromator is scanned from high energy (UV) to lower energy (red) to minimize photobleaching of the sample by the UV radiation. The sample compartment contains the focusing optics, polarizing elements (prisms or films), filter holders, and sometimes front-face viewing optical elements. Usually sample compartment focusing optics are either mirror based (as shown here) or lens based, the former being preferred to eliminate chromatic aberrations and facilitate focusing over a wider spectral range. Following the light path beyond the exit slit of the excitation monochromator into the sample compartment is the best location for an order-sorting filter holder or filter-changing apparatus (filter wheel). Following the primary beam path into the sample compartment, a beam splitter is encountered that reflects a small fraction (on the order of a few percent) onto a reference detector diode (or quantum counter cell plus reference detector) to monitor the intensity of the excitation light and correct for lamp spectral intensity or drift on the emission and absorbance detector signals. An absorbance detector can be provided as shown to monitor the transmission spectrum of light through the solvent blank and sample to calculate the samples absorbance and/or transmittance. The emission paths normally comprise at right angles to the excitation path emission monochromators (L-side and T-side) that are tuned via the "blaze" of the grating(s) and detector spectral response towards redder "Stokes-shifted" wavelengths compared to the excitation monochromator. Figure 1 also highlights the key components functionally associated with the major spectral correction factors including the effective relevant excitation correction factor elements and beam path (in green) and the emission monochromator-detector correction factors (L-side in red and T-side in black), the details of which are described below.

Figure 2 exemplifies another typical commercial unitized absorbance-fluorescence spectrophotometer, namely, the Aqualog® from Horiba Instruments Inc. [4]. The Aqualog® optical bench includes a special aberration-corrected double-grating excitation monochromator, a reference detector and absorbance detector (both Si photodiodes), and a unique emission detector comprised of a thermoelectrically cooled back-illuminated CCD and spectrograph. The system incrementally scans excitation from high energy to low energy and can collect the full emission spectrum at each excitation increment to rapidly and simultaneously generate absorbance spectra and fully corrected excitation-emission spectral maps or matrices (EEMs).

The chapter is organized as follows to relate to the typical optical benches described above: first there are descriptions of the protocols and data treatments associated with the NIST traceable spectral wavelength, band-pass, and absorbance accuracy

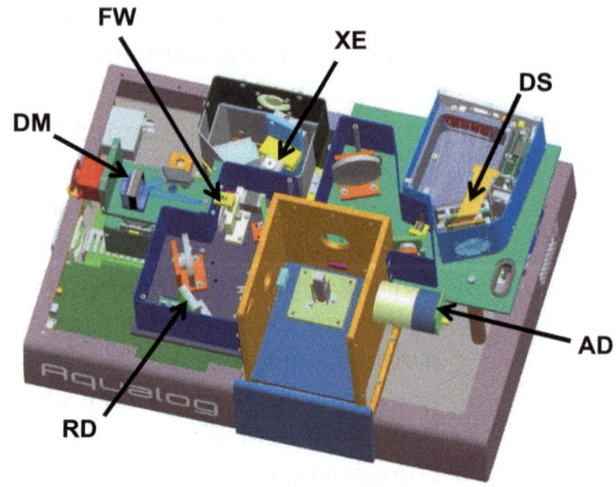

Fig. 2 Optical bench layout of the Aqualog®, a commercial absorbance-fluorescence spectrophotometer manufactured by HORIBA Instruments Inc. The major components labeled in the legend include the vertically mounted 150 W xenon arc source (XE), the subtractive 10-cm double-grating excitation monochromator (DM), the order-sorting filter wheel (FW), the silicon diode reference detector (RD), the silicon diode absorbance detector (AD), and the thermoelectrically cooled CCD spectrograph (DS)

calibration procedures. Following the protocols for calibration of the optical elements, the NIST traceable spectral shape correction factor protocols are explained for the excitation and emission beam paths. Then secondary correction protocols and algorithms for the spectral data are defined including (1) dealing with and/or removing overlapping scatter signals associated with the grating orders of the monochromators (spectrographs) and Raman bands from the solvent(s), (2) subtracting solvent background emission signals, and (3) correcting for sample concentration-dependent absorbance of the excitation and emission beams in the sample cuvette (inner-filter effects). The chapter concludes with an example analysis implementing the recommended best practices for fluorescence NIST traceable spectral data collection and treatment as they can facilitate validation of qualitative and quantitative analyses for intra- and interlaboratory comparisons.

2 Methods

2.1 Spectrofluorimeter Excitation and Emission Optical Component Calibration Procedures

Accurate spectral correction requires accurate calibration of all aspects of the instrument including wavelength accuracy, slit-band-pass accuracy, and optical beam alignment including the instrument throughput and signal to noise. The optical alignment and slit calibration are normally performed only by qualified personnel trained by the manufacturer. Most fluorescence instruments are calibrated using software tools provided by the manufacturer to

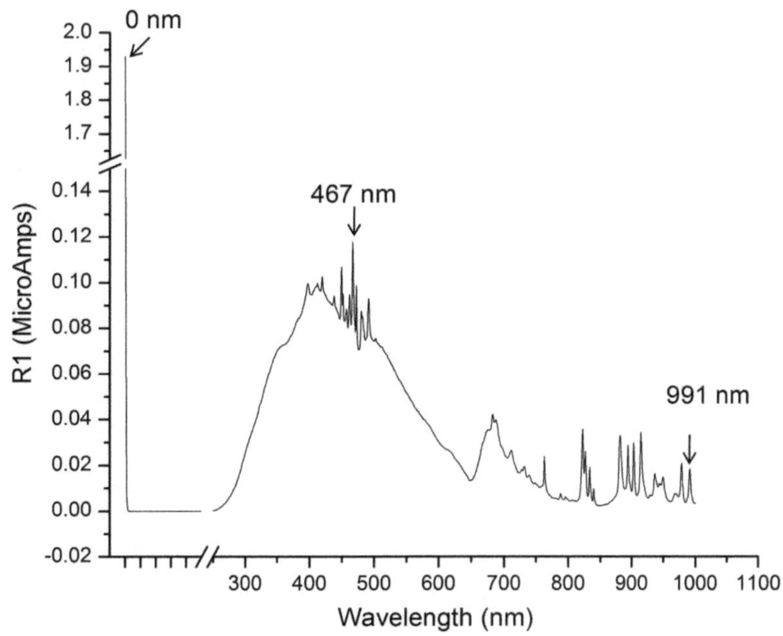

Fig. 3 A typical reference diode detector signal (in microamp units) of the xenon lamp spectrum obtained by scanning the excitation monochromator from 0 to 1,000 nm in 1-nm steps, 1-nm band-pass, and 0.1-s integration time. Three landmark xenon peaks are identified for excitation monochromator calibration purposes including the 0-nm peak, the 467-nm peak, and the 991-nm peak

within their factory specified wavelength accuracy, nominally 0.5–1 nm. It is logical and conventional to first calibrate the excitation optical path from the source up to the reference detector. This can be done by referring to the distinctive emission lines of gas-filled xenon or mercury pen lamps passed through the excitation monochromator [2, 3]. Figure 3 shows a typical xenon lamp spectrum suitable for wavelength calibration illustrating three landmark calibration features: (1) the 0-nm peak associated the zero-order emission of the grating, (2) the first-order peak at 467 nm, and (3) another peak at 991 nm commonly used to calibrate the NIR region. Often times it is a good idea to verify the full range of the monochromator versus the calibration lamp data by beginning the scan at 0 nm to visualize the 0th-order peak (white light); this helps to ascertain the monochromator is calibrated in the first order up to and including the referenced gas emission lines (*see* **Note 1**). Figure 4 shows the results of two standard reference material sample-dependent methods commonly used to verify calibration of the excitation optical path, respectively, for absorbance (photometric) and wavelength accuracy. The samples are included in the NIST traceable Starna RM-06HLKI United States Pharmacopeia reference set. Figure 4a shows a typical absorbance scan of the

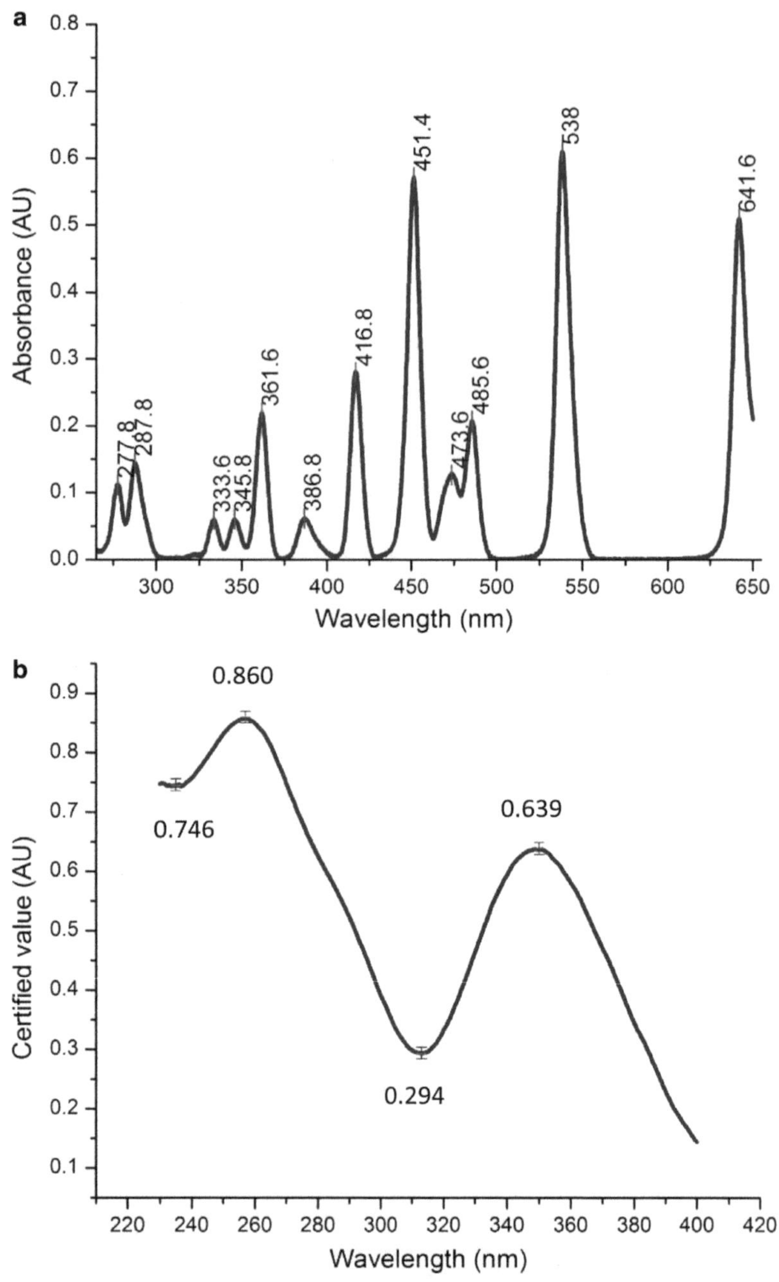

Fig. 4 Panel (**a**) shows holmium oxide excitation monochromator spectral calibration standard data from Starna RM-06HLKI set sample traceable to NIST SRM 2034, the peaks are identified within wavelength specification as shown, and the absorbance data were acquired with the absorbance detector and reference detector signals ratioed as $A1_c/R1_c$ using a blank solution cuvette of 0.105 N $HClO_4$. Panel (**b**) shows the Starna potassium dichromate photometric (absorbance) accuracy standard sample from the Starna RM-06HLKI set traceable to NIST SRM 935. The calibration data with absorbance values identified within the specifications as shown were measured according to the manufacturer's certificate instructions for Starna RM-06HLKI

holmium oxide dissolved in 0.1-N perchloric acid measured against the 0.105-N perchloric acid blank. The RM certificate for the holmium spectrum contains the values for the peaks as shown and as defined for the representative spectral band-pass (in this case 5 nm for a typical Aqualog® instrument). Figure 4b shows the typical absorbance spectrum of the potassium dichromate standard solution measured relative to its solvent blank. The values and symbols in the figure refer to the four major calibration values for the RM, namely, the absorbance values at 235 nm, 257 nm, 313 nm, and 350 nm, respectively. Again both the holmium and dichromate RM data provide verification of the excitation path wavelength accuracy and just as importantly the photometric accuracy which itself is critical to correct for inner-filter effect corrections as explained later.

Emission path wavelength calibration for UV-visible spectrofluorimeters usually is achieved by one or more of the following three methods: (1) by analyzing either the peak Raman scattering signal of a purified water sample, (2) measuring the emission peaks of known reference materials under specified excitation-emission wavelengths, or (3) measuring the peak coordinates of the lamp lines used for excitation (*see* Fig. 3 above) from a diffuse reflector material such as Spectralon™ placed in the sample compartment (*see* **Note 2**). Notably accurate water Raman calibration is the most popular method and is most effective when one uses highly purified water in pristine scratch-free cuvettes devoid of scattering particles and importantly with very low total dissolved organic carbon (TOC) at <10-ppb concentrations. The typical water Raman scan as shown in Fig. 5a is performed using 5-nm band-pass on excitation and emission with excitation at 350 nm and emission scanned from 365 to 450 nm (0.5-nm increments). As shown in Fig. 5a, the water Raman (OH stretch) peak signal used for emission calibration is at 3,382 cm^{-1} from the excitation energy such that 350-nm excitation yields a peak at 397 ± 0.5 nm.

In addition to the emission wavelength calibration, the water Raman sample analysis conditions above are used to calibrate the instrument throughput and signal-to-noise ratio. These data are often instrumental in diagnosing issues with the instrument primarily those associated with the age and output of the excitation source to indicate a need for lamp servicing. The typical Horiba Instrument Inc. water Raman analysis includes in addition to the scan in Fig. 5a a second kinetics scan to define the peak-to-peak and root-mean-square (*RMS*) value of the noise of the Raman sample signal at 450 nm, excited at 350 nm under 5-nm band-pass conditions. The kinetics are measured for 30 s at 1-s intervals and 1-s integration. From these data, the average peak-to-peak noise value divided by 5 is used to estimate the *RMS*. Two signal-to-noise ratio (*SNR*) figures of merit are computed from the data as shown in the legend of Fig. 5b, namely, a first standard deviation

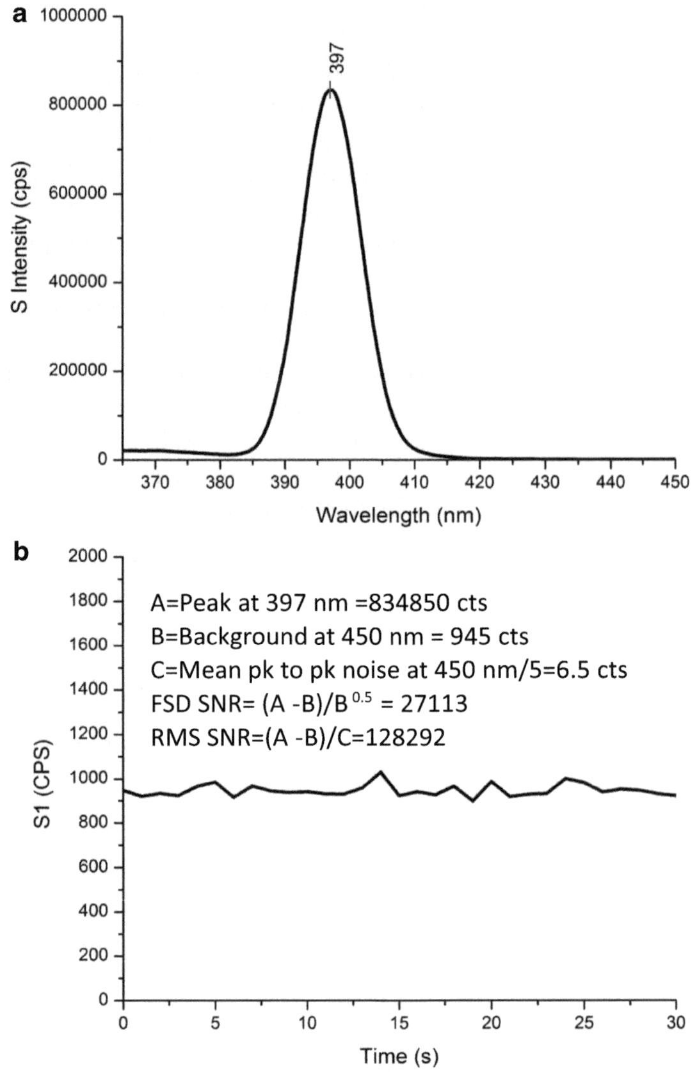

Fig. 5 Typical data used to calculate the water Raman signal-to-noise ratios using both a first standard deviation (FSD) and root-mean-square method (RMS). Panel (**a**) shows a water Raman spectrum at 5-nm band-pass obtained by exciting the Starna 3Q-10 water sample at 350 nm and scanning the emission in 0.5-nm steps with 1-s integration from 265 to 450 nm. The peak calibration wavelength is labeled at 397 nm. Panel (**b**) shows the kinetics scan used to evaluate the peak-to-peak noise and RMS of the water Raman sample; the Starna 3Q-10 sample was excited at 5-nm band-pass at 350 nm, and the kinetics of the signal at 450 nm were monitored for 30 s at 1-s intervals. The legend in panel (**b**) illustrates the parameters from panels (**a**) and (**b**) and the calculations of the FSD and RMS

($FSD\ SNR$) and a root mean square ($RMS\ SNR$). In addition to the throughput information indicated by the peak signal, the two figures of merit each provide unique independent information about the instrument's performance. The *FSD SNR* indicates the

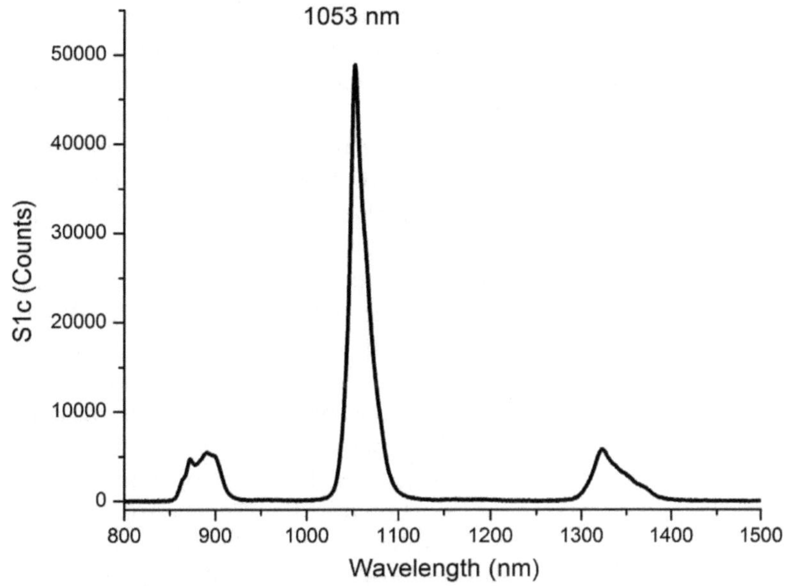

Fig. 6 A typical emission scan of a 2 % Nd-doped laser phosphate glass material (Schott Glass) measured at 5-nm band-pass, excitation at 575 nm, and emission at 1-nm increments from 800 to 1,550 nm with a 150-ms integration time. The instrument was a Nanolog® from Horiba Instruments Inc., equipped with a double-grating excitation monochromator with gratings at blazed at 500 nm with 1,200 lines/mm. The emission was measured with a Symphony InGaAs array detector (512 × 1 array, 50-μm pixels, −170 °C) mounted on an iHR320 imaging spectrograph using an emission grating blazed at 800 nm with 100 lines/mm

stray-light rejection of the system as indicated by the average intensity of the background signal at 450 nm. The *RMS SNR* indicates the electronic noise associated with the instrument, which influences the background intensity readings at 450 nm (*see* **Note 3**).

Near-infrared emission path wavelength calibration and throughput are commonly achieved using excitation-emission lines from rare-earth-doped laser glass materials (Schott Laser Glass 2 % Nd) in addition to measuring at characteristic reflected lamp emission lines. The data in Fig. 6 show the results of an emission scan of the 2 % Nd glass sample measured at an angle of 30° from normal excitation. The band-pass was 5 nm, excitation was at 575 nm, and the emission was measured with an InGaAs array and an emission spectrograph over the range of 800–1,500 nm with an integration time of 30 ms for the entire emission scan. The landmark emission peak at 1,053 nm is used to verify wavelength calibration, and the intensity of the peak can be used to judge the total instrument throughput under the defined sample position and instrument conditions.

2.2 Excitation Intensity and Spectral Correction Factors

Experimentally the excitation intensity can vary widely as a function of the light source, grating efficiency, selected wavelength, chosen band-pass, and placement of filters before the sample excitation. As explained above, the excitation intensity and spectral response are typically monitored with a reference detector that receives a small fraction (on the order of a few percent) of the beam reflected from a "pick-off" beam splitter. The beam splitter is ideally made of quartz material to effectively transmit light from the deep ultraviolet (UV) into the near infrared. The beam splitter should be angled to minimize polarization artifacts. Whereas the reference detector signal itself accounts for the vast majority of the excitation correction, there is still a substantial fraction of the ideal response that requires correction that is associated with the combined wavelength dependence of the reference detector and beam-splitter transmission-reflection properties [2, 3, 5, 6].

As shown in Fig. 7a, the spectral response of the detector and beam splitter is typically corrected by measuring the comparative responses of a certified calibrated diode (or quantum counter with a known quantum yield) placed in the sample position and the reference diode during a scan of the xenon lamp spectrum. The NIST traceable calibrated diode was obtained from International Light Technologies Inc., of Peabody, MA, USA. Importantly, both the reference and calibrated detector responses must first be corrected by subtracting their dark current-based signals. The "observed" response of the uncalibrated reference diode is ratioed to yield the "ideal" response of the calibrated diode using the information from the calibration file data (Fig. 7b) provided by the diode manufacturer. The multiplicative excitation correction factor displayed in Fig. 7c is equivalent to the (ideal/observed) response ratio such that subsequent multiplication of observed reference diode data yields the ideal spectral response. Most spectrofluorimeter software packages implement detector signal algebra tools to facilitate implementation of the stored excitation correction factor files. The following reference detector signal convention is used to yield the fully corrected reference diode signal as

$$R1_c = (R1 - R_{\text{dark}}) \times X_{\text{correct}} \tag{1}$$

where $R1$ is the raw detector signal, R_{dark} is the dark current, and X_{correct} is the excitation correction factor file.

Care should be noted that the accuracy of the X_{correct} file depends on matching the experimental band-pass and monochromator wavelength stepping increments closely to those used for X_{correct} file generation. The corrected reference diode signal is particularly important for fluorescence excitation scans, absorbance/transmission scans, and excitation-emission matrices. However, it is recommended to always apply the $R1_c$ signal correction for emission data be it a scan or single point to account for lamp

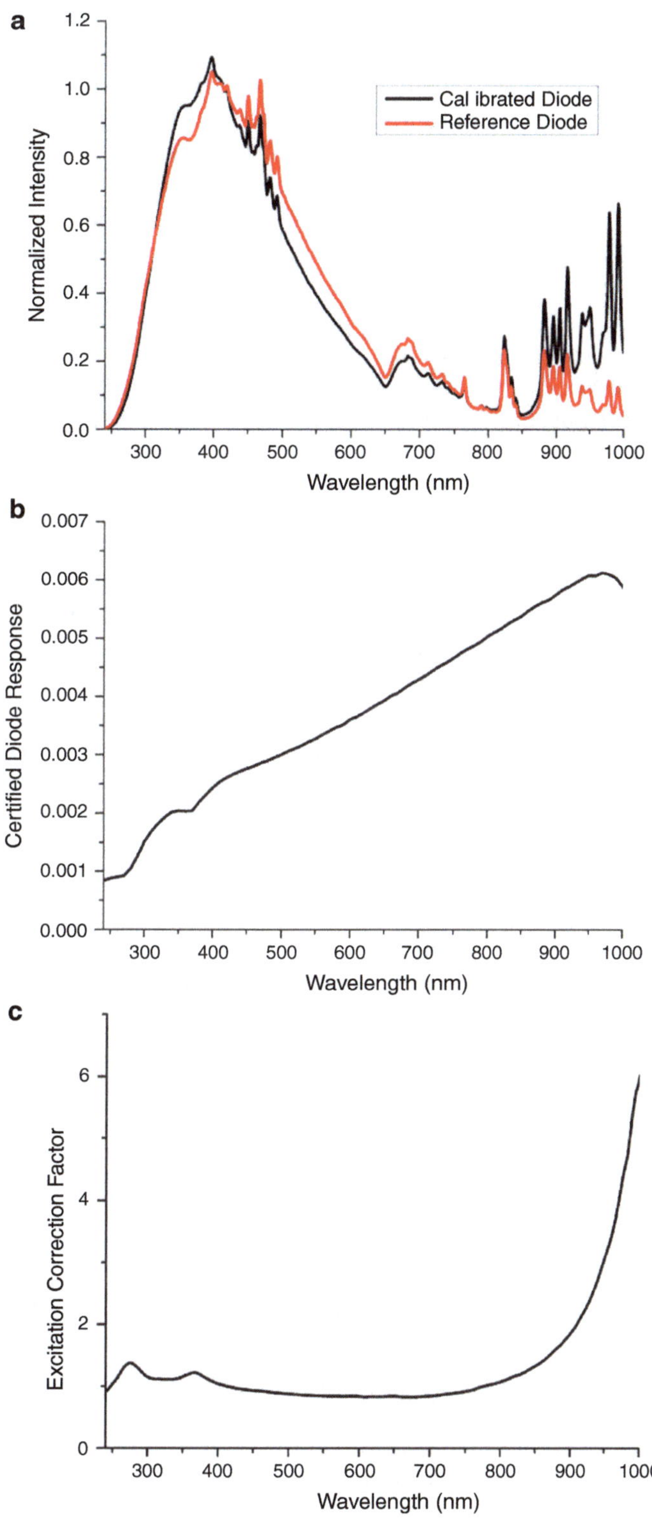

Fig. 7 Typical calibrated and reference detector signal data (**a**) and diode calibration data (**b**) used to calculate the excitation correction factor (**c**) to account for the spectral dependence of the reference diode and beam splitter. The *y*-scale units for panel (**a**) originally measured in Amps units were normalized at 414 nm, panel (**b**) calibration file units are Amps/W, and panel (**c**) units were normalized at 414 nm

drift and day-to-day changes in lamp intensity. Importantly, it follows that use of optical filters in the excitation path for order sorting, band-pass selection, or neutral intensity attenuation is generally recommended, as discussed above for Figs. 1 and 2, to be after the excitation mono exit slit and before the beam splitter and reference diode in order to not interfere with the excitation correction factor's spectral accuracy which will lead to artifacts in both the excitation and absorbance spectra (*see* **Note 4**).

2.3 Emission Intensity and Spectral Correction Factors

The measurement of the polychromatic fluorescence sample emission as mentioned above is strongly influenced by the optical and detector elements of the instrument downstream from sample excitation. The two key factors are the monochromator grating efficiency curve and the spectral response of the detector material, be that the cathode material for a photomultiplier tube (PMT) or semiconductor material of a photodiode, diode array, or CCD device. However, mirror reflective materials and sample compartment/monochromator component coatings with luminescence backgrounds may also contribute.

To generate emission spectral correction factors, it is important to correct first for the detector's dark current signal; this is because the correction being usually multiplicative as explained will generate artifacts when used to multiply the dark current signal of the emission detector. It is important to consider the band-pass, stepping increment, and optical elements like polarizers or filters in the light path relative to the experimental data the factor is intended to correct. The spectral resolution is especially critical for resolving the polarization artifacts known as Wood's anomalies common to ruled gratings and observed as sharp transitions in reflected intensity as a function of wavelength. Again it must be reemphasized that the polarizing elements in the light path will directly influence the resolution of Wood's anomalies.

There are two popular materials and protocols for generating the emission correction factors, namely, (1) use of a calibrated light source or (2) use of reference material samples both with calibrated spectral emission shape data [2, 3, 7–9]. In both cases, the procedure is similar, namely, to record the "observed" emission spectra of the calibrated reference material (or lamp) and use this information to generate as a function of wavelength the (ideal/observed) ratio factor, typically referred to as the emission correction factor. The emission correction factor is used to multiply (or divide) the observed emission detector response to generate the ideal detector response. The signal algebra for emission correction parallels that of the excitation correction above where the fully corrected emission detector signal is calculated as

$$S1_c = (S1 - S_{dark}) \times M_{correct} \quad (2)$$

where *S1* is the raw detector signal, S_{dark} is the detectors dark current signal, and $M_{correct}$ is the emission correction factor file.

Figure 8 shows example data measured on a Fluorolog® 3 system equipped with Hamamatsu R928P PMT detector used to generate an $M_{correct}$ file using the ideal NIST sourced certified reference material spectra from SRMs 2940, 2942, and 2943 [7–9] (Panel a) and the observed emission spectra (Panel b) to calculate the (ideal/observed) emission correction ratio factor (Panel c). The observed data in Fig. 8b were collected under the conditions defined in the SRM certificates, and the results from the three spectra in Fig. 8a and Fig. 8b were normalized at the overlapping wavelength regions for the certified spectra to generate the $M_{correct}$ factor (Fig. 8c).

Here again the proper practical use and accuracy of the emission correction factor file are subject to several caveats, the most important being the cautious use of optical filters after the sample and before the emission monochromator and detector. Any optical filter or polarizing element in this location that was not present during the emission correction factor generation will cause a deviation in the ideal spectral response; this is particularly true for optical band-pass filters. However, it must also be noted that even the so-called neutral density filters in this location rarely have spectrally flat responses and will cause significant spectral deviations (*see* **Note 5**). Secondly, it is recommended that the band-pass resolution and stepping increments used for the emission correction should match those of the experiment to best resolve spectral features associated with the grating efficiency and the Wood's anomalies (polarization artifacts). In summary, it follows that whenever the corrected response of the emission optical path is desired, all experimental optical elements (*see* **Note 6**) and resolution settings match those used to generate the emission correction factor files.

2.4 Typical Spectral Resolution, Throughput, and Detector Saturation Issues

The spectral resolution of a typical absorbance-fluorescence spectrophotometer is limited by a few key factors, primarily, the focal length of the monochromators, the grating groove density, and the slit width as well as the mechanical stepping interval precision of the monochromators. Generally, the band-pass that determines spectral resolution is limited, for scanning monochromator systems with ruled gratings, by the factor known as the reciprocal linear dispersion. As an example, a single grating monochromator with a 180-mm focal length and grating groove density of 1,200 lines/mm exhibits a reciprocal linear dispersion of 4.2 nm/mm. Thus with slits that vary from 0 to 7 mm in physical width, the band-pass ranges from 0 to 29.4 nm. Resolution of spectral features for a scanning system with a single-channel detector like a PMT or diode is primarily limited by the band-pass and secondarily limited by the number of stepping increments within the band-pass; a general rule of thumb is that the stepping increments required to resolve two

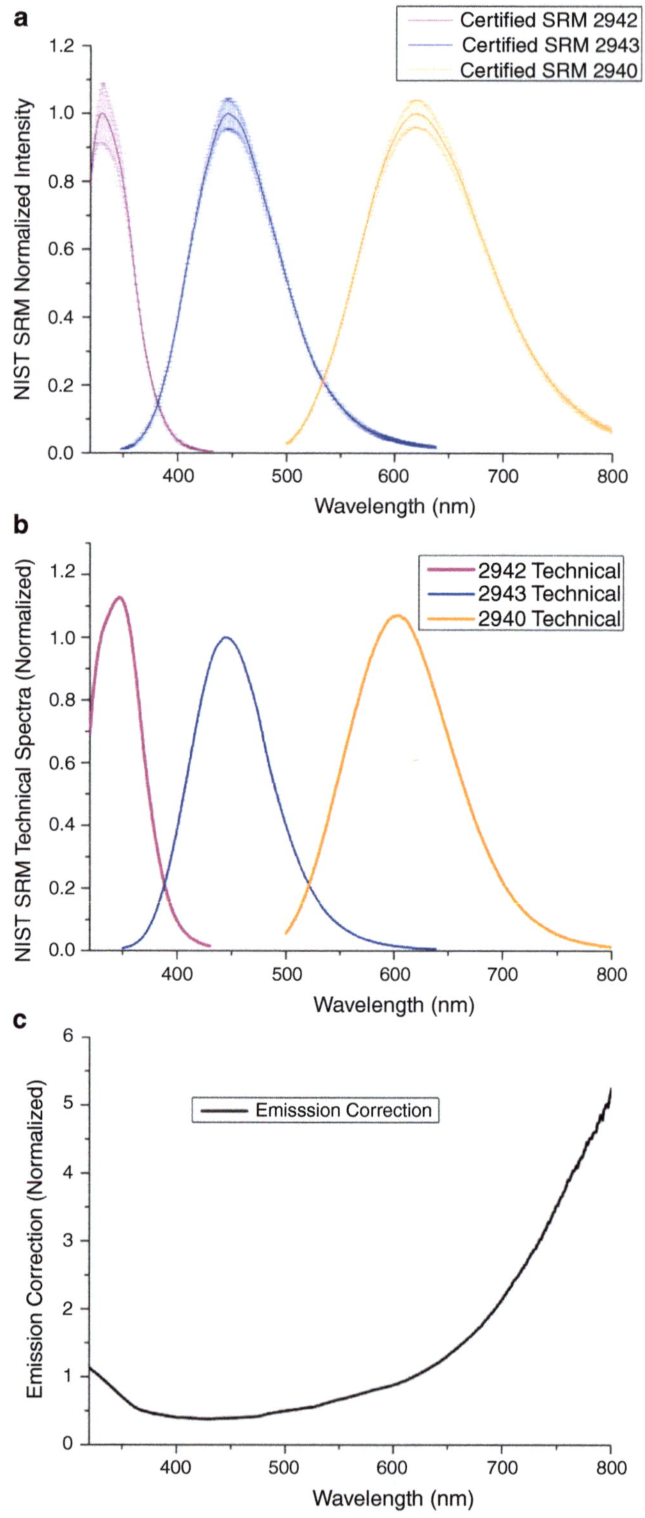

Fig. 8 Panel (**a**) shows the certified NIST SRM spectral data with certified uncertainties (U95) for the SRM 2942 (*magenta*), SRM 2943 (*blue*), and SRM 2940 (*orange*) fluorescence spectral shape standards. Panel (**b**) shows the

spectral peaks within the band-pass limit are 3–5 steps within the band-pass. It is generally advisable for purposes of reducing scan time and sample bleaching/photochemistry to use the maximum band-pass required to resolve the spectral features of interest. Another aspect of the resolution relates to light throughput where for typical bilateral slits the intensity is proportional to the square of the band-pass. Hence, resolution adjustments through slit width attenuation act counter to the throughput intensity.

Importantly a similar limitation as above for scanning monochromators applies to multichannel detectors like diode arrays or CCD detectors when used with imaging spectrographs, as in the Aqualog® example shown in Fig. 2. In this case, one must be concerned with the reciprocal linear dispersion as it limits the "coverage" of the dispersed spectral region imaged across the detectors element array. The center wavelength is determined by the angle of the grating in the spectrograph. Here one should take into consideration, in addition to the band-pass, the detectors elemental density (number of pixels or pixel columns) in the spectral region of interest. Generally the resolution of spectral features within a given band-pass is potentially limited by the pixel density, and it is again recommended 3–5 pixels per band-pass interval be measured at a minimum. In most cases, multichannel detector pixel density can be controlled with electronic or software binning operations, and gain adjustments can be made to optimize the data collection time for a given spectral resolution condition. Following the above spectral resolution issues, another key issue is to avoid detector saturation which obviously can distort spectral data. Most multichannel detectors are limited to 16-bit resolution which means effectively $2^{16} = 65{,}536$ counts can be collected per pixel bin. Here again, the appropriate gain settings should be determined to avoid saturation and minimize the cumulative influence of the dark current and array readout noises on the final spectrum as measured by a multichannel detector.

In the case of PMT detectors, the signal saturation conditions are not as straightforward, and in this chapter, the limitations are described only in terms of single-photon counting (SPC) PMT applications with behavior defined by Poissonian statistics [10]. An SPC PMT exhibits potential signal saturation when the photon

Fig. 8 (continued) measured technical spectra for the same reference materials measured according to the instructions in the NIST certificate for band-pass and excitation and emission wavelength scan conditions. Panel (**c**) shows the emission correction factor calculated using the data in panel (**a**) and (**b**) using an algorithm that first normalizes the certified spectra at their overlapping wavelengths collection rate exceeds the limits of its amplifier-discriminator circuitry. This type of saturation is known as "pulse pileup" and it occurs when the photon collection rate exceeds the "pulse-pair-resolution time" (PPRT) limit. As an example, a typical SPC R928P PMT used in a Horiba Fluorolog® fluorometer has a PPRT of around 12 ns which in practice limits the linearity of the measured photon count rate only up to four million photon counts per second (cps). It is in fact recommended for optimal counting linearity the limit is between one and two million cps. Note the rate of photon collection imposes one limit but the real signal-to-noise ratio for an SPC PMT is in fact determined further by the Poissonian statistics that dictate the standard deviation (σ) of the signal that itself is equal to the square root of the number of counts measured. It follows that for SPC one is advised to choose an integration time sufficient to collect the desired number counts to determine the *SNR*. As an example, if the signal cps is 10,000, then for 1 s integration time, the $\sigma = 10{,}000^{0.5} = 100$ and the $SNR = 10{,}000{:}100 = 100$. Often times it is advisable to first pre-determine the maximum count rate for the sample under chosen excitation-emission conditions as limited by the pulse pileup and then collect data with the integration time as a function of counts until the desired *SNR* value is attained. Typically most SPC instrument software allows one to choose between cps and counts modality for this purpose.

2.5 Combined Excitation and Emission Signal Algebra Correction Considerations and Masking of Scatter Signals

As noted above, the culminated advantage of the spectrofluorimeter compared to an absorbance spectrophotometer is the sensitivity and added selectivity of the fluorescence emission information. Duly noted of course are the fluorimeter's complex spectral response influence, signal saturation issues, and correction requirements. Hence of particular concern here are the beneficial end results of having both the excitation and emission optical elements fully corrected and NIST traceable [2, 3]. The first being that a fully corrected excitation scan (measured as $S1_c/R1_c$) yields, given one is analyzing a pure fluorescent compound at a concentration below the recognized inner-filter effect threshold, the equivalent of the absorbance spectrum for that compound. Likewise a fully correct emission scan for a pure component provides unique molecular-specific information not available from either the absorbance or excitation spectra. Most importantly the emission intensity provides information about the molecular environment of the fluorophore(s). Yet arguably the most powerful and complete steady-state fluorescence sample information comes when one combines the corrected excitation and emission scans with absorbance scan information to generate a fully corrected excitation-emission map (EEM) or matrix [2, 3]. In this manner, one can simultaneously generate the individual excitation and emission spectra for all fluorescent sample components while also providing information on

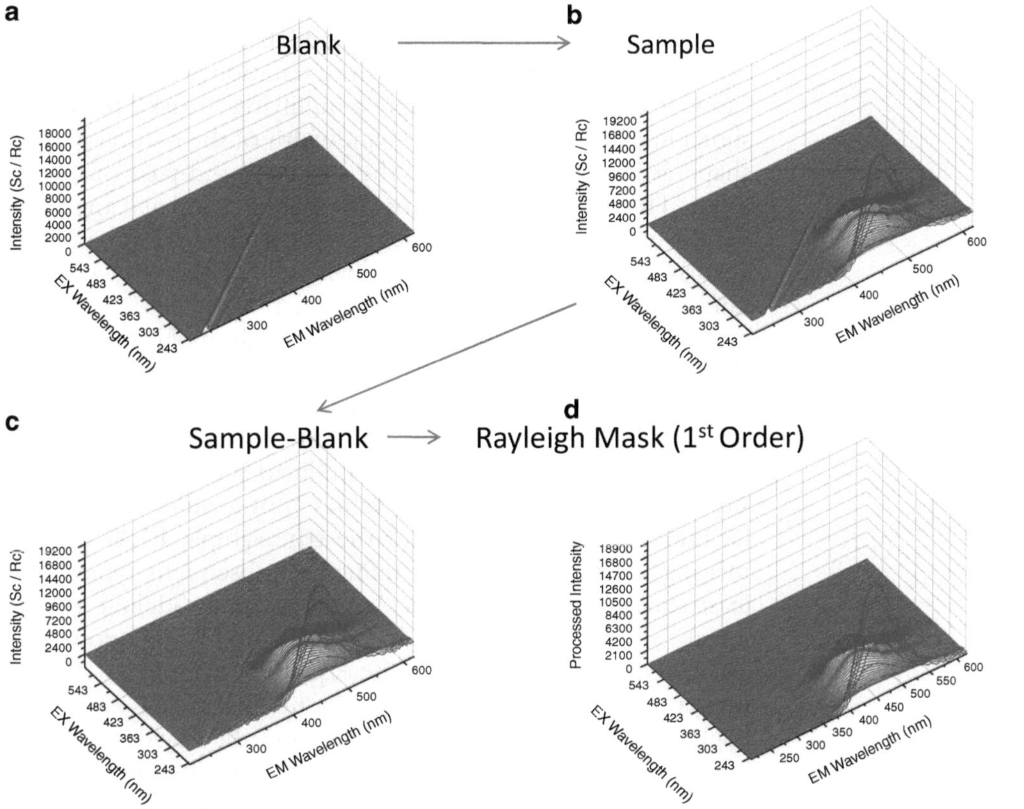

Fig. 9 The sequence of (**a**) blank, (**b**) sample, (**c**) sample-blank, and (**d**) Rayleigh masking operations for a typical excitation-emission map of a sample of 1-ppm quinine sulfate (Starna RM-QS) measured relative to its 0.105-N $HClO_4$ solvent blank cell (**a**)

components that absorbed light but did not fluoresce as well as information needed to correct the fluorescence spectra for the sample concentration-dependent inner-filter effects as described below.

However, preliminary to the inner-filter effect correction, a sequence of "secondary corrections" is required. As an example, Fig. 9 shows a sequence of EEMs collected for the Starna RM-QS kit which includes a blank 0.105-N $HClO_4$ solvent cuvette (a) and a solution of 1 μg/l (1.28×10^{-4} mol/l) quinine sulfate sourced from the NIST SRM 936a material diluted in 0.105-N $HClO_4$ (b). The first recommended part of the sequence involves subtraction of the corrected (SI_c/RI_c) emission signal of the blank EEM (a) (represented as a function of wavelength) from the sample emission EEM (b) to yield (c). The blank subtraction removes any common solvent emission and subtracts the majority of the solvent's Raman spectrum (in this case, the water Raman signals). It is important to note however that in the case of solvent Raman signals, the sample absorbance will lead to a concomitant absorbance of the Raman lines where the absorbance spectrum overlaps the Raman lines' signals which correlate linearly with a 3,382-cm^{-1} energy shift from the excitation wavelength. In such cases, it is advised to also apply a nullifying "mask" to the Raman signal line areas equivalent to bandwidth of the natural Raman line shape as it is also limited by the spectral band-pass of the instrument. This is important because the Raman band will appear, in the blank-subtracted EEM, as a negative or attenuated signal line in the absorbance overlap regions, which will distort the overlapping sample EEM signals after blank subtraction.

It is also customary to apply a Rayleigh masking algorithm for the first- (as shown in d) and/or second-order Rayleigh scattering lines prior to application of the inner-filter effect correction as detailed in the next section. Notably the masked Rayleigh and Raman areas are generally either nullified set (equal to zero intensity) or converted to missing values (not-a-number solutions) to facilitate spectral modeling of the true sample EEM signals; the missing values will thus not influence subsequent typical classical least-squares fitting or other multivariate modeling operations which effectively fit through the missing data regions for the simulated data. In the case of the Aqualog® from HORIBA Instruments Inc., the Rayleigh masking nullification is automatically applied in the instrument's software, and the Rayleigh and Raman mask areas are converted to missing values by the Solo + MIA software package from Eigenvector Research Inc. as part of the file import filter(s) specific to the Aqualog® EEM data files.

2.6 Correction for Sample Concentration-Dependent Inner-Filter Effects

As noted above, fluorescence spectra are subject to significant and predictable distortions caused by concentration-dependent inner-filter effects [1–3, 10–13]. The primary inner-filter effect (PIF) as illustrated graphically in Fig. 10 occurs when excitation light is absorbed by components in the sample before it reaches the area of intersection of the excitation and emission beam paths where fluorescence can be excited and hence detected. The secondary inner-filter effect (SIF) occurs as shown (Fig. 10) when the fluorescence excited in the intersected regions of the excitation and emission beam paths is absorbed by sample components before it can exit the cell to reach the emission detector. Both SIF and PIF can be caused by sample components that absorb light independent of whether these components fluoresce. The concentration dependence of the PIF and SIFs has been extensively studied as have their dependence on the geometry of the measuring cell and light beam paths. In general it is recognized that when the optical density of the solution in the fluorescence excitation and emission region is below 0.05, the inner-filter effects are superfluous; note, however, these effects are always present at any OD value in the overlapping absorbance and fluorescence emission regions.

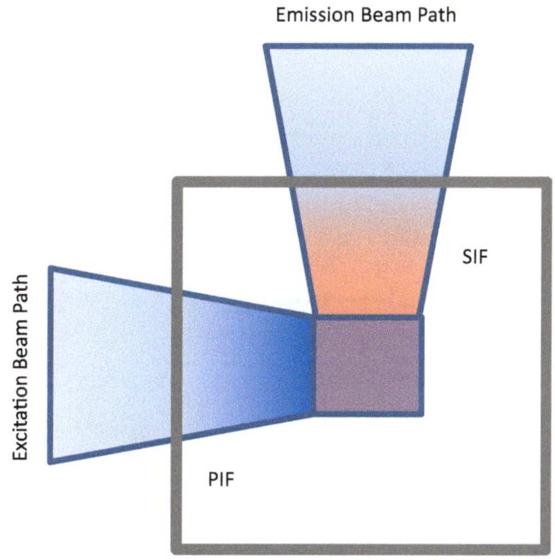

Fig. 10 A schematic illustrating the primary and secondary inner-filter effects (PIF and SIF) relative to the excitation and emission optical beam paths in a hypothetical 1-cm pathlength cell. The view is meant to represent the top view of the cross section of the meridian of the beam paths, i.e., the optical center of the cuvette

The literature cites many studies of the inner-filter effect corrections, all of which relate to formulating and testing equations dealing with the optical extinction coefficients, cell path lengths, cell positions, and beam geometry as the latter are defined by the focal properties (f-numbers) of the excitation and emission beam paths. One of the most commonly applied equation systems applied to the data below in Fig. 11 was derived by MacDonald et al. in ref. [13]. In practice one normally measures the absorbance spectrum of the sample where this overlaps the fluorescence excitation and/or emission spectral regions of interest. Ideally this is done simultaneously with the fluorescence measurements in the same cell and instrument to minimize differences in temperature, photobleaching, and any other chemistry or photochemistry occurring in the sample or associated with the instrument or sample cells that may quench or change the fluorescence spectrum or intensity when comparing the absorbance and fluorescence data. One key factor for inner-filter effect corrections is that they are in practice only at most as accurate as the absorbance data with respect to concentration dependence. Hence it is recommended practice and the user's responsibility to evaluate the linearity of the absorbance versus concentration of the sample system to ascertain the Beer-Lambert linearity requirement [4]. If the concentration is beyond linearity (high OD, usually >2.0), then the fluorescence inner-filter correction will reflect this inaccuracy relating to concentration-dependent

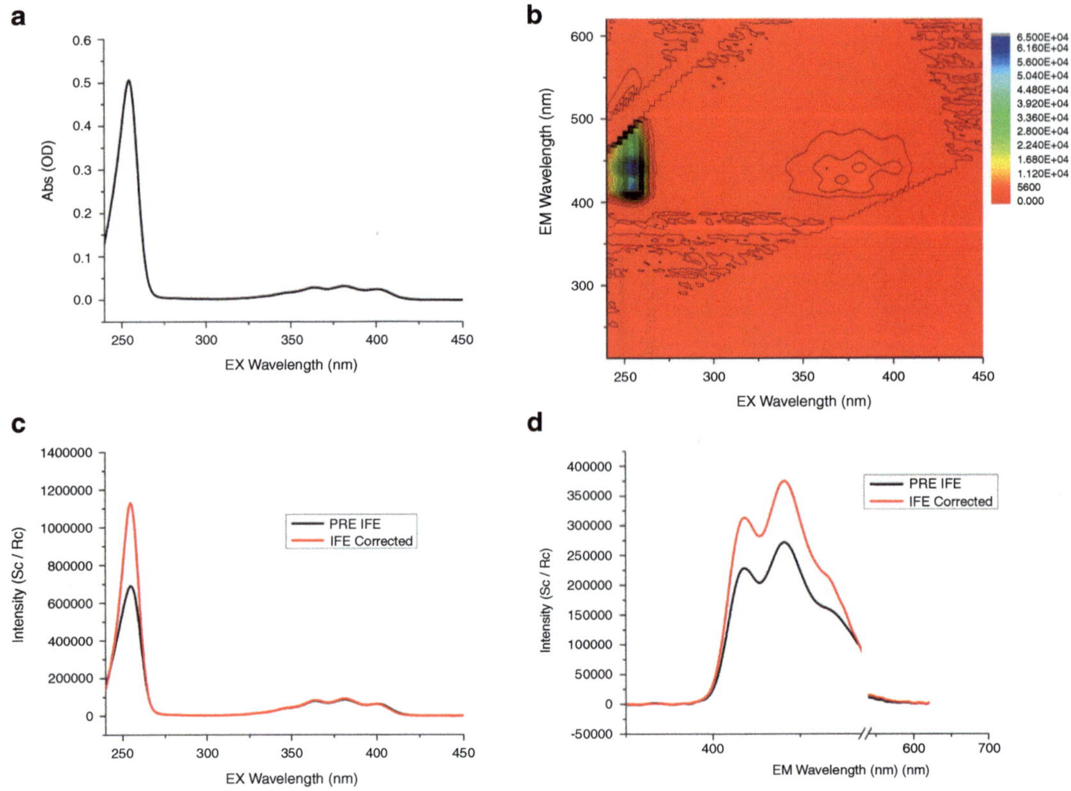

Fig. 11 The effects of inner-filter effect correction on a typical ethanolic solution of 9-cyanoanthracene. Panel (**a**) shows the absorbance spectrum that was measured simultaneous to the excitation-emission matrix data in panel (**b**). Note the data in panel (**b**) were corrected for both PIF and SIF using the absorbance data in panel (**a**). Panels (**b**) and (**c**), respectively, show the integral profiles of the excitation and emission axes from panel (**a**) (IFE corrected) and from the same data before IFE (PRE-IFE)

intensity. In conclusion, it must be emphasized that user evaluation of the linearity of the NIST traceable absorbance and fluorescence data as a function of sample concentration or systematic dilution is a key user responsibility especially to ascertain the accuracy of the inner-effect algorithm employed.

Figure 11 illustrates simultaneously acquired absorbance (a) and EEM data (b) collected with an ethanolic solution of 9-cyanoanthracene using an Aqualog®. The absorbance is quite strong in the deeper UV absorbance band region as indicated in the spectrum (a) and corrected (blank subtracted, Rayleigh masked (first and second order), and inner-filter effect corrected) EEM contour plot (b). Panels c and d show the significance of the inner-filter effect correction effect by comparing the integrated EEM data profiles from the excitation axis (c) and emission axis (d) before (black) and after (red) IFE correction. Clearly the IFE correction restores the correct excitation and emission profiles corresponding to the expected profiles under so-called infinite dilution conditions. Especially clear was the strong PIF influence on the excitation spectrum associated with the stronger absorbance in the deeper UV absorbance band.

2.7 Conclusion: Example Spectral Modeling and Component Validation for NIST Traceable Intra- and Interlaboratory Comparison

The end game for the fluorescence spectral corrections procedures described above is to provide NIST traceable spectral data that can be used for valid intra- and interlaboratory comparisons. Most notably, the EEM methods and spectral correction protocols explained above can be used for effective standard compound calibration and mixture decomposition using a number of modeling tools including classical least-squares regression (CLS) and parallel factor analysis (PARAFAC) methods inter alia. In this chapter example, we present a data case illustrating the ability to calibrate a model with known pure standard compounds using CLS and then to apply the model to effectively quantify the components in an unknown mixture. The CLS protocol is operated within the Solo + MIA software package from Eigenvector Research Inc. The protocol as it relates to the data in Fig. 12 includes importing the EEM data sets from the pure compounds, 9-cyanoanthracene (9-CA, Panels a and b), anthracene (ANTH, Panels c and d), and POPOP (Panels e and f) into the calibration model field and then reshaping the 3D EEM data into a 2D array. Next the concentration data for each member of the calibration set is entered, and the calibration model is fit to the pure compounds. Then the unknown mixture (MIX, Panels g and h) data set is entered into the validation entry field and tested against the calibration model assuming some unknown composition of the pure compound. Figure 13 shows the results of the calibration and validation model predictions in terms of their relative concentration loadings for the standards (black symbols) compared to the unknown mixture (red symbols). The calibrated model described the unknown validation model with >99.99 % of the variance accounted for by the three components. Note similar results can also be obtained with 3-way PARAFAC calibration and validation modeling (also available in the Solo + MIA package from Eigenvector Research Inc.), although the CLS modeling routine is better suited for concentration-dependent calibration and validation with these types of samples [13–16].

The most significant feature of the model success illustrated in Fig. 13 is clearly that the EEM spectra correction methods described above facilitate resolution of the exact spectral components in the mixture despite their highly overlapping absorbance and emission spectra. Again the nontriviality of this separation should not be underemphasized since as shown in Fig. 11 the absorbances of 9-cyanoanthracene and anthracene are very strong and overlap significantly in the analyzed region of interest. Thus in conclusion it is clear that the NIST traceable spectral calibration and correction methods described here can generate corrected spectra from pure samples and/or complex mixtures that are well suited for validating intra- and interlaboratory fluorescence studies.

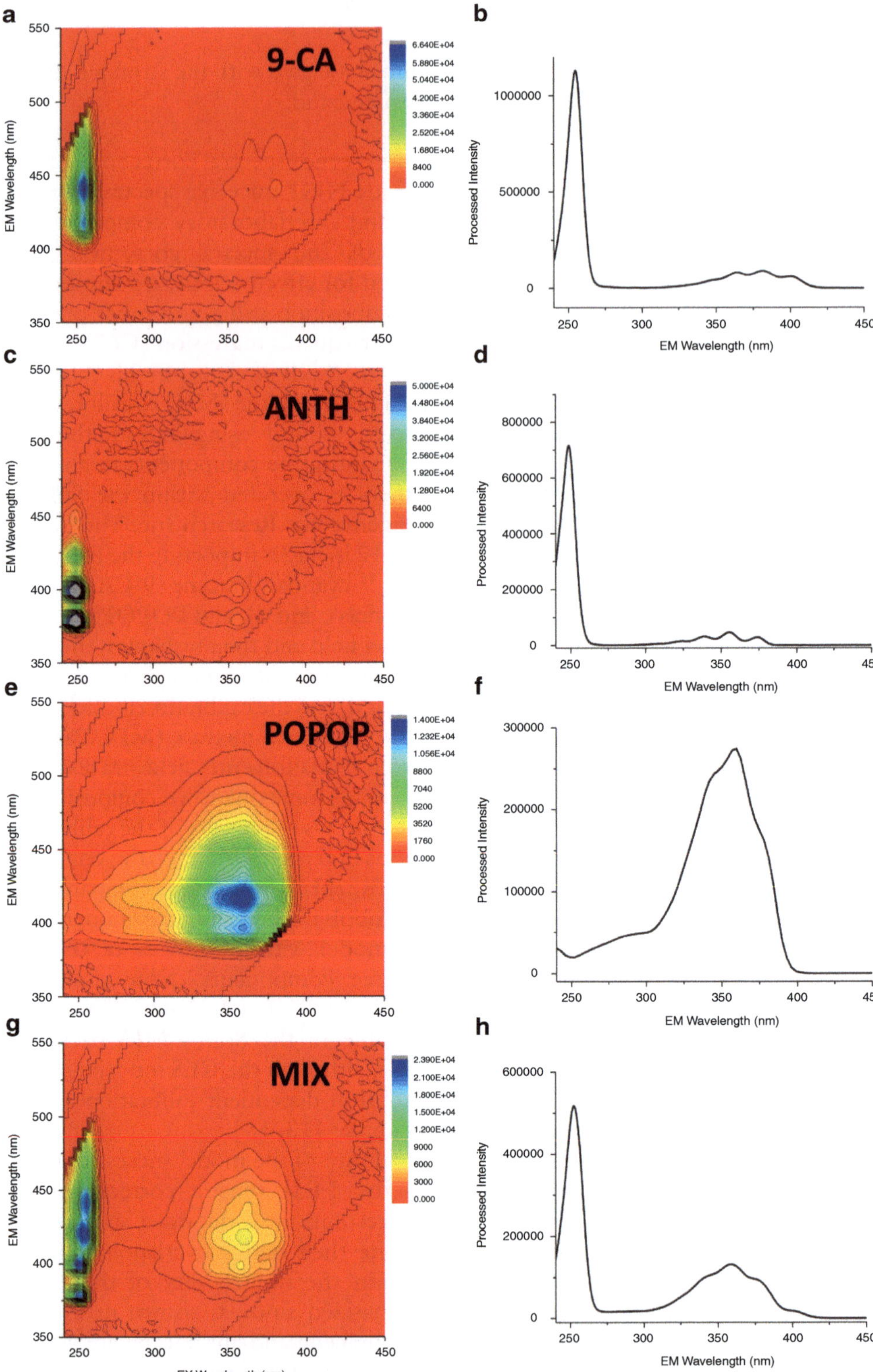

Fig. 12 EEM (*left column*) and integrated excitation profile data (*right column*) from a set of ethanolic solutions representing three pure calibration samples, namely, 9-cyanoanthracene (9-CA, panels **a** and **b**), anthracene (ANTH, panels **c** and **d**), and POPOP (POPOP, panels **e** and **f**) and a test mixture of the three components (MIX, panels **g** and **h**). All four sample EEMs were corrected using the protocols defined for the 9-CA sample shown in Fig. 11 panels (**a**) and (**b**)

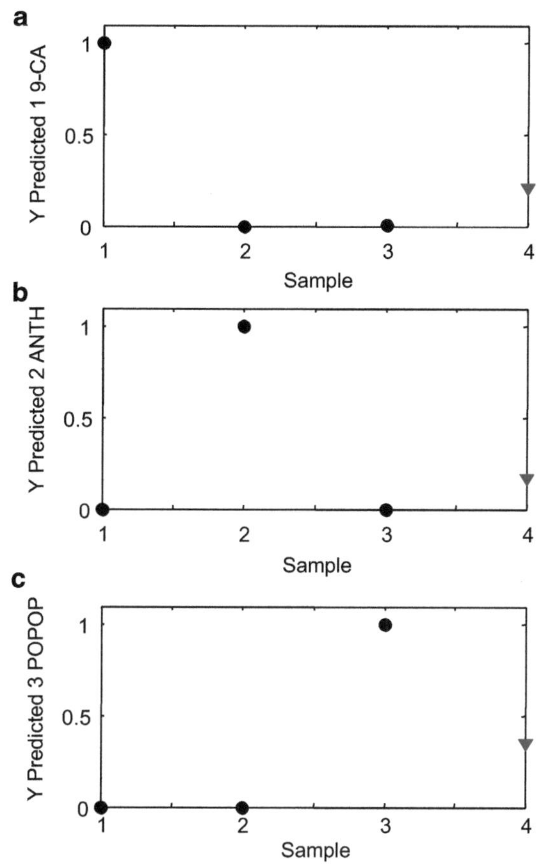

Fig. 13 The results of a classical least-squares spectral analysis using the Solo + MIA package from Eigenvector Research Inc. for the three calibration samples (*black circles*) and test mixture (*triangles*) from the EEM data shown in Fig. 12. Each panel refers to the four samples on the *x*-axis with the concentration loadings for the calibration samples (1, 2, and 3) noted on the *y*-axis and the loading of the respective component in mixture on the *right side* (sample 4). Panel (**a**) represents the 9-CA, panel (**b**) represents ANTH, and panel (**c**) represents POPOP. The model mixture model fit accounted for >99.99 % of the model variance with 9-CA accounting for 56.39 %, ANTH accounting for 31.75 %, and POPOP representing 11.85 % of the total model variance

3 Notes

1. It is not uncommon after a computer crash or power failure during data collection or inadvertent miscalibration of the excitation monochromator (by erroneously adjusting the emission monochromator settings instead) to lose track of the landmark peaks of the xenon lamp under the normal scan conditions. It is thus advised to make a complete survey scan of the excitation monochromator range to help identify the

zero-order and first-order emission bands and peaks. Noting there is usually an area of a few hundred nanometers between the zero-order peak and the first-order emission band with no detectable signal due to the low grating efficiency and lack of emission from the lamp itself (*see* Fig. 3).

2. As explained in **Note 1**, commonly the emission monochromator can also lose calibration well outside the recognized landmarks of the water Raman lines and reference sample spectra. In many cases as with the excitation monochromator, it can be helpful to make a full emission range of the water Raman sample under the conditions described above except scanning from 0 to 800 nm to identify the following landmark peak structures: first the zero-order peak, second the first-order Rayleigh peak (which should be 350 nm up from the zero-order peak), and third the water Raman peak (which should be 47 nm up from the first-order Rayleigh peak). It follows that sometimes the second-order peak positions are also valuable to verify that the monochromator has not been calibrated to second order; thus, the second-order Rayleigh will be 700 nm from the zero-order and 350 nm from the first-order Rayleigh and the second-order Raman will be 94 nm from the second-order Rayleigh.
3. The FSD water Raman SNR is particularly susceptible to two sources of noise associated with sample preparation. The first being the ubiquitous organic matter contamination common to aged water samples and/or water that has not been treated to remove the organic matter initially. The organic matter results in a broad fluorescence spectral signature peaking under the 350-nm excitation in the 400–450-nm emission region. The second major source or noise is from scattered excitation light caused by particulates in the sample, or scratches or fingerprints, etc. on the cuvette surfaces; this scatter usually elicits as a broad sloping signal superimposed on the water Raman spectrum with a power function roughly equivalent to the wavelength of energy to the fourth power, thus being strongest in the UV region.
4. It is not advised to place any filter capable of altering the spectrum received by the reference detector between the detector and the beam splitter since this alters the recorded spectrum of light from that impinging on the sample position leading to anomalous excitation spectra.
5. When using neutral density filters for signal attenuation (to avoid detector saturation), careful placement of the reflective side(s) is advised to ensure proper reproducibility of the attenuation factor. Likewise careful selection of UV transmitting (e.g., fused quartz) filters is advised.
6. The use of "low-fluorescence" order-sorting filters is strongly advised.

References

1. Holbrook RD, DeRose PC, Leigh SD et al (2006) Excitation–emission matrix fluorescence spectroscopy for natural organic matter characterization: a quantitative evaluation of calibration and spectral correction procedures. Appl Spectrosc 60:791–799
2. DeRose PC, Early EA, Kramer GW (2007) Qualification of a fluorescence spectrometer for measuring true fluorescence spectra. Rev Sci Instrum 78:033107–033112
3. DeRose PC, Resch-Genger U (2010) Recommendations for fluorescence instrument qualification: the new ASTM standard guide. Anal Chem 82:2129–2133
4. Gilmore AM (2011) Water quality measurements with the Aqualog®. Spectroscopyonline.com http://reg.accelacomm.com/servlet/Frs.FrsGetContent?id=50198696
5. Larason TC, Bruce SS, Parr AC (1998) Spectroradiometric detector measurements: part I—ultraviolet detectors and part II—visible to near-infrared detectors. NIST Special Publication 250-41, Washington, DC
6. Taylor DG, Demas JN (1979) Light intensity measurements I: large area bolometers with microwatt sensitivities and absolute calibration of the Rhodamine B quantum counter. Anal Chem 51:712–722
7. DeRose PC, Smith MV, Mielenz KD et al (2008) Characterization of standard reference material 2941, uranyl-ion-doped glass, spectral correction standard for fluorescence. J Lumin 128:257–266
8. DeRose PC, Smith MV, Mielenz KD et al (2009) Characterization of standard reference material 2940, Mn-ion-doped glass, spectral correction standard for fluorescence. J Lumin 129:349–355
9. DeRose PC, Smith MV, Mielenz KD et al (2011) Characterization of standard reference material 2943, Cu-ion-doped glass, spectral correction standard for blue fluorescence. J Lumin 131:2509–2514
10. Lakowicz JR (2006) Principles of fluorescence spectroscopy, 3rd edn. Springer Science and Business Media, LLC, New York
11. Gu Q, Kenny JE (2009) Improvement of inner filter effect correction based on determination of effective geometric parameters using a conventional fluorimeter. Anal Chem 81: 420–426
12. Larsson T, Wedborg M, Turner D (2007) Correction of inner-filter effect in fluorescence excitation-emission matrix spectrometry using Raman scatter. Anal Chim Acta 583:357–363
13. MacDonald BC, Lvin SJ, Patterson H (1997) Correction of fluorescence inner filter effects and the partitioning of pyrene to dissolved organic carbon. Anal Chim Acta 338:155–162
14. Bro R (1997) PARAFAC. Tutorial and applications. Chemom Intell Lab Syst 38:149–171
15. Bro R (2006) Review on multiway analysis in chemistry—2000–2005. Crit Rev Anal Chem 36:279–293
16. Ravna C, Skibsted E, Bro R (2008) Near-infrared chemical imaging (NIR-CI) on pharmaceutical solid dosage forms—comparing common calibration approaches. J Pharm Biomed Anal 48:554–561

Chapter 2

Steady-State Fluorescence Polarization/Anisotropy for the Study of Protein Interactions

Nicholas G. James and David M. Jameson

Abstract

Fluorescence methods are often employed for the characterization of molecular interactions. In particular, polarization/anisotropy studies are widely utilized in the life sciences as they allow quantification of protein interactions in the micro- and nanomolar concentration range. Herein we shall briefly describe the theoretical aspects of polarization/anisotropy and outline an experiment for determination of the dissociation constant for a protein–ligand complex.

Key words Polarization, Anisotropy, Protein interaction, Protein–ligand binding

1 Introduction

Fluorescence methodologies have been used in the biological and biochemical fields since the first half of the last century [1, 2]. Fluorescence studies offer unique advantages over traditional biochemical approaches to ligand–protein interactions, such as column chromatography or radiolabeling, by providing information down to subnanomolar concentrations. The characterization and understanding of the forces driving these interactions are of critical importance since biological specificity, essential to life, is largely based on protein–ligand interactions. In many cases, the intrinsic protein fluorescence, largely due to tryptophan and to a lesser extent tyrosine residues [3], can be used to monitor the protein interaction. In other cases, an extrinsic fluorescence probe can be attached to the protein or ligand. Although direct intensity measurements can be valuable for detecting protein interactions, in some cases, the fluorescence intensity change upon ligand binding is negligible [4, 5]. In these cases, protein interactions can often be monitored using polarization, which relies on the differences in rotational mobility of bound vs. free ligand; hence a change in fluorescence yield or spectral shift is not required. Polarization studies have

Yves Engelborghs and Antonie J.W.G. Visser (eds.), *Fluorescence Spectroscopy and Microscopy: Methods and Protocols*, Methods in Molecular Biology, vol. 1076, DOI 10.1007/978-1-62703-649-8_2,

been used by biochemists for studying protein interactions for more than 50 years as the theory had been thoroughly developed and tested by Francis Perrin and Gregorio Weber [6, 7]. However, the introduction of the TDx polarization instrument for fluorescence immunoassays by Abbott Laboratories in the early 1980s [8] as well as the development of polarization plate readers soon thereafter significantly increased the use of this method because it introduced a broader community to the rapid screening of a large number of protein–ligand combinations. Polarization has since become one of the most prominent methods for investigating protein interaction as polarization measurements are done at equilibrium (no need to separate bound and free) and can be easily automated [9].

Herein we demonstrate how to use fluorescence polarization to determine the fraction of bound and free ligand when monitoring protein–ligand interaction. We start with a large excess (relative to the K_d) of the species (protein or ligand) that is not being monitored, such that all of the fluorescent ligand or protein is bound. Successive dilution of the sample not being monitored (e.g., ligand), while maintaining the concentration of the monitored substance (e.g., protein), provides a complete binding isotherm. In case the emission intensity changes during the outlined experiment, this change is taken into account, as outlined in the theory section, when determining the fraction bound. We note that in the case of significant intensity changes, there is no need to measure fluorescence polarization/anisotropy to follow the binding. But we also note that intensity is an "extensive property," i.e., it depends on the amount of material present, while polarization/anisotropy is an "intensive property," which is independent of the amount of material (the fluorescence lifetime is also an intensive parameter). There are often advantages to the use of intensive parameters, for example, the fact that they are independent of the instrument platform being used for the measurements. From the plot of fraction bound and protein concentration, we quantify the dissociation constant (K_d) for this protein–ligand pair. An example, from the literature, is also provided in which the polarization change of the ligand is monitored, as well as an example determining the strength of protein–protein interaction [10, 11]. A similar procedure, along with a rigorous treatment of attendant errors, has been described previously [12] and has been used to determine dissociation constants in a number of reports [13, 14].

1.1 Theory

For a molecule to fluoresce, an energy source capable of promoting the molecule to an excited state must be utilized. The most common way of achieving these specific energies is through a xenon arc lamp and a monochromator, which allows for the selection of wavelengths in the range of 200–1,000 nm. To achieve polarized emission from the sample, a specific electric vector of the excitation light must be isolated as the excitation source typically has electronic radiation aligned in all directions within the plane perpendicular to the propagation of the light. This isolation is traditionally done by placing a polarizer orientated parallel to the laboratory axis just after the light source. Such polarization of light allows for the specific excitation of fluorophores that have excitation vectors aligned properly with the plane-polarized excitation light. This phenomenon is termed photoselection. The polarization (P) is then determined by measuring emission through a polarizer parallel ($I_{||}$) and perpendicular ($I_{\perp}$) to the excitation light:

$$P = \left(\frac{I_{||} - I_{\perp}}{I_{||} + I_{\perp}}\right) \tag{1}$$

Another frequently used expression, in the context of polarization, is anisotropy and is defined as:

$$r = \left(\frac{I_{||} - I_{\perp}}{I_{||} + 2I_{\perp}}\right) \tag{2}$$

The information content from polarization and anisotropy is essentially identical [15] and can be interconverted using the expression:

$$r = \frac{2P}{(3 - P)} \tag{3}$$

Some researchers fervently adhere to the use of the polarization function, while others cling just as fervently to the anisotropy function. We simply note that anisotropy is probably the more popular function in biophysical studies, while polarization is almost universally used in clinical chemistry application. The limits of polarization, in a completely orientated system like a crystal, are +1 or −1 (+1 or −0.5 for anisotropy) as defined by the equations above. However, in the case of a solution of fluorophores, these theoretical limits are not reached. This partial depolarization is due to the random orientation of the electric vector of each molecule in an ensemble of randomly oriented fluorophores. The limits for these cases have been derived, based on photoselection rules, as +0.5 and −0.33 for polarization or +0.4 and −0.2 for anisotropy [7]. A number of factors contribute to the observed polarization of a fluorophore. One of the factors, which is a measure of polarization of the fluorophore in the absence of rotation, is termed the limiting polarization (P_0) and is determined:

$$\frac{1}{P_0} - \frac{1}{3} = \frac{5}{3}\left(\frac{2}{3\cos^2\theta - 1}\right) \tag{4}$$

where θ is the angle between absorption and emission oscillator.

Polarization measures the rotational mobility of the fluorescent molecule, which makes it more appealing for monitoring interactions than intensity measurements. In 1925–1926, Francis Perrin derived an expression for obtaining the polarization based on the rotational diffusion and excited state lifetime [6]:

$$\frac{1}{P} - \frac{1}{3} = \left(\frac{1}{P_0} - \frac{1}{3}\right)\left(1 + \left(\frac{RT}{\eta V}\right)\tau\right) \tag{5}$$

where P is the polarization observed, P_0 is the limiting polarization of the fluorophore, R is the gas constant, T is the absolute temperature, V is the molar volume of the fluorescent molecule, η is the solvent viscosity, and τ is the excited state lifetime. This equation is often rewritten to contain the Debye rotational relaxation time (ρ):

$$\frac{1}{P} - \frac{1}{3} = \left(\frac{1}{P_0} - \frac{1}{3}\right)\left(1 + \frac{3\tau}{\rho}\right) \tag{6}$$

The major advantage of polarization, compared to the other methods, is its ability to quantify the extent of bound ligand in a sample without recourse to a separation step, as, for example, is required with radioisotope labeling. This mathematical resolution of free from bound is accomplished through the additivity of polarization of different species, which was derived by Gregorio Weber in 1952 [7]:

$$\left(\frac{1}{P_{\text{obs}}} - \frac{1}{3}\right)^{-1} = \sum f_i \left(\frac{1}{P_i} - \frac{1}{3}\right)^{-1} \tag{7}$$

where P_{obs} is the actual observed polarization coming from all the components, f_i is the fractional contribution of the ith component to the photocurrent, and P_i is the polarization of the ith component. Additivity for anisotropy can also be expressed as:

$$r_{\text{obs}} = \sum f_i r_i \tag{8}$$

As can be deduced from the previous equations, a change in fractional contribution to the photocurrent of the bound or free fluorophore during the experiment will alter their contributions in the additivity equation. This bias can be corrected, when solving for the fraction bound (f_b), by taking into account the change in intensity of the bound fluorescent ligand, termed g, namely,

$$f_b = \frac{(3 - P_b)(P_{\text{obs}} - P_f)}{(3 - P_{\text{obs}})(P_b - P_f) + (g - 1)(3 - P_f)(P_b - P_{\text{obs}})} \tag{9}$$

$$f_b = \frac{(r_{\text{obs}} - r_f)}{(r_b - r_f) + (g - 1)(r_b - r_{\text{obs}})} \tag{10}$$

where P_{obs} is the observed polarization, P_f is the polarization of the free ligand, and P_b is the polarization of the bound ligand. An expression for anisotropy, which takes into account changes in intensity (g), is also shown above where r_{obs} is the observed anisotropy, r_f is the anisotropy of the free ligand, and r_b of the bound one.

The dissociation constant corresponding to a reversible equilibrium between a protein (P), a ligand (L), and a protein–ligand complex (PL) is given by:

$$K_d = \frac{[P][L]}{[PL]} \tag{11}$$

Determination of this dissociation constant requires determination of the concentration of bound ligand (PL) as a function of the total protein (P_T) and ligand (L_T) concentrations:

$$[PL] = \frac{(K_d + L_T + P_T) - \sqrt{\left[(K_d + L_T + P_T)^2 - 4P_TL_T\right]}}{2} \tag{12}$$

Dissociation constants for the protein–ligand system can then be determined by fitting a plot of f_b vs. concentration to the following equation:

$$f_b = \frac{[\text{Protein}]}{[\text{Protein}] + K_d} \tag{13}$$

2 Materials

All solutions were made up using ultrapure water (specific resistance 17.5 MΩ) collected from a Barnstead Nanopure II at 25 °C and, when required, filtered through a 0.22 μm filter from Millipore. Buffer and protein stocks were stored at 4 °C. Stock solutions of furosemide were stored at room temperature:

1. Monomeric HSA (mHSA): ~1 mM stock solution isolated from lyophilized, 99 % agarose gel electrophoresis, HSA from Sigma (St. Louis, MO, USA). *See* Subheading 3.1 for purification procedure. (We note that lyophilized HSA typically contains about 10–15 % of a covalent dimer and higher oligomers due to exchange of disulfide cross-links within the protein population [16].)
2. Furosemide: 850 μM (0.28 g/L) stock solution was made in 0.05 M HEPES, pH 7.0 (Sigma, St. Louis, MO, USA).
3. Sephacryl S300 buffer: 0.05 M HEPES, pH 7.5, and 0.05 M NaCl. Weigh out 11.9 g of HEPES powder and 2.9 g NaCl. Dissolve in ~900 mL. Adjust the pH to 7.5 with 2 M NaOH. Add ~100 mL of water to make 1 L. Filter through a 0.22 μm and store at 4 °C.
4. Sephacryl S300 column: Contained in a Pharmacia XK16 column with ~120 mL of S300 resin. Column stored in S300 buffer or, for long-term storage, in 70 % ethanol.

5. Amicon Ultra centrifugal filters: Ultracel—10 kDa MWCO from Millipore (Bellirica, MA, USA).
6. Phosphate buffer (PB): 0.05 M phosphate buffer, pH 7.0. Dissolve 0.17 g of potassium phosphate, monobasic, (=KH_2PO_4) and 0.22 g of potassium phosphate, diabasic, (=K_2HPO_4) in 50 mL of water. The pH should be near 7; however, small adjustments with 1 N HCl were done to place it exactly at 7.
7. Fluorescent cuvettes: Suprasil quartz, 10 × 4 mm.

3 Methods

Carry out all experiments at room temperature.

3.1 Isolation of Pure Monomeric HSA from a Lyophilized Source

1. Start by equilibrating the Sephacryl S300 column, which is typically stored in 17 % ethanol, with 5 column volumes of S300 buffer. Our column is ~120 mL and was equilibrated by running 600 mL of S300 buffer through.
2. Dissolve ~50 mg of lyophilized HSA into 1 mL of Sephacryl S300 buffer. Run this solution over the S300 column (*see* **Note 1**).
3. Collect the fractions and record the absorbance at 280 nm (the main protein absorbance wavelength). The absorbance values for the fractions collected from our S300, run at 1 mL/min with 600 μL fractions, are shown in Fig. 1a. Run out fractions around the main peak on a native gel, Fig. 1b, to isolate the fractions containing pure mHSA (*see* **Note 2**).

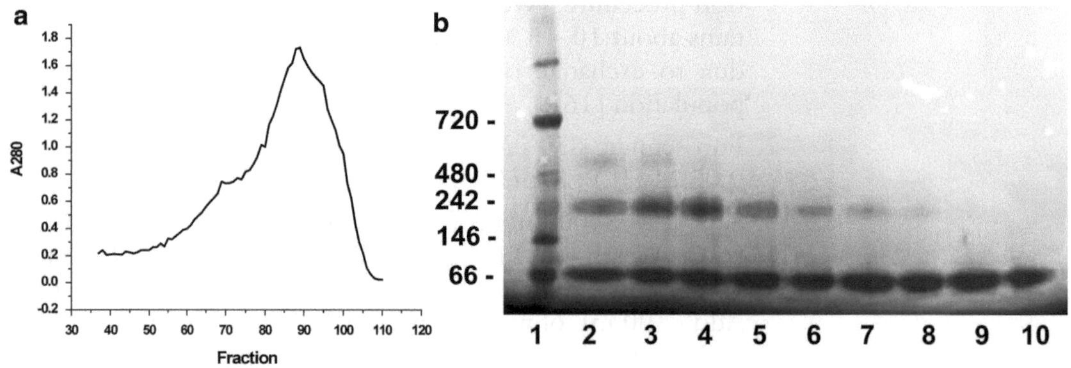

Fig. 1 Isolation of monomeric HSA from a lyophilized source. (**a**) Protein absorbance (A_{280}) of 600 μL fractions collected off a Sephacryl S300 column. (**b**) A native, non-denaturing gel of fractions 59 (*Lane 2*), 65 (*Lane 3*), 71 (*Lane 4*), 77 (*Lane 5*), 80 (*Lane 6*), 84 (*Lane 7*), 89 (*Lane 8*), 94 (*Lane 9*), and 99 (*Lane 10*). Around 3 μg, determined by the A_{280}, of each fraction was loaded. *Lane 1* contained 5 μL of NativeMark (Invitrogen)

4. Pool the fractions containing mHSA and concentrate them in an Ultracel 10 kDa. Once all the fractions have been added and concentrated, exchange the buffer to PB by adding in 4 mL and concentrating the sample. This step needs to be done at least three times. In our case, the stock solution of mHSA was ~1 mM in PB. This monomeric sample is used throughout these experiments.

3.2 Polarization of mHSA During Furosemide Binding

1. Make a 1 μM solution of mHSA in a fluorescence cuvette by diluting a small amount of the concentrated mHSA with PB (near 800 μL). The concentration can be known precisely by measuring the absorbance at 280 nm and using the absorption coefficient of 35,121 M^{-1} cm^{-1} [17].
2. For these experiments, we will be varying the amount of ligand, furosemide, while maintaining a constant concentration of mHSA (*see* **Note 3**). This procedure is done by diluting the excess furosemide with volumes of 1 μM mHSA. To keep things identical throughout the experiment, a 10 mL solution of 1 μM mHSA is made in PB by adding 0.010 mL of the stock mHSA to 9.990 mL of PB.
3. Place the cuvette into the fluorimeter, with polarizers on the excitation and emission side, and measure the polarization of mHSA using 300 nm excitation (*see* **Note 4**). This polarization value is P_f (Eq. 9) and the intensity will be used to calculate g. The emission can be collected either through a monochromator or directly through a filter (*see* **Note 5**).
4. Make a solution containing ~30 μM of furosemide by adding enough ligand from the stock solution to the cuvette and ensure that the final volume is 800 μL (*see* **Note 6**). Measure the polarization and use this value for P_b, assuming all mHSA is bound (Eq. 9). The g value for Eq. 9 can now be determined by dividing the intensity of the bound mHSA (~45,000 counts) by the free mHSA (~420,000 counts). This gives a g value of 0.11.
5. Dilute the sample to 20 μM furosemide by removing 265 μL from the cuvette and adding back 265 μL of the 1 μM solution of mHSA and measure the polarization. Repeat this step to 15, 10, 7.5, 5, 2.5, 1.25, 0.625, 0.530, 0.430, and 0.215 μM.
6. Figure 2a shows the plot of polarization vs. total ligand concentration (M), on a log scale.
7. Using the polarization values and the g value from above, the fraction of bound ligand at each concentration can be determined from Eq. 9. The plot of f_b vs. total ligand concentration (M) is shown in Fig. 2b, along with the best fit using Eq. 13. This fit gives a K_d of 0.46 ± 0.03 μM for furosemide binding to mHSA (*see* **Note 7**).

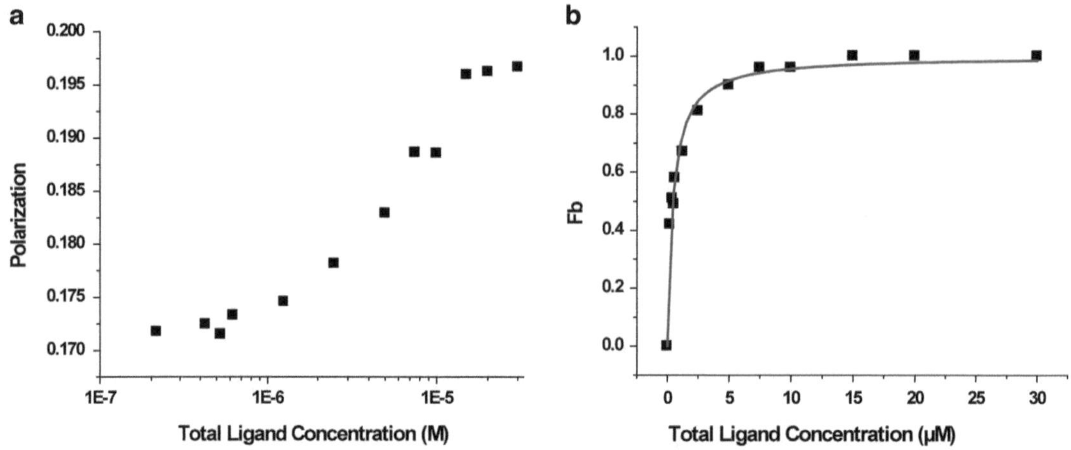

Fig. 2 Steady-state polarization of the intrinsic tryptophan of mHSA during furosemide binding. (**a**) Polarization values of mHSA at varying concentrations of furosemide, plotted on a logarithmic scale, while maintaining 1 μM mHSA. (**b**) Plot of fraction bound furosemide, corrected for changes in intensity, at each concentration of furosemide. Line is the best-fit binding isotherm, Eq. 13, with a K_d of 0.4 μM

3.3 Monitoring the Polarization Change of a Fluorescent Ligand for Determination of the K_d Between Protein–Ligand Interactions

1. In this example, which has been published previously [10], the change in polarization of the fluorescent ligand, mant-GTPγS, binding to a large protein, dynamin 2 (Dyn2), is followed.
2. A stock solution of mant-GTPγS (a fluorescent, slowly hydrolyzable GTP analog) was made in 20 mM HEPES, pH 7.4, 300 mM NaCl, 5 mM $MgCl_2$, 1 mM EDTA, 0.2 mM PMSF, and 0.5 mM DTT [18], and the concentration was adjusted to 1 μM by measuring the absorbance at 354 nm and using the absorption coefficient of 5,700 M^{-1} cm^{-1} [19].
3. The anisotropy of the 1 μM solution was determined by excitation at 354 nm and measuring the polarized emission through a 430 nm long-pass filter.
4. Anisotropy measurements were initially made with 1 μM mant-GTPγS in solution with the GTP-binding protein Dyn2 at 38 μM. The Dyn2 concentration was lowered to ~35 μM by removing a fraction of the sample in the cuvette and adding back an identical volume of 1 μM mant-GTPγS. This step was repeated until the concentration of Dyn2 was 2.5 μM.
5. The plot of anisotropy vs. protein concentration (μM) is shown in Fig. 3 along with the best fit to the anisotropy function (Eq. 11).

3.4 Protein Dimer–Monomer Dissociation Constant Determined by Steady-State Polarization

1. One can also study the strength of association of a protein that forms higher-order oligomers in vitro by polarization. In this case, the ligand bound is the unlabeled protein.
2. We will describe the procedure outlined by Ross et al. [11] who investigated the monomer–dimer equilibrium of endophilin.

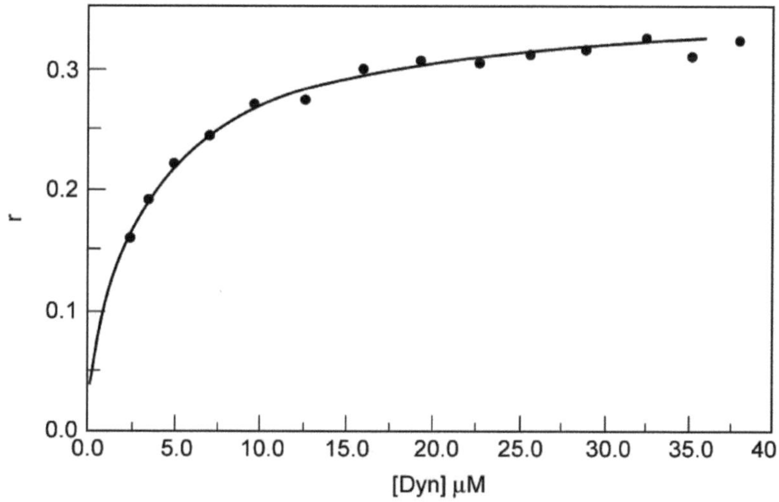

Fig. 3 In this experiment, 1 μM mant-GTPγS (a fluorescent, non-hydrolyzable GTP analog) was present, and the concentration of the GTP-binding protein, Dyn2, was varied by starting at high concentrations followed by dilution. The binding curve was fit to the anisotropy equation (in this case, the yield of the fluorophore increased about twofold upon binding). A K_d of 8.3 μM was found

3. 500 μL of endophilin A2 (4 μM) was reacted with 10 μL of 6 mM Alexa Fluor 488 succinimidyl ester (*see* Fig. 4). The reaction took place in 20 mM HEPES, pH 7.5 with 100 mM NaCl, and was allowed to proceed overnight at 4 °C. The solution was dialyzed (6–8,000 MWCO) against the reaction buffer, and labeling was determined to be 2:1, protein to fluorophore (*see* **Note 8**).
4. The labeled protein was seeded with unlabeled protein to achieve a starting concentration of 20 μM. Anisotropy was recorded by excitation with 488 nm light with the polarized emission being collected directly through a 525 nm long-pass filter (no. 3484; Pyrex) (*see* **Note 9**).
5. The sample was sequentially diluted, with 20 mM HEPES, pH 7.5, and 100 mM NaCl, down to 100 nM protein concentration.
6. Figure 3 shows the polarization as a function of protein concentration and fit to the dimer–monomer binding isotherm with a K_d of 15 μM.

4 Notes

1. The Sephacryl S300 resin is not optimal for separation of 132 and 66 kDa (dimer vs. monomer HSA). Therefore, there will be some overlap of the higher-order oligomers (trimer and dimer)

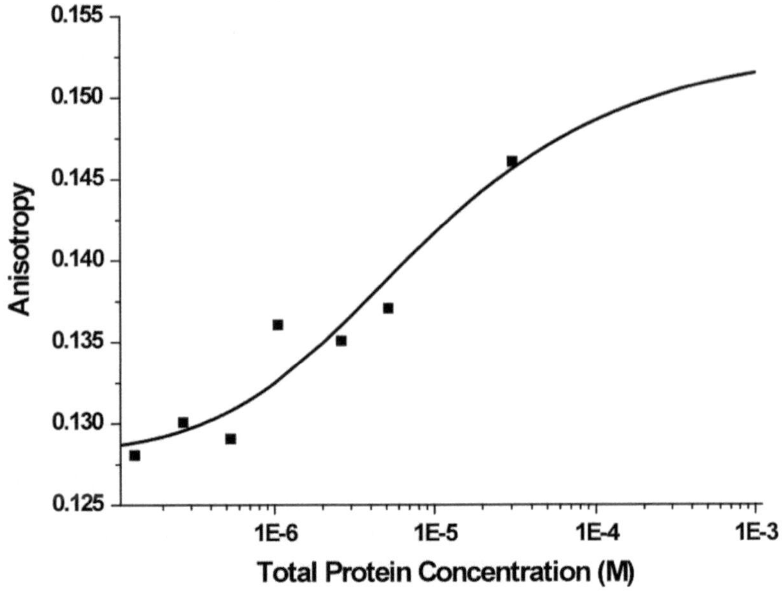

Fig. 4 Steady-state anisotropy of extrinsically labeled endophilin. Anisotropy values, along with the best fit to the binding isotherm, of varying concentrations of Alexa Fluor 488 labeled endophilin. This figure is a modification from a previous publication [10]

with the monomer, which is clearly visible from Fig. 1. To minimize this, run the S300 column at 1 mL/min (or slower) and collect 500 μL fractions. Better separation can be accomplished by using a resin with higher separation in this area, such as an S200 or lower.

2. Lyophilization of HSA causes oligomerization through disulfide linkage exchange [16]. Therefore, a native, non-denaturing gel is needed to see the higher-order oligomers. Fraction 89 (highest point on the main peak) still contains a faint dimer band, and, as such, fractions 90–110 were pooled and concentrated.

3. An issue arises with this ligand as furosemide has an absorption band starting from ~310 nm and fluorescence emission from 350 to 500 nm [20]. The polarized emission from mHSA can be isolated by collecting the emission through a monochromator set at 330 nm or through a band-pass filter centered in the 330 nm region. It is absolutely critical to avoid the emission from furosemide because, without directly exciting furosemide, the emission will result from energy transfer between tryptophan and furosemide. This transfer will cause a decrease in polarization, when in actuality the polarization should increase. And, while it is possible to collect the data through the monochromator, it is highly recommended that an interference filter

be used. This point is stressed as the emission of mHSA is quenched ~90 % at the start of the experiment and more counts results in improved precision of the measurement [21].

4. At this wavelength, we are only exciting tryptophan residues, thus avoiding any potential tyrosine–tryptophan energy transfer, which could lower the observed polarization. We recommend using a 300 nm narrow band-pass filter (ours is FF01-295/15-25 from Semrock) after the excitation monochromator to ensure no excitation light above 305 nm impinges on the sample as the ligand, furosemide, can be excited by wavelengths above 310 nm [20].

5. There are advantages and disadvantages to measuring polarization through a monochromator vs. a filter (long-pass or band-pass). The major advantage is the ability to select a specific emission wavelength. However, the major disadvantage is the dramatic loss in intensity. Another factor that needs to be considered critically when using a monochromator is the *G*-factor, which takes into account the dependency of the transmission efficiency of the monochromator on the polarization direction of the light. This value should be measured for each individual system; however, in the case of directly measuring polarization through a filter, the value is typically around 1 (in our system it is 1.03 ± 0.02). Calculation of this value requires emission intensity collected through excitation and emission polarizers set at horizontal and horizontal (I_{HH}) and horizontal and vertical (I_{HV}):

$$G = \frac{I_{HV}}{I_{HH}}$$

which can then be added to modify Eqs. 1 and 2 to:

$$P = \frac{(I_{\parallel} - GI_{\perp})}{(I_{\parallel} + GI_{\perp})}$$

$$r = \frac{(I_{\parallel} - GI_{\perp})}{(I_{\parallel} + 2GI_{\perp})}$$

The *G*-factor, when measuring through the monochromator at 300 nm, in our system is 2.28 ± 0.03.

6. Ensure that the stock solution of furosemide is concentrated such that volume added is low, minimizing the dilution effect on the mHSA fluorescence. The final volume of the sample is critical because the meniscus needs to be out of the path of the excitation light and the observation volume.

7. Voelker et al. [4] did a study of furosemide binding to HSA using the fluorescence quenching upon binding. Their reported K_d, under similar conditions as our study, was

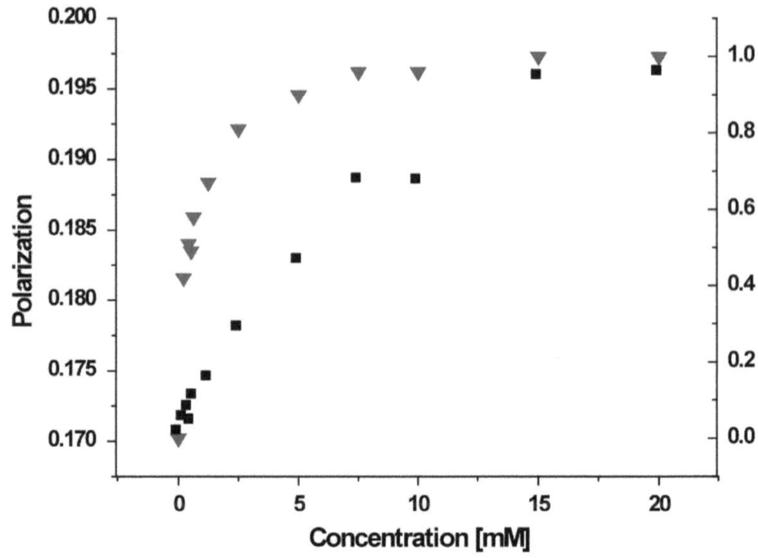

Fig. 5 Corrected and uncorrected f_b of furosemide to mHSA. If the f_b were not corrected for the large difference between bound and free forms from Subheading 3.2, the data would look like the *black squares* with a K_d of ~3.5 μM

0.65 ± 0.05 μM, which is almost identical to the value being reported here. With regard to the intensity change, if we did not take this change into account when calculating f_b (Eq. 9), the plot of f_b vs. total ligand concentration (M) would look like Fig. 5. The gray squares are the corrected f_b values. A K_d of 3.5 μM is obtained when the uncorrected values are fit to Eq. 13. This value is nearly ten times higher than the corrected K_d, 0.4 μM, demonstrating the need for correction.

8. The optimal conditions for labeling a protein must be worked out for each individual case, i.e., what probe concentration, what protein concentration, what pH, how long, and what temperature. After the labeled protein is isolated from the unreactive probe (either by gel-filtration chromatography, spin columns, or dialysis), the extent of labeling must be determined. Clearly, if one knows the extinction coefficient of the protein and the probe at 280 nm and if one knows the extinction coefficient of the probe at a higher wavelength, where the protein does not absorb light, then the labeling ration calculation is straightforward. However, one may not readily know the contribution of the probe at 280 nm. In this case, one should utilize a classic protein determination assay, such as the Bradford method, to determine the protein concentration. If one knows the probe's extinction coefficient at 280 nm, then the proteins coefficient may be estimated from its amino acid sequence using the formula worked out empirically by Pace et al. [17], namely,

$$\varepsilon_{280\mathrm{nm}}(\mathrm{M}^{-1}\mathrm{cm}^{-1}) = (\#\mathrm{Trp})\,(5{,}500) + (\#\mathrm{Tyr})\,(1{,}490) + (\#\mathrm{Cystines})\,(125)$$

9. The Raman peak will be located at 585 nm with 488 nm excitation light. As such, for these experiments, the Raman peak will begin to show at lower and lower intensities due to the long-pass filter allowing all light above 525 nm. Because Ross et al. [11] were not able to maintain a constant intensity, this experiment was only able to go down to ~100 nM (due to Raman contamination). A more complete binding isotherm can be obtained with a higher labeling ratio.

Acknowledgements

This work was supported in part by funding from Allergan, Inc. and also by NIH grant RO1GM076665.

References

1. Valeur B, Berberan-Santos MN (2012) Molecular fluorescence: principles and applications, 2nd edn. Wiley-VCN, Weinheim, Germany
2. Jameson DM, Croney JC, Moens PDJ (2003) Fluorescence: basic concepts, practical aspects and some anecdotes. In: Marriott G, Parker I (ed) Methods in enzymology, vol 360, Chapter 1, pp 1–42
3. Ross JA, Jameson DM (2008) Time-resolved methods in biophysics. 8. Frequency domain fluorometry: applications to intrinsic protein fluorescence. Photochem Photobiol Sci 7:1301–1312
4. Voelker JR, Jameson DM, Brater DC (1989) In vitro evidence that Urine composition affects the fraction of active furosemide in the Nephrotic Syndrome. J Pharmacol Exp Ther 250:772–778
5. Dandlinker WB, Feijen GA (1961) Quantification of the antigen-antibody reaction by the polarization of fluorescence. Biochem Biophys Res Commun 5:299–304
6. Perrin F (1926) Polarisation de la lumière de fluorescence. Vie moyenne des molécules dans l'état excité. J Phys Radium 7:390–401
7. Weber G (1952) Polarization of the fluorescence of macromolecules. I. Theory and experimental method. Biochem J 5:145–155
8. Jolley ME, Stroupe SD, Wang C-HJ et al (1981) Fluorescence polarization immunoassay I. Monitoring aminoglycoside antibiotics in serum and plasma. Clin Chem 27: 1190–1197
9. Jameson DM, Ross JA (2010) Fluorescence polarization/anisotropy in diagnostics and imaging. Chem Rev 110:2685–2708
10. Eccleston JF, Hutchinson JP, Jameson DM (2005) Fluorescence-based assays. In: King FD, Lawton G (ed) Progress in Medical Chemistry, vol 43. Chapter 2, pp 20–47
11. Ross JA, Chen Y, Mueller JD et al (2011) Dimeric Endophilin A2 stimulates assembly and GTPase activity of dynamin 2. Biophys J 100:729–737
12. Jameson DM, Mocz G (2005) Fluorescence polarization/anisotropy approaches to study protein-ligand interaction: effects of errors and uncertainties. In: Nienhaus GU (ed) Methods in Molecular Biology vol 305. Chapter 15, pp 301–322
13. Montecinos-Franjola F, Ross JA, Sanchez SA et al (2012) Studies on the dissociation and Urea-induced unfolding of FtZ supports the dimer nucleus polymerization mechanism. Biophys J 102:2176–2185
14. Farris M, Lague A, Manuelyan Z et al (2012) Altered nuclear cofactor switching in retinoic-resistant variants of the PML-RARα oncoprotein of acute promyelocytic leukemia. Proteins 80:1095–1109
15. Mocz G (2006) Information content of fluorescence polarization and anisotropy. J Fluoresc 16:511–524
16. Costantino HR, Shieh L, Klibanov AM, Langer R (1997) Heterogeneity of serum albumin samples with respect to solid-state aggregation

via thiol-disulfide interchange: implications for sustained release from polymers. J Control Release 44:255–261

17. Pace CN, Vajdos F, Fee L, Grimsley G, Gray T (1995) How to measure and predict the molar absorption coefficient of a protein. Protein Sci 4:2411–2423
18. Binns DD, Helms MK, Barylko B et al (2000) The mechanism of GTP hydrolysis by Dynamin II: a transient kinetic study. Biochemistry 39:7188–7196
19. Hiratsuka T (1983) New ribose-modified fluorescent analogs of adenine and guanine nucleotides available as substrates for various enzymes. Biochim Biophys Acta 742: 496–508
20. Forray AW, Kimpel B, Blair AD, Culter RE (1974) Furosemide concentrations in serum and urine, and its binding by serum proteins as measured fluorometrically. Clin Chem 20: 152–158
21. Jameson DM, Weber G, Spencer RD, Mitchell G (1978) Fluorescence polarization: measurements with a photon-counting photometer. Rev Sci Instrum 49:510–514

Chapter 3

Quantitative Fluorescence Spectral Analysis of Protein Denaturation

Ivo H.M. van Stokkum and Sergey P. Laptenok

Abstract

This chapter describes a procedure of global analysis of the steady-state spectra measured with different concentrations of the denaturant to quantitatively study protein denaturation. With the help of physicochemical models, relevant spectral parameters that characterize the folding intermediate and thermodynamic parameters that describe a three-state model N $\Leftrightarrow$ I $\Leftrightarrow$ U can be estimated.

Key words Global analysis, Protein denaturation, Singular value decomposition, Steady-state fluorescence

1 Introduction

During denaturant-induced equilibrium (un)folding of a particular protein, a molten globule-like folding intermediate is formed [1]. Here we describe how the steady-state fluorescence spectrum monitored as a function of the denaturant concentration can be used to infer the properties of the folding intermediate with the help of global analysis [2].

2 Materials

Materials, Protein Expression, and Purification. All chemicals used were of the highest purity available. The concentration guanidine hydrochloride (GuHCl) was determined by measuring the refractive index of the sample used, as described previously [3]. A variant of apoflavodoxin from *Azotobacter vinelandii*, which contains two tryptophan residues (i.e., W74-W128-F167 (WWF)), was obtained and purified as described [3]. In all experiments protein concentration was 4 μM in 100-mM potassium pyrophosphate buffer, pH = 6.0. Temperature was set to 25 °C.

Yves Engelborghs and Antonie J.W.G. Visser (eds.), *Fluorescence Spectroscopy and Microscopy: Methods and Protocols*, Methods in Molecular Biology, vol. 1076, DOI 10.1007/978-1-62703-649-8_3,

Steady-State Fluorescence Spectra. Steady-state fluorescence spectra were obtained with a Fluorolog 3.2.2 spectrofluorometer (Horiba, Jobin Yvon, Optilas, Alphen aan den Rijn, the Netherlands), as described previously [3]. The excitation wavelength was 300 nm, excitation and emission slit widths were 2 nm, and emission spectra were recorded between 305 and 400 nm with 1-nm steps. All spectra were corrected for wavelength-dependent instrumental response characteristics. Background fluorescence emission was measured under the same circumstances, except that now no protein is present in the samples, and was subsequently subtracted from the corresponding fluorescence spectra of samples with protein.

3 Methods

3.1 Determination of the Number of Components Contributing to the Steady-State Fluorescence Spectra

The steady-state spectra measured at n_d denaturant concentrations can be collated in a matrix Ψ where a column of the $n_\lambda \times n_d$ matrix Ψ contains a spectrum measured at n_λ wavelengths at a particular denaturant concentration, whereas a row contains the emission measured at a particular wavelength at n_d denaturant concentrations. The rank of this matrix Ψ can be estimated with the help of the singular value decomposition (SVD) [4–11] (*see* **Note 1**).

Figure 1 depicts the results from the singular value decomposition (SVD) analysis of the data matrix Ψ obtained from denaturant-induced unfolding of WWF apoflavodoxin. The scree plot of the singular values (Fig. 1b) shows a kink after $i = 3$ indicating the presence of at least three significant components [12]. The first three LSVs and RSVs (black, red, and blue in Fig. 1c, d) show clear structure. The fourth LSV and RSV (light green) are noise-like traces. In conclusion, SVD indicates that three species are present in the data matrix. There are several methods to resolve these species; most well known are soft modeling (e.g., Multivariate Curve Resolution (MCR) [13–15]) and hard modeling with the help of models that incorporate physicochemical knowledge and aim for the estimation of parameters that are physicochemically interpretable. The latter approach is termed global analysis [11, 16]. Advantages of hard over soft modeling have been demonstrated [13, 17]. A prerequisite for global analysis (*see* **Note 2**) is the availability of suitable physicochemical models.

3.2 Global Analysis of the Steady-State Fluorescence Spectra with the Help of a Spectral Model

At each GuHCl denaturant concentration, the observed emission spectrum $\psi(\lambda)$ is described as a linear combination of spectra arising from native protein (n), folding intermediate (i), and unfolded protein (u):

$$\psi(\lambda) = c_n f_n(\lambda) + c_i f_i(\lambda) + c_u f_u(\lambda) \tag{1}$$

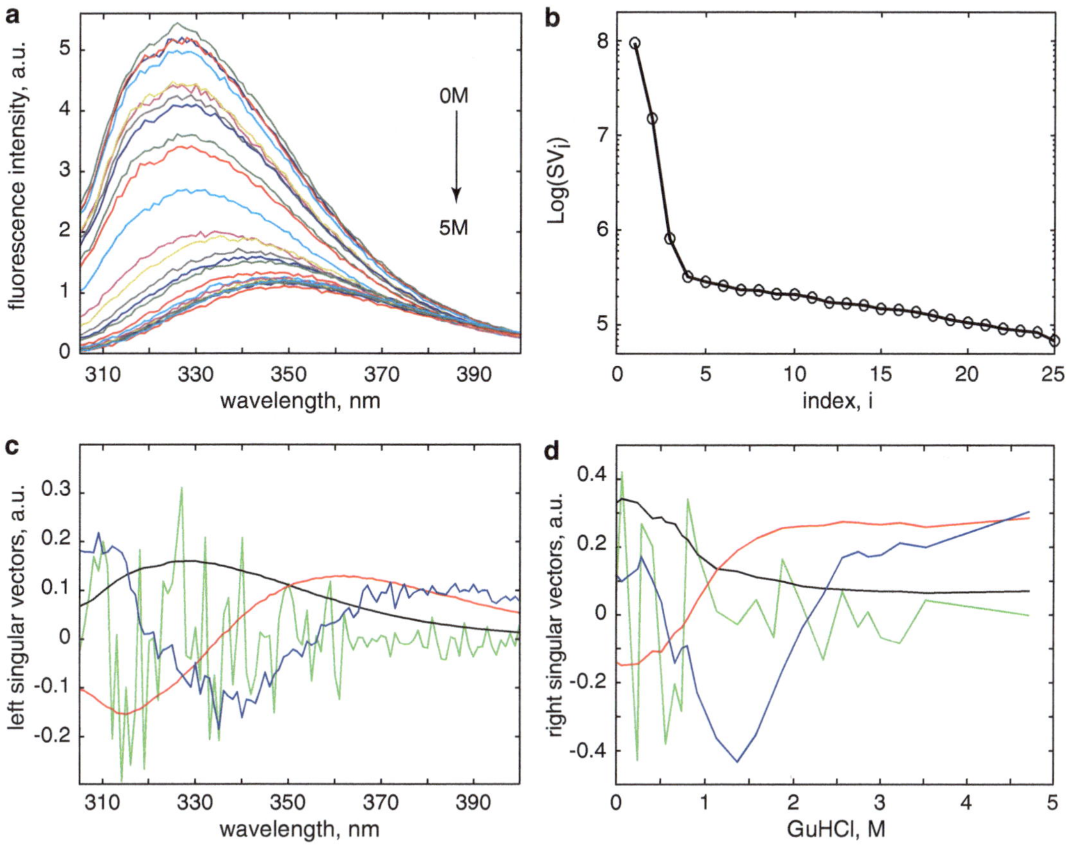

Fig. 1 Results from the singular value decomposition (SVD) analysis of the data matrix collated from steady-state fluorescence data obtained from denaturant-induced unfolding of WWF apoflavodoxin. (**a**) Denaturation trajectory of steady-state fluorescence spectra obtained at increasing concentrations of denaturant. (**b**) Scree plot of the singular values shows a kink after $i = 3$ indicating the presence of at least three significant components. (**c**) The first four left singular vectors (LSVs) (colored *black*, *red*, *blue*, and *light green*, respectively). (**d**) The accompanying first four right singular vectors (RSVs)

Steady-state fluorescence spectra obtained of protein at 0 and 4.72 M GuHCl are used as reference spectra that characterize the native and unfolded protein, respectively. The steady-state fluorescence spectrum of the folding intermediate was modeled as a skewed Gaussian in the energy domain ($\bar{\nu} = 1/\lambda$) and is described by three parameters: peak location $\bar{\nu}_{\max}$, width $\Delta\bar{\nu}$, and skewness b [16, 11]:

$$f_i(\bar{\nu}) = \bar{\nu}^5 \exp(-\ln(2)\{\ln(1 + 2b(\bar{\nu} - \bar{\nu}_{\max})/\Delta\bar{\nu})/b\}^2) \qquad (2)$$

where the parameter $\bar{\nu}_{\max}$ is the Franck-Condon wave number of maximum emission. The full width at half maximum (FWHM) is given by $\Delta\bar{\nu}_{1/2} = \Delta\bar{\nu}\sinh(b)/b$ (*see* **Note 3**).

Now all spectra are globally analyzed as a linear combination of spectra arising from native protein, folding intermediate, and unfolded protein. The three parameters that describe the shape of

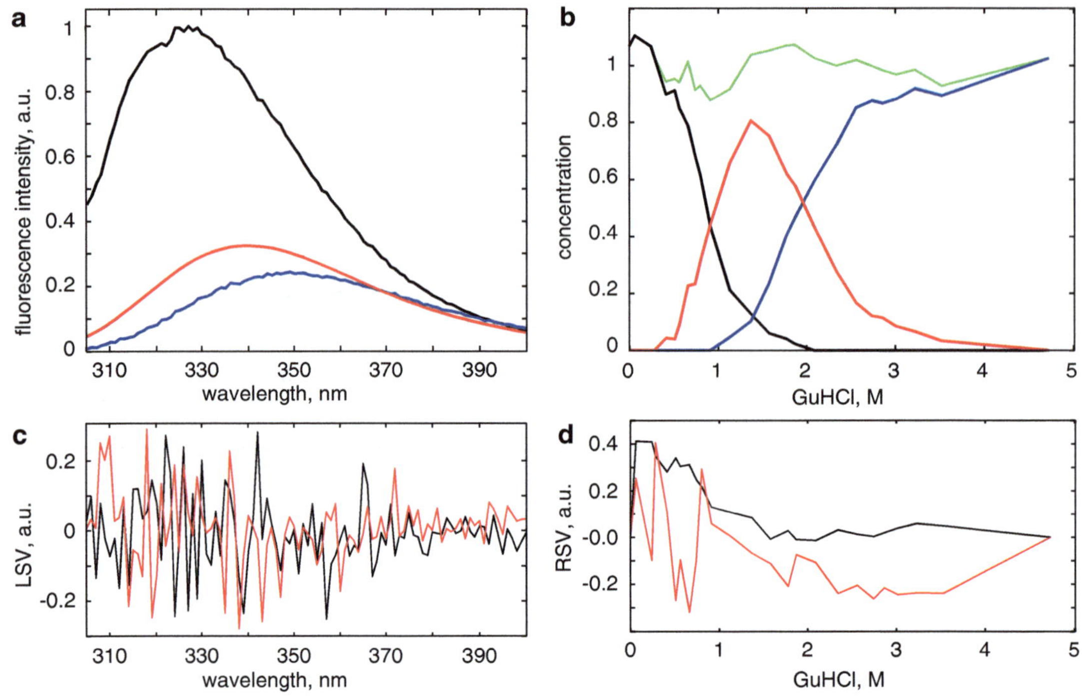

Fig. 2 Global analysis of steady-state fluorescence data matrix with the help of a spectral model. (**a**) Steady-state fluorescence spectra of WWF apoflavodoxin in 0 M GuHCl (native protein, *black*) and in 4.72 M GuHCl (unfolded protein, *blue*), respectively. The steady-state fluorescence spectrum of the folding intermediate (*red*) is modeled as a skewed Gaussian and is estimated from the global analysis of all unfolding data. (**b**) Estimated concentrations of the different folding species (colored *black*, *red*, and *blue*) as a function of denaturant concentration. The sum of the three concentrations is shown as a thin *light green line*. (**c**, **d**) Results from the singular value decomposition of the residual matrix. (**c**) The first (*black*) and second (*red*) LSV. (**d**) The first (*black*) and second (*red*) RSV

the spectrum of the folding intermediate and the concentrations of each folding species are the unknown parameters that need to be estimated from a global fit of all data. The $n_\lambda \times n_d$ matrix Ψ can be written as a matrix product.

$$\Psi = F(\bar{\nu}_{\max}, \Delta\bar{\nu}, b)C^{\mathrm{T}} \tag{3}$$

where the F matrix contains three columns $\left[f_n f_i(\bar{\nu}_{\max}, \Delta\bar{\nu}, b) f_u\right]$ and n_λ rows and the C matrix contains three columns $\left[c_n c_i c_u\right]$ and n_d rows. The rank of the C, F, or Ψ matrix is three, consistent with the SVD. The concentration parameters were constrained to be nonnegative (*see* **Note 4**). The estimated spectrum of the folding intermediate is depicted in red in Fig. 2a. Note that it is blue shifted relative to that of the unfolded protein (blue). The estimated spectral parameters are $\bar{\nu}_{\max} = 29,290 \pm 40/\mathrm{cm}$, $\Delta\bar{\nu} = 4,830 \pm 30$, and $b = -0.209 \pm 0.006$. The root mean square error of the fit was 0.41 % of the maximum of the data. The matrix of residuals resulting from the global analysis can best be diagnosed with the help of its SVD. Shortcomings of the model used show up as trends in the most important left or right singular vectors. No such trends are present in the first (black) and second (red) LSV or RSV (Fig. 2c, d). Therefore, the fit can be accepted.

In this way the product $c_i f_i$ can be estimated, and thus, we can determine the shape of c_i as a function of the GuHCl denaturant concentration. In order to estimate the concentration c_i relative to the other concentrations, we use the constraint that the sum $c_n + c_i + c_u$ should be close to one at all GuHCl concentrations. This is estimated by means of a subsequent linear regression. All estimated concentrations are depicted in Fig. 2b, as well as this sum $c_n + c_i + c_u$ (depicted in light green). The small deviations of this sum from one are considered acceptable.

3.3 Global Analysis of the Fractions of the Different Folding Species with the Help of a Thermodynamic Model

As shown by Bollen et al. [1], denaturant-induced equilibrium unfolding of apoflavodoxin is described by a three-state model: N ⇔ I ⇔ U, in which N represents native, U represents unfolded molecules, and I is a folding intermediate. Consequently, the two corresponding equilibrium constants (i.e., K_{IN} and K_{UI}) and associated free energy differences (i.e., ΔG^0_{IN} and ΔG^0_{UN}) are

$$
\begin{aligned}
K_{\mathrm{IN}} &= \frac{[\mathrm{N}]}{[\mathrm{I}]} = exp\left[-(\Delta G^0_{\mathrm{IN}} + m_{\mathrm{IN}} \times [\mathrm{D}])/0.59\right] \\
K_{\mathrm{UI}} &= \frac{[\mathrm{I}]}{[\mathrm{U}]} = exp\left[-(\Delta G^0_{\mathrm{UI}} + m_{\mathrm{UI}} \times [\mathrm{D}])/0.59\right]
\end{aligned}
\qquad (4)
$$

where m_{IN} and m_{UI} describe the denaturant dependence of ΔG^0_{IN} and ΔG^0_{UI}. The number 0.59 in this equation equals the gas constant R times temperature T (=298 K) and is in kcal/mol. The fractional populations of each folding state (c_{U}, c_{I}, c_{N}) follow from

$$
\begin{aligned}
c_{\mathrm{U}} &= \frac{1}{1 + K_{\mathrm{UI}} + K_{\mathrm{IN}} \times K_{\mathrm{UI}}} \\
c_{\mathrm{I}} &= \frac{K_{\mathrm{UI}}}{1 + K_{\mathrm{UI}} + K_{\mathrm{IN}} \times K_{\mathrm{UI}}} \\
c_{\mathrm{N}} &= \frac{K_{\mathrm{IN}} \times K_{\mathrm{UI}}}{1 + K_{\mathrm{UI}} + K_{\mathrm{IN}} \times K_{\mathrm{UI}}}
\end{aligned}
\qquad (5)
$$

Global analysis yielded the fractional populations (Fig. 2b) of the folding species at a particular denaturant concentration, and these fractions were subsequently used to estimate ΔG^0_{IN}, m_{IN}, ΔG^0_{UI}, and m_{UI}.

A least absolute values (LAV, *see* **Note 5**) approach was used during global analysis, because this approach is more robust against outliers than the least squares (LS) method (*see* **Note 6**). The LAV criterion that is minimized as a function of the thermodynamic parameters is

$$
\min\left(\sum_i \left|c^i_{\mathrm{U}^*} - c^i_{\mathrm{U}}\right| + \left|c^i_{\mathrm{I}^*} - c^i_{\mathrm{I}}\right| + \left|c^i_{\mathrm{N}^*} - c^i_{\mathrm{N}}\right|\right) \qquad (6)
$$

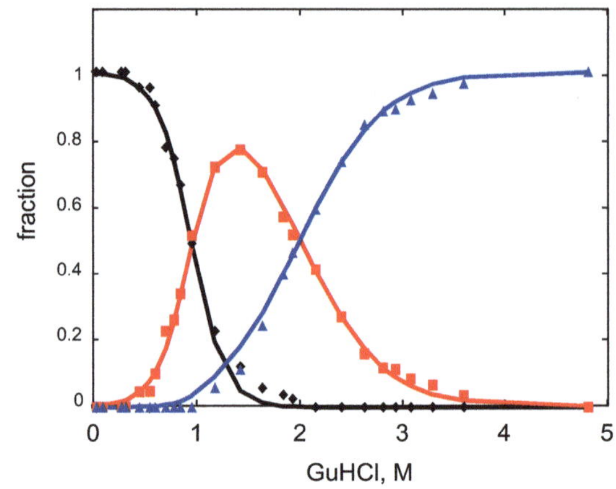

Fig. 3 Global analysis of the fractions of the different folding species with the help of a thermodynamic model. Symbols indicate relative concentrations of the three species (colored *black*, *red*, and *blue*) as a function of the denaturant concentration estimated in Fig. 2b (*diamonds*: native; *squares*: intermediate; *triangles*: unfolded). *Solid lines* depict global LAV fit

Table 1
Thermodynamic parameters estimated from GuHCl-induced equilibrium unfolding of WWF apoflavodoxin (*see* Note 8)

Criterion	ΔG^0_{UI} (kcal/mol)	m_{UI} (kcal/mol M^{-1})	ΔG^0_{IN} (kcal/mol)	m_{IN} (kcal/mol M^{-1})	ΔG^0_{UN} (kcal/mol)	m_{UN} (kcal/mol M^{-1})
LAV	−2.9	1.5	−3.3	3.6	−6.2	5.1
LS	−3.0	1.5	−3.1	3.4	−6.1	4.9

where c_U, c_I, and c_N are calculated with Eq. 5 and c_{U*}, c_{I*}, and c_{N*} are the *normalized* concentrations (*see* **Note** 7) from the global analysis (Fig. 2b), with i the summation index, which corresponds to the different concentrations of denaturant used. The results from the global analysis of the fractions of the different folding species, with the help of a thermodynamic model, are depicted in Fig. 3. The fit is considered satisfactory. The thermodynamic parameters estimated with LS or LAV listed in Table 1 are well interpretable [2]. The difference in the estimated values between the LAV and LS criteria is within 10 % relative precision and thus not significant. This is no surprise, since there are no large outliers here. However, in general it is advisable to use the LAV method [18, 19] when available.

4 Notes

1. The singular value decomposition (SVD) is a model-free matrix factorization technique, which decomposes the data into a sum of orthonormal vector products scaled by singular values. Here, the left singular vector (LSV) represents spectral dimension and the right singular vector (RSV) represent denaturant concentration dimension. The contribution to the data is the product of the *n*th left singular vector and right singular vector scaled by the *n*th singular value. The singular vectors are ordered based on their contribution to the data as represented by the magnitude of the singular values as shown in the scree plot. The ordinate of this scree plot is logarithmic. Ideally, the transition between data and noise appears as a kink in the scree plot. The technique can be used to explore the number of independent components in the data matrix, which is an important aspect of defining an initial model.
2. Public domain software for global analysis is available [20, 21].
3. Note that the expression for the skewed Gaussian contains a term $\ln(1 + 2b(\bar{\nu} - \bar{\nu}_{\max})/\Delta\bar{\nu})/b$ for which a limit exists when skewness parameter b approaches zero. Since $\lim_{b\to 0} \ln(1 + bx)/b = x$ the expression simplifies to $f_i(\bar{\nu}) = \bar{\nu}^5 \exp\left(-\ln(2)\{2(\bar{\nu} - \bar{\nu}_{\max})/\Delta\bar{\nu})\}^2\right)$ which is a normal Gaussian with FWHM $\Delta\bar{\nu}$. In practical computations with nonzero b, the argument of the natural logarithm has to be tested first. If it is positive, the amplitude $f_i(\bar{\nu})$ can be computed, else $f_i(\bar{\nu}) \equiv 0$.

   For the actual computation of $f_i(\bar{\nu})$ at a particular wavelength λ, one substitutes $\bar{\nu} = 1/\lambda$ in Eq. 2. The conversion from wavelength to wave number [22], $f(\bar{\nu}) = \lambda^2 f(\lambda)$, is already taken into account in Eq. 2 [11].
4. There are several ways to ensure nonnegativity of the concentration parameters. Firstly, one can use unconstrained least squares and estimate the nonlinear spectral shape parameters (peak location $\bar{\nu}_{\max}$, width $\Delta\bar{\nu}$, and skewness b) and the matrix of conditionally linear parameters C with the help of the variable projection algorithm [16, 23]. When some of the estimated concentration parameters become negative, they can be constrained to zero, which means that at that denaturant concentration a certain component does not contribute. After imposing the constraint, the data have to be refitted. This process can be automated with the help of the nonnegative least squares algorithm [24] in combination with the variable projection algorithm [25].
5. Least absolute values minimization can most easily be done using the Excel Solver function or with dedicated algorithms [26].

6. Outliers are often present with this small number (typically 15–25 denaturant concentrations) of data points. A disadvantage of LAV analysis is that it does not report standard errors. An estimate of the relative standard errors is 10 %.
7. The small deviations of the sum $c_n + c_i + c_u$ from 1 (indicated in light green in Fig. 2b) that were present are removed when dividing by this sum. The normalized concentrations are defined as $c_{N^*} = c_n/(c_n + c_i + c_u)$, $c_{I^*} = c_i/(c_n + c_i + c_u)$, and $c_{U^*} = c_u/(c_n + c_i + c_u)$ at each denaturant concentration.
8. Note that ΔG^0_{UN} and m_{UN} which describe the thermodynamic stability against unfolding have been computed using $\Delta G^0_{UN} = \Delta G^0_{UI} + \Delta G^0_{IN}$ and $m_{UN} = m_{UI} + m_{IN}$.

References

1. Bollen YJM, Sanchez IE, van Mierlo CPM (2004) Formation of on- and off-pathway intermediates in the folding kinetics of Azotobacter vinelandii apoflavodoxin. Biochemistry 43 (32):10475–10489. doi:10.1021/bi049545m
2. Laptenok SP, Visser NV, Engel R et al (2011) A general approach for detecting folding intermediates from steady-state and time-resolved fluorescence of single-tryptophan-containing proteins. Biochemistry 50(17):3441–3450. doi:10.1021/bi101965d
3. Visser NV, Westphal AH, van Hoek A et al (2008) Tryptophan-tryptophan energy migration as a tool to follow apoflavodoxin folding. Biophys J 95(5):2462–2469. doi:10.1529/biophysj.108.132001
4. Golub GH (1996) Matrix computations. The Johns Hopkins University Press, Baltimore
5. Henry ER (1997) The use of matrix methods in the modeling of spectroscopic data sets. Biophys J 72(2):652–673
6. Henry ER, Hofrichter J (1992) Singular value decomposition—application to analysis of experimental-data. Methods Enzymol 210: 129–192
7. Hendler RW, Shrager RI (1994) Deconvolutions based on singular-value decomposition and the pseudoinverse—a guide for beginners. J Biochem Biophys Methods 28(1):1–33. doi:10.1016/0165-022x(94)90061-2
8. Shrager RI (1986) Chemical-transitions measured by spectra and resolved using singular value decomposition. Chemom Intell Lab Syst 1(1):59–70. doi:10.1016/0169-7439(86)80026-0
9. Shrager RI, Hendler RW (1982) Titration of individual components in a mixture with resolution of difference spectra, pKs, and redox transitions. Anal Chem 54(7):1147–1152. doi:10.1021/ac00244a031
10. Satzger H, Zinth W (2003) Visualization of transient absorption dynamics—towards a qualitative view of complex reaction kinetics. Chem Phys 295(3):287–295. doi:10.1016/j.chemphys.2003.08.012
11. van Stokkum IHM, Larsen DS, van Grondelle R (2004) Global and target analysis of time-resolved spectra. Biochim Biophys Acta 1657:82–104. doi:10.1016/j.bbabio.2004.04.011
12. Catell RB (1966) The scree test for the number of factors. Multivar Behav Res 1:245–276
13. de Juan A, Maeder M, Martinez M, Tauler R (2000) Combining hard- and soft-modelling to solve kinetic problems. Chemom Intell Lab Syst 54(2):123–141. doi:10.1016/s0169-7439(00)00112-x
14. de Juan A, Tauler R (2003) Chemometrics applied to unravel multicomponent processes and mixtures—revisiting latest trends in multivariate resolution. Anal Chim Acta 500(1–2): 195–210. doi:10.1016/s0003-2670(03)00724-4
15. de Juan A, Tauler R (2006) Multivariate curve resolution (MCR) from 2000: progress in concepts and applications. Crit Rev Anal Chem 36(3–4):163–176. doi:10.1080/10408340600970005
16. van Stokkum IHM, Linsdell H, Hadden JM et al (1995) Temperature-induced changes in protein structures studied by fourier transform infrared spectroscopy and global analysis. Biochemistry 34(33):10508–10518
17. van Stokkum IHM, Mullen KM, Mihaleva VV (2009) Global analysis of multiple gas chromatography-mass spectrometry (GC/MS)

data sets: a method for resolution of co-eluting components with comparison to MCR-ALS. Chemom Intell Lab Syst 95(2):150–163. doi:10.1016/j.chemolab.2008.10.004

18. Vandenbos A (1988) Nonlinear least-absolute-values and minimax model-fitting. Automatica 24(6):803–808. doi:10.1016/0005-1098(88)90056-8
19. Seber GAF, Wild CJ (1989) Nonlinear regression. Wiley, New York
20. Mullen KM, van Stokkum IHM (2007) TIMP: an R package for modeling multi-way spectroscopic measurements. J Stat Softw 18(3)
21. Snellenburg JJ, Laptenok SP, Seger R et al (2012) Glotaran: a JAVA-based graphical user interface for the R-package TIMP. J Stat Softw 49:1–22
22. Lakowicz JR (2006) Principles of fluorescence spectroscopy, 3rd edn. Springer, Science and Business Media, LLC, New York
23. Golub GH, LeVeque RJ (1979) Extensions and uses of the variable projection algorithm for solving nonlinear least squares problems. In: Proceeding of the 1979 Army Numerical Analysis and Computer Conferences, ARO Report 79–3
24. Lawson CL, Hanson RJ (1974) Solving least squares problems. Prentice Hall, Englewood Cliffs
25. Mullen KM, van Stokkum IHM (2009) The variable projection algorithm in time-resolved spectroscopy, microscopy and mass spectrometry applications. Numer Algorithms 51 (3):319–340. doi:10.1007/s11075-008-9235-2
26. Schlossmacher EJ (1973) Iterative technique for absolute deviations curve fitting. J Am Stat Assoc 68(344):857–859. doi:10.2307/2284512

Chapter 4

High-Pressure Fluorescence Applications

Mariano Dellarole and Catherine A. Royer

Abstract

Fluorescence is the most widely used technique to study the effect of pressure on biochemical systems. The use of pressure as a physical variable sheds light into volumetric characteristics of reactions. Here we focus on the effect of pressure on protein solutions using a simple unfolding example in order to illustrate the applications of the methodology. Topics covered in this review include the relationships between practical aspects and technical limitations; the effect of pressure and the study of protein cavities; the interpretation of thermodynamic and relaxation kinetics; and the study of relaxation amplitudes. Finally, we discuss the insights available from the combination of fluorescence and other methods adapted to high pressure, such as SAXS or NMR. Because of the simplicity and accessibility of high-pressure fluorescence, the technique is a starting point that complements appropriately multi-methodological approaches related to understanding protein function, disfunction, and folding from the volumetric point of view.

Key words High-pressure fluorescence, Protein volumetric properties, Protein folding, Protein cavities

1 Introduction

> The purpose of this note is to state a fact of possible biological interest which I have discovered incidentally in the course of other work. If the white of an egg is subjected to hydrostatic pressure at room temperature*, it becomes coagulated, presenting an appearance much like that of a hard-boiled egg.
>
> P. W. BRIDGMAN, 1914 J Biol Chem 19, 511–512 (*N.B. ~7 kbar during 30 min)

Beyond the accepted interest for the understanding of life in extreme habitats [1–3], the utilization of high pressure (HP) to study biochemical equilibria yields precise insights into reaction volumes:

$$\left(\frac{\partial \ln K}{\partial p}\right)_T = -\frac{\Delta V}{RT} \tag{1}$$

where K is the equilibrium constant, p the pressure, ΔV the difference of the final and initial partial molar volumes of the reactants (final minus initial), and RT the product of the molar gas constant

Yves Engelborghs and Antonie J.W.G. Visser (eds.), *Fluorescence Spectroscopy and Microscopy: Methods and Protocols*, Methods in Molecular Biology, vol. 1076, DOI 10.1007/978-1-62703-649-8_4,

and the temperature. HP studies can make a significant contribution to the general understanding of protein folding [4] and protein function (*see* ref. 5 for a textbook review). Since the observation of HP-induced *boiling* of the *white of an egg* by a hundred years ago [6], numerous apparatus have been adapted to work under pressure in order to assess diverse reaction coordinates of protein processes [5]. These innovations include turbidity spectrometers [7], fluorometers and absorbance spectrophotometers [8], fluorescence correlation (FCS) microscopes [9], circular dichroism [10, 11] and Fourier-transform infrared (FTIR) [12, 13] instruments, electron paramagnetic resonance (EPR) [14] and nuclear magnetic resonance (NMR) [15–17] spectrometers, X-ray diffractometers [18], small-angle X-ray scattering (SAXS) [19, 20] and neutron small-angle scattering (SANS) [21] devices, electrospray ionization mass spectrometers [22], electrophoresis apparatus [23], and calorimeters [24]. Due to this continual technical HP development, novel insights into protein physical chemistry have been forthcoming regarding folding [17, 25], refolding from aggregates [26–28], miss-folding [29], subunit dissociation, e.g., [30–40], dynamics [41], phase diagrams [37, 42–46], and interactions with lipids [47], DNA, e.g., [48, 49] or chemical compounds [38, 49–54].

Today the combination of HP-fluorescence with other techniques is no longer carried out by distinct groups on different systems. Rather collective research efforts in the HP field have become the norm, as opposed to the exception [55, 56]. For example, the combination of HP-FTIR, HP-SAXS, and HP-crystallography ensured the volumetric characterization attained by HP-fluorescence on the effect of single mutants perturbing cavities [20, 57–59]. Contrary to protein perturbations by temperature, pH, or denaturants, pressure is considered a more gentle perturbation, leading to fully reversible events even at elevated protein concentrations (>1 mM) [7, 15, 42, 60–68]. This is because pressure acts non-thermally [69] by eliminating internal solvent-excluded void volumes at subunit interfaces or internal cavities, thus leading to pressure-induced dissociation [29], or unfolding [21, 70, 71]. Because it is technically and theoretically rather straightforward, in addition to being rapid and sensitive, fluorescence spectroscopy is widely used in the field of protein biophysics [72]. These attributes greatly facilitated the extension of fluorescence to the high-pressure field, and account for its early use.

The use of protein fluorescence spectroscopy in combination with pressure as a physical variable has been reviewed in detail elsewhere [23, 25, 63, 69, 73–78]. Given the recent renewed interest in pressure perturbation, our present aim is to revisit HP-fluorescence methodology in order to render the approach accessible to researchers and students from diverse academic backgrounds. For this purpose, we describe the potentiality and limitations of the technique by illustrating pressure denaturation by

HP-fluorescence using the model monomeric and globular protein system, staphylococcal nuclease (SNase) [12, 21, 44, 79–89]. This example will allow us to consider thermodynamic (Subheading 3.1) and kinetic (Subheading 3.2) parameterization of the volumetric differences between the folded state, the transition state ensemble (TSE), and unfolded state via equilibrium pressure-induced unfolding and pressure-jump (*P*-jumps) relaxation studies, respectively. Additionally we will introduce to HP-fluorescence the linking between thermodynamics and kinetics by analysis of relaxation kinetics amplitudes [90, 91]. Our current understanding of the effect of pressure on proteins allows extension of HP-fluorescence information to more complex systems and/or higher resolution methods, e.g., single-molecule HP FCS or residue-specific HP NMR.

2 Materials

HP-fluorescence experiments involve similar materials as standard fluorescence, e.g., ultrapure reagents for buffers, pure protein samples, faultless quartz cuvettes, and so on. However specific technical care must be taken regarding the effect of pressure itself and the HP-equipment setup. In this section, we describe in detail the architecture and operation of the HP setup (Fig. 1) and we refer to Subheading 4 where technical issues and or tips are given. We note that the pressure limitation of the equipment is that of its lowest pressure component. In our system, the pump and lines and valves are limited to 60,000 psi (~4.5 kbar) whereas the cell itself is rated at 7 kbar.

1. *The equipment.* A standard fluorometer is adapted with an HP setup as described in Fig. 1. In comparison to conventional fluorescence, the light path in HP systems, colored in orange, enters into the cell body in order to excite the sample cell immersed into the pressure-transmitting fluid, colored in green, and the emitted fluorescence is collected at 90°. It is important to limit light loss both in the excitation and emission path. Consequently, the pressure-transmitting fluid, usually water, ethanol, or isopropanol, must be of the highest spectroscopic grade possible to avoid signal interferences (*see* **Note 1**) to address possible interference on the fluorescence yield. The pressure-generating fluid must have a low compressibility and an index of refraction close to that of the solution under study. Since we work on aqueous protein solutions, water is the preferred pressure-transducing liquid. However, it can only be used if the high-pressure cell and the pressure lines, pump, and gauge are made of stainless steel. Excitation is carried out using a lamp and monochromator coupled to a multimode optical fiber that is focused into the rear window of the sample

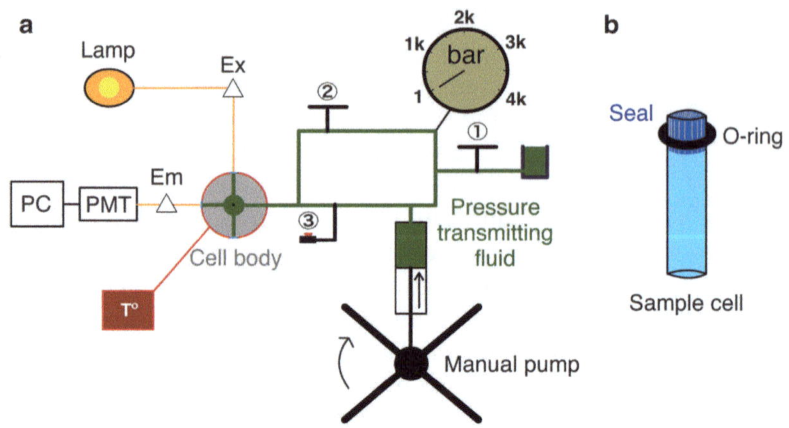

Fig. 1 Schematic sketch of HP-fluorescence apparatus. (**a**) Pressure is generated when the HP system is hermetic, manual valve (1) closed, by compression of the pressure-transmitting fluid (colored in *green*) via the action of the manual pump. (**b**) The sample cell is located at the center of the cell body and perceives pressure through compression of the impermeable seal. When the manual valve (2) is open, pressure is transmitted to the sample concomitantly with pump compression. When the manual valve (2) is closed, the automatic piston valve (3) is used for small rise time *P*-jumps. A barometer located between valves (1) and (2) records the pressure of the system. The temperature of the sample is controlled by a thermal bath (colored in *red*) linked to the cell body. Excitation of the sample is carried out using a lamp and monochromator (Ex Δ). A monochromator in the emission path (Em Δ) precedes the photomultiplier tubes (PMT) and photon counting electronics module connected to a computer (PC)

cell using a lens of appropriate focal length. A detector system is placed at one of the windows of the sample cell that is at 90° to the excitation window. This limits detection of scattered exciting light. Nonetheless, since the inside of the cell is made of stainless steel, scattered light cannot be entirely eliminated. For some applications (in the presence of scattering molecules such as micelles for example), a cut-on filter is placed directly after the emission collection lens. There is also a monochromator in the detection path and the light is focused onto the entrance slit using a lens of appropriate focal length. The detectors used are generally photomultiplier tubes, operating in photon counting mode. The digital output of the PMT is sent to a photon counting electronics module for discrimination and amplification prior to being sent to an interface card in a PC. The hardware used in our system is from ISS Inc (Champaign, IL, USA).

2. *The sample cell.* The sample cell is located in the center of the cell body and surrounded by pressure-transmitting fluid. Commercially available cuvettes adapted to a rigid Teflon stopper

(ISS, Champaign, IL, USA) and artisanal cuvettes adapted to an impermeable seal with an o-ring will properly transmit pressure to the sample by compression. We address loading issues in **Note 2**. The sample cell as well as the cell body structural design vary according to the setup and have been described with different degrees of detail elsewhere [8, 23, 29, 57, 73–76, 92].

3. *The high-pressure cell.* The high-pressure cell used in our setup is made of heavy stainless steel alloy and was designed in-house and built by SOFOP, La Broussine, Olemps, France [93]. The temperature of the sample is controlled through a thermostat circulating water bath externally coupled to the cell body. The stainless steel cell body acts as a good-temperature buffer, but thermal equilibration is slow. As volumetric properties are sensitive to temperature thermal control must be ensured; *see* **Note 3**. The cell is equipped with three sapphire windows, one for excitation, one for emission, and one for transmission (used in optical alignment). The use of sapphire windows precludes fluorescence anisotropy measurements, which require fused silica quartz windows, because sapphire is naturally birefringent. However, quartz is not as pressure resistant as sapphire, and we have found it more advantageous to use the sapphire. The sample cell is loaded through an opening in the top of the cell, and the pressure line is connected in place of the fourth window on the side of the vessel. We have found it very useful to separate the sample loading input from the pressure input, which then is almost never disconnected. This greatly reduces issues of leaks at the seals between the line and the plug.
4. *Manual pump.* Pressure is induced by either a manual or a motorized pump acting on the transmitting fluid through a piston and monitored via a pressure gauge. The pump in our system was purchased from SITEC, Hi-Pro Pressure Products Ltd., Aylesbury, England. The HP pump will be more resistant at elevated pressures (>2 kbar); thus if the pump strength is the same, the time to increase pressure will be longer. A compressed air-controlled piston valve (Autoclave France, Nogent-sur-Oise, France) is inserted into the system to allow rapid and kinetically identical pressure rise. The manual pump is increased to the desired pressure with the valve closed and then the pressure increment to the sample cell can be triggered with dead times of ~1 s. Care must be taken while incrementing pressure in order to avoid thermal fluctuations; *see* **Note 4**. It is important to include valves to isolate each component of the system in order to identify the sources of eventual leaks.
5. *Buffer.* Under HP, the choice of the sample buffer is limited. Even buffers from the Good's list [94] can be bad pH

controllers [95]. Electrostriction of solvent around charges is accompanied by a decrease in volume. Hence, it is key to use buffer systems in which the unprotonated and protonated forms carry the same number of charges and/or small ionization volume [61, 96]. In particular care must be taken to avoid buffers of high ionization volume such as phosphate buffer essential. For example the following buffers are suitable for HP measurements: CAPS for pH range 9.5–11.5, Tris–HCl for pH range 7–9, MES or Bis-Tris for pH range 5.5–7.5, and sodium acetate for pH range 3.8–5.8. Once the buffer is defined, combinations with other additives have been widely implemented (*see* **Note 5**).

6. *Fluorophores.* The quality of the sample, i.e., purity and solubility, is crucial to avoid contaminating fluorescence, scattering, or other signal distortions. Recombinant or purchased polypeptides must be at least 95 % pure; mass spectrometry or polyacrylamide gels can be used to ensure the purity of the sample. The presence, quantity, and position (if structure is known) of intrinsic fluorescent residues must be known. The same information must be known if extrinsic covalently bonded fluorophores are used, e.g., FITC or other organic dyes. Often, the intrinsic stability of a protein or a complex and/or a small difference in volume can preclude the study of the desired phenomenon (unfolding, dissociation, etc.). In these cases, one can use stabilizing or destabilizing additives such as osmolytes or chemical denaturants or an acid or a base, as needed to "tune" the transition into the available pressure window. In such cases, care must be taken to use highly pure additives as well. Moreover, as presented below volumetric properties on proteins are sensitive to the presence of internal void volumes; additional information of the protein sample like cavity density and distribution as well as global stability may also be valuable before starting an HP experiment; *see* **Note 6**.

3 Methods

3.1 Equilibrium Unfolding

As described in the previous section and in comparison with atmospheric fluorescence experiments, the time required for cell loading, thermal equilibration, and equipment maintenance complicates but does not preclude the performance of titration experiments (*see* **Note 7**). For that reason we chose to describe a simple unfolding study on SNase protein (Fig. 2) that can be extrapolated to other experimental approaches. The high quantum yield and solvatochromic properties of tryptophan residue allow monitoring accurately the folded and unfolded states of proteins [72]. The protein

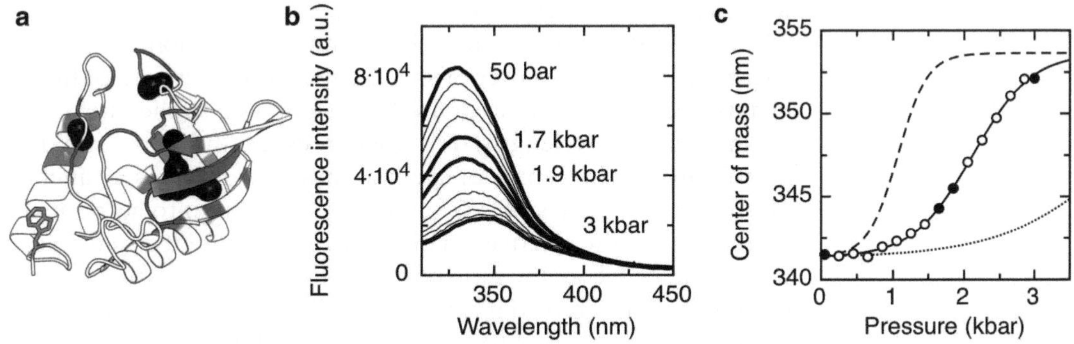

Fig. 2 Equilibrium HP unfolding of SNase protein. (**a**) Cartoon representation of the crystallographic structure of SNase (PDB ID 1SNC), the single tryptophan residue is represented with sticks, the molecular surface of cavities is represented in *black-filled circles*, and residues surrounding cavities are colored in *grey*. (**b**) Fluorescence emission spectra of SNase protein carried out at different pressures. The initial and the final emission spectra as well as the spectra at 1.7 and 1.9 kbar are highlighted in *black*. (**c**) Center of spectral mass calculated from the emission spectra recorded at different pressures. *Solid line* corresponds to data fitted to a two-state unfolding model and leads to a ΔG of 2.9 $\pm$ 0.1 kcal/mol and a ΔV of 60 $\pm$ 6 ml/mol. *Dashed* and *dotted lines* correspond to simulations of a two-state unfolding model by only changing ΔV to 120 or 30 ml/mol, respectively. *Filled circles* correspond to highlighted spectra of (**b**). Cavities were calculated using HOLLOW [139] and represented using PyMOL [148]. Unfolding was carried out at 10 mM Bis-Tris pH 5.5 at 20 °C, protein concentration of 50 μM, and excitation wavelength of 290 nm

intrinsic fluorophore signal can be followed in function of pressure by observing the following:

1. The emission intensity (Fig. 2) at a given wavelength or the entire spectral intensity over the emission band which decreases for SNase protein upon unfolding as a result of both static and dynamic quenching effects [79, 97, 98].
2. The shift to higher wavelengths of the maximum of the emission intensity upon denaturation as a consequence of becoming more exposed to solvent [72] (Fig. 2); *see* **Note 8**.
3. Anisotropy: Upon unfolding the anisotropy value decreases due to an increment of rotational mobility [72]. Note that this is only possible if the windows of the high-pressure cell are in quartz, and even then correction factors must be determined and applied to correct for pressure-induced birefringence in the quartz windows [92, 99]; *see* **Note 8**.

In Fig. 2b we show the spectra of emission of SNase protein as function of pressure. The equilibration times after each pressure increment must be complete and well defined (*see* **Note 9**). In Fig. 2c we plot the center of spectral mass of the emission spectra, an accurate method to quantify spectral shifts [31, 32]:

$$\langle \lambda \rangle_j = \frac{\sum_j F_j \lambda_j}{\sum_j F_j} \tag{2}$$

with F_j fluorescence intensity at wavelength $\lambda_j = 320, 321, 322, \ldots, 450$ nm. One can also do the computation in terms of energy (cm^{-1}). The center of spectral mass provides data of higher quality because it is an integral value, as opposed to the intensity at an individual wavelength. For each pressure-induced unfolding measurement, the center of spectral mass was fitted as a function of pressure for values of $\Delta G^{\text{o}}_{\text{u}}$ and $\Delta V^{\text{o}}_{\text{u}}$ taking into account the correction by the relative quantum yield of unfolded and folded states obtained from the ratio of the high- and low-pressure spectra. For pressure-induced equilibrium unfolding analysis, the free energy of unfolding is assumed to evolve linearly with the pressure, p:

$$\Delta G^{\text{o}}_{\text{u}p} = \Delta G^{\text{o}}_{\text{u}} + p\Delta V^{\text{o}}_{\text{u}} \tag{3}$$

where

$$\Delta G^{\text{o}}_{\text{u}p} = -RT \ln K_{\text{u}p} \tag{4}$$

and

$$K_{\text{u}i} = \frac{\left(\langle\lambda\rangle_{\text{f}} - \langle\lambda\rangle_{p}\right)}{\left(\langle\lambda\rangle_{p} - \langle\lambda\rangle_{\text{u}}\right)} \tag{5}$$

To account for differences in quantum yield between the folded and unfolded state in calculating the observable from the free energy for the nonlinear least squares fit, we assume that the observable ($<\lambda>_p$) at pressure p is a linear weighted sum of the observable value for F and U ($<\lambda>_{\text{f}}$ and $<\lambda>_{\text{u}}$), weighted for both their fractional population at that pressure, $\alpha_{i,p}$ (obtained from the value of free energy), and their fractional quantum yield calculated, f_Q:

$$\langle\lambda\rangle_p = \sum_i \alpha_{ip}\lambda_i f_Q, \quad \text{with} \quad f_Q = \frac{\alpha_i Q_i}{\sum_i \alpha_i Q_i} \tag{6}$$

where Q is the normalized quantum yield of species, i (in this case there are only two species, F and U). For deconvolution details refer to [23, 25, 63, 69, 73–77]. Relative to the partial molar volume of a folded protein (volume/mole), the magnitude of the partial molar volume decrease upon unfolding ΔV^{o} is rather small, approximately two orders lower, and in the range of 50–150 ml/mol or 0.5–2 % of the total [23, 25, 63, 69, 73–75]. Nonetheless, these small differences of volumes, $\Delta V^{\text{o}} > 5$ ml/mol, can be detected in pressure experiments, since the pressures used are rather high (2–4 kbar). As extensively demonstrated by the effect of cavity creating mutants on the hyperstable variant SNase, Δ + PHS [87], but also perceived earlier on T4 lysozyme [58] and other proteins [71, 82, 100–102], picturing the tridimensional location of cavities (Fig. 2a) can be helpful for data interpretation (refer also to **Note 6**).

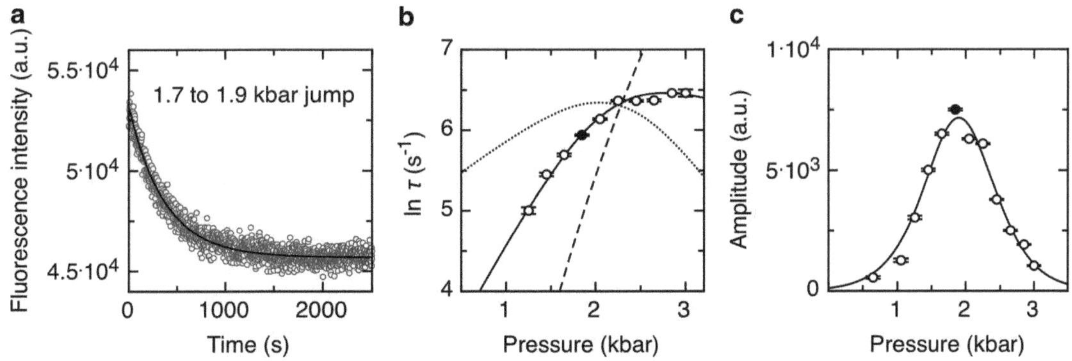

Fig. 3 HP relaxation kinetics of SNase protein. (**a**) Fluorescence intensity kinetics of a 0.2 kbar typical P-jump to 1.9 kbar. *Solid line* corresponds to data fitted to a mono-exponential function, leading to a macroscopic τ of 380 s and an amplitude of fluorescent signal of 7,500 $\pm$ 50 a.u. (**b**) Dependence of the lnτ(s) with pressure. *Solid line* corresponds to data fitted to a two-state unfolding model and leads to an activation volume of 52 $\pm$ 0.1 ml/mol and a k_f of 0.077 $\pm$ 0.011. *Dashed* and *dotted lines* correspond to simulations of the two-state unfolding model by setting the activation volume to 100 or 25 ml/mol, respectively. (**c**) Dependence of the amplitude with pressure. *Solid line* corresponds to data fitted to a two-state unfolding model (see text) and leads to a ΔG of 3.1 $\pm$ 0.1 kcal/mol, a ΔV of 70 $\pm$ 1 ml/mol, and a $p_{1/2}$ of 2.00 $\pm$ 0.01 kbar in agreement with results presented in Fig. 2. *Filled circles* correspond to data from the example shown in panel A. *P*-jumps were carried in between each emission spectrum presented in Fig. 2, the excitation wavelength was 290 nm, and emitted intensity was recorded at 340 nm

The larger the amount of solvent-excluded volume in the folded state which is eliminated in the unfolded state, the higher the absolute value of the volume change of unfolding, ΔV°, sharpening the curves, as simulated in Fig. 2c. Temperature modulates significantly the magnitude of ΔV° because the thermal expansivity of folded states is much smaller than that of unfolded states (refer to **Note 3**). This phenomenon, along with the ΔC_p of protein folding, is largely responsible for ellipticity and tilt of protein folding pressure–temperature-phase diagrams [46], and can lead to a change in sign of the volume change at high temperature [103].

3.2 Relaxation Kinetics

It is of interest to use pressure jump relaxation to investigate the volumetric properties of the TSE for two-state protein folding. In Fig. 3a we describe an example of the 0.2 kbar *P*-jump from 1.7 to 1.9 kbar of SNase protein. Here we simply monitor the intensity at a given wavelength, as this observable does not need to be quantum yield corrected. Intensity versus time profiles were fitted to a single exponential decay:

$$I_{(t)} = I_0 \exp^{-(t/\tau)} + I_f \qquad (7)$$

where I_0 is the intensity at time 0 before the jump, τ is the apparent macroscopic relaxation time or the time at which the population of the assembly is reduced to $1/e$ times its initial value, and I_f is the final fluorescence intensity. As revealed by random distribution of residuals, the absolute difference between experimental data and the fitted function, the decay of fluorescence intensity upon incrementing pressure fits well to the single exponential. It is important to ensure the end of the transition (*see* **Note 9**).

From the solution of the set of differential equations defining the change in population of the folded and unfolded states after a pressure jump,

$$\frac{\delta[F]}{\delta t} = -k_u[F] \quad \text{and} \quad \frac{\delta[U]}{\delta t} = -k_u[U] \tag{8}$$

we find the eigenvalue

$$\tau(p) = 1/(k_f(p) + k_u(p)) \tag{9}$$

where $\tau(p)$ is the relaxation time after a pressure jump to pressure p. This value depends only upon the final pressure, not the direction or the size of the jump, in the limit of relaxation kinetics [104]. In formal analogy with Eq. 1, the dependence of the apparent rate constant τ on the pressure can be derived from the Eyring theory [105] and can be written as

$$\left(\frac{\partial \ln \tau}{\partial p}\right)_T = -\frac{\Delta V^\ddagger}{RT} \tag{10}$$

Plots of the natural logarithm of the relaxation time vs. pressure can be fit for the folding and unfolding rate constants at atmospheric pressure, k_f and k_u, and the activation volumes for the folding and unfolding reactions, $\Delta V^\ddagger{}_f$ and $\Delta V^\ddagger{}_u$. The ratio k_f/k_u corresponds to the equilibrium constant K_f, and the sum of the activation volumes, $\Delta V_f^\ddagger$ and $\Delta V_u{}^\ddagger$, is equal to the equilibrium volume change of folding ΔV_f. The natural logarithm dependence of τ with pressure for SNase, Fig. 3b, presents profiles that are analogous to the chevron plots obtained with chemical denaturants [106, 107], with the exception that the sign of the activation volumes is not necessarily known. For example, denaturant always decreases the folding rate and increases the unfolding rate, although of course this occurs at varying degrees for different proteins. The activation volumes, on the other hand, can both be positive [108]. Hence, the plots do not necessarily present a chevron shape [90]. The basic principles of relaxation kinetics can be found in order to improve the robustness of the fitting, it is suitable to restrict the quantity of parameters to the smallest amount possible; the simpler (as in simpler not simplest) the better [109]. Instead of fitting to four parameters, we restricted to two parameters $\Delta V^\ddagger{}_f$ and k_f by constraining the k_u and $\Delta V_u{}^\ddagger$ values to the equilibrium constant and volume change [89, 110].

The ratio between $\Delta V^\ddagger$ and ΔV, or V_{TSE}-value [111], informs about the volumetric properties of the TSE relative to the folded and unfolded states. As simulated in Fig. 3b, the V_{TSE}-value relationship modulates the shape of the inverted chevron-plot. A value of 1 or higher implies that the rate-limiting step of the reaction is rather native-like presenting solvent excluded cavities, as it is the case for the presented data with SNase, observed in different conditions [5, 12, 112]. Volumetrically native-like TSEs have been reported for a number of other proteins [59, 67, 78, 113–117]. On the contrary, a low *V*-value, < (as in much much less) <1, involves an unfolded like TSE, i.e., the cavities are not formed and this volume occupied by solvent [118]. Moreover, in analogy to the *Phi*-value analysis [119], the V_{TSE}-value analysis [111] emerges as a tool to provide volumetric information on the effect of mutations on the TSE and thereby the structural properties of the protein at the folding barrier [120].

In Fig. 3c we plot the dependence of relaxation amplitudes with pressure. Relaxation amplitudes were calculated from the mono-exponential decays (Fig. 3a). The width of the Gaussian distribution of the relaxation amplitudes reflects the sharpness of the sigmoid of the equilibrium-unfolding curve (Fig. 2c); the maximum amplitude value corresponds to the largest difference between the initial and the final states. In two-state systems, the position near the maximum amplitude (Fig. 3c) corresponds to the pressure unfolding midpoint, analogous to the melting point, i.e., the $p_{1/2}$ value. This $p_{1/2}$ corresponds to the equilibrium situation where $K = 1$; thus from Eq. 1 $p_{1/2} = \Delta V/\Delta G$ for simple two-state systems. In some cases as in *p*-jumps, *K* and ΔV can be determined from relaxation amplitudes; refer to [90]. We analyzed the dependence of *p*-jump relaxation amplitudes as described for thermal relaxation amplitudes by the Gruebele laboratory [91] (Fig. 3c). Because amplitude signal focuses on the transition of the reaction, the fit is robust allowing bona fide parameterization of curves with incomplete baseline signal, underlining the usefulness of the method in experiments with low signal/noise in the folded or the unfolded states (*see* **Note 10**).

The combination of X-ray crystallography, FTIR, NMR, calorimetry, SAXS, and fluorescence carried out as a function of pressure, plus the determination of the volumetric reaction coordinates that can be assessed by them, provides a comprehensive framework to address fundamental unresolved questions such as the mechanism of protein folding or adaptation to extreme habitats [4, 121] (*see* **Note 11**).

4 Notes

1. In order to obtain reasonable signal-to-noise ratio (S/N) some considerations must be taken regarding the architecture of the HP setup (Fig. 1). Because of the differences in the light path of HP-fluorescence in comparison to standard fluorescence, in particular due to the thick sapphire windows and the

internal scattering of the stainless steel high-pressure cell, a concentration increment of approximately one order of magnitude is required in order to obtain similar S/N. Moreover, if room temperature differs significantly from the experimental one, condensation may occur at the exterior of the cell body windows. External blowing of $N_{2(g)}$ gas on the windows is recommended to avoid condensation. Another artifact that may affect the S/N is the presence of air bubbles on the transmitting fluid in the light path of the cell body. Removal of bubbles should be done before each experiment by homogenizing the transmitting fluid of the light path tubing of the cell body using a plastic Pasteur pipette for example. As air is more compressible than liquid, performing a fluorescence emission scan before and after applying a soft pressure (~0.05 kbar) can be enough to identify and dissolve annoying bubbles.

2. Upon loading the sample into the cuvette, the sample cell must be filled completely in order to allow an appropriate transmission of pressure. The presence of bubbles in the sample cell must be avoided; thus it is recommended to leave a convex meniscus on the top of the cuvette before placing the impermeable seal. According to Steam Tables [122], the sample will be compressed an ~20 % of its volume at 4 kbar, thus leaving a couple of millimeters of seal after the o-ring will avoid sample leaking upon o-ring rolling up or down. Consequently, foresee additional sample as loading may encompass some sample loss.

3. Considering the large mass of the cell body, thermo-equilibration time should be long enough, at least ~20 min. This is particularly relevant because protein volumetric properties are extremely sensitive to temperature due to expansivity effects:

$$E = \left(\frac{\partial V}{\partial T}\right)_p \quad (11)$$

where E is the partial molar expansivity. The effect of temperature ΔV has been shown on SNase protein [44], among others, to present a strong thermal dependence, >1 ml/mol per degree. Different proteins seem to respond differently to temperature [25], stimulating interest in understanding the origin of such changes, which most certainly contain information about protein structure-energetic relationships. Even more interesting, similar expansivity effects have been also observed at the TSE [44], by studying the temperature dependence of the activation volume of the folding reaction.

4. We consider the nonaggressive effect of HP on proteins, a leading advantage allowing the stabilization and thus accurate characterization of protein subunits and rare protein

intermediates of functional relevance, e.g., [31, 36, 39, 40, 51, 64, 115, 116, 123]. In order to guaranty this characteristic, progressive and small pressure-jumps (*P*-jumps), of at least 0.2 kbar, should be applied while incrementing or diminishing pressure. Abrupt *P*-jumps will cause uncontrolled thermal effects (adiabatic heating which is about 0.3°/100 bar) that may lead to irreversible situations or amorphous aggregations. If the desired equilibrium pressure change is superior to 0.2 kbar, slow rate compressions to reach them are recommended, e.g., 0.1 kbar/min. Triggered pressure jumps will ensure low experimental dead time but the pressure jump will not be controlled a priori. For example if constant 0.2 kbar jump is desired, the increment in pressure before triggering should be higher, ~50 bar, in order to reach a final jump of 0.2 kbar. Even then it will hardly be the same final pressure jump magnitude all over the curve. If relaxation times are slow enough, we recommend increasing pressure without the use of the automatic piston valve; with experience, the user may reach dead times of ~7 s. Beyond the collateral increase of biceps strength, constant and small pressure-jumps are important when relaxation amplitude analysis is pursued; *see* Subheading 3.

5. Controlling properly the pH is a minor but central limitation of HP; once the proper buffer is defined, combinations without HP side effects have been performed: for example sugars, e.g., sucrose and sorbitol [50], IPTG [49], xylose [54], or with osmolites, e.g., glycerol [38, 51], TMAO [52], urea [26, 39, 40, 59, 121, 123, 124], or crowding agents, e.g., PEG, dextran [53], or salts, e.g., K_2SO_4, $CaCl_2$ [50], guanidinium chloride [26–28, 87, 117, 125], or oxido-reducing agents, e.g., glutathione [27], DTT [28], ethanol [126], or benzene [57]. Moreover, the use of external dyes like ANS [39, 113, 127–132], thioflavin T [125, 126, 132], fluorescein [133, 134], and 2-amimopurine [57] has been implemented in protein systems without intrinsic fluorescence or to explore specific protein interactions or binding to external dyes under HP. Regarding the characterization of rare conformers, fluorescent probes will preferentially reveal macroscopic ones, i.e., intermediate species from multimers [30–38] or oligomeric intermediates from amyloid polymerizations [29]. Subtle conformational fluctuations like low-lying conformers belonging to the native manyfold of a protein energetic landscape can be assessed by more resolutive HP techniques such as HP NMR [135]. Beyond the classic use of intrinsic fluorescence properties of proteins, the use of other fluorescent dyes [39, 113, 125, 126, 128–132, 134] may enrich the research scope of HP-fluorescence spectroscopy.
6. The main technical limitation of HP-fluorescence apparatus corresponds to the maximal attainable pressure; depending on

the HP setup it can range from 3 to 7 kbar (1 kbar ≈ 100 MPa) [136], 3–4 kbar being the limitation of commercially attainable apparatus (e.g., ISS, Champaign, IL, USA). And the pressure range that proteins require to unfold or dissociate is large, 2–11 kbar [25]. The density of protein cavities predominates the effect of pressure determining the magnitude of ΔV as empirically observed by [20, 58, 71, 82, 100–102], confirmed by [87], and in agreement with low contributions of protein surface hydration, size, and sequence to ΔV [79, 108]. As at atmospheric pressure the presence of water in nonpolar cavities is *vanishingly small* [137], in order to evaluate the sensitivity of a protein system to HP and if the chosen protein system has known structures it is suitable to analyze the protein interior spaces. Several open-access software packages have been developed for quantifying and representing protein cavities, via molecular grid screening methods, in short, tools for the study of "biomacromolecular speleology" or *speleoproteinology*, e.g., CAVER, sphgen, CASTp, VOIDOO, or HOLLOW [138–142]. Moreover, combinations with small amounts of guanidinium chloride [26–28, 87, 117, 125] and urea [26, 39, 40, 59, 121, 123, 124] or varying the pH or the temperature are viable options to *modify* the stability of the folded state of proteins in order to complete a pressure denaturation curve on the pressure limits of the apparatus. Similar strategies can be implemented for interaction studies by varying the ionic strength.

7. Overall HP-fluorescence experiments require from the beginning to the end preparation and equilibration times longer than standard fluorescence. This is due to the positive activation volumes for protein folding. Indeed many examples of hysteresis in reversibility are simply a manifestation of the fact that the system was not given the time to equilibrate prior to changing the pressure again. Moreover the technical difficulty to use automated titrating modules or to add new reagents without opening the HP system reduces the feasibility to perform accurate titrations [143]. That is, to perform titration under HP, the user usually must load each sample one after the other, verify the presence of bubbles on the system, and equilibrate well the desired temperature. These steps may induce considerable experimental errors and sample duplicates should be considered. Even though it is not impossible, it is tedious to perform titrations at HP in order to evaluate the effect of pressure on an interaction [5]. Moreover, while the HP system is robust, minimal maintenance should be considered in controlling valve leaking and oxidation processes of the cell body stainless steel alloy, especially if water is the chosen pressure-transmitting fluid. In this case, for long periods

without usage it is recommended to replace all water traces of the cell body with pure ethanol, then remove the ethanol, and ensure dryness using compressed air prior to use.

8. Pressure profiles of the shift to higher wavelengths of the maximum of the emission intensity upon denaturation and of anisotropy must be fit with a quantum yield-weighting factor if the unfolded (or dissociated) form exhibits a different quantum yield than the folded (or associated) form. This is because the anisotropy and center of mass are linearly weighted for the fractional population of each species and their contribution to the total fluorescence intensity [99].

9. In order to ensure equilibrium conditions after pressure perturbation it is recommended to follow fluorescence intensity as function of time until its stabilization whether kinetic characterization is desired or not. At least five times the value of τ should be recorded in order to ensure the end of the decay and thus an equilibrated situation. Pressure relaxation times can be extremely long or short depending on the protein system and conditions [23, 25, 63, 69, 73–77] and under-estimations of the time required to reach the equilibrium may result in overestimations on the magnitude of the ΔV [79, 112]. As some proteins can present extremely long equilibration times, >8 h, increasing temperature or using small amounts of chemical denaturant can modulate the magnitude of the relaxation to convenient times.

10. Relaxation amplitude analysis is relevant for the HP field where the physical limitation to reach high pressures and the elevated unfolding pressure required to unfold proteins are usually combined. In addition, going back from kinetics data to thermodynamics appears as a validation method. For example, unpublished NMR data from our laboratory on single residue resolution *P*-jump studies using NMR SoFast HMQC spectroscopy [144] present a lower S/N than in HP-fluorescence, diminishing the quantity of interpretable kinetic experiments, i.e., the baselines. Despite the lower S/N, amplitude analysis validated well the data obtained on both techniques (unpublished results). Thus, by using the same sample one can explore thermodynamic and kinetic characterization of the unfolding reaction, thus increasing the confidence by analysis of relaxation amplitudes, with special interest on protein reactions presenting more than two states, e.g., [118, 145].

11. Together with intrinsic protein fluorescence, the use of external additives and/or alternative signals such as following light scattering by setting the excitation and emission wavelengths to 320 nm completes the characterization of oligomeric states on large protein polymers [29] or even on the study of viral

capsides [127]. Different methods have specific characteristics, advantages, and disadvantages. Regarding sample preparation, HP-NMR as well as HP-SAXS require elevated amount of protein and the apparatus are found to be usually shared with several laboratories, being overall expensive methods. In this context, it is advantagous to carry out HP fluorescence studies to characterize the system thoroughly prior to moving on to these more complex approaches. Moreover, combined research provides new point of views, giving robustness to the conclusions of the work through global analysis. The combination of HP-NMR with HP-fluorescence has shed light in many cases, such as conformational fluctuations and local unfolding observations [146]; for example characterization of a globular highly hydrated protein intermediate by fluorescence and HP-NMR [121], insight into amyloid dissociations [147] or the effect of cavities on the contributions to ΔV [87].

References

1. Somero GN (1992) Adaptations to high hydrostatic pressure. Annu Rev Physiol 54: 557–577
2. Abe F, Horikoshi K (2001) The biotechnological potential of piezophiles. Trends Biotechnol 19(3):102–108
3. Daniel I, Oger P, Winter R (2006) Origins of life and biochemistry under high-pressure conditions. Chem Soc Rev 35(10):858–875
4. Matthews BW (2012) Proteins under pressure. Proc Natl Acad Sci USA 109(18): 6792–6793
5. Markley JL, Northrop DB, Royer CA (eds) (1996) High pressure effects in biophysics and enzymology. Oxford University Press, Oxford, p 381. ISBN: 9780195097221
6. Bridgman PW (1914) The coagulation of albumen by pressure *J.* Biol Chem 19:511–512
7. Suzuki K, Miyosawa Y, Suzuki C (1963) Protein denaturation by high pressure. Measurements of turbidity of isoelectric ovalbumin and horse serum albumin under high pressure. Arch Biochem Biophys 101:225–228
8. Weber G, Tanaka F, Okamoto BY, Drickamer HG (1974) The effect of pressure on the molecular complex of isoalloxazine and adenine. Proc Natl Acad Sci USA 71(4):1264–1266
9. Mueller JD, Gratton E (2003) High-pressure fluorescence correlation spectroscopy. Biophys J 85(4):2711–2719
10. Harris RD, Jacobs M, Long MM, Urry DW (1976) A high-pressure sample cell for circular dichroism studies. Anal Biochem 73(2): 363–368
11. Hayashi R, Kakehi Y, Kato M, Tanimizu N, Ozawa S, Matsumoto M, Kawai S, Pudney P (2002) Circular dichroism under high pressure. In: Rikimaru H (ed) Trends in high pressure bioscience and technology, vol. 19. Elsevier Science B.V., Amsterdam, p 579–586. http://www.sciencedirect.com/science/article/pii/S0921042302801575
12. Panick G, Malessa R, Winter R et al (1998) Structural characterization of the pressure-denatured state and unfolding/refolding kinetics of staphylococcal nuclease by synchrotron small-angle X-ray scattering and Fourier-transform infrared spectroscopy. J Mol Biol 275(2):389–402
13. Panick G, Malessa R, Winter R (1999) Differences between the pressure- and temperature-induced denaturation and aggregation of beta-lactoglobulin A, B, and AB monitored by FT-IR spectroscopy and small-angle X-ray scattering. Biochemistry 38(20):6512–6519
14. McCoy J, Hubbell WL (2011) High-pressure EPR reveals conformational equilibria and volumetric properties of spin-labeled proteins. Proc Natl Acad Sci USA 108(4): 1331–1336
15. Jonas J, Ballard L, Nash D (1998) High-resolution, high-pressure NMR studies of proteins. Biophys J 75(1):445–452
16. Kamatari YO, Kitahara R, Yamada H, Yokoyama S et al (2004) High-pressure NMR spectroscopy for characterizing folding intermediates and denatured states of proteins. Methods 34(1):133–143

17. Lassalle MW, Akasaka K (2007) The use of high-pressure nuclear magnetic resonance to study protein folding. Methods Mol Biol 350:21–38
18. Collins MD, Kim CU, Gruner SM (2011) High-pressure protein crystallography and NMR to explore protein conformations. Annu Rev Biophys 40:81–98
19. Woenckhaus J, Kohling R, Winter R, Thiyagarajan P et al (2000) High pressure-jump apparatus for kinetic studies of protein folding reactions using the small-angle synchrotron x-ray scattering technique. Rev Sci Instrum 71(10):3895–3899
20. Ando N, Chenevier P, Novak M et al (2008) High hydrostatic pressure small-angle X-ray scattering cell for protein solution studies featuring diamond windows and disposable sample cells. J Appl Crystallogr 41(1):167–175
21. Paliwal A, Asthagiri D, Bossev DP, Paulaitis ME (2004) Pressure denaturation of staphylococcal nuclease studied by neutron small-angle scattering and molecular simulation. Biophys J 87(5):3479–3492
22. Stefanowicz P, Petry-Podgorska I, Kowalewska K et al (2009) Electrospray ionization mass spectrometry as a method for studying the high-pressure denaturation of proteins. Biosci Rep 30(2):91–99
23. Balny C, Masson P, Travers F (1989) Some recent aspects of the use of high-pressure for protein investigations in solution. High Pressure Res 2(1):1–28
24. Lin LN, Brandts JF, Brandts JM, Plotnikov V (2002) Determination of the volumetric properties of proteins and other solutes using pressure perturbation calorimetry. Anal Biochem 302(1):144–160
25. Royer CA (2002) Revisiting volume changes in pressure-induced protein unfolding. Biochim Biophys Acta 1595(1–2):201–209
26. St John RJ, Carpenter JF, Balny C, Randolph TW (2001) High pressure refolding of recombinant human growth hormone from insoluble aggregates. Structural transformations, kinetic barriers, and energetics. J Biol Chem 276(50):46856–46863
27. St John RJ, Carpenter JF, Randolph TW (2002) High-pressure refolding of disulfide-cross-linked lysozyme aggregates: thermodynamics and optimization. Biotechnol Prog 18(3):565–571
28. Seefeldt MB, Ouyang J, Froland WA et al (2004) High-pressure refolding of bikunin: efficacy and thermodynamics. Protein Sci 13(10):2639–2650
29. Kim YS, Randolph TW, Seefeldt MB, Carpenter JF (2006) High-pressure studies on protein aggregates and amyloid fibrils. Methods Enzymol 413:237–253
30. Weber G (1986) Phenomenological description of the association of protein subunits subjected to conformational drift. Effects of dilution and of hydrostatic pressure. Biochemistry 25(12):3626–3631
31. Royer CA, Weber G, Daly TJ, Matthews KS (1986) Dissociation of the lactose repressor protein tetramer using high hydrostatic pressure. Biochemistry 25(25):8308–8315
32. Silva JL, Miles EW, Weber G (1986) Pressure dissociation and conformational drift of the beta dimer of tryptophan synthase. Biochemistry 25(19):5780–5786
33. Scarlata SF, Ropp T, Royer CA (1989) Histone subunit interactions as investigated by high pressure. Biochemistry 28(16): 6637–6641
34. Pin S, Royer CA, Gratton E et al (1990) Subunit interactions in hemoglobin probed by fluorescence and high-pressure techniques. Biochemistry 29(39):9194–9202
35. Silva JL, Silveira CF, Correia Junior A, Pontes L (1992) Dissociation of a native dimer to a molten globule monomer. Effects of pressure and dilution on the association equilibrium of arc repressor. J Mol Biol 223 (2):545–555
36. Pin S, Royer CA (1994) High-pressure fluorescence methods for observing subunit dissociation in hemoglobin. Methods Enzymol 232:42–55
37. Herberhold H, Marchal S, Lange R et al (2003) Characterization of the pressure-induced intermediate and unfolded state of red-shifted green fluorescent protein–a static and kinetic FTIR, UV/VIS and fluorescence spectroscopy study. J Mol Biol 330(5): 1153–1164
38. Valente-Mesquita VL, Botelho MM, Ferreira ST (1998) Pressure-induced subunit dissociation and unfolding of dimeric beta-lactoglobulin. Biophys J 75(1):471–476
39. Foguel D, Silva JL, de Prat-Gay G (1998) Characterization of a partially folded monomer of the DNA-binding domain of human papillomavirus E2 protein obtained at high pressure. J Biol Chem 273(15):9050–9057
40. Gorovits B, Raman CS, Horowitz PM (1995) High hydrostatic pressure induces the dissociation of cpn60 tetradecamers and reveals a plasticity of the monomers. J Biol Chem 270(5):2061–2066

41. Heremans K, Smeller L (1998) Protein structure and dynamics at high pressure. Biochim Biophys Acta 1386(2):353–370
42. Zhang J, Peng X, Jonas A, Jonas J (1995) NMR study of the cold, heat, and pressure unfolding of ribonuclease A. Biochemistry 34(27):8631–8641
43. Herberhold H, Winter R (2002) Temperature- and pressure-induced unfolding and refolding of ubiquitin: a static and kinetic Fourier transform infrared spectroscopy study. Biochemistry 41(7):2396–2401
44. Panick G, Vidugiris GJ, Malessa R et al (1999) Exploring the temperature-pressure phase diagram of staphylococcal nuclease. Biochemistry 38(13):4157–4164
45. Ravindra R, Winter R (2003) On the temperature–pressure free-energy landscape of proteins. Chemphyschem 4(4):359–365
46. Smeller L (2002) Pressure-temperature phase diagrams of biomolecules. Biochim Biophys Acta 1595(1–2):11–29
47. Heremans K (1980) Biophysical chemistry at high pressure. Rev Phys Chem Jpn 50: 259–273
48. Lima LM, Foguel D, Silva JL (2000) DNA tightens the dimeric DNA-binding domain of human papillomavirus E2 protein without changes in volume. Proc Natl Acad Sci USA 97(26):14289–14294
49. Royer CA, Chakerian AE, Matthews KS (1990) Macromolecular binding equilibria in the lac repressor system: studies using high-pressure fluorescence spectroscopy. Biochemistry 29(20):4959–4966
50. Herberhold H, Royer CA, Winter R (2004) Effects of chaotropic and kosmotropic cosolvents on the pressure-induced unfolding and denaturation of proteins: an FT-IR study on staphylococcal nuclease. Biochemistry 43 (12):3336–3345
51. Oliveira AC, Gaspar LP, Da Poian AT, Silva JL (1994) Arc repressor will not denature under pressure in the absence of water. J Mol Biol 240(3):184–187
52. Krywka C, Sternemann C, Paulus M et al (2008) Effect of osmolytes on pressure-induced unfolding of proteins: a high-pressure SAXS study. Chemphyschem 9(18):2809–2815
53. Wang S, Tate MW, Gruner SM (2012) Protein crowding impedes pressure-induced unfolding of staphylococcal nuclease. Biochim Biophys Acta 1820(7):957–961
54. Frye KJ, Royer CA (1997) The kinetic basis for the stabilization of staphylococcal nuclease by xylose. Protein Sci 6(4):789–793
55. King L, Weber G (1986) Conformational drift of dissociated lactate dehydrogenases. Biochemistry 25(12):3632–3637
56. Fujisawa T, Kato M, Inoko Y (1999) Structural characterization of lactate dehydrogenase dissociation under high pressure studied by synchrotron high-pressure small-angle X-ray scattering. Biochemistry 38(20): 6411–6418
57. Ando N (2009) Biomacromolecules under high hydrostatic pressure. In: Department of Physics. Cornell University, Ithaca, NY, pp 221 http://www.google.fr/url?sa=t&rct=j&q=&esrc=s&source=web&cd=1&ved=0CF4QFjAA&url=http%3A%2F%2Fbigbro.biophys.cornell.edu%2Fpublications%2Ff26%2520Ando%2520PhD%2520thesis.pdf&ei=5wUsUJWROcq50QXl5YG4Aw&usg=AFQjCNEH7ffkJeYKGtdisUpi9JXZvs9i7Q
58. Ando N, Barstow B, Baase WA, Fields A, Matthews BW, Gruner SM (2008) Structural and thermodynamic characterization of T4 lysozyme mutants and the contribution of internal cavities to pressure denaturation. Biochemistry 47(42):11097–11109
59. Rouget JB, Schroer MA, Jeworrek C et al (2010) Unique features of the folding landscape of a repeat protein revealed by pressure perturbation. Biophys J 98(11):2712–2721
60. Hawley SA (1971) Reversible pressure–temperature denaturation of chymotrypsinogen. Biochemistry 10(13):2436–2442
61. Zipp A, Kauzmann W (1973) Pressure denaturation of metmyoglobin. Biochemistry 12(21):4217–4228
62. Schade BC, Rudolph R, Ludemann HD, Jaenicke R (1980) Reversible high-pressure dissociation of lactic dehydrogenase from pig muscle. Biochemistry 19(6):1121–1126
63. Royer CA (1995) Application of pressure to biochemical equilibria: the other thermodynamic variable. Methods Enzymol 259: 357–377
64. Mohana-Borges R, Lima Silva J, de Prat-Gay G (1999) Protein folding in the absence of chemical denaturants. Reversible pressure denaturation of the noncovalent complex formed by the association of two protein fragments. J Biol Chem 274(12):7732–7740
65. Torrent J, Connelly JP, Coll MG et al (1999) Pressure versus heat-induced unfolding of ribonuclease A: the case of hydrophobic interactions within a chain-folding initiation site. Biochemistry 38(48):15952–15961
66. Ishimaru D, Lima LM, Maia LF et al (2004) Reversible aggregation plays a crucial role on

the folding landscape of p53 core domain. Biophys J 87(4):2691–2700

67. Tan CY, Xu CH, Wong J et al (2005) Presssure equilibrium and jump study on unfolding of 23-kDa protein from spinach photosystem II. Biophys J 88(2):1264–1275
68. Brandts JF, Oliveira RJ, Westort C (1970) Thermodynamics of protein denaturation. Effect of pressure on the denaturation of ribonuclease A. Biochemistry 9(4):1038–1047
69. Weber G, Drickamer HG (1983) The effect of high pressure upon proteins and other biomolecules. Q Rev Biophys 16(1):89–112
70. Hummer G, Garde S, Garcia AE et al (1998) The pressure dependence of hydrophobic interactions is consistent with the observed pressure denaturation of proteins. Proc Natl Acad Sci USA 95:1552–1555
71. Collins MD, Hummer G, Quillin ML et al (2005) Cooperative water filling of a nonpolar protein cavity observed by high-pressure crystallography and simulation. Proc Natl Acad Sci USA 102(46):16668–16671
72. Lakowicz JR (1983) Principles of fluorescence spectroscopy. Plenum Press, New York
73. Gross M, Jaenicke R (1994) Proteins under pressure. The influence of high hydrostatic pressure on structure, function and assembly of proteins and protein complexes. Eur J Biochem 221(2):617–630
74. Molina-Garcia AD (2002) The effect of hydrostatic pressure on biological systems. Biotechnol Genet Eng Rev 19:3–54
75. Ando N (2006) High hydrostatic pressure effect on proteins: fluorescence studies. In: Encyclopedia of analytical chemistry, Wiley. http://dx.doi.org/10.1002/9780470027318.a9246
76. Torrent J, Font J, Herberhold H et al (2006) The use of pressure-jump relaxation kinetics to study protein folding landscapes. Biochim Biophys Acta 1764(3):489–496
77. Ruan K, Balny C (2002) High pressure static fluorescence to study macromolecular structure-function. Biochim Biophys Acta 1595(1–2):94–102
78. Silva JL, Foguel D, Royer CA (2001) Pressure provides new insights into protein folding, dynamics and structure. Trends Biochem Sci 26(10):612–618
79. Royer CA, Hinck AP, Loh SN et al (1993) Effects of amino acid substitutions on the pressure denaturation of staphylococcal nuclease as monitored by fluorescence and nuclear magnetic resonance spectroscopy. Biochemistry 32(19):5222–5232
80. Vidugiris GJA, Truckses DM, Markley JL, Royer CA (1996) High-pressure denaturation of staphylococcal nuclease proline-to-glycine substitution mutants. Biochemistry 35:3857–3864
81. Frye KJ, Perman CS, Royer CA (1996) Testing the correlation between delta A and delta V of protein unfolding using m value mutants of staphylococcal nuclease. Biochemistry 35(31):10234–10239
82. Frye KJ, Royer CA (1998) Probing the contribution of internal cavities to the volume chnage of protein unfolding under pressure. Protein Sci 7:2217–2222
83. Seemann H, Winter R, Royer CA (2001) Volume, expansivity and isothermal compressibility changes associated with temperature and pressure unfolding of Staphylococcal nuclease. J Mol Biol 307(4):1091–1102
84. Woenckhaus J, Kohling R, Thiyagarajan P et al (2001) Pressure-jump small-angle x-ray scattering detected kinetics of staphylococcal nuclease folding. Biophys J 80(3):1518–1523
85. Mitra L, Rouget JB, Garcia-Moreno B et al (2008) Towards a quantitative understanding of protein hydration and volumetric properties. Chemphyschem 9(18):2715–2721
86. Kitahara R, Hata K, Maeno A et al (2011) Structural plasticity of staphylococcal nuclease probed by perturbation with pressure and pH. Proteins 79(4):1293–1305
87. Roche J, Caro JA, Norberto DR et al (2012) Cavities determine the pressure unfolding of proteins. Proc Natl Acad Sci USA 109 (18):6945–6950
88. Lassalle MW, Yamada H, Akasaka K (2000) The pressure-temperature free energy-landscape of staphylococcal nuclease monitored by (1)H NMR. J Mol Biol 298(2):293–302
89. Brun L, Isom DG, Velu P et al (2006) Hydration of the folding transition state ensemble of a protein. Biochemistry 45(11):3473–3480
90. Bernasconi CF (1976) Relaxation kinetics. In: Claude F (ed) Bernasconi. Academic, New York
91. Girdhar K, Scott G, Chemla YR, Gruebele M (2011) Better biomolecule thermodynamics from kinetics. J Chem Phys 135(1):015102
92. Paladini AA, Weber G (1981) Absolute measurements of fluorescence polarization at high pressures. Rev Sci Instrum 52(3):419–427

93. Balny C, Saldana JL, Dahan N (1987) High-pressure stopped-flow fluorometry at subzero temperatures: application to kinetics of the binding of NADH to liver alcohol dehydrogenase. Anal Biochem 163(2):309–315
94. Good NE, Winget GD, Winter W et al (1966) Hydrogen ion buffers for biological research. Biochemistry 5(2):467–477
95. Quinlan RJ, Reinhart GD (2005) Baroresistant buffer mixtures for biochemical analyses. Anal Biochem 341(1):69–76
96. Distèche A (1972) Effects of pressure on the dissociation of weak acids. Symp Soc Exp Biol 26:27–60
97. Shortle D (1986) Guanidine hydrochloride denaturation studies of mutant forms of staphylococcal nuclease. J Cell Biochem 30(4):281–289
98. Eftink MR, Gryczynski I, Wiczk W et al (1991) Effects of temperature on the fluorescence intensity and anisotropy decays of staphylococcal nuclease and the less stable nuclease-conA-SG28 mutant. Biochemistry 30(37):8945–8953
99. Royer CA (1985) The use of high pressure fluorescence spectroscopy to investigate subunit interactions in oligomeric proteins. University of Illinois at Urbana-Champaign, Urbana, IL, pp 364
100. Cioni P (2006) Role of protein cavities on unfolding volume change and on internal dynamics under pressure. Biophys J 91(9): 3390–3396
101. Botelho MM, Valente-Mesquita VL, Oliveira KM et al (2000) Pressure denaturation of beta-lactoglobulin. Different stabilities of isoforms A and B, and an investigation of the Tanford transition. Eur J Biochem 267(8):2235–2241
102. Lassalle MW, Yamada H, Morii H et al (2001) Filling a cavity dramatically increases pressure stability of the c-Myb R2 subdomain. Proteins 45(1):96–101
103. Royer C, Winter R (2011) Protein hydration and volumetric properties. Curr Opin Colloid Inter Sci 16(6):568–571
104. Eigen M, de Maeyer L (1963) Techniques of organic chemistry, In: Weissberger A (ed), vol 8. Interscience Publishers, New York, p. 865. http://books.google.fr/books?id=A58-hAQAAMAAJ
105. Eyring H (1935) The activated complex in chemical reactions. J Chem Phys 3(2): 107–115
106. Matthews CR, Crisanti MM, Manz JT, Gepner GL (1983) Effect of a single amino acid substitution on the folding of the alpha subunit of tryptophan synthase. Biochemistry 22(6):1445–1452
107. Chan HS, Dill KA (1998) Protein folding in the landscape perspective: chevron plots and non-Arrhenius kinetics. Proteins 30(1):2–33
108. Rouget JB, Aksel T, Roche J et al (2011) Size and sequence and the volume change of protein folding. J Am Chem Soc 133 (15):6020–6027
109. Bachmann A, Kiefhaber T (2008) Kinetic mechanisms in protein folding. In: Protein folding Handbook. Wiley, Weinheim, pp 377–410. http://dx.doi.org/10.1002/9783527619498.ch12a
110. Stites WE, Gittis AG, Lattman EE, Shortle D (1991) In a staphylococcal nuclease mutant the side-chain of a lysine replacing valine 66 is fully buried in the hydrophobic core. J Mol Biol 221(1):7–14
111. Mitra L, Hata K, Kono R et al (2007) V(i)-value analysis: a pressure-based method for mapping the folding transition state ensemble of proteins. J Am Chem Soc 129(46):14108–14109
112. Vidugiris GJA, Markley JL, Royer CA (1995) Evidence for a molten globule-like transition state in protein folding from determination of activation volumes. Biochemistry 34: 4909–4912
113. Desai G, Panick G, Zein M et al (1999) Pressure-jump studies of the folding/unfolding of trp repressor. J Mol Biol 288(3): 461–475
114. Jacob MH, Saudan C, Holtermann G et al (2002) Water contributes actively to the rapid crossing of a protein unfolding barrier. J Mol Biol 318(3):837–845
115. Kitahara R, Royer C, Yamada H et al (2002) Equilibrium and pressure-jump relaxation studies of the conformational transitions of P13MTCP1. J Mol Biol 320(3):609–628
116. Mohana-Borges R, Silva JL, Ruiz-Sanz J, Gd PG (1999) Folding of a pressure-denatured model protein. Proc Natl Acad Sci USA 96(14):7888–7893
117. Pappenberger G, Saudan C, Becker M et al (2000) Denaturant-induced movement of the transition state of protein folding revealed by high pressure stopped-flow measurements. Proc Natl Acad Sci USA 97:17–22
118. Osvath S, Quynh LM, Smeller L (2009) Thermodynamics and kinetics of the pressure unfolding of phosphoglycerate kinase. Biochemistry 48(42):10146–10150
119. Fersht AR, Sato S (2004) Phi-value analysis and the nature of protein-folding transition states. Proc Natl Acad Sci USA 101(21): 7976–7981
120. Fersht AR (1985) Enzyme structures and mechanisms. W. H. Freeman & Co., New York

121. Ohmae E, Murakami C, Tate S et al (2012) Pressure dependence of activity and stability of dihydrofolate reductases of the deep-sea bacterium Moritella profunda and Escherichia coli. Biochim Biophys Acta 1824(3):511–519
122. Keenan JH, Moore JG, Hill PG, Keyes FG (1969) Steam tables: thermodynamic properties of water including vapor, liquid, and solid phases (english units). In: Wiley JS (ed) Incorporated, http://books.google.fr/books/about/Steam_tables.html?id=HgV-RAAAAMAAJ&redir_esc=y
123. Rene-Trouillefou M, Benzaria A, Marchal S et al (2010) Staphylococcal enterotoxin A: partial unfolding caused by high pressure or denaturing agents enhances superantigenicity. Biochim Biophys Acta 1804(6):1322–1333
124. Goncalves RB, Sanches D, Souza TL et al (2008) The proapoptotic protein Smac/DIABLO dimer has the highest stability as measured by pressure and urea denaturation. Biochemistry 47(12):3832–3841
125. Chatani E, Kato M, Kawai T et al (2005) Main-chain dominated amyloid structures demonstrated by the effect of high pressure. J Mol Biol 352(4):941–951
126. Grudzielanek S, Smirnovas V, Winter R (2006) Solvation-assisted pressure tuning of insulin fibrillation: from novel aggregation pathways to biotechnological applications. J Mol Biol 356(2):497–509
127. Goncalves RB, Mendes YS, Soares MR et al (2007) VP4 protein from human rhinovirus 14 is released by pressure and locked in the capsid by the antiviral compound WIN. J Mol Biol 366(1):295–306
128. Li TM, Hook JW 3rd, Drickamer HG, Weber G (1976) Plurality of pressure-denatured forms in chymotrypsinogen and lysozyme. Biochemistry 15(25):5571–5580
129. Gaspar LP, Terezan AF, Pinheiro AS et al (2001) The metastable state of nucleocapsids of enveloped viruses as probed by high hydrostatic pressure. J Biol Chem 276(10):7415–7421
130. Ruan K, Xu C, Yu Y et al (2001) Pressure-exploration of the 33-kDa protein from the spinach photosystem II particle. Eur J Biochem 268(9):2742–2750
131. Souza MO, Creczynski-Pasa TB, Scofano HM et al (2004) High hydrostatic pressure perturbs the interactions between CF(0)F(1) subunits and induces a dual effect on activity. Int J Biochem Cell Biol 36(5):920–930
132. Marchal S, Shehi E, Harricane MC et al (2003) Structural instability and fibrillar aggregation of non-expanded human ataxin-3 revealed under high pressure and temperature. J Biol Chem 278 (34):31554–31563
133. Yamamoto S, Otsuka Y, Borjigin G et al (2005) Effects of a high-pressure treatment on the activity and structure of rabbit muscle proteasome. Biosci Biotechnol Biochem 69(7):1239–1247
134. Rietveld AW, Ferreira ST (1996) Deterministic pressure dissociation and unfolding of triose phosphate isomerase: persistent heterogeneity of a protein dimer. Biochemistry 35(24):7743–7751
135. Akasaka K, Li H (2001) Low-lying excited states of proteins revealed from nonlinear pressure shifts in 1H and 15N NMR. Biochemistry 40(30):8665–8671
136. Ruan K, Lange R, Meersman F et al (1999) Fluorescence and FTIR study of the pressure-induced denaturation of bovine pancreas trypsin. Eur J Biochem 265(1):79–85
137. Wolfenden R, Radzicka A (1994) On the probability of finding a water molecule in a nonpolar cavity. Science 265(5174):936–937
138. Hendrix DK, Kuntz ID (1998) Surface solid angle-based site points for molecular docking. Pac Symp Biocomput 317–326. http://www.ncbi.nlm.nih.gov/pubmed/9697192
139. Ho BK, Gruswitz F (2008) HOLLOW: generating accurate representations of channel and interior surfaces in molecular structures. BMC Struct Biol 8:49
140. Till MS, Ullmann GM (2010) McVol—a program for calculating protein volumes and identifying cavities by a Monte Carlo algorithm. J Mol Model 16(3):419–429
141. Kleywegt GJ, Jones TA (1994) Detection, delineation, measurement and display of cavities in macromolecular structures. Acta Crystallogr D Biol Crystallogr 50(Pt 2):178–185
142. Petrek M, Otyepka M, Banas P et al (2006) CAVER: a new tool to explore routes from protein clefts, pockets and cavities. BMC Bioinformatics 7:316
143. Bugnon P, Laurenczy G, Ducommun Y et al (1996) High-pressure stopped-flow spectrometer for kinetic studies of fast reactions by absorbance anbd fluorescence detection. Analyt Chem 68:3045–3049
144. Schanda P, Brutscher B (2006) Hadamard frequency-encoded SOFAST-HMQC for ultrafast two-dimensional protein NMR. J Magn Reson 178(2):334–339

145. Vidugiris GJ, Royer CA (1998) Determination of the volume changes for pressure-induced transitions of apomyoglobin between the native, molten globule, and unfolded states. Biophys J 75(1):463–470
146. Kitahara R, Simorellis AK, Hata K, Maeno A, Yokoyama S, Koide S, Akasaka K (2012) A delicate interplay of structure, dynamics, and thermodynamics for function: a high pressure NMR study of outer surface protein A. Biophys J 102(4):916–926
147. Shah BR, Maeno A, Matsuo H, Tachibana H, Akasaka K (2012) Pressure-accelerated dissociation of amyloid fibrils in wild-type hen lysozyme. Biophys J 102(1):121–126
148. DeLano W (2002) The PyMOL molecular graphics system. DeLano Scientific, Palo Alto, Ca, USA

Part II

Time-Resolved Fluorescence Spectroscopy (TRFS)

Chapter 5

Frequency Domain Fluorometry: Theory and Application

Carissa M. Vetromile and David M. Jameson

Abstract

Frequency domain fluorometry is a widely utilized tool in the physical, chemical, and biological sciences. This chapter focuses on the theory of the method and the practical aspects required to carry out intensity decay, i.e., lifetime measurements on a modern frequency domain fluorometer. Several chemical/biological systems are utilized to illustrate data acquisition protocols. Data analysis procedures and methodologies are also discussed.

Key words Frequency domain fluorometry, Multifrequency phase-modulation fluorometry, Ethidium bromide, tRNA, Phasor plot, Human serum albumin (HSA)

1 Introduction

Fluorescence spectroscopy has proven to be an extremely valuable tool in a wide variety of fields ranging from biology to chemistry as well as medical diagnostics and material sciences. Its usefulness stems from the detailed information available from multiple fluorescent parameters, i.e., emission spectra, excitation spectra, polarizations/anisotropies, lifetimes, and quantum yields. While knowledge of all these fluorescent properties is important to fully understand the processes being investigated, the focus of this review is on time-resolved measurements, in particular intensity decay or lifetime measurements.

The absorption of light generates an excited state within femtoseconds and deactivation typically occurs on the order of picoseconds to tens of nanoseconds depending on the fluorophore and its surroundings. The newly generated population of photoexcited molecules decays exponentially from the S_1 electronic excited state (usually, although some exceptions exist) through emission of a photon, though other non-radiative processes usually deactivate the excited state also. The time it takes for the number of excited molecules to decay to $1/e$ of the excited state population is defined

Yves Engelborghs and Antonie J.W.G. Visser (eds.), *Fluorescence Spectroscopy and Microscopy: Methods and Protocols*, Methods in Molecular Biology, vol. 1076, DOI 10.1007/978-1-62703-649-8_5,

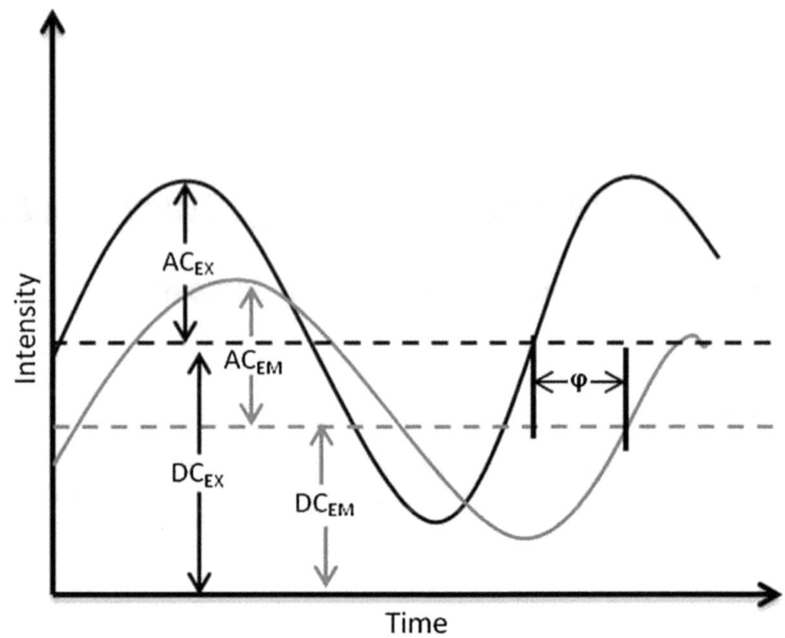

Fig. 1 Schematic representation of sinusoidally modulated exciting light (*black*) and the resulting sinusoidally modulated fluorescence (*grey*) delayed in phase and demodulated

as the excited state lifetime, τ, and is unique to a given molecule in a given environment [1]. This environmental sensitivity is one reason why the excited state lifetime is an important fluorescent property. Two principle methods, namely, frequency domain and time domain, are presently used to obtain time-resolved information. This chapter discusses the frequency domain method only and focuses specifically on data acquisition and analysis of intensity decay, i.e., lifetime data. We also describe practical aspects of the frequency domain method, also known as multifrequency phase and modulation fluorometry, by describing measurements on a simple fluorophore in solution (fluorescein), protein fluorescence (human albumin serum), and a ligand binding system (ethidium bromide-tRNA).

The first accurate frequency domain lifetime measurements were realized by Enrique Gaviola in 1926 [2, 3], while the mathematical theory behind frequency domain fluorometry was first described by Dushinsky in 1933 [4]. Dushinsky determined that a fluorophore, with a single exponential excited state decay time, τ, excited by light sinusoidally modulated at high frequencies, $E(\mathrm{t})$, emits light, $F(\mathrm{t})$, sinusoidally modulated at the same angular frequency, ω (equal to $2\pi f$, where f is the linear light modulation frequency), as the exciting light, but delayed in phase (φ) and demodulated as depicted in Fig. 1 and expressed as:

$$F(t) = F(0)[1 + M_F \cos(\omega\tau + \varphi) \tag{1}$$

The relative modulation of the emission, M, is given by the ratio of the AC/DC components for the exciting, M_E, and emitted light, M_F:

$$M_E = \left(\frac{AC_E}{DC_E}\right) \quad \text{and} \quad M_F = \left(\frac{AC_F}{DC_F}\right) \tag{2}$$

$$M = \left(\frac{M_F}{M_E}\right) \tag{3}$$

The correlation between the phase delay, modulation, and excited state lifetime is given by:

$$\tan\varphi = \omega\tau_p \tag{4}$$

$$M = \frac{1}{\sqrt{1 + (\omega\tau_m)^2}} \tag{5}$$

where τ_p and τ_m are independently determined and represent the phase and modulation lifetimes, respectively. For systems characterized by a single exponential decay, the phase and modulation lifetimes are identical and are independent of the light modulation frequency. For more complex systems with multiple fluorescing species, τ_p is less than τ_m and both are dependent upon the modulation frequency. In 1969, Weber and Spencer [5] described a frequency domain fluorometer which utilized the principle of cross-correlation, wherein the light source (in their case a xenon arc passing through a Debye-Sears light modulation tank) was modulated at one frequency, while the detector was modulated at a slightly different frequency. The "cross-correlation" frequency was typically low (about 40 Hz) such that a low-frequency signal could be isolated, which was far easier to work with than the original high frequencies. In this instrument only two light modulation frequencies were available. In 1983, Enrico Gratton introduced a true multifrequency fluorometer by passing laser light (originally an argon ion laser) through a Pockels cell [6]. Table 1 gives an example of τ_p and τ_m determined for a solution containing two components of 12.08 and 1.38 ns in the proportions of 53 and 47 % (fractional intensities, i.e., contributions to the total intensity), respectively. These differences between τ_p and τ_m, and their frequency dependence, permit one to extract the underlying lifetime components.

Weber [7] defined the terms S and G as

$$\varphi = \tan^{-1}(S/G) \tag{6}$$

$$M = (S^2 + G^2)^{-1/2} \tag{7}$$

Table 1
τ_p and τ_m values obtained for a fluorophore mixture (details in text)

Frequency (MHz)	τ_p (ns)	τ_m (ns)
5	6.76	10.24
10	6.02	9.70
30	3.17	6.87
70	1.93	4.27

where

$$S = \sum_i f_i M_i \sin \varphi_i \quad (8)$$

$$G = \sum_i f_i M_i \cos \varphi_i \quad (9)$$

Resolution of the underlying lifetime components in a mixture, and assignment of their relative contributions to the signal, requires fitting the frequency dependence of the phase delay and demodulation to a specific decay scheme (*see* **Note 1**). Two commercially available programs for analysis, Vinci Multidimensional Spectroscopy software and Globals for Spectroscopy, will be discussed later in the Analysis section.

The relationship between S and G provides another way to treat the data known as the phasor method [8]. This model-less approach offers immediate graphical evidence of lifetime heterogeneity as well as possible excited state reactions present in the sample. In recent years, the phasor method has gained in popularity as a tool used in fluorescence lifetime image microscopy (FLIM), but it is also useful in cuvette studies [9–11]. We should also note that this approach was introduced for FLIM studies by several groups who designated the graphs as phasor plots [12], AB plots [13], and polar plots [14]. A plot of S versus G values for a particular modulation frequency is characterized by a vector of length equal to M with an angle from the x-axis equal to φ as shown in Fig. 2. For a system with a single exponential lifetime, the frequency dependence of the vector forms a semicircle with a radius of ½ known as the universal circle. For systems that are multi-exponential, the phasor point is restricted to the inside of the universal circle [9, 10]. Thus, a phasor plot immediately indicates lifetime heterogeneity.

2 Materials

The frequency domain time-resolved measurements described in this chapter were carried out on a Chronos (ISS, Champaign, IL) frequency domain fluorometer. This instrument is designed to

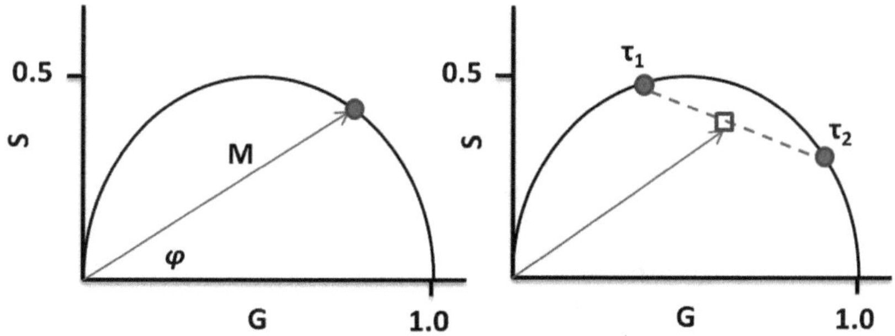

Fig. 2 Phasor plot illustrating the basic concept of a phasor point at one frequency for a system with a single exponential lifetime (*left*) and for a system with a double exponential lifetime (*right*)

accommodate light-emitting diodes (LEDs), laser diodes (LDs), as well as CW and mode locked lasers. Clearly, the sample under investigation, be it a protein or a simple fluorophore, must be as pure as possible. Equally important, though, is that the reagents used for the lifetime measurements, both sample and reference solvents, must be of the highest possible purity to eliminate errors due to background fluorescence. Solvent impurities should always be checked before measurements (*see* **Note 2**). Also, one must keep in mind that additional sample purification may be required, and one should not necessarily trust the purity of commercial compounds. It is important to follow proper protocol for storage, usage, and disposal for all reagents. It is also very important to know the optical transparency of the cuvettes being used. Plastic or glass cuvettes typically absorb light in the near-UV (<330 nm) wavelength range. For UV-visible range experiments, quartz cuvettes are recommended and were utilized for the lifetime experiments described here.

3 Methods

All multifrequency phase and modulation lifetime measurements involve choosing an excitation light source and optical filters, sample preparation, reference preparation, selecting an optimal frequency range, data acquisition, and analysis [8, 15, 16]. Below we have separated each of these required steps for additional discussion and to provide experimental details for the acquisition of lifetimes for three systems, namely, fluorescein in 0.01 M NaOH, intrinsic protein fluorescence (HSA), and EB-tRNA binding. Finally, the notes section provides additional information and nuances one learns with experience.

3.1 Experimental Considerations

One of the first tasks in any lifetime measurement is choosing the excitation wavelength. The probability of a species to absorb a photon of an appropriate energy to generate an excited state is proportional to the molecular absorption coefficient (extinction

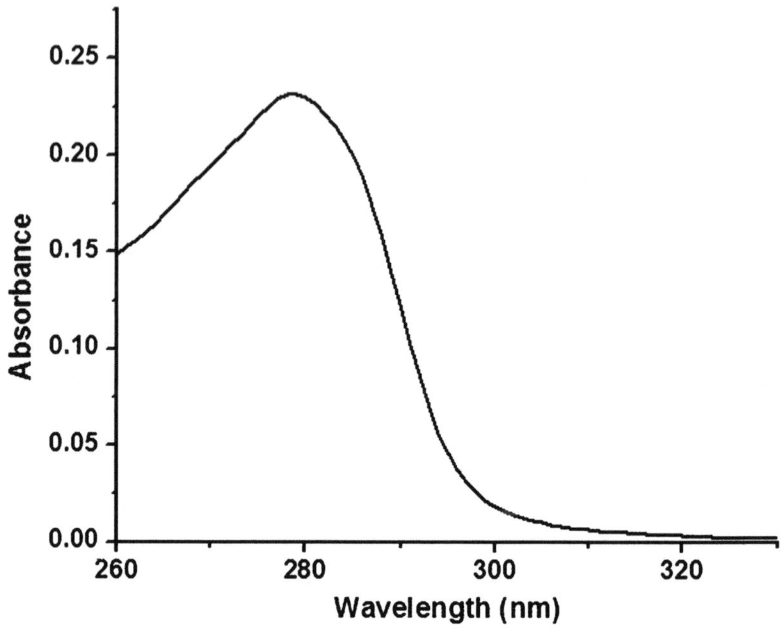

Fig. 3 UV-vis absorption spectrum of 15 μM HSA in 20 mM HEPES, pH 7.2

coefficient), and hence the absorption spectrum of the sample should be known. For instance, the absorption spectra of HSA in buffered solution displays a prominent absorption band centered near 278 nm, with absorption out to about 315 nm as shown in Fig. 3 (*see* **Note 3**). Therefore, an excitation source anywhere from 250 to 315 nm may theoretically be used (*see* **Note 4**). Fortunately, with the emergence of light-emitting diodes (LEDs) and laser diodes (LD), the choice of available wavelengths has increased and the costs have decreased. Together, LEDs and LDs provide a wavelength range that spans the entire UV and visible range of the spectrum (*see* **Note 5**). Unlike xenon arc lamps and CW lasers implemented in earlier instruments, the input voltage of the LEDs and LDs can be directly modulated [8, 15, 16]. While there are advantages to these types of light sources, LDs and LEDs often emit over a small range of wavelengths and require additional optical components to increase wavelength selectivity. The best types of optical filters to achieve a narrow excitation wavelength range of a few nanometers are interference filters. The narrow wavelength range of interference filters are specified by its wavelength plus the FWHM (full width at half max), i.e., a 300 nm interference filter with FWHM of 5 nm passes 50 % of the laser intensity at 295 and 305 nm as it does at 300 nm. The excitation wavelength, interference filter, and emission filter for the three systems are provided in Table 2 (*see* **Note 6**). The "magic angle" conditions [17, 18] for the excitation and emission polarizers are also summarized in Table 2 (*see* **Note 7**).

Table 2
Summary of the light source and instrument configuration for the three example systems

Sample	λ_{ex} (type/wavelength)	Filter type	Magic angle configuration (excitation/emission)	Frequency range (MHz)
Fl in 0.1 M NaOH	LED/471 nm	LP 515 nm	0°/55°	10–100
HSA 20 mM HEPES, pH 7	LD/300 nm	LP 305 nm	–/35°	10–150
EB-tRNA	LED/471 nm	LP 525 nm	0°/55°	1–100

Table 3
Summary of the sample preparations for the three example systems

Sample	Solvent	Temperature (°C)
500 nM Fl	0.01 aqueous NaOH	25
60 μM HSA	20 mM HEPES, pH 7.0	20
1 μM EB	20 mM HEPES, 2 mM $MgCl_2$, 0.1 mM EDTA, 100 mM KCl, pH 7.0	25
1 μM EB, 4.0 μM tRNA	20 mM HEPES, 2 mM $MgCl_2$, 0.1 mM EDTA, 100 mM KCl, pH 7.0	25

3.1.1 Sample Preparation

As already mentioned, optical grade solvents and highest purity compounds are required for reliable lifetime data. All experiments described here were performed in a 1 cm cuvette with a final volume of 2 mL. Fluorescein was purchased from Invitrogen and used without further purification. HSA was purchased from Sigma and purified according to the protocol described in the review by James et al. [10]. Yeast tRNA and ethidium bromide solution (10 mg/mL) were purchased from Invitrogen. Table 3 is a complete summary of the sample preparations.

3.1.2 Reference Compound

After the excitation wavelength has been determined and the sample prepared, a reference compound can be chosen. In the early days of phase-modulation fluorometry [5], a light-scattering solution, such as glycogen in water, was typically utilized to provide a reference signal from which to measure the phase shift and excitation light modulation. As higher frequencies became available, it was apparent that the PMTs being utilized could demonstrate considerable "color effects," which distorted the phase lifetime [19]. Even though better PMTs became available, the use of reference standards became common, i.e., lifetime standards with emission characteristics similar to those of the sample such that the PMT could always observe the same wavelength.

Table 4
Common fluorescence lifetime standards used in frequency domain spectroscopy: from refs. 1 and 20

Compound	Solvent	Lifetime (ns)	λ_{ex} (nm)	λ_{em} (nm)
2-aminopurine	Water	11.34	290	380
Anthracene	Methanol	5.1	295–360	375–442
Coumarin 153	Methanol	4.3	295–442	495–550
Dimethyl POPOP	Ethanol	1.45	300–400	390–560
Erythrosine B	Water	0.089	488–568	550–580
Fluorescein	NaOH/water	4.1	400	490–520
Indole	Water	4.49	290	360
NATA	Water	3.1	295–309	330–410
POPOP	Cyclohexane	1.12	295–360	380–450
p-Terphenyl	Ethanol	1.05	284–315	330–380
Rhodamine B	Water	1.74	488–575	560–630

Virtually, any fluorescent species with monoexponential decay can be deemed a standard; however, there are a few other qualities that make particular fluorophores preferred lifetime standards (*see* **Notes 8** and **9**) [20]. Over the years a detailed list of dependable fluorescence lifetime standards has been compiled. Some of those standards are summarized in Table 4 with the excitation wavelength, emission range, and lifetime.

One must choose a standard that displays sufficient emission intensity under the optimal instrument configuration designed for the sample, i.e., excitation wavelength, optical filters, and PMT voltage. With that being said, the reference emission intensity should be comparable to the sample intensity. Comparable counts on the detector will reduce inconsistencies in gain settings and signal to noise between the reference and sample which can be a source of error [1]. Table 5 lists the references utilized for the example systems.

3.1.3 Frequency Range

The frequency range of a multifrequency instrument should scale from kilohertz to hundreds of megahertz. The LD and LEDs available from ISS for the ChronosFD have a frequency response up to 600 MHz—although the frequency range of the various light sources differs. The idea, however, is not to utilize the entire frequency range of the instrument but rather to focus on the optimal range. The ideal multifrequency range for determining decay lifetimes by analysis of both phase delay and demodulation requires multiple data points before and after the optimal frequency. For a known lifetime, the optimal frequency is easy to establish, that is, when $\omega\tau = 1$, keeping in mind that $\omega = 2\pi f$ and

Table 5
Summary of the reference solutions used for each of the three example systems

Sample system	Reference	Solvent	Reference lifetime (ns)
Fl	Fluorescein	0.01 aqueous NaOH	4.1
HSA	p-Terphenyl	Ethanol	1.05
EB-tRNA	Ethidium bromide	DI H_2O	1.71

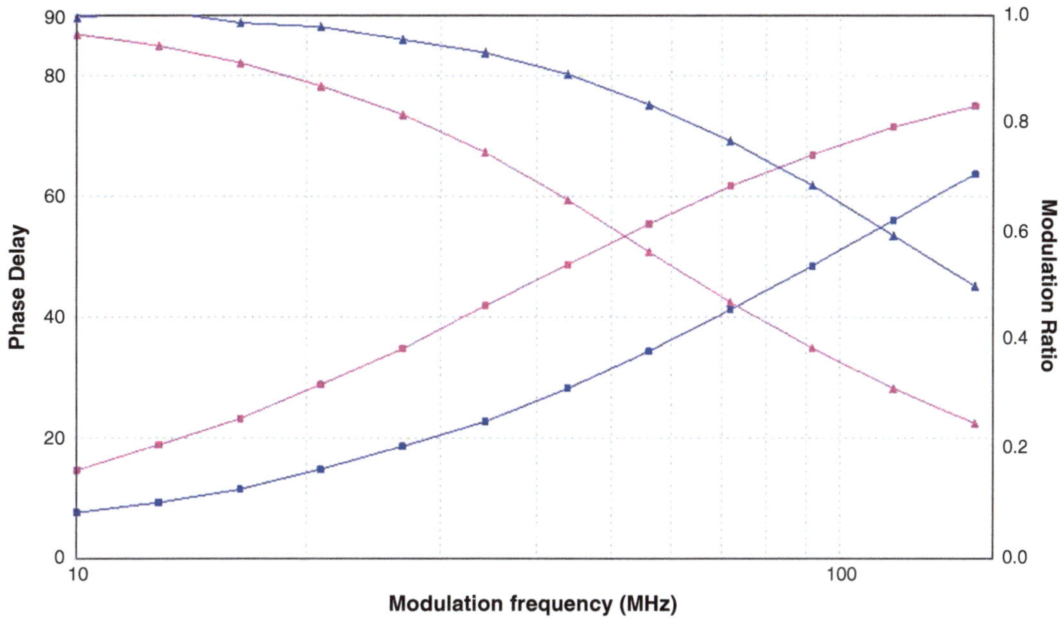

Fig. 4 Phase (*square*) and modulation (*triangle*) plot of fluorescein (*blue*) and ethidium bromide (*pink*)

$f = 1/2\pi\tau$, where τ is the excited state lifetime [1]. For example, the optimal frequency range for a lifetime like that of fluorescein (4.1 ns) is 10–100 MHz (optimal frequency of 40 MHz). On the other hand, that is not an optimal range for ethidium bromide (1.71 ns), which would have an optimal frequency of ~100 MHz, as shown in Fig. 4. We usually prefer to use a range which encompasses around 10–15° of phase shift up to around 70° of phase shift. More often than not, the general range of the experimental lifetime is known and choosing the frequency range becomes more apparent with experience. Table 2 presents the frequency range used for each of the example systems.

3.2 Data Analysis and Applications

Data Analysis is the most challenging aspect of any lifetime measurement. It involves selecting the appropriate decay scheme, evaluating the fit, and deciding whether or not the results are reasonable. Gregorio Weber solved the exact analytical solution to extract N lifetimes and fractional intensities from phase and modulation data at N frequencies [7]. However, the algorithm became impractical for data with increasing number of frequencies due to the necessity for increased precision [8]. With development of true multifrequency instruments [6], lifetime values are extracted from phase and modulation data using a nonlinear least square method [8]. The goodness of fit is determined by reduced chi-squared (χ^2) value using

$$\chi^2 = \sum \left([(P_c - P_m)/\sigma_p]^2 + [(M_c - M_m)/\sigma_m]^2 \right) / (2n - fp - 1) \tag{10}$$

where P and M are the phase and modulation data, the subscripts c and m indicate the calculated and measured values, and the σ_p and σ_m refer to the standard deviations. The number of frequencies is n, and fp is the number of free parameters. The sum in Eq. 10 is over n frequencies. Two commercial programs available for data analysis and phasor plots are Globals for Spectroscopy and Vinci2 (*see* **Note 10**). Analysis using these programs first requires the choice of a decay scheme (single, multi-exponential, discrete or distributions, etc.) followed by the selection of estimated parameters (lifetimes, fractional intensities, etc.). Satisfactory fits will have a reduced χ^2 value close to 1 [1]. The standard deviations generally used for phase shift and modulation ratio are typically 0.2° and 0.004, respectively, but of course those values may be changed to reflect the particular instrumentation and conditions. The data analysis for each of the three systems is described below.

3.2.1 Fluorescein, 0.01 M NaOH

To begin, data analysis for fluorescein in solution is demonstrated in Vinci2-Analysis as shown in Fig. 5a. The ISS data files can be opened in this program without any conversion of the file type (we note that with Globals the file type must be modified as discussed further below). Once the file is open, one selects the Fitting tab, and chooses lifetimes, which will open the options for the possible decay schemes. For fluorescein in basic solution, the decay scheme should be discrete monoexponential (in this case one will find that adding additional exponential terms does not improve the fit). After the scheme is chosen, the fit is carried out and the results, including residuals, are displayed (Fig. 5b). To view a complete summary of the results, select the Fit Report tab (Fig. 5c). Next, to convert data to a phasor plot, select View followed by phasor plot. The phasor plot of this sample is an ideal example of how one can immediately ascertain that the sample is single exponential, namely, all of the phasor points lie directly on the universal circle as predicted for a single exponential decay (Fig. 6).

3.2.2 HSA, 20 mM HEPES, pH 7

HSA demonstrates a slightly more complex system, namely, tryptophan fluorescence in a protein matrix rather than a single fluorophore free in an isotropic environment, i.e., in solution. The analysis described is performed in Globals for Spectroscopy. In order to use Globals for fitting ISS data files, they must be

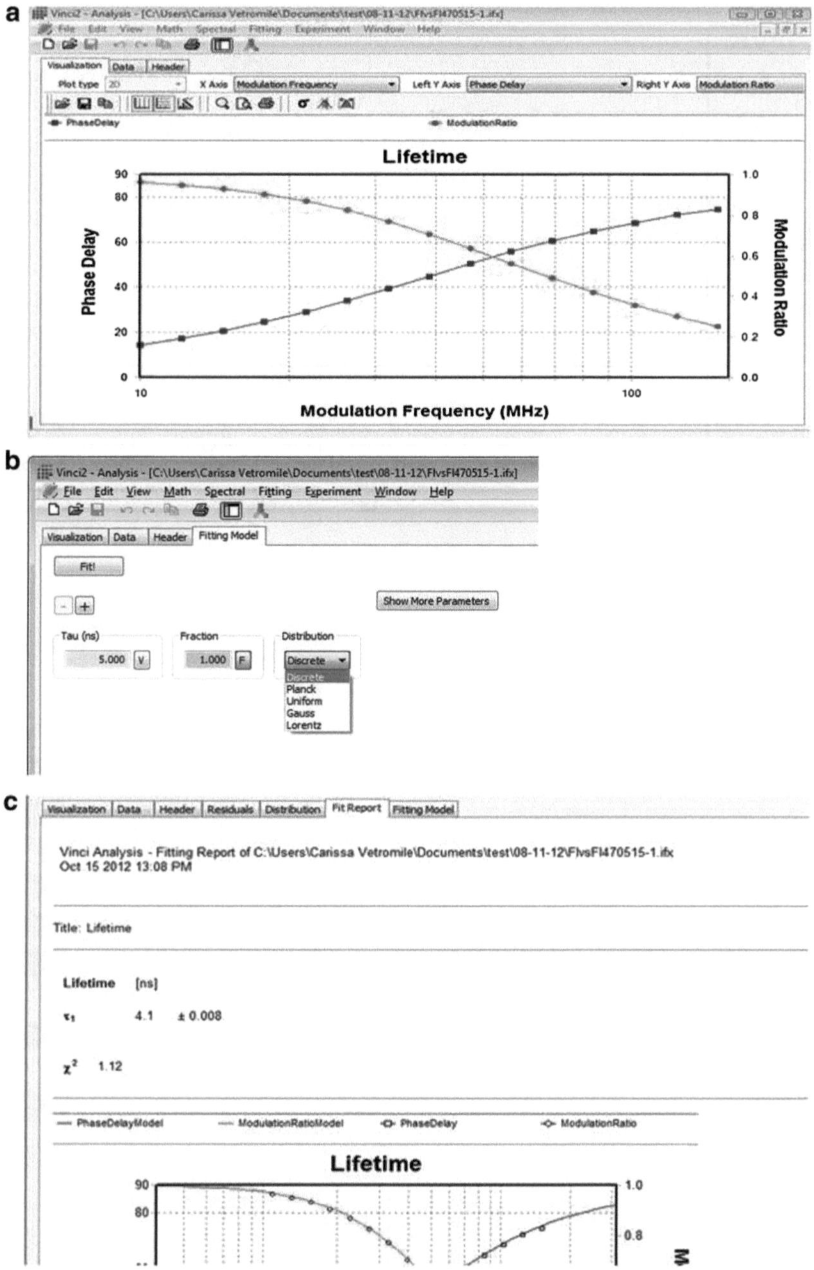

Fig. 5 Phase and modulation plot of fluorescein (**a**), page where one selects a decay model (**b**), and the fit report for a single exponential decay model (**c**) shown using Vinci2-Analysis (ISS)

converted to ISS Lif format. To do this one opens the file in Vinci2, then selects file, export, ISS Lif format (binary), and save. One then opens Globals and inserts the recently saved file. Before selecting the model, be sure to change Rec number to 1 as shown in Fig. 7. It is recommended that the reader refers to the tutorials on model selecting provided by Globals to become familiar with the program,

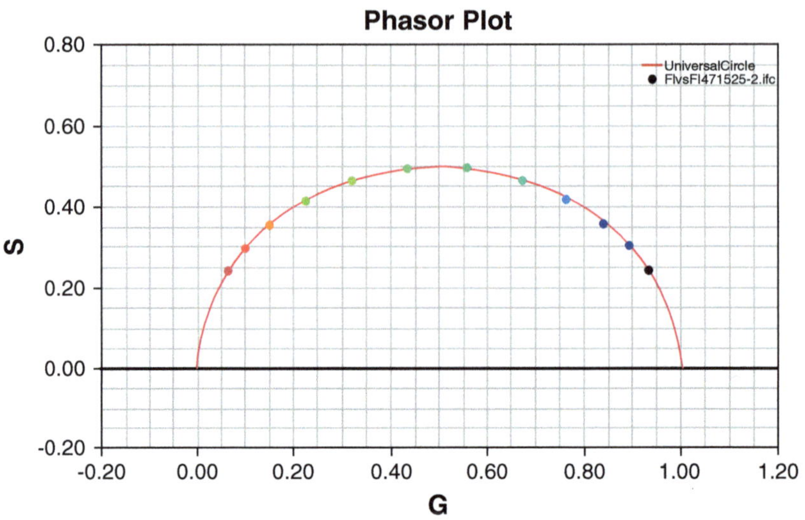

Fig. 6 Phasor plot of fluorescein in 0.01 M NaOH

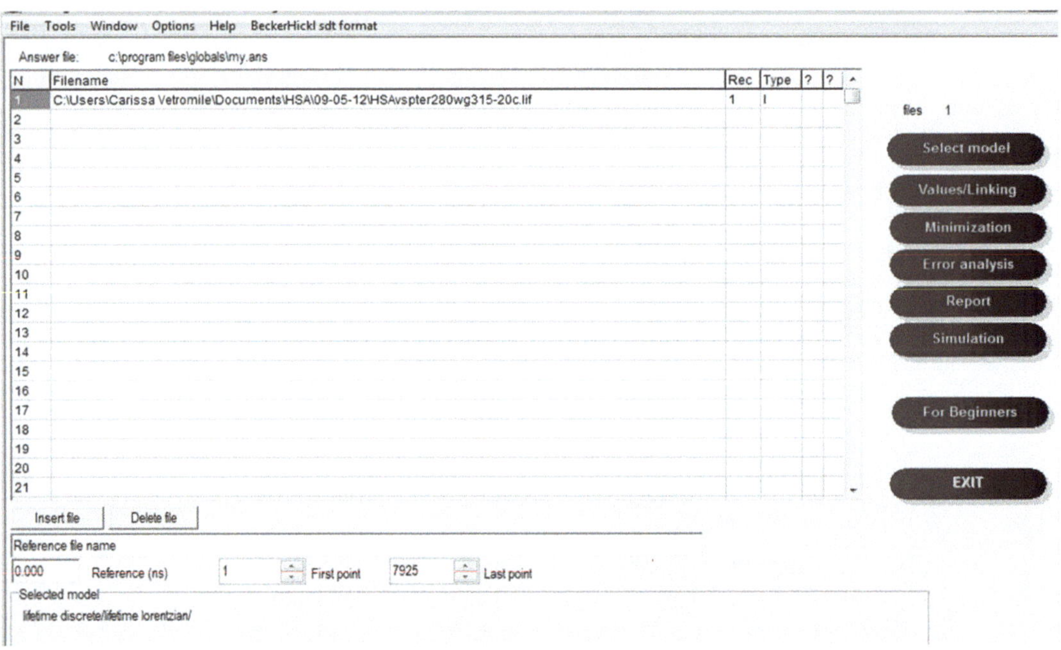

Fig. 7 Image of Globals for Spectroscopy workstation where one inserts and analyzes data files

which can be found under the help tab on the program or at http://www.lfd.uci.edu/globals/tutorials/.

In cases where the lifetime is unknown, we prefer to begin the analysis with a simple single exponential decay scheme. From Fig. 8, it is clear that HSA tryptophan fluorescence does not fit to a single exponential, as is also evident from the phasor plot in Fig. 8 ($\chi^2 = 63.09$). One nice feature of Globals is that the phasor plot

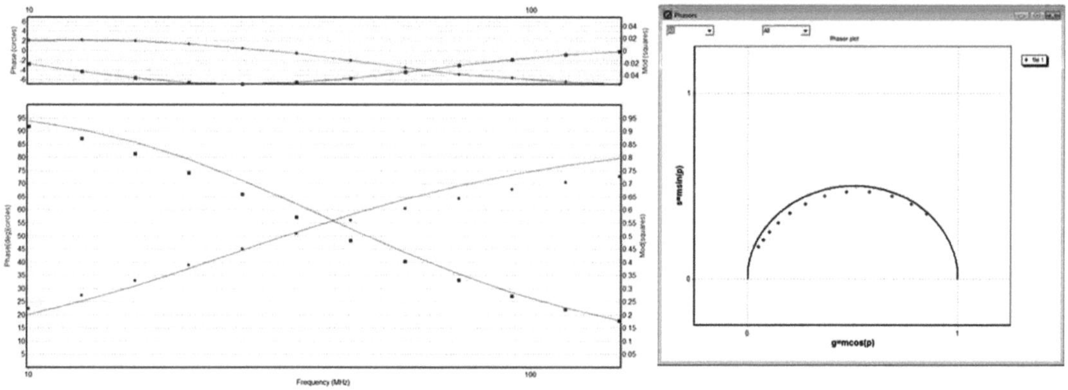

Fig. 8 Phase and modulation plot of HSA in 20 mM HEPES, pH 7.2 fit to a single exponential decay model (*left panel*) and the resulting phasor plot (*right panel*). $\chi^2 = 63.03$

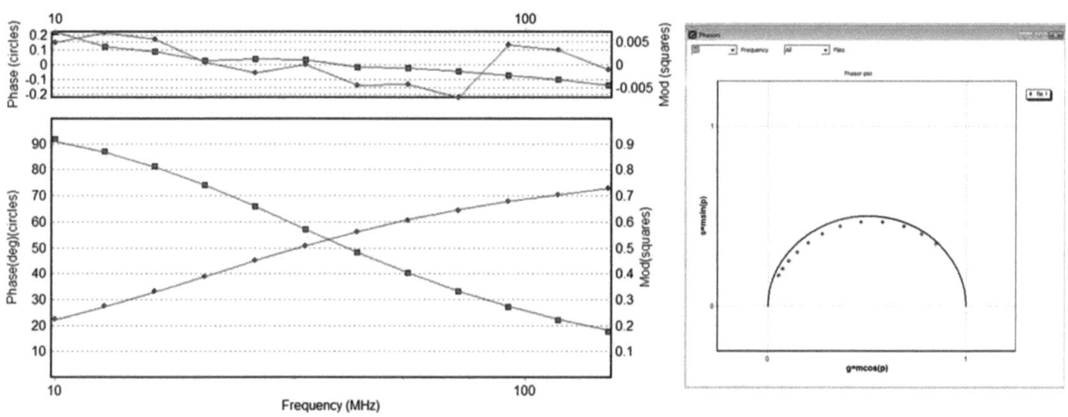

Fig. 9 Phase and modulation plot of HSA in 20 mM HEPES, pH 7.2 fit to heterogeneous decay model described by a fast discrete exponential lifetime and a longer lifetime with a Lorentzian distribution (*left panel*) and the resulting phasor plot (*right panel*). $\chi^2 = 0.61$

also displays the fit on the plot (+). When dealing with a sample that is unknown, it can be beneficial to check the phasor plot first before fitting the data. If the phasor plot reveals the lifetime is not single exponential, then the next step is to systematically try fitting the phase and modulation data to different decay schemes, i.e., a single lifetime distribution, two discrete lifetimes, one discrete on distribution, and so on.

Fluorescence lifetime measurements of the single tryptophan in HSA have been carried out by many labs, and the general agreement is that the tryptophan exhibits a heterogeneous lifetime [21]. We found that under our conditions the data fits best to two components, a fast discrete lifetime (1.03 ns) accompanied with a longer lifetime (6.73 ns) with a Lorentzian distribution (width = 1.03) as shown in Fig. 9 ($\chi^2 = 0.63$). Lifetime distribution models for protein fluorescence were discussed in detail by Alcala et al. [22–24].

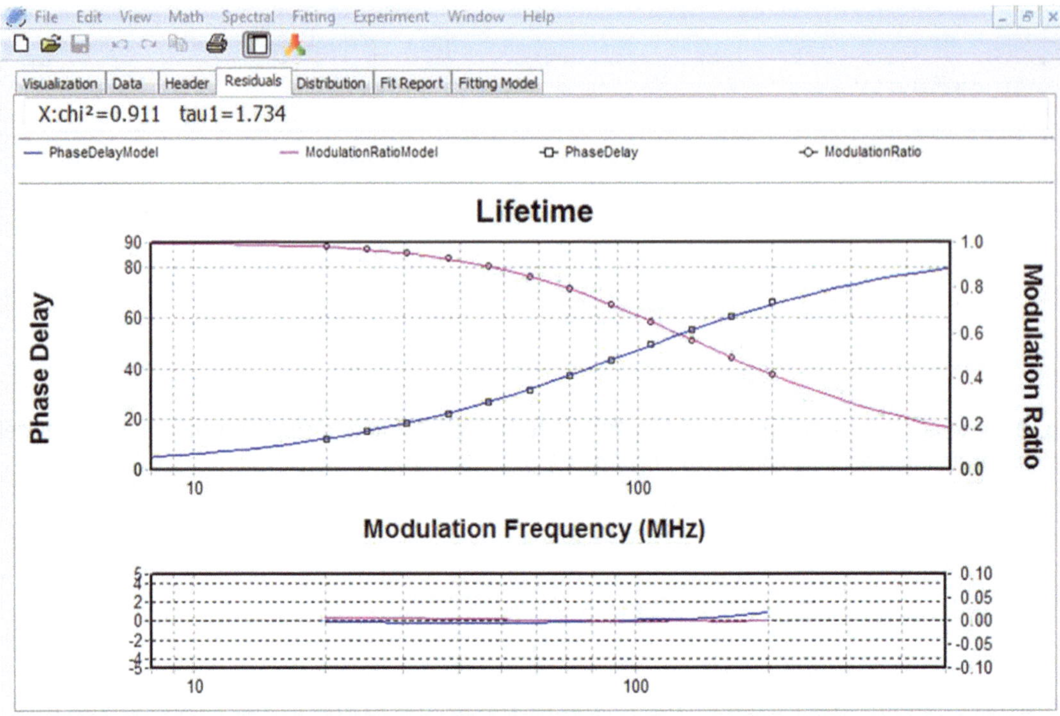

Fig. 10 Phase and modulation plot of ethidium bromide fit to a single exponential decay model

3.2.3 EB-tRNA

Another application of frequency domain fluorometry is toward the study of ligand binding. The final example system, ethidium bromide binding to yeast tRNA, is a great instance of this due to the significant increase in the fluorescence lifetime of ethidium bromide upon binding to tRNA. Previously reported fluorescence lifetime measurements of ethidium bromide binding to tRNA suggest the existence of a strong binding site along with some weaker binding sites [27]. To keep this exercise from becoming too complex, we will investigate ethidium bromide in the absence and presence of excess tRNA, to ensure only the strong binding site is occupied.

To begin one examines ethidium bromide in the absence of tRNA under the conditions described in Table 3. The measured phase and modulation data for ethidium bromide fits to a single exponential decay scheme with a fluorescence lifetime of 1.73 ns as shown in Fig. 10. Next, to investigate the effects of the strong binding site, one must fit the phase and modulation data of ethidium bromide with excess tRNA. Using conditions described above, the data should also fit to a single exponential decay model (all ethidium bromide molecules bound) but with a significantly longer lifetime (Fig. 11). Once again, the phasor plot of both ethidium bromide in the absence and presence of excess tRNA provides another excellent example of how one can instantly observe the differences between samples and distinguish the heterogeneity of an unknown sample (Fig. 12).

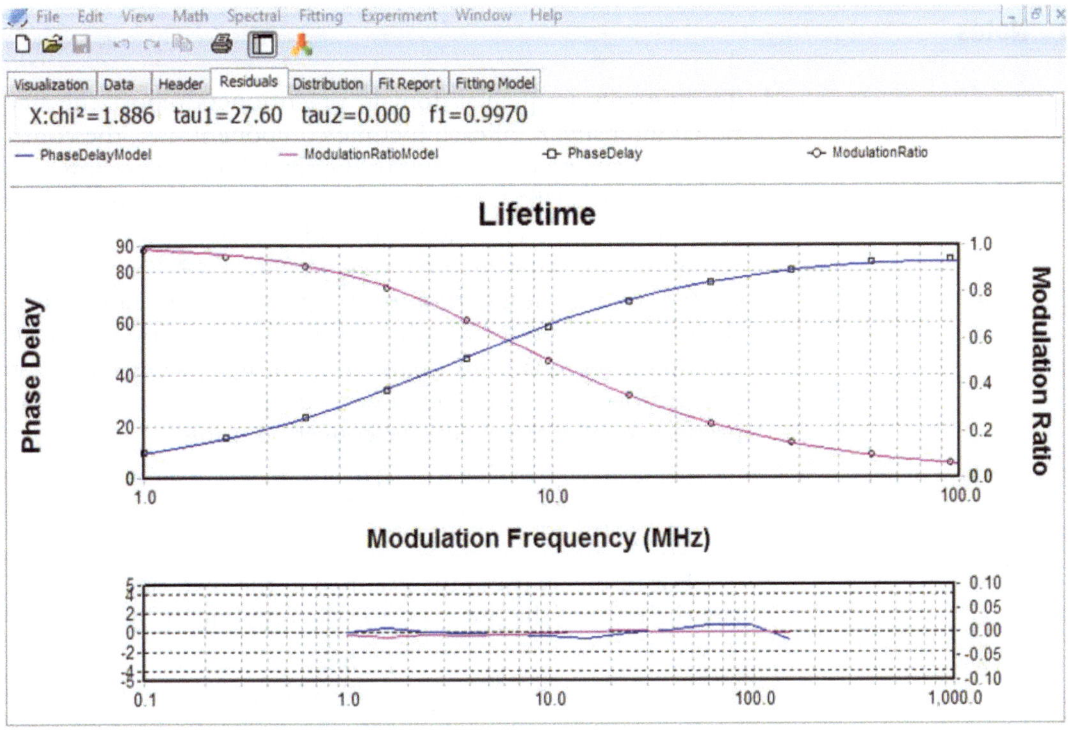

Fig. 11 Phase and modulation plot of ethidium bromide in the presence of excess yeast tRNA fit to a single exponential decay model

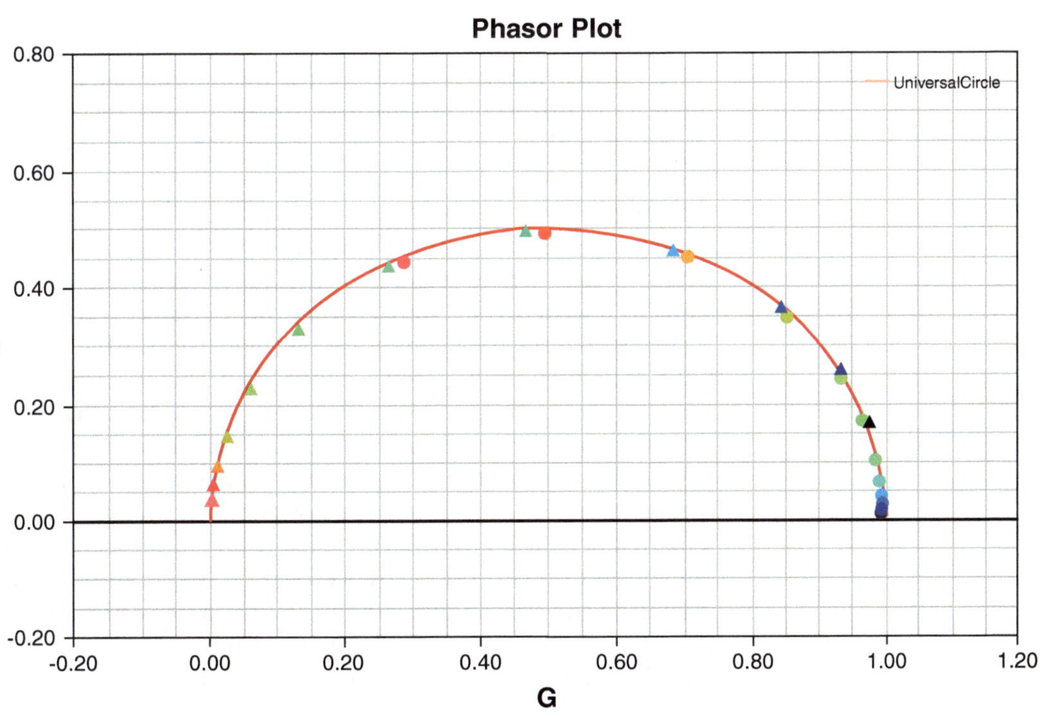

Fig. 12 Phasor plot of ethidium bromide in the absence (*circles*) and presence (*triangles*) of excess yeast tRNA

4 Notes

1. Traditionally frequency domain practitioners report the fractional intensity, f_i, of each lifetime component, i.e., the contribution of the *i*th component to the total fluorescence intensity. The sum of all fractional intensities equals one. In contrast, time domain practitioners generally report pre-exponential terms, α_i, for each lifetime component. These α_i values are related to the number of emitting molecules, not directly their contributions to the observed emission intensity. Assuming the quantum yields for each fluorescing species are proportional to their lifetimes, fractional intensities and pre-exponential factors can be converted using

$$f_i = \frac{\alpha_i \tau_i}{\sum_i \alpha_i \tau_i} \tag{11}$$

For instance, using the example in the text, one can calculate the ratio of the pre-exponential components by

$$0.53 = \frac{12.08\alpha_1}{12.08\alpha_1 + 1.38\alpha_2}$$

$$\frac{\alpha_1}{\alpha_2} = 0.13 \cong 1/8$$

Thus, for this example there are ~8 times as many molecules with a lifetime of 1.38 ns as there are molecules with a lifetime of 12.08 ns.

2. After the instrument is set up for the measurements, i.e., exciting wavelength and filters are in position, a cuvette with sample is placed in the sample holder and the PMT voltages and/or gain settings are adjusted to give an appropriate signal. Then, this sample cuvette should be momentarily replaced by a cuvette containing only the solvent. The shutters to the light source and detectors are opened using the Vinci2 software and the solvent signal is checked. This signal should be the same as (or at least very close to) the dark configuration (detector shutter open but light source shutter closed); if not then the solvent may be contaminated. In our experience, for example, ethanol frequently contains contaminants that exhibit fluorescence upon UV or mid-UV excitation. Even if the ethanol is only being used for the reference compound, spurious fluorescence will be problematic since the lifetime of this background fluorescence will be convoluted in with the reference lifetime. We note that even a very clean solvent can give rise to a Raman peak, which may pass the emission filter, but usually the Raman peak is very weak and is seldom a problem with lifetime measurements.

3. It is not always necessary to excite at the maximum absorption band since the emission of a photon is from the lowest excited

electronic state, regardless of which excited state is initially populated (there are a few exceptions). That being said, the more probable the transition, the larger the population of excited state molecules, and the greater the florescence intensity. In addition to the excitation wavelength, the sample concentration is another way to adjust the observed fluorescence intensity; however, one must be careful not to introduce inner filter effects to the sample. Inner filter effects can occur in samples sufficiently concentrated such that reabsorption and reemission of photons can occur [1]. It is best to work with solutions having optical densities less than 0.1 at the excitation wavelength.

4. This choice of 280 versus 300 nm excitation is always in question when investigating intrinsic protein fluorescence [23]. Intrinsic protein fluorescence originates from the three aromatic amino acids: tryptophan, tyrosine, and phenylalanine. In 1957, Weber and Teale published the first photophysical characterization of these aromatic amino acids and influenced the foundation for investigating proteins with fluorescence techniques [24]. Tryptophan, tyrosine, and phenylalanine have different molar extinction coefficients and quantum yields [15]; nevertheless, excitation at 280 nm results in significant emission from all tyrosine and tryptophan (for a more detailed discussion of this topic, *see* ref. 15). The contribution to the total intensity from each is not equal, with tryptophan fluorescence typically dominating. It was also established by Gregorio Weber that energy transfer from tyrosine to tryptophan can occur in proteins, a process which clearly complicates the excited state decay kinetics. Fortunately, tryptophan is the only aromatic amino acid with significant absorption at 300 nm, and excitation at this wavelength can be used to remove the effects of energy transfer from tyrosine and promote tryptophan selectivity [25, 26]. Unfortunately, in the vast majority of single tryptophan protein studies, the excited state lifetime is not characterized by a single exponential decay. The complexity of tryptophan fluorescence in proteins is beyond the scope of this chapter and readers are referred to ref 15 and references therein for more information.
5. The ISS website provides a list of available light sources for ChronosFD. LED wavelengths range from 280 to 520 nm, while LD wavelengths range from 405 to 830 nm.
6. Optical filters are often used in lifetime measurements to ensure that the detected light is from the fluorophore under investigation and not parasitic or scattered light. There are many types of optical filters including bandpass, interference, and longpass (cut-on) filters. Both bandpass and interference filters allow a specific range of wavelengths to pass through. The wavelength range of bandpass filters are generally 50–100 nm, while

interference filters have a more narrow range, typically 5–10 nm. Longpass filters block light at lower wavelengths until a specific range is reached and then allow longer wavelengths to pass. Before running a measurement, it is always a good idea to test whether a chosen emission filter is blocking all the excitation light. To test the filter, it should be placed in the excitation path with the emission and excitation shutters opened. If the filter is good, the observed signal should be on the same order as the dark signal.

7. When fluorescence is monitored at 90° to the excitation, as in most instruments, the vertically polarized molecules are favored over the horizontally polarized molecules since one of the horizontally polarized populations cannot be observed and as a result the measured lifetime is affected if the emission is polarized. To reduce this effect, "magic angle" configurations of the polarizers in the excitation and emission pathway can be used [17]. There are four configurations which yield equivalent results: (1) 35° excitation, no emission polarizer; (2) no excitation polarizer, 35° emission polarizer; (3) 55° excitation polarizer, 0° emission polarizer; and (4) 0° excitation polarizer, 55° emission polarizer. The configurations can be interchanged depending on what works best for the users' instrument and sample.
8. Nine laboratories tested the quality of multiple fluorescence standards; before starting they summarized a list of characteristics all reliable standards should possess: (1) display single exponential decay regardless of excitation or emission wavelength, (2) good quantum yield, (3) considerable Stokes shift, (4) characteristic lifetime should be on the same order as sample, and (5) must be chemically stable and photostable [20].
9. Lifetime standards may also be used to test for systematic errors. It is a good rule of thumb to test the accuracy of the instrument using a well-known fluorescence standard as a sample and reference before starting an unknown sample to insure the instrument configuration will yield accurate results.
10. Globals for Spectroscopy was developed by the Laboratory for Fluorescence Dynamics (LFD). The software is an extremely valuable tool for fluorescence data analysis including a number of methods not mentioned here. Globals can be downloaded from the LFD website at http://www.lfd.uci.edu/globals/. In addition, tutorials can be found on the LFD website to assist in using the program.

Acknowledgments

This work was partially supported by funding from Allergan, Inc.

References

1. Valeur B, Berberan-Santos MN (2012) Molecular fluorescence principles and applications, 2nd edn. Weinheim, Germany
2. Gaviola E (1926) Die Abklingungszeiten der Fluoreszenz von Farbstofflösungen. Ann Phys 386:681–710
3. Gaviola E (1927) Ein Fluorometer. Apparat zur Messung von Fluoreszenzabklingungszeiten. Z Phys 42:853–861
4. Dushinsky F (1933) Der zeitliche Intensitätsverlauf von intermittierend angeregter Resonanzstrahlung. Z Phys 81:7–22
5. Spencer RD, Weber G (1969) Measurements of subnanosecond fluorescence lifetimes with a cross-correlation phase fluorometer. Ann N Y Acad Sci 158:361–376
6. Gratton E, Limkeman M (1983) A continuously variable frequency cross-correlation phase fluorometer with picosecond resolution. Biophys J 44:315–324
7. Weber G (1981) Resolution of the fluorescence lifetimes in a heterogeneous system by phase and modulation measurements. J Phys Chem 85:949–953
8. Jameson DM, Gratton E, Hall RD (1984) The measurement of heterogeneous emission by multifrequency phase and modulation fluorometry. Appl Spectrosc Rev 20:55–106
9. Štefl M, James NG, Ross JA et al (2011) Applications of phasors to in vitro time-resolved fluorescence measurements. Anal Biochem 410:62–69
10. James NG, Ross JA, Štefl M et al (2011) Applications of phasor plots to in vitro protein studies. Anal Biochem 410:70–76
11. Buscaglia R, Jameson DM, Chaires JB (2012) G-Quadruplex structure and stability illuminated by 2-aminopurine phasor plots. Nucleic Acids Res 40:4203–4215
12. Digman MA et al (2008) The phasor approach to fluorescence lifetime imaging analysis. Biophys J 94:L14–L16
13. Hanley QS, Clayton AHA (2005) AB-plot assisted determination of fluorophore mixtures in a fluorescence microscope using spectra quenchers. J Microsc 218:62–67
14. Redford GI, Clegg RM (2005) Polar plot representation for frequency-domain analysis of fluorescence lifetimes. J Fluoresc 15:805–815
15. Ross JA, Jameson DM (2008) Time-resolved methods in biophysics. Frequency domain fluorometry: applications to intrinsic protein fluorescence. Photochem Photobiol Sci 7:1301–1312
16. Gratton E, Jameson DM, Hall RD (1984) Multifrequency phase and modulation fluorometry. Annu Rev Biophys Bioeng 13:105–124
17. Spencer RD, Weber G (1970) Influence of Brownian rotations and energy transfer upon the measurement of fluorescence lifetimes. J Chem Phys 52:1654–1663
18. Teale FWJ (1983) Phase and modulation fluorometry. In: Cundell RB, Dale RE (eds) Time–resolved fluorescence spectroscopy in biochemistry and biology. Plenum Press, New York, pp 59–80
19. Jameson DM, Weber G (1981) Resolution of the pH-dependent heterogeneous fluorescence decay of tryptophan by phase and modulation measurements. J Phys Chem 85:953–958
20. Boens N, Qin W, Basarić N et al (2007) Fluorescence lifetime standards for time and frequency domain fluorescence spectroscopy. Anal Chem 79:2137–2149
21. Helms MK, Peterson CE, Bhagavan NV et al (1997) Time-resolved fluorescence studies on site-directed mutants of human serum albumin. FEBS Lett 408:67–70
22. Alcala R, Gratton E, Prendergast FG (1987) Resolvability of fluorescence lifetime distributions using phase fluorometry. Biophys J 51:587–596
23. Alcala JR, Gratton E, Prendergast FG (1987) Fluorescence lifetime distributions in proteins. Biophys J 51:597–604
24. Alcala R, Gratton E, Prendergast FG (1987) Interpretation of fluorescence decays in proteins using continuous lifetime distributions. Biophys J 51:925–936
25. Barbieri B, Terpetschnig E, Jameson DM (2005) Frequency-domain fluorescence spectroscopy using 280-nm and 300-nm light-emitting diodes: measurement of proteins and protein-related fluorophores. Anal Biochem 344:298–300
26. Weber G, Teale FWJ (1957) Ultraviolet fluorescence of aromatic amino acids. Biochem J 65:476–482
27. Tramier M, Holub O, Croney JC, Ishii T, Seifried SE, Jameson DM (2002) Binding of ethidium to Yeast tRNA^{Phe}: a new perspective on an old bromide. In: Kraayenhof R, Visser AJWG, Gerritsen HC (eds) Fluorescence spectroscopy, Imaging and Probes. Springer, Berlin, pp 111–120

Chapter 6

Polar Plot Representation of Time-Resolved Fluorescence

John Paul Eichorst, Kai Wen Teng, and Robert M. Clegg

Abstract

Measuring changes in a molecule's fluorescence emission is a common technique to study complex biological systems such as cells and tissues. Although the steady-state fluorescence intensity is frequently used, measuring the average amount of time that a molecule spends in the excited state (the fluorescence lifetime) reveals more detailed information about its local environment. The lifetime is measured in the time domain by detecting directly the decay of fluorescence following excitation by short pulse of light. The lifetime can also be measured in the frequency domain by recording the phase and amplitude of oscillation in the emitted fluorescence of the sample in response to repetitively modulated excitation light. In either the time or frequency domain, the analysis of data to extract lifetimes can be computationally intensive. For example, a variety of iterative fitting algorithms already exist to determine lifetimes from samples that contain multiple fluorescing species. However, recently a method of analysis referred to as the polar plot (or phasor plot) is a graphical tool that projects the time-dependent features of the sample's fluorescence in either the time or frequency domain into the Cartesian plane to characterize the sample's lifetime. The coordinate transformations of the polar plot require only the raw data, and hence, there are no uncertainties from extensive corrections or time-consuming fitting in this analysis. In this chapter, the history and mathematical background of the polar plot will be presented along with examples that highlight how it can be used in both cuvette-based and imaging applications.

Key words Polar plot, Phasor plot, Fluorescence lifetime, Frequency domain

1 Introduction and History of the Polar Plot

The polar plot arose as a convenient method of analysis to study the response of systems such as electric impedance [1, 2], dielectric relaxation [3], fluorescence [4–7], and relaxation kinetics [8]. The signals of these systems decay exponentially (relax) in response to a sudden small external stimulus. For example, in the case of chemical relaxation kinetics, the measured emission of a fluorescence probe, which is thermodynamically coupled to a chemical reaction of interest, will decay exponentially in response to a sudden change in pressure or temperature of the system. Instead of reconstructing the exponential relaxations in the time domain, it is often more convenient to subject the system to repetitive perturbations,

Yves Engelborghs and Antonie J.W.G. Visser (eds.), *Fluorescence Spectroscopy and Microscopy: Methods and Protocols*, Methods in Molecular Biology, vol. 1076, DOI 10.1007/978-1-62703-649-8_6,

and examine the signals in the frequency domain (as sets of sine and cosine functions; see below). Both methods yield equivalent information concerning the underlying physical parameters. The polar plot simplifies the analysis of these types of measured signals by displaying the measured data graphically on the Cartesian plane.

Any repetitive signal can be decomposed into a sum of sine and cosine functions oscillating at a set of frequencies by Fourier series analysis [9, 10]. The coefficients that weight the sine and cosine functions at each frequency indicate the contribution that the specific frequency has to the original signal. These coefficients, which can span a wide range of frequencies depending on the waveform of the repetitive signal, are referred to as the signal's frequency response. When summed over all frequencies and combined with a DC (or steady-state) term (long time average), the original signal can be reconstructed in its entirety. At each frequency, a particular amplitude of oscillation and phase delay characterize the system.

When studying relaxations in the frequency domain, the amplitude of oscillation and phase delay are dependent on both the repetitive frequency and the decay constants of the different components contributing to the relaxation. (See the later section discussing the measurement of lifetimes for derivations related to the amplitude of oscillation (modulation ratio) and phase delay to fluorescence lifetime measurements.) The polar plot applies a coordinate transformation to translate the modulation ratio and phase delay information into a polar vector on the Cartesian plane. All relaxations described by a single exponential decay will reside on a semicircle centered at (0.5, 0) with a radius of 0.5. Relaxations with multiple decaying components will be projected inside the semicircle. The relative contribution of independently decaying components to a measured decay can be assessed by simple vector addition in this representation. As a result, simply by observing changes in the position of a vector on the polar plot, quantitative information can be extracted to describe the relaxations of various systems.

The application of polar plots to study the lifetimes of fluorescent molecules (fluorescent relaxations) began in the early 1980s [6]. At that time, measuring lifetimes with large sets of repetitive frequencies was beginning to be used in order to examine the contribution of multiple distinct decaying components in cuvette-based fluorescence measurements [11–13]. The polar plot was used to verify the relationship between decay constants derived from the modulation ratio and phase delay for samples with multi-component decays. When measuring lifetimes in images became more widely available, the simple graphical presentation and computational ease of the polar plot shortened data acquisition and analysis times and made the interpretations of complex systems more transparent. In parallel, three labs introduced the polar plot for such applications: the AB plot [4], polar plot [5], and phasor plot [7].

Currently, this graphical representation of fluorescence lifetimes is standard in most software packages for the acquisition and analysis of lifetime measurements.

In this chapter, we describe the basic steps necessary to generate a polar plot and extract physical information about a sample using the polar coordinate. The fundamentals of the polar plot can be equivalently applied to data collected from either cuvette-based measurements or to images taken with FLIM (fluorescence lifetime-resolved imaging microscopy). Specific examples from a variety of chemical and biological systems are also given to demonstrate the usefulness and graphical ease of the polar plot.

2 Methods

2.1 Fundamental Fluorescence Response

When a fluorescent molecule (fluorophore) absorbs light, fluorescence is observed as the excited fluorophore loses energy by the emission of a photon and returns to the ground state (de-excitation). The lifetime of the fluorophore is the average amount of time that the fluorophore spends in the excited state. If more avenues of de-excitation are available other than fluorescence, such as energy transfer, the fluorophore leaves the excited state more rapidly resulting in a shorter lifetime. The process of depleting the excited state can be described with first order rate kinetics where each pathway of de-excitation contributes a rate specific to its probability of occurring. Therefore, the probability that a fluorophore will remain in the excited state following a brief excitation decays exponentially in time [14–17].

Experimentally, the lifetime can be measured directly by exciting the molecule with a very brief (with respect to the lifetime) pulse of light and recording the fluorescent decay. The measured intensity at any point along the decay is proportional to the number of molecules remaining in the excited state at that time. The general equation for the fluorescent decay of a sample containing a set of fluorophores with different lifetimes is written in Eq. 1,

$$F_\delta(t) = \sum_i a_i \, e^{-t/\tau_i} \tag{1}$$

In Eq. 1, the factor a_i is proportional to the normalized concentration (species fraction) of fluorophore i with lifetime τ_i.

2.2 Frequency Response

Measuring a sample's lifetime in the frequency domain requires an excitation light that has its intensity repetitively modulated. Any repetitive waveform, including a pulsed excitation source, can be used to modulate the intensity of the exciting light to measure lifetimes in the frequency domain. This is because any sinusoidal constituent of a Fourier expansion of a repetitively intensity-modulated excitation source can be analyzed independently as a purely sinusoidal waveform exciting the fluorescence. It is also possible to transform data taken in the time domain to the frequency domain, such that a polar plot-based analysis can be applied. Thus, by analyzing the results from measurements taken either in the frequency domain or in the time domain, the polar plot can be constructed to graphically examine the frequency response of a fluorescing system.

In the frequency domain, the fluorescence from the sample is excited by a repetitively intensity-modulated source. The sample's measured fluorescence $S(t)$ is analyzed to extract the average fluorescence, phase delay, and amplitude of oscillation (modulation ratio); from these parameters, a polar plot is constructed. If the fluorescence response to a repetitively intensity-modulated exciting light is defined over a definite period of repetition T, a Fourier series analysis expands the measured repetitive fluorescent signal into a basis set of sine and cosine functions oscillating at a set of modulation frequencies $\{\omega_n\}$ [9, 10], $\omega_n = 2\pi n/T$. The contribution of a specific repetitive frequency to the measured fluorescence (referred to as $S_n(t)$) is defined by the coefficients a_n, b_n weighting the sine and cosine basis functions at each frequency ω_n as shown in Eq. 2:

$$S_n(t) = \mathrm{DC} + \sum_{n=0}^{\infty} (a_n \cos(\omega_n t) + b_n \sin(\omega_n t)) \tag{2}$$

where a_n and b_n are defined in Eqs. 3 and 4.

$$a_n = \frac{2}{T} \int_0^T S(t) \cos(\omega_n t) dt \tag{3}$$

$$b_n = \frac{2}{T} \int_0^T S(t) \sin(\omega_n t) dt \tag{4}$$

The nth component from the sum in Eq. 2 is an independent sinusoidal response representing the sample's emitted fluorescence modulated at ω_n. This response is defined in terms of a modulation ratio M_{Meas} and phase delay φ_{Meas}. For instance,

$$a_n \cos(\omega_n t) + b_n \sin(\omega_n t) = \sqrt{a_n^2 + b_n^2} \cos(\omega_n t + \varphi_{\mathrm{Meas}}) \tag{5}$$

where

$$\varphi_{\mathrm{Meas}} = \tan^{-1}\left(\frac{a_n}{b_n}\right) \tag{6}$$

and M_{Meas} is defined as

$$M_{\mathrm{Meas}} = \frac{\sqrt{a_n^2 + b_n^2}}{\mathrm{DC}}. \tag{7}$$

DC is the long time fluorescence intensity. The measured amplitude of oscillation (modulation ratio) M_{Meas} and phase delay φ_{Meas} from a specific sample will depend on the instrument used for acquisition. The measurements taken on a given instrument depend on the excitation source, the sample holder (e.g., type/size of cuvette) and, in some cases, the gain of the detector. Therefore, the values for M_{Meas} and φ_{Meas} need to be corrected by analyzing the excitation light in order to relate them to the lifetime of the sample.

The waveform of the excitation light is decomposed with a similar Fourier analysis, and analogous values for the modulation ratio M_{Exc} and phase delay φ_{Exc} are defined. The variables M_{Meas} and φ_{Meas} are then corrected in order to extract the modulation ratio M_{Fluor} and phase delay φ_{Fluor} dependent on the lifetimes of the sample, Eqs. 8 and 9.

$$M_{\mathrm{Fluor}} = \frac{M_{\mathrm{Meas}}}{M_{\mathrm{Exc}}} \tag{8}$$

$$\varphi_{\mathrm{Fluor}} = \varphi_{\mathrm{Exc}} - \varphi_{\mathrm{Meas}} \tag{9}$$

Values for M_{Exc} and φ_{Exc} are commonly determined by measuring a sample with known modulation ratio and phase decay.

If the sample is characterized by only a single exponential decay (lifetime), M_{Fluor} and φ_{Fluor} can be expressed as Eqs. 10 and 11,

$$M_{\mathrm{Fluor}} = \frac{1}{\sqrt{1 + (\omega_n \tau_{\mathrm{MOD}})^2}} \tag{10}$$

$$\varphi_{\mathrm{Fluor}} = \tan^{-1}(\omega_n \tau_\varphi) \tag{11}$$

where τ_{MOD} and τ_φ are both equal to the lifetime of the sample. If the sample is characterized by multiple exponential decays, τ_{MOD} and τ_φ do not refer to any particular constituent decay.

Experimentally, instruments that measure lifetimes in the frequency domain directly output values for the oscillation (modulation) frequency ω_n, the modulation ratio M_{Fluor}, and phase delay of the sample φ_{Fluor}. Equivalently, Fourier analysis can be applied to the exponential decays produced by instruments operating in the time domain to find M_{Fluor} and φ_{Fluor}.

2.3 Polar Plot

The (x, y) coordinates of the polar plot are constructed from M_{Fluor} and φ_{Fluor} as shown in Eqs. 12 and 13.

$$x = M_{\mathrm{Fluor}} \cos(\varphi_{\mathrm{Fluor}}) \tag{12}$$

$$y = M_{\mathrm{Fluor}} \sin(\varphi_{\mathrm{Fluor}}) \tag{13}$$

As demonstrated in Fig. 1a, the polar coordinates of any fluorescing species characterized by a single lifetime (single exponential decay) will reside on a semicircle with a radius of 0.5 centered at (0.5,0). If the sample contains multiple fluorescing species with different

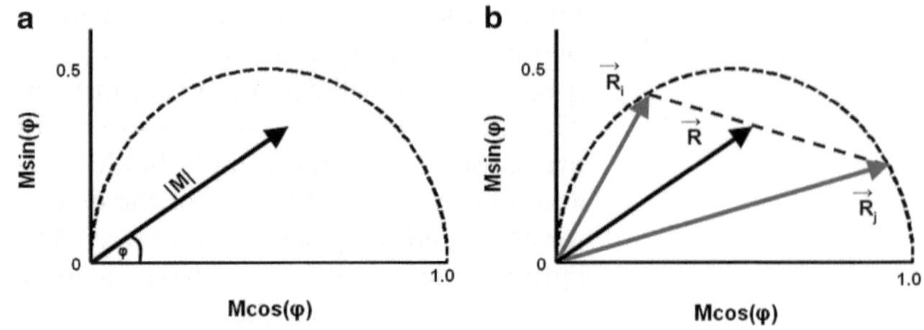

Fig. 1 Basics of the polar plot. (**a**) The polar coordinate of any fluorophore can be equivalently described as a vector with a length equal to the modulation ratio making an angle relative to the *x*-axis equal to the phase delay. (**b**) The polar coordinates of two different fluorophores defined by the vectors $\vec{R_i}$ and $\vec{R_j}$ contribute intensity to the measured polar coordinate $\vec{R}$. All possible measured combinations of the two fluorophores *i* and *j* lie on the *dashed line* projected within the semicircle

lifetimes, the (x, y) coordinates of the polar plot reflect the contributions of the constituent lifetimes,

$$x = \sum_i \alpha_i M_i \cos(\varphi_i) \tag{14}$$

$$y = \sum_i \alpha_i M_i \sin(\varphi_i) \tag{15}$$

The factors α_i in Eqs. 14 and 15 are the contribution of emitted fluorescence from the species with lifetime τ_i to the measured steady-state intensity. The terms α_i are referred to as the intensity fraction of species i. The intensity fraction is related to the species fraction by the following relation,

$$\alpha_i = \frac{a_i \tau_i}{\sum_i a_i \tau_i} \tag{16}$$

(Please *see* Eq. 1 for the definition of the species fractions a_i). Therefore, the resulting polar coordinate of a sample containing fluorophores with many lifetimes is simply the sum of the polar coordinates of each constituent lifetime component weighted by the corresponding intensity fraction. The factors M_i and φ_i are the M_{Fluor} and φ_{Fluor} that would be recorded if each constituent fluorescing species were to be measured independently.

The graphical ease of the polar plot is demonstrated in Fig. 1b, where the polar coordinate $\vec{R}$ of a sample containing a mixture of two fluorophores with single component lifetimes is shown. The polar coordinates of the two constituent fluorophores are $\vec{R_i}$ and $\vec{R_j}$. The locations of the polar coordinates representing all linear combinations of fluorescing components i and j are projected as a dashed line between the coordinates of $\vec{R_i}$ and $\vec{R_j}$. The polar coordinate $\vec{R}$ is a weighted sum of $\vec{R_i}$ and $\vec{R_j}$; therefore, the relative distance between $\vec{R}$ and $\vec{R_i}$ or $\vec{R_j}$ along the dashed line can be used to determine the corresponding intensity fractions. Experimentally, the modulation frequency must be selected carefully to sufficiently resolve and distinguish the variable contributions of fluorophores with different lifetimes (*see* **Note 1**).

The polar coordinate calculated from M_{Fluor} and φ_{Fluor} uniquely characterizes any fluorescent sample from those with a single component lifetime to those described by complex distributions of lifetimes. Hence, changes between different samples or among different fluorescent populations in the sample can be visualized simply by monitoring changes in polar coordinate. Without the need for iterative fitting or intensive computations, calculations of distance on the polar plot can also be used to quantitatively calculate both intensity and species fractions of contributing fluorescent components.

3 Illustrative Applications

3.1 Dynamic Quenching

Generally, fluorescence quenching refers to any process that decreases a sample's measured fluorescence [16]. Dynamic quenching takes place while the fluorophore is in the excited state and therefore shortens the measured lifetime. An example of dynamic quenching is shown in Fig. 2a where fluorescein was quenched by increasing concentrations of iodide [18]. In this case, the experiment took place in a cuvette; therefore, all molecules of iodide have an equal chance of accessing fluorescein. The polar (x,y) coordinates were directly computed from the measured modulation ratio and phase delay and are presented on the polar plot (Fig. 2a). All of the measured polar coordinates lie on the semicircle, with the exception of a few points that possibly underwent photobleaching during acquisition (*see* **Note 2**). As the quencher concentration increases, the measured polar coordinate moves along the semicircle in a clockwise manner indicating that the lifetime of the sample decreases homogenously as single component lifetime. Without the need acquire data at large sets of modulation frequencies and iteratively fit, the process of quenching can be monitored quantitatively with a measurement taken at a single modulation frequency when using the polar plot.

3.2 Spectrally Resolved Polar Plot

Fluorescent samples can often contain sets of fluorophores, which absorb and emit light at similar sets of wavelengths. As a result, measuring the sample's emission and lifetime through filters may not be sufficient to characterize the individual fluorescent species

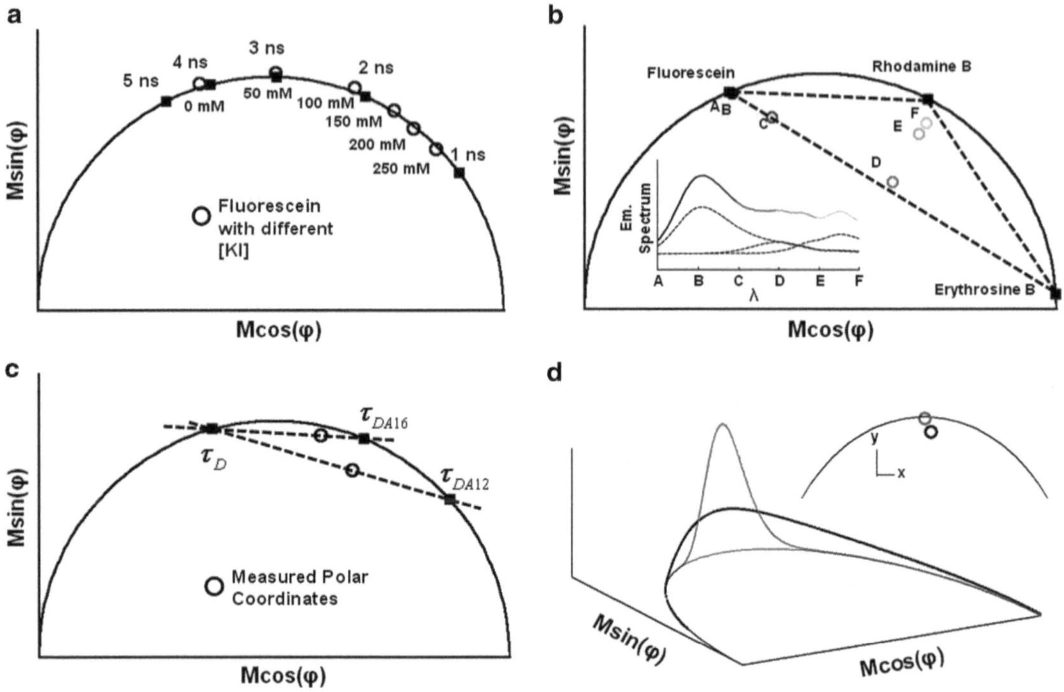

Fig. 2 Applications of polar plot. (**a**) The measured lifetime of fluorescein decreases by dynamic quenching as the concentration of iodide is increased. The resulting polar coordinates plotted at each concentration of iodide span the perimeter of the universal semicircle. (**b**) The modulation ratio and phase delay were simulated at small intervals of wavelengths along the emission spectrum (*inset*) of a sample containing a mixture of fluorescein, erythrosine B, and rhodamine B. The movement of the polar plot trajectory reveals information on the dye composition of the sample. (**c**) The polar coordinates of two samples containing fluorescein (donor) and tetramethyl-rhodamine (acceptor) separated by either 12- or 16-base pairs of DNA are shown (*gray circles*). A background of DNA labeled only with the donor was added to these samples to move these polar coordinates into the universal semicircle. By drawing a straight line between the polar coordinate of the noninteracting donor τ_D and the measured points (*gray circles*), the lifetimes of the donor quenched on the doubly labeled strands by FRET (τ_{DA12} and τ_{DA16}) can be determined. (**d**) The measured polar coordinates of ensembles containing two different distributions of lifetimes are plotted on the polar plot. The projected lines above the polar plot indicate the intensity fraction of the component lifetimes along the universal semicircle for each distribution. The polar coordinate for the wider distribution of lifetime lies further inside the semicircle compared to the narrower distribution (inset) (color figure online)

in these types of samples. Decomposing the spectra of a sample containing many fluorophores in order to examine the contributions of each specific fluorophore to the emitted intensity is commonly applied to analyze steady-state measurements [19, 20]. When combined with lifetime measurements and analysis on the polar plot, decomposing spectra (spectral unmixing) can define trajectories on the polar plot that would not otherwise be easily accessible to determine the composition and changes in fluorescent species [21–24]. Unlike measurements taken with filters that pass large sets of wavelengths to the detector, the data collected for eventual unmixing is sampled at small intervals of wavelengths spanning the emission spectrum of the sample. With sufficient signal, spectral unmixing has been useful for studying FRET as well as examining contributions from dimly fluorescing components in samples [23].

For example, the simulation presented in Fig. 2b describes a sample that contains the fluorescent dyes fluorescein (lifetime = 4.3 ns [25]), erythrosine B (lifetime = 89 ps [26]), and rhodamine B (lifetime = 1.74 ns [26]). The steady-state fluorescence intensity, modulation ratio, and phase delay were incrementally sampled at a set of emission wavelengths from 490 to 600 nm. These are labeled as A–F in the inset of Fig. 2b. Within the steady-state spectrum (inset in Fig. 2b), the three contributions of each component spectrum derived from the unmixing are shown.

The polar coordinates computed from the fluorescence at each wavelength (A–F) along with the polar coordinates of pure fluorescein, erythrosine B, and rhodamine B are plotted in Fig. 2b. As the emission wavelengths were moved from region A to region D, the simulated polar coordinate moved from a position near fluorescein's polar coordinate towards the coordinate of erythrosine B. In addition to the movement of the polar coordinate, the unmixing at these emission wavelengths also reveals the spectral contribution of erythrosine B, which is not clearly visible in the emission spectrum (peak near region D). At longer emission wavelengths (regions E and F in Fig. 2b), the collected intensity as estimated by both the spectral unmixing (peak near region F) and the movement of the polar coordinate has definite contributions from rhodamine B's fluorescence. Therefore, the spectral unmixing when combined with the trajectories on the polar plot can specifically assign intensities to fluorophores and characterize complicated samples. However, in this example even without unmixing or any knowledge about the constituent fluorophores, these changes in trajectory of the polar coordinates clearly indicate that the fluorescence signal is derived from multiple fluorophores.

As a general comment, any measured polar coordinate that contains fluorescence from fluorophores with three distinct lifetimes will lie inside the area of a triangle formed by the polar coordinates computed from the lifetimes of each of the three constituent fluorophores (dashed triangle in Fig. 2b).

The calculation for spectral unmixing is briefly described below Eq. 17. The measured spectrum containing the mixture of fluorophores $I_{\mathrm{Meas}}(\lambda_{\mathrm{Em}})$ equals the sum of reference spectra (from the constituent fluorophores) $F_{\mathrm{ref},i}(\lambda_{\mathrm{Em}})$ each weighted by a specific contribution (A_i).

$$I_{\mathrm{Meas}}(\lambda_{\mathrm{Em}}) = A_1 F_{\mathrm{ref},1}(\lambda_{\mathrm{Em}}) + A_2 F_{\mathrm{ref},2}(\lambda_{\mathrm{Em}}) + A_3 F_{\mathrm{ref},3}(\lambda_{\mathrm{Em}}) + \cdots + \tag{17}$$

In addition to iterative fitting, a matrix calculation can also be carried out to extract the weighted contributions (A_i) from each reference spectrum by multiplying both sides of the above equation by the inverse of the matrix containing the reference spectra Eq. 18.

$$\begin{bmatrix} A_1 \\ A_2 \\ \vdots \end{bmatrix} = \begin{bmatrix} F_{\mathrm{ref},1}\left(\lambda_{\mathrm{Em}}^1\right) & F_{\mathrm{ref},2}\left(\lambda_{\mathrm{Em}}^1\right) & \cdots \\ F_{\mathrm{ref},1}\left(\lambda_{\mathrm{Em}}^2\right) & F_{\mathrm{ref},2}\left(\lambda_{\mathrm{Em}}^2\right) & \cdots \\ \vdots & \vdots & \vdots \end{bmatrix}^{-1} \begin{bmatrix} I_{\mathrm{Meas}}\left(\lambda_{\mathrm{Em}}^1\right) \\ I_{\mathrm{Meas}}\left(\lambda_{\mathrm{Em}}^2\right) \\ \vdots \end{bmatrix} \tag{18}$$

Although the reference matrix is not square, a pseudo-inverse can be found using singular value decomposition (SVD) [27].

The intensity contributions from the component fluorophores extracted from unmixing are presented initially relative to the measured (reference) spectra of each component fluorophore. This set of reference spectra includes the emission spectrum of each constituent fluorophore measured independently. The conditions of these measurements such as the local environment, solvent, and instrument must closely parallel the conditions of the measurement of interest (containing the mixture of fluorophores). In the case of biological samples such as live cells or tissues containing a set of fluorophores, the reference spectra should be measured in equivalent cells or tissues singly labeled with the constituent fluorophores. Very often, pH and the concentration of ions can cause spectral shifts and changes in the shape of measured spectra [28–30].

3.3 Förster Resonance Energy Transfer

Förster Resonance Energy Transfer (FRET) is the non-radiative transfer of energy between two fluorescent molecules referred to as the donor and acceptor [31–34]. The primary considerations for efficient transfer are the distance between the donor and acceptor and the overlap between the absorbance spectrum of the acceptor and the emission spectrum of the donor. Other parameters such as the mutual orientation of the donor and acceptor molecules and the range of wavelengths being applied can also alter the amount of FRET taking place. The FRET efficiency E can be measured with both steady-state and time-dependent methods. The Förster radius is the distance where the efficiency of energy transfer is 50 %. The FRET efficiency can be determined by the following relation Eq. 19 if the lifetime of the free donor and the lifetime of the donor undergoing FRET are known.

$$E = 1 - \frac{\tau_{\mathrm{DA}}}{\tau_{\mathrm{D}}} \tag{19}$$

Many biological samples such as cells and macromolecules studied using FRET often contain a mixture of paired and unpaired donors and acceptors. In other words, there is frequently a background of unpaired (noninteracting) donors and acceptors in addition to the donors, which are conjugated (or near) to acceptors and are consequently undergoing FRET. In these cases, the measured FRET efficiencies can be lowered simply by the presence of noninteracting donors and acceptors. As a result, quantitatively relating the measured FRET efficiency to the distance between the donors and acceptors undergoing FRET is complicated, but can be greatly simplified by using the polar plot (Fig. 2c).

In the example presented in Fig. 2c, the FRET efficiency between a fluorescein derivative (donor) and tetramethylrhodamine (acceptor) was measured on opposing ends of a 12-base-pair DNA strand and separately on the opposite ends of a 16-base-pair strand of DNA. In these two solutions of DNA, a background of unpaired donors is present in addition to the doubly labeled DNA strands. The polar coordinates calculated from the measured phase delay and modulation ratio of the donor in both of these sets of DNA are projected on the polar plot (gray circles in Fig. 2c). The lifetime of the unpaired donor was measured from a separate solution of single-stranded DNA labeled only with the donor (τ_D in Fig. 2c).

The polar coordinates describing the two solutions of 12- and 16-base-pair doubly labeled DNA (along with some singly labeled DNA) are composed of multicomponent lifetimes and as a result lie inside the universal semicircle. However, fluorescein has a single lifetime; therefore, any quenched lifetime of the fluorescein as a result of FRET will also have only one component. Without any fitting, the polar plot can be used to find the single component lifetimes of the donor undergoing FRET on the 12-base-pair strand of DNA and the 16-base-pair strand of DNA even with the background of singly labeled DNA (donor alone) in the solutions. By plotting a straight line that passes through the polar coordinate of donor alone (τ_D in Fig. 2c) and the measured polar coordinate of either one of the solutions containing the doubly labeled DNA (gray circles in Fig. 2c), extrapolation of the straight line intersects the semicircle revealing the polar coordinate of donor undergoing FRET. As shown by the two intersections on the universal semicircle (τ_{DA12} and τ_{DA16} in Fig. 2c), the donors that were only 12-base pairs apart from their corresponding acceptors had a shorter lifetime than the donors that were separated by 16-base pairs from their acceptors. Both of these lifetimes (τ_{DA12} and τ_{DA16}) can be determined quantitatively from the polar plot for more rigorous calculations of FRET efficiencies and distances between donors and acceptors.

3.4 Distributions of Lifetimes

Variations in the local environment of a fluorophore can result in a continuous distribution of lifetimes for samples containing even a single fluorescent species. However, in a pure solution with a single fluorescent species, this distribution is often very narrow and can be analyzed as a discrete lifetime. Examples of samples from biology that can exhibit wider distributions of lifetimes include tryptophan residues continuously fluctuating between substates in proteins [35] and DNA labeled at multiple sites (unpublished measurements on Acridine Orange attached to DNA). When the measured modulation ratio and phase delay from these types of samples are projected on the polar plot, the measured polar coordinate will be inside the semicircle, but near to the average lifetime of the distribution. In the example shown in Fig. 2d, the movement of the polar coordinate into the semicircle can be observed as a function of the width of the distribution of lifetimes. In Fig. 2d, two Gaussian distributions of single lifetime are plotted. The plots located above the universal semicircle (3D projection) describe the intensity fraction that each single component lifetime contributes to the distribution. As the width of the Gaussian distribution widens, the contribution of the wings of the distribution will pull the measured polar coordinate farther inside the semicircle. A polar coordinate located inside the semicircle can also be indicative of a small number of discrete lifetimes or the result of additional background in the measurement (*see* **Note 3**). Additional knowledge of systems with many lifetimes is needed in order to more reliably interpret the measured polar coordinate as a distribution or a small number of lifetimes.

4 Notes

1. Optimal Modulation Frequency
   The separation on the polar plot between fluorescing species with distinct lifetimes can change dramatically depending on the frequency ω_n that the polar plot refers to. For example, in Fig. 3, the polar coordinates defining 0.5 and 2 ns are presented at the following modulation frequencies 10, 20, 40, and 80 MHz (Fig. 3). Measuring a difference between the two polar coordinates at 10 MHz (Fig. 3a) would yield only a small change in distance that likely would not be sufficient to resolve these two species experimentally, especially if the data is from a FLIM experiment. However, the two points are easily resolved at 80 MHz (Fig. 3d). The optimal modulation frequency can be calculated by optimizing the expression for distance on the polar plot as a function of frequency as described in [5]. However, in the laboratory, a simple rule is to select a modulation frequency near $1/\tau$ (where τ is close to the lifetimes of interest) in order to maximize the resolution between different fluorescent species on the polar plot.

2. Photobleaching
   Photobleaching (photodestruction of the fluorophore) causes a reduction in, or complete loss of, the emitted fluorescence intensity from the sample during the time of data acquisition (illumination of the sample) [16]. On the polar plot, the most

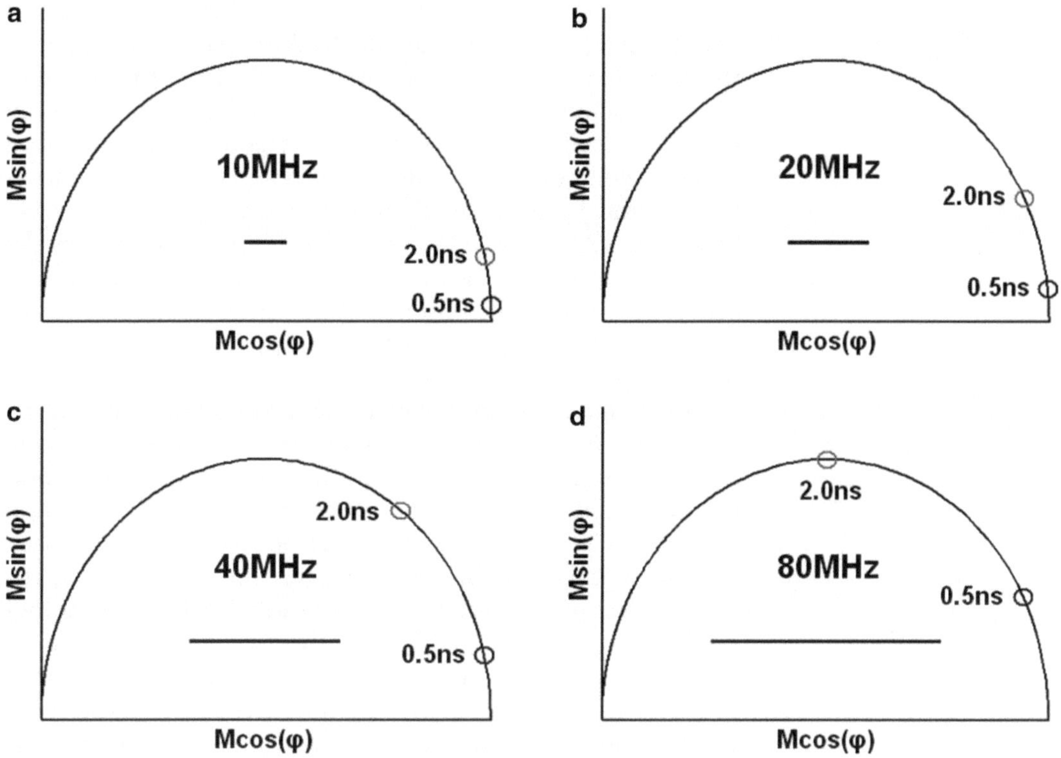

Fig. 3 Selecting the optimal modulation frequency (**a–d**). The polar coordinates corresponding to 0.5 ns (*black open circle*) and 2 ns (*gray open circle*) are plotted at modulation frequencies of 10, 20, 40, and 80 MHz. On each plot, the distance between the two polar coordinates is presented (*black line* in center of polar plot) indicating that the resolution between the two points improves as a function of the modulation frequency (color figure online)

noticeable effect of photobleaching is an artificial increase in the modulation ratio that can move the polar coordinate outside of the universal semicircle [5]. Changes in the phase delay can also occur, which when coupled with corresponding changes in the modulation ratio can create curved trajectories on the polar plot. The precise effects of the bleaching manifested in these trajectories also depends on the rate of bleaching, instrumental parameters related to the excitation source, as well as the lifetime of the sample.

Although several computational strategies exist to correct for photobleaching [36], maintaining low levels of exciting light and using anti-fade agents remain the most direct means to avoid artifacts from loss of signal. However, lower excitation light means fewer detected photons, lower signal-to-noise, and possibly longer data acquisition times.

3. Background or Extraneous Fluorescence

   The polar plot also graphically relates the relative contribution in intensity of different fluorophores and is, therefore, very useful for identifying unwanted light or extraneous fluorescence. Examples include background from media in the case of live cell imaging, scattered light from the excitation source reaching the detector, fluorescence from mounting media in fixed samples, and also the autofluorescence from organic compounds in biological samples. Un-modulated background from low levels of dark current can almost never be avoided. All of these sources of background have themselves a polar coordinate indicative of their time-dependent fluorescence. As a result, the removal of background requires simple vector subtraction provided that the polar coordinate of the pure sample of interest is known.

Complications arise when the polar coordinate of the sample or fluorophore of interest is not known (or could be changing); in that case, the effects of background cannot be accurately determined using simple vector addition. Recently, a technique has been introduced to remove background in this case that incorporates the analysis of spatial frequencies using wavelets to distinguish specific morphological features in images [37]. In other words, images can be reconstructed with only morphological features relevant to the sample while those features characteristic of background can be disregarded. Unlike Fourier analysis that presents the spatial frequencies in the image globally, wavelets can uniquely indicate the spatial frequencies present at specific locations throughout the image [38–40].

Wavelets are oscillatory functions used to decompose images based on the spatial frequencies contained in the wavelet decomposition. In order to decompose an entire image, a set of stretched (scaled) versions of a specific wavelet are used to highlight a range of frequencies throughout the image. At each pixel of the image, a coefficient can be calculated for a specific scaled version of the wavelet indicating the similarity between that region of the image and the scaled wavelet. As described, this process (referred to as the continuous wavelet transform (CWT)) is computationally expensive. The discrete wavelet transform (DWT) is a more economical approach that samples both the scales of the wavelet and the positions at which it samples by powers of two. Unlike the CWT, the DWT outputs a "details matrix" generated from a low-pass filter (scaling function) and an approximation of the image resulting from a scaled version of the wavelet (high frequencies). Therefore, at each level of the decomposition describing the image, each approximation contains the details while the low-pass filter (scaling function) creates low-resolution renderings of the original image.

In a process known as multiresolution analysis [41], background can then be removed or suppressed by reconstructing the original image only with a narrow bandpass of spatial frequencies defined by the difference between two approximations derived by the DWT. As a result, only morphological features in the images described by that bandpass of spatial frequencies defined by the two approximations are retained. When this technique was applied to study lifetime images, the time-dependent fluorescence (phase delay and modulation ratio) associated with the morphological feature highlighted by the multiresolution analysis is also specifically retained without the addition of background elements.

Removing background is also complicated because the data collected contains noise. As a consequence of noise, the measured polar coordinate of a solution containing a single fluorophore will actually be a distribution because of the associated uncertainty in the phase delay and the modulation ratio. Therefore, removing background, on a pixel-wise basis is more complicated than a simple addition of vectors. Binning in the image as well as finding the average polar coordinate in selected regions of the image can be done to simplify the process of background removal. However, binning or the use of common low-pass filters such as median or mean filters to remove noise can significantly degrade the spatial resolution in the image. In order to avoid the loss of fine features in an image, these strategies should only be applied with sufficient knowledge of the pixel size and resolution of the imaging system. Recently, methods to remove noise while retaining spatial resolution have been introduced for both Poisson noise and the Gaussian noise inherent to lifetime imaging systems with image intensifiers [25, 42].

References

1. Fatt P (1964) An analysis of the transverse electrical impedance of striated muscle. Proc R Soc Lond B Biol Sci 159:606–651
2. Rigaud B, Hamzaoui L, Frikha MR et al (1995) In vitro tissue characterization and modeling using electrical impedance measurements in the 100 Hz–10 MHz frequency range. Physiol Meas 16:A15–A28
3. Cole KS, Cole RH (1941) Dispersion and absorption in dielectrics. I. Alternating current characteristics. J Chem Phys 9:341–351
4. Hanley QS, Clayton AH (2005) AB-plot assisted determination of fluorophore mixtures in a fluorescence lifetime microscope using spectra or quenchers. J Microsc 218:62–67
5. Redford GI, Clegg RM (2005) Polar plot representation for frequency-domain analysis of fluorescence lifetimes. J Fluoresc 15:805–815
6. Jameson DM, Gratton E, Hall RD (1984) The measurement and analysis of heterogeneous emissions by multifrequency phase and modulation fluorometry. Appl Spectros Rev 20: 55–106
7. Digman MA, Caiolfa VR, Zamai M, Gratton E (2008) The phasor approach to fluorescence lifetime imaging analysis. Biophys J 94: L14–L16
8. Halvorson HR (1979) Relaxation kinetics of glutamate dehydrogenase self-association by pressure perturbation. Biochemistry 18: 2480–2487
9. Brigham EO (1988) The fast Fourier transform and its application. Prentice Hall, Englewood Cliffs, NJ
10. Bracewell RA (1978) The Fourier transformation and its applications. McGraw-Hill Kogakusha, Ltd., Hamburg
11. Gratton E, Jameson DM, Hall RD (1984) Multifrequency phase and modulation fluorometry. Annu Rev Biophys Bioeng 13:105–124

12. Gratton E, Limkeman M, Lakowicz JR et al (1984) Resolution of mixtures of fluorophores using variable-frequency phase and modulation data. Biophys J 46:479–486
13. Gratton E, Limkeman M (1983) A continuously variable frequency cross-correlation phase fluorometer with picosecond resolution. Biophys J 44:315–324
14. Noomnarm U, Clegg RM (2009) Fluorescence lifetimes: fundamentals and interpretations. Photosynth Res 101:181–194
15. Birks J (1970) Photophysics of aromatic molecules. Wiley-Interscience, New York
16. Lakowicz J (2006) Principles of fluorescence spectroscopy. Springer, New York, NY
17. Valeur B (2002) Molecular fluorescence: principles and applications. Wiley-VCH, Weinheim, Germany
18. Kasha M (1952) Collisional perturbation of spin-orbital coupling and the mechanism of fluorescence quenching. A visual demonstration of the perturbation. J Chem Phys 20:71–74
19. Zimmermann T (2005) Spectral imaging and linear unmixing in light microscopy. Adv Biochem Eng Biotechnol 95:245–265
20. Haraguchi T, Shimi T, Koujin T et al (2002) Spectral imaging fluorescence microscopy. Genes Cells 7:881–887
21. Rueck A, Huelshoff C, Kinzler I et al (2007) SLIM: a new method for molecular imaging. Microsc Res Tech 70:485–492
22. Hanley QS, Arndt-Jovin DJ, Jovin TM (2002) Spectrally resolved fluorescence lifetime imaging microscopy. Appl Spectrosc 56:155–166
23. Chen YC, Clegg RM (2011) Spectral resolution in conjunction with polar plots improves the accuracy and reliability of FLIM measurements and estimates of FRET efficiencies. J Microsc 244:21–37
24. Strat D, Dolp F, von Einem B et al (2011) Spectrally resolved fluorescence lifetime imaging microscopy: Förster resonant energy transfer global analysis with a one- and two-exponential donor model. J Biomed Opt 16:026002
25. Spring BQ, Clegg RM (2009) Image analysis for denoising full-field frequency-domain fluorescence lifetime images. J Microsc 235:221–237
26. Boens N, Qin W, Basaric N et al (2007) Fluorescence lifetime standards for time and frequency domain fluorescence spectroscopy. Anal Chem 79:2137–2149
27. Tsurui H, Nishimura H, Hattori S et al (2000) Seven-color fluorescence imaging of tissue samples based on Fourier spectroscopy and singular value decomposition. J Histochem Cytochem 48:653–662
28. Sjoback R, Nygren J, Kubista M (1995) Absorption and fluorescence properties of fluorescein. Spectrochimica Acta A 51:L7–L21
29. Elsliger MA, Wachter RM, Hanson GT et al (1999) Structural and spectral response of green fluorescent protein variants to changes in pH. Biochemistry 38:5296–5301
30. Hawe A, Sutter M, Jiskoot W (2007) Extrinsic fluorescent dyes as tools for protein characterization. Pharm Res 25:1487–1499
31. Clegg RM (1992) Fluorescence resonance energy transfer and nucleic acids. Methods Enzymol 211:353–388
32. Förster T (1951) Fluoreszenz organischer Verbindungen. Vandenhoeck & Ruprecht, Göttingen, Germany
33. Förster T (1948) Zwischenmolekulare Energiewanderung und Fluoreszenz. Ann Phys 437:55–75
34. Förster T (1993) Intermolecular energy migration and fluorescence. In: Mielczarek EV, Greenbaum E, Knox RS (eds) Biological physics. American Institute of Physics, New York, pp 148–160
35. Gratton E, Silva N, Mei G et al (1992) Fluorescence lifetime distribution of folded and unfolded proteins. Int J Quantum Chem 42:1479–1489
36. Malachowski G, Clegg RM, Redford GI (2007) Analytic solutions to modeling exponential and harmonic functions using Chebyshev polynomials: fitting frequency-domain lifetime images with photobleaching. J Microsc 228:282–295
37. Buranachai C, Kamiyama D, Chiba A et al (2008) Rapid frequency-domain FLIM spinning disk confocal microscope: lifetime resolution, image improvement and wavelet analysis. J Fluoresc 18:929–942
38. Hubbard BB (1998) In: Peters AK (ed) The world according to wavelets. The story of a mathematical technique in the making. 2nd edn. Wellesley, Natick, MA, pp 111–259
39. Rao RM, Bopardikar AS (1998) Wavelet transforms. Introduction to theory and applications. Addison-Wesley, Reading, MA
40. Walker JS (1997) Fourier analysis and wavelet analysis. Not Am Math Soc 44:658–670
41. Mallat SG (1989) A theory for multiresolution signal decomposition: the wavelet representation. IEEE Trans Pattern Anal Mach Intell 11:674–692
42. Willet RM, Nowak RD (2004) Fast multiresolution photon-limited image reconstruction. IEEE Int Symp Biomed Imaging 2: 1192–1195

Chapter 7

Ensemble and Single-Molecule Detected Time-Resolved FRET Methods in Studies of Protein Conformations and Dynamics

Tomer Orevi, Eitan Lerner, Gil Rahamim, Dan Amir, and Elisha Haas

Abstract

Most proteins are nanomachines that are selected to execute specific functions and therefore should have some degree of flexibility. The driving force that excites specific motions of domains and smaller chain elements is the thermal fluctuations of the solvent bath which are channeled to selected modes of motions by the structural constraints. Consequently characterization of the ensembles of conformers of proteins and their dynamics should be expressed in statistical terms, i.e., determination of probability distributions of the various conformers. This can be achieved by measurements of time-resolved dynamic non-radiative excitation energy transfer (trFRET) within ensembles of site specifically labeled protein molecules. Distributions of intramolecular segmental end-to-end distances and their fast fluctuations can be determined, and fast and slow conformational transitions within selected sections of the molecule can be monitored and analyzed. Both ensemble and single-molecule detection methods can be applied for data collection. In combination with synchronization methods, time-resolved FRET was also used for studies of fast conformational transitions, in particular the folding/unfolding transitions.

Key words Protein conformational dynamics, Time-resolved FRET, Single-molecule detection, Site-specific labeling, Distance distributions, Intramolecular diffusion coefficient, Fast fluctuations and conformational transitions

1 Introduction

Measurements of dynamic non-radiative excitation energy transfer (FRET) [1–7], which is based on the distance dependent interactions between excited state dipoles, can be applied for the determination of long-range distances between aminoacyl residues in protein molecules. This method has several advantages which makes it uniquely suited to the study of protein conformation and dynamics: close to ideal time resolution; very high sensitivity, up to single-molecule detection; accuracy at distances representing molecular dimensions (10–100 Å); measurements of intermolecular interactions; detection

Yves Engelborghs and Antonie J.W.G. Visser (eds.), *Fluorescence Spectroscopy and Microscopy: Methods and Protocols*, Methods in Molecular Biology, vol. 1076, DOI 10.1007/978-1-62703-649-8_7,

of protein–protein interactions; and oligomerization and the ability to recover distributions of intramolecular distances of fast fluctuating disordered protein molecules.

A series of parameters can be chosen to characterize the partially folded states of protein molecules. These include the mean, the width, the number of subpopulations, and the shape of the distributions of the intramolecular distances. Ensemble time-resolved FRET (trFRET) and single-molecule detected FRET (smFRET) measurements of double-labeled protein samples can provide all these parameters.

In their disordered states and in most cases also in the folded state, protein molecules undergo rapid conformational fluctuations. The lifetime of the excited states of donor probes in trFRET experiments (nanoseconds) defines time windows that enable the detection of rapid fluctuations of intramolecular distances both in the ensemble and in the single-molecule modes. Slower fluctuations can be detected by single-molecule FRET spectroscopy or autocorrelation analysis of intensity fluctuations of double-labeled protein samples [8, 9]. Various modes of trFRET measurements, either in the ensemble of the single-molecule detection, are particularly powerful in studies of flexible protein molecules, of fully or partially unfolded states of globular proteins, and of the intrinsically disordered proteins (IDPs).

FRET measurements are also useful for detection and analysis of intermolecular interactions. A common characteristic of many proteins is the tendency to form amyloid aggregates, protein–protein or protein–ligand interactions. The protein/protein interactions can be monitored by various modes of intermolecular FRET measurements.

We shall first briefly review the theoretical basis of the FRET-based methods and then describe protocols for preparation of labeled samples of proteins suitable for FRET experiments. This will be followed by sections describing the various analysis methods that were developed.

1.1 Theoretical Background

1.1.1 Non-radiative Excitation Energy Transfer

Non-radiative transfer of excitation energy requires some interaction between a donor molecule and an acceptor molecule (Fig. 1). This transfer can occur if at least some vibronic levels in the donor have the same energy as corresponding transitions in the acceptor. Energy transfer can result from different interaction mechanisms. The interactions may be Coulombic and/or due to intermolecular orbital overlap. The Coulombic interactions consist of long-range dipole–dipole interactions (Förster's mechanism) and short-range multipolar interactions. In the energy transfer process, the initially excited electron on the donor, D, returns to its ground state orbital on D, while simultaneously, an electron on the acceptor, A, is promoted to an excited state. For permitted transitions on D and A, the Coulombic interaction is predominant, even at short distances. This is the case of the singlet–singlet transfer:

$$^1D^* + {}^1A \rightarrow {}^1D + {}^1A^*$$

which is effective over a long interaction range (up to 80–100 Å).

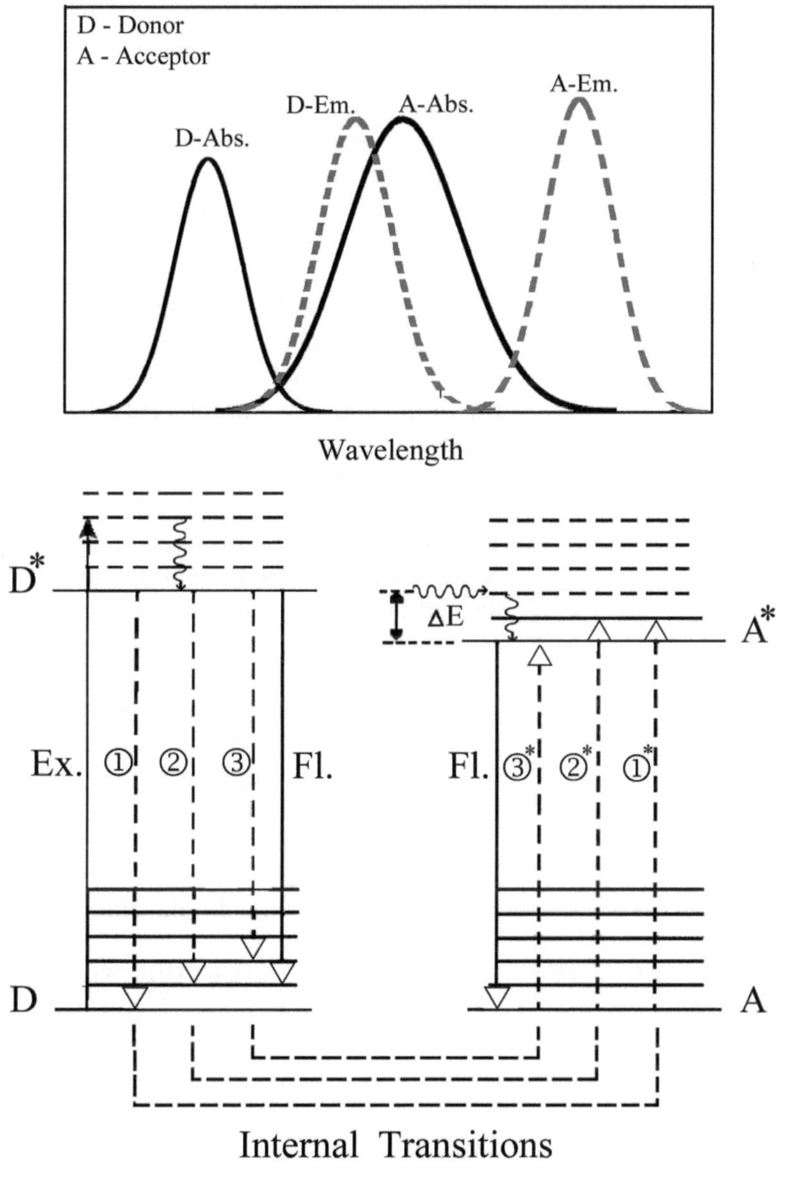

Fig. 1 Energy level diagram demonstrating the mechanism of resonance coupling of the non-radiative transitions in the donor and acceptor probes under conditions where the vibrational relaxation is faster than the energy transfer (very weak coupling). The non-radiative transitions are shown in dashed lines and the radiative transitions are shown as *continuous vertical lines*. The upper inset shows a scheme of the corresponding absorption and emission spectra of the probes and the overlap integral

Two main mechanisms are relevant in the context of the application of energy transfer for IDP research: the Förster's mechanism for long-range interactions and the triplet–triplet transfer for the detection of shorter-range interactions.

The term fluorescence resonance energy transfer (FRET) is commonly used to describe singlet–singlet energy transfer via a mechanism based on long-range dipole–dipole resonance coupling[1] [1].

1.1.2 Rate of Non-radiative Excitation Energy Transfer Within a Donor: Acceptor Pair

A molecule in an electronically excited state (donor) can transfer its excitation energy to another molecule (acceptor) [1] provided that the donor-acceptor pair fulfills several conditions: (a) The energy of the donor must be luminescent; (b) the emission spectrum of the donor should have some overlap with the absorption spectrum of the acceptor; and (c) the distance between the two probes must not exceed an upper limit (usually within the range of up to 100 Å); (d) orientation dependence, see further. The transfer is readily observed when two different probes are involved, but transfer within a homogeneous population of chromophores can occur, as well. Such transfer was first detected in 1927 by Perrin, who detected the loss of polarization of the emitted light [10].

The dipole–dipole interaction which leads to the transfer of excitation energy is very weak, usually of the order of ~2–4 cm^{-1}, while the spectroscopic energies that are transferred are much higher, ~15,000–40,000 cm^{-1}.

Förster's theory describes the rate of energy transfer for an isolated pair of chromophores, which fulfill the requirements for energy transfer by the dipole–dipole interaction to be:

$$k_{\mathrm{T}} = \frac{9(\ln 10)\kappa^2 \Phi_{\mathrm{D}}^{\mathrm{o}}}{128\pi^5 N_{\mathrm{A}} r^6 \tau_{\mathrm{D}}^{\mathrm{o}}} \int_0^{\infty} f(\lambda)\varepsilon(\lambda)\lambda^4 \mathrm{d}\lambda \tag{1}$$

$$k_{\mathrm{T}} = \frac{1}{\tau_{\mathrm{D}}^{\mathrm{o}}} \left(\frac{R_{\mathrm{o}}}{r}\right)^6 \tag{1a}$$

where $\kappa^2 = \cos\theta_{\mathrm{DA}} - 3\cos\theta_{\mathrm{D}}\cos\theta_{\mathrm{A}} = \sin\theta_{\mathrm{D}}\sin\theta_{\mathrm{A}}\cos\varphi - 2\cos\theta_{\mathrm{D}}\cos\theta_{\mathrm{A}}$ (θ_{DA} is the angle between the donor and the acceptor dipoles, θ_{D} and θ_{A} are the angles between the donor and the acceptor dipoles and the line joining their centers, and φ is the angle between the projections of the transition moments on the plane perpendicular to the line through the centers). κ^2 can in principle assume values from 0 (perpendicular transition moments) to 4 (collinear transition moments) or 1 (when the transition moments are parallel); Φ_{D}^{0} and

[1]The acronym FRET denoting fluorescence resonance energy transfer is inaccurate because the transfer does not involve any fluorescence but non-radiative transfer of the electronic excitation energy of the donor. This mechanism does not depend on the fluorescence properties of the acceptor. The correct term should be EET representing *electronic energy transfer* or *excitation energy transfer* or RET for *resonance energy transfer*. However, since the term FRET has been accepted in the literature, we will continue to use it here (Robert Clegg proposed in 2006 to reserve the F in FRET for Förster).

τ_D^0 are the quantum yield and the fluorescence lifetime of the donor in the absence of an acceptor; r is the distance between the centers of the two dipoles (donor and acceptor); N_A is Avogadro's number per mmole; $f(\lambda)$ is the fluorescence intensity of the donor in the range of λ to $\lambda + d\lambda$ normalized so that $\int_0^\infty f(\lambda)d\lambda = 1$; and $\varepsilon(\lambda)$ is the absorption coefficient of the acceptor at the wavelength λ. In Eq. 1, r and λ are in cm units, $\varepsilon(\lambda)$ is in cm^2/mol, and $J(\lambda)$ in units of cm^6/mol^{-1}. When those units are used R_o is defined by Eq. 2 and is given by:

$$R_o^6 = 8.8 \times 10^{-28} \Phi_D^o \kappa^2 n^{-2} J \tag{2}$$

where J is the overlap integral in Eq. 1. The energy transfer process competes with the spontaneous decay of the excited state of the donor, characterized by the rate constant $k_D^o = 1/\tau_D^o$. If $\varepsilon(\lambda)$ is given in $cm^{-1}\ mol^{-1}$ units, then $R_o^6 = 8.8 \times 10^{-25} \Phi_D^o \kappa^2 n^{-4} J$. Thus, the probability ρ for the donor to retain its excitation energy at the time t after excitation is given by:

$$-\left(\frac{1}{\rho}\right)\frac{d\rho}{dt} = \frac{1}{\tau_D^o} + \frac{1}{\tau_D^o}\left(\frac{R_o}{r}\right)^6 \tag{3}$$

and the efficiency E of energy transfer is expressed by:

$$E = \frac{R_o^6}{R_o^6 + r^6} \tag{4}$$

R_o is thus the inter-dipole distance at which the transfer efficiency (and the donor lifetime) is reduced to 50 % and has the strongest dependence on changes of that distance (Fig. 2). (At very short interprobe distances where quadrupole interactions become efficient, the point dipole approximation does not hold neither the inverse six power relation between distance and transfer efficiency.)

There is a strong experimental support for Förster's Eqs. 1, 2, 3, and 4 (reviewed by Steinberg [4]). Weber and Teale [11] and Latt et al. [12] showed the dependence on the overlap integral. The R^{-6} dependence was demonstrated by Stryer and Haugland [13].

Equation 4 shows that the distance between a donor and an acceptor can be determined by measuring the efficiency of transfer, provided that r is not too different from R_o and that all molecules in the sampled ensemble share the same intramolecular distance between the labeled residues. If this is not the case, an average distance r_{av} that would be extracted from the measured transfer efficiency does not correspond to any simple "average" distance.

The efficiency of energy transfer is independent of the value of the lifetime of the excited state of the donor. Efficient energy transfer can occur even for "long life" excited states (e.g., for phosphorescence emission) provided that the quantum yield of emission is reasonably high. Such transfer phenomena have been observed. Equations 3 and 4 thus define the characteristic time and

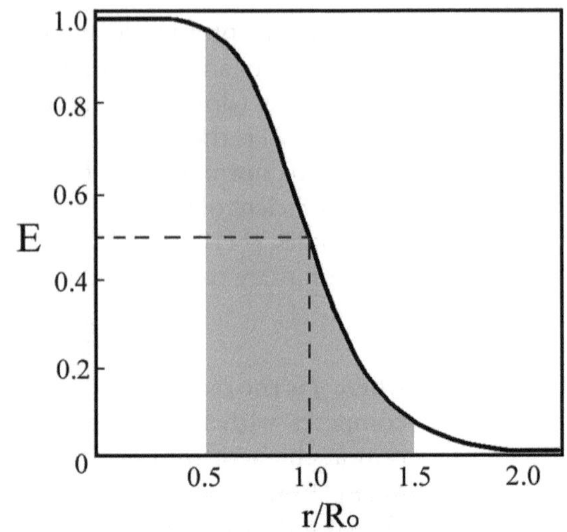

Fig. 2 Variation of the transfer efficiency, *E*, as a function of the distance between the donor and the acceptor probes. The shadowed range represents the limits of the distance range where reliable measurements of distances are possible

distance ranges ("windows") in which the transfer efficiency, *E*, is most sensitive to conformational transitions and their rates (Fig. 2). Typical values available for probes that are used in protein chemistry are in the range of 10–80 Å and down to picosecond time intervals. Many conformational changes and processes in folded and unfolded proteins occur on these time scales. Hence, the researcher can design experiments with the time and distance resolutions most suitable for each molecular question.

1.1.3 The Orientation Factor

Distance determination by measurements of transfer efficiencies is complicated by the strong dependence of the probability for energy transfer on the orientation of the interacting dipoles. Therefore, knowledge of the orientation factor, κ^2, is essential for applications of FRET in studies of protein conformations. When the two dipoles undergo rapid orientational averaging during the lifetime of the excited state of the donor, the orientation factor can be averaged to yield a numerical value of 2/3. The large span of possible κ^2 values, between 0 and 4, makes it the primary source of uncertainty in distance determination by FRET measurements (*see* **Notes 8** and **9**).

2 Materials

Prepare all solutions using ultrapure water and analytical grade reagents. Use spectroscopic grade solvents for fluorescence studies. All materials used should be checked for fluorescence emission in the relevant spectral range.

3 Methods

3.1 Strategies for Site-Specific Double Labeling of Proteins

Despite major biotechnological developments, preparation of homogeneous well-defined samples is still a major biochemical challenge.

Site specific labeling can be achieved by at least nine basic strategies: (a) Use of natural probes including the tyrosine or tryptophan residues and a few prosthetic groups. Site-directed mutagenesis can be applied for the preparation of single tryptophan and single tyrosine mutants [14, 15]. (b) Nonselective chemical modification of reactive side chains and chain terminal reactive groups (amine, carboxyl, and mainly sulfhydryl groups) combined with high-resolution or affinity chromatography-based separation methods [16–20]. (c) Solid-phase synthesis of protein fragments and ligation of synthetic peptide fragments or of a genetically truncated protein fragment and a synthetic, labeled complementing fragment [21]. Total synthesis of labeled protein molecules is applicable for medium-sized proteins, e.g., the total synthesis of Aβ fragments [22, 23]. (d) Production of edited mutants with no more than one pair of exposed reactive side chains, preferably cysteine residues [24, 25] or one cysteine and one tryptophan [15, 26–28] at the desired sites, followed by chemical modification and chromatographic separation. Nitration of tyrosine residues has also been successfully applied [29]. (e) Cell-free protein synthesis and the use of edited t-RNAs and edited genetic code for the incorporation of non-natural amino acids into selected sites in the chain [30]. (f) Enzyme-catalyzed insertion of fluorescent substrates, taking advantage of the specificity and mild conditions of the enzymatic reaction. (g) Use of a protein genetically fused with one or two fluorescent proteins [31]. (h) Exploitation of natural mechanisms for posttranslational chemical modification for either attachment of a fluorescence probe to a selected side chain or production of a nonnative reactive site chain followed by chemical modification for attachment of a reactive probe group [32]. (i) Use of labeled ligands, which can be selective when applied in vivo [33], e.g., tagged antibodies, SNAP tags, and chelating tails that chelate rare-earth ions (*see* **Notes 1–3**).

Strategy (a) avoids the potential artifacts caused by labeling procedures and uncertainties such as the labeling ratio. On the other hand, the main limitations of using the natural probes are the limited arsenal of available probes, their less than ideal spectral characteristics, the limited range of R_o values that can be achieved by using them, and the fact that most proteins contain several Trp and Tyr residues. Applications of strategy (b) are limited to small proteins or proteins where a limited number of the reactive side chains can be found [20]. Nitration of tyrosine residues was also successfully applied in few cases [34, 35] but that is not a "clean" reaction. Strategy (c) is the ultimate one, which achieves full freedom for the designer of protein derivatives best suited for testing well-defined working hypotheses. This strategy will most probably become the method of choice in the field of spectroscopic research of protein folding [30, 36]. The limited scope of application of this approach at the present time is mainly due to the very sophisticated and multistep operations involved and the low yields. These yields are limiting only in the case of measurements of folding kinetics, where high amounts of samples are employed (e.g., the stopped-flow double-kinetics experiment). In most FRET experiments, the sensitivity of the fluorometric detection systems makes this method satisfactory and promising. Strategy (e) is theoretically the ideal strategy; using such techniques, any protein with any number of Cys or Trp residues can be labeled. But, even though the technique was developed more than 20 years ago, it is still not widely used due to its sophistication and low yields.

Strategy (f) is not of general use due to the limited number of known suitable enzymatic reactions and the multiplicity of reactive side chains. The well-documented catalysis of peptide bond synthesis by proteolytic enzymes was described (for C-terminal labeling of RNase A), but it is not easy to apply as a general method. Strategy (h) is attractive but still has the problem of multiple sites that can be modified. Transglutaminase catalysis of insertion of primary amino groups attached to fluorescent probes was also attempted [18, 37, 38], but the development of strategy (d) offers simpler route. Strategy (g) is good for in vivo experiments, but not so useful for structural studies in vitro due to the very large size of the "probes."

Strategy (d) is currently the most practical approach for several reasons: first, exposed cysteine residues are rare, and in many proteins these residues can be replaced by Ser or Ala residues without major perturbations of structure and function. Thus, most proteins can be engineered to have two exposed sulfhydryl side chains. Second, the arsenal of sulfhydryl reactive pairs of probes covering a wide range of spectroscopic characteristics and R_o values is abundant, and the experimental design can be optimized to meet a wide variety of distances and labeling conditions. Third, these reagents can be selectively targeted at the cysteine residues, and huge amounts of labeled mutants suitable for kinetic experiments can be produced [15, 39–42].

3.1.1 Preparation of Double-Labeled Mutants Using Engineered Cysteine Residues (Strategy D)

Although reagents for selective alkylation of sulfhydryl residues are available, the challenge here is to develop methods for selective modification of each one of the two reactive sulfhydryl groups in the same protein molecule.

Sub-milligram amounts of labeled proteins can suffice for a full set of FRET experiments under equilibrium conditions. Such minute amounts of site-specific double-labeled protein samples can be prepared, using high-resolution chromatography. This chromatography is a mixture of the possible products, following a nonselective reaction of two Cys mutants with a fluorescent reagent. Homogeneous double-labeled products can be obtained in a second cycle of reaction and separation. Hundred-fold larger preparations of labeled proteins are needed for stopped-flow kinetic experiments. In this case, it is desirable to develop methods in which the selectivity is achieved by differential reactivity of selected sites on the protein molecule. A general systematic procedure was developed for high-yield selective labeling of each one of two cysteine residues in mutant protein molecules (*see* **Notes 4–7**). The determination of individual reaction rate constants of the engineered SH groups and the optimization of reaction conditions enabled high yields (25 %) of the preparation of pure, site-specific double-labeled *E. coli* adenylate kinase (AK) mutant. These high yields make some otherwise difficult experiments feasible, such as the double-kinetics time-resolved FRET measurements [43].

Jacob et al. [44] improved the prediction of the rates of reactions of engineered cysteine residues at selected sites in a protein. They based this on the observed correlation between reactivity of engineered sulfhydryl groups and the effect of neighboring charged groups (due to dependence of its pK_a) and the freedom of rotation of the side chain (Fig. 3). The selectivity of the labeling

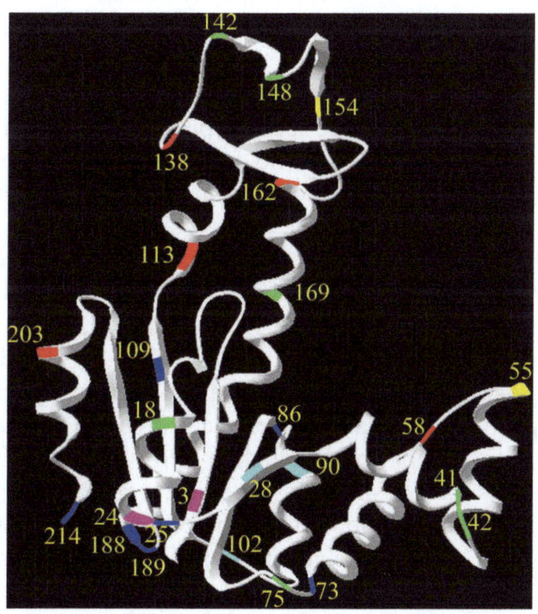

Fig. 3 Color-coded AK sequence positions for which cysteine–DTNB reaction rates were determined: five cysteines reacted extremely quickly (*red*, 50–200 ms), seven very fast (*green*, 200–1,000 ms), two fast (*yellow*, 2 and 3 s), three slowly (*cyan*, 5–30 s), seven very slowly (*blue*, 30–200 s), and two extremely slowly (*magenta*, 20 min and 15 h). These differences can be the basis for selectivity in the reaction of fluorescent probes with one of two cysteine residues in a protein molecule (reprinted with permission from Jacob et al. [44])

reaction can be further enhanced by local increase or reduction of reactivity of selected side chains by means of conformational effects. Solvent composition (organic solvents, salt components, and pH) and temperature are the most common variables that can differentially affect the reaction rate constants of selected side chains. In this work, the ratio of the reaction rate constants was examined only under a range of temperatures and salt concentrations. More drastic effects can be obtained by addition of ligand (substrates, inhibitors, or protein–protein interactions) [17] (*see* **Notes 4** and **5**).

The level of homogeneity required for spectroscopic measurements makes the chromatographic separation steps indispensable. Separation of the reaction products can be enhanced when one of the labels is either charged or very hydrophobic [20]. Yet, either hydrophobicity or extra electrostatic charges can affect the folding pathway of proteins and can be a disadvantage when the modifications are made for structural studies. Affinity chromatography was also used for separating multiple-labeled protein derivatives. It is also possible to use a mixed disulphide resin for separation of fractions with free cysteine residues in the process of labeling two-cysteine mutants. The single-labeled mutants which are used for the reference measurements can be prepared by straightforward labeling of single cysteine mutants. Caution must be taken since in several cases it was shown that the spectroscopic properties of a probe attached at one site on a protein are significantly modified when the second cysteine is inserted at the second site for the double labeling. In this case, it is desirable to use the corresponding single-labeled protein sample of the two-cysteine mutant in which the second cysteine residue is blocked by a non-fluorescent alkylation reagent such as iodoacetamide (*see* **Notes 1** and **4–7**).

3.2 Determination of Intramolecular Distances in Double-Labeled Protein Molecules Using FRET Measurements

Clear presentation of theoretical background for the methods described below may be found in the monographs by B. Valeur [45], J.R. Lakowicz [46], and B.W. van der Meer [6] and numerous reviews such as [4, 26, 47–49, 50] published over the past 40 years.

The first pioneering application of FRET investigation of a natural flexible polypeptide was reported by Edelhoch, Brand, and Wilchek in 1963 [51, 52]. Edelhoch and his colleagues applied strategy (a), using a pair of intrinsic probes (Tyr to Trp transfer). Since then, the technology of both site-specific labeling and time-resolved fluorescence measurements has advanced; together with the methodological developments, the number and power of the applications of FRET measurements in characterization of disordered peptides and proteins proliferated.

3.2.1 Single Distance Between Donor and Acceptor

Determination of distances by determination of FRET efficiency, E, is possible by measuring the decrease of donor emission or an enhancement of the acceptor emission by steady-state and time-resolved methods:

$$r = \left(\frac{1}{E} - 1\right)^{1/6} R_o \quad (5)$$

Due to the comparative nature of this mode of determination of E, the concentrations of the probes (and hence the labeled protein molecules) and their microenvironments must be the same for each set of measurements.

3.3 Time-Resolved FRET Detection Methods

Time-resolved emission of the donor or the acceptor fluorescence provides direct information regarding the transfer rates independent of the concentrations, as this information is provided by the shape of the fluorescence decay curves and not the amplitudes. The donor and the acceptor fluorescence decay curves contain additional information not available from the steady-state measurements. This information, which can be extracted directly by proper analytical procedures, is of particular interest for studies of protein dynamics, unfolded states of proteins, protein folding, and intrinsically disordered proteins (IDPs). Analysis of trFRET experiments can resolve conformational subpopulations, distributions of distances, and fast conformational fluctuations on nanosecond time scales.

3.3.1 Determination of Donor Fluorescence Decay Rates

When an acceptor is located close enough to the donor, two competing processes affect τ_D, the lifetime of the excited state of the donor, and the transfer rate:

$$1/\tau_D = 1/\tau_D^o + k_{ret} \quad (6)$$

The ensemble average transfer efficiency can be determined using the fluorescence lifetimes of the donor:

$$E = 1 - \frac{\tau_D}{\tau_D^o} \quad (7)$$

and the apparent mean distance is calculated by:

$$r = \frac{R_o}{(\tau_D/\tau_D^o - 1)^{1/6}} \quad (8)$$

When a non-exponential decay of the donor alone (τ_D^o) is observed, a common situation due to heterogeneity of microenvironments, average lifetime values can be used to calculate E, provided that the deviation from monoexponentiality is moderate:

$$E = 1 - \frac{\langle \tau_D \rangle}{\langle \tau_D^o \rangle} \quad (9)$$

where $\langle \tau \rangle = \sum_i \alpha_i \tau_i \Big/ \sum_i \alpha_i$

When the heterogeneity of the decay rates is enhanced by the FRET effect, distribution analysis accompanied by appropriate control experiments should be applied.

3.3.2 Determination of Intramolecular Distance Distributions

The distance between two selected residues in a protein is by definition nonuniform, and hence, Eqs. 6 or 9 should be replaced by distance analysis, assuming a probability distribution of the intramolecular distance between the labeled sites, $N_o(r)$, with finite full width at half maximum (FWHM) will be determined. In that case, the FRET efficiency determined by steady-state measurements is averaged over the efficiencies of all fractions in the distribution:

$$\langle E \rangle = \int_0^\infty N_o(r) \frac{R_o^6}{R_o^6 + r^6} \mathrm{d}r \tag{10}$$

where $N_o(r)\mathrm{d}r$ is the probability of the fraction of molecules with D–A pairs at distance r to $r + \mathrm{d}r$. $N_o(r)$ can be resolved when the experiment is repeated with several pairs of probes of different R_o values [53]. A single time-resolved experiment contains all the information needed for the determination of $N_o(r)$.

When the fast conformational exchange within the ensemble of labeled protein molecules is slow relative to the lifetime of the excited state of the donor, $N_o(r)$ can be recovered by modeling $i(t)$ using the expression:

$$i(t) = k \int_0^\infty N_o(r) \exp\left[-\frac{t}{\tau_D^0}\left(1 + \left(\frac{R_o}{r^6}\right)^6\right)\right] \mathrm{d}r \tag{11}$$

where k is a proportionality factor. Knowledge of R_o and τ_D^0 enables the extraction of $N_o(r)$ from the multiexponential decay curve, $i(t)$.

3.3.3 Evaluation of the Effect of Fast Conformational Fluctuations and Determination of Intramolecular Diffusion Coefficients

When exchange of conformers occurs between fractions of an ensemble of protein molecules under equilibrium conditions (where the equilibrium probability distribution of the distances is unchanged), the transfer efficiency is enhanced. For analysis of trFRET experiments this enhancement is modeled as an intramolecular diffusion of the labeled sites on the chain, which acts towards restoration of the equilibrium probability distribution. This phenomenon can be viewed as a process of diffusion under the conformational force field which governs the equilibrium probability distribution of the intramolecular distance, $N_o(r)$ [54, 55]. Equation 12 considers both the energy transfer reaction term, $k(r)$, and the restoration force by the diffusion term. Equation 12 is the basis for the experimental data analysis [54–57] for determination of the parameters of the distribution, $N_o(r)$:

$$\frac{\partial \overline{N}_i(r,t)}{\partial t} = \frac{D}{N_0(r)} \times \frac{\partial}{\partial r}\left[N_0(r)\frac{\partial}{\partial r}\overline{N}_i(r,t)\right] - k_i(r)\overline{N}_i(r,t) \tag{12}$$

where i denotes the ith species of the donor fluorescence decay (in the absence of an acceptor), $N^*_i(r,o) = N_0(r)$ (the equilibrium distance distribution), $N^*i(r,t)\mathrm{d}r$ is the probability for molecules with an excited state species i of the donor to have an intramolecular distance in the range r to $r + \mathrm{d}r$, and $\overline{N}_i(r,t) = N^*_i(r,t) / N_0(r)$. D is defined as the intramolecular diffusion coefficient of the segments carrying the two probes. In the following equations τ_i and α_i are the lifetime and the normalized pre-exponential factor, respectively, of the excited state of the ith species of the donor in the absence of an acceptor, and $k_i(r)$, the reaction term, includes the spontaneous emission rate and the Förster energy transfer rate:

$$k_i(r) = \frac{1}{\tau_i} + \frac{1}{\tau_\mathrm{r}}\left(\frac{8.79 \times 10^{-25}\mathrm{n}^{-4}\kappa^2\mathrm{J}}{r^6}\right) \tag{13}$$

τ_r is the radiative lifetime of the donor, and the other terms were defined in Eqs. 1 and 1a. Equation 13 considers the variation of R_o due to variations of the donor quantum yield Φ^0_D. This is represented by the ratio of the donor lifetime components and the radiative lifetime, which is assumed to be independent of local environmental effects.

In the case of ensemble measurements the solution of Eq. 12 is obtained by numerical methods. Since the boundaries are chosen in such a way that the distribution function is very small at both the upper and lower extreme distances, reflective boundary conditions are arbitrarily used [58]:

$$\left[\partial\overline{N}(r,t)\partial/r\right]_r = r\,\mathrm{min},\ r\,\mathrm{max} = 0 \tag{14}$$

$i^\mathrm{D}_\mathrm{c}(t)$, the calculated fluorescence decay curve of the donor, is readily obtained from $\overline{N}(r,t)$ by the relationship [56]:

$$i^\mathrm{D}_\mathrm{c} = m\int_{\mathrm{r_{max}}}^{\mathrm{r_{max}}} N_0(r)\left[\sum_i \alpha_i\overline{N}_i(r,t)\right]\mathrm{d}r \tag{15}$$

where m is a constant and α_i is the pre-exponential factor of the ith excited state species of the donor in the absence of an acceptor.

The ideal case is such that the fluorescence decay of the donor in the absence of an acceptor is mono-exponential. In such a case, all molecules share a single value of R_o, the sum in Eq. 15 reduces to one term, and the reactive term, $k_i(r)$ Eq. 13, can be replaced by a simpler expression:

$$k(r) = -1/\tau\left[1 + (R_\mathrm{o}/r)^6\right] \tag{16}$$

In a case where the diffusion rate is known to be negligible on the time scale of the lifetime of the excited state of the donor, the Fick term in Eq. 12 is zero and $N_\mathrm{o}(r)$ can be derived directly from an analysis of the donor decay curve. In such a case, Eq. 12 can be replaced by [59] (*see* **Notes 10** and **11**):

$$i_c^D(t) = k \int_0^{\infty} N_o(r)(\exp -k(r))\mathrm{d}r \quad (17)$$

3.4 Experimental Methods of trFRET Applications in Characterization of Protein Conformation and Dynamics and Intermolecular Interactions

The transfer efficiency in an ensemble of protein molecules cannot be used for calculation of mean interprobe distances due to its nonlinear distance dependence and the intrinsic heterogeneity of conformations. Only quasi-mean distances can be determined, and changes of FRET efficiency can be related to changes in the distances. This limitation highlights the strength of methods of time-resolved FRET measurements in the ensemble and the single-molecule detection modes.

3.4.1 Time-Resolved Measurements

Equations 6, 7, 8, and 9 are used for the determination of the average FRET efficiency by time-resolved measurements. The main technical advantage of this measurement is that there is no need to know the concentrations of the proteins, and the accuracy of the lifetime measurements surpasses that of the steady-state intensity measurements.

The fluorescence decay curves of the probes in trFRET experiments contain detailed information on heterogeneity of the ensembles of proteins (Fig. 4).

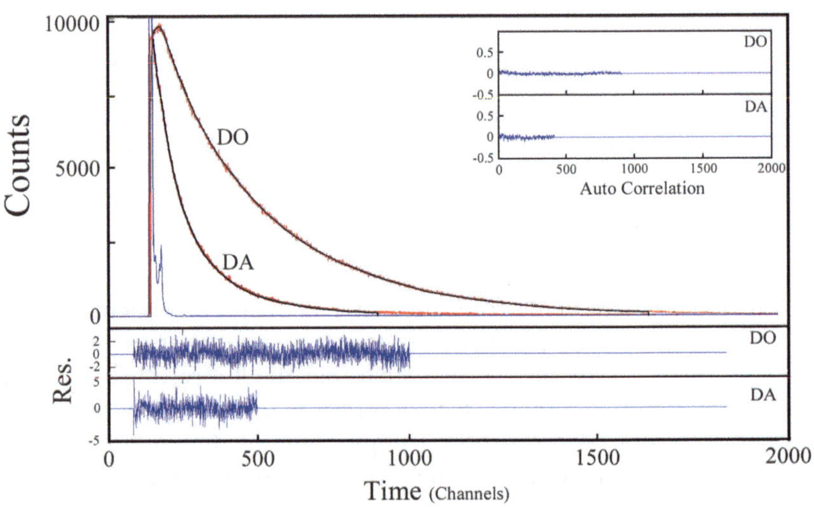

Fig. 4 Fluorescence decay curves of tryptophan in mutants of α-synuclein, one labeled with the donor-only (the DO experiment) and the other labeled by both probes (the DA experiment). The associated residuals and the autocorrelation function are shown in the lower and upper insets. The decay rate of the tryptophan residue in the DA mutant is accelerated due to the FRET effect. The residuals (*bottom*) and the corresponding autocorrelation functions (*upper right*) served as indicators of the quality of fit of the experimental data to a multiexponential function. The average lifetime of tryptophan for the DO mutant was 2.2 ns. The time equivalent of each channel is 12.5 ns and the lifetime components were $\tau_1 \approx 4$ ns (30–55 %), $\tau_2 \approx 1.8$–2.3 ns (30–45 %), and $\tau_3 \approx 0.8$ ns (15–30 %)

The technology of time-resolved fluorescence measurements matured during the last quarter of the twentieth century. Fluorescence decay can be detected either by the single-photon counting method or by the phase modulation method, and in principle, both methods are equivalent. While phase experiments are relatively easy to perform, single-photon counting techniques offer more straightforward methods for incorporation of acceptor decay curves in the analysis and easier control of background subtraction and other corrections.

3.4.2 Data Analysis

The basic procedure used for analysis of time-resolved FRET experiments is based on curve fitting methods. In this procedure, the parameters of the distance distribution function, the diffusion coefficient, the fluorescence lifetimes, and their associated pre-exponents are searched [60, 61].

The calculated fluorescence decay profile, $F_{\text{calc}}(t, \lambda_{\text{ex}}, \lambda_{\text{em}})$, is composed of both the donor and the acceptor decay curves, $F_{\text{D}}(t)$and $F_{\text{A}}(t)$, respectively, which are calculated using the Förster theory applied to the relevant model distribution:

$$F_{\text{calc}}(t, \lambda_{\text{ex}}, \lambda_{\text{em}}) = F_{\text{A}}(t) + F_{\text{D}}(t) \tag{18}$$

Several computational methods were developed to search for the set of free parameters. The maximum entropy method (MEM) is also applied to the analysis of time-resolved fluorescence and FRET experiments. These methods yield the theoretical decay curve best fitted to the experimentally recorded fluorescence decay curves $F_{\text{exp}}(t, \lambda_{\text{ex}}, \lambda_{\text{em}})$ [62, 63]. This procedure is inherently difficult since the analysis of any multiexponential decay curve is a mathematically "ill defined" problem [64, 65]. The quality of fit of the experimental and calculated curves is judged by the minimization to the χ^2 values and the tests for randomness of the deviations [60, 65].

The differential equation 15 does not have a simple general analytical solution (except for specific cases). Consequently, a numerical solution based on the finite difference (FA) method [66, 67] was applied. Therefore, the most common current practice is the use of model distribution functions for $N_o(r)$ with a small number of free parameters (Fig. 5). The model used in the experiments described below is a Gaussian distribution ($N_o(r) = C\{f4\pi r^2 \exp[-a_1(r-\mu_1)]^2+(1-f)\,4\pi r^2 \exp[-a_2(r-\mu_2)]^2\}$), where a_i and μ_i are free parameters which determine the width and the mean of the distributions, respectively, f is the equilibrium fraction of subpopulation 1, and C is a normalization factor. (When two subpopulations are assumed, Eq. 15 is solved for each subpopulation, and the two computed decay curves are then summed up and fitted to the experimental curve.) A model including the contribution of intramolecular diffusion can also be used in this context.

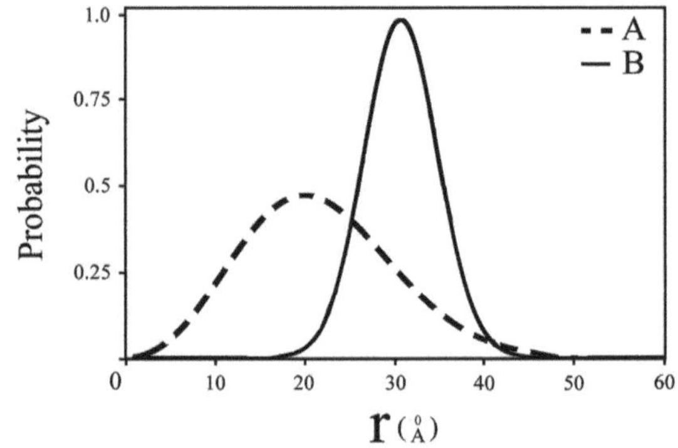

Fig. 5 Intramolecular end-to-end distance distributions which were obtained by analysis of fluorescence decay curves of the probes attached to a protein molecule which exist in two conformations in solution. The analysis is based on solution of Eq. 12 and the curve fitting algorithm. The parameters of interest are the mean and the width of the distributions and the ratio of the two subpopulations

Uncertainty ranges in the value of each parameter obtained by the fitting procedure are enhanced by the random noise level and correlations between the parameters. These ranges are routinely reduced by computing each set of parameters by global analysis of multiple independent measurements monitoring one system. In particular, coupled analysis of the fluorescence decay of both the donor and the acceptor probes contributes to overdetermination of the free parameters. Whatever methods are used for data analysis, the best fit must be tested by the statistical tests and fine details of the distribution of the residuals (*see* **Note 11**).

3.5 Single-Molecule Detection-Based trFRET Methods

Single-molecule detection (SMD) technology enables determination of FRET efficiency and time-resolved (tr)FRET determination of one molecule at a time in an ensemble of double-labeled molecules. The main advantages of using this technology are (a) resolution of subpopulations in ensembles of FRET active molecules (since the characteristics of the ensemble are determined by summing up the individual systems in the ensemble, free of any assumed model) and (b) resolution of dynamic changes between and within subpopulations; in contrast, ensemble detection methods (which depend on population synchronization) are limited in both these respects. General rules can be derived only from knowledge of characteristics of ensembles of molecules, due to the stochastic nature of the elementary steps of all molecular processes. Yet, an ensemble of labeled protein molecules can be assumed to be an ergodic system that will yield the same description by measuring either an infinite number of individual molecules for a brief time period or a single system for infinite time. Current SMD technology allows us to track each photon emitted from a labeled protein, along with several of its properties: the time of emission, the energy (wavelength range), and polarization.

The main factors that limit the applicability of trFRET measurements under SMD measurements are the inherent limits of the flux of fluorescence photons from any single molecule and the finite number of excitation/emission cycles' that organic molecules can undergo before photobleaching. The rate of accumulation of a statistically significant number of photons in such experiments is usually much lower than the fundamental rates of protein conformational dynamics. Therefore, even in the SMD mode of measurement, the results represent an average of a structural characteristic of the labeled protein molecule over a finite time interval. In an ensemble measurement, a high number of molecules from the ensemble contribute to the signal. Methods were developed to maximize the quality and types of data that can be obtained in SMD-trFRET experiments.

3.6 Typical Single-Molecule FRET (smFRET) Experiment Setup and Measurements

A drop (generally at picomolar concentration) of single- or double-labeled protein (using probes with high extinction coefficient and quantum yield) with FRET donor and acceptors is placed on top of a cover slip on a confocal microscope stage. The solution is excited by either continuous wave (cw) or pulsed laser illumination tightly focused inside the solution. The excitation power is weak (tens of microwatts) so that only at the focal volume will the flux of photons be sufficiently high to yield high local excitation intensity inside a Gaussian illumination profile, with a typical volume of 1 femtoliter. (This volume usually contains none or one molecule at a time and with low probability even two or more.) When a single molecule is in the focal volume it will be excited; photons are emitted either from donor fluorescence or from acceptor fluorescence after excitation energy is transferred due to FRET. These photons are then collected by at least two sensitive detectors (typically avalanche photodiodes). The wavelength range detected by each detector is selected by optical filters specific for either the donor or the acceptor fluorescence spectra. Each photon detected by each detector is registered with an absolute arrival time relative to the beginning of the measurement, called the *macrotime*. The photon arrival macrotimes are recorded in the donor/acceptor detection channels. The detection channels register all detection events including shot-noise-related events. Those are events that do not originate from fluorescence but from scattering leakage, from thermal electrons, or from detector afterpulsing [68, 69]. The shot-noise signals are weak but frequent in comparison to the signal events originating from fluorescence. A signal event that exceeds a certain predefined threshold of photon flux within a predetermined time window is called a photon *burst*.

When performing smFRET measurements, the detection channels register photon bursts originating from the donor fluorescence

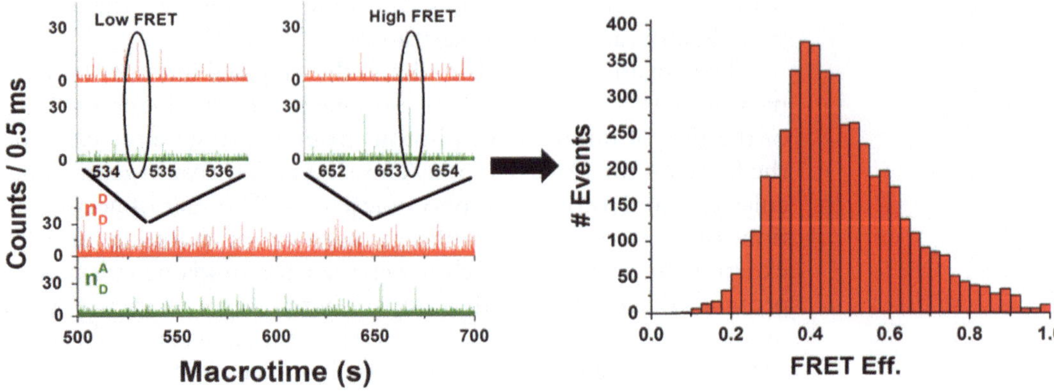

Fig. 6 An example of the mode of accumulation of FRET-related single-molecule bursts towards formation of a FRET histogram. A sample of AK molecules labeled at the CORE domain as well as at the ATP-binding (LID) domain was used. The plot on the top focuses on typical smFRET bursts out of a long train of bursts (*middle*). FRET efficiency is calculated for each burst that had a combined number of photons above a given threshold (here, 25 photons). The FRET efficiency values are then accumulated into a FRET histogram. The skewed histogram suggests two subpopulations

and acceptor bursts generated from acceptor fluorescence. The Proximity Ratio (PR) Eq. 19, used for FRET determination, is defined as the ratio of the number of single-photon events registered at the acceptor channel (under excitation at the donor excitation wavelength) in one burst, n_D^A, divided by the total events (in all channels and also under excitation at the donor excitation wavelength) registered for that given burst, $n_D^D + n_D^A$:

$$PR = \frac{n_D^A}{n_D^D + n_D^A} \quad (19)$$

Figure 6 illustrates how data is collected and accumulated to a FRET histogram. Photon counts (including donor and acceptor bursts) are accumulated in predetermined time intervals, called *time bins* (usually in the ms range, comparable to typical diffusion times of macromolecules through the focal volume) for both the donor and the acceptor detection channels. For those time bins having total number of photons higher than a predetermined threshold (e.g., 25 photons), the *PR* value is calculated. A histogram is then constructed showing the number of events (bursts) as a function of the calculated *PR* values (*see* **Notes 13** and **14** for experimental details). Various modes of data analysis were developed for extracting subpopulation and dynamic characteristics of the molecules as described below.

The calculated *PR* values are affected by several sources of non-FRET-related photons that might affect the FRET efficiency calculation: (a) The photons detected in the acceptor channel, n_D^A, include photons resulting from direct excitation of the acceptor [70–72]. (b) The red edge of the donor emission may be passed through the filter selecting for the acceptor emission ("leakage" problem) [72]. (c) The two detectors may have different detection efficiencies, and the two dyes (donor and acceptor labels) often have different fluorescence quantum yields. A correction factor, the γ *factor*, is therefore used. It is proportional to the ratio of the quantum yields of the acceptor and the donor and the ratio of the corresponding channel detection efficiencies. Corrections for direct acceptor excitation (a), donor fluorescence leakage to the acceptor channel (b), and the quantum yield and detection efficiency ratios (c) should be taken into account to yield the energy transfer efficiency, E (*see* **Note 15**).

The E values of each burst are accumulated to yield an energy transfer efficiency histogram such as that shown in Fig. 6. Such a histogram is the main mode of data presentation for smFRET experiments. It produces "model-free distribution," which is a distance-related characteristic of conformational subpopulations in ensembles of labeled protein molecules. A minimal width of the FRET efficiency distribution is contributed by the shot noise, defined by Eq. 20 [73]:

$$\sigma(E) = \sqrt{\frac{E(1-E)}{n_D^D + n_D^A}} \tag{20}$$

It has also been established that energy transfer efficiency histograms with widths higher than the shot-noise-limited width qualitatively report on the existence of more than one subpopulation of molecules and possible dynamic exchange between subpopulations [69, 74]. In millisecond time intervals (a typical time bin), a single molecule might undergo transitions between different conformations (*see* **Note 16**). Using Eq. 20 one can clearly understand that the apparent skewness of the FRET histogram shown in Fig. 6 is actually two FRET efficiency populations combined and could not be explained by a shot-noise-limited single population.

The mean and the width of the FRET efficiency histograms are a function of both the equilibrium distribution of conformations in the ensemble and the rates of their interconversion. The dynamic and time-independent characteristics can be resolved either by FRET fluorescence cross-correlation spectroscopy (FRET–FCCS) [8, 75–77] or by analyzing the rates of decay of the fluorescence of the donor and the acceptor probes. All the information needed for either modes of analysis is available in the registered photon arrival times. In the FRET-related FCCS analysis, the data originating from donor and acceptor detected bursts of a given E subpopulation are cross correlated. If an exponential anticorrelated trend appears in the cross-correlation function, this represents the interconversion relaxation time of the molecules between different distance-related states for times faster than typical diffusion time through the focal volume. Analysis of single-molecule trFRET data can further resolve the number of fast interconverting populations and the structural characteristics of each such one of them.

3.7 The Strength of Time-Resolved smFRET Analysis

In time-correlated single-photon counting (TCSPC) with a single-molecule fluorescence detection system, the time elapsed between the moment of excitation and the moment of detection of each detected photon event is registered. This is the photon's *microtime*. A histogram graphing the number of photon counts vs. their *microtime* values yields the donor and acceptor fluorescence decays. That histogram can be accumulated by adding all photon events collected over extended period of time yielding fluorescence decays identical to the ones achieved by ensemble-level measurements.

Microtime histograms can be accumulated per single bursts by a method called burst-integrated fluorescence lifetime (BIFL) [78]. Best fits to single lifetime exponential functions yield an average fluorescence lifetime for a given burst. The burst-related fluorescence lifetimes as well as E values are accumulated in two-dimensional histograms, such as that shown in Fig. 7, in a method called multiparameter fluorescence detection (MFD) [70, 79–84]. In the example shown, the E–τ correlogram is obtained by analysis of the single-molecule trFRET experiment using double-labeled AK molecule labeled at its CORE domain as well as at its ATP-binding LID domain. On a millisecond time scale, it has two

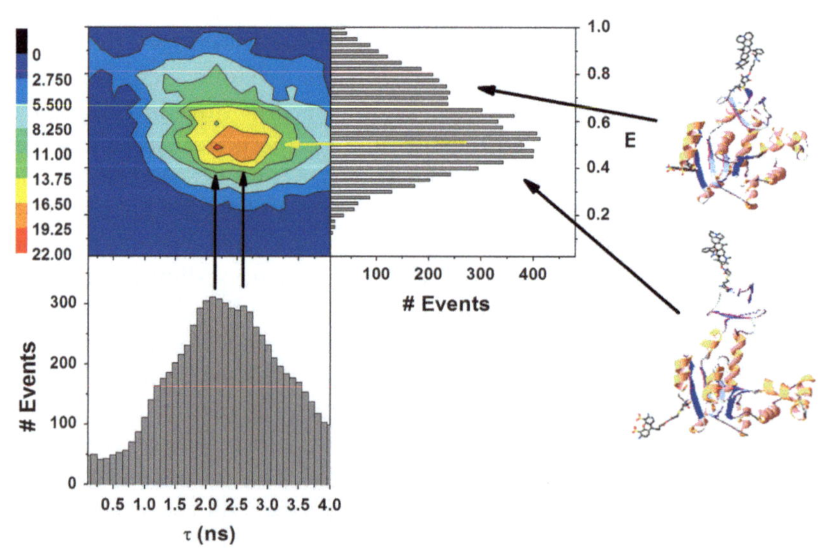

Fig. 7 Multiparameter fluorescence detection (MFD) FRET. Average donor lifetimes (*x* axis) as well as *E* values per burst (y) are accumulated into an E–τ correlogram, for analysis of the registered photon events obtained from smFRET measurement of double-labeled AK molecules (labeled at the CORE and the LID domains). Here, one can see that in the apo-AK, the molecule has two conformations characterized by two states of the interface between the LID and CORE domains; these *open*- and *closed*-like states interconvert with a time constant in the range of the burst bin time (1 ms) or slower. The *open* conformation has two characteristic average fluorescence lifetimes, which suggest additional dynamics occurring faster than the burst time scale

shot-noise-limited E subpopulations and at least four donor average lifetime components. When correlating lifetime components and FRET populations, it is evident that the low FRET population has two characteristic donor lifetimes. Since donor fluorescence lifetime is directly linked to FRET efficiency, such results indicate that there are two states represented by two different distance distributions, which interconvert at least ten times faster than the 1 ms time bin used. This interpretation relies on the assumption that each state distance distribution is represented by a single fluorescence lifetime. An E–τ correlogram can be quantitatively analyzed on the basis of models that distinguish static from dynamic phenomena [70, 81] (*see* **Notes 17** and **18**).

In a further analysis, microtime histograms can also be accumulated vs. the smFRET E values of the bursts [85], as shown in Fig. 8. A fluorescence decay curve obtained from a single burst is affected by and, hence, contains information regarding the average rates of conformational fluctuations in a single molecule. These occur on a time scale orders of magnitude faster than the burst duration. Each fluorescence decay component resolved in analysis of molecules sharing the same E value describes a subensemble of all molecules having a given averaged FRET efficiency,

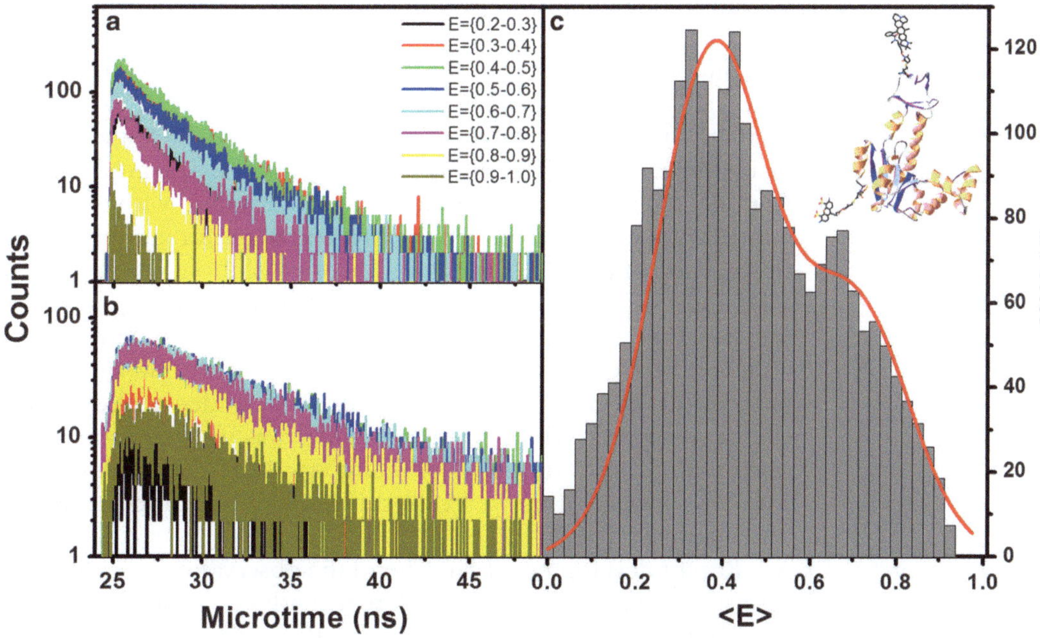

Fig. 8 Fluorescence decays per E values obtained from analysis of apo-*E. coli* adenylate kinase labeled at the LID and CORE domains. Donor fluorescence decay is single exponential for E values of the low FRET population with no overlap with the high FRET population ($E = (0.2–0.43)$); when overlap starts, a bi-exponential decay emerges. At high E values when population overlap ends (0.7–1.0), the bi-exponential trend ends and a short lifetime monoexponential decay is found. FRET in acceptor fluorescence decays is recognized by the grow-in phase of the intensity vs. time curve

i.e., most probably sharing a similar mean intramolecular distance. The example presented in Fig. 8 shows the smFRET histogram as well as its donor and acceptor fluorescence decays at different *E* value intervals obtained for the same AK sample shown in Fig. 7. Two distinct smFRET subpopulations are resolved. The donor fluorescence decays are a single exponential for *E* value ranges in which the overlap between FRET populations is low (*E* is 0.2–0.3; *E* is 0.7–1) and exhibit a clear deviation from a single exponent for *E* value ranges where the two FRET populations overlap (*E* is 0.3–0.7). Acceptor fluorescence decay curves can also be similarly analyzed (*see* **Note 18**).

Thus, fluorescence decay kinetics as a function of either a burst or an *E* value resolve subpopulations in the ensemble. Therefore, as in ensemble-level trFRET, such data should be fitted to a model that takes into account a distance distribution and an interchromophoric diffusion coefficient for each fast interconverting population. There are many advantages but also caveats for this approach (*see* **Notes 18** and **19**). The combination of smFRET to resolve slow interconverting conformational populations, FCCS to detect the existence of faster dynamics, and trFRET in both the ensemble and single-molecule modes is generally most effective for characterization of conformational dynamics over the nanosecond to seconds time regimes.

3.8 Two Modes of Measurement: Freely Diffusing and Immobilized Molecules

Continuous monitoring of smFRET from a single molecule over time can be achieved by limiting molecular mobility by several methods, tethering a molecule to a surface [86–88], and caging in a tethered vesicle [89, 90] or in agar gel pores. The implementation of such methods should be accompanied with rigorous controls to eliminate effects of structural perturbations. Rapid observation of freely diffusing molecules in solution, which traverse a confocal effective excitation volume (ms time scale), is free of structural perturbations (except for those contributed by the attached probes) but enables collection of very few bursts per pass. In this mode of measurement no direct evidence of dynamics can be observed, yet an smFRET model was very recently developed that includes the effect of interconversion dynamics between FRET subpopulations in freely diffusing single-molecule FRET measurements [91].

Recent developments can extend the observation time of freely diffusing molecules by three different methods, by analysis of the inter-burst time intervals [92], by a three-dimensional tracking system feedback loop [93–96], and by new excitation/detection arrays [97, 98].

3.9 The Need for Pure Samples

The minute amount (few nanomoles) of double-labeled protein molecules used in smFRET detected experiments is much smaller than the samples that must be prepared for rigorous characterization of the protein samples (*see* **Notes 1–5**). Insufficient characterization might be a source of misinterpretation and artifacts. The large probes attached to protein molecules in preparations for time-resolved smFRET experiments might affect the distribution of conformers and their thermal stability. This effect can lead to misinterpretation of smFRET data and wrong conclusions. This emphasizes the importance of exhaustive biochemical characterization of prepared samples in preparation for smFRET measurements (*see* **Note 19**).

4 Notes

1. *Possible Pitfalls Associated with the Preparation of Labeled Protein Samples for trFRET Experiments*

   (a) *Incomplete specificity* of the labeling might prohibit meaningful interpretation of FRET data, particularly for flexible systems. Distributions of labeling reactions (e.g., mixtures of species with different distances between the labeled sites) might, when used in FRET experiments, be thought to represent distributions of a specific distance. In principle, under conditions where the extent of mislabeling is known, it should be possible to adjust the model in the analysis algorithm and account for subpopulations of a distance distribution and relate them to the labeled species.

   (b) *Impurities:* Two types of impurities are of concern. The first is the presence of fluorescent impurities and background emission, and the second is incomplete labeling. As long as the background emission is not too high (ca. <20 % of the signal), and the emission can be determined independently (control experiments), reliable distance determination is possible by careful subtraction of the background emission.

   Incomplete labeling of the single-labeled samples should not interfere with analysis of trFRET data provided that the extent of the partially labeled fraction is independently known [99] and does not exceed ca. 20 % of the population.

   (c) *Conformational bias due to the labeling procedure*: The ensembles of conformers of disordered proteins are particularly sensitive to effects of solvent [100] and chemical or mutational modifications. Structural perturbations can be minimized by the proper choice of labeling sites and probes. Rigorous control experiments for determination of the extent of perturbation due to the modifications are mandatory. Control experiments can include the determination of biological activities, aggregation curves, and

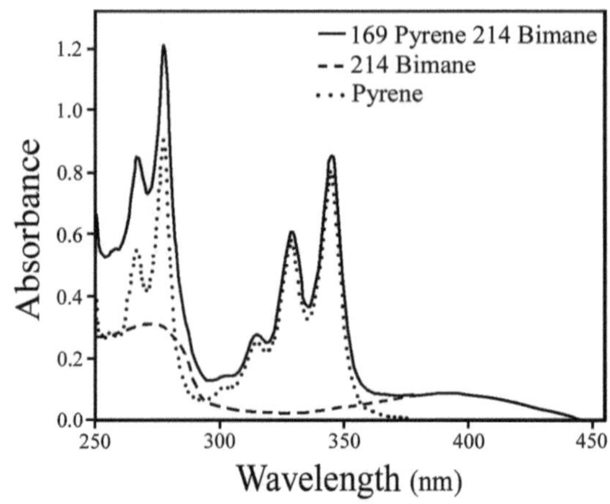

Fig. 9 Test for extent of double labeling of a globular protein. In this example the protein was AK with two engineered Cys residues (169 and 214), the donor was N-1-acetamide-methyl-pyrene (attached to Cys 169), and the acceptor was bimane (attached to Cys 214). In a perfectly double-labeled protein, the spectrum (—) should be a sum of the spectra of the single-labeled mutants

detection of structural properties using CD, IR, and other spectral characteristics (Fig. 9).

2. *Experimental Guidelines for Preparation of Double-Labeled Proteins Designed for Time-Resolved FRET Studies*
   1. The most practical strategy to achieve specific labeling of a defined site with a dye of choice is based on site-specific mutation of the chosen residue to cysteine. The combination of high specificity of the cysteine sulfhydryl group with thiol-reactive probes, the vast availability of commercial thiol-reactive dyes, and the relative ease of sequential manipulation with site-specific mutagenesis makes this strategy the method of choice.
   2. As a general rule, modification of glycine, proline, and charged or bulky hydrophobic residues should be avoided if possible. Alanine and serine are the best candidates for cysteine substitutions. For many protein systems, Ala or Ser to Cys substitution does not significantly alter protein structure or function.
   3. In order to minimize packing perturbation, increase solubility, and allow free rotation of the fluorophore, it is advised to choose highly solvent exposed residues for labeling. Solvent accessibility can be evaluated based on the solved X-ray crystallographic structure (where available). Estimation concerning the predicted dyes rotational degree of freedom can be done by inspection of the number of

allowed rotamers at a specific position (e.g., using *Swiss PDB viewer* or molecular dynamics simulation).

4. The best product yields are obtained when there is a considerable (>2 orders of magnitude) difference between the reactivities of the selected cysteines. Therefore, information regarding the cysteine reactivates is essential and attainable using freely available software (*see* **Note 4**).

3. *Protein Sample Preparation*

1. Preparation of double-labeled protein samples is a low-yield process. Typically, one can expect a yield of 20 % (of the final product) when there is significant specificity of the cysteine reactivities and only 2 % when there is no selectivity. Therefore, in order to have at least 1 mg of pure, well-characterized and analyzed, double-labeled protein sample, at least 5–50 mg of purified unlabeled mutant protein should be prepared.

2. When planning the required quantities of labeled protein for a study, one should take into account the need for control experiments and characterization of the labeled protein (e.g., CD, mass spectroscopy, activity assay, stability test (e.g., determination of thermal or denaturation transition curve), as well as spectroscopic controls such as quantum yield determination and anisotropy decay measurements). In many instances, most of the labeled sample is consumed in these control experiments.

3. Accordingly, even for single-molecule studies that require picomolar sample concentration, the effective requirement for labeled protein (including ensemble control experiments) is still demanding.

4. Finally, to maximize the final product yields, it is recommended that the reactants (unlabeled protein) should be of high purity (> 90 %).

4. *Cysteine Reactivity as a Tool Towards Selectivity*

1. The tendency of a particular cysteine to take part in a nucleophilic substitution determines its reactivity. Under given conditions, a difference between the reactivities of two cysteine residues gives rise to labeling selectivity. This specificity directly determines the distribution of products at the end of a labeling reaction of a dye with mutant protein that contains two cysteines (or more). Hence, selectivity is a powerful tool to consider when planning a labeling scheme.

2. Consider a mutant protein with two reduced cysteine residues available for modification at sites α and β. We represent the protein as

$$C^*_\alpha - C_\beta$$

where α or β denote the position of the cysteine and "*" indicates a reactive Cys. Subsequently, a cysteine can be labeled with a donor, acceptor, or acetamide to give "D," "A," or "AA," respectively (as discussed below).

3. Labeling of a protein that bears two cysteines that lack selectivity ($C_\alpha^*-C_\beta^*$) with a molar ratio of 1:1 between the protein and thiol-reactive dye (D) is predicted to give the following product distribution:
   (1) 25% unlabeled protein ($C_\alpha^*-C_\beta^*$)
   (2) 25% dye attached to position α ($D_\alpha-C_\beta^*$)
   (3) 25% dye attached to position β ($C_\alpha^*-D_\beta$)
   (4) 25% double-labeled protein ($D_\alpha-D_\beta$)
4. In cases where cysteine selectivity is present (reactivity difference of two orders of magnitude, or more, between both cysteines), an enrichment of one product can be accomplished. For example, if position α is much more reactive than position β, the main product of a labeling reaction (using a 1:1 M ratio between protein and dye) would be $D_\alpha-C_\beta$. Such selectivity can be very useful for increasing the yields of a desired product. Furthermore, selectivity can be used to place the donor or acceptor at a specific and predetermined position. After the first labeling step, the preferred labeled species should be purified before a subsequent labeling step on an additional unlabeled cysteine (*see* **Note 5**).

   The intrinsic reactivity of a cysteine residue is context dependent. Factors such as pH, solvent accessibility, charge, and dipoles of nearby side chains can directly affect cysteine reactivity. An algorithm that predicts cysteine reactivity was developed [44] and is now available for free use (CYSPKA see below). With an input of PDB coordinates, the software provides the predicted reactivity of a desired position as an output. These data can be very useful for a rational design of labeling sites. Thus, it is possible to design a protein that bears only one reactive cysteine under certain conditions, for example, folding conditions or inhibitor binding. This position can be preferably labeled in an initial labeling step with relatively high yields. The second position can then be labeled under different conditions that are tuned for optimal reactivity at the second position. For example, denaturing conditions allow the reactivity of the unreacted cysteine or removal of a bound inhibitor.

CYSPKA is a Python program that runs in PyMOL. Instructions for the installation of Python environment and PyMOL – http://www.pymolwiki.org/index.php/Windows_Install.

CYSPKA downloadable code is available at http://www.pymolwiki.org/index.php/Cyspka.

5. *Labeling Cysteine with Thiol-Reactive Dyes*

   1. The most efficient labeling is obtained when the protein concentration in the reaction vessel is at the range of 2–20 mg/ml (lower concentrations can be used for proteins that tend to aggregate). Prior to labeling, the protein sample is incubated for 1 h with 10–20 mM of DTT in order to activate the cysteines. After activation, the DTT is removed by gel filtration or dialysis against the reaction buffer (HEPES 50 mM adjusted to pH-7.2). At this point, it is recommended to insure that the molar ratio between activated Cys and the labeling reagent is close to 1:1 or 1:2 in the case of a single or double Cys sample, respectively (protocols for measuring the concentration of activated Cys may be found in ref. [101]). The labeling reaction with the first dye is carried out in a glass vessel under reducing environment (TCEP 50 μM). The dye is added to the vessel to give a molar ratio of 1:1 between the protein and the dye. Insoluble dyes can be first dissolved in DMSO and then added to the reaction (a final DMSO concentration of 50 % v/v is acceptable). The reaction is allowed to proceed for 3–24 h at room temperature under gentle stirring and protected from light. The reaction is terminated by addition of DTT to a final concentration of 10–20 mM. Unbound dye should be removed by gel filtration or dialysis.
   2. As mentioned above, nonreactive Cys can sometimes be activated when the protein loses its native structure. In such cases, it is possible to perform the labeling reaction, with the second dye, under unfolding conditions, repeating the same steps as in the first labeling reaction (a high concentration of Gnd–HCl can be used for unfolding).
   3. Occasionally, dilution of the Gnd–HCl at the end of the labeling reaction (to facilitate refolding) will lead to sample aggregation. This phenomenon can be eliminated by gradual dilution of the reaction volume back to folding conditions. We developed a flow dialysis protocol that proved to be very effective for avoiding aggregation. In this protocol, the reaction volume is transferred to a dialysis membrane soaked in a TRIS buffer containing moderate concentrations of Gnd–HCl. The surrounding Gnd–HCl is slowly titrated with TRIS buffer without changing the dialysis volume.

6. *Use of Donor-Only Labeled Protein as a Reference for the Double-Labeled Protein*

   Time-resolved FRET measurements require preparation of a double-labeled sample (donor and acceptor) along with a donor-only labeled protein that is used as a spectroscopic reference. For some purposes, an acceptor-only control is also necessary (e.g., diffusion during the donor fluorescence decay). As a general approach, we recommend using one mutant with two sites available for modification both for the single-labeled controls and double-labeled samples (Fig. 10). In some cases, it is possible to use another mutant with a single reactive site available for labeling with either the donor or the acceptor.

   This approach is valid when (1) the double- and single-labeled proteins share the same structural characteristics that are relevant for the specific study (e.g., folding mechanism, conformational dynamics during catalysis). (2) In the absence of FRET, the spectroscopic properties of the donor in both samples are identical. When a two-cysteine mutant is used both for single and double labeling, the only difference between the two samples is the presence of an acceptor in the double-labeled sample and a free cysteine in the single-labeled one. The difference between the samples is further reduced by blocking the acceptor site with acetamide. Finally, the donor-only and double-labeled sample should be subjected to rigorous tests listed above to confirm that a joint analysis between them will be valid.

7. *General Strategy for Targeted Labeling of a Double Cysteine Protein*

   In light of all the considerations described above, a general guideline for the double-labeling strategy is presented. The strategies are summarized in the flowchart (Fig. 11).

   Strategy A

   1. Preparation of one or three protein samples (purity level of >90 %):

      (a) A mutant with two reactive cysteine residues ($C_\alpha - C_\beta$)

      Optional:

      (b) Mutant with a single substitution to Cys at the donor position ($C_\alpha - N_\beta$)

      (c) Mutant with a single substitution to Cys at the acceptor position ($N_\alpha - C_\beta$)

      where N represents the native residue. The last two mutants are useful as markers for chromatographic separation of the products of the labeling of the two-Cys mutant and in some cases will serve as a reference for the DO and AO experiments.

2. Small-scale labeling of both single mutants with the donor probe. Labeling under unfolding conditions is advisable to insure high reactivity. The expected products are ($D_\alpha - N_\beta$) and (N_α-D_β).
3. Analytical run of the products (separately) with a high-resolution ion exchange column, under identical conditions (Fig. 12). If a reasonable chromatographic separation

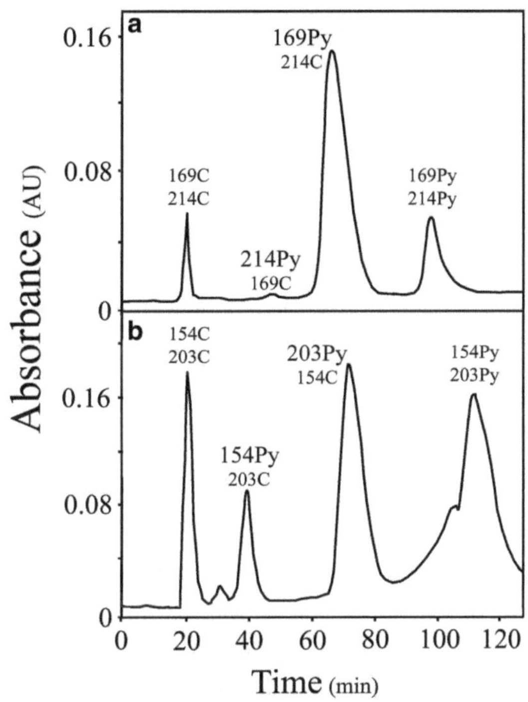

Fig. 10 Chromatographic analysis of the products of site-specific labeling of a two-cysteine mutant based on difference in the reactivity of the sulfhydryl groups. (**a**) A 5 mg sample of AK mutant with two engineered cys at residues 214 (poor reactivity) and 169 (high reactivity) was reacted with N-1-iodo-acetamido-methyl-pyrene. The reaction mixture was loaded on a phenyl sepharose column (diameter 0.5 cm and length 6 cm) and eluted by a 60 min gradient from 0.5 M to 0.0 M ammonium sulfate (in 20 mM tris buffer pH 7.8, 2 Mm EDTA) at a flow rate of 1 ml/min. The 280 nm detector trace shows the unlabeled mutant (UL) at 20 min, trace amount of protein labeled at residue 214 (40 min), large amount of protein labeled at residue 169 (70 min), and smaller amount of protein labeled at both cysteine residues (DL, 100 min). This was a case of large difference in the reactivity where the yield of 214cys169pyrene-AK was in preparative amount. (**b**) A similar procedure was applied to a mutant in which the difference in the reactivity of the two Cys residues (203 and 154) was not large enough and the enrichment of the product singly labeled at the more reactive residue (203pyrene154cys eluted at 70 min) was poor. A major fraction of the protein was lost in the DL fraction. The second single-labeled "isomer," 154 pyrene 203 cys, eluted at ~40 min

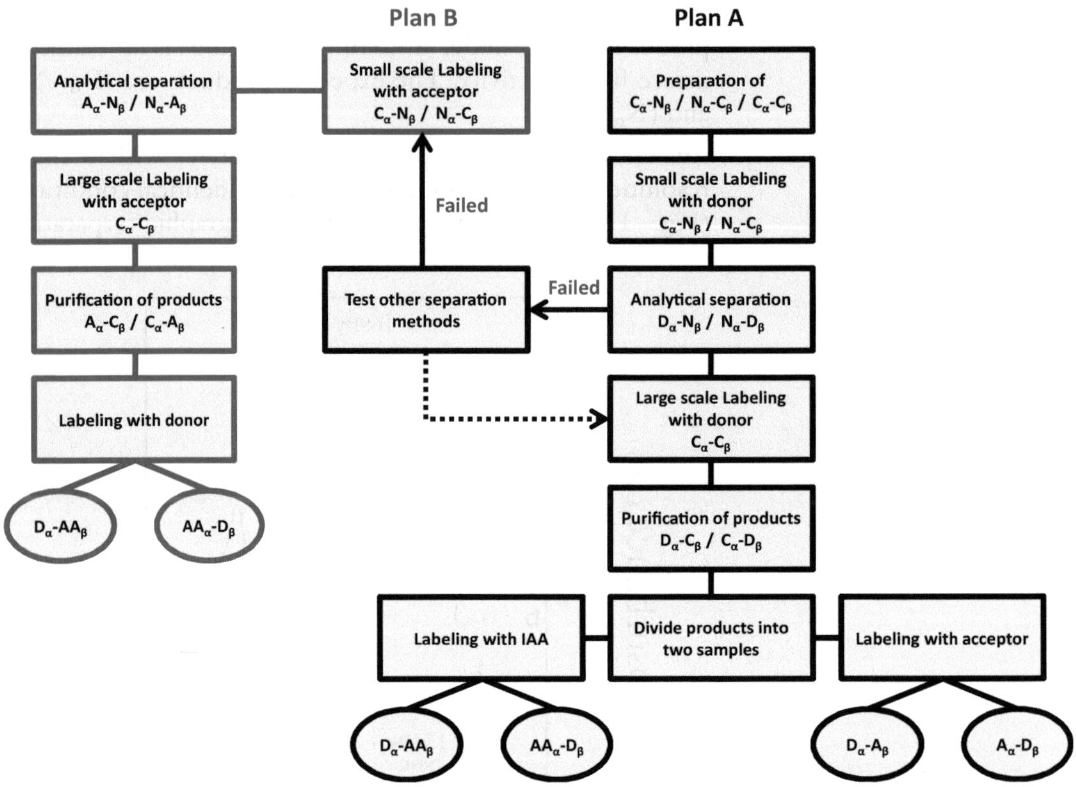

Fig. 11 Flow chart that represents strategies A and B for labeling a two-cysteine mutant

between the two labeled single cysteine mutants (isomers) can be attained, move to **step 4**. If not, additional separation columns (HPLC, Phenyl Sepharose, etc.) or other separation conditions should be tested. If separation between the two protein isomers is impossible, move to strategy "B."

4. Labeling the double Cys mutant with the donor dye (1:1 M ratio between protein and dye). If there is no selectivity, then an expected yield of 25 % for each single-labeled product should be expected. If there is selectivity, the more highly reactive species will be enriched.
5. Preparative chromatographic run under conditions determined as described in Subheading 3. Both single-labeled products should be collected for further use (Fig. 12).
6. Split each single-labeled fraction into two samples (duplicate): (a) Label one duplicate with iodoacetamide to prepare a donor-only mutant. (b) Label the second duplicate with the acceptor dye to prepare a double-labeled mutant. Reaction "a" and "b" can be performed under denaturing conditions with excess dye (1:5).

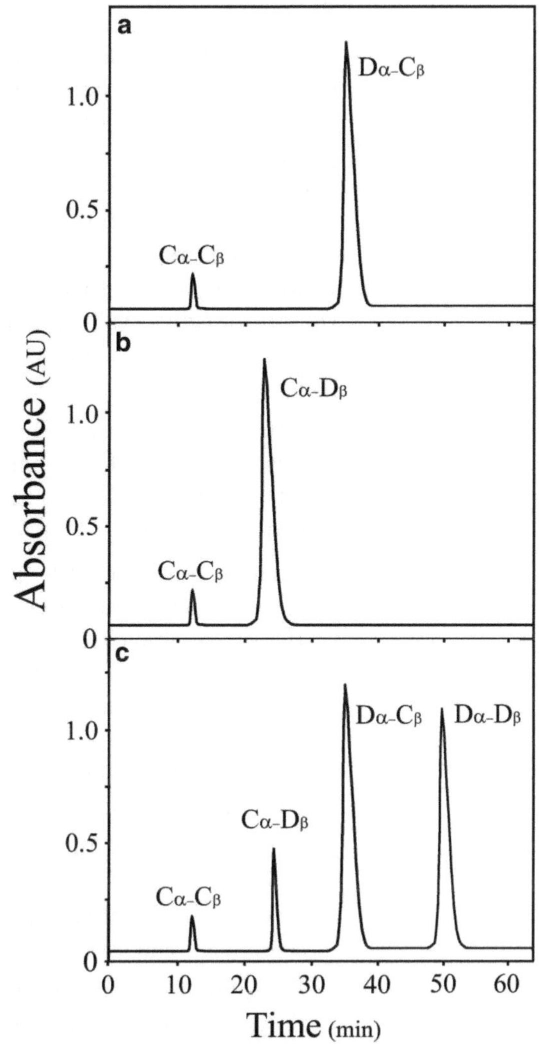

Fig. 12 Identification of the single- and double-labeled "isomers" of the labeled two-cysteine mutant by HPLC analysis. (**a**) and (**b**) products of the single cysteine mutants were analyzed. (**c**) When a reaction mixture of two-cysteine mutant with the same reagent was analyzed, the three reaction products can be identified according the results shown in (**a**) and (**b**)

7. Preparative chromatographic run of sample "b" to remove unlabeled fractions (in most cases, chromatographic separation between a donor-only and donor–acceptor-labeled protein can be performed without difficulty). Chromatographic separation between (D_α-C_β) and (D_α-AA_β) (where AA represents acetamide) is very difficult to attain (in order to remove unlabeled protein from reaction "a"). An Elman test can be performed to insure that the labeling reaction reaches completion.

Strategy B

When chromatographic separation between the donor-labeled singles is not possible, preparation of a donor-only sample with iodoacetamide in the acceptor position is very difficult to achieve. In this case the single donor-labeled mutants (D_α–N_β) and (N_α–D_β) (prepared as described in Strategy A, Subheading 2) could be used as references, rather than (D_α–AA_β) and (AA_α–D_β) (as described for Strategy A, Subheading 6). Here, special care should be taken to ensure that the donor only is valid as a reference (*see* **Note 4**). Nevertheless, preparation of a double-labeled protein is still possible, as follows:

1. Small-scale labeling of both single mutants with the **acceptor** dye (*see* **Note 4**). Labeling under unfolding conditions is advisable in order to insure high reactivity. The expected products are (A_α–N_β) and (N_α–A_β).
2. Analytical run of the products (separately) with a high-resolution ion exchange column, under identical conditions. If a reasonable chromatographic separation between the two single mutants (isomers) can be found, move to **Step 3**. If not, other separation columns (HPLC, Phenyl Sepharose) or separation conditions should be tested. If separation between the isomers is impossible, changing the FRET dye pairs should be considered.
3. Labeling the double Cys mutant with acceptor dye (1:1 M ratio between protein and dye). If there is no selectivity, 25 % of each single-labeled product is to be expected. If there is selectivity, the more highly reactive species will be enriched.
4. Preparative chromatographic run under conditions described in Subheading 2. Both single-labeled products should be collected for further use.
5. Label each of the products with the donor dye to prepare a double-labeled mutant. The reaction can be performed under denaturing conditions with excess of dye (1:5).
6. Preparative chromatographic run in order to remove unlabeled fractions.

8. *Avoiding Errors due to the Effect of R_o.*

The main uncertainties that should be eliminated in FRET-based experiments are (a) the possible bias of the Förster critical distance used in the calculation of a distance by incomplete averaging of the orientations of the probes, (b) multi-exponential decay of the donor emission that can result from the distribution of either the distances or the orientations in trFRET experiments, and (c) variation of parameters that affect the Förster critical distance, R_o, as a result of change of environment of the probe due to change of solvent or any conformational transition (such as ligand binding, folding/ unfolding, or chemical modification).

Overcoming the uncertainty due to the problem of the orientation factor, κ^2.

Various approaches were proposed for the reduction of these uncertainties and the determination of the range of possible errors in distance determination. Dale et al. [102–104] reported the most extensive studies and developed methods for estimation of the range of the orientation factor, $<\kappa^2>_{min}$, and $<\kappa^2>_{max}$, using fluorescence anisotropy measurements. Rapid reorientation during the lifetime of the excited state of the probes is reflected in the anisotropy measurements and therefore can be estimated experimentally for each preparation (for details of all the angles taken into account see Dale et al. [103, 104] and Van der Meer [6] pp. 55–83). An ideal case, where full averaging is an intrinsic characteristic of the system, is achieved with single-atom probes, such as lanthanides and chelated transitions metal ions [5, 105–107]. An alternative approach to this problem is based on the fact that many chromophores show mixed polarizations in their spectral transitions, i.e., their absorption and emission across the relevant spectral range of overlap involves a combination of two or more incoherent dipole moments [108]. Haas et al. [107] showed that the occurrence of mixed polarizations in the energy donor and acceptor may markedly limit the range of possible values that κ^2 can assume and, thus, under favorable conditions may alleviate the problem of the orientation factor in FRET experiments. This is the case for many common probes which have high levels of mixed polarization, such as aromatic probes and derivatives of naphthalene. This effect, when considered in planning FRET experiments by using probes that show a high degree of internal depolarization, can reduce the κ^2-related uncertainties to a range in which they do not exceed the other common experimental uncertainties. An improved approximation of the value of κ^2 is obtained when the contribution of the rotation of the whole molecule to the emission anisotropy of the probes is first subtracted from the observed anisotropy. This procedure can be performed when the fluorescence anisotropy decay is determined and the analysis shows the relative amplitude associated with the decay component corresponding to the molecular rotational diffusion. The uncertainty in r can be further reduced by employing two independent FRET experiments. In these experiments, different pairs of probes are used for FRET determination of the distance between a single pair of sites in a protein molecule.

9. *Approaches to Determine and Control the Value of R_o*

How to determine and control the value of R_o?
The distance range for the statistically significant determination of distances for each pair is defined by the characteristics of the probes and is given by the R_o value. Equation 4 shows that the limits of that range are $r = R_o \pm 0.5\ R_o$ (Fig. 2). Maximal sensitivity of the FRET effect to small changes in r is observed when $r = R_o$. Therefore, optimal design of FRET experiments includes a selection of a pair of probes that have an R_o value close to the expected r value.

R_o depends on four parameters which are characteristic of each pair of probes in Eq. 6. These parameters were discussed critically by Eisinger, Feuer, and Lamola [109]. Modulation of the value of R_o of any specific pair in order to adjust it to the expected r value is limited to a narrow range due to the sixth power relation of R_o and the spectroscopic constants of the probes. Equation 6 shows that the donor quantum yield Φ_D^o and the overlap integral parameters are the most readily available for manipulation to modulate the R_o value.

Index of refraction n.
The values used for n to calculate R_o for a pair of probes attached to a protein molecule vary widely, from 1.33 to 1.6. Eisinger, Feuer, and Lamola [109] recommended using the value $n = 1.5$ in the region of spectral overlap of the aromatic amino acids in proteins.

The donor quantum yield Φ_D^o
Φ_D^o should be determined for each probe when attached to the protein molecule at the same site used for the FRET experiment. Many experiments showed that the Φ_D^o is affected by both the site of attachment of the probes and by modifications at other sites in the protein molecule, by attachment of an acceptor at second sites, by changes of solvent conditions, and by ligand binding or folding/unfolding transitions. Therefore, R_o should be determined repeatedly after changes in the conditions to avoid misinterpretation of changes of R_o as changes of r. The R_o depends on the sixth root of Φ_D^o and is therefore not very sensitive to uncertainties in Φ_D^o. Nevertheless, this same dependence limits the range that R_o can be manipulated. For instance, a twofold decrease Φ_D^o would result in ~9 % decrease of R_o. Increasing Φ_D^o without a change of probes usually requires change of solvent and therefore is not very useful in folding studies; but Φ_D^o can be decreased by adding quenching solute without a change of solvent.

Bound donor probes may have few modes of interactions with the macromolecular environment in the ground or in the electronically excited state. In such a case, each subpopulation which has a different lifetime also has a different R_o, as indicated in Eqs. 1 and 2.

The spectral overlap integral J.

The most effective variation of R_o is available by selecting probes according to the overlap of their spectra. The value of *J* depends on the local environment of the probes. As a rule, absorption spectra vary relatively slightly with a change of solvent, temperature, or local environment, but emission spectra may be very sensitive to such changes of the local environment of the probes. An example of this dependence is the fluorescence of indole derivatives [110]. The selection of probes with narrow emission and absorption spectra and with high extinction coefficients of the acceptor can yield high R_o values.

10. *Experimental Challenges in the Implementation of trFRET in Protein Research*

Practical implementation of the FRET approach in the characterization of protein ensembles either in the ensemble or the single-molecule modes involves four very different steps: design and preparation of pure, homogenous, structurally and functionally intact double-labeled protein samples; spectroscopic determination of transfer efficiencies and fluorescence decay profiles of the probes; control experiments; and data analysis.

Optimized Design and Preparation of Labeled Protein Samples for Time-Resolved FRET Experiments

A minimum of two labeled protein species should be prepared for each well-controlled, time-resolved FRET experiment. These are (a) a site specifically double-labeled (by both donor and acceptor) species and (b) a single donor-labeled species (at the same site as in (a)). The best results can be obtained when the following requirements are satisfied: (a) the mean distance between the two selected sites should be close to the R_o characteristic for the pair of probes used; (b) the ensemble characteristics of the labeled protein derivatives should be minimally perturbed; (c) the probes should be of small size and with minimal potential interactions with the backbone or the side chains; and (d) the probes should have ideal spectroscopic characteristics to enable detection and data analysis free of correction factors. Proteins are rarely totally unaffected by chemical or mutational modifications, and the spectroscopically ideal probes are usually also very large. Therefore, the design process is a matter of optimization, not maximization.

The extent of perturbation is an important factor in the evaluation of the FRET results. Occasionally, modified protein molecules are found to be unsuitable for planned experiments due to very strong perturbations. Consequently, an alternative modification design should be attempted. The concern for purity cannot be overemphasized, since background fluorescence or fluorescence of impurities can drastically distort the results of distance determinations.

Incomplete labeling of the reference single-labeled mutant in each set should not prevent the use of the preparation in FRET experiments, provided that the extent of labeling is known. Even incomplete labeling of the double-labeled mutant can be tolerated in ensemble steady-state or time-resolved detected experiments, provided that the extent of labeling is known and (in some cases) the partially labeled fractions do not exceed ~25 % of the total population of labeled molecules. The strict requirements for complete labeling and accurate knowledge of the extent of labeling are relaxed in single-molecule FRET experiments. In these measurements, it is possible to discard records of single-molecule emission intensity that do not show both donor and the acceptor emission during data processing.

For optimal accuracy of distance determination in proteins by steady-state or time-resolved FRET, the following spectral characteristics of the probes are desired: (a) R_o of the selected pair should be as close as possible to the expected mean distance between the two labeled sites (including the linker of the probes); (b) the orientation of the probes should be averaged during the lifetime of the excited state of the donor; (c) for time-resolved ensemble experiments, the ratio of the donor and acceptor mean fluorescence lifetimes should be $\tau_D/\tau_A \geq 3$ for effective global analysis; (d) photostability; (e) minimal environmental sensitivity of the spectral characteristics of the probes; (f) large Stokes shifts of the acceptor (to achieve minimal overlap of the emission spectra of the donor and the acceptor); and (g) a high extinction coefficient to enable measurements of dilute solutions (where aggregation problems are minimized or the presence of only one molecule at a time can be detected in the smFRET mode of measurement).

It is always important to show that the photons which appear as enhancement of the acceptor emission are accountable by reduction of the photon emission by the donor and vice versa. Unlike intensity measurements where correct calculation of the changes in probes' quantum yields strongly depends on knowledge of the exact concentrations, average lifetime changes can be obtained regardless of the concentrations provided that the sample is uniform [7].

11. *Data Analysis*

The interprobe distance distribution (IDD) function is obtained from global analysis of two or four experimental fluorescence decay curves. Analysis of only two experiments, the DO and DA decay curves, can yield the IDD without need for additional independent decay curves. Yet, for reliable determination of the intramolecular diffusion coefficient, inclusion of the acceptor decay curves (experiments AO and DA) is recommended. To compare the calculated fluorescence decay curves with those obtained experimentally, the calculated curves are convolved with the excitation pulse profile, $G(t)$:

$$F_{\mathrm{c}}(t) = \int_{0}^{t} G(t) I_{\mathrm{c}}(t - s) \mathrm{d}s \qquad (21)$$

Curve fitting procedures are used for determination of the free parameters of the model functions. Analysis of the fluorescence decay curves is performed using the nonlinear least-squares program, GLOBAL (a Levenberg–Marquardt algorithm) [55, 112, 113]. This software is available commercially or by request from the Haas lab. Several other methods for curve fitting and judgment of the quality of fit are available; the most commonly used is the maximum entropy method (MEM) [29, 114]. The calculated decay curves of the two probes in the double-labeled (donor and acceptor) samples are prepared by numerical solution of the second-order differential Eqs. 12 and 13.

The quality of fit of the calculated $F_{\mathrm{c}}(t)$ and the corresponding experimental intensity, at each time interval, $F(t)$, is evaluated using four criteria: (a) the global and local χ^2 values, (b) the distributions of the residuals, (c) the autocorrelation of the residuals, and (d) the error intervals of the calculated parameters [60] (Fig. 13).

The result of the best fit using distance distribution model (black, Fig.13) is compared to the best fit results obtained by fitting the same dataset to model of weighted sum of three exponent function (red), $I(t) = \Sigma_i I_{0i} \exp(-t/\tau_i)$, where I_{0i} and τ_i are the amplitude and the lifetime, respectively, for the ith decay component. This reference curve serves as a representation of the noise level of the specific experiment. Shown in Fig. 13 are the computed distance distributions (D, G), the fitting residuals (B, E), and their autocorrelation (C, F). The distribution of the residuals in panel B (single population model) shows that there is more information in the data, which is not accounted for by the single population distribution model. This is sharply demonstrated by the corresponding autocorrelation of the residuals. The normal distribution of the residuals (panel E) and the zero autocorrelation (panel F) obtained by fitting the two subpopulations support the conclusion that the ensemble of the labeled

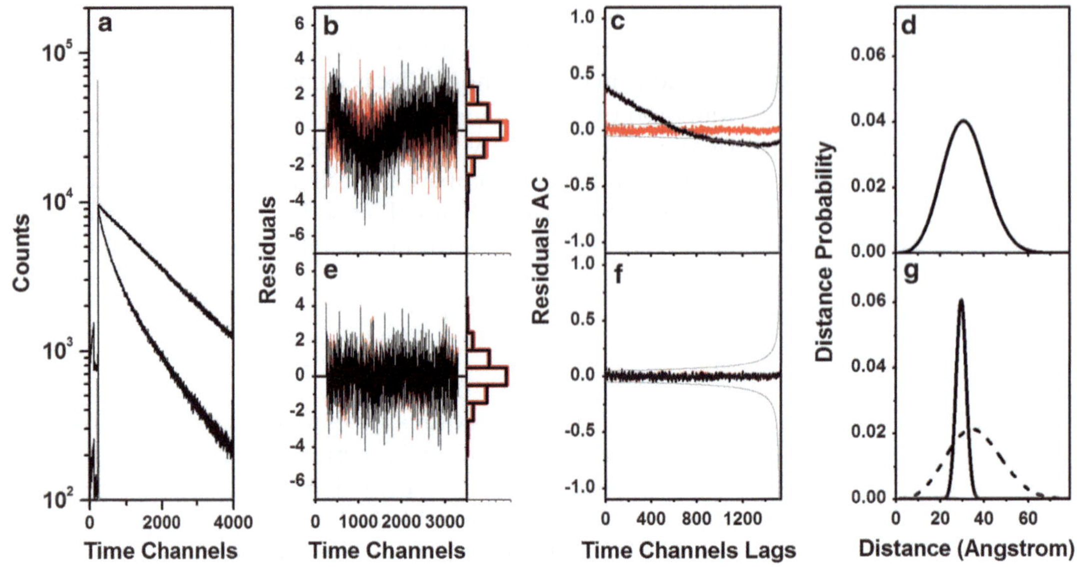

Fig. 13 Ensemble mode trFRET measurements can resolve two subpopulations characterized by their intramolecular distance distribution. In the example shown the decay of the probes attached to AK labeled with a donor, N-1-acetamide-methyl-pyrene (Pyr), at residue 58 and an acceptor, maleimide(1-(3-(succinimidyloxycarbonyl)benzyl)-4-(5-(4-ethoxyphenyl)oxazol-2-yl)pyridinium bromide) (PyMPO), at residue 188 was measured in the presence of 0.67 M guanidinium hydrochloride, the chemical folding midpoint. The decay curves (panel **a**, upper curve is the donor-only decay and the lower curve, the donor in the presence of acceptor decay; the channel width is 16 ps) were analyzed (based on Eqs. 12, 13, 14, 15, 16, and 17) in search for best fit parameters of the intramolecular distance distribution assuming either one (panels **b–d**) or two subpopulations (panels **e–g**)

protein contains at least two subpopulations. The third criterion for discrimination between the two models is the corresponding χ^2 values (1.5 and 1.2 for the single and the two subpopulations respectively) also shows improved fit by the two subpopulations model. In addition, the χ^2 values obtained in the fitting to three exponential decay for each experiment enable evaluation of the significance of the χ^2 values obtained for each set of experiments in the analyses based on the solution of Eq. 12 or any other model.

Averaged energy transfer efficiency is determined from the time-resolved measurements using Eq. 7.

In MEM analysis [115], it is assumed that the decay originates from a distribution of fluorescence lifetimes in the range of 10 ps to 10 ns, or in a similar range, with all lifetimes initially assumed to have equal probability weights (amplitudes). Subsequently, the distribution is modified in each iteration of the analysis, leading to a maximization of the Shannon–Jaynes entropy, $S = \Sigma - P_i \log P_i$, where P_i is the probability (amplitude) of the *i*th lifetime, and to a minimization of the residuals (χ^2). For a particular value of χ^2, there could be many possible values of P_i. MEM analysis identifies the distribution for which S is maximal. The calculation is terminated when successive iterations do not change the values of χ^2, S, and the distribution profile. Thus, MEM analysis enables the best (least biased) fit of the data using combination of several Gaussian distributions without imposing the envelope distribution. A software package to perform these calculations is commercially available from Maximum Entropy Solutions, Ely, UK.

An alternative method of analysis which might have much use in the future was introduced by Schröder and Grubmüller [116]. In this method, the FRET parameters (fluorescence decay curves) are calculated from a simulated ensemble of conformers using atomistic molecular dynamics.

Ultrafast dynamics studied by upconversion laser spectroscopy and intramolecular trFRET measurements can reveal the nonequilibrium dynamics of subdomain structures in the protein. In this case, the donor fluorescence decay is fit with stretched exponentials and not translated into the distance distributions. This method is of much interest in studies of the elementary steps of the action of globular proteins [117].

Analysis of the acceptor fluorescence decay.
The FRET contribution to the acceptor fluorescence decay is most effective in the rise time (grow-in) section of the decay. Time-resolved fluorescence experiments using tryptophan excitation have shown that FRET is manifested by a distinct growing in of acceptor fluorescence at a rate characteristic for this process [118]. Here, the problem is the leakage of the donor emission and the direct excitation of the acceptor.

11a. *Global Analysis.*
The routine global analysis procedure includes joint analysis of at least one set containing at least two or four fluorescence decay curves of the probes attached to the protein: (a) The "DD experiment," determination of τ_D the fluorescence decay of the donor in the presence of FRET (excitation at the donor excitation wavelength and detection at the donor emission wavelength). (b) The "DA experiment," determination of τ_A, the acceptor emission, in the presence of FRET, with excitation at the donor excitation range, and detection of the acceptor emission at the long wavelength range. Two internal reference measurements are included for determination of the probes' lifetime when attached to the same corresponding sites as in the double-labeled mutant but in the absence of FRET. (c) The "DO experiment," a measurement of τ_D^o, the emission of the donor in the absence of an acceptor, using protein molecule labeled by the donor alone. (d) The "AO experiment," measurement of τ_A^o, the decay of

the acceptor emission in the double-labeled protein, excited directly at a wavelength where the donor absorption is negligible or using a protein preparation labeled by the acceptor only. When collected under the same experimental conditions (optical geometry, state of the instrument, time calibration, linearity of the response, and solution parameters) used for the first two experiments, the DO and AO experiments serve as "internal standards" for the time calibration of this experiment and the dependence of the spectroscopic characteristics of the probes.

Rigorous error analysis.

The statistical significance of the value of any one of the free parameters that is determined by any curve fitting method can be evaluated by a rigorous analysis procedure. This procedure, if repeated for every free parameter, actually generates projections of the χ^2 hypersurface on specific axes in the parameter space. Error limits at any required significance level can then be determined using an *F-test* [55]. This procedure considers all correlations between the parameters. The evaluation of each analysis and of the significance of the parameters is based on four indicators: (1) the global χ^2 values, (2) the distributions of the residuals, (3) the autocorrelation of the residuals, and (4) the error intervals of the calculated parameters obtained by the rigorous analysis procedure. In addition, each "DD experiment" is analyzed by a multi-exponential (three or four exponentials) decay model. The χ^2 values obtained in these analyses serve for evaluation of the significance of the χ^2 values obtained for each set of experiments in the global analysis, based on the solution of Eqs. 15, 16, 17, and 18.

12. *Detection of Protein–Protein and Protein–Ligand Association by trFRET Methods*

Questions of intermolecular interactions, e.g., protein– protein interactions and protein–ligand interaction, can be studied in two levels by trFRET measurements: (a) the question of proximity at nm scale in vitro and in vivo and (b) the question of the mode of interaction, the dimensions of the complex, the orientation of the interacting molecules, the fraction of bound molecules, etc.

The main problem in such experiments is that unlike the case of intramolecular FRET, where the population of pairs is known, here one has to deal with at least three components mixture: the donor and the acceptor monomers and the donor and acceptor complex. In case of homodimers or oligomers the mixture is even more complex. trFRET enables a solution to this problem. The solution can be based either on measurement of the donor lifetimes or the acceptor.

In the first case the procedure should be as follows:

First determine the fluorescence lifetime of the donor-labeled monomer τ_D^o. Then measure the lifetime of the donor in series of solutions of fixed concentration of donor-labeled monomers and variable concentrations of the acceptor-labeled monomer or ligand. Then use global analysis assuming two subpopulations, one characterized by $\tau_D^o$and the other either single of distribution of lifetimes. Determine the decreasing fraction of the contribution of the τ_D^o subpopulation to the fluorescence decay of the donor in the donor/acceptor mixtures.

The second type of experiments (in case of heterocomplexes) would be as follows:

a protein labeled by a donor with long fluorescence lifetime and the second protein or the ligand labeled with short lifetime acceptor. Measure the fluorescence decay in three experiments: DO and AO under excitation at the donor absorption spectral range. Then measure the fluorescence decay of the acceptor in solutions of mixtures of donor-labeled and acceptor-labeled molecules, which is designated the "A–D" experiment (A–D means the emission of the acceptor under excitation at the donor absorption spectral range in the presence of free-donor- and free-acceptor-labeled monomers).

Repeat the acceptor fluorescence decay measurement (the A–D experiment) using mixtures of the two monomers at constant donor-labeled protein concentration and series of higher concentrations of the acceptor-labeled protein/ligand. Measure the donor fluorescence decay under the same conditions as above.

Use global analysis of the three experiments including the A–D experiments of the series of acceptor-labeled monomer, when limiting the analysis of the A–D experiments to the section of the fluorescence decay that starts at five times the lifetime of the acceptor in the AO experiment. In this manner the contribution of the FRET and its amplitude can be determined on the background of large directly excited acceptor (free and in the complex) emission and unpaired donor emission which is invisible.

The fraction of long lifetime donor can be determined, and it represents the fraction of unbound donor by determination of distribution of the donor lifetimes.

The advantage of this strategy is that by monitoring the FRET via the acceptor emission decay only bound donor-labeled protein molecules are detected via the acceptor fluorescence decay. The mean FRET efficiency can be determined, and in optimal cases, the distribution of the distance between the two labeled sites can also be determined.

The parameters that should be considered in a design of such experiment are:

(a) The ratio of fluorescence lifetime of the donor and the acceptor (should be larger than 8).

(b) The expected distance between the labeled sites and accordingly the mean transfer efficiency in the bound pair of molecules, optimal value $E = 0.5$. Higher values would shorten the long lifetime component in the acceptor emission and lower values would reduce the amplitude of the long lifetime component in the acceptor decay. Durroux and coworkers [119] used the long-lived chelated lanthanide ions emission as donor in search for protein–protein interactions in cells.

13. *Count Thresholds for Burst Detection in Single-Molecule Detection Experiments*
Both signals originating from photon detection as well as non-photon detection events can be described as Poisson processes. Therefore, to distinguish between the important photon and the irrelevant non-photon events, one needs to make sure that the mean of the event count histogram of the fluorescence photons detected from measured single molecules has a higher number of counts than the non-photon background histogram, the shot noise. Usually, in a well-calibrated system, shot noise is characterized by low counts in any given time bin, though consists of the majority of detected events. Therefore, it results in an event count histogram characterized by a Poisson distribution with a very low mean count rate, explaining why the mean fluorescence count rate should be higher. An optimal count rate threshold should be applied to select bursts with high-enough event count to yield a sufficient signal to background (S/B) ratio, in this generally low S/B system.

An assessment of the shot-noise statistics can be attained by performing the same measurements on a sample containing solvent only. Its event count histogram represents the contributions of background signals, including the non-filtered leaked scattering signals as well as signals resulting from impurities.

The experimental setup and the measured sample should be planned so that the background signals will be minimal. Signals originating from Rayleigh and Raman scattering should be carefully filtered out. Band-pass filters for selection of emission wavelengths for fluorescence detection should be selected so that, on the one hand, they will include large spectral range around the maximal fluorescence signal of the donor/acceptor but, on the other hand, will exclude wavelengths close to the excitation wavelengths and wavelengths corresponding to the Raman scattering of water. It is highly recommended to measure the sample with the focal volume

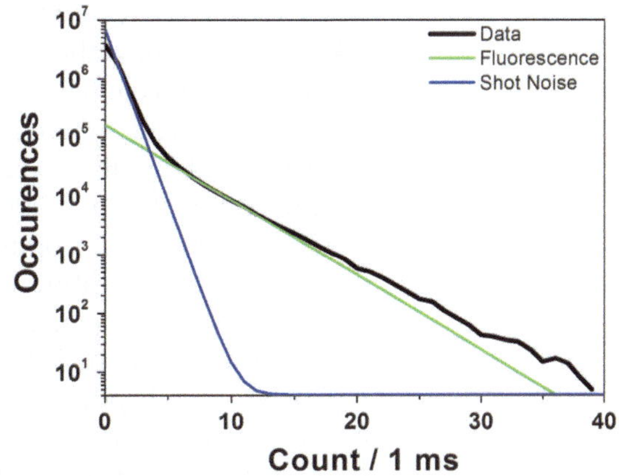

Fig. 14 An example of how to assess a minimal counting threshold. The total photon count histogram for 1 ms time bins is presented in *black*. It was fitted with a two process function, here using a bi-exponential function. The high count process in *green* represents fluorescence, while the low count process in *blue* represents shot noise

aimed deep inside the sample and far away from the cover slip surface to reduce auto-fluorescence and scattering from the glass cover slip. When epifluorescence is utilized, the excitation should be carefully separated from fluorescence by use of a good dichroic mirror. Emitted fluorescence light should be tightly focused to the center of the APD detectors.

A typical freely diffusing single-molecule FRET experiment is presented in Fig. 14. Photon counts are accumulated into 1 ms time bins in both donor and acceptor detection channels. Then, a photon counting histogram is calculated. One can clearly distinguish the low count but frequent shot-noise process from the high count infrequent detection of fluorescence. One way of calculating count thresholds above which 1 ms counts are considered as bursts is to extrapolate the shot-noise process in the counting histogram to zero, and the photon count value at which there are no shot-noise contributions will be set as the minimal counting threshold.

An example is given in Fig. 14 showing the total detected events per 1 ms time bins, and minimal threshold is assessed to be ~12 events. Since in low FRET a low number of counts will be detected in the acceptor detection channel, and in high FRET a low number of counts will be detected in the donor detection channel, hence, a threshold is placed on the total of both detection channels.

Yet, a total count threshold of 12 detected photons is insufficient, as illustrated by the following example: a single-molecule traversing the excitation volume could yield a total of 12 photons and a FRET value of 0.5. This means that 6 of the photons were detected as donor fluorescence, while the other 6 were detected as acceptor fluorescence. Now, if we take into account the uncertainty regarding whether each detected event is in fact due to fluorescence or to background noise, the "real" FRET value taking only fluorescence photons could easily be due to 4 donor photons and 6 acceptor photons yielding a FRET value of 0.6. This is the manifestation of the FRET population shot-noise width in Eq. 20. The higher the counting threshold, the lower the probability of signal contributions not from fluorescence photons, and the more accurate the FRET value calculation will be, yielding narrower FRET populations. Nevertheless, the higher the counting thresholds are, fewer events will pass these thresholds, and the acquisition times would be extended.

Time-resolved capabilities can also be useful in reducing background contributions. As discussed here, the photon and the non-photon arrival macrotimes are Poissonian distributed. However, these processes are distributed totally differently in means of microtimes. After shot-noise is minimized, most of the remaining noise originates from detector afterpulsing. Afterpulsing is random within the fluorescence lifetime time regime and hence is distributed uniformly, while photon arrival microtimes are exponentially distributed due to the nature of the fluorescence process. On that basis, signals with a short microtime will have a higher probability of originating from fluorescence rather than shot noise. The higher an event microtime, the lower the fluorescence decay at that time, and therefore, according to the inverse square root law of S/N in fluorescence decays, the lower the probability that it originated from fluorescence. The time-gating filtering method is based on the accumulation of bursts, not after count thresholding, but rather by selection of detection events with low microtime. This methodology yields smFRET histograms with lower widths and higher information content. This result in lower count rates and longer acquisition times, but the smFRET histograms are biased towards higher FRET efficiency values since the higher microtimes in the donor fluorescence decays are a signature of low FRET rate contributions.

14. *The Choice of Time Bin*

The time bin chosen for data accumulation is very fundamental. Theoretically, the shorter the time bin taken, the less averaging of dynamic data would occur. Yet, due to the inherently low rate of photon emission and detection, there is a practical lower limit for time bin duration. In smFRET of freely diffusing molecules, the time bin is directly related to the duration that single molecules remain inside the effective excitation volume.

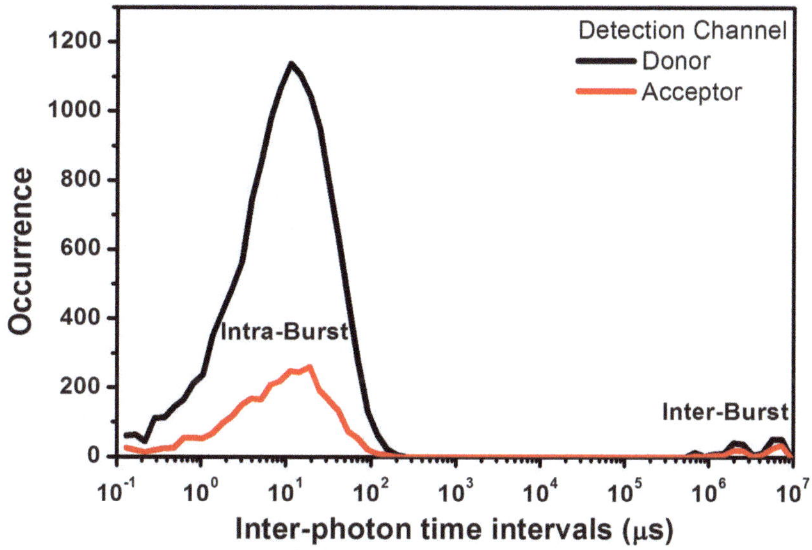

Fig. 15 An example of an inter-photon histogram in smFRET of freely diffusing molecules

Another way of assessing the optimal time bin duration is by considering the inter-photon histogram (IPH) [120]. The IPH is calculated from time intervals between consecutive photon arrival macrotimes. In smFRET of freely diffusing molecules one should expect an IPH with two peaks. The first occurs at short times belonging to intervals between photon detection of an individual single molecule, and the second is observed at long times representing the waiting times between the exit of a single molecule from the excitation volume and the entry of a new single molecule.

Figure 15 illustrates an inter-photon time histogram in an smFRET measurement of freely diffusing molecules. One can clearly distinguish the population centered in microseconds from the population centered in seconds. The peak value of the microsecond population for both donor and acceptor fluorescence detection is around 20 μs. Thus, to attain an average of 100 detected fluorescence photons, one needs a time bin of 2 ms. This approach represents one possible way of assessing a suitable time bin.

General vs. time binned bursts.

Burst durations vary due to transitions to dark state and variation of the duration of the diffusion of a labeled molecule through the excitation beam. Therefore, for the sake of data acquisition, each burst should be detected and collected, irrespective of the time bin in which it was collected.

Rates of conformational changes can be estimated by analyses of series of smFRET histograms obtained with different time bin duration. Copich and Szabo [121] have shown that

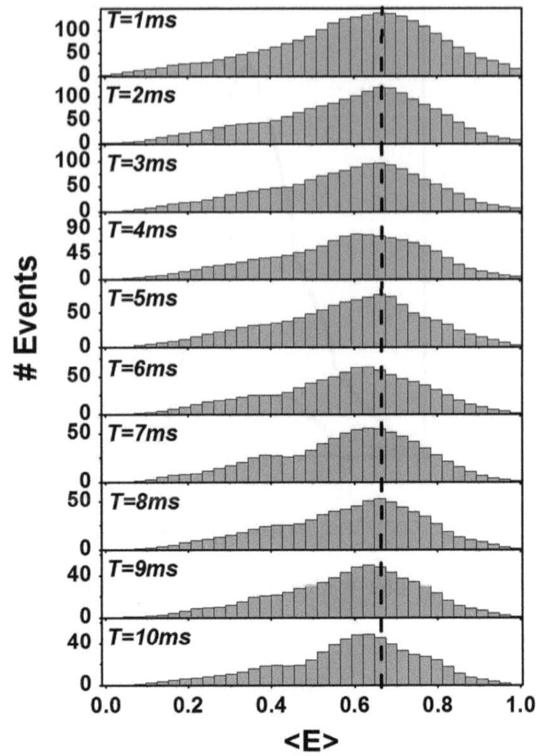

Fig. 16 smFRET of freely diffusing double-labeled AK molecules and the detection of conformational dynamics. An example of FRET histograms collected using various time bins and the interconversion averaging effect. A *dashed vertical line* is positioned at the FRET efficiency value of the high FRET population mean value, to emphasize the two populations' dynamical mixing effect

smFRET histograms accumulated using a 0.1 ms time bin may exhibit the two distinct FRET populations with widths limited by shot noise, while using a 10 ms time bin resolved only a single FRET population. This single population was the average of the FRET values of the two conformations weighted by their equilibrium fractions. Using a 1 ms time bin, the FRET histogram revealed the two FRET populations as well as another intermediate population, accounting for cases in which a single molecule entered the effective excitation volume in one state and exited it after interconverting to the second state. Without time binning, this important dynamic data might be missed. Figure 16 shows that as the time bin duration grows longer, the major FRET population shifts towards shorter FRET values, while the minor FRET population shifts towards longer FRET values, gradually creating a weighted averaged FRET population.

15. *Corrections Needed to Transform Proximity Ratios to FRET Efficiencies*
The equation for correctly calculating smFRET efficiencies is [122]:

$$E = \frac{F_{\mathrm{FRET}}}{\gamma n_{\mathrm{D}}^{\mathrm{D}} + F_{\mathrm{FRET}}};\ F_{\mathrm{FRET}} = n_{\mathrm{D}}^{\mathrm{A}} - \mathrm{Lk} - \mathrm{Dir} \quad (22)$$

where γ was defined above, $n_{\mathrm{D}}^{\mathrm{D}}$ and $n_{\mathrm{D}}^{\mathrm{A}}$ are the number of donor and acceptor photons per given time bin after donor excitation, Lk is a correction factor for donor fluorescence leakage into the acceptor detection channel, and Dir is a correction factor accounting for the acceptor direct excitation.

Lk can be calculated as shown in Eq. 23 [122]:

$$\mathrm{Lk} = l \cdot n_{\mathrm{D}}^{\mathrm{D}} \quad (23)$$

With l being the leakage coefficient and can be easily determined by the measurement of $n_{\mathrm{D}}^{\mathrm{A}}/n_{\mathrm{D}}^{\mathrm{D}}$ for a donor-only (DO) sample,

Dir can be calculated as shown in Eq. 24:

$$\mathrm{Dir} = d \cdot n_{\mathrm{D}}^{\mathrm{A}} \quad (24)$$

where d can be calculated from the ratio of acceptor to donor extinction coefficients at the wavelength of excitation. It can also be directly measured as the ratio of $n_{\mathrm{D}}^{\mathrm{A}}$values of an acceptor-only (AO) sample with a donor and acceptor (DA) sample measured under the same concentrations.

The DO and AO control measurements can be done by using ensemble fluorescence decays in order to get accurate correction factors with minimal error due to background signals.

The γ factor can be calculated in many ways. One of the ways is to measure smFRET with time-resolved capabilities and use the relation:

$$E = \frac{n_{\mathrm{A}}}{\gamma n_{\mathrm{D}} + n_{\mathrm{A}}} = 1 - \frac{\tau_{\mathrm{DA}}(E)}{\tau_{\mathrm{DO}}} \quad (25)$$

By measuring the average donor fluorescence lifetime per each E range in the FRET histogram, $\tau_{\mathrm{DA}}(E)$, and using the known fluorescence lifetime in the absence of an acceptor, τ_{DO}, the photon count-related FRET can be compared to the absolute values of trFRET, and the gamma factor can be resolved as is exemplified in Fig. 17. A donor fluorescence decay was accumulated from microtimes at each FRET efficiency range. The average donor fluorescence lifetimes were elucidated out of bi-exponential fits and were plotted against their corresponding photon count-related FRET values. According to the method presented here, by using Eq. 25, a linear dependence should be obtained as described by Eq. 26:

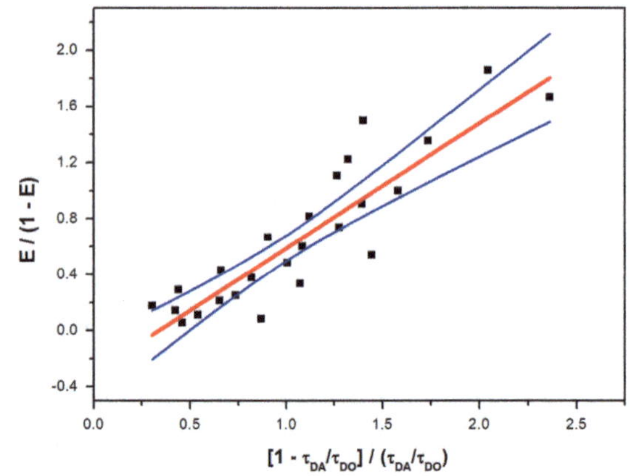

Fig. 17 An example of the calculation of gamma factor from photon count FRET and donor fluorescence lifetimes with a DO donor lifetime of 4 ns

$$\frac{E}{1-E} = \frac{1 - \frac{\tau_{\mathrm{DA}}(A)}{\tau_{\mathrm{DO}}}}{\frac{\tau_{\mathrm{DA}}(A)}{\tau_{\mathrm{DO}}}} \gamma \tag{26}$$

Therefore, these quantities were calculated and were fit to a linear equation. The data points were weighted by the total event count per each E range. The resulting slope of the linear function is the gamma factor.

Other methods of extracting the gamma factor directly out of a smFRET measurement by the utilization of the alternating laser excitation approach (ALEX) [123–125] were developed (*see* **Note 15**).

16. *Models for Fitting FRET Histograms*

smFRET histograms are fitted to a sum of Gaussian function (Eq. 27) in order to find the FRET population characteristics:

$$f(E) = \sum_{i=1}^{n} \frac{a_i}{\sigma_i \sqrt{2\pi}} e^{-\frac{(E-\mu_i)^2}{2\sigma_i^2}}; \forall i \exists \sigma_i = \sqrt{\frac{\mu_i(1-\mu_i)}{n_{\mathrm{D}}^{\mathrm{D}} + n_{\mathrm{D}}^{\mathrm{A}}}} \tag{27}$$

where a_i is the amplitude of the *i*th population; μ_i, the mean FRET value; and σ_i, the shot-noise limited width. For $n = 2$, this model covers a two-population realization. Depending on the interconversion times between the two populations, there will be occasions in which a molecule at state A has entered the excitation volume and somewhere in the middle of the duration will interconvert to state B. If states A and B have different mean FRET efficiency values, then such contributions, if frequent enough, will yield values which are averages of the means of states A and B. These contributions actually affect the FRET histogram at interconversion times up to ten times faster than the time bin. There were qualitative attempts to deal with such dynamics in smFRET by analysis of deviations from shot-noise population widths [69, 81]. It was recently found that FRET population broadening is not always caused by distance dynamics but rather may be due photophysical properties of the dye [69]. Szabo and Gopich developed a model that takes this into account and makes use of the change of smFRET histogram as a function of time bin to elucidate not only FRET populations but also their interconversion rate constants.

An accurate analysis of smFRET histograms is also important as a preparative step before analysis of fluorescence decays per a given E range. Returning to the two-population example (Fig. 8), after elucidating the donor fluorescence decays for the FRET efficiency, E, at the range 0.2–0.3, it may be analyzed with the Haas–Steinberg equation (Eq. 12) using a two-population distance distribution. This is very important in order to determine the equilibrium fraction to be used in the two-population distance distribution.

17. *Multiparameter Fluorescence Detection FRET: Accumulating Burst-Related Fluorescence Lifetimes*

Bursts contain few photon detection events, and their microtimes are usually insufficient for obtaining of a fluorescence decay that can be properly fitted with a multiexponential function. A maximum likelihood estimator was used to obtain average lifetime in the course of the burst time bin [126].

A system was developed [81] that enables the parallel measurement of FRET photon count, fluorescence lifetimes, and fluorescence anisotropies per a burst using a system with four detectors. The emitted signal is split chromatically once to account for donor/acceptor fluorescence and again with a polarizer to account for parallel and perpendicularly polarized photons. Since FRET, fluorescence lifetimes, and fluorescence anisotropies are interdependent and since these measurements are determined in parallel for each burst, the values can be accumulated in FRET-lifetime-anisotropy correlograms. The technique is called multiparameter fluorescence detection [81, 84].

18. *Analysis of Fluorescence Decays at Different FRET Efficiency Ranges*

Another time-resolved single-molecule FRET approach is to build fluorescence decays by accumulating donor/acceptor microtimes for a large number of bursts that fall within a given photon count FRET efficiency range. This yields donor/acceptor fluorescence decays associated with different FRET efficiency bins in the smFRET histogram. Such an approach has been taken to study structural heterogeneity and fluctuations in nucleic acids and denatured protein [85]. In this analysis, microtime histograms include events from a myriad of bursts, and hence, the resulting decay curves are more reliable and can be analyzed using functions containing more than just a single exponent.

In order to understand how distances are distributed in each FRET population, one can also globally analyze all of the donor/acceptor fluorescence decays using Haas–Steinberg equations (Eq. 12). The two FRET populations in the smFRET histogram provide a model-free justification for the use of a two-population distance distribution.
Using this method, one can quite accurately attain a quantitative assessment of how distances are distributed in each FRET population and the segmental mobility in each FRET population.

19. *Using Time-Resolved Fluorescence Data*
Even if one uses pulsed laser excitation at an adequate repetition rate, photobleaching occurs due to the high excitation photon flux during the pulse duration (up to hundreds of picoseconds). It has been shown that in FRET, the acceptor preferentially undergoes photobleaching, yielding the well-known zero peak in the smFRET E histogram [72, 123–125, 127]. In double-labeled protein, the probability of photobleaching of the acceptor is significantly higher than that of the donor. Therefore, bursts with only donor fluorescence and E values of 0 accumulate during data collection. Methods based on synchronized pulsed two-laser excitation (donor and acceptor) were developed for distinguishing between events associated with "true" very low FRET in double-labeled molecules and events with no FRET due to acceptor photobleaching. This way, the bursts resulting from molecules with active (i.e., not bleached) acceptor can be identified. Yet, it is still possible for the non-fluorescent photobleached acceptor probe to maintain spectral overlap and FRET. Using the method of pulsed interleaved excitation (PIE) [72, 127], donor and acceptor detection channels are referenced not only from their corresponding detection channels but also by their microtimes. Photons having microtimes that belong to the microtime range of the donor fluorescence signal are assumed to originate from donor excitation, while those having microtimes in the range of the direct excitation of the acceptor are assumed to originate from acceptor excitation, as demonstrated in Fig. 18.

Accordingly, we can calculate not only n_D^D and n_D^A but also n_A^A, the number of directly excited acceptor fluorescence photons per burst. Then we can find a suitable event count threshold for n_A^A. Only bursts with n_A^A above its minimal threshold indicate a fluorescently active acceptor, i.e., one that yields fluorescence upon excitation. What about bursts lacking a fluorescently active donor? If the samples are dye labeled so that when exciting the donor there is no direct excitation of the

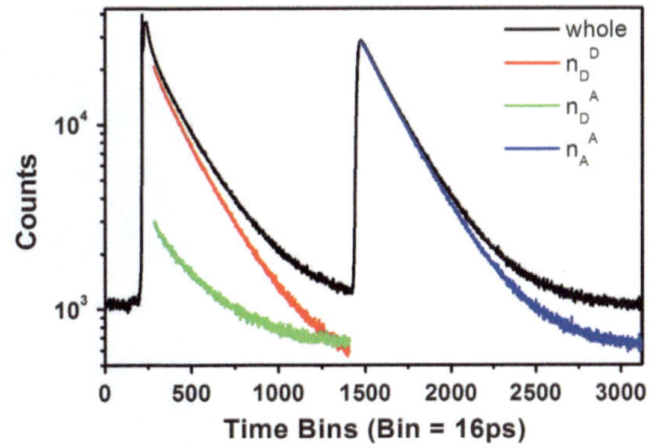

Fig. 18 The microtime histogram from a pulsed interleaved excitation (PIE) measurement (*black*) and its components, donor fluorescence after donor excitation (*red*), acceptor fluorescence after donor excitation (FRET; *green*), and acceptor fluorescence after direct acceptor excitation (*blue*)

acceptor and then if the donor is in a dark state, it will not transfer energy to the acceptor, and thus no photons will be detected in either donor or acceptor detection channels. Nevertheless, in cases with direct acceptor excitation, there may be bursts with no donor fluorescence and with acceptor fluorescence resembling a 100 % FRET situation. A method analogous in nature to PIE but using a different methodology and terminology has been invented and dubbed "alternating laser excitation" (ALEX) [123–125]. In ALEX, besides the FRET efficiency per burst defined in Eq. 22, another measure, the stoichiometry per burst is calculated by using the data from the acceptor excitation Eq. 28.

$$S = \frac{n_D^A + \gamma n_D^D}{n_A^A + n_D^A + \gamma n_D^D} \tag{28}$$

This simply translates to the ratio of photons originating from donor excitation to the total amount of photons. Note that when there is no acceptor, $n_A^A = 0$, which will yield stoichiometry values of 1 and an artificial FRET efficiency value of zero. Whenever there is no donor, $n_D^A = 0$; $n_D^D = 0$, which will yield stoichiometry values of close to zero and an artificial FRET efficiency value of 1. But when both donor and acceptor are present and fluorescently active, S will always have values in the range $0 < S < 1$. Therefore, a two-dimensional histogram of E and S values can be accumulated, plotted, and used to filter out non-FRET events. Nanosecond ALEX is an approach in which alternating excitation is achieved in a time-resolved fashion [85]. Figure 19 illustrates an E–S correlogram.

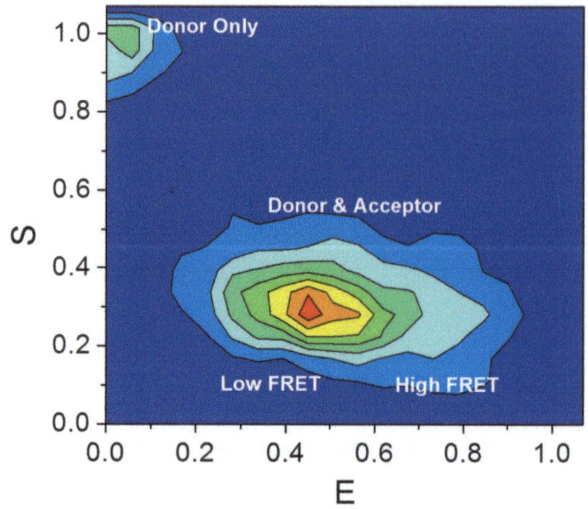

Fig. 19 An example of an *E–S* correlogram using alternating laser excitation (nsALEX) smFRET

Since this method is based on time-resolved data, one can calculate the donor-only fluorescence lifetime from the $E = 0$ and $S = 1$ population, the acceptor intrinsic lifetime per each burst and donor and acceptor fluorescence decays after donor excitations per different E and S values in the E–S correlogram. The slope of $1/S$ vs. E for the donor and acceptor populations has also been proven useful in the determination of the γ-factor as the slope directly related to it [128]. The ALEX approach is also most useful in intermolecular FRET due to the possibility of elucidating both FRET populations of interacting labeled subunits from non-interacting subunits by the use of the $S = 0$ and $S = 1$ populations. Since in FRET, it is the acceptor which preferentially photobleaches, a totally pure double-labeled sample should not show any $S = 0$ acceptor-only population.

References

1. Förster T (1948) Zwischen molekulare Energie wanderung und Fluoreszenz. Ann Phys (Leipzig) 2:55–75
2. Förster T (1959) Transfer mechanisms of electronic excitation. Discussions of the Faraday Soc 27:7–17
3. Förster T (1965) Delocalized excitation and excitation transfer. In: Sinaonglu O (ed) Modern quantum chemistry, Istanbul lectures part III: action of light and organic crystals. Academic, New York, pp 93–137
4. Steinberg IZ (1971) Long-range nonradiative transfer of electronic excitation energy in proteins and polypeptides. Annu Rev Biochem 40:83–114
5. Stryer L, Thomas DD, Meares CF (1982) Diffusion-enhanced fluorescence energy transfer. Annu Rev Biophys Bioeng 11:203–222
6. van der Meer BW, Coker G III, Chen SYS (1994) Resonance energy transfer theory and data. VCH, New York
7. Haas E (2004) Fluorescence resonance energy transfer (FRET) and single molecule fluorescence detection studies of the mechanism of protein folding and unfolding. In: Kiefhaber T (ed) Protein folding handbook. Part I. WILEY-VCH Verlag GmbH &Co, Weinheim, pp 573–633
8. Haas E, Steinberg IZ (1984) Intramolecular dynamics of chain molecules monitored by

fluctuations in efficiency of excitation energy transfer. A theoretical study. Biophys J 46:429–437

9. Nettels D, Gopich IV, Hoffmann A, Schuler B (2007) Ultrafast dynamics of protein collapse from single-molecule photon statistics. Proc Natl Acad Sci USA 104:2655–2660
10. Perrin J (1927) Fluorescence et induction moléculaire par resonance. C R Hebd Seances Acad Sci 184:1097–1100
11. Weber G, Teale FJW (1959) Discuss Faraday Soc 27:134–151
12. Latt SA, Cheung HT, Blout ER (1965) Energy transfer. A system with relatively fixed donor-acceptor separation. J Am Chem Soc 87:995–1003
13. Stryer L, Haugland RP (1967) Energy transfer: a spectroscopic ruler. Proc Natl Acad Sci USA 58:719–726
14. Orevi T, Ben Ishay E, Pirchi M et al (2009) Early closure of a long loop in the refolding of adenylate kinase: a possible key role of non-local interactions in the initial folding steps. J Mol Biol 385:1230–1242
15. Navon A, Ittah V, Landsman P et al (2001) Distributions of intramolecular distances in the reduced and denatured states of bovine pancreatic ribonuclease A. Folding initiation structures in the C-terminal portions of the reduced protein. Biochemistry 40:105–118
16. Jager M, Nir E, Weiss S (2006) Site-specific labeling of proteins for single-molecule FRET by combining chemical and enzymatic modification. Protein Sci 15:640–646
17. Jager M, Michalet X, Weiss S (2005) Protein–protein interactions as a tool for site-specific labeling of proteins. Protein Sci 14:2059–2068
18. Folk JE, Chung SI (1985) Transglutaminases. Methods Enzymol 113:358–375
19. Amir D, Levy DP, Levin Y, Haas E (1986) Selective fluorescent labeling of amino groups of bovine pancreatic trypsin inhibitor by reductive alkylation. Biopolymers 25:1645–1658
20. Amir D, Haas E (1987) Estimation of intramolecular distance distributions in bovine pancreatic trypsin inhibitor by site-specific labeling and nonradiative excitation energy-transfer measurements. Biochemistry 26:2162–2175
21. Deniz AA, Laurence TA, Beligere GS et al (2000) Single-molecule protein folding: diffusion fluorescence resonance energy transfer studies of the denaturation of chymotrypsin inhibitor 2. Proc Natl Acad Sci USA 97:5179–5184
22. Kim J, Lee M (2004) Observation of multi-step conformation switching in beta-amyloid peptide aggregation by fluorescence resonance energy transfer. Biochem Biophys Res Commun 316:393–397
23. Gustiananda M, Liggins JR, Cummins PL, Gready JE (2004) Conformation of prion protein repeat peptides probed by FRET measurements and molecular dynamics simulations. Biophys J 86:2467–2483
24. Kim Y, Ho SO, Gassman NR et al (2008) Efficient site-specific labeling of proteins via cysteines. Bioconjug Chem 19:786–791
25. Garzon-Rodriguez W, Sepulveda-Becerra M, Milton S, Glabe CG (1997) Soluble amyloid Abeta-(1-40) exists as a stable dimer at low concentrations. J Biol Chem 272: 21037–21044
26. Wu P, Brand L (1994) Resonance energy transfer: methods and applications. Anal Biochem 218:1–13
27. Lakowicz JR, Kusba J, Szmacinski H et al (1991) Resolution of end-to-end diffusion coefficients and distance distributions of flexible molecules using fluorescent donor-acceptor and donor-quencher pairs. Biopolymers 31:1363–1378
28. Grupi A, Haas E (2011) Segmental conformational disorder and dynamics in the intrinsically disordered protein alpha-synuclein and its chain length dependence. J Mol Biol 405:1267–1283
29. Lee JC, Langen R, Hummel PA et al (2004) Alpha-synuclein structures from fluorescence energy-transfer kinetics: implications for the role of the protein in Parkinson's disease. Proc Natl Acad Sci USA 101:16466–16471
30. Liu DR, Magliery TJ, Pastrnak M, Schultz PG (1997) Engineering a tRNA and aminoacyl-tRNA synthetase for the site-specific incorporation of unnatural amino acids into proteins in vivo. Proc Natl Acad Sci USA 94: 10092–10097
31. Szilvay GR, Blenner MA, Shur O et al (2009) A FRET-based method for probing the conformational behavior of an intrinsically disordered repeat domain from Bordetella pertussis adenylate cyclase. Biochemistry 48:11273–11282
32. Buckler DR, Haas E, Scheraga HA (1993) C-terminal labeling of ribonuclease A with an extrinsic fluorescent probe by carboxypeptidase Y-catalyzed transpeptidation in the presence of urea. Anal Biochem 209:20–31
33. Masi A, Cicchi R, Carloni A et al (2010) Optical methods in the study of

protein–protein interactions. Adv Exp Med Biol 674:33–42

34. Tcherkasskaya O, Ptitsyn OB, Knutson JR (2000) Nanosecond dynamics of tryptophans in different conformational states of apomyoglobin proteins. Biochemistry 39:1879–1889
35. Rischel C, Thyberg P, Rigler F, Poulsen FM (1996) Time-resolved fluorescence studies of the molten globule state of apomyoglobin. J Mol Biol 257:877–885
36. Santoro SW, Anderson JC, Lakshman V, Schultz PG (2003) An archaebacteria-derived glutamyl-tRNA synthetase and tRNA pair for unnatural amino acid mutagenesis of proteins in Escherichia coli. Nucleic Acids Res 31: 6700–6709
37. Takashi R (1988) A novel actin label: a fluorescent probe at glutamine-41 and its consequences. Biochemistry 27:938–943
38. Fink ML, Chung SI, Folk JE (1980) gamma-Glutamylamine cyclotransferase: specificity toward epsilon-(L-gamma-glutamyl)-L-lysine and related compounds. Proc Natl Acad Sci USA 77:4564–4568
39. Sinev M, Landsmann P, Sineva E et al (2000) Design consideration and probes for fluorescence resonance energy transfer studies. Bioconjug Chem 11:352–362
40. Ortiz JO, Bubis J (2001) Effects of differential sulfhydryl group-specific labeling on the rhodopsin and guanine nucleotide binding activities of transducin. Arch Biochem Biophys 387:233–242
41. Ratner V, Kahana E, Eichler M, Haas E (2002) A general strategy for site-specific double labeling of globular proteins for kinetic FRET studies. Bioconjug Chem 13:1163–1170
42. Lymperopoulos K, Kiel A, Seefeld A et al (2010) Fluorescent probes and delivery methods for single-molecule experiments. Chemphyschem 11:43–53
43. Ben Ishay E, Rahamim G, Orevi T et al (2012) Fast subdomain folding prior to the global refolding transition of E. coli adenylate kinase: a double kinetics study. J Mol Biol 423:613–623
44. Jacob MH, Amir D, Ratner V et al (2005) Predicting reactivities of protein surface cysteines as part of a strategy for selective multiple labeling. Biochemistry 44:13664–13672
45. Valeur B (2002) Molecular fluorescence principles and applications. Weinheim, Wiley-VCH Verlag GmbH
46. Lakowicz JR (1999) Principles of fluorescence spectroscopy, 2nd edn. Kluwer Academic/ Plenum, New York
47. Stryer L, Thomas DD, Carlsen WF (1982) Fluorescence energy transfer measurements of distances in rhodopsin and the purple membrane protein. Methods Enzymol 81:668–678
48. Clegg RM (1992) Fluorescence resonance energy transfer and nucleic acids. Methods Enzymol 211:353–388
49. Clegg RM (1996) Fluorescence resonance energy transfer. In: Wang XF, Herman B (eds) Fluorescence imaging spectroscopy and microscopy. John Wiley & Sons Inc., New York, pp 179–252
50. Edidin M (2003) Fluorescence resonance energy transfer: techniques for measuring molecular conformation and molecular proximity. Curr Protoc Immunol Chapter 18:Unit 18 10
51. Edelhoch H, Brand L, Wilcheck M (1963) Fluorescence studies of tryptophyl peptides. Israel J Chem 1:1–2
52. Edelhoch H, Brand L, Wilchek M (1967) Fluorescence studies with tryptophyl peptides. Biochemistry 6:547–559
53. Cantor CR, Pechukas P (1971) Determination of distance distribution functions by singlet-singlet energy transfer. Proc Natl Acad Sci USA 68:2099–2101
54. Haas E, Katchalsky-Katzir E, Steinberg IZ (1978) Brownian motion of the ends of oligopeptide chains in solution as estimated by energy transfer between the chain ends. Biopolymers 17:11–31
55. Beechem JM, Haas E (1989) Simultaneous determination of intramolecular distance distributions and conformational dynamics by global analysis of energy transfer measurements. Biophys J 55:1225–1236
56. Haran G, Haas E, Szpikowska BK, Mas MT (1992) Domain motions in phosphoglycerate kinase: determination of interdomain distance distributions by site-specific labeling and time-resolved fluorescence energy transfer. Proc Natl Acad Sci USA 89:11764–11768
57. Huang F, Lerner E, Sato S et al (2009) Time-resolved fluorescence resonance energy transfer study shows a compact denatured state of the B domain of protein A. Biochemistry 48:3468–3476
58. Chandrasekhar S (1943) Stochastic problems in physics and astronomy. Rev Mod Phys 15:1–89
59. Steinberg IZ, Haas E, Katchalsky-Katzir E (1983) Long-range non-radiative transfer of electronic excitation energy. In: Cundal RB, Dale RE (eds) Time resolved spectroscopy in biochemistry. Plenum Publ. Corp, New York, pp 411–451
60. Grinvald A, Steinberg IZ (1974) On the analysis of fluorescence decay kinetics by the

method of least-squares. Anal Biochem 59:583–598
61. Beechem JM (1989) A second generation global analysis program for the recovery of complex inhomogeneous fluorescence decay kinetics. Chem Phys Lipids 50:237–251
62. Beechem JM, Brand L (1985) Time-resolved fluorescence of proteins. Annu Rev Biochem 54:43–71
63. Ameloot M, Beechem JM, Brand L (1986) Simultaneous analysis of multiple fluorescence decay curves by Laplace transforms. Deconvolution with reference or excitation profiles. Biophys Chem 23:155–171
64. Lanczos C (1956) Applied analysis. Prentice Hall Inc, Engelwood Cliffs, pp 272–304
65. Bevington PR (1969) Data reduction and error analysis for the physical sciences. McGraw-Hill, New York
66. Press WH, Flannery BP, Teukolski SA, Vetterling WF (1989) Numerical recipes: The Art of Scientific Computing. Cambridge University Press, Cambridge
67. Ames WF (1977) Numerical methods for partial differential equations 2nd ed. Academic, New York
68. Cherny DI, Eperon IC, Bagshaw CR (2009) Probing complexes with single fluorophores: factors contributing to dispersion of FRET in DNA/RNA duplexes. Eur Biophys J 38:395–405
69. Chung HS, Louis JM, Eaton WA (2010) Distinguishing between protein dynamics and dye photophysics in single-molecule FRET experiments. Biophys J 98:696–706
70. Kudryavtsev V, Sikor M, Kalinin S et al (2012) Combining MFD and PIE for accurate single-pair Förster resonance energy transfer measurements. Chemphyschem 13:1060–1078
71. Orte A, Clarke RW, Klenerman D (2008) Fluorescence coincidence spectroscopy for single-molecule fluorescence resonance energy-transfer measurements. Anal Chem 80:8389–8397
72. Ruttinger S, Macdonald R, Kramer B et al (2006) Accurate single-pair Förster resonant energy transfer through combination of pulsed interleaved excitation, time correlated single-photon counting, and fluorescence correlation spectroscopy. J Biomed Opt 11:024012
73. Nir E, Michalet X, Hamadani KM et al (2006) Shot-noise limited single-molecule FRET histograms: comparison between theory and experiments. J Phys Chem B 110: 22103–22124
74. Hoefling M, Lima N, Haenni D et al (2011) Structural heterogeneity and quantitative FRET efficiency distributions of polyprolines through a hybrid atomistic simulation and Monte Carlo approach. PLoS One 6:e19791
75. Choi J, Kim S, Tachikawa T et al (2011) pH-induced intramolecular folding dynamics of i-motif DNA. J Am Chem Soc 133: 16146–16153
76. Mukhopadhyay S, Krishnan R, Lemke EA et al (2007) A natively unfolded yeast prion monomer adopts an ensemble of collapsed and rapidly fluctuating structures. Proc Natl Acad Sci USA 104:2649–2654
77. Nettels D, Hoffmann A, Schuler B (2008) Unfolded protein and peptide dynamics investigated with single-molecule FRET and correlation spectroscopy from picoseconds to seconds. J Phys Chem B 112:6137–6146
78. Eggeling C, Berger S, Brand L, Fries JR et al (2001) Data registration and selective single-molecule analysis using multi-parameter fluorescence detection. J Biotechnol 86:163–180
79. Eggeling C, Widengren J, Brand L et al (2006) Analysis of photobleaching in single-molecule multicolor excitation and förster resonance energy transfer measurements. J Phys Chem A 110:2979–2995
80. Felekyan S, Kalinin S, Sanabria H et al (2012) Filtered FCS: species auto- and cross-correlation functions highlight binding and dynamics in biomolecules. Chemphyschem 13:1036–1053
81. Kalinin S, Sisamakis E, Magennis SW et al (2010) On the origin of broadening of single-molecule FRET efficiency distributions beyond shot noise limits. J Phys Chem B 114:6197–6206
82. Neubauer H, Gaiko N, Berger S et al (2007) Orientational and dynamical heterogeneity of rhodamine 6G terminally attached to a DNA helix revealed by NMR and single-molecule fluorescence spectroscopy. J Am Chem Soc 129:12746–12755
83. Sabir T, Schroder GF, Toulmin A et al (2011) Global structure of forked DNA in solution revealed by high-resolution single-molecule FRET. J Am Chem Soc 133:1188–1191
84. Weidtkamp-Peters S, Felekyan S, Bleckmann A et al (2009) Multiparameter fluorescence image spectroscopy to study molecular interactions. Photochem Photobiol Sci 8:470–480
85. Laurence TA, Kong X, Jager M, Weiss S (2005) Probing structural heterogeneities and fluctuations of nucleic acids and denatured proteins. Proc Natl Acad Sci USA 102:17348–17353
86. Allen MW, Urbauer RJ, Zaidi A et al (2004) Fluorescence labeling, purification, and

immobilization of a double cysteine mutant calmodulin fusion protein for single-molecule experiments. Anal Biochem 325:273–284

87. Foldes-Papp Z, Baumann G, Demel U, Tilz GP (2004) Counting and behavior of an individual fluorescent molecule without hydrodynamic flow, immobilization, or photon count statistics. Curr Pharm Biotechnol 5:163–172
88. Lamichhane R, Solem A, Black W, Rueda D (2010) Single-molecule FRET of protein–nucleic acid and protein–protein complexes: surface passivation and immobilization. Methods 52:192–200
89. Liu B, Mazouchi A, Gradinaru CC (2010) Trapping single molecules in liposomes: surface interactions and freeze-thaw effects. J Phys Chem B 114:15191–15198
90. Okumus B, Wilson TJ, Lilley DM, Ha T (2004) Vesicle encapsulation studies reveal that single molecule ribozyme heterogeneities are intrinsic. Biophys J 87:2798–2806
91. Gopich IV, Szabo A (2012) Theory of the energy transfer efficiency and fluorescence lifetime distribution in single-molecule FRET. Proc Natl Acad Sci USA 109:7747–7752
92. Hoffmann A, Nettels D, Clark J et al (2011) Quantifying heterogeneity and conformational dynamics from single molecule FRET of diffusing molecules: recurrence analysis of single particles (RASP). Phys Chem Chem Phys 13:1857–1871
93. Bronstein I, Israel Y, Kepten E et al (2009) Transient anomalous diffusion of telomeres in the nucleus of mammalian cells. Phys Rev Lett 103:018102
94. Madl J, Weghuber J, Fritsch R et al (2010) Resting state Orai1 diffuses as homotetramer in the plasma membrane of live mammalian cells. J Biol Chem 285:41135–41142
95. Rolfe DJ, McLachlan CI, Hirsch M et al (2011) Automated multidimensional single molecule fluorescence microscopy feature detection and tracking. Eur Biophys J 40:1167–1186
96. Sakon JJ, Weninger KR (2010) Detecting the conformation of individual proteins in live cells. Nat Methods 7:203–205
97. Colyer RA, Scalia G, Rech I et al (2010) High-throughput FCS using an LCOS spatial light modulator and an 8 x 1 SPAD array. Biomed Opt Express 1:1408–1431
98. Michalet X, Colyer RA, Scalia G et al (2010) High-throughput single-molecule fluorescence spectroscopy using parallel detection. Proc Soc Photo Opt Instrum Eng 7608
99. Mansoor SE, Farrens DL (2004) High-throughput protein structural analysis using site-directed fluorescence labeling and the bimane derivative (2-pyridyl)dithiobimane. Biochemistry 43:9426–9438
100. McNulty BC, Tripathy A, Young GB et al (2006) Temperature-induced reversible conformational change in the first 100 residues of alpha-synuclein. Protein Sci 15:602–608
101. Riddles PW, Blakeley RL, Zerner B (1979) Ellman's reagent: 5,5′-dithiobis(2-nitrobenzoic acid)–a reexamination. Anal Biochem 94:75–81
102. Eisinger J, Dale RE (1974) Letter: interpretation of intramolecular energy transfer experiments. J Mol Biol 84:643–647
103. Dale RE, Eisinger J (1976) Intramolecular energy transfer and molecular conformation. Proc Natl Acad Sci USA 73:271–273
104. Dale RE, Eisinger J, Blumberg WE (1979) The orientational freedom of molecular probes. The orientation factor in intramolecular energy transfer. Biophys J 26:161–193
105. Selvin PR, Hearst JE (1994) Luminescence energy transfer using a terbium chelate: improvements on fluorescence energy transfer. Proc Natl Acad Sci USA 91:10024–10028
106. Mersol JV, Wang H, Gafni A, Steel DG (1992) Consideration of dipole orientation angles yields accurate rate equations for energy transfer in the rapid diffusion limit. Biophys J 61:1647–1655
107. Haas E, Katchalski-Katzir E, Steinberg IZ (1978) Effect of the orientation of donor and acceptor on the probability of energy transfer involving electronic transitions of mixed polarization. Biochemistry 17:5064–5070
108. Albrecht AC (1960) Forbidden characterizing allowed electronic transitions. J Chem Phys 33:156
109. Eisinger J, Feuer B, Lamola AA (1969) Intramolecular singlet excitation transfer. Applications to polypeptides. Biochemistry 8: 3908–3915
110. Cortijo M, Steinberg IZ, Shaltiel S (1971) Fluorescence of glycogen phosphorylase b. Structural transitions and energy transfer. J Biol Chem 246:933–938
111. Hawe A, Poole R, Jiskoot W (2010) Misconceptions over Förster resonance energy transfer between proteins and ANS/bis-ANS: Direct excitation dominates dye fluorescence. Anal Biochem 401:99–106
112. Beechem JM, Brand L (1986) Global analysis of fluorescence decay: applications to some unusual experimental and theoretical studies. Photochem Photobiol 44:323–329
113. Beechem JM (1992) Global analysis of biochemical and biophysical data. Methods Enzymol 210:37–54

114. Jha SK, Dhar D, Krishnamoorthy G, Udgaonkar JB (2009) Continuous dissolution of structure during the unfolding of a small protein. Proc Natl Acad Sci USA 106:11113–11118
115. Brochon JC (1994) Maximum entropy method of data analysis in time-resolved spectroscopy. Methods Enzymol 240:262–311
116. Schroder GF, Grubmüller H (2003) Maximum likelihood trajectories from single molecule fluorescence resonance energy transfer experiments. J Chem Phys 119:9920–9924
117. Stevens JA, Link JJ, Zang C et al (2011) Ultrafast dynamics of nonequilibrium resonance energy transfer and probing globular protein flexibility of myoglobin. J Phys Chem A 116:2610–2619
118. Visser NV, Borst JW, Hink MA et al (2005) Direct observation of resonance tryptophan-to-chromophore energy transfer in visible fluorescent proteins. Biophys Chem 116:207–212
119. Cottet M, Faklaris O, Maurel D et al (2012) BRET and Time-resolved FRET strategy to study GPCR oligomerization: from cell lines toward native tissues. Front Endocrinol (Lausanne) 3:92
120. Gopich IV, Szabo A (2007) Single-molecule FRET with diffusion and conformational dynamics. J Phys Chem B 111:12925–12932
121. Chung HS, Louis JM, Eaton WA (2009) Experimental determination of upper bound for transition path times in protein folding from single-molecule photon-by-photon trajectories. Proc Natl Acad Sci USA 106:11837–11844
122. Lee NK, Kapanidis AN, Wang Y, Michalet X, Mukhopadhyay J, Ebright RH, Weiss S (2005) Accurate FRET measurements within single diffusing biomolecules using alternating-laser excitation. Biophys J 88:2939–2953
123. Kapanidis AN, Laurence TA, Lee NK et al (2005) Alternating-laser excitation of single molecules. Acc Chem Res 38:523–533
124. Hirschfeld V, Hubner CG (2010) A sensitive and versatile laser scanning confocal optical microscope for single-molecule fluorescence at 77 K. Rev Sci Instrum 81:113705
125. Sobhy MA, Elshenawy MM, Takahashi M et al (2011) Versatile single-molecule multicolor excitation and detection fluorescence setup for studying biomolecular dynamics. Rev Sci Instrum 82:113702
126. Edel JB, Eid JS, Meller A (2007) Accurate single molecule FRET efficiency determination for surface immobilized DNA using maximum likelihood calculated lifetimes. J Phys Chem B 111:2986–2990
127. Fore S, Yuen Y, Hesselink L, Huser T (2007) Pulsed-interleaved excitation FRET measurements on single duplex DNA molecules inside C-shaped nanoapertures. Nano Lett 7: 1749–1756
128. Lee BC, Zuckermann RN, Dill KA (2005) Folding a nonbiological polymer into a compact multihelical structure. J Am Chem Soc 127:10999–11009

Chapter 8

MD + QM Correlations with Tryptophan Fluorescence Spectral Shifts and Lifetimes

Patrik R. Callis and Jose R. Tusell

Abstract

Principles behind quenching of tryptophan (Trp) fluorescence are updated and extended in light of recent 100-ns and 1-μs molecular dynamics (MD) trajectories augmented with quantum mechanical (QM) calculations that consider electrostatic contributions to wavelength shifts and quenching. Four studies are summarized, including (1) new insight into the single exponential decay of NATA, (2) a study revealing how unsuspected rotamer transitions affect quenching of Trp when used as a probe of protein folding, (3) advances in understanding the origin of nonexponential decay from 100-ns simulations on 19 Trps in 16 proteins, and (4) the correlation of wavelength with lifetime for decay-associated spectra (DAS). Each study strongly reinforces the concept that—for Trp—electron transfer-based quenching is controlled much more by environment electrostatic factors affecting the charge transfer (CT) state energy than by distance dependence of electronic coupling. In each case, water plays a large role in unexpected ways.

Key words Tryptophan, Fluorescence, Quantum yield, Lifetime, Protein, Molecular dynamics, Quantum mechanics, QM-MM, MD + QM, Electrostatics

1 Introduction and Overview

The use of fluorescence-based methods to study cellular biology at the molecular level of detail has become a rapidly growing billion dollar industry. Most of these methods depend upon changes in the intensity and/or wavelength of light emitted from a probe molecule when a protein or nucleic acid undergoes a functional process.

The aims of this chapter—which involve a unique ability to compute fluorescence intensity and wavelength from first principles—are strongly linked by a common theme: *the extraordinary ruggedness of the spatial and temporal electrostatic landscape within proteins is mirrored in the ruggedness of the fluorescence intensity and wavelength landscape of environment-sensitive fluorophores.* This is significant because the diversity and strength of electric fields in these biopolymers is increasingly recognized as key to their function as enzymes [1–4]. The ability of enzymes to

Yves Engelborghs and Antonie J.W.G. Visser (eds.), *Fluorescence Spectroscopy and Microscopy: Methods and Protocols*, Methods in Molecular Biology, vol. 1076, DOI 10.1007/978-1-62703-649-8_8,

undergo rapid changes between defined conformations is almost certainly an important part of their overall catalytic activity, because fluctuations to specific reactive conformations are likely at the heart of enzyme specificity and reactivity [5].

Fluorescence generates enormous amount of data that correlates with protein motions. Clearly the ability to compute fluorescence behavior from molecular dynamics (MD) simulations not only points the way towards understanding of the underlying processes but also provides confidence for the same types of computations that can predict enzymatic efficiency.

Our computations over several years have clearly shown that the electrostatic landscape—including solvent—is absolutely the key to predicting fluorescence behavior of environment-sensitive probes [6–25]. Thus, the fluorescence of Trp (and a few other fluorophores) is well poised to help reveal the emerging view of protein dynamic disorder that involves transitions between multiple minima in a rough energy landscape, leading to *nonexponential* correlation functions [26]. Continued effort in these areas is encouraged by the emerging view that the catalytic power of enzymes is largely due to a specifically oriented, *preorganized* electrostatic environment, whose energy may come from reduction in folding energy [27]. We have noted on several occasions that an ordered electrostatic environment coupled with large fluctuations is precisely what determines whether fluorescence will be strong or weak, and whether its average wavelength will be short or long. This meshes perfectly with the exciting recent observation by Marcus and others that the nonexponential autocorrelation time behavior of fluctuations of electrostatic interaction energy on time scales as long as milliseconds to seconds at a local site in enzymes is in agreement with measurable dielectric dispersion of enzymes and is in common with single molecule experimental observations of catalysis rate, fluorescence lifetime, and spectral diffusion—all occurring over the time scale of biological importance (milliseconds to seconds) [28, 29].

Such a time scale is mostly unreachable by real-time computational trajectory simulations today, but the millisecond barrier has been broken in one laboratory [30]. Simulations are increasingly able to reveal fine details of biological system function by generating microscopic descriptions of behavior that are experimentally testable [5]. Vendruscolo and Dobson [31] have provided a recent analysis based on Moore's Law that suggests that within less than 10 years MD simulations will routinely reach the time scale of biological importance.

The ubiquitous nonexponential decay of Trp fluorescence is a direct consequence of its sensitivity to the electrostatic environment. Though often lamented, nonexponential decay may be considered a great asset by virtue of providing a unique window on protein and solvent dynamics that are relevant to the basic function of proteins and is routinely exploited with the aim of gaining insight into solvation dynamics near proteins [32]. The main theme of this chapter is, however, to provide a microscopic view of underlying mechanisms for nonexponential fluorescence decay from Trp in proteins arising from heterogeneity in excited state population decay due to quenching by electron transfer. This is based on much longer (100 ns to 1 μs) MD + QM simulations than have typically been reported in our previous studies. With 2–3 orders of magnitude longer simulations, discrete subpopulations become much more evident, and the variable nature of these populations begins to be seen. This has also led to a new emphasis on the interesting intermediate case in which discrete populations interconvert on the same time scale as the excited state lifetime. In such a case, decay curves may be fit well to two or three exponentials, but the decay constants cannot be attributed to directly as belonging to a physical species.

1.1 Outline

Subheading 2 is a summary of the underlying principles.

In Subheading 3, we present the enigmatic case of what is *not* nonexponential decay: the pure mono-exponential decay of the fluorescence from *N*-acetyltryptophanamide (NATA), the most commonly used model for Trp behavior in proteins. A major frustration for several years has been our inability to fathom why NATA exhibits almost purely exponential decay. We introduce a new concept in which quantum yield and lifetime reduction comes about by rare, short-lived, specially solvated forms that have little to do with rotamer state [33, 34].

In Subheading 4, we describe how 1-μs simulations of Trp fluorescence quenching during the folding of the extensively studied, small, fast-folding villin headpiece, lend insight into the efficacy of local Trp fluorescence quenching for reporting the global folding rate [25].

Subheading 5 takes advantage of the recent advances in the speed of MD simulations that allow us to examine 100-ns simulations of fluorescence dynamics from 19 Trps in 16 selected proteins. On this time scale, the proteins begin to exhibit a range of tendencies for transitions between discrete states that affect fluorescence lifetimes and quantum yields [33, 34].

Subheading 6 presents an MD-QM study of cyclic peptides that reveals a universal understanding of how heterogeneity in Trp fluorescence lifetimes almost always gives rise to what we call pseudo-time-dependent fluorescence Stokes shifts (pseudo-TDFSS), a phenomenon arising from what has been a mystifying strong correlation of the wavelength and lifetimes of DAS [17, 24].

Subheading 7 is a synopsis of methods relevant to this chapter.

The long simulations underlying all these topics will help understand how to quantitatively compute fluorescence lifetimes and intensities for proteins that display the type of fluctuations between conformations that are also important in enzyme catalytic activity. A common theme is the importance of water and how Trp fluorescence quenching is almost always controlled by the electrostatically controlled energy gap between the charge transfer (CT) and fluorescing state (^1L_a).

2 Some General Principles Behind Trp Fluorescence Quenching and Spectra

The change in electronic state of environment-sensitive fluorophores accompanying emission of a photon always involves a large instantaneous change in electronic charge distribution. Within a protein or solvent, there are strong Coulombic interactions between the fluorophore electrons and all the atoms of the protein, which means that the energy of the emitted photon will depend sensitively upon the changed positions of the electrons in the fluorophore relative to those of the protein. Several quantum mechanical studies unanimously predict that the ground → ^1L_a state involves a large change in permanent dipole moment arising from electron density shifting from the pyrrole ring to the benzene ring of the indole moiety [8]. QM-MD simulations found good correlation of predicted and experimental Trp fluorescence wavelengths, assuming that ^1L_a was always the emitting state [10]. Subsequent fluorescence anisotropy experiments demonstrated that even for the extremely blue-shifted, structured fluorescence from Trp 48 of azurin, the emission is 100 % from ^1L_a [35]. Only Trp 72 of the M97V mutant of transhydrogenase has been reported to emit partly from the ^1L_b state [36].

Much the same may be said about the energy difference between the ^1L_a state and the CT state. Quenching rate and wavelength shift are caused by the *same* phenomenon acting on *different* electronic states: wavelength is determined by the position of charges acting on the electron density difference between the ground and fluorescing state; quenching is determined by the position of charges acting on the electron density difference between the fluorescing state and a CT state. The most common CT states active in quantum yield reduction of Trp in proteins involve electron transfer from the indole ring to one of two nearby backbone amides. While this is the focus of this chapter, quenching by a nearby histidine cation is featured in Subheading 4.

No two single-Trp proteins will have exactly the same fluorescence λ_{max}, quantum yield (Φ_f), and decay profile. That is because the average surrounding atomic configuration for a Trp in a given protein will be unique to that protein. We have constructed a relatively simple scheme that is reasonably successful for understanding and predicting the main properties of Trp fluorescence quantum yields and lifetimes [9, 11–16, 18–25]. The fluorescence quantum yield is estimated with

$$\Phi_f = \langle k_r / (k_r + k_{nr} + k_{et,i(t)}) \rangle_{i,t} \tag{1}$$

where k_r is the radiative rate constant, k_{nr} is the nonradiative rate constant in the absence of electron transfer. The electron transfer rate constant, $k_{et,i}(t)$, includes time dependence and the possibility of heterogeneity with the subscript i indicating "species" i.

Typically $k_r = 4 \times 10^7\ s^{-1}$ and $k_{nr} = 9 \times 10^7\ s^{-1}$for 3-methylindole (3MI) in aqueous solution [37]. The corresponding fluorescence decay times ($\tau_{f,i}$) are estimated as

$$\tau_{f,i} = (k_r + k_{nr} + k_{et,i}(t))^{-1} \tag{2}$$

All variability in τ_{fi} and Φ_{fi} comes from $k_{et,i}(t)$. k_r and k_{nr} are considered effectively constant in comparison to $k_{et,i}(t)$, which is found to fluctuate rapidly with high amplitude, and is the primary source of quantum yield variability and lifetime heterogeneity. Although proton transfer quenching may be a possibility, there has been little convincing evidence for this outside of free Trp in solution and has not been considered in our predictions to date.

Most generally $k_{et}(t)$ is computed from the Fermi golden rule expression (*see* Subheading 7), which is proportional to the product of the square of the electronic coupling matrix element, V_{et}^2, and the Franck–Condon-weighted density of states (FCDS). The FCDS (i.e., number of states in resonance with the fluorescing state), $\rho(\Delta E_{00})$, a function of the separation of CT and ^1L_a state zero point energies, ΔE_{00}, (which will be often referred to here as the "energy gap"). Details of how these are calculated are addressed further in Subheading 7 and elsewhere [23–25].

We note that the golden rule, which is derived from first-order time-dependent quantum mechanics under certain approximations that are usually reasonable, underlies a vast number of phenomena besides electron transfer, e.g., absorption of light by atoms and molecules, and Förster resonant energy transfer (FRET).

A primary thesis of this chapter is that quantum yield and lifetime for Trp in proteins are almost always dictated by the CT-^1L_a energy gap. This is so for three reasons: (1) only the two nearest backbone amides are significant quenchers, (2) the amide CT states are almost always at considerably higher energy than the ^1L_a state, and (3) V_{et} is so large for the two local amides that the quenching is very high during brief periods of resonance. In other words, the rate limiting step in quenching is waiting for a fluctuation that creates resonance, not the speed of the V_{et} controlled rate of quenching during the times of resonance. The reason quenching is dominated by the two nearest amides is due to the ~1 eV greater electrostatic stabilization due to the attraction of the negatively charged amide and positively charged indole ring in the CT state compared to more distant amide CT states.

The energy gap is essentially the difference in electric potential energy for an electron in the excited ring and one in the empty orbital localized on the C of the amide. This is another way of saying that the fluorescing state and the CT state must have the same energy or electron transfer cannot happen.

Alternatively, this may be stated in terms of the electric field direction. If an electron experiences a field that forces it from the ring to the amide, electron transfer will require less energy. For example, a positive charge near the amide of Trp and/or a negative charge near the Trp ring will create a smaller energy gap than will the opposite case (a negative charge near the amide and/or a positive charge near the ring). Therefore, hydrogen bonding to the oxygen of the Trp amide nearly always reduces the energy gap (whether by participation in a helix or beta sheet or by water accessibility) [14, 20, 21, 23]. Although hydrogen bonds are particularly effective, this principle applies quite generally to all surrounding atoms with a significant partial charge. In particular, oxygen atoms of carbonyl, carboxylate groups, and water, when close to the Trp ring, stabilize the CT state.

Given these electrostatic guidelines, there can be little doubt that different rotameric configurations of Trp could plausibly have different energy gaps and therefore different quenching rates by the local amide, but we will see that rotamer variation is not the only mechanism to achieve a low-energy gap.

2.1 Interaction

Ab initio calculations have provided quite large values for the electronic coupling, V_{et}, for electron transfer between the Trp ring and amide [18]. When $\chi_1 \approx -60°$, V_{et} is always close to 900 cm^{-1} and never reaches zero. In contrast, a χ_1 of either 60° or ±180° yields a $|V_{et}|$ that fluctuates from 0 to 400 cm^{-1}. Either of these couplings are so strong that at any distance possible for the two amides, the quenching rate would be ~10^{12} s^{-1} for an activationless (resonant) process. This is obviously not the case because it would lead to a quantum yield ~1×10^{-4}. If the energy gap did not matter, no Trp would fluoresce in proteins. Figure 1 shows the average V_{et} for 19 different single-Trp proteins in 16 different protein environments. The magnitude of V_{et} is determined for the most part by the χ_1 dihedral angle. It is seen that there is no correlation of quantum yield with V_{et}.

3 Why NATA Has Single Exponential Decay: It Is the Water

A main theme of this chapter is to provide a view of underlying mechanisms for nonexponential fluorescence decay from Trp in proteins, based on MD + QM simulations. We begin with a spectacular case of what is *pure mono-exponential decay*: the fluorescence decay from *N*-acetyltryptophanamide (NATA). A major frustration

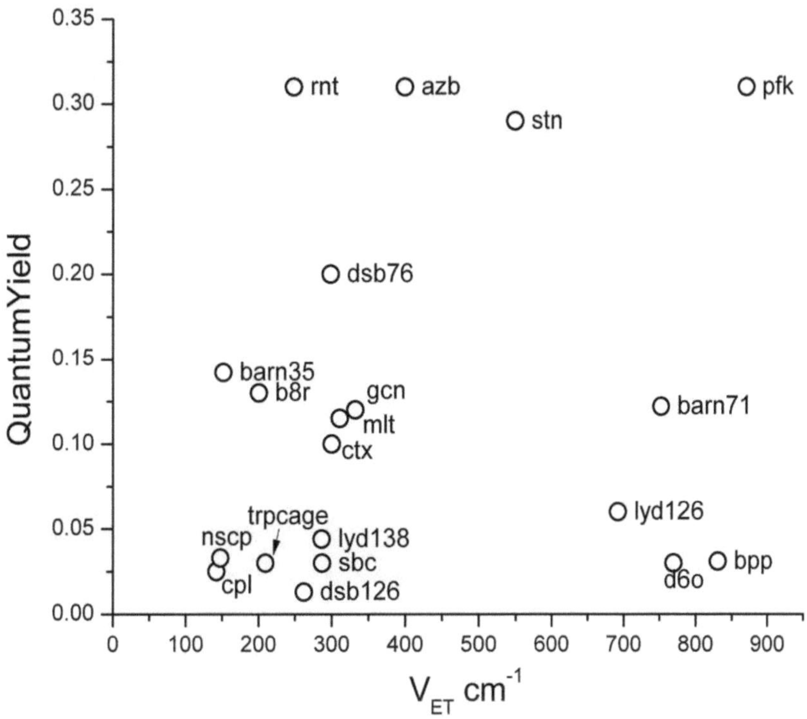

Fig. 1 Plot of observed quantum yield vs. average root mean square of electron transfer coupling element, illustrating the lack of correlation. This is because the electrostatically modulated CT-^1L_a energy gap is usually the controlling factor

for several years has been an inability to understand why NATA exhibits almost purely exponential decay using MD + QM methods. Recent near microsecond simulations have confirmed earlier observations that the short lifetime and low quantum yield relative to the unquenched 3MI is caused by rare solvation events [25, 34].

Figure 2 shows QM-computed amide CT (black) and ^1L_a (red) transition energies during a 600-ns MD simulation. The main feature to notice from this figure is the rare outlying excursions of the CT energy to exceptionally low values. These appear to lie outside the normal Gaussian distribution. The top panel of Fig. 3 shows how the all important energy gap (the difference between the two curves of Fig. 2) fluctuates in time during this 600-ns simulation. The second panel shows how the excited state lifetime varies with time, as computed by the Fermi rule, in which the energy gap is the controlling factor. The most extreme downward fluctuations of the energy gap due to large stabilization by water are seen to correspond exactly to high quenching rates that compete with the unquenched rate (~1.2×10^8 s^{-1}). This is illustrated graphically in Fig. 4. Similar simulations in which coordinates are saved at 2 fs intervals show that these excursions typically form and

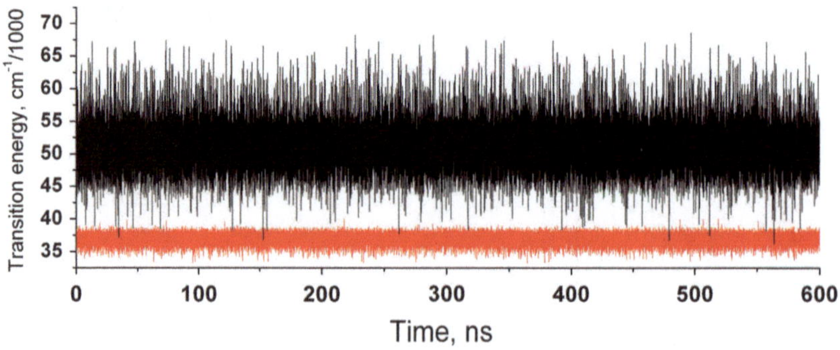

Fig. 2 600-ns MD + QM trajectories of ${}^{1}L_a$ (*red*) and lowest CT state (*black*) transition energies of NATA in water showing extreme downward fluctuations (small energy gap) of the CT state caused by super stabilization by water that are primarily responsible for the lower fluorescence quantum yield and lifetime of NATA relative to 3-methylindole

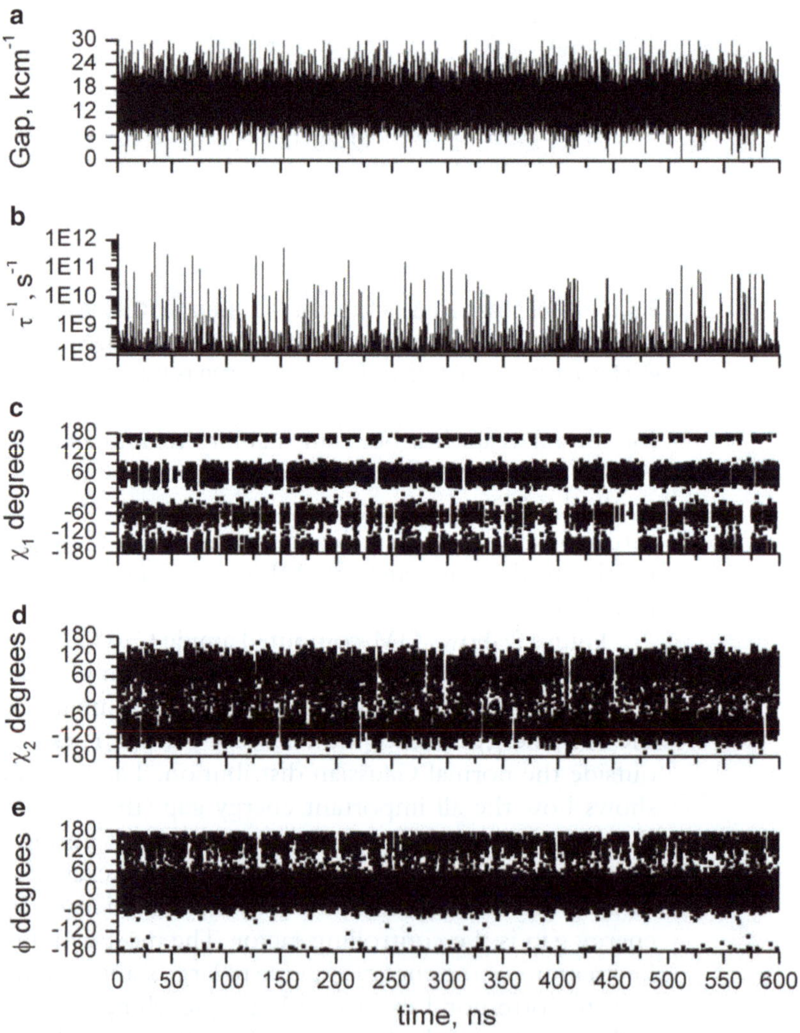

Fig. 3 600-ns MD + QM trajectories for NATA in water of properties related to the single exponential decay. (**a**) CT-${}^{1}L_a$ energy gap in thousands of cm^{-1}. (**b**) Instantaneous fluorescence decay rate, showing the rare extreme decay rates from quenching by electron transfer to the amide CT state. (**c–e**) Trajectories of the χ_1, χ_2, and Ψ dihedral angles, respectively, showing numerous rotamer state transitions. The high quenching rates seen in panel **b** do not correlate with rotamer state

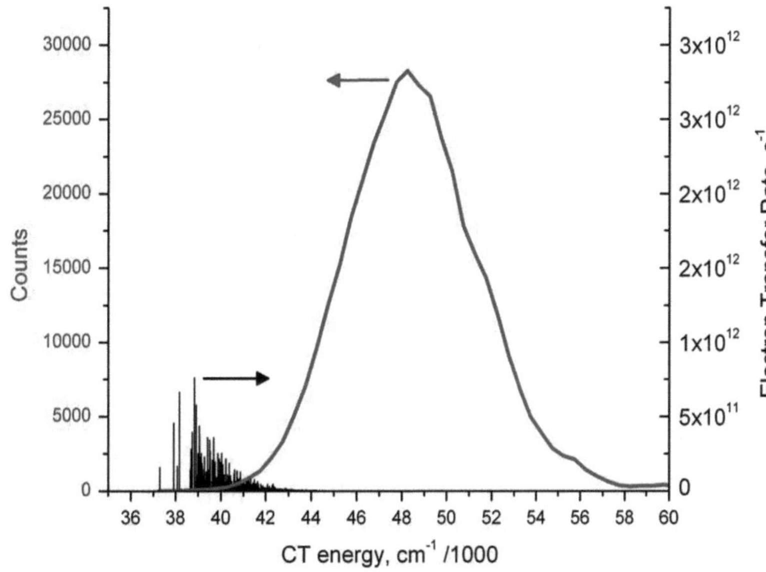

Fig. 4 Distribution of *instantaneous* CT state energies (in cm^{-1}) calculated from QM, and the corresponding *instantaneous* electron transfer rate constants (*vertical black lines*) derived from a 600-ns MD + QM trajectory. Most of the electron transfer happens during fluctuations in solvation stabilization that shift the instantaneous CT energy to low energies close to that of ^1L_a. This reveals the basic explanation for the single exponential decay fluorescence decay for NATA

dissolve in ~50 fs, i.e., the inertial response time of water molecules in liquid water [38]. The lower three panels of Fig. 3 plot the three dihedral angles of importance showing that all rotamers are significantly populated. Careful examination shows virtually no correlation with the extreme quenching events. The C-terminal amide is overwhelmingly the site of electron transfer, regardless of rotamer.

These simple observations suggest a picture that more naturally explains the observed single exponential decay than the traditional rotamer-based correlation with distance of the amide from the ring. To better understand how this new picture robustly predicts single exponential decay, we summarize the pertinent aspects of a generalized solution of the kinetic equations describing fluorescence decay [25, 39–41].

3.1 Theory

In general, the nature of the heterogeneity predicted from the simulations has a profound effect on how quantum yield and lifetime is to be calculated. Two "species" are identified, one designated "F" in which the energies of the CT and ^1L_a states are too *f*ar apart for effective quenching and one designated "C" in which the energies of the CT and ^1L_a states are *c*lose, for which fluorescence is strongly quenched by electron transfer to an amide. There are two obvious extremes. When interconversion between the rapidly quenched C species and slowly quenched F species is much slower than the fluorescence lifetime of either species, the quantum yield is given simply by the average of the individual quantum yields

$$\Phi_{\text{slow exchange}} = f_C \Phi_C + f_F \Phi_F \tag{3}$$

where f_C and f_F are the fractions of time spent in the close and far conformations. At the other extreme, when the two species exchange at rates much faster than the fluorescence decay rate, the quantum yield becomes

$$\Phi_{\text{fast exchange}} = \frac{k_r}{k_r + k_{nr} + f_C k_{ET,C} + f_F k_{ET,F}} \tag{4}$$

Even when $f_C << f_F$, if $k_{et,C}$ is large enough, the quantum yield can become very small according to this equation.

Our simulations on NATA in water show a fast mean exchange rate between the two populations because the stabilized CT states persist typically for only ~100 fs. This means that the environment of the excited molecule will be a hybrid of the two pure species and that the equilibrium between the two species will be perturbed because the rapidly depleted close species will be replenished by the fast exchange process. This effect, which is not given by Eqs. 3 and 4, represents a condition under which the two limits just mentioned will be in considerable error. In this intermediate region, we expect to see two decay constants that are not the pure lifetimes of each of the species, a consequence of the environment changing during the excited state lifetime. This can be seen from a simple compartmental analysis consisting of the F and C species, connected by rate constants shown in Fig. 5. In this scheme, it is not necessary to include the ground states, because the excitation process is assumed to have already occurred at time zero. It is implicit that the disappearance of the excited species means return to the ground state, whose population has no effect on the excited

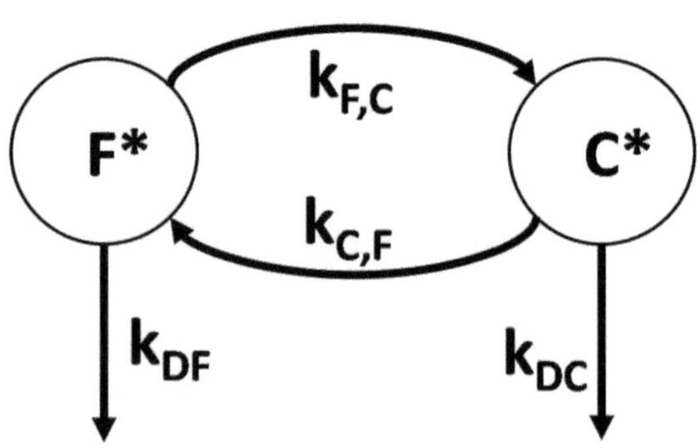

Fig. 5 Simple compartmental scheme consisting of the slow decaying, far-apart energy gap (F) species and fast decaying, close-together energy gap (C) species, which exchange by rate constants k_{CF} (C → F, slow), k_{FC} (F→C, fast), k_{DC} (fast), and k_{DF} (slow)

state dynamics in our simplified scheme, wherein the exchange rates between the two species is independent of electronic state.

This type of analysis has been used in numerous applications to multiple fluorescent species with somewhat different focus and emphasis [40, 42, 43]. The decay rates out of the two excited species are

$$\begin{aligned} k_{\mathrm{DF}} &= k_{\mathrm{r}} + k_{\mathrm{nr}} + k_{\mathrm{ET,F}} \\ k_{\mathrm{DC}} &= k_{\mathrm{r}} + k_{\mathrm{nr}} + k_{\mathrm{ET,C}} \end{aligned} \tag{5}$$

where k_{r}, k_{nr}, $k_{\mathrm{et,F}}$, and $k_{\mathrm{et,C}}$ are the radiative, nonradiative, and electron transfer rate to amide in the far and close states. The coupled rate equations generated from the scheme in Fig. 5 are shown in Eq. 6.

$$\begin{bmatrix} -\dfrac{\mathrm{dC}}{\mathrm{d}t} \\ -\dfrac{\mathrm{dF}}{\mathrm{d}t} \end{bmatrix} = \begin{bmatrix} (k_{\mathrm{C,F}} + k_{\mathrm{DC}}) & -k_{\mathrm{F,C}} \\ -k_{\mathrm{C,F}} & (k_{\mathrm{F,C}} + k_{\mathrm{DF}}) \end{bmatrix} \begin{bmatrix} \mathrm{C} \\ \mathrm{F} \end{bmatrix} \tag{6}$$

These coupled equations are solved by standard procedures [39–41] after finding the eigenvalues of the matrix. The reciprocals of the two eigenvalues are the two exponential decay times describing the relaxation and decay of the coupled excited species and are given by

$$\lambda_{\pm} = \frac{(k_{\mathrm{F,C}} + k_{\mathrm{DF}}) + (k_{\mathrm{C,F}} + k_{\mathrm{DC}})}{2} \pm \sqrt{\left[\frac{(k_{\mathrm{F,C}} + k_{\mathrm{DF}}) - (k_{\mathrm{C,F}} + k_{\mathrm{DC}})}{2}\right]^2 + k_{\mathrm{F,C}} k_{\mathrm{C,F}}} \tag{7}$$

In the slow exchange limit, λ_+ and λ_- are simply the fast and slow decay components, k_{DC} and k_{DF} of the true species. In the fast exchange limit, λ_+ approaches $k_{\mathrm{F,C}} + k_{\mathrm{C,F}}$, i.e., the relaxation rate constant for equilibration between the two species while λ_- approaches the equilibrium average of k_{DF} and k_{DC}. These limiting rates give the quantum yields expressed in Eqs. 3 and 4 as a function of the relative populations. In the intermediate cases, the averaging is stochastic. The eigenvectors represent deviations from equilibrium concentrations that will decay exponentially. The initial concentrations may be decomposed into a linear combination of the eigenvectors, depicted here as $c_{\mathrm{F}+}$, $c_{\mathrm{F}-}$, $c_{\mathrm{C}+}$, and $c_{\mathrm{C}-}$, which are the projections of the eigenvectors on the ground-state equilibrium concentrations established by the equilibrium constant given by the ratio of the exchange rates. We assume for simplicity that the exchange rates between F and C are the same in the ground and excited states.

$$F(0) = \frac{k_{C,F}}{k_{F,C} + k_{C,F}}; \quad C(0) = \frac{k_{F,C}}{k_{F,C} + k_{C,F}} \tag{8}$$

$$F(t) = c_{F+} \exp(-\lambda_+ t) + c_{F-} \exp(-\lambda_- t) = c_{F+} \exp\left(\frac{-t}{\tau_+}\right) + c_{F-} \exp\left(\frac{-t}{\tau_-}\right)$$
$$C(t) = c_{C+} \exp(-\lambda_+ t) + c_{C-} \exp(-\lambda_- t) = c_{C+} \exp\left(\frac{-t}{\tau_+}\right) + c_{C-} \exp\left(\frac{-t}{\tau_-}\right) \tag{9}$$

The quantum yield is then equal to

$$\begin{aligned}\Phi &= k_r \int_0^\infty [(c_{F+} + c_{C+}) \exp(-\lambda_+ t) + (c_{F-} + c_{C-}) \exp(-\lambda_- t)] \mathrm{d}t \\ &= k_r \left\{ \frac{(c_{F+} + c_{C+})}{\lambda_+} + \frac{(c_{F-} + c_{C-})}{\lambda_-} \right\} \\ &= k_r \{(c_{F+} + c_{C+})\tau_+ + (c_{F-} + c_{C-})\}\tau_-\end{aligned} \tag{10}$$

3.2 Results

Keying on the MD + QM trajectories shown in Fig. 2, we discovered that treating the smaller, rapidly quenched population as "species" C and the remainder as F gives a robustly single exponential decay, for a model wherein $k_{F,C}$ remains constant at ~10^9 s^{-1}, providing $k_{DC} \cong k_{C,F}$ is $\geq k_{F,C}$. This is shown in Fig. 6 for the case in which $k_{F,C} = 0.4 \times 10^9$ s^{-1}, a choice that gives $\Phi_f = 0.135$ and a nearly single exponential decay with $\tau_f = 3$ ns over the range of $k_{DC} = k_{C,F} = 1 \times 10^{10}$ to 1×10^{12} s^{-1} with essentially undetectable amounts of the shorter lifetime component, i.e., in close agreement with experiment.

3.3 Discussion

We now discuss the considerable difference between the longstanding, well-reasoned hypothesis by Petrich et al. [44] for the single exponential decay observed for NATA. Petrich et al. argue from a common point of view in which the electron transfer rate is controlled entirely by the electronic coupling matrix element V_{et}, i.e., ignoring entirely consideration of the density of states (resonance factor) of the Fermi rule. Then, because V_{et} is known to fall off exponentially with distance, the qualitative argument becomes one of comparing distance from the indole ring to the two amides. The basis of the argument was that in NATA, the three χ_1 rotamers all have the carbonyl carbon equidistant from the indole nitrogen and would therefore not display heterogeneity.

Our simulations, however, paint a completely different picture—one in which distance plays no role at all, and the resonance (density of final states) is the *only* variable that matters. This is because of the weak electron accepting nature of the amide (recall that formamide and acetamide do not quench the indole ring in

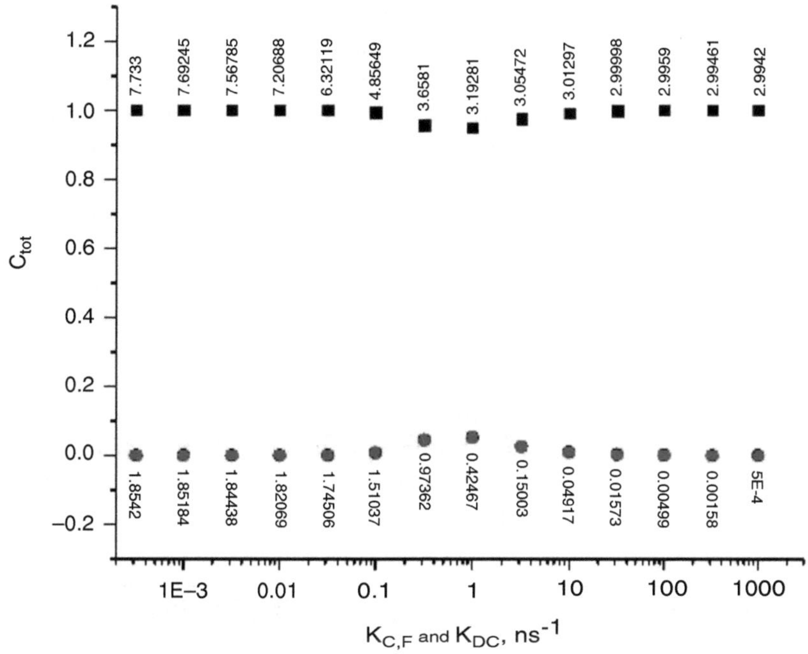

Fig. 6 Transition from slow exchange limit (pure heterogeneity) to the fast exchange limit as function of $\log_{10}(k_{C,F})$. $k_{C,F}$ is the rate of exchange from the "C" to the "F" species and is held equal to $k_{DC,}$ the rate of decay of the excited state for the "C" species to the ground state, as shown in Fig. 5. The vertical axis gives the fractional amplitudes of the long-lifetime (*upper curve*, *filled squares*) and short-lifetime (*lower curve*, *circles*) exponentials, whose corresponding lifetimes in nanosecond are shown as *numbers* attached to the *points*. Note that in the presence of interconversion between the species, there is no direct correspondence between the "species" and the "lifetimes." As the exchange rate increases, the two lifetimes evolve ultimately to a single average lifetime that applies to both species and an extremely fast transient corresponding to establishing a steady state following a delta function excitation pulse

water) [45–48]. In fact, as noted above, the coupling is so strong (200–800 cm^{-1}) that at any distance possible for the two amides the quenching rate would be ~10^{12} s^{-1}, corresponding to a quantum yield of ~1×10^{-4}.

In our view, strong quenching comes about only during transient random solvation events that greatly stabilize the CT state of one amide or the other through favorable electrostatic interaction with water. These rare (~1 %) fluctuations are essentially independent of rotamer and are usually characterized by having two or three water molecules donate hydrogen bonds to the carbonyl O of the electron accepting amide. Other waters add stabilization by having their dipoles point in a direction that stabilizes the CT state. In these solvation events, on the order of 100 water molecules within 10 Å of the NATA molecule typically contribute a net stabilization of $-8{,}000$ to $-16{,}000$ cm^{-1} (1–2 eV), about two thirds of which are stabilizing. Figure 4 shows a superposition of the distribution of CT-^1L_a energy gaps and a histogram of corresponding k_{et} values, demonstrating how only the low-energy tail of the distribution contributes to the quantum yield reduction.

Rolinski et al. have reported that a single exponential fit for NATA fluorescence gives unacceptably high χ^2 values (>1.4) [49] and find a reasonable fit from a maximum entropy method using gamma function distributions peaking at 3.0 with ~0.6-ns FWHM at 300 K, which they attribute to a single rotamer. We find that there is a mixture of rotamers that interchange on a time scale similar to the excited state lifetime. A narrow distribution of exponentials seems consistent with our simulations. Our choice to arbitrarily assume only two species in our analysis is obviously over simplified, and dividing the classes of rates in many different species could plausibly lead to such a distribution.

4 Simulations of Trp Fluorescence During Folding of the Villin Headpiece

Both Trp fluorescence and molecular dynamics have been increasingly exploited to study the rate of protein folding/unfolding [50, 51]. One of the purposes of developing the ability to predict quenching rates has been to help verify or disprove the many hypotheses that are generated from the results of experimental and theoretical studies on protein folding. Such an application was recently carried out [25] for the highly studied, fast-folding C-terminal domain of the villin headpiece mutant HP35 N27H (villin), wherein Trp is stacked on a histidine cation (a potent quencher by electron transfer [37, 52, 53]) in a helix when folded. This mutant was designed by Eaton and coworkers [54, 55] to follow the folding by the quenching of the Trp fluorescence in the folded state. The left panel of Fig. 7 shows the stacked structure as it appears in the X-ray structure. In this section, we describe how several 1-μs simulations on villin unexpectedly revealed that only a transient, closely stacked structure shown in the middle panel of Fig. 7 appears to quench significantly enough to produce the changes in quantum yield seen experimentally during folding [25].

In this seemingly very different setting, there is a striking parallel to the behavior of NATA; a minor species dominates the quenching, and it is solvation energy affecting energy gap that makes the minor species so easily quenched.

Because protein folding naturally would take a Trp in the protein through an astronomical number of environments, one can legitimately wonder how dynamic changes in fluorescence quantum yield could be a faithful measure of folding dynamics. Indeed, viability of Trp fluorescence as an indicator of global folding rate was recently brought into question in the comparison of

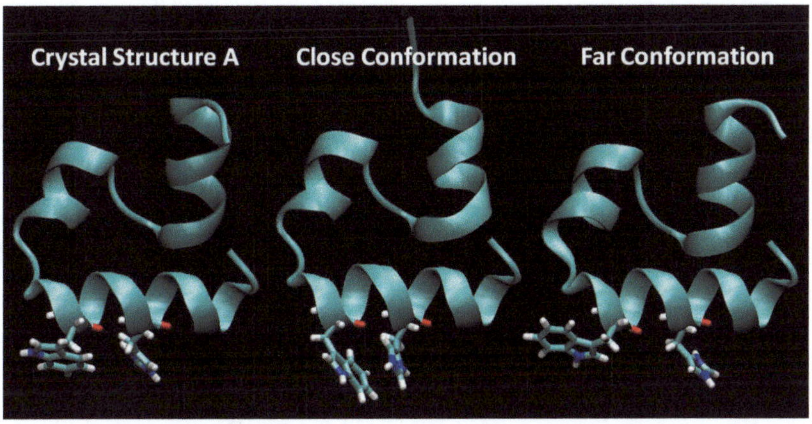

Fig. 7 Comparison of the X-ray structure (1YRF "A") with representative "close" and "far" conformations showing Trp23 and His27 relative conformations taken from a 1-μs MD simulation of villin HP-35 at 300 K in explicit TIP3P water using the AMBER99SB-ILDN force field. The X-ray structure on the *left* and the structure on the *right* are the two main "far" conformations, neither of which we believe allows significant quenching. The "close" (*center*) conformation is reached only transiently when χ_1 is ~180° and χ_2 is ~+90° and is the only conformation for which we believe quenching is possible. Most of the trajectory is spent in the X-ray conformation, for which we believe quenching is not possible or is minor

simulated and experimental folding rates wherein classical MD simulations predict folding times that are several microseconds longer [56, 57] than experiment [51, 58, 59]. MD simulations suggested that the helix containing the Trp-His^+ is formed much before the final folded structure is achieved [56, 57]. On the other hand, recent alternative experiments supported the ability of Trp fluorescence to report the global folding rate [59, 60].

4.1 Results

Figures 8 and 9 show trajectories of properties computed from two of four ~1-μs MD simulations performed on the villin head piece with explicit water involving an eclectic sample of mutants, force fields, and temperatures [25]. Figures 8a and 9a show trajectories of the distance between the Trp23 and His^+27 centroids of mass (COM) derived from simulations on two different mutants. This distance is naturally considered the best single indicator of the quenching rate of Trp fluorescence by His^+ [56, 57]. All trajectories exhibited a similar behavior, marked by several regions of discrete inter-residue distance. Most striking and significant is the small percentage of time for which the closest distance (~4 Å) is visited during the trajectories of folded villin—*while the Trp-His^+ local structure is helical 100 % of the time according to backbone RMSD and inspection of structures.* We find that in all cases, the close distance state arises solely through a concerted χ_1 and χ_2 transition, an example of which is shown in Figs. 9b, c. The close conformation is achieved only if χ_1 is ~180° and χ_2 is ~+90°. The vast majority of the time χ_1 is ~+60° and χ_2 is ~−90° (left panel of Fig. 7).

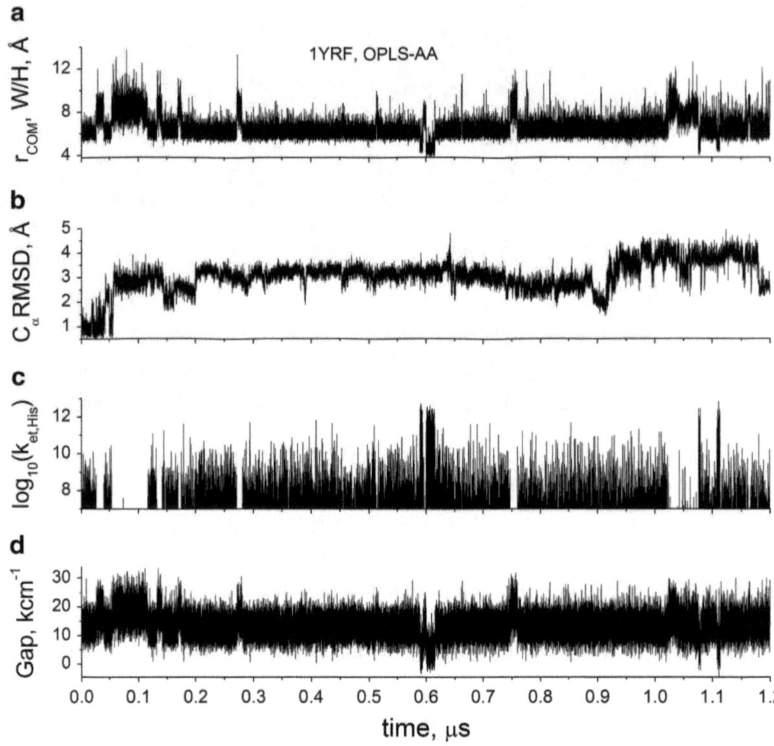

Fig. 8 1 μs simulation of HP35(N27H$^+$) with OPLS-AA force field. (**a**) Center of mass (COM) distance between Trp (W) and His$^+$ (H) rings, (**b**) C_α RMSD relative to crystal structure (1YRF), (**c**) calculated electron transfer rate, s^{-1}, for quenching by His$^+$, and (**d**) CT(His$^+$)-^1L_a energy gap from MD + QM. Note the significant correlation between distance, rate, and energy gap, but that there is no correlation between RMSD and quenching ability for His$^+$. In this simulation, quenching occurs only during the brief periods in which a rotamer switch brings the center of mass of the rings within <5.5 Å (the center picture in Fig. 7)

Figure 8c shows how the computed quenching rate constant varies with time during the trajectory. Note the strong correlation of Trp-His$^+$ distance with quenching rate. For reference, if the quenching rate drops below 1×10^7 s^{-1}, there is negligible effect on the quantum yield. Figure 8d shows that the important energy gap between the CT and ^1L_a states has an unexpected strong correlation with Trp-His$^+$ distance. Further analysis revealed that the source of the strong correlation lies primarily in a large change in solvation energy at short distance compared to longer distances. At large separation, the vertical CT energy immediately after electron transfer is highly *destabilized* by the polarization of the water oriented by His$^+$ prior to electron transfer. The negative oxygen atoms of water are mostly in close contact with the His$^+$. This destabilization decreases sharply when the separation is small

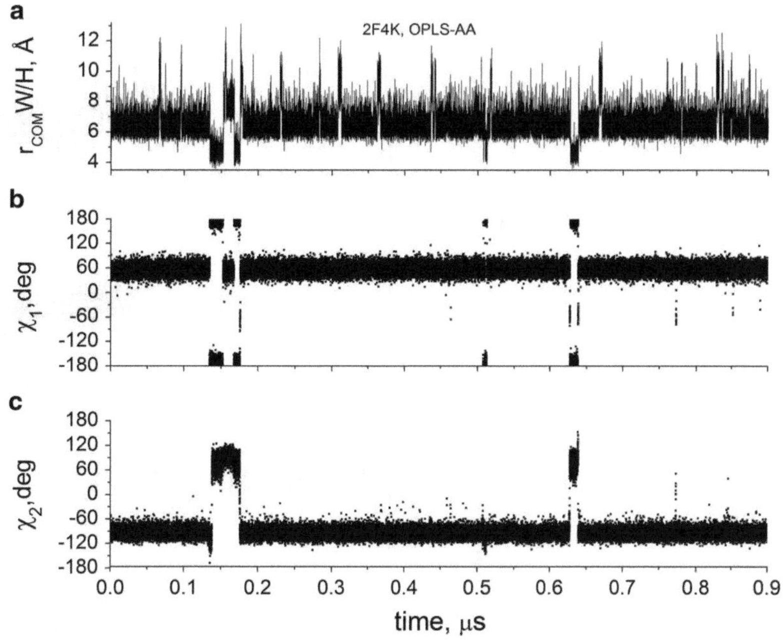

Fig. 9 0.9 μs simulation of HP35(N27H$^+$,K24Nle,K29Nle) with OPLS-AA force field showing correlation with Trp dihedral angles, χ_1 and χ_2. As in Fig. 8, quenching occurs only during the brief periods in which a rotamer switch brings the center of mass of the rings within <5.5 Å. (**a**) Center of mass (COM) distance between Trp (W) and His$^+$ (H) rings, (**b**) Trp23 χ_1 vs. time, (**c**) Trp23 χ_2 vs. time

enough to exclude water in the region between the donor and acceptor. The energy gap is the key variable in the Franck–Condon-weighted density of states (FCDS) (the overlap of the photoelectron spectrum for removal of an electron from the ring and that for adding an electron to the amide, in units of states per cm^{-1}) which in turn is a major factor in the Fermi rule for the electron transfer rate and is a measure of the extent of resonance of the CT state with the emitting state. (This is precisely analogous to the overlap of acceptor absorption spectrum with donor emission spectrum in FRET.) Figure 10 shows how the FCDS and, accordingly, the Fermi rule-computed electron transfer rate constant drop sharply by 3 orders of magnitude as the Trp-His$^+$ distance increases from 4 to 5.5 Å. Figure 10 also shows that in this distance range, the average square of electronic coupling in the Fermi rule was computed to change very little, contrary to usual assumptions and experimental findings at larger distances. Similar deviation at short distance has been reported previously [61, 62].

The picture that emerges from the results presented above is that quenching by His$^+$ happens almost exclusively at the closest distance. The duration of the close state was found to be on the

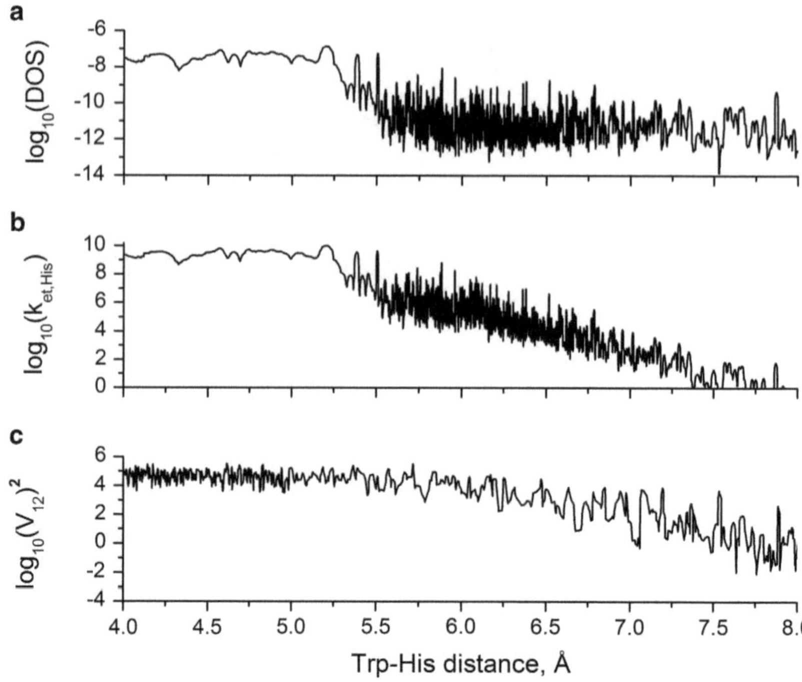

Fig. 10 Log_{10} plot vs. Trp-His^+ center of mass distance of (**a**) Franck-Condon-weighted density of states, in units of states per cm^{-1} (FCDS), (**b**) Fermi rule-calculated electron transfer rate constant for Trp → His^+ in units of s^{-1}, and (**c**) square of electron transfer Hamiltonian matrix element (V_{et}) for coupling fluorescing and histidine CT state in units of $(cm^{-1})^2$

order of the excited state lifetime of Trp, meaning that fluorescence intensity will exhibit heterogeneity. In other words, there are in effect two species: as in NATA, one which is not quenched and for which the CT state is far above the fluorescing state (which happens to be when Trp and His are far apart) and one which is highly quenched in which the CT state energy is close to that of the fluorescing state (which happens to be when the Trp and His are close in distance). This is seen in Fig. 10 by realizing that when the distance exceeds ~5.2 Å, the effect of His^+ quenching drops below $10^7\ s^{-1}$ and becomes negligible compared to amide quenching and the non-electron transfer decay channels.

The current assumption by experimentalists and computationalists alike is that when the region encompassing these two residues is helical, there will be significant quenching, but no quenching otherwise. Because of the isomorphism with the NATA fluorescence decay kinetics, Eqs. 5–10 may be applied directly to the villin case. Using a combination of the MD simulations and experimental fluorescence behavior of Trp in barnase, which is also quenched by His^+, it was possible to show that a model consistent with Fig. 10 accounts for the small increase in quantum yield observed upon unfolding. Unlike the NATA case, the exchange dynamics between the F and C species is in the slow limit, and it is reasonable to consider them as true species. The ratio of the exchange rate constants provides the equilibrium constant for the F–C process and cannot be accurately determined from the simulations.

4.2 Summary

Contrary to consensus expectations, simulations predict that most of time His^+27 does not quench fluorescence from Trp23, even while helix III is formed. Calculations (as calibrated with Trp94 of barnase) say that quenching happens only during certain periods in the folded helical state when specific rotamer conformations bring Trp and His^+ into close contact, a contact distance substantially closer than in the crystal structure. At the crystal distance or greater, quenching is not possible because the energy of the CT state responsible for quenching by His^+ is greatly ***destabilized*** by greater solvation of His^+, taking the CT state out of resonance with the fluorescing state. The resulting heterogeneity regarding the fluorescence signal, in which there are effectively two species, means that small changes in the fraction of the highly quenched species (i.e., the ratio of the exchange rates, corresponding to ~1 kcal/mol in ΔG^0) will have a measurable effect on the average quantum yield. This in turn means that intermediate unfolded structures with helix III fully intact could easily have significantly different quantum yields due to tertiary contacts with the Trp-His region, thereby providing a mechanism by which Trp fluorescence should be able to reliably report the global folding rate, despite early folding of helix III. Furthermore, the 70-ns relaxation time reported by Kubelka et al. [63] is consistent with the rate of exchange between the C and F conformations in the simulations, when a temperature dependence is imposed on the equilibrium between the C and F conformations of the helical structure arising from a small negative ΔH^0 associated with forming the C conformation.

5 Proteins and Discrete States: 100-ns Simulations

The unprecedented 1-ms MD simulation of BPTI [30] by Shaw and coworkers has graphically revealed what had been increasingly inferred from experimental studies that functional proteins exhibit transitions among multiple distinct structural states. Such major "basin hopping" was found to occur on the 10–100 μs time scales, typically involving backbone conformational changes, with a large separation of time scales from local side chain dynamics occurring on the nanosecond scale.

That multiple basins ("species") are populated in the ground-state ensemble of a protein was already widely believed to be the general origin of nonexponential decay of Trp fluorescence when the basin hopping rate is on the order of the excited state lifetime or

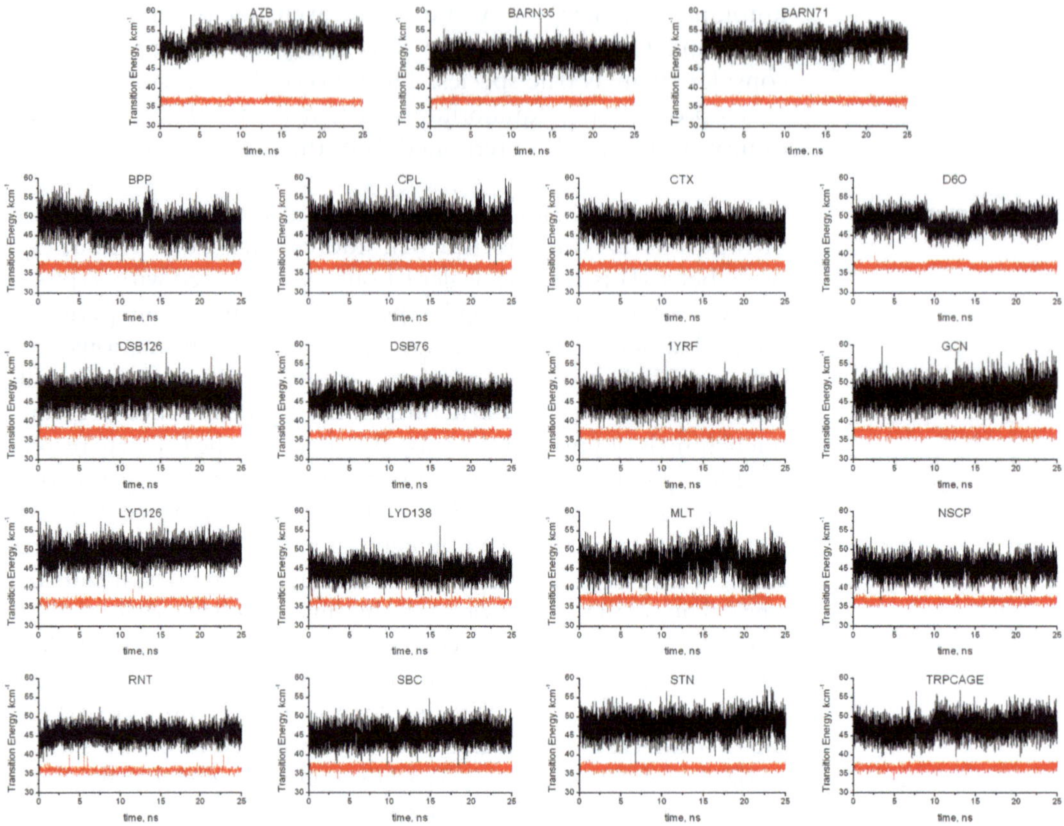

Fig. 11 ZINDO vertical transition energies for ground to ^1L_a state (*red*) and ground to amide CT state (*black*) during 25-ns simulations for 20 single-Trp proteins using the OPLS-AA force field

longer. The idea that these basins are confined only to different rotamer states has attained almost dogma status, albeit primarily because of the absence of easily grasped alternatives. In this section, we show that Trp rotamer transitions do occur during 100-ns simulations, that nonexponential decay does not always correlate with rotamer transitions, and that some proteins display considerable heterogeneity with predicted nonexponential decay *in the complete absence of rotamer transitions.*

Figure 11 presents a time trace of the ^1L_a (red) and CT (black) Zindo-calculated energies from 25,000 calculations for each of 19 single-Trp proteins (described in Table 1) during 25-ns MD trajectories. Because these simulations are 5–20 times longer than typical fluorescence lifetimes of Trp in proteins, heterogeneity is clearly revealed by the variations in the CT-^1L_a energy gaps. The CT-^1L_a energy gaps give indications of the density of states used in the Fermi golden rule; the smaller the gap, the more often there are significantly high values of the Franck–Condon factors, i.e., the more resonance of the fluorescing and CT states (energy must be conserved for the electron transfer to happen).

Table 1
Experimental fluorescence quantum yields and pdb codes for proteins

Exp.	Abbreviation	Description	PDB code	Reference[a]
0.013	dsb126	dsBa W126	1dsb	[65]
0.025	cpl	Human cyclophilin A	2cpl	[79]
0.03	trpcage	TrpCage (nmr structure) 1 and 2	1l2y	[102][b]
0.03	d6o	fkb506-binding protein	1d6o	[103][c]
0.03	sbc	Subtilisin C	1sbc	[104][d]
0.031	bpp	Phospholypase A2	2bpp	[104]
0.033	nscp	NSCP W57 W4F W170F	2scp	[67]
0.044	lyd138	T4 lysozyme W138	1lyd	[105]
0.06	lyd126	T4 lysozyme W126	1lyd	[105]
0.1	ctx	Cobra toxin	1ctx	[104]
0.115	mlt	Melittin	2mlt	[104]
0.12	gcn	Glucagon	1gcn	[104]
0.122	barn71	Barnase W71	1a2p	[78]
0.142	barn35	Barnase W35	1a2p	[78]
0.2	dsb76	dsba W76	1dsb	[65]
0.29	stn	staph. nuclease	1stn	[104]
0.0555	1yrf	Villin	1yrf	[106]
0.31	rnt	Ribnuclease T1	9rnt	[104]
0.31	azb	Apo-azurin W48	1azb	[104]

[a]Reference for experimental quantum yield.
[b]Estimated relative to NATA.
[c]Upper limit based on 12-fold increase upon denaturation.
[d]Estimated from lifetimes.

A previously used scheme was used to predict quantum yields from these trajectories [12], in which an individual trajectory is assigned a single Gaussian distribution for the energy gap based on the mean and standard deviations from the trajectory. A constant interaction element is also used. The results were not significantly better than when only much shorter (150 ps) simulations were used, as seen in Fig. 12. This scheme might be applicable for some proteins that do not contain multiple energetic basins for the CT state, but significant deviation might be expected if multiple energy basins exist. Obviously, this procedure cannot treat nonexponential fluorescence decay because the single Gaussian distribution combined with the single V_{et} yields only a single average lifetime. Ab initio computations of V_{et} as a function of rotamer

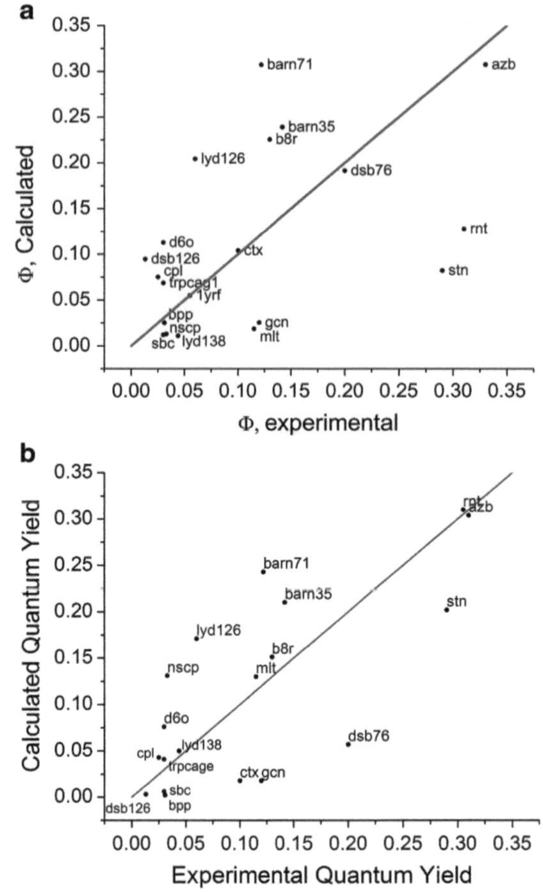

Fig. 12 (**a**) Theoretical vs. experimental fit using Eqs. 1 and 11. The added constant to obtain the fit was 240 cm^{-1}. (**b**) Theoretical vs. experimental fit using Eqs. 1 and 11 from ref. 17

conformation [18] suggest that this assumption would not hold for proteins that exhibit tryptophan χ_1, χ_2, or ψ dihedral transitions. Furthermore, the results from Subheadings 3 and 4 above demonstrate that when the interconversion between species is on the order of the fluorescence lifetime, one cannot expect the average rate or average quantum yield to be accurate. Finally, there is the question of whether the choice of MD force fields used plays a role in the results.

To partially explore the influence of force field, Fig. 13 shows 100-ns simulations for the same proteins as Fig. 11 but using an improved force field, AMBER99SB-ILDN, which has been shown to be more effective in protein-folding simulations [64]. The remainder of this section revolves around the simulations shown in Fig. 13, from which a more realistic and thorough procedure for capturing quantum yields and decay time heterogeneity was applied. Addressing some of the obvious problems with the first approach,

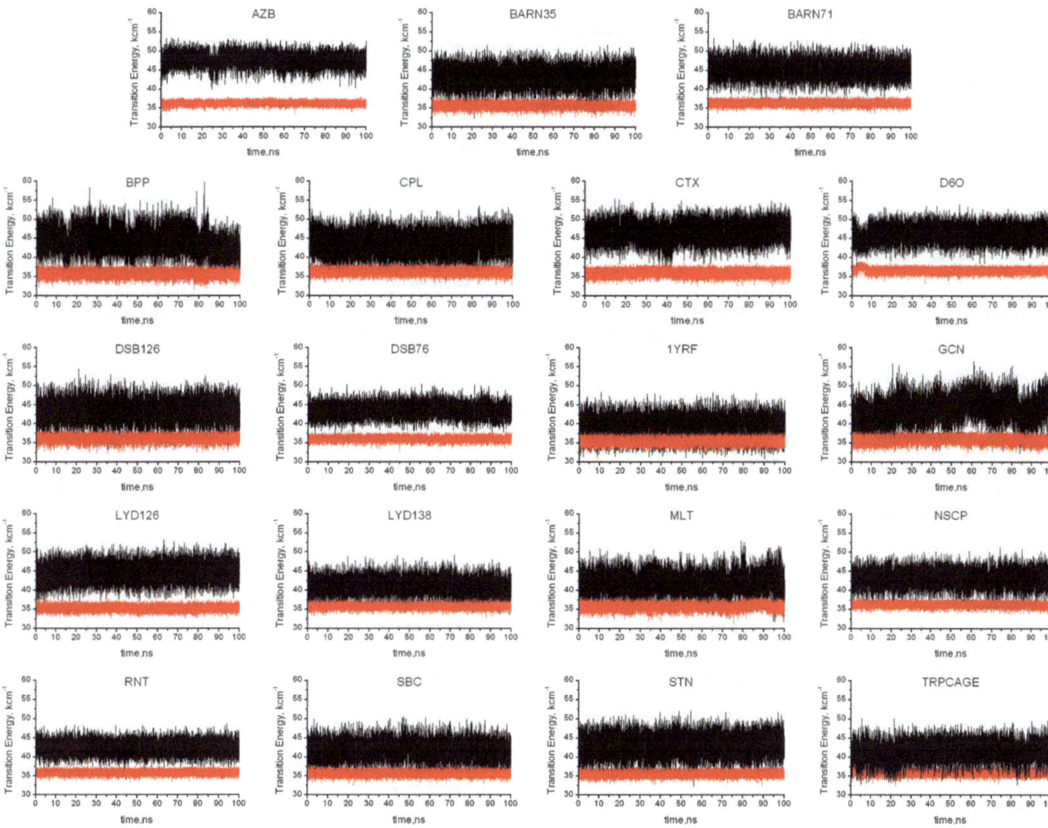

Fig. 13 ZINDO vertical transition energies for ground to ^1L_a state (*red*) and ground to amide CT state (*black*) during 100-ns simulations for 19 Trps in 16 proteins using the AMBER99SB-ILDN force field

the time dependence of both the energy gap and V_{et} are taken into consideration. An instantaneous τ^{-1} was calculated as a function of time for each of the proteins. The instantaneous τ^{-1} included a reasonable approximation to an instantaneous V_{et} by interpolation from a table of ab initio computed V_{et} values [18] as a function of the three dihedrals χ_1, χ_2, and ψ. Figure 14 shows for each of the 19 Trps in 16 proteins the Trp χ_1, χ_2, and ψ dihedrals. V_{et} as a function of these dihedral angles is shown in the lower right panels in Fig. 14. The trajectories of instantaneous τ^{-1} values for each protein are presented in Fig. 15. From the time traces for τ^{-1} the kinetic equation is integrated to obtain average survival curves, i.e., decays of the fluorescent state for each protein. The results are presented for all 19 Trps in Fig. 16. Integrating these survival curves and multiplying them by k_r gives the fluorescence quantum yield for each of the Trps, which are plotted vs. experiment in Fig. 17.

5.1 Results

A comparison between AMBER99SB-ILDN transition energies (Fig. 13) and OPLS-AA transition energies (Fig. 11) shows that the transition energies to the ^1L_a state are not significantly different

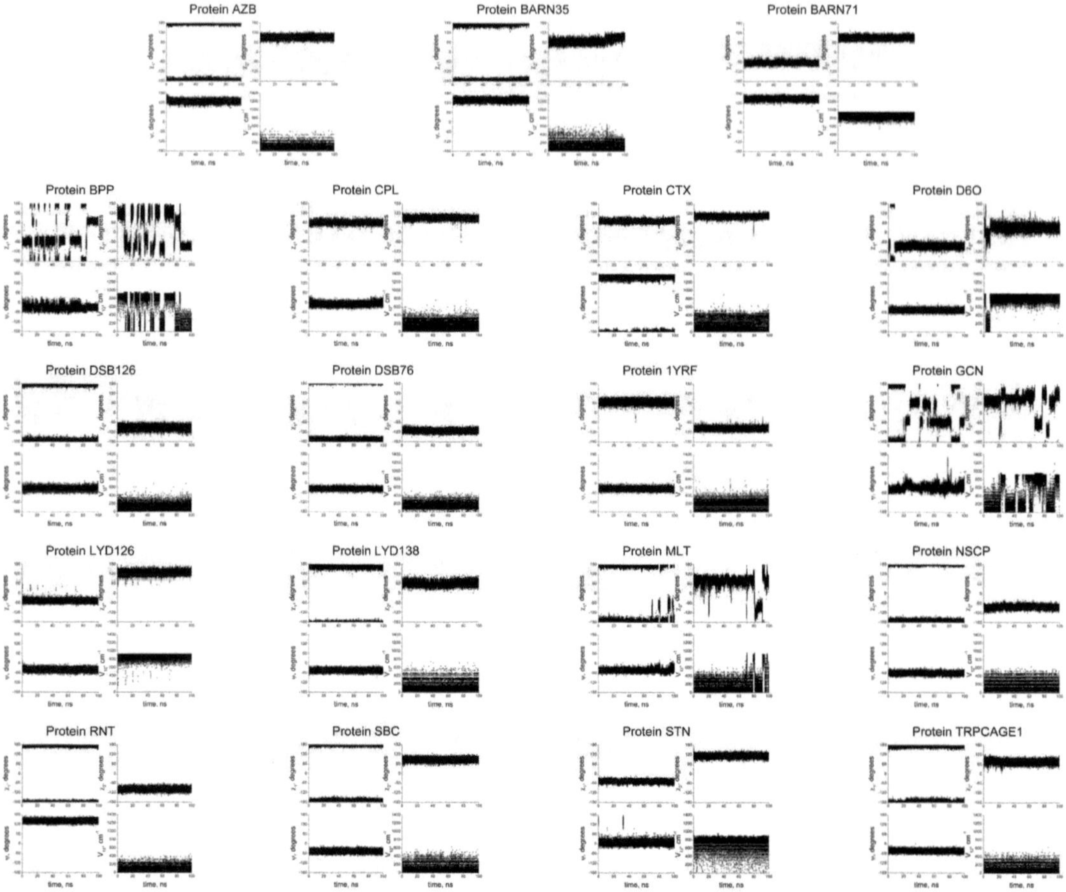

Fig. 14 χ_1, χ_2, ψ, and the calculated V_{et} as a function of time during 100-ns simulations using the AMBER99SB-ILDN force field shown in Fig. 13

between the force fields whereas the CT states show significant differences. These differences are apparently due to differences in assigned charges of each residue in the two force fields, which, interestingly, lead to average CT state energies that are consistently lower for AMBER99SB-ILDN than for OPLS-AA.

Regardless of differences in the force fields, comparing Figs. 11 and 13 shows the existence of some consistency in the general behavior of the proteins portrayed by the two force fields. The uniquely short-wavelength fluorescing Trp 48 of azurin (AZB) is seen in both cases to have by far the largest energy gap and smallest standard deviation, a clear reflection of the uniquely hydrocarbon local environment and complete insulation from water. Two other cases of distinctly nonpolar environment giving rise to quite blue-shifted experimental fluorescence are RNT and DSB76. In both force fields, these have noticeably smaller standard deviations of transition energy for both the fluorescing and CT states.

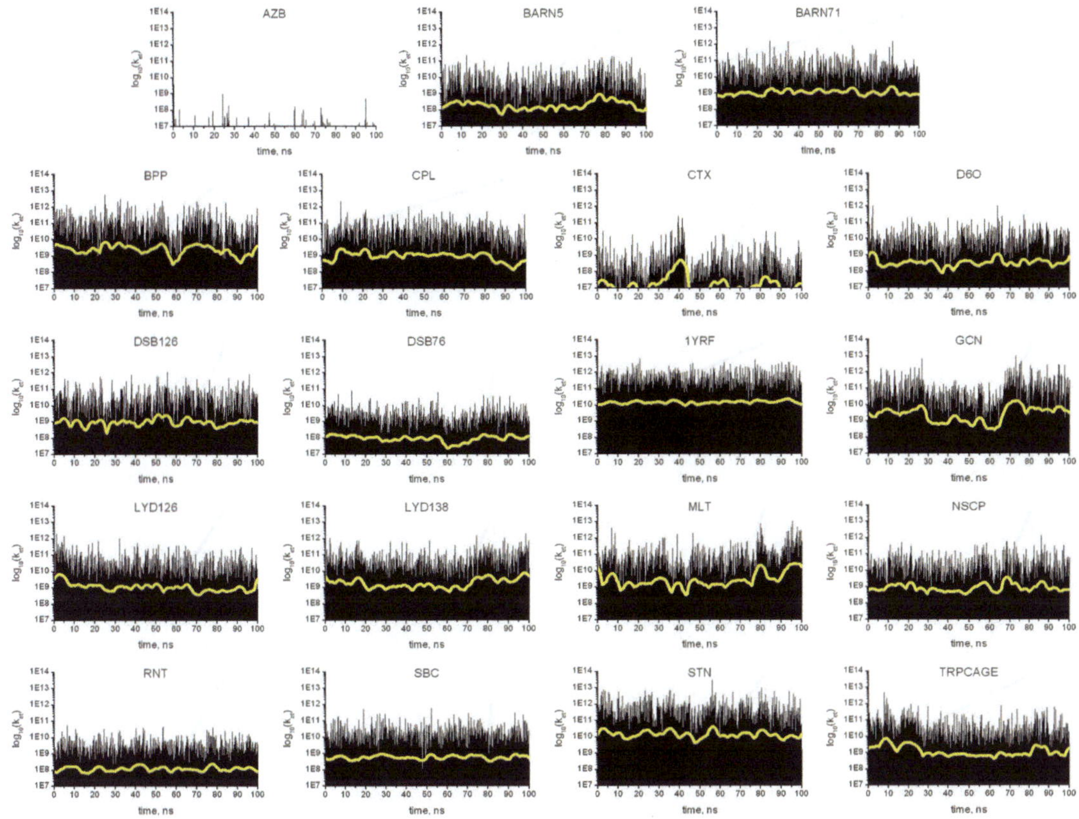

Fig. 15 Time traces of the Fermi rule-computed instantaneous sum of all rate constants, $k_r + k_{nr} + k_{et,amide}$, from the 500,000 Zindo-calculated energies for the 100-ns trajectories of Fig. 13 using the AMBER99SB-ILDN force field. The *yellow line* shows the smoothed curves using an average window of 5 ns

5.2 Rotamers and Heterogeneity

More than any other variable, rotamer states have been invoked as the probable reason behind heterogeneity of Trp fluorescence lifetimes. The electron transfer rate is basically determined by the product of two largely independent factors in the Fermi rule expression, (Eq. 7.1) V_{et}^2 and ρ. Clearly different rotamers could have different values for these factors: V_{et} is well known to decrease exponentially with distance, although this is not so clearly the case at the short distances that matter here [61, 62]. For the specific case of Trp ring electron transfer to the nearby amides, there is theoretical evidence that through bond effects may be more importantly modulated by dihedral angle than by distance alone [18]. Similarly, it is easy to imagine that for different rotamers states, Trp ring and amides would have quite different electrostatic interactions with the surrounding protein and solvent, which in turn could have strong effects on the CT-^1L_a energy gap, i.e., on ρ.

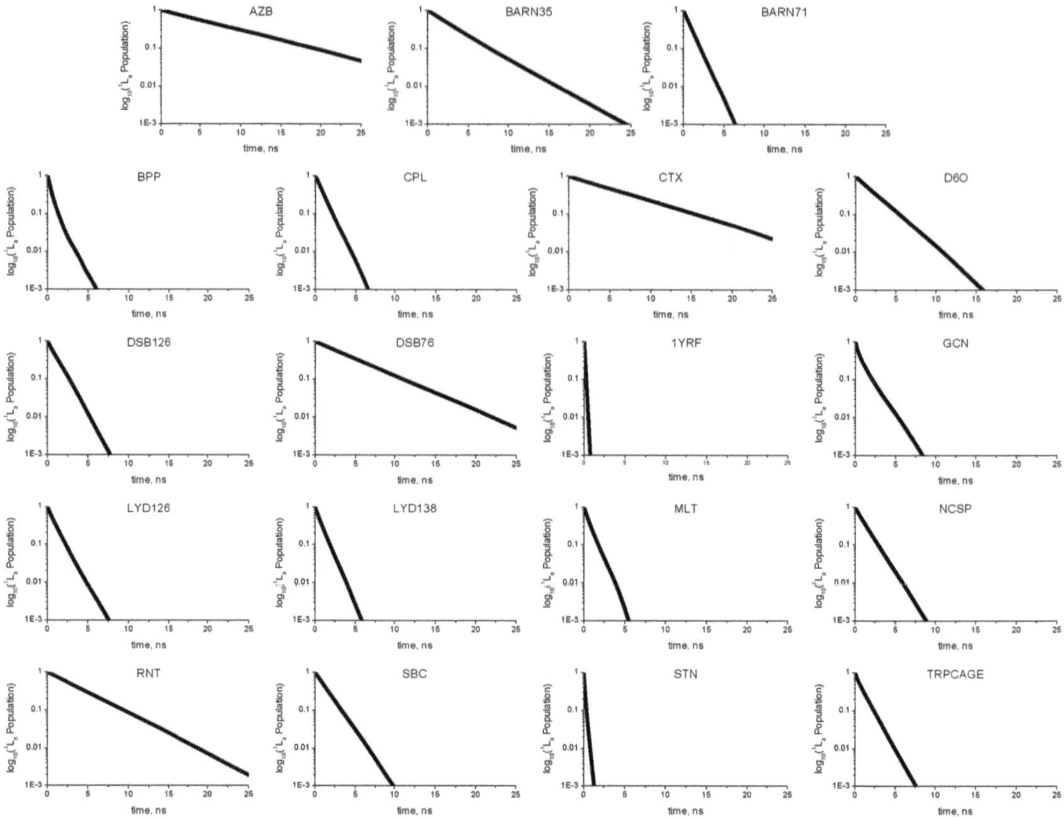

Fig. 16 Average survival curves calculated using the time traces for the sum of the rate constants for each protein in Fig. 13. A 25-ns window was used to calculate each survival curve, as described in Subheading 7. Of the 19 Trps, BPP, GCN, and MLT show significant multiexponential kinetics as expected from their heterogeneous behavior

Figure 14 shows time traces of the Trp χ_1, χ_2, and ψ dihedrals for each of the 19 Trps, with the corresponding V_{et} shown as a function of these dihedral angles in the lower right panels. We see that rotamer transitions are frequent only for BPP and GCN. Rotamer transitions are occasional for D6O and MLT and essentially absent for the other 15 cases. These simulations therefore provide an opportunity to examine the extent that rotamers contribute to heterogeneity and to also see whether heterogeneity may be present without different rotamer state populations.

Figure 15 shows the trajectory for $\log_{10}$ of *instantaneous* excited state decay constants $\tau^{-1}(t)$—the true measure of lifetime heterogeneity. $\tau^{-1}(t)$ is defined as $(k_r + k_{nr} + k_{et}(t))^{-1}$. These were obtained by combining results from Figs. 13 and 14. A smoothed average is superimposed to be able to follow the average behavior of τ^{-1}, using a 5-ns window, chosen because in this time the fraction left of a completely unquenched tryptophan would be about 0.6. One sees that most of the proteins exhibit several time periods in which k_{et} is different by an order of magnitude and which persist for more than 5 ns. By the criteria set forth in Subheadings 3 and 4 above, and just by inspection, we should expect that these proteins will exhibit heterogeneity in their lifetimes. Notable are GCN, BPP, CTX, D6O, and MLT.

Because we believe that V_{et} and ρ operate independently, it is useful therefore to speak of two types of heterogeneity, V_{et} heterogeneity and ρ or energy gap heterogeneity. Below, we discuss these separately. The extent to which energy gap heterogeneity, qualitatively seen in Fig. 13, and V_{et} heterogeneity, seen in Fig. 14, correlate with k_{et} and with Trp rotamer transitions may be inferred by comparison with k_{et} plotted in Fig. 15.

5.3 V_{et} Heterogeneity

At the short distances between the Trp ring and the two closest amides, V_{et}^2 was found to be controlled by the χ_1 dihedral, whereas ρ is determined by the energy gap, which may or may not be related to the dihedrals. Proteins that have a χ_1 value near $-60°$ maintain a large V_{et} (~800–900 cm^{-1}) that does not go to zero, whereas proteins that have nominal values of χ_1 of $\pm180°$ or $60°$ have much smaller values approximately (fluctuating between 0 and ~200 cm^{-1}). Examples of V_{et} heterogeneity include BPP, which shows χ_1 transitions from $-60°$ to $+60°$ or $\pm180°$ that change the magnitude of V_{et}. Other proteins that show this behavior are D6O, GCN, and MLT. χ_2 undergoes transitions from $+90°$ to $-90°$ with little effect on V_{et}.

The roughly 16-fold variations in V_{et}^2 due to rotamer transitions of χ_1 in and out of $-60°$ do not correlate well with the instantaneous rates of Fig. 15 because the energy gap often is seen here to be large when V_{et} is high. The larger energy gap effectively cancels rate enhancement from V_{et}. For BPP, when $\chi_1 = -60°$ and V_{et} ~800 cm^{-1}, there is a reasonably large energy gap, but when $\chi_1 = 180°$, the energy gap is much smaller. Only at 60 ns in the trajectory does the energy gap stay large when χ_1 is 180°, creating the large dip in the quenching rate. Much the same happens for GCN. In most regions where $\chi_1 = -60°$, the energy gap is large, negating the large V_{et} at that angle. The most prominent feature is the almost 2 order-of-magnitude increase in rate between 65 and 75 ns, which is caused entirely by a dip in the energy gap while $\chi_1 = -60°$. A few of the blips can be traced to changes in V_{et}, while the energy gap is constant. For MLT, the high rates at 80 and 95 ns correlate with short periods in which $\chi_1 = -60°$.

We believe that the anti-correlation of rate contributions from V_{et} and ρ is not entirely random. When $\chi_1 = -60°$, the C of the Trp amide is considerably farther from ring atoms than when $\chi_1 = +60°$ or 180°, which place the C almost over the ring. In the CT state, C is quite negative after electron transfer, and the ring is positive. At the increased distance for $\chi_1 = -60°$, the electrostatic stabilization of the CT χ state is smaller by a few thousand cm^{-1} than for

+60° or 180°, meaning that the energy gap will always be larger when V_{et} is largest in vacuum, a circumstance that will increase the probability of this type of correlation in the protein environment. This type of long-term water trapping at the electron acceptor carbonyl is the major reason for conserved low fluorescence quantum yield of the γ-crystallins (eye lens proteins) [14, 20, 21]

5.4 Energy Gap Heterogeneity

We have just seen that energy gap heterogeneity can apparently be correlated with V_{et} heterogeneity in a way to reduce the impact of large increases in V_{et}. For the many cases in which there are no rotamer transitions, there are several proteins for which significant modulation of k_{et} comes from heterogeneity arising from energy gap variations alone. A spectacular case is that of CTX, in which the large increase followed by a steep drop in rate seen from 40 to 45 ns is caused by trapping of a water in a small pocket between the carboxyl of Asp27 such that it H-bonds to the O of the Trp amide, stabilizing the CT state. This H-bond breaks when Asp27 moves away, and the water molecule reverses its orientation to a destabilizing orientation, which will cause k_{et} to decrease.

5.5 Survival Curves

Figure 16 shows semilog plots of the corresponding fluorescence decay curves derived from average survival curves computed from the instantaneous rates by methods shown in Subheading 7, using a 25-ns averaging window. These reveal significant nonexponential decay for BPP, GCN, and MLT, proteins which exhibit distinct basins that persist for ~25 ns. Other proteins for which nonexponential decay has been experimentally observed include D6O, DSB126, and NSCP. These show considerable energy gap heterogeneity in Fig. 15 on the 5-ns scale, but do not produce significant nonexponential decay in the survival curves of Fig. 16.

The fluorescence quantum yield for each of the proteins was estimated by integrating the survival curves and multiplying them by the radiative rate constant. In Fig. 17, the predicted quantum yields deduced from these survival curves that take into account 25-ns heterogeneity on the 100-ns time scale are compared to quantum yields predicted by a previous procedure based on average energy gaps and constant individual average V_{et} for each protein (Fig. 12b).

5.6 Discussion

The value of the montage of exploratory results presented in this section lies not so much in improved predictive power but in its portrayal of the diversity that may be expected in how the ubiquitous nonexponential decay of Trp fluorescence comes about. This study demonstrates a certain robustness of the general procedure with regard to different force fields and geometry of the quantum molecule, while at the same time, revealing large effects due to choice of force field that seem specific to certain proteins. Tools are introduced which can be refined to perhaps better capture nonexponential decay.

Proteins such as DSB126 and NSCP are known to exhibit considerable lifetime heterogeneity [65–67]. A case was made in those studies for different rotamers as the source of heterogeneity, but in our study, no rotamer transitions were observed in 100-ns simulations for these proteins. Considerable rate constant modulations on the 5-ns scale seen in Fig. 13 for both proteins, however, could in principle, be responsible for this heterogeneity. For DSB126, the heterogeneity on the 5–10-ns scale seems to be due to global motion in which water at long range contributes to increasing the gap, possibly slaved by the Lys132. Survival curves based on a smaller averaging window may reveal nonexponential decay. Alternatively, 100-ns may simply be insufficient time to observe the rotamer transitions in these proteins and that an equilibrium population of different rotamers may take several orders of magnitude longer to be established. The 1-ms simulations of Shaw and coworkers suggest this possibility [30].

5.7 Conclusions

The slight improvement gained by using 100-ns simulations combined with survival curve analysis raises the possibility that further improvement may result from even longer simulations. There is, of course, no guarantee that in some cases the interconversion rate between important conformations is so slow that milliseconds or longer would be required to capture the correct picture.

Regardless of whether these trajectories are sufficiently long to provide significantly more accurate predictions, the results provide a view of the diversity of mechanisms by which the ubiquitous nonexponential decay of Trp fluorescence may come about.

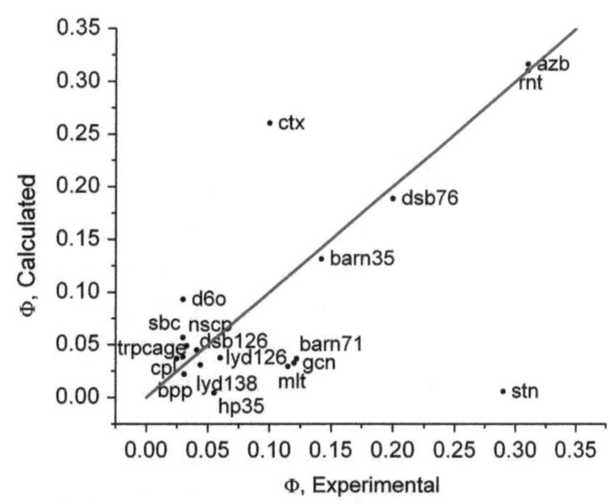

Fig. 17 Fit of calculated vs. experimental quantum yields obtained using AMBER99SB-ILDN and the survival curve procedure

6 DAS Wavelength vs. Lifetime Correlation and Solvation Dynamics

In time-resolved studies of Trp fluorescence, λ_{max} has always been found to shift to longer wavelengths following excitation. This is easily anticipated given that the large increase in permanent dipole upon excitation requires solvent reorganization that will lower the free energy of the excited state more than the ground state. Ultrafast time-resolved experiments on solvent-exposed Trp in proteins always exhibit such relaxation on a 1–2-ps time scale, due to the fast reorientation of water [38]. Red-shifting relaxation has even been seen on the scale of several nanoseconds for Trps with essentially no electron transfer quenching [68, 69]. More recently, an experimental window has opened on the intermediate times of 10–100 ps, in which dozens of Trps in proteins have exhibited a time-dependent red shift [32, 70, 71]. Some of these have also exhibited population decays indicative of electron transfer-based quenching. Given the considerable discussion of lifetime heterogeneity in the preceding sections of this chapter, it is natural to address the question of why there has not been reported a case in which λ_{max} was seen to shift to the blue in time. This is, in principle, possible if, for example, there are two species in which the population with the more blue-shifted fluorescence has a much longer lifetime than the population with the more red-shifted fluorescence. In such a seemingly possible scenario, the red-shifted population would decay away first leaving the blue-shifted population, giving the appearance of a blue-shifting time-dependent Stokes shift. This should be called a pseudo-time-dependent Stokes shift (pseudo-TDFSS), because it is not related to solvation dynamics. Fortunately, just such a phenomenon has been observed by Boxer and coworkers for the protein GB1 containing the artificial amino acid aladan [72], whose side chain is a well-known solvent-sensitive dye with a very large dipole change upon excitation. In most sites, the fluorescence was observed to initially shift to the red, but during the latter half of the decay, λ_{max} was seen to blue shift with increasing time. The authors gave the explanation just described above. Below, we describe a study in which a rationale is given for why no such reverse Stokes shifting has been reported for Trp.

As is evident from the above sections, nonexponential decay of Trp fluorescence is an inescapable consequence of its great sensitivity to the electrostatic protein environment. More often than not a single Trp in a protein will display fluorescence decays requiring a linear combination of two to four exponential terms and whose relative weights will depend on the wavelength at which the fluorescence is detected. It is common for investigators to mathematically fit such changing decay profiles by a linear combination of different decay-associated spectra (DAS) [73–77], each of which globally decays with a best fit decay time. In the vast majority of cases, there is a strong correlation between the lifetimes and λ_{max} of the DAS: the short-lifetime component has a blue-shifted DAS and the long-lifetime component has a red-shifted DAS. Therefore, almost always λ_{max} shifts to longer wavelengths with time. For a single species in an environment that relaxes in response to a large increase in permanent dipole upon excitation, this correlation is as it must be, because the excited state energy must move downward as the solvent aligns with the new dipole. The hallmark of such pure relaxations is the fitting by DAS with negative amplitudes on their red edge with decay times corresponding to the time constants of the rate of red shift of λ_{max}. This signature, however, has been observed on the ns time scale for only a few cases [68, 69]. In the vast majority of cases for single Trps in room temperature proteins, ns time scale red shifts are reported *without* negative amplitude DAS, which implicates multiple discrete species (subpopulations) with distinct environments (and therefore different λ_{max} and population decay values). Clearly a strong physical principle is dictating this correlation, a phenomenon that has been commonly observed in proteins over many years [53, 66, 67, 78–89]. This correlation was not understood, and the absence of plausible hypotheses for the nature of this principle had repeatedly been used as evidence against the subpopulation models [90].

Barkley and coworkers have recently conducted a revealing Trp fluorescence lifetime study on a set of seven single-Trp cyclic hexapeptides, consisting only of amino acids whose side chains are nonquenching in water [17, 24, 91, 92]. This work was designed specifically to examine fluorescence lifetime heterogeneity that arises plausibly only from different rates of quenching by electron transfer from the excited Trp indole ring to a nearby backbone amide due to different rotameric states of the Trp. There is no other source of quenching nor can the highly solvent-exposed Trps plausibly exhibit nonexponential decay due to solvent relaxation on the ns time scale for such small structures. The peptides exhibit variations in fluorescence λ_{max} of 350–355 nm, quantum yields of 0.04–0.24, and triple exponential fluorescence decays with lifetimes of 4.4–6.6, 1.4–3.2, and 0.2–1.0 ns with relative amplitudes always varying with emission wavelength.

The most striking result of the DAS analysis on the seven cyclic hexapeptides is the extremely strong correlation of DAS λ_{max} and τ_f: in every case but one, ***the shortest wavelength DAS is associated with the shortest*** τ_f ***and the long-wavelength DAS with the longest*** τ_f [17, 91]. The probability of the singular result of perfect correlation for each of the seven peptides is $(1/3!)^7 = 4 \times 10^{-6}$, yet this result was nearly realized. Figure 18 shows this highly improbable correlation, the explanation of which is discussed in this section [24]. A relaxation model is ruled out because the required negative amplitudes in the decay kinetics were not observed in the cyclic peptides [17]. In addition, a single rotamer of the solvent-exposed

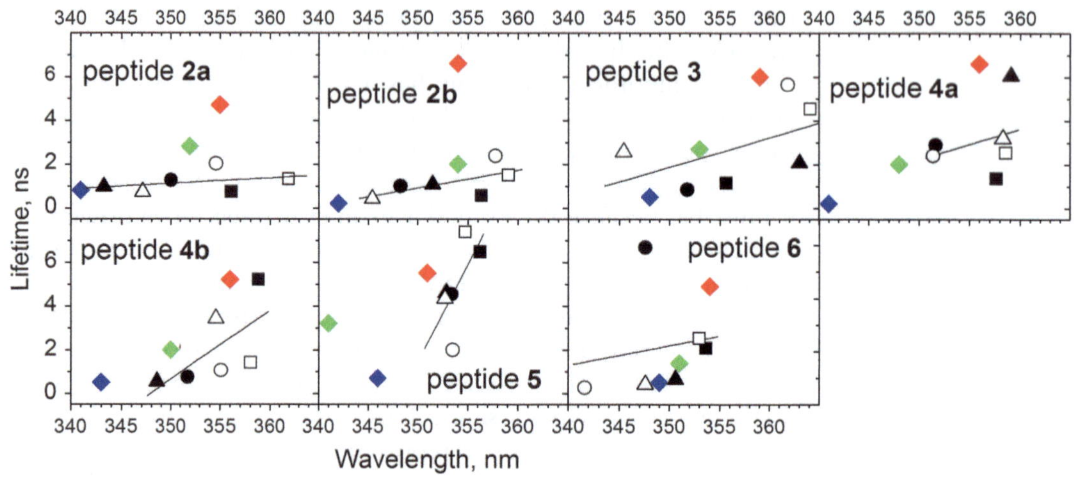

Fig. 18 Wavelength–lifetime correlation in cyclic hexapeptides. Experimental data [23]: τ_f vs. DAS λ_{max} of (*filled red diamond*) long-, (*filled green diamond*) intermediate-, and (*filled blue diamond*) short-lifetime components. Calculated values of τ_{fi} vs. λ_{max} for six Trp rotamers: (*filled square*) $(\chi_1, \chi_2) = (-60°, -90°)$, (*open square*) $(\chi_1, \chi_2) = (-60°, 90°)$; (*filled circle*) $(\chi_1, \chi_2) = (180°, -90°)$, (*open circle*) $(\chi_1, \chi_2) = (180°, 90°)$; and (*filled triangle*) $(\chi_1, \chi_2) = (60°, -90°)$, (*open triangle*) $(\chi_1, \chi_2) = (60°, 90°)$. (*Solid line*) Regression line through calculated points

tryptophan in the cyclic peptides of this study is not expected to exhibit nonexponential decay due to dielectric relaxation of solvent alone in response to the excitation-induced dipole change, because this process experimentally is known to have a time constant of 1.2 ps [16, 93].

The results of these experiments were striking, the evidence for an underlying physical principle so compelling, and the systems so simple that one could not help but believe the physical basis could be discovered. This section describes the results of MD-QM simulations on each of the six canonical rotamers for the seven cyclic peptides that revealed the physical basis.

Early attempts to find a hint of the correlation in simulations met with failure, because the method at first used a small electronic coupling [11, 12]. However, when the much stronger coupling was found from ab initio calculations [18], the correlation emerged strikingly. The stronger coupling, with concomitant greatly increased sensitivity of the electron transfer quenching rate to the energy separation between the CT and fluorescing (^1L_a) states, led naturally to a picture in which those rotamers having shorter λ_{max} (blue shifted) have an increased probability for exhibiting shorter fluorescence lifetimes.

6.1 Results

A typical broad fluorescence spectrum for an aqueous solvent-exposed Trp has an emission peak (λ_{max}) near 350 nm. Simulations show that this is an ensemble average wherein each single molecular λ_{max} fluctuates—in sub-ps times—over a range of 3,000–4,000 cm^{-1} (fwhw) or 40 nm. We find that parameters

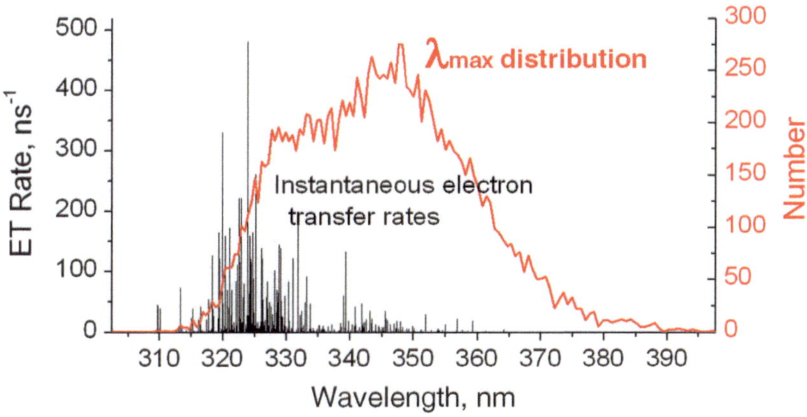

Fig. 19 Typical (*noisy red curve*) distribution of *instantaneous* λ_{max} values calculated from ^1L_a energy and corresponding (*vertical black lines*) calculated *instantaneous* ET rates both derived from a 150-ps QM-MM trajectory. Most of the electron transfer happens during fluctuations in solvation that shift the instantaneous λ_{max} far to the blue. This reveals the basic explanation for the λ_{max}-lifetime correlation observed for the cyclic hexapeptides

providing the observed range of wavelengths and lifetimes predict that only the short-wavelength half of the distribution contributes to quenching by electron transfer from the indole ring to a local backbone amide, with typically only the bluest 25 % contributing 90 % of the electron transfer, as illustrated in Fig. 19. Thus, rotamers with blue-shifted steady-state spectra tend to have short lifetimes. Exceptions occur for rotamers with high lying charge transfer states, which have long lifetimes. This mechanism is expected to be operative in proteins that also show multiexponential decay, even in the absence of rotamer heterogeneity.

The results of the MD + QM studies for 36 of the 42 rotamers are summarized in Fig. 20, in which simulated fluorescence spectra created directly from the distribution of vertical ground → ^1L_a transition energies are shown along with calculated lifetimes for each.

Overall, the simulation results collected in Fig. 20 give a surprisingly accurate account of the experimental picture. Computed rotamer λ_{max} and lifetime values span the observed range, and the specific characteristics exhibited by the individual peptides are captured within the uncertainty of the method and temperature differences between experiment (5 °C) and simulation (27 °C). Major points of agreement are listed below:

- Peptides **2a** and **2b** are quite similar in not having any long-lifetime components and having lower quantum yields when compared to the other peptides, which do have a long-lifetime component.
- Peptide **4a** has a much longer average lifetime than peptide **4b**.

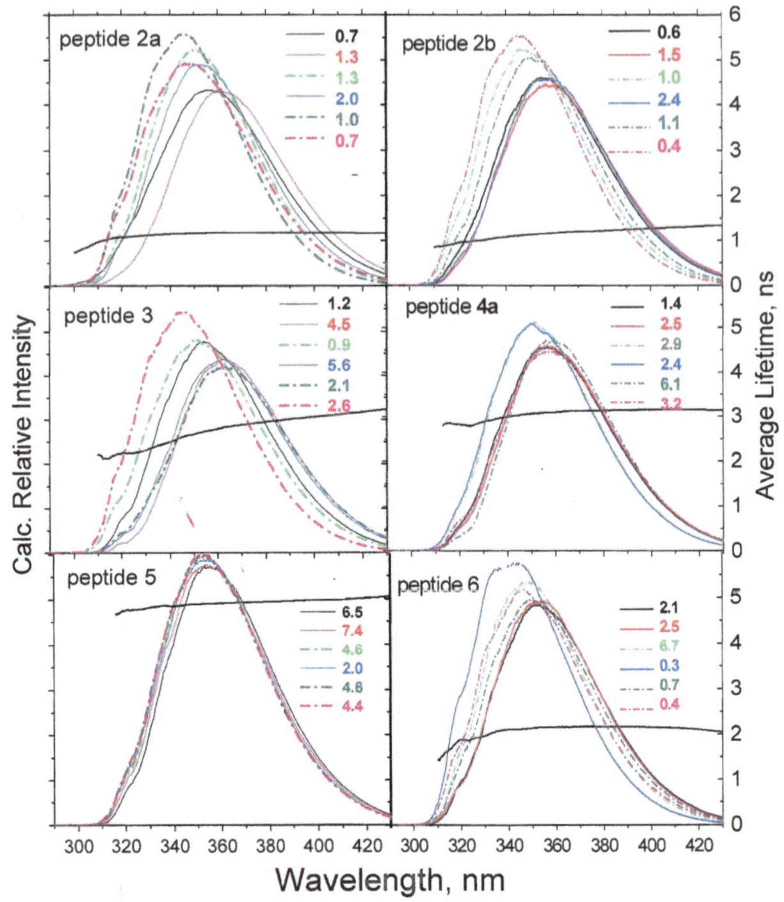

Fig. 20 Calculated fluorescence spectra from QM-MD simulations for each of the six canonical rotamers for six cyclic hexapeptides. The legend gives calculated fluorescence lifetime in nanosecond for the six Trp (χ_1, χ_2) rotamers in descending order: (*black solid line*) (−60°, −90°), (*red solid line*) (−60°, 90°); (*green dash-dot line*) (180°, −90°), (*blue solid line*) (180°, 90°); and (*blue dash-dot line*) (60°, −90°), (*pink dash-dot line*) (60°, 90°). (*Black solid line*) Average lifetime calculated for equal populations of rotamers increases monotonically with wavelength

- All but peptide **5** show large dispersion of DAS λ_{max}.
- Peptide **3** has uniquely the longest DAS λ_{max}.
- The domination of peptide **5** by long λ_{max} values is consistent with its uniquely high quantum yield and long average lifetime.
- Agreement with experiment that peptide **5** is the one exception for which the short-wavelength DAS does not have the shortest lifetime. Only the $(\chi_1, \chi_2) = (180°, 90°)$ rotamer has a somewhat short lifetime with neither the shortest nor longest λ_{max}.

- Peptide **6** uniquely has three rotamers with $\tau_{fi} < 1$ ns and only one rotamer with predicted lifetime >2.5 ns, in harmony experimentally with the major lifetime component of 0.5 ns and a uniquely low quantum yield of 0.04.
- Except for peptides **2a** and **2b**, predicted quantum yields are in the range of experimental values.
- The average lifetime of each peptide, calculated for the arbitrary assumption that the six rotamers are equally populated, increases with increasing wavelength (Fig. 20). This guarantees that decomposition into DAS will show the λ_{max}–τ_f correlation found experimentally.

6.2 Summary

Because of the strong electronic coupling, the lifetime and quantum yield for a given rotamer of a given peptide is extremely sensitive to the energy gap (as it is for any Trp in any environment). The energy gap fluctuates due to essentially independent fluctuations in both CT and ^1L_a state energies. Typically, quenching happens only when downward fluctuations of CT coincide with blue-shifted fluctuations of ^1L_a. The probability of such occurrences is considerably enhanced for those rotamers that have a blue-shifted average λ_{max}, but for a given peptide, the rotamer with the most blue-shifted λ_{max} will not necessarily have the shortest lifetime if it happens to have a high CT state energy.

We conclude that whatever the cause of the heterogeneity, there is a strong propensity for fluorophores whose quantum yield depends upon a variable energy gap, such as Trp, to exhibit red-shifting pseudo-TDFSS, which many times could be difficult to distinguish from TDFSS caused by solvent relaxation.

7 Methods

In this section, we sketch the evolution of our methods for predicting quantum yields and lifetimes over the period 2003 to present. Despite what might appear to be large differences in parameters and procedures, the predictive power and considerable insight level has remained nearly constant. Only the procedure used in Subheading 5 is capable of explicitly revealing—at least partially—nonexponential decay caused by lifetime heterogeneity. In all cases, quantum yield variation is assumed to come only from variations in the quenching by electron transfer to one of the two local backbone amides and is given by Eq. 1. In all cases, the quenching rate constant begins with the Fermi golden rule:

$$k_{et,i}(t) = \frac{2\pi}{\hbar} V_{et,i}^2(t)\rho_i(\Delta E_{00}(t)) = 4\pi^2 c V_{et,i}^2(t)\rho_i(\Delta E_{00}(t)) \quad (11)$$

Table 2
Evolution of method for quenching by local amides

k_{et} equation	V_{et} (cm^{-1})	Geom.	D_{vert} (cm^{-1})	MD engine Sim. length	Force field	Reference
$4\pi^2 c V_{et}{}^2 \rho_i(\langle \Delta E_{vert} + D_{vert} \rangle_t)$	20	CT	−4,000	CHARMM 50 ps, water sphere	CHARMM 22	2003–2004 [11, 12]
$4\pi^2 c \langle V_{et,i}{}^2 \rangle \langle \rho_i(\Delta E_{vert}(t) + D_{vert}) \rangle$	140–800	CT	+4,700	CHARMM 150 ps, water sphere	CHARMM 22	2007 [18]
$4\pi^2 c V_{et}{}^2 \langle \rho_i(\langle \Delta E_{vert,i} \rangle_t + D_{vert}) \rangle$	200	CT	+6,500	CHARMM 150 ps, water sphere	CHARMM 22	2011 [24]
$4\pi^2 c \langle V_{et,i}{}^2 \rangle \langle \rho_i(\langle \Delta E_{vert} \rangle_t + D_{vert}) \rangle$	200–800	G	+240	Gromacs 4.0.775 25 ns, PBC	OPLS	2012 [33]
$4\pi^2 c V_{et,i}{}^2(t) \rho(\Delta E_{vert}(t) + D_{vert})$	200–800	G	+1,940	Gromacs 4.5.4 100 ns, PBC	AMBER99SB-ILDN	2012 [33]

where the index, i, specifies the protein in which the Trp resides or, in the most recent procedure, may refer to a particular subpopulation within a protein. In the convenient right-hand version of Eq. 11, which applies specifically to the use of cm^{-1} units for all energy terms, c is the speed of light in cm s^{-1}, V_{et} is the quantum mechanical electronic coupling matrix element connecting the initial and final states in cm^{-1}, and $\rho(\Delta E_{00})$ is the density of final states in states per cm^{-1}. $\rho(\Delta E_{00})$ is a measure of the overlap integral of the photoelectron spectrum and therefore is a function of the separation of the zeropoint energies of the ^1L_a and CT states, ΔE_{00}. If ΔE_{00} is $>>kT$ (i.e., ~1,000 cm^{-1}), electron transfer is not possible.

In all cases, the important ΔE_{00} variable is assessed with Zerner's semiempirical, spectroscopically calibrated INDO/S-CIS program ("Zindo") which has been quite successful in a variety of applications [94–96], and was modified to shift transition energies by inputting electric fields and potentials at each atom [97, 98]. An additional, essential modification required for reasonable estimation of amide CT energies was the use of new oxygen parameters [99].

Typically, the QM-part of the calculations was performed on a fragment of the protein including capped version of tryptophan plus the preceding amide, making the QM molecule *N*-formyltryptophanamide. The electrostatic environment is introduced from the potential at each QM atom caused by all non-QM atoms. In early versions (2001–2007), we labeled the method as QM-MM or QM-MD. MD was always used, and QM was driving the MD only by providing QM charges to the non-QM atoms. CHARMM [100] was the MD engine, and the protein was embedded in a water sphere. Since 2010, Gromacs [101] was the MD engine, and the protein was embedded in a water cube with periodic boundary conditions. The method was called MD + QM because charges on the Trp were kept constant, usually the default force-field ground-state values. QM was then run asynchronously after the MD coordinate frames were stored on disk.

Table 2 summarizes the major developments for estimating quenching by local amides, which is responsible for the bulk of quantum yield variations. The first column indicates the various approaches to evaluating V_{et} and $\rho(\Delta E_{vert} + D_{vert})$. ΔE_{vert} is the energy gap given by the vertical transition energies determined by the quantum mechanical procedure, and D_{vert} is the shift to convert ΔE_{vert} to ΔE_{00}, the difference of the two zero point energies. The D_{vert} adjustment is necessary because we run the MD trajectories at one geometry only, meaning that ΔE_{vert} contains terms due to stretching or compressing bonds away from the minimum of state potential surfaces. In the early work, MD was run with the QM (Zindo) part in the CT geometry, meaning that ground and ^1L_a energies are raised. The difference between the ground → CT and ground → ^1L_a vertical transition energies are therefore much smaller than the true ΔE_{00} energy gap. It is reasonable therefore to add an appropriate constant to the QM ΔE_{vert}. In the latest simulations, ground-state geometry was used to avoid the awkward task of creating special topology files for the excited states. In these cases, the CT state is raised in energy much more than the ^1L_a state, and a constant should be subtracted from the vertical energy gap. D_{vert} is always used as a fitting constant because both the underlying quantum and MD procedure methods providing ΔE_{vert} are semiempirical and will always depend on the choice of program and force field.

In the original method [11, 12], both V_{et} and D_{vert} were strictly empirical parameters that gave the best fit, and their unphysical nature was pointed out. The small ad hoc V_{et} of 20 cm^{-1} was compensated by subtracting 4,000 cm^{-1} from ΔE_{vert}, thereby creating an artificially small energy gap, even though it was recognized that because CT state geometry was used, D_{vert} should have been added. This was corrected in the next version following computation of ab initio values for V_{et} as a function of the dihedral angles χ_1, χ_2, and ψ for a model tryptamide [18]. Despite an immense difference in the values of V_{et} used, the fit of calculated vs. experimental quantum yield was not significantly different. This constancy of usefulness seems to lie in an invariance in the product of the V^2 and FCDS achieved by shifting the FCDS function of ΔE_{00}.

Common to all procedures, a crystal structure for a protein was taken directly from the protein data bank. For the GROMACS-based procedures, each protein was solvated in a water cube so at least 1 nm of tip3p water separated the surface of the protein and the edge of the cube. Either sodium or chloride ions were usually added to neutralize the overall charge of the protein. An energy minimization was carried out until the force on all atoms was less than 1,000 kJ/mol nm. NVT and NPT equilibrations where carried out for 100 ps each. In these equilibrations, the time step was set to 2 fs, and the positions of all protein atoms were constrained to their respective minimized coordinates with a force constant of 1,000 kJ/mol nm. Electrostatics were calculated using particle mesh Ewald. The temperature was set at 300 K, and the protein and solvent were coupled separately to a velocity rescale thermostat with a time constant of 0.1 ps. Pressure was set at 1 bar using a Parrinello-Rahman barostat with a 2 ps time constant and a compressibility of 4.5×10^{-5} bar^{-1}. Periodic boundary conditions were used with a cubic box as the unit cell. After equilibration, production simulations were carried out under NPT conditions, with pressure and temperature again set at 1.0 bar and 300 K, respectively, using velocity rescaling and a Parrinello-Rahman barostat using the same time constants and compressibility. Coordinates were typically saved every 2 ps. All amino acids had ground-state charges given by the force field used, including tryptophan.

For the 100-ns simulations, the AMBER99SB-ILDN force field was used in conjunction with a more detailed analysis that featured updating all aspects of the Fermi rule expression at each time frame. A table in which V_{et} was listed as a function of the dihedral angles χ_1, χ_2, and ψ from ab initio calculations for a model tryptamide [18]. The GROMACS utility g_angle was used to calculate the dihedral angles χ_1, χ_2, and ψ as a function of time, which gave $V_{\mathrm{et}}(t)$ by interpolation. Along with instantaneous Franck-Condon densities, the quenching rate as a function of time was obtained.

7.1 Survival Curves

The time series of instantaneous rates was used to construct a series of survival

$$\begin{aligned}
&\frac{\mathrm{d}P(t)}{\mathrm{d}t} = -k(t)P(t) \\
&\int_{^1\mathrm{L_a}(0)}^{^1\mathrm{L_a}(t)} \frac{\mathrm{d}P(t')}{P(t')} = -\int_0^t [k_0 + k_{\mathrm{et}}(t')]\mathrm{d}t' = -\left(k_0 t + \int_0^t k_{\mathrm{et}}(t')\mathrm{d}t'\right) \\
&\ln\left(\frac{P(t)}{P(0)}\right) = -\left(k_0 t + \int_0^t k_{\mathrm{et}}(t')\mathrm{d}t'\right) \\
&P(0) = 1 \\
&P(t) = P(0)\exp\left(-\left(k_0 t + \int_0^t k_{\mathrm{et}}(t')\mathrm{d}t'\right)\right)
\end{aligned} \tag{12}$$

curves by integration of the first-order rate equation: where $k_0 = k_{\mathrm{r}} + k_{\mathrm{nr}}$ which are the usual radiative and nonradiative decay constants for 3-methylindole and $k_{\mathrm{et}}(t)$ is the instantaneous electron transfer rate constant at time t. The window for each survival curve was 25 ns, which is three times the lifetime of a completely unquenched tryptophan. After calculating and saving a survival curve, the window was moved 0.05 ns in time, and a new survival curve was calculated and saved. This was done until the last frame of the window reached the end of the simulation. With this procedure, a total of 1,601 survival curves were calculated. All of these survival curves were used to calculate an average survival curve by averaging all the populations at time t. The quantum yield is then calculated using the average survival curve using the following:

$$\Phi = \frac{\int_0^{t'} \langle P(t)\rangle \mathrm{d}t}{\int_0^{t'} \exp(-k_{\mathrm{r}})\mathrm{d}t} = \frac{\int_0^{t'} \langle P(t)\rangle \mathrm{d}t}{\frac{1}{k_{\mathrm{r}}}} = k_{\mathrm{r}} \int_0^{t'} \langle P(t)\rangle \mathrm{d}t \tag{13}$$

Acknowledgements

We gratefully acknowledge support from NSF Grants MCB-0446542 and MCB-0847047, and crucial supercomputer resource TRAC grants MCB090176 from Teragrid/XSEDE for the years 2009–2012. We also acknowledge contributions to the work presented here by Drs. James Vivian, Tiqing Liu, Alexander Petrenko, Pedro Muino, and Chia-Pin Pan. We thank Drs. Ludwig Brand, Bruce Hudson, Mary Barkley, Jay Knutson, Dimitri Toptygin, Robert Woody, Andy Albrecht, Jonathan King, Linda Kurz, Jaap Broos, Dongping Zhong, Jan Kubelka, William Eaton, and J. Michael Schurr for collaboration, inspiration, and helpful conversations.

References

1. Warshel A (1991) Computer modeling of chemical reactions in enzymes and solutions. Wiley-Interscience, New York
2. Johnson ET, Parson WW (2002) Electrostatic interactions in an integral membrane protein. Biochemistry 41:6483–6494
3. Warshel A, Sharma PK, Kato M et al (2006) Modeling electrostatic effects in proteins. Biochim Biophys Acta 1764:1647–1676
4. Olsson MHM, Parson WW, Warshel A (2006) Dynamical contributions to enzyme catalysis: critical tests of a popular hypothesis. Chem Rev 106:1737–1756
5. McGeagh JD, Ranaghan KE, Mulholland AJ (2011) Protein dynamics and enzyme catalysis: insights from simulations. Biochim Biophys Acta 1814:1077–1092
6. Muino PL, Harris D, Berryhill J et al (1992) Simulation of solvent dynamics effects on the fluorescence of 3-methylindole in water. Proc SPIE Int Soc Opt Eng 1640:240–250
7. Muiño PL, Callis PR (1994) Hybrid simulations of solvation effects on electronic spectra: indoles in water. J Chem Phys 100:4093–4109
8. Callis PR (1997) 1La and 1Lb transitions of tryptophan: applications of theory and experimental observations to fluorescence of proteins. Methods Enzymol 278:113–150
9. Callis PR, Burgess BK (1997) Tryptophan fluorescence shifts in proteins from hybrid simulations: an electrostatic approach. J Phys Chem 101:9429–9432
10. Vivian JT, Callis PR (2001) Mechanisms of tryptophan fluorescence shifts in proteins. Biophys J 80:2093–2109
11. Callis PR, Vivian JT (2003) Understanding the variable fluorescence quantum yield of tryptophan in proteins using QM MM simulations. Quenching by charge transfer to the peptide backbone. Chem Phys Lett 369:409–414
12. Callis PR, Liu T (2004) Quantitative prediction of fluorescence quantum yields for tryptophan in proteins. J Phys Chem B 108:4248–4259
13. Kurz LC, Fite B, Jean J et al (2005) Photophysics of tryptophan fluorescence: link with the catalytic strategy of the citrate synthase from Thermoplasma acidophilum. Biochemistry 44:1394–1413
14. Chen JJ, Flaugh SL, Callis PR et al (2006) Mechanism of the highly efficient quenching of tryptophan fluorescence in human gamma D-crystallin. Biochemistry 45:11552–11563
15. Callis PR, Liu TQ (2006) Short range photoinduced electron transfer in proteins: QM-MM simulations of tryptophan and flavin fluorescence quenching in proteins. Chem Phys 326:230–239
16. Xu JH, Toptygin D, Graver KJ et al (2006) Ultrafast fluorescence dynamics of tryptophan in the proteins monellin and IIA(Glc). J Am Chem Soc 128:1214–1221
17. Pan CP, Callis PR, Barkley MD (2006) Dependence of tryptophan emission wavelength on conformation in cyclic hexapeptides. J Phys Chem B 110:7009–7016
18. Callis PR, Petrenko A, Muino PL et al (2007) Ab initio prediction of tryptophan fluorescence quenching by protein electric field enabled electron transfer. J Phys Chem B 111:10335–10339
19. Muiño PL, Callis PR (2009) Solvent effects on the fluorescence quenching of tryptophan

by amides via electron transfer. Experimental and computational studies. J Phys Chem B 113:2572–2577

20. Xu J, Chen J, Toptygin D et al (2009) Femtosecond fluorescence spectra of tryptophan in human gamma-crystallin mutants: site-dependent ultrafast quenching. J Am Chem Soc 131:16751–16757
21. Chen J, Callis PR, King J (2009) Mechanism of the very efficient quenching of tryptophan fluorescence in human gamma D- and gamma S-crystallins: the gamma-crystallin fold may have evolved to protect tryptophan residues from ultraviolet photodamage. Biochemistry 48:3708–3716
22. Callis PR (2009) Exploring the electrostatic landscape of proteins with tryptophan fluorescence. In: Geddes CD (ed) Reviews in fluorescence 2007, vol 4. Springer, New York, pp 199–248
23. Callis PR (2011) Predicting fluorescence lifetimes and spectra of biopolymers. Methods Enzymol 487:1–38
24. Pan CP, Muino PL, Barkley MD et al (2011) Correlation of tryptophan fluorescence spectral shifts and lifetimes arising directly from heterogeneous environment. J Phys Chem B 115:3245–3253
25. Tusell JR, Callis PR (2012) Simulations of tryptophan fluorescence dynamics during folding of the villin headpiece. J Phys Chem B 116:2586–2594
26. Luo GB, Andricioaei I, Xie XS et al (2006) Dynamic distance disorder in proteins is caused by trapping. J Phys Chem B 110:9363–9367
27. Warshel A, Sharma PK, Kato M et al (2006) Electrostatic basis for enzyme catalysis. Chem Rev 106:3210–3235
28. Prakash MK, Marcus RA (2007) An interpretation of fluctuations in enzyme catalysis rate, spectral diffusion, and radiative component of lifetimes in terms of electric field fluctuations. Proc Natl Acad Sci USA 104:15982–15987
29. Prakash MK, Marcus RA (2008) Dielectric dispersion interpretation of single enzyme dynamic disorder, spectral diffusion, and radiative fluorescence lifetime. J Phys Chem B 112:399–404
30. Shaw DE, Maragakis P, Lindorff-Larsen K et al (2010) Atomic-level characterization of the structural dynamics of proteins. Science 330:341–346
31. Vendruscolo M, Dobson CM (2011) Protein dynamics: Moore's law in molecular biology. Curr Biol 21:R68–R70
32. Zhong DP, Pal SK, Zewail AH (2011) Biological water: a critique. Chem Phys Lett 503:1–11
33. Tusell JR. Computation of fluorescence quenching by amide and histidine. Ph.D. Dissertation, Montana State University, Bozeman; 2011.
34. Tusell JR, Callis PR (2012) Computational predictions of exponential and non-exponential tryptophan fluorescence decay in NATA, the villin headpiece subdomain, and other proteins. Biophys J 102:217a
35. Broos J, Tveen-Jensen K, de Waal E et al (2007) The emitting state of tryptophan in proteins with highly blue-shifted fluorescence. Angew Chem Int Ed Engl 46:5137–5139
36. Jensen TJ, Strambini G, Gonnelli G et al (2008) Mutations in transhydrogenase change the fluorescence emission state of trp-72 from ^1L_a to ^1L_b. Biophys J 95 (7):3419–3428
37. Chen Y, Barkley MD (1998) Toward understanding tryptophan fluorescence in proteins. Biochemistry 37:9976–9982
38. Jimenez R, Fleming GR, Kumar PV et al (1994) Femtosecond solvation dynamics of water. Nature 369:471–473
39. Boens N, Andriessen R, Ameloot M et al (1992) Kinetics and identifiability of intramolecular 2-state excited-state processes—global compartmental analysis of the fluorescence decay surface. J Phys Chem 96:6331–6342
40. Vandommelen L, Boens N, Ameloot M et al (1993) Species-associated spectra and upper and lower bounds on the rate constants of reversible intramolecular 2-state excited-state processs with added quencher—global compartmental analysis of the fluorescence decay surface. J Phys Chem 97:11738–11753
41. Van Dommelen L, Boens N, Ameloot M (1998) Time-resolved fluorescence of glucagon studied by global compartmental analysis. J Phys Chem B 102:3287–3294
42. Ameloot M, Beechem JM, Brand L (1986) Compartmental modeling of excited-state reactions—identifiability of the rate constants from fluorescence decay surfaces. Chem Phys Lett 129:211–219
43. Lofroth JE (1986) Time-resolved emission-spectra, decay-associated spectra, and species-associated spectra. J Phys Chem 90:1160–1168
44. Petrich JW, Chang MC, McDonald DB et al (1983) On the origin of nonexponential fluorescence decay in tryptophan and its derivatives. J Am Chem Soc 105:3824–3832
45. Bushueva TL, Busel EP, Burstein EA (1975) The interaction of protein functional groups with indole chromophore III. Amine, amide, and thiol groups. Stud Biophys Berlin 52:41–52

46. Ricci RW, Nesta JM (1976) Inter- and intramolecular quenching of indole fluorescence by carbonyl compounds. J Phys Chem 80:974–980
47. Froehlich PM, Nelson K (1978) Fluorescence quenching of indoles by amides. J Phys Chem 82:2401–2403
48. Chen Y, Liu B, Yu H-T et al (1996) The peptide bond quenches indole fluorescence. J Am Chem Soc 118:9271–9278
49. Rolinski OJ, Scobie K, Birch DJS (2009) Protein fluorescence decay: A gamma function description of thermally induced interconversion of amino acid rotamers. Phys Rev E 79(5 Pt 1):050901
50. Buchner GS, Murphy RD, Buchete NV et al (2011) Dynamics of protein folding: probing the kinetic network of folding-unfolding transitions with experiment and theory. Biochim Biophys Acta 1814:1001–1020
51. Kubelka J, Henry ER, Cellmer T et al (2008) Chemical, physical, and theoretical kinetics of an ultrafast folding protein. Proc Natl Acad Sci USA 105:18655–18662
52. Willaert K, Engelborghs Y (1991) The quenching of tryptophan fluorescence by protonated and unprotonated imidazole. Eur Biophys J 20:177–182
53. Willaert K, Loewenthal R, Sancho J et al (1992) Determination of the excited-state lifetimes of the tryptophan residues in barnase, via multifrequency phase fluorometry of tryptophan mutants. Biochemistry 31:711–716
54. Eaton WA, Munoz V, Thompson PA et al (1998) Kinetics and dynamics of loops, alpha-helices, beta-hairpins, and fast-folding proteins. Acc Chem Res 31:745–753
55. Thompson PA, Munoz V, Jas GS et al (1999) Using tryptophan fluorescence to probe helix-coil kinetics. Biophys J 76:A175
56. Freddolino PL, Schulten K (2009) Common structural transitions in explicit-solvent simulations of villin headpiece folding. Biophys J 97:2338–2347
57. Ensign DL, Kasson PM, Pande VS (2007) Heterogeneity even at the speed limit of folding: large-scale molecular dynamics study of a fast-folding variant of the villin headpiece. J Mol Biol 374:806–816
58. Kubelka J, Chiu TK, Davies DR et al (2006) Sub-microsecond protein folding. J Mol Biol 359:546–553
59. Cellmer T, Buscaglia M, Henry ER et al (2011) Making connections between ultrafast protein folding kinetics and molecular dynamics simulations. Proc Natl Acad Sci USA 108:6103–6108
60. Reiner A, Henklein P, Kiefhaber T (2010) An unlocking/relocking barrier in conformational fluctuations of villin headpiece subdomain. Proc Natl Acad Sci USA 107:4955–4960
61. Beratan DN, Betts JN, Onuchic JN (1991) Protein electron transfer rates set by the bridging secondary and tertiary structure. Science 252:1285–1288
62. Risser SM, Beratan DN, Meade TJ (1993) Electron-transfer in DNA—predictions of exponential-growth and decay of coupling with donor-acceptor distance. J Am Chem Soc 115:2508–2510
63. Kubelka J, Eaton WA, Hofrichter J (2003) Experimental tests of villin subdomain folding simulations. J Mol Biol 329:625–630
64. Lindorff-Larsen K, Maragakis P, Piana S et al (2012) Systematic validation of protein force fields against experimental data. PLoS One 7: e32131
65. Hennecke J, Sillen A, Huber WM et al (1997) Quenching of tryptophan fluorescence by the active-site disulfide bridge in the DsbA protein from Escherichia coli. Biochemistry 36:6391–6400
66. Sillen A, Hennecke J, Roethlisberger D et al (1999) Fluorescence quenching in the DsbA protein from Escherichia coli: the complete picture of the excited-state energy pathway and evidence for the reshuffling dynamics of the microstates of tryptophan. Proteins 37:253–263
67. Sillen A, Diaz JF, Engelborghs Y (2000) A step toward the prediction of the fluorescence lifetimes of tryptophan residues in proteins based on structural and spectral data. Protein Sci 9:158–169
68. Toptygin D, Savtchenko RS, Meadow ND et al (2001) Homogeneous spectrally- and time-resolved fluorescence emission from single-tryptophan mutants of IIA^{Glc}. J Phys Chem B 105:2043–2055
69. Toptygin D, Gronenborn AM, Brand L (2006) Nanosecond relaxation dynamics of protein GB1 identified by the time-dependent red shift in the fluorescence of tryptophan and 5-fluorotryptophan. J Phys Chem B 110:26292–26302
70. Peon J, Pal SK, Zewail AH (2002) Hydration at the surface of the protein Monellin: dynamics with femtosecond resolution. Proc Natl Acad Sci USA 99:10964–10969
71. Zhang LY, Yang Y, Kao YT et al (2009) Protein hydration dynamics and molecular mechanism of coupled water-protein fluctuations. J Am Chem Soc 131:10677–10691

72. Abbyad P, Shi XH, Childs W et al (2007) Measurement of solvation responses at multiple sites in a globular protein. J Phys Chem B 111:8269–8276
73. Knutson JR, Beechem JM, Brand L (1983) Simultaneous analysis of multiple fluorescence decay curves—a global approach. Chem Phys Lett 102:501–507
74. Beechem JM, Knutson JR, Ross JBA et al (1983) Global resolution of heterogeneous decay by phase modulation fluorometry—mixtures and proteins. Biochemistry 22:6054–6058
75. Knutson JR, Walbridge DG, Brand L (1983) Studies of ligand-binding to alcohol-dehydrogenase with decay-associated fluorescence spectroscopy. Biophys J 41:A268
76. Davenport L, Knutson JR, Brand L (1983) Fluorescence studies of the interaction of equilenin with liposomes. Biophys J 41:A373
77. Knutson JR, Walbridge DG, Brand L (1982) Decay-associated fluorescence-spectra and the heterogeneous emission of alcohol-dehydrogenase. Biochemistry 21:4671–4679
78. De-Beuckeleer K, Volckaert G, Engelborghs Y (1999) Time resolved fluorescence and phosphorescence properties of the individual tryptophan residues of barnase: evidence for protein-protein interactions. Proteins 36:42–53
79. Gastmans M, Volckaert G, Engelborghs Y (1999) Tryptophan microstate reshuffling upon the binding of cyclosporin A to human cyclophilin A. Proteins 35:464–474
80. Hogue CWV, Doublie S, Xue H et al (1996) A concerted tryptophanyl-adenylate-dependent conformational change in Bacillus subtilis tryptophanyl-tRNA synthetase revealed by the fluorescence of Trp92. J Mol Biol 260:446–466
81. Dahms TES, Willis KJ, Szabo AG (1995) Conformational heterogeneity of tryptophan in a protein crystal. J Am Chem Soc 117:2321–2326
82. Szabo AG, Rayner DM (1980) Fluorescence decay of tryptophan conformers in aqueous solution. J Am Chem Soc 102:554–563
83. Hudson BS, Huston JM, Soto-Campos G (1999) A reversible "dark state" mechanism for complexity of the fluorescence of tryptophan in proteins. J Phys Chem A 103:2227–2234
84. Merrill AR, Steer BA, Prentice GA et al (1997) Identification of a chameleon-like pH-sensitive segment within the colicin E1 channel domain that may serve as the pH-activated trigger for membrane bilayer association. Biochemistry 36:6874–6884
85. Willis KJ, Szabo AG (1992) Conformation of parathyroid-hormone—time-resolved fluorescence studies. Biochemistry 31: 8924–8931
86. Alberti P, Bombarda E, Kintrup M et al (1997) Structural investigation of Tet repressor loop 154–167: a time-resolved fluorescence study of three single Trp mutants. Arch Biochem Biophys 346:230–240
87. Neyroz P, Menna C, Polverini E et al (1996) Intrinsic fluorescence properties and structural analysis of p13(suc1) from *Schizosaccharomyces pombe*. J Biol Chem 271:27249–27258
88. Kim SJ, Chowdhury FN, Stryjewski W et al (1993) Time-resolved fluorescence of the single tryptophan of *Bacillus stearothermophilus phosphofructokinase*. Biophys J 65:215–226
89. Lakowicz J (2000) On spectral relaxation in proteins. Photochem Photobiol 72:421–437
90. Lakowicz JR (2006) Principles of fluorescence spectroscopy, 3rd edn. Springer, New York
91. Pan CP, Barkley MD (2004) Conformational effects on tryptophan fluorescence in cyclic hexapeptides. Biophys J 86:3828–3835
92. Adams PD, Chen Y, Ma K et al (2002) Intramolecular quenching of tryptophan fluorescence by the peptide bond in cyclic hexapeptides. J Am Chem Soc 124:9278–9286
93. Pal SK, Peon J, Zewail AH (2002) Biological water at the protein surface: dynamical solvation probed directly with femtosecond resolution. Proc Natl Acad Sci USA 99:1763–1768
94. Ridley J, Zerner M (1973) Intermediate neglect of differential overlap (INDO) technique for spectroscopy: pyrrole and the azines. Theor Chim Acta (Berl) 32:111–134
95. Thompson MA, Zerner MC (1991) A theoretical examination of the electronic structure and spectroscopy of the photosynthetic reaction center from *Rhodopseudomonas viridis*. J Am Chem Soc 113:8210–8215
96. Cory MG, Zerner MC, Xu XC et al (1998) Electronic excitations in aggregates of bacteriochlorophylls. J Phys Chem B 102:7640–7650
97. Theiste D, Callis PR, Woody RW (1991) Effects of the crystal field on transition moments in 9-ethylguanine. J Am Chem Soc 113:3260–3267
98. Sreerama N, Woody RW, Callis PR (1994) Theoretical study of the crystal field effects on

the transition dipole moments in methylated adenines. J Phys Chem 98:10397–10407

99. Li J, Williams B, Cramer CJ et al (1999) A class IV charge model for molecular excited states. J Phys Chem 110:724–733
100. MacKerell AD Jr, Bashford D, Bellott M et al (1998) All atom empirical potential for molecular modeling and dynamics studies of proteins. J Phys Chem B 102: 3586–3616
101. Hess B, Kutzner C, van der Spoel D et al (2008) Algorithms for highly efficient, load-balanced, and scalable molecular simulation. J Chem Theory Comput 4:435–447
102. Qiu L, Pabit SA, Roitberg AE et al (2002) Smaller and faster: the 20-residue Trp-cage protein folds in 4μs. J Am Chem Soc 124:12952–12953
103. Egan DA, Logan TM, Liang H et al (1993) Equilibrium denaturation of recombinant human FK binding protein in urea. Biochemistry 32:1920–1927
104. Eftink MR (1991) Fluorescence techniques for studying protein structure. Methods Biochem Anal 35:127–205
105. Harris DL, Hudson BS (1991) Fluorescence and molecular dynamics study of the internal motion of the buried tryptophan in bacteriophage T4 lysozyme: effects of temperature and alteration of nonbonded networks. Chem Phys 158:353–382
106. Kubelka J. Unpublished work. 2011.

Chapter 9

Analysis of Time-Dependent Red Shifts in Fluorescence Emission from Tryptophan Residues in Proteins

Dmitri Toptygin

Abstract

Instantaneous fluorescence emission spectra measured at different times after excitation often shift to the red as the delay between the excitation pulse and fluorescence detection is increased. In the case of Trp fluorescence in proteins, the time-dependent red shift (TDRS) may have its origins in relaxation, heterogeneity, or a mixture of the two. In those cases where it is possible to rule out the contribution of heterogeneity, the TDRS can be used to study nonequilibrium relaxation dynamics of the protein matrix and the solvent on the picosecond and nanosecond time scales. Here we describe the experimental and computational procedures involved in recording spectrally and time-resolved fluorescence, detecting heterogeneity, and extracting information about protein/solvent relaxation dynamics.

Key words Tryptophan fluorescence, Time-correlated single-photon counting, Solvent relaxation, Dynamic Stokes shift, Time-dependent red shift, Heterogeneous fluorescence

1 Introduction

Fluorescence emission intensity from a large ensemble of fluorophores excited at time zero by a very short pulse of radiation (δ-excitation) can be described by a function of two variables: a spectral variable and the time t after excitation. Two alternative spectral variables will be used here: λ, the emission wavelength, and ν, the emission wavenumber ($\nu = 1/\lambda$). Depending on the choice of the spectral variable, the function of two variables will be denoted either $F_\lambda(\lambda,t)$ or $F_\nu(\nu,t)$. These functions can be thought of as surfaces in 3D space, where the x coordinate is the spectral variable (λ or ν), the y coordinate is the time (t), and the z coordinate is the fluorescence emission intensity. Slices of these surfaces by the planes parallel to the x and z axes represent instantaneous fluorescence emission spectra at different times after excitation. Slices of these surfaces by the planes parallel to the y and z axes

Yves Engelborghs and Antonie J.W.G. Visser (eds.), *Fluorescence Spectroscopy and Microscopy: Methods and Protocols*, Methods in Molecular Biology, vol. 1076, DOI 10.1007/978-1-62703-649-8_9,

resemble fluorescence decay curves measured at different emission wavelengths or wavenumbers.

The functions $F_\lambda(\lambda,t)$ and $F_\nu(\nu,t)$ contain information about the heterogeneity of the excited fluorophore ensemble and about the nonequilibrium relaxation dynamics triggered by the excitation. Theoretically, four different combinations are possible. First, we will consider the case where the ensemble is homogeneous and there is no relaxation in the excited state. In this case each of the functions $F_\lambda(\lambda,t)$ and $F_\nu(\nu,t)$ is a product of one function of the spectral variable and the other function of time, i.e.,

$$F_\lambda(\lambda,t) = f_\lambda(\lambda)\cdot\frac{1}{\tau}\exp(-t/\tau) \tag{1}$$

$$F_\nu(\nu,t) = f_\nu(\nu)\cdot\frac{1}{\tau}\exp(-t/\tau) \tag{2}$$

Here the functions $f_\lambda(\lambda)$ and $f_\nu(\nu)$ represent the spectral densities of the emitted photons on the wavelength scale and on the wavenumber scale, and τ is the lifetime of the excited state. Note that the decay curves are monoexponential and the lifetime has the same value at all emission wavelengths (or wavenumbers). If these conditions are not met, then the functions $F_\lambda(\lambda,t)$ and $F_\nu(\nu,t)$ cannot be represented in the form 1 and 2, and therefore the ensemble is heterogeneous and/or relaxation is present.

The second case to be considered involves a heterogeneous fluorophore ensemble without relaxation. For the sake of simplicity, we will assume that the heterogeneous ensemble consists of a finite number N_S of homogeneous sub-ensembles, the emission of which can be described by Eqs. 1 or 2 but with different spectral distributions $f_{\lambda n}(\lambda)$ and $f_{\nu n}(\nu)$ and different lifetimes τ_n, where the index $n = 1,2,\ldots,N_S$ denotes different populations. Since fluorescence emission intensities are additive, the functions $F_\lambda(\lambda,t)$ and $F_\nu(\nu,t)$ are merely the sums over all sub-ensembles:

$$F_\lambda(\lambda,t) = \sum_{n=1}^{N_S} f_{\lambda n}(\lambda)\cdot\frac{1}{\tau_n}\exp(-t/\tau_n) \tag{3}$$

$$F_\nu(\nu,t) = \sum_{n=1}^{N_S} f_{\nu n}(\nu)\cdot\frac{1}{\tau_n}\exp(-t/\tau_n) \tag{4}$$

The third case to be considered involves a homogeneous ensemble with relaxation. It has been shown [1–3] that the emission from a homogeneous ensemble of fluorophores in a relaxing host matrix can be represented in the form

$$F_\nu(\nu,t) = \nu^3 S[\nu - \nu_c(t)]D(t) \tag{5}$$

Here the factor ν^3 takes care of the fact that under all circumstances the spontaneous emission rate is proportional to the cube of the emission frequency. The function $S(x)$ represents the vibrational envelope of the electronic transition responsible for fluorescence emission; by definition [1, 2] this function has the following properties: $\int S(x)\mathrm{d}x = 1$, $\int xS(x)\mathrm{d}x = 0$, where the integration is carried out from $-\infty$ to $+\infty$. The function $\nu_c(t)$ represents the time variation of the spectral center of gravity of $F_\nu(\nu,t)/\nu^3$,

$$\nu_c(t) = \frac{\int_0^\infty \nu^{-2} F_\nu(\nu, t)\mathrm{d}\nu}{\int_0^\infty \nu^{-3} F_\nu(\nu, t)\mathrm{d}\nu} \quad (6)$$

This function describes the dynamics of the time-dependent red shift (TDRS) and contains information about the relaxation of the protein matrix and the solvent. The function $D(t)$ represents the time variation of the product of the squared magnitude of the electronic transition dipole moment and the size of the excited-state population [1–3]. It is also known as the damping function [4–7]. If we disregard the time variation of the electronic transition dipole moment, then $D(t)$ will also represent the decay of the excited-state population [3]. Whether the transition dipole moment is constant or varies with time, the damping function can be expressed directly in terms of $F_\nu(\nu,t)$:

$$D(t) = \int_0^\infty \nu^{-3} F_\nu(\nu, t)\mathrm{d}\nu \quad (7)$$

The functions $\nu_c(t)$ and $D(t)$ can be parametrized using linear combinations of exponential terms. It has been shown [1, 2] that such parametrization together with the expansion of the function $S(x)$ in Taylor series makes it possible to transform Eq. 5 to the following form:

$$F_\nu(\nu, t) = \sum_{n=1}^{N_E} \alpha_{\nu n}(\nu) \cdot \exp(-t/\tau_n) \quad (8)$$

The first theoretical derivation of an expression equivalent to Eq. 8 for an ensemble of fluorophores exhibiting a time-dependent spectral shift can be found in the work of Gakamsky et al. [8, 9]. Theoretically, the number of exponential terms N_E in Eq. 8 is infinite, which is a direct consequence of the infinite number of terms in the Taylor expansion of $S(x)$. However, in the Taylor series the term becomes smaller and smaller with an increase in the term power, which makes it possible to truncate the series after just a few terms. Even if we do not neglect very small terms in the Taylor series, all but a finite number of the terms in the sum Eq. 8 will have their time constants τ_n shorter than the time resolution of the experimental setup that will be used to measure spectrally and time-resolved fluorescence. This means that for all practical purposes, N_E is a finite number and usually quite small (does not exceed 7).

Equations 4 and 8 are remarkably similar; if we set $N_S = N_E$ and $f_{\nu n}(\nu)/\tau_n = \alpha_{\nu n}(\nu)$, then they become completely identical. Yet, Eq. 4 describes the emission from a heterogeneous ensemble of fluorophores without relaxation, while Eq. 8 describes the emission from a homogeneous ensemble of fluorophores in the presence of relaxation. This shows that it may be difficult to experimentally distinguish between the two cases. From the conceptual point of view, the main difference between the two cases is that in Eq. 4 N_S is the number of different fluorescent species with different lifetimes τ_n, whereas in Eq. 8 N_E is not the number of species and τ_n are not even lifetimes. Another important difference is that since the functions $f_{\nu n}(\nu)$ represent the steady-state emission spectra of different species, the values of $f_{\nu n}(\nu)$ are not allowed to be negative in any part of the emission spectrum. On the contrary, the functions $\alpha_{\nu n}(\nu)$ are, generally, some linear combinations of the function $\nu^3 S(\nu-\nu_c)$ with its derivatives with respect to ν_c. The derivatives have both positive and negative values; therefore the functions $\alpha_{\nu n}(\nu)$ often have negative values. Thus, the presence of negative values in some of $\alpha_{\nu n}(\nu)$ can be used as the evidence that relaxation takes place in the fluorophore ensemble. Unfortunately, the absence of negative values in $\alpha_{\nu n}(\nu)$ does not rule out relaxation, as well as the presence of negative values in $\alpha_{\nu n}(\nu)$ does not rule out heterogeneity.

If the fluorophore ensemble is permanently heterogeneous, then the TDRS calculated in accordance with Eq. 6 contains contributions from both relaxation and heterogeneity, which are impossible to separate; this TDRS is not an accurate representation of protein and solvent relaxation dynamics. For this reason it is important to have a criterion to distinguish between a homogeneous and a heterogeneous fluorophore ensemble. One such criterion [2] uses the fact that for a homogeneous ensemble, the full width at half maximum FWHM (along the variable ν) of the function $F_\nu(\nu,t)/\nu^3$ does not change with time. Furthermore, when $F_\nu(\nu,t)/\nu^3$ (considered as a function of ν for $t = \text{const}$) is peak normalized, the shape of the resulting spectrum does not change as it shifts to the red (parallel translation) with increasing t. Empirically the shape conservation of the normalized spectra on the wavenumber scale during a red shift was first observed with a solvatochromic fluorescent dye 2-p-toluidinonaphthalene-6-sulfonate adsorbed to apomyoglobin or membranes [5, 6], then with indole in glycerol [1], and later with tryptophan residues in proteins [2, 3]. Theoretically, the shape conservation follows directly from Eq. 5 that was first derived in [1] under the assumption that the Franck–Condon factor envelope does not change as the spectrum shifts to the red. It must be emphasized that the width and shape conservation is characteristic of fluorophore ensembles without permanent or long-term heterogeneity. In a fluorophore ensemble where the difference between two or more sub-ensembles is conserved for the entire life of the excited state, the instantaneous emission spectrum is expected to shrink at late times, when one of the sub-ensembles outlives all the others.

This chapter describes experimental and computational procedures that are critical for obtaining the corrected (for instrumental artifacts) and smooth function $F_\nu(\nu,t)$, which is suitable for applying the shape conservation criterion and obtaining the TDRS for tryptophan residues in proteins.

2 Experimental Data Collection

Three kinds of measurements are required to generate spectrally and time-resolved fluorescence intensity surfaces $F_\lambda(\lambda,t)$ and $F_\nu(\nu,t)$ for Trp residues in proteins. First, a time-correlated single-photon counting (TCSPC) instrument equipped with monochromators and polarizers records the "decay curves" at multiple emission wavelengths. Second, a steady-state instrument with a known spectral sensitivity curve measures the steady-state emission spectrum of the same protein sample. Third, an absorption spectrum of the same protein sample is obtained using a spectrophotometer (it will be used for the inner-filter correction). The order of the measurements is important, since each of these measurements increases the concentration of photobleached Trp residues, the emission from which contaminates the fluorescence signal (*see* **Note 1**). The contamination has the most detrimental effect on the results of the time-resolved measurements, an insignificant effect on the steady-state spectra, and virtually no effect on the absorption spectra.

2.1 TCSPC Measurements

1. *Wavelengths.* TCSPC decay curves are measured through a monochromator (in the fluorescence emission path) at multiple equally spaced emission wavelengths between 300 and 450 nm, at 5-nm intervals. For those proteins that also contain tyrosine residues, the optimum exciting wavelength is 296 nm; for proteins without tyrosines, it may be better to excite near 289 nm (this increases the amplitude of the TDRS). The optimum spectral resolution of the monochromator in the emission path equals 8 nm; however, for weakly fluorescent or low-concentration protein samples, it can be increased to 16 nm to effectively quadruple the sensitivity. If the excitation is at 296 nm and the protein solution scatters light too much, then the shortest emission wavelength is set to 305 nm rather than 300 nm. The decay curves are measured starting from the longest wavelength (450 nm) and ending with the shortest wavelength (300 nm), since the concentration of photobleached Trp keeps increasing during the experiment and the

contamination from photobleached Trp emission has the most detrimental effect at wavelengths above 380 nm (*see* **Note 1**).

2. *Photobleaching.* To avoid severe local photobleaching at the focus of the exciting beam, the beam is defocused so that its diameter is 1.5 mm at the center of the sample cell and attenuated (using neutral density filters) to 5.0 μW or less in the case of 296-nm excitation (2.0 μW or less at 289-nm excitation). The protein solution is stirred naturally every time when the protein cell is swapped with the scatterer cell (see below); therefore no magnetic stirring is needed to refresh the protein solution within the active volume. To avoid photobleaching the entire volume of the cell, the protein solution is replaced after H hours of measurements, where $H = C \cdot V/P$, $C = 10$ μW h/mL for 296-nm excitation, C=4 μW h/mL for 289-nm excitation, V is the volume of the protein solution in mL, and P is the mean power of the exciting beam in μW. The minimum volume of protein solution equals 1.5 mL. All measurements are performed in 10 mm by 10 mm quartz cell with all sides polished. Cells of smaller size increase photobleaching beyond acceptable levels, probably due to restricted convection in a narrow cell.
3. *Polarization.* Fluorescence is excited with polarized radiation, and fluorescence emission path contains a polarizer before the monochromator. The "magic angle" configuration is used to cancel all effects of rotational diffusion on the recorded data. The "magic angle" configuration is achieved when $\cos^2\theta_{exc}\cdot\cos^2\theta_{em} = 1/3$, where θ_{exc} and θ_{em} are the angles between the electric vector of the exciting and emission radiation, respectively, and the direction (usually vertical) of the normal to the plane containing the optical axes of both the exciting beam and the emission beam. For example, if the optical axes of the exciting and emission beams are horizontal and the exciting radiation is vertically polarized ($\theta_{exc} = 0$), then the polarizer in the emission path is set at $\theta_{em} = 55°$.
4. *Time resolution.* The overall time resolution of the TCSPC instrument depends on the exciting pulse width, the optical pulse broadening in the monochromator, the electron transit time dispersion in the microchannel plate photomultiplier, the bandwidths and noise levels of electronic preamplifiers, and the quality of the discriminators (either constant-fraction or pico-Timing discriminators). The overall time resolution in Eq. 20can be characterized by the FWHM of the impulse response function (IRF). An IRF FWHM of 55–70 ps is desirable; 100 ps or longer is unacceptable.
5. *IRF recording.* A cell of the same material and shape as the cell containing the protein solution is filled with a scatterer solution

and also placed in the motorized slide (or turret) for the TCSPC measurement. The scatterer cell is used to measure the IRF quasi-simultaneously with the decay curve: out of each minute of photon counting, photons from the protein solution are counted for 50 s with the monochromator set to the emission wavelength, and photons from the scatterer solution are counted for 10 s with the monochromator set to the excitation wavelength (these are the real times rather than the live times; *see* **Note 2**). Photon counts from the protein and from the scatterer are stored in different memory segments of the multichannel analyzer (MCA), since transferring the data from the MCA to a computer takes longer than swapping the two cells and moving the monochromators between the two wavelengths (*see* **Note 3**). The mechanics of the slide or turret must have enough precision to ensure that the physical positions of different cells (when they are in the beam) differ by no more than 0.1 mm. The group velocity of light in the scatterer solution must match that in the protein sample, i.e., if the protein solvent is water, then the scatterer must be also in water; however, if the protein solvent contains 80 % glycerol, then the scatterer solution must also contain 80 % glycerol. Millimolar concentrations of ions that are commonly present in pH buffers, such as Na^+ or $H_2PO_4^-$, have insignificant effect on the group velocity and do not need to be duplicated in the scatterer solution. Ludox colloidal silica is used to make water-based scatterer solutions. The concentration of Ludox is adjusted to match within 10 % the number of scattered photons per second with the number of protein fluorescence photons per second at the peak emission wavelength.

6. *Duration*. The goal is to register approximately 250 millions of fluorescence emission photon counts at all emission wavelengths together. The decay curves measured near the emission maximum usually have about 16 million counts per curve; those measured at 300 nm and at 450 nm have less than one million. To avoid pileup problems, the mean rate at which emission photons are detected must not exceed 1/200 of the excitation pulse rate, which equals 4.1 MHz in our experimental setup and cannot be made much higher than that because there must be sufficient time for the excited-state population to decay completely to zero before the next excitation pulse arrives. It takes 800 s to register 16×10^6 photon counts at $20 \times 10^3\ s^{-1}$ mean counting rate; however, if the dead time and quasi-simultaneous IRF recording are taken into account, then 800 s becomes about 20 min. With 20 min per wavelength and 31 wavelengths, the duration of the entire TCSPC experiment is over 10 h. This time can be cut in half if two identical fluorescence emission wings are used in parallel

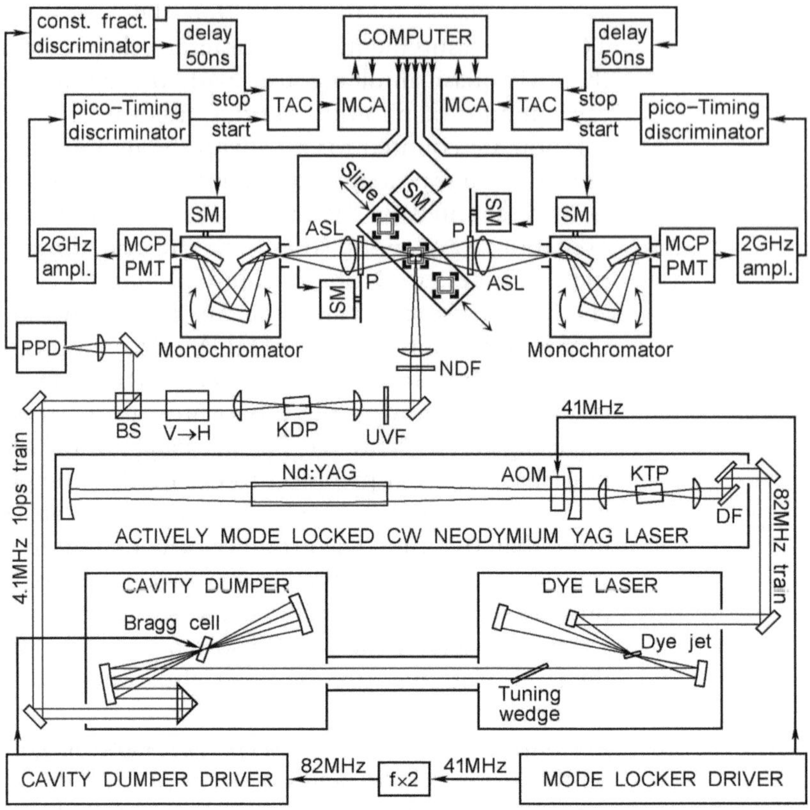

Fig. 1 Schematic diagram of the experimental setup used to collect TCSPC data. Acousto-optical modulator (AOM), aspheric lens (ASL), beam splitter (BS), dichroic filter (DF), potassium dihydrophosphate nonlinear optical crystal (KDP), potassium thallium phosphate nonlinear optical crystal (KTP), multichannel analyzer (MCA), microchannel plate photomultiplier (MCP-PMT), neutral density filter (NDF), neodymium-doped yttrium aluminum garnet (Nd:YAG), dichroic film polarizer (P), picosecond photodiode (PPD), synchronous motor (SM), time-to-amplitude converter (TAC), vertical-to-horizontal polarization rotator (V $\rightarrow$ H), ultraviolet-transmitting visible absorbing filter (UVF)

(T-format) as shown in Fig. 1 (*see* **Note 4**). TCSPC data from each wing are saved in separate computer files and later combined together using the program *comb_tcp*, which will be described in Subheading 3.2. The program shifts the decay curves and corresponding IRF before the addition to ensure the minimum possible width of the IRF for the combined data. Whether one or two wings are used to measure fluorescence emission, if the duration of the measurement at one wavelength exceeds 300 s, then the measurement is split in two or more parts of equal duration, and the results are saved in separate files to minimize the broadening of the IRF due to slow drifts in the instrument. The data are later combined using the program *comb_tcp*.

7. *Experimental setup.* A schematic diagram of the experimental setup that satisfies these requirements and was actually used to collect TCSPC data in references [1–3] is shown in Fig. 1. The motorized slide carries three jacketed cell holders, two of which are hooked up to a water bath thermostat by flexible PFA tubing and can be used in the temperature range from −50 °C to +100 °C (at $T < +5$ °C ethanol or methanol is added to water). The third cell holder always carries the scatterer cell and works at room temperature.
8. *Calibration.* Wavelength is automatically calibrated by the computer using a low-pressure Hg discharge lamp; wavelength accuracy is 0.4 nm and reproducibility 0.2 nm. Polarizers are motorized and automatically calibrated by the computer using highly polarized scattered light from a Ludox scatterer cell; calibration accuracy is 0.6° and reproducibility 0.2°. The timing calibration is achieved by manually replacing a short RG-58 cable between a pico-Timing discriminator output and the start input of a TAC by a longer cable of the same type. This introduces a calibrated delay of 19.960 ns and shifts the IRF peak to a different MCA channel. The TACs in both wings are adjusted so that the shift of the IRF peak equals 1,500 channels (*see* **Note 5**). This defines the time scale of 13.307 ps per channel. For different wings the IRF peaks can be in different channels, and for each wing the IRF peak can shift by as much as ten channels during 5 h; this is considered acceptable. The position of the peak is not as important as the time scale, which is kept constant and equal for both wings. A typical IRF is shown in Fig. 2, on the linear Y scale (top panel) and on the logarithmic Y scale (bottom panel).

2.2 Steady-State Emission Spectrum

1. *Wavelengths.* The steady-state emission spectrum is measured using a SLM-48000S spectrofluorometer (or a comparable instrument) by scanning the emission monochromator over the wavelength range from 280 to 480 nm and recording fluorescence emission intensity at 1-nm wavelength steps. The slits of the emission monochromator are set to 4-nm spectral resolution. The slits of the excitation monochromator are set to 2-nm spectral resolution. The excitation monochromator is set to the same wavelength that was used to excite fluorescence in the TCSPC measurement, and the protein solution is thermostated at the same temperature as in the TCSPC measurement. The emission spectrum of the protein solution is recorded four times. The emission of a "blank" cell containing the same solvent without the protein is also recorded four times. The mean blank spectrum is then subtracted from the mean protein spectrum.

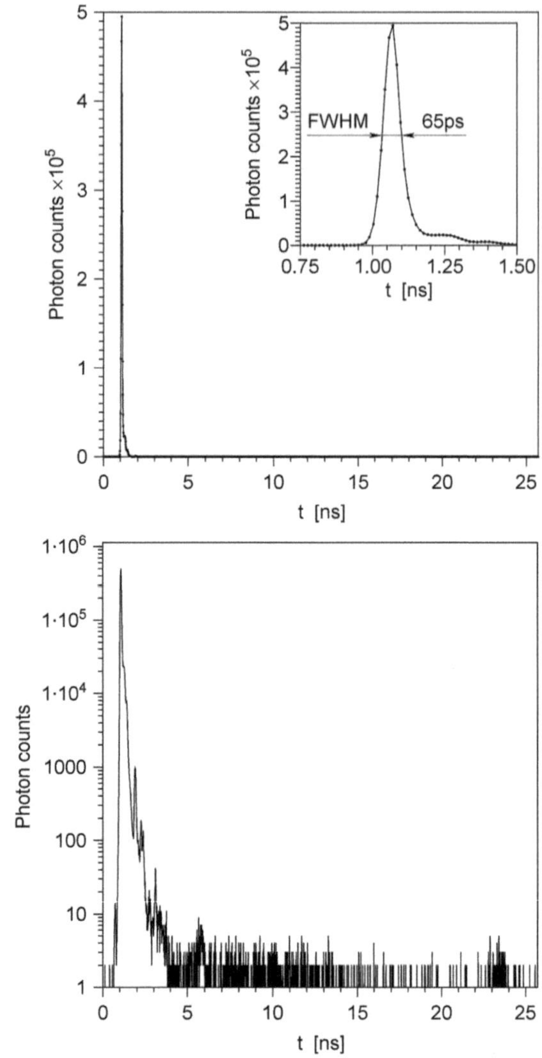

Fig. 2 Typical impulse response function (IRF) of the instrument shown in Fig. 1, plotted on the linear intensity scale (*top panel*) and on the logarithmic intensity scale (*bottom panel*). The inset in the top panel depicts the IRF on an expanded time scale and reveals its full width at half maximum (FWHM). The plot on the logarithmic scale (*bottom panel*) reveals several small-intensity satellite peaks resulting from the reflections of light by miscellaneous optical elements. The highest satellite peak has about 1/500 of the intensity of the main peak; on the linear intensity scale (*top panel*), this peak is invisible

2. *Photobleaching*. It was empirically found that in the case of the SLM-48000S spectrofluorometer, using 2-nm slits on the excitation monochromator and continuous stirring with a small magnetic stir bar is sufficient to reduce photobleaching to insignificant levels during a typical 20-min measurement. As a precaution, the spectra of the four individual scans are

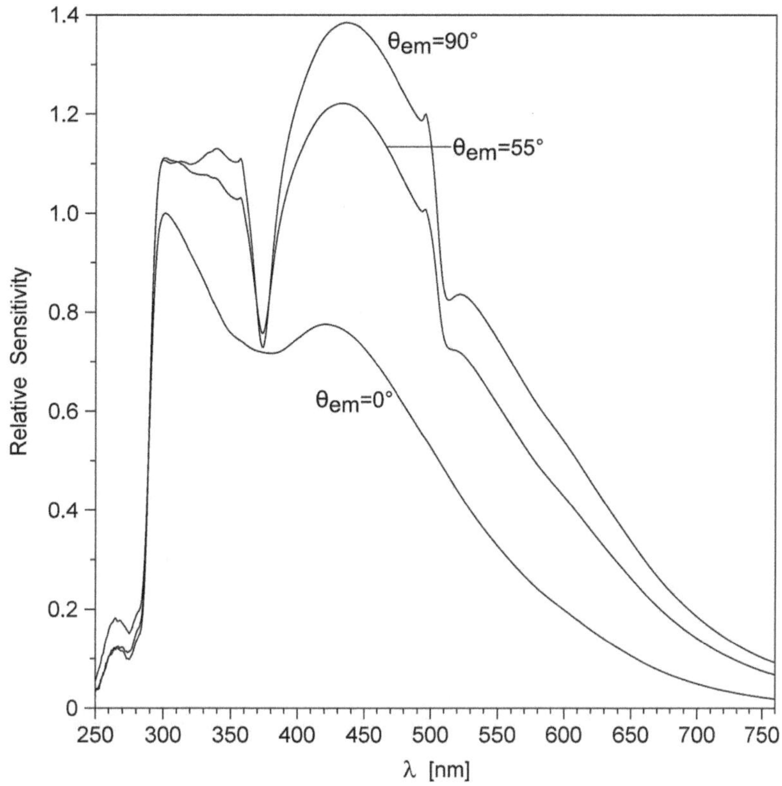

Fig. 3 Spectral sensitivity variation for SLM-48000S spectrofluorometer equipped with Glan–Thompson polarizers, MC-200 monochromators, and R-926 photomultipliers in the emission light path. θ_{em} is the angle of the polarizer, measured from the vertical. Wood's anomalies are observed near $\lambda = 373$ nm and $\lambda = 505$ nm with $\theta_{em} = 90°$ and $\theta_{em} = 55°$, but not with $\theta_{em} = 0°$, because in the latter case the polarization (electric vector) is parallel to the grooves on the diffraction grating

monitored during the measurement; a systematic decrease in the peak intensity from the first scan to the second, third, and fourth serves as a good indicator of photobleaching (with 2-nm slits and magnetic stirring, the decrease is not observed).

3. *Polarization*. The use of polarizers is vital for accurate spectral correction, since the spectral sensitivity curve of every fluorescence spectrometer equipped with diffraction grating monochromators is significantly different for the polarizations parallel and normal to the direction of the grooves on the diffraction grating (*see* Fig. 3). Without the polarizers the spectral sensitivity would depend on the degree of polarization of the fluorescence emitted from the sample, which would make it impossible to correct emission spectra of different

samples using the same spectral sensitivity curve. The SLM-48000S spectrofluorometer is equipped with calcite prism polarizers in both the excitation (Glan–Taylor) and the emission (Glan–Thompson) light path. The "magic angle" configuration is used to cancel the effect of rotational diffusion on the recorded steady-state spectrum. The "magic angle" configuration is achieved when $\cos^2\theta_{exc}\cdot\cos^2\theta_{em} = 1/3$, where θ_{exc} and θ_{em} are the angles that the polarizers in the excitation and emission path make with the vertical direction (*see* Subheading 2.1, **item 3**). To avoid Wood's anomalies in the emission spectrum (*see* Fig. 3), the polarizer in the emission path is set parallel to the direction of the grooves on the diffraction grating ($\theta_{em} = 0$), and the polarizer in the excitation path is set at $\theta_{exc} = 55°$.

4. *Spectral correction*. It must be emphasized that the spectral sensitivity curve of every fluorescence spectrometer depends on the orientation of the polarizer in the emission light path and on the width of the emission monochromator slits. The spectral sensitivity curve for $\theta_{em} = 0$ and 4-nm monochromator slits is obtained using thermal radiation with the spectral distribution close to that of a blackbody of a known color temperature near 3,000 K. The source of the thermal radiation must have a quartz bulb. The best choice is a tungsten band lamp calibrated at the National Bureau of Standards; if this is not available or does not fit in the cell compartment, then the second-best choice is a miniature 12 V 20 W halogen lamp with a G-4 base, RadioShack catalog number 272-1177, powered by a regulated 16.5 V DC power supply. The lamp is placed in the well that normally houses a fluorescence cell. The band lamp is oriented so that the tungsten band is normal to the axis of the emission optical path. The miniature halogen lamp is oriented so that the axis of the spiral tungsten filament is parallel to the axis of the emission optical path. The lamp holder is sturdy and allows translational alignment in two directions normal to the axis of the emission optical path. The lamp is aligned to maximize the signal from the photomultiplier that normally measures fluorescence emission intensity. A technical spectrum of the lamp is recorded from 250 to 800 nm at 1-nm steps. The technical spectrum is then divided by the blackbody radiation spectrum to obtain the spectral sensitivity curve. The blackbody radiation spectrum is calculated as

$$I_{bb}(\lambda) = \frac{2\pi c\lambda^{-4}}{\exp\left(\frac{hc}{\lambda k_B T}\right) - 1}, \quad (9)$$

where I_{bb} is the spectral density of photons emitted from a unit surface area per unit time per unit wavelength range, h is the

Planck constant, c is the speed of light, λ is the emission wavelength, k_B is the Boltzmann constant, and T is the color temperature. In contrast to Eq. 9, the most well-known form of Planck radiation law contains the numerator $2\pi hc^2\lambda^{-5}$ rather than $2\pi c\lambda^{-4}$ and gives the spectral energy density rather than the spectral density of photons. Throughout this chapter fluorescence intensity is expressed as photon density rather than energy density per unit spectral interval; therefore we must use $2\pi c\lambda^{-4}$ in the numerator of Eq. 9 to obtain the correct spectral sensitivity curve. The accuracy of the spectral sensitivity curve obtained in this manner depends on the accuracy of the color temperature T. For a tungsten band lamp, the temperature is obtained from its passport. For the miniature 12 V 20 W halogen lamp powered by 16.5 V DC, $T = 3{,}120$ K. For the same lamp powered by 13.6 V DC, $T = 2{,}860$ K; however, at $T < 3{,}000$ K the lamp has insufficient intensity in the UV part of the spectrum, which creates technical problems. The accuracy of the spectral sensitivity curve is tested by taking the fluorescence emission spectrum of quinine sulfate dihydrate under the conditions (pH, temperature, exciting wavelength, concentration) described in [10], correcting it using the spectral sensitivity curve obtained as described here and comparing the corrected spectrum with the one tabulated in [10].

2.3 Absorption Spectrum

The absorption spectrum is scanned in the wavelength range from 280 to 480 nm or wider. The wavelength step should be 1 nm or a fraction of 1 nm, so that the computer file would contain the value of the absorbance (a.k.a. optical density) at every integer wavelength from 280 to 480 nm. Any UV–VIS spectrophotometer can be used for this purpose. If the spectrophotometer contains two cell holders, one for the "sample" cell and one for the "reference" cell, then the "reference" cell is left empty, because the absorption spectrum is to be used for the inner-filter correction (*see* **Note 6**). The absorption spectrum is usually measured after all time-resolved and steady-state fluorescence measurements. If there is a suspicion that the protein molecules may aggregate during the lengthy fluorescence measurements, then the absorption spectrum of the protein cell is measured twice, both before and after fluorescence measurements. With aggregation the absorption spectrum acquires a tail at $\lambda > 315$ nm, where none of the 20 protein amino acids absorb radiation. The tail represents the radiation losses due to scattering, and its spectral shape can be mathematically described by the function $1/\lambda^4$. Fluorescence emission from aggregated proteins is always heterogeneous, as well as fluorescence emission from proteins having more than one Trp residue in the amino acid sequence.

3 Computational Procedures

The experimental data obtained as described above need to be processed by a number of computer programs, with each program using a specific algorithm and intended to perform a specific task. The computer programs can be obtained from the author of this chapter. All programs run under Linux operating system on computers with Intel i686 processors and require four shared object libraries, *libm.so.6*, *libgcc_s.so.1*, *libc.so.6*, and *libgfortran.so.3*, the former three of which are standard components of any Linux operating system, while the fourth one can be easily installed from the Internet. The programs, Linux operating system, and the additional libraries are free. Below each program is described together with the underlying math.

3.1 Correcting and Smoothing Steady-State Spectra

1. *Correction*. Technical steady-state emission spectrum should be corrected for the spectral sensitivity variation of the instrument and for the secondary inner-filter effect (the primary inner-filter effect changes the absolute intensity but not the shape of the emission spectrum). The program *correct* corrects for the spectral sensitivity variation and, if necessary, may also correct for the secondary inner-filter effect using the following recipe:

$$I_{\mathrm{corr}}(\lambda) = \frac{10^{A(\lambda)/2}}{S(\lambda)} I_{\mathrm{tech}}(\lambda) \tag{10}$$

Here $I_{\mathrm{tech}}(\lambda)$ and $I_{\mathrm{corr}}(\lambda)$ are the technical and the corrected spectrum, $A(\lambda)$ is the gross absorbance (optical density) spectrum of the same solution in the same cell, and $S(\lambda)$ is the spectral sensitivity curve of the steady-state fluorescence spectrometer obtained for $\theta_{\mathrm{em}} = 0$ and 4-nm monochromator slits.

2. *Smoothing*. An experimental steady-state emission spectrum always contains random noise. High-frequency random noise, if it is not removed from the spectrum, will translate into "fine structure" in the functions $F_\lambda(\lambda,t)$ and $F_\nu(\nu,t)$, which will hinder the calculation of the FWHM, peak position, etc., and make the results useless. Removal of the high-frequency noise is accomplished by fitting a smooth analytical function to the corrected experimental spectrum. Before the fitting, it is necessary to visually inspect the corrected experimental spectrum and to pick a cutoff wavelength to the right of the scattered excitation light peak but to the left of the shortest wavelength at which TCSPC data were collected (usually 300 nm). The optimum cutoff wavelength is shown by an arrow in Fig. 4. The data at wavelengths shorter than the cutoff wavelength are not used in the fitting. The fitting function must have a number of fitting parameters that can be varied from just 3 (Gaussian) or 5

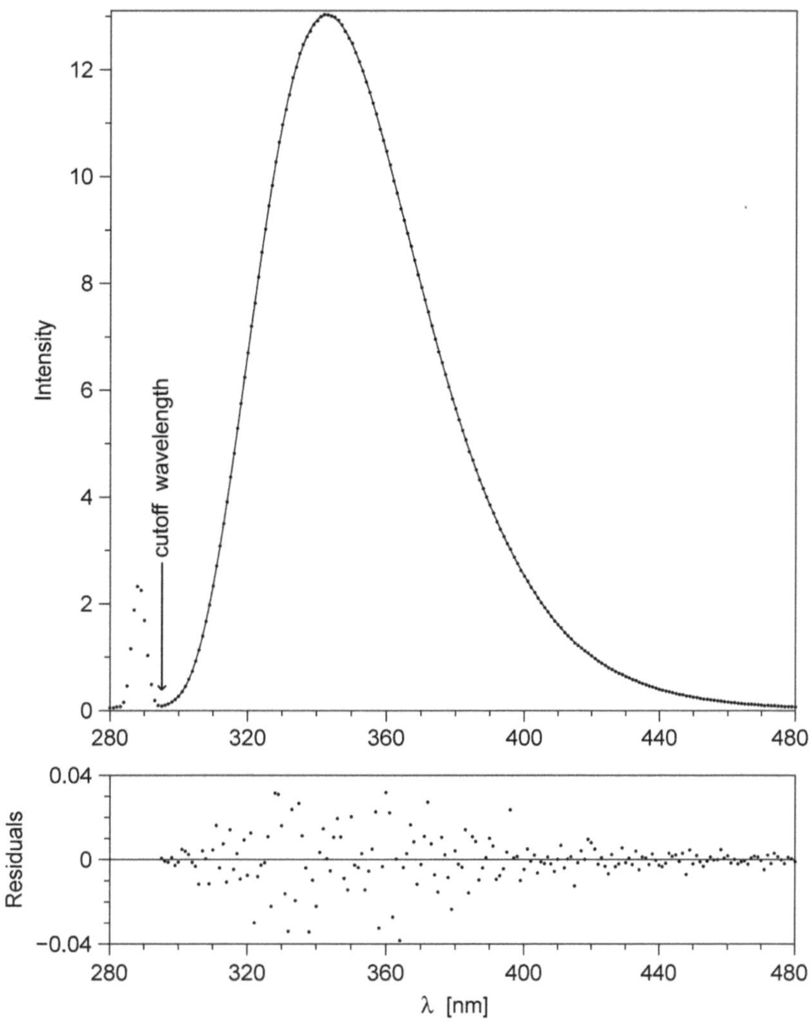

Fig. 4 Corrected spectrum of fluorescence emission from Trp residues in E21W variant of IIA^{Glc} protein at +5 °C, excited at 288.5 nm; this spectrum was used to generate the time-resolved emission spectra published in [2]. This figure illustrates an intermediate step in the data processing, and it was not included in [2]. Main panel: corrected experimental data are depicted by *dots*; the *solid line* represents the best fit with the model function in Eq. 11 and $m = 10$. The fitting was done by the program *exppolnu*. The *arrow* points to the cutoff wavelength, which is selected to the right of the scattered light peak and to the left of λ=300 nm, the shortest wavelength at which TCSPC data are collected. The data to the left of the cutoff wavelength are not used in the fitting. The bottom panel shows the residuals (the differences between the data and the fit) on a 40-fold expanded vertical scale. If the residuals are not random (the *dots* seem to form a wave), then the value of *m* needs to be increased

(skewed Gaussian) to 20 or more depending on the complexity of the spectral shape. The most general fitting function of this type is

$$f_\lambda(\lambda) = \exp\left(\sum_{n=0}^{m} a_n \lambda^{-n}\right) \tag{11}$$

This function is actually an exponential of a polynomial of ν, where $\nu = \lambda^{-1}$. The choice of ν rather than λ for the argument of the polynomial function is dictated by the fact that the vibrational structure in electronic spectra has a constant period on the ν scale, but not on the λ scale. The degree of the polynomial equals m. The number of free fitting parameters a_n equals $m + 1$. For $m = 2$ the function in Eq. 11 represents a symmetrical Gaussian on the wavenumber scale. For $m = 4$ it represents a skewed Gaussian. The fitting of the model function in Eq. 11 to the corrected fluorescence emission spectrum is done using the program *exppolnu*. For Trp residues in proteins, the value of $m = 10$ always gives a perfect fit. In Fig. 4 the corrected emission spectrum is shown by dots, the best fit by the model function in Eq. 11 with $m = 10$ is shown by the solid line. The magnified residuals are shown in the bottom panel.

3.2 Combining TCSPC Data

1. *Combining multiple TCSPC data sets.* TCSPC data from the right and the left wing of the instrument shown in Fig. 1 need to be added together before the data analysis. The time scale (the time per one MCA channel) is always kept the same for both wings, but the position of the IRF peak may be different, which needs to be taken into account when the TCSPC data are combined. Furthermore, the IRF peak in each wing may shift by as much as ten MCA channels during a 5-h experiment. For this reason, long data collections are always split in parts not exceeding 300 s in duration and the data are saved in different files. The data obtained back-to-back from the same wing, the data from different wings, and the data from similar experiments conducted on different days are combined using the program *comb_tcp*. The underlying algorithm combines two TCSPC data sets each time it is invoked. If the number of TCSPC data sets exceeds two, then the algorithm is invoked more than once (the algorithm can be invoked any number of times during one run of the program *comb_tcp*). During the first invocation, data sets 1 and 2 are combined. During the second invocation, the combination of sets 1 and 2 is combined with set 3. During the third invocation, the combination of sets 1, 2, and 3 is combined with set 4. This is repeated until all data sets are combined. The final result is independent of the order in which the data sets are combined.

2. *Algorithm for combining two TCSPC data sets.* A TCSPC data set consists of a decay curve, which will be denoted $D_m(i)$, and the IRF, which will be denoted $E_m(i)$. The IRF is measured quasi-simultaneously with the corresponding decay curve using a Ludox scatterer cell. Both $D_m(i)$ and $E_m(i)$ are integer-valued functions of the integer variable i, which represents MCA channel number, whereas m denotes the number of the set. For notation simplicity it will be assumed that two sets with $m = 1$ and $m = 2$ are being added. The algorithm starts by calculating the correlation function:

$$C(k) = \sum_i E_1(i)E_2(i + k) \tag{12}$$

The summation is carried out over all i for which the expression under the summation sign is defined. The optimum shift k_O is determined as the value of the argument at which the correlation function $C(k)$ reaches its absolute maximum. The minimum possible width of the IRF for the combined data is achieved when the shift equals k_O. The decay curves are shifted and added in the proportion one to one:

$$D_{comb}(i) = D_1(i) + D_2(i + k_O) \tag{13}$$

Since the random errors in $D_1(i)$ and $D_2(i)$ are statistically independent and the photon counts strictly follow Poissonian statistics, the result $D_{comb}(i)$ also strictly follows Poissonian statistics, which is critical for applying the data analysis programs described below. Whether the IRFs $E_1(i)$ and $E_2(i)$ can be added in the proportion one to one is an important question. If the data sets 1 and 2 are from different wings that have slightly different spectral sensitivity curves or of if the data were obtained on different days, when the Ludox concentration was not exactly the same, then adding $E_1(i)$ and $E_2(i)$ in the proportion one to one will result in a serious systematic error. Alternatively, the IRFs can be combined with coefficients a_1 and a_2:

$$E_{comb}(i) = a_1E_1(i) + a_2E_2(i + k_O) \tag{14}$$

To define the optimum values of a_1 and a_2, consider the following. If $D_1(i)$ is a numerical convolution of $a_1E_1(i)$ with a function $F(i)$ that describes a δ-excitation decay law, while $D_2(i)$ is a numerical convolution of $a_2E_2(i)$ with the very same function $F(i)$, then no systematic errors will arise from the operation in Eqs. 13 and 14, regardless of how different the IRFs $E_1(i)$ and $E_2(i)$ are and also regardless of the value of k_O. The integral of a convolution of two functions equals the product of the integrals of these functions, which helps to determine the following ratio between a_1 and a_2 that must be observed in order to avoid systematic errors:

$$\frac{a_1}{a_2} = \frac{\left(\sum_i D_1(i)\right) \times \left(\sum_i E_2(i)\right)}{\left(\sum_i E_1(i)\right) \times \left(\sum_i D_2(i)\right)} \tag{15}$$

One Eq. 15 is not sufficient to uniquely define the values of two unknowns, a_1 and a_2. An additional requirement that $E_{comb}(i)$ obeys near-Poissonian statistics is sufficient to remove the ambiguity. Poissonian statistics implies that the variance of $E_1(i)$ equals the ensemble mean of $E_1(i)$, and the variance of $E_2(i)$ equals the ensemble mean of $E_2(i)$. If we also require that the variance of $E_{comb}(i)$, averaged over all MCA channels, equals the ensemble mean of $E_{comb}(i)$, averaged over all MCA channels, then this requirement results in the following constraint:

$$\sum_i \left(a_1^2 - a_1\right) E_1(i) + \left(a_2^2 - a_2\right) E_2(i) = 0 \tag{16}$$

The algorithm on which the program *comb_tcp* is based solves the system of Eqs. 15 and 16 with respect to the unknowns a_1 and a_2 and then uses the values of a_1, a_2, and k_O to calculate $E_{comb}(i)$ in accordance with Eq. 14. It must be emphasized that while $D_{comb}(i)$ exactly follows Poissonian statistics, and this is strictly required by the data analysis programs described below, $E_{comb}(i)$ only resembles Poissonian statistics in terms of its variance, and in fact it is not even an integer-valued function. The data analysis programs do not care whether $E_{comb}(i)$ is an integer-valued function and whether it follows Poissonian statistics. The main role of the additional constraint in Eq. 16 is to make sure that after a very large number of successive invocations of the algorithm the final $E_{comb}(i)$ does not vanish or approach infinity, but keeps growing at roughly the same pace as $D_{comb}(i)$.

3.3 TCSPC Data Analysis

1. *Convolution*. A TCSPC data set consists of a decay curve $D_m(i)$ and the IRF $E_m(i)$ measured quasi-simultaneously; i is the number of MCA channel, and m is the serial number of the data set. Note that from here different m values represent the data sets obtained at different emission wavelengths, whereas in Subheading 3.2, where the program *comb_tcp* was described, different m values represented the data sets obtained at identical emission wavelengths but either at different times or from different wings of the instrument. As soon as TCSPC data are loaded into a data analysis program, each $E_m(i)$ is renormalized to unit area, i.e.,

$$\sum_{i=1}^{N_C} E_m(i) = 1 \tag{17}$$

Here N_C is the number of MCA channels used for the data collection (*see* **Note 5**). Then every decay curve $D_m(i)$ is fitted by a numerical convolution of the corresponding IRF $E_m(i)$ with the discretized model function $F_m(i)$:

$$G_m(i) = \sum_{j=1}^{N_C} E_m(j) F_m(i-j) \tag{18}$$

The model parameters are adjusted to achieve the best fit between the numerical convolution $G_m(i)$ and the experimental decay curve $D_m(i)$. The discretized model function $F_m(i)$ is obtained from the continuous mathematical model function $F_m(t)$ using the following integral that expresses the essence of the centered delta-function convolution method:

$$F_m(i) = \int_{(i-1/2)\Delta t}^{(i+1/2)\Delta t} F_m(t)\mathrm{d}t \tag{19}$$

Here Δt is the time per one MCA channel, which is calibrated as described in Subheading 2.1, **item 8**.

2. *Model Function.* The following mathematical model function is used by the data analysis programs *l_global* and *tcphoton* (*see* **Note 7**):

$$F_m(t) = b_m + s_m \delta(t) + \theta(t) \sum_{n=1}^{N_E} \frac{f_{mn}}{\tau_n} \exp(-t/\tau_n) \tag{20}$$

Here t is time measured from δ-excitation; the index m is relevant to the global analysis only, representing the serial number of the wavelength at which the emission was collected; $\delta(t)$ and $\theta(t)$ are Dirac delta and Heaviside step functions, respectively; N_E denotes the total number of exponentials; the model parameters b_m, s_m, f_{mn}, and τ_n represent the photomultiplier background (dark counts), the intensity of scattered exciting radiation (including Raman), the intensities associated with exponential terms (allowed to be both positive and negative), and the time constants. The same set of time constants τ_n applies to all emission wavelengths, which is consistent with the models in Eqs. 3, 4, and 8.

3. *Weighted least squares.* The programs *l_global* and *tcphoton* use a slightly improved version of the Gauss–Newton algorithm to find those values of the model parameters b_m, s_m, f_{mn}, and τ_n that minimize the sum of weighted least squares:

$$S = \sum_{m=1}^{N_W} \sum_{i=1}^{N_C} W_m(i) \left(D_m(i) - G_m(i)\right)^2 \tag{21}$$

Here N_W is the number of wavelengths (for single-curve analysis $N_W = 1$), N_C is the number of used MCA channels (*see* **Note 5**), $D_m(i)$ is the experimental decay curve, $G_m(i)$ is the convolution from Eq. 18, and $W_m(i)$ are the weights for the experimental data $D_m(i)$. It has been shown [11] that in the method of weighted least squares, the most accurate values of the model parameters are obtained when the weights $W_m(i)$ are taken equal to the inverses of the variances of the corresponding data $D_m(i)$. The data $D_m(i)$ strictly follow Poissonian statistics. For a random variable that follows Poissonian statistics the variance equals its ensemble mean. Of course, there is no way of knowing the ensemble means for $D_m(i)$, since this would imply that one can obtain experimental data completely free from random noise. The programs *l_global* and *tcphoton* calculate the weights using the method of successive approximations. For the zeroth approximation, they take $W_m(i) = 1/D_m(i)$, where division by zero is avoided by using unit weights for those MCA channels where the photon count equals zero. The weights obtained in the $(N-1)$th approximation are used for the Nth round of weighted least-square fitting with the model function in Eq. 20 containing a very large number of exponentials with τ_n equally spaced on the logarithmic scale and fixed (this is known as the exponential series model function; it fits any TCSPC data set regardless of the underlying physical processes). The best-fit $G_m(i)$ is used to calculate the new weights $W_m(i) = 1/G_m(i)$, which are then used as the next approximation. In practice the weights $W_m(i)$ stabilize after just two rounds of fitting with the exponential series model; however, the data analysis programs described here perform eight rounds to ensure that the weights are calculated with the greatest possible precision.

4. *Chi square*. If the weights $W_m(i)$ are taken equal to the inverses of the variances of the corresponding data $D_m(i)$, then the definition of the quantity S in Eq. 21 coincides with the definition of the random variable χ^2 (not reduced), which is well known because of the χ^2 goodness-of-fit test [11]. The reduced χ^2, which is most commonly reported in literature, is defined as the ratio

$$\text{reduced}\,\chi^2 = \frac{S}{N_D - N_P} \tag{22}$$

Here N_D is the total number of data points and N_P is the total number of model parameters, the values of which were adjusted by the program to minimize S. In the case of the global analysis (program *l_global*), $N_D = N_W \cdot N_C$ and $N_P = 2\,N_w + N_W \cdot N_E + N_E$, where N_W is the number of wavelengths, N_C is the number of used MCA channels (*see* **Note 5**), and N_E

Table 1
5 and 0.5 % thresholds for the χ^2 distribution versus the number of degrees of freedom

$N_D - N_P$	χ^2 (5 %)	χ^2 (0.5 %)
100	1.243	1.402
200	1.164	1.258
500	1.104	1.163
1,000	1.074	1.115
2,000	1.052	1.081
5,000	1.033	1.052
10,000	1.023	1.036
20,000	1.016	1.026
50,000	1.010	1.016
100,000	1.007	1.012

is the number of exponentials. In the case of the single-curve analysis (program *tcphoton*), $N_D = N_C$ and $N_P = 2 + 2N_E$. The difference $N_D - N_P$ is usually called the number of degrees of freedom. In a typical single-curve analysis, $N_D - N_P$ is close to 1916; in a global analysis of the data from 31 wavelengths, it is about 59,516. The upper limits for the reduced χ^2 can be found in Table 1, which gives two thresholds for selected numbers of degrees of freedom. The probability that the reduced χ^2 exceeds the first threshold equals 5 %; the probability that it exceeds the second threshold equals 0.5 %. If it is found that the reduced χ^2 is greater than the 5 % threshold for the given number of degrees of freedom, then this may imply that N_E needs to be increased. If the reduced χ^2 is greater than the 0.5 % threshold and the increase in N_E does not lower it, then this is a clear indication of systematic errors (technical problems with the instrument or bad fluorescent sample, e.g., aggregated protein and clumps in the solution).

5. *Global analysis.* The global analysis implies that a number $N_W > 1$ of TCSPC data sets are analyzed simultaneously and some of the model parameters are relevant to all data sets. In the model function in Eq. 20, the model parameters τ_n are relevant to all data sets; these are called the global parameters. The model parameters b_m, s_m, and f_{mn} are each relevant to only one TCSPC data set; these are called the local parameters. It is important that the model function in Eq. 20 is linear in all local parameters, but it is not linear in the global parameters. The algorithm in the basis of the program *l_global* was specifically

designed to be very efficient in the case where the model is linear in all local parameters. The program does not require initial guesses for the local parameters, and it tolerates the initial guesses for the global parameters that can be very far from the best-fit values. For example, fluorescence emission from the single Trp residue of the B1 domain of streptococcal protein G (GB1) at +5 °C and 296-nm excitation was found to contain four exponentials with $\tau_1 = 73.3$ ps, $\tau_2 = 579$ ps, $\tau_3 = 3.09$ ns, and $\tau_4 = 7.15$ ns [3]. When the 31 TCSPC data sets from the experiments described in [3] are analyzed using the program *l_global* with the initial guesses $\tau_1 = 20.0$ ps, $\tau_2 = 200$ ps, $\tau_3 = 5.00$ ns, and $\tau_4 = 50.0$ ns, the program finds the best fit and the best values of all parameters in less than 4 s (benchmarked on HP 510 laptop with 2.13 GHz Pentium M processor); it takes about half of this time to read all TCSPC data sets from the disk. The final values of the fitting parameters found with different initial guesses typically differ by no more than one part in one million (the first six significant digits coincide). The program always outputs the best-fit values of all global fitting parameters with standard deviations, the best-fit values of all local fitting parameters with standard deviations, and the reduced χ^2. If requested, it also generates a color map of weighted residuals (*see* Fig. 5 for typical examples) and the autocorrelation of global unweighted residuals in a numerical format.

6. *Single-curve analysis.* The program *tcphoton* can be used to analyze one TCSPC data set at a time. Equations 17–22 describe the work of this program if the index m is dropped from all functions and parameters and the summation from $m = 1$ to N_W in Eq. 21 is also dropped. Unfortunately, in the case of single-curve analysis, the algorithm is more sensitive to the initial guesses for τ_n, especially when the number of exponentials is greater than two (if the initial guesses for τ_n are not good, then the program asks to choose different initial guesses). The initial guesses for b, s, and f_n are not important and can be all set to zero, which is recommended for faster convergence. The program works significantly faster than *l_global* (typically the fitting takes a few milliseconds) and it outputs not only the values of all model parameters with confidence intervals (*see* **Note 8**) and the reduced χ^2 but also a combination plot of the data with fit, weighted residuals and autocorrelation of unweighted residuals (*see* two panels of Figs. 6 and 7 for typical examples). The combination plot is useful in finding the sources of systematic errors: the errors resulting from insufficient number of exponentials are shown in the bottom panel of Fig. 6, whereas Fig. 7 depicts the errors resulting from a common hardware defect.

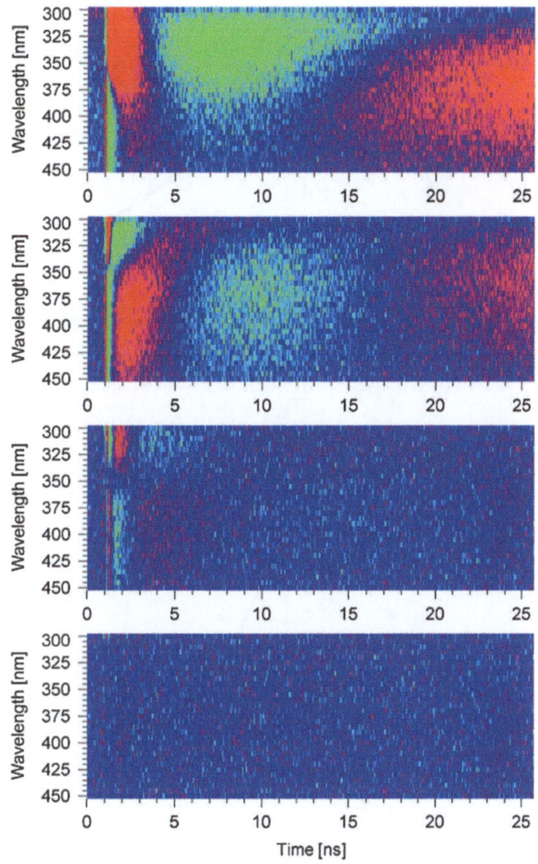

Fig. 5 Four examples of weighted residuals color maps generated by the program *l_global*. In all four cases the program simultaneously analyzed the same 31 TCSPC data sets obtained at emission wavelengths from 300 to 450 nm at 5-nm wavelength intervals; these 31 TCSPC data sets were used to generate the time-resolved emission spectra published in [3]. This figure illustrates an intermediate step in the data processing and it was not included in [3]. Fluorescence emission was from the single Trp residue of the B1 domain of streptococcal protein G (GB1) at +5 °C and 296-nm excitation. The four panels (top to bottom) represent the fits to the data with the models of one, two, three, and four exponentials, respectively. Corresponding reduced χ^2 values are 4.956, 2.134, 1.075, and 1.004. The weighted residuals (WR) are calculated as $[D_m(i) - G_m(i)]\sqrt{W_m(i)}$ and depicted using the following color scheme: *blue* for near-zero WR, *red* for positive WR, and *green* for negative WR

3.4 Reconstruction of Amplitude Spectra and Time-Resolved Spectra

1. *Amplitude spectra on the wavelength scale.* The values of the parameters f_{mn} and τ_n of the model function in Eq. 20 are determined from experimental TCSPC data by the program *l_global* as described in Subheading 3.3, **steps 1** through **5**. Note that the sum of f_{mn} over all n represents the total emission intensity at the emission wavelength λ_m. If the TCSPC

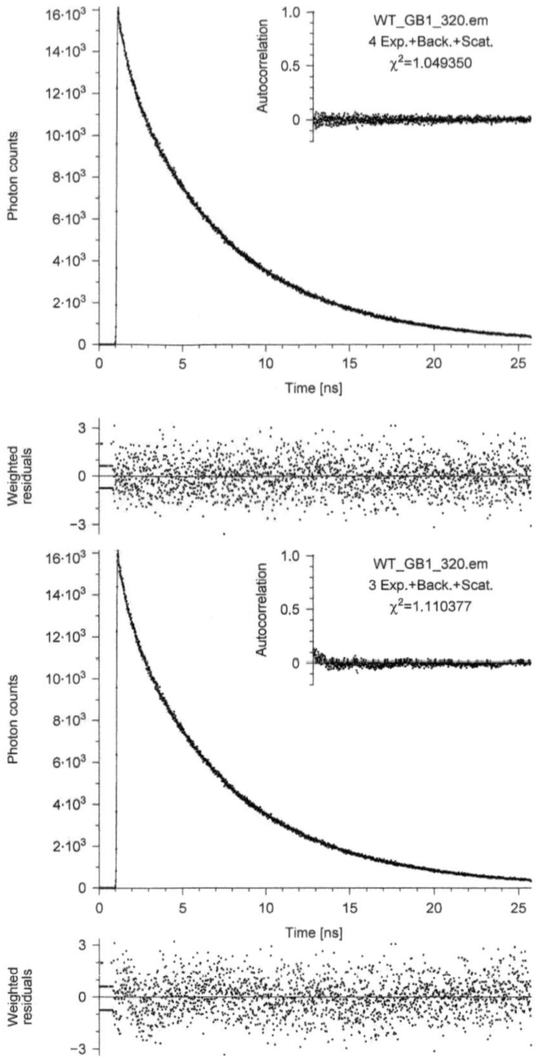

Fig. 6 The *top half* and the *bottom half* of this figure each represents an example of a combination plot generated by the program *tcphoton*. The program analyzed a single TCSPC data set obtained at the emission wavelength 320 nm; fluorescence emission was from the single Trp residue of the B1 domain of streptococcal protein G (GB1) at +5 °C and 296-nm excitation; this is one of the 31 TCSPC data sets that were used to generate the time-resolved emission spectra published in [3]; however, no single-curve analysis of this data was included in [3]. The main panel shows the experimental photon counts $D(i)$ (*dots*) and the convolution $G(i)$ (*solid line*), which represents the best fit with the model of four exponentials ($N_E = 4$) for the combination plot at the *top* and three exponentials ($N_E = 3$) for the combination plot at the bottom. The strip at the bottom of each combination plot shows the weighted residuals defined as $[D(i) - G(i)]\sqrt{W(i)}$. The time axis of the main panel applies also to the weighted residual strip. The inset in each combination plot shows the peak-normalized autocorrelation of unweighted residuals; the autocorrelation is defined as $A(k) = \sum_i [D(i) - G(i)] \times [D(i+k) - G(i+k)]$ and peak normalized by dividing $A(k)$ by $A(0)$ for every k. The time axes of the main panel and the autocorrelation plot are shifted (their origins do not coincide); however, the time scales (numbers of nanoseconds per inch) are the same. The total time interval for which the autocorrelation is plotted equals about one half of that for the data in the main panel. Note that with the model of three exponentials (bottom half of the figure), the autocorrelation inset and the strip of weighted residuals reveal small systematic differences between the data and the model. Also note that the reduced χ^2 value obtained with the model of three exponentials is higher than that obtained with the model of four exponentials and that it exceeds the 0.5 % threshold from Table 1

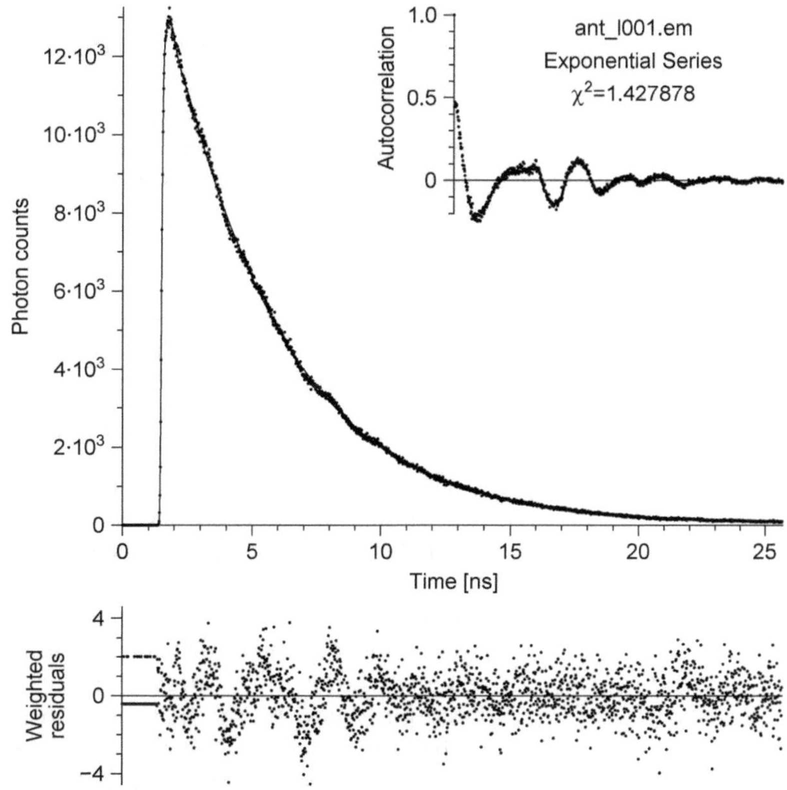

Fig. 7 The combination plot (generated by the program *tcphoton*) that clearly shows a technical problem (a cross talk between the start and the stop signals). The format of the combination plot is described in the legend of Fig. 6

instrument was free from the spectral sensitivity variation, the dead time, and uncontrolled variations in the intensity of the exciting laser beam, then the sum of f_{mn} over all n would reproduce $f_\lambda(\lambda_m)$ within a constant factor, where $f_\lambda(\lambda)$ is the steady-state emission spectrum, corrected and smoothed as described in Subheading 3.1. The instrumental artifacts result in a significant difference between the sum of f_{mn} over all n and $f_\lambda(\lambda_m)$, which shows that the absolute f_{mn} values cannot be used in reconstructing the spectra associated with the exponential terms. The relative values of these parameters are obtained by renormalizing the absolute values to unit sums at each wavelength:

$$f'_{mn} = \frac{f_{mn}}{\sum_{k=1}^{N_E} f_{mk}} \tag{23}$$

The renormalized parameters are here denoted with a prime; they are free from the artifacts associated with the spectral sensitivity variation, the dead time, and the uncontrolled variations in exciting laser power. The products of the renormalized parameters and the corrected steady-state emission spectrum give the intensity spectra associated with the exponential terms:

$$f_{\lambda n}(\lambda_m) = f'_{mn} \cdot f_{\lambda}(\lambda_m) \tag{24}$$

The amplitude spectra associated with the exponential terms are obtained by dividing the intensity spectra by the corresponding time constants:

$$\alpha_{\lambda n}(\lambda_m) = f_{\lambda n}(\lambda_m)/\tau_n \tag{25}$$

The amplitudes $\alpha_{\lambda n}(\lambda_m)$ are depicted by circles in Fig. 8. The error bars in Fig. 8 are calculated from the confidence intervals for the parameters f_{mn} (*see* **Note 8**) by performing operations similar to those in Eqs. 23–25. The next task is to draw smooth continuous lines that pass as close as possible to the circles and cross most of the error bars. Again, we will use weighted least-square fitting by a smooth model function to generate the smooth lines. By the method of trial and error, it was found that the functions suitable for fitting the spectra $\alpha_{\lambda n}(\lambda_m)$ or $f_{\lambda n}(\lambda_m)$ are considerably more complex and need more fitting parameters as compared to the functions suitable for fitting the renormalized parameters f'_{mn} (considered as functions of λ_m). The parameters f'_{mn} can be successfully fit by the polynomial functions:

$$p_n(\lambda) = \sum_{l=0}^{L} c_{ln}\lambda^{-l} \tag{26}$$

The right-hand side of Eq. 26 is a polynomial function of ν, where $\nu = 1/\lambda$. It was found empirically that when two polynomial functions of the same degree L are compared, the polynomial function of ν always gives a better fit than the polynomial function of λ. In Subheading 3.1, **step 2**, a polynomial function of ν was also involved in fitting the steady-state spectrum. The calculation of f'_{mn}, $f_{\lambda n}(\lambda_m)$, $\alpha_{\lambda n}(\lambda_m)$, the error bars for $\alpha_{\lambda n}(\lambda_m)$, and fitting of f'_{mn} with the polynomials from Eq. 26 is done by the program *trspectr*. The best-fit polynomials $p_n(\lambda)$ are then used by the program to generate the continuous amplitude spectra:

$$\alpha_{\lambda n}(\lambda) = p_n(\lambda) \cdot f_{\lambda}(\lambda)/\tau_n \tag{27}$$

The program *trspectr* automatically generates a plot similar to the one in Fig. 8. The plot should be visually inspected by the user of the program. If the TCSPC data are very accurate (long

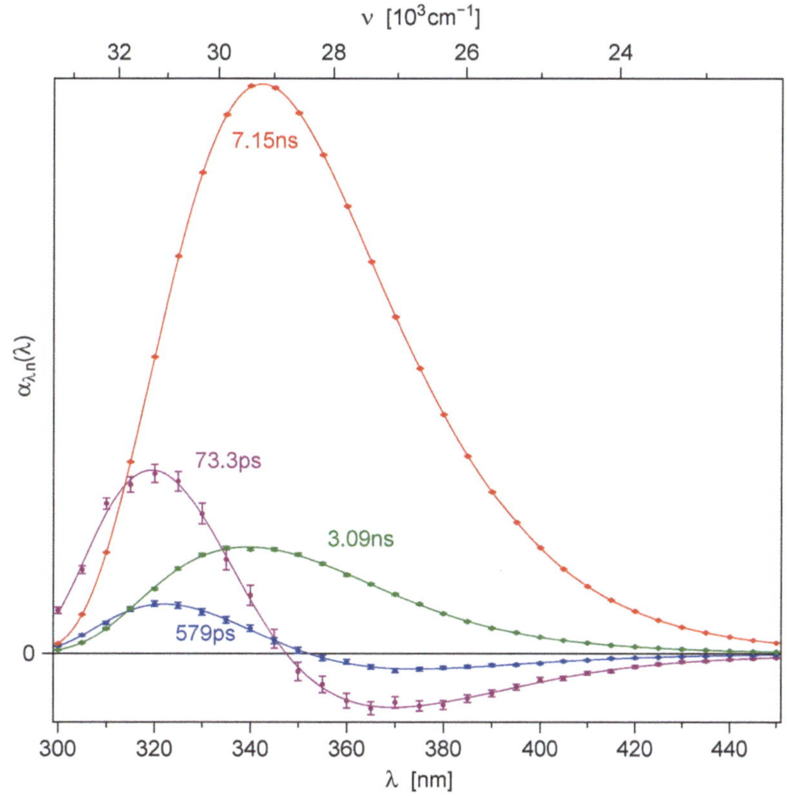

Fig. 8 Amplitude spectra on the linear wavelength scale corresponding to the four exponential terms in the fluorescence emission from the single Trp residue of the B1 domain of streptococcal protein G (GB1) at +5 °C and 296-nm excitation; these spectra represent an intermediate step in the generation of the time-resolved emission spectra published in [3]; however, no data on the linear wavelength scale was included in [3]. Dots: $\alpha_{\lambda n}(\lambda_m)$ obtained in accordance with Eqs. 23–25. Error bars: 95 % confidence intervals (*see* **Note 8**). *Solid lines*: continuous amplitude spectra $\alpha_{\lambda n}(\lambda)$ obtained in accordance with Eq. 27 using the fitting polynomials from Eq. 26 of the degree $L = 8$. The figure was generated by the program *trspectr*

data collection, narrow confidence intervals) or the degree L of the fitting polynomials is chosen too low, then the continuous amplitude spectra will miss too many error bars; *see* the top panel of Fig. 9. The error bars represent 95 % confidence intervals (*see* **Note 8**); therefore a continuous spectrum is allowed to miss about 1 error bar out of 20. If the number of missed error bars is significantly greater than 5 %, then this is a sign that the degree L is too low. On the other hand, if the TCSPC data are not sufficiently accurate or the degree L is chosen too high, then the continuous spectra will develop bumpy structure; *see* the bottom panel of Fig. 9. The degree

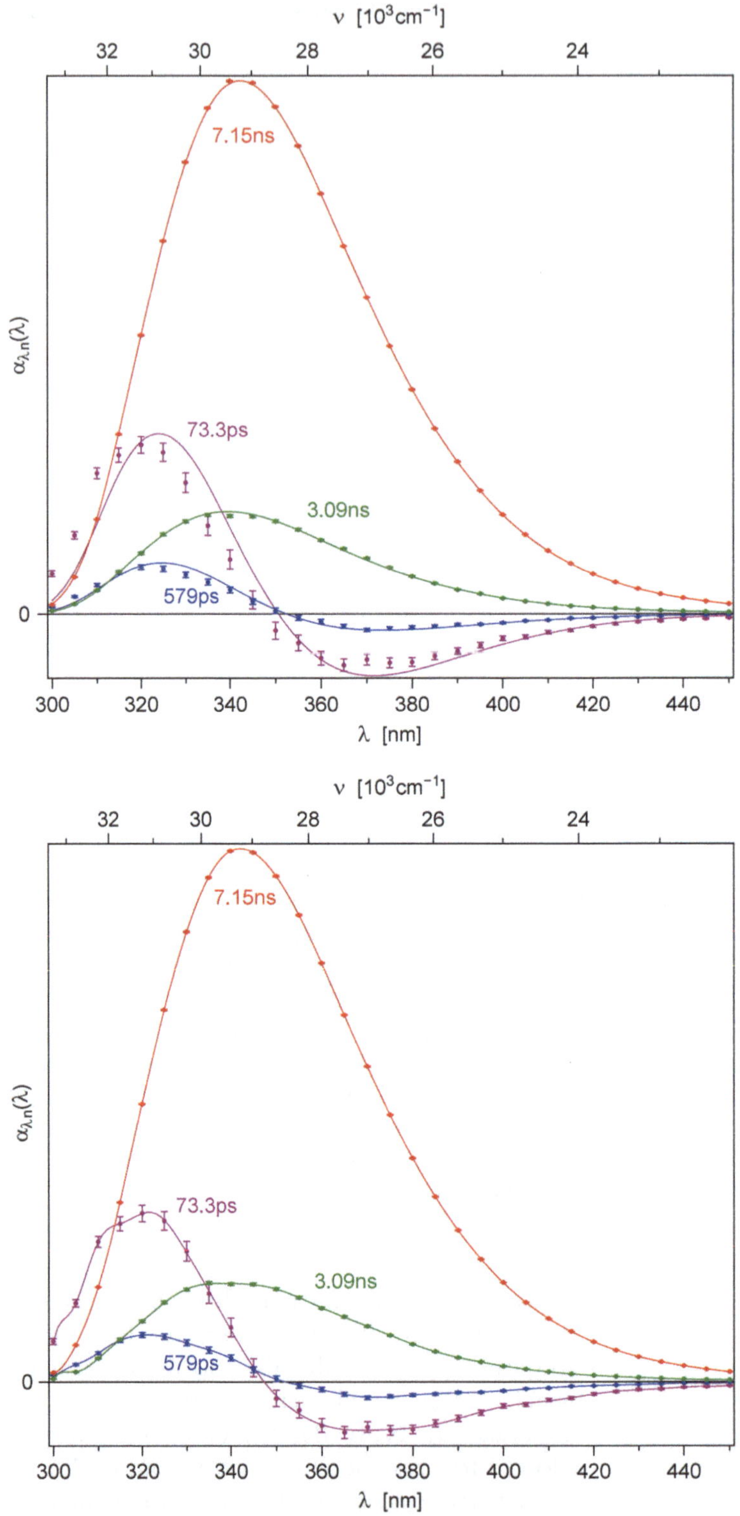

Fig. 9 Same as Fig. 8, except that the fitting polynomials of the degrees $L = 2$ and $L = 20$ were used to generate the continuous amplitude spectra $\alpha_{\lambda n}(\lambda)$ in the *top panel* and in the *bottom panel*, respectively. Note that with $L = 2$ (*top panel*) the continuous amplitude spectra miss a large number of error bars, whereas with $L = 20$ (*bottom panel*) the continuous amplitude spectra are bumpy. Both panels were generated by the program *trspectr*

L can be changed and the plot will be quickly regenerated by the program *trspectr*. With the right choice of the degree L, the plot looks like the one in Fig. 8. There is usually a range of acceptable L values. $L = 8$ was used for Fig. 8; however, with $L = 6$ and $L = 10$, the result is almost indistinguishable from that in Fig. 8. At $L = 4$ the number of missed error bars increases, and at $L = 2$ the continuous lines miss most of the error bars; *see* the top panel of Fig. 9. The shapes of the continuous lines become slightly bumpy at $L = 12$ and very bumpy at $L = 20$; *see* the bottom panel of Fig. 9.

2. *Conversion from wavelength scale to wavenumber scale.* A spectrum on the wavelength scale represents the density of photons per unit wavelength. A spectrum on the wavenumber scale represents the density of photons per unit wavenumber. Since $d\lambda/d\nu = \nu^{-2}$, the general rule requires that the spectrum is multiplied by ν^{-2} when it is converted to the wavenumber scale. For example, the steady-state emission spectrum is converted using the following equation:

$$f_\nu(\nu) = \nu^{-2} f_\lambda(\nu^{-1}) \tag{28}$$

This also applies to the amplitude spectra:

$$\alpha_{\nu n}(\nu) = \nu^{-2} \alpha_{\lambda n}(\nu^{-1}) \tag{29}$$

The program *trspectr* converts amplitude spectra in accordance with Eq. 29 and generates a plot on the linear wavenumber scale; *see* the top panel of Fig. 10.

3. *Time-resolved spectra.* Time-resolved spectra on the wavenumber scale are generated by the program *trspectr* from the continuous amplitude spectra on the wavenumber scale in accordance with the following equation:

$$\nu^{-3} \cdot F_\nu(\nu, t) = \nu^{-3} \sum_{n=1}^{N_E} \alpha_{\nu n}(\nu) \cdot \exp(-t/\tau_n) \tag{30}$$

Then the program peak normalizes the $\nu^{-3} \cdot F_\nu(\nu,t)$ spectra (for fixed values of t) and generates a plot on the linear wavenumber scale; *see* the bottom panel of Fig. 10. This plot is directly relevant to the shape conservation criterion described in the Introduction.

3.5 Calculation of the Width and the Center of Gravity of Time-Resolved Spectra

1. *Width calculation.* The program *trspectr*, in addition to generating graphical output shown in Figs. 8, 9, and 10, also saves the smooth amplitude spectra $\alpha_{\nu n}(\nu)$ and the time constants τ_n in numerical format. The program *nu_FWHM* reads the amplitude spectra $\alpha_{\nu n}(\nu)$ and the time constants τ_n, generates $\nu^{-3} \cdot F_\nu(\nu,t)$ for a large number of t values according to Eq. 30, calculates the FWHM of the spectrum for every t value, and

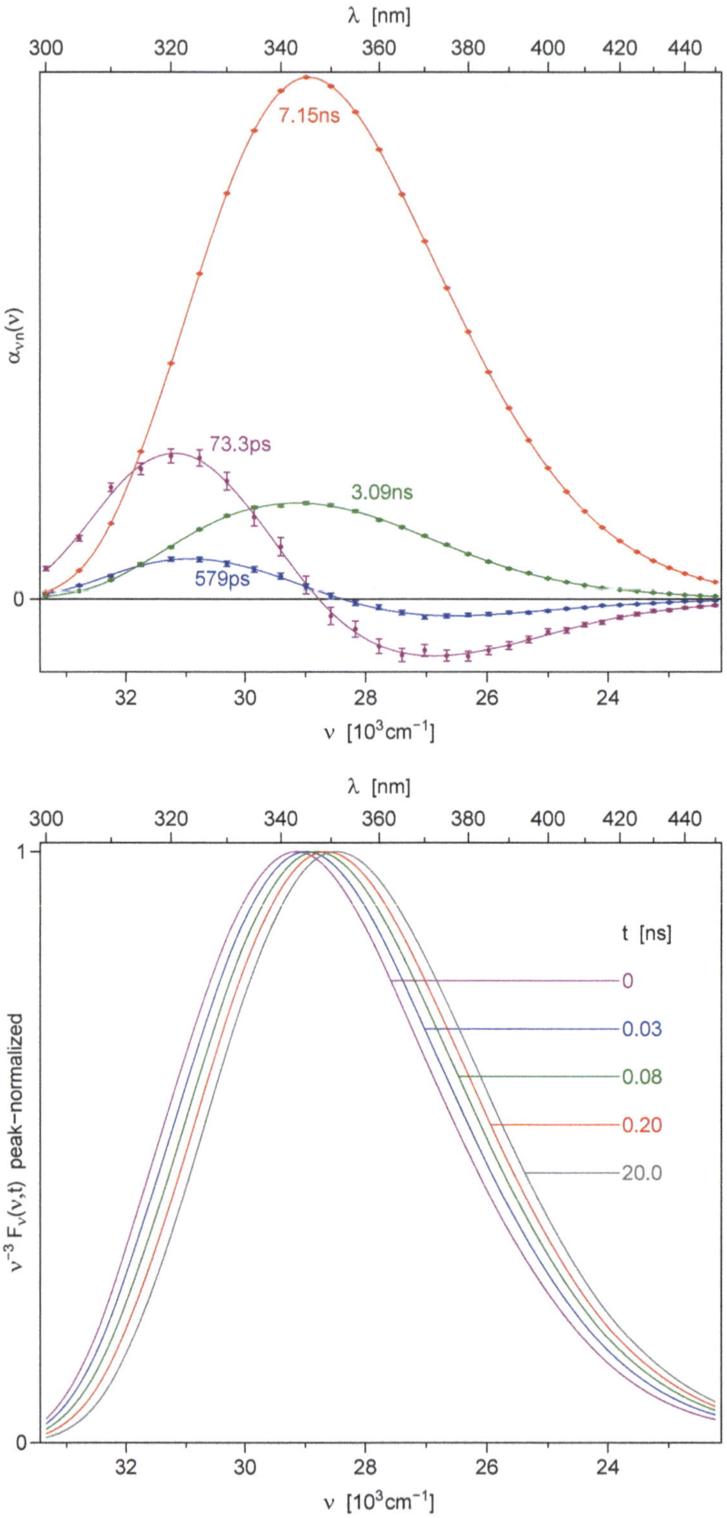

Fig. 10 *Top panel*: amplitude spectra on the linear wavenumber scale, obtained from the $\alpha_{\lambda n}(\lambda)$ spectra in Fig.8 using Eq. 29. *Bottom panel*: Time-resolved emission spectra $\nu^{-3} \cdot F_\nu(\nu, t)$ obtained from the amplitude spectra in the *top panel* using Eq. 30. Both panels were generated by the program *trspectr*. Reprinted with permission from [3]. Copyright (2006) American Chemical Society

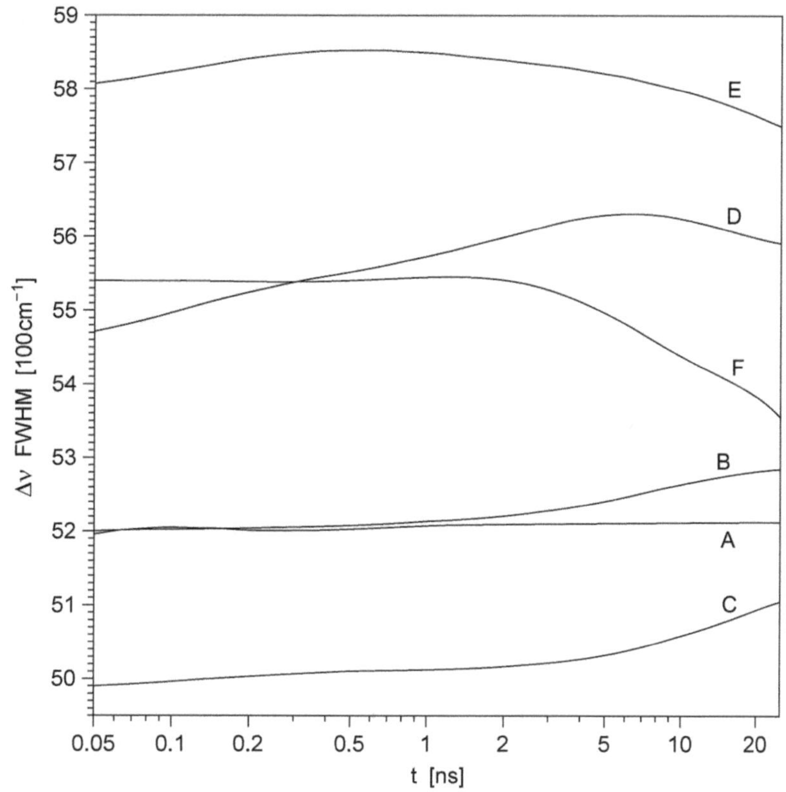

Fig. 11 Full width at half maximum of $\nu^{-3}\cdot F_{\nu}(\nu,t)$ spectra as a function of time, plotted on a logarithmic time scale. *A*, *B*, *C*—single-Trp globular proteins with well-defined structures. *D*, *E*—single-Trp proteins, intrinsically unstructured. F—a protein with four Trp residues. *A*: the $\nu^{-3}\cdot F_{\nu}(\nu,t)$ spectra from the *bottom panel* of Fig. 10, adapted with permission from [3]. Copyright (2006) American Chemical Society. *B*: adapted with permission from [2]. Copyright (2001) American Chemical Society. *C–F*: this data was originally published in Biochemical Society Transactions [12]. Copyright (2008) the Biochemical Society

saves the result in numerical format. Any plotting program can be used to plot the data. In Fig. 11 the FWHM for six proteins is plotted versus *t* on a logarithmic time scale. In accordance with the width conservation criterion (*see* Subheading 1), a decrease in the width at long times after excitation is characteristic of heterogeneous systems. The decrease is seen in curves D, E, and F. Curves D and E represent single-Trp proteins with the Trp residues in intrinsically unstructured regions, where heterogeneity is expected. Curve F represents a protein with four Trp residues in the amino acid sequence; fluorescence emission of all proteins with more than one Trp residue in the amino acid sequence must be heterogeneous. Curve A shows neither an increase nor a decrease in the width at long times after excitation. This is the classical case consistent with the theory of homogeneous relaxation [1, 2]. Curves B and C show an increase in the width at late times, which cannot be explained in terms of heterogeneity. It is most likely that at late times water molecules penetrate in a hydrophobic pocket containing the Trp residues in proteins B and C; Brownian motion of polar molecules generates fast oscillations in the electric field that homogeneously broadens the emission spectrum. It is worth mentioning that not only the spectral width in heterogeneous systems D, E, and F decreases with time at long times but also it is greater than the spectral width in homogeneous systems A, B, C at any time.

2. *Center of gravity calculation*. The program *nu_centr* reads the amplitude spectra $\alpha_{\nu n}(\nu)$ and the time constants τ_n, generates $\nu^{-3}\cdot F_\nu(\nu,t)$ according to Eq. 30 for a large number of t values, calculates the center of gravity of the spectrum for every t value,

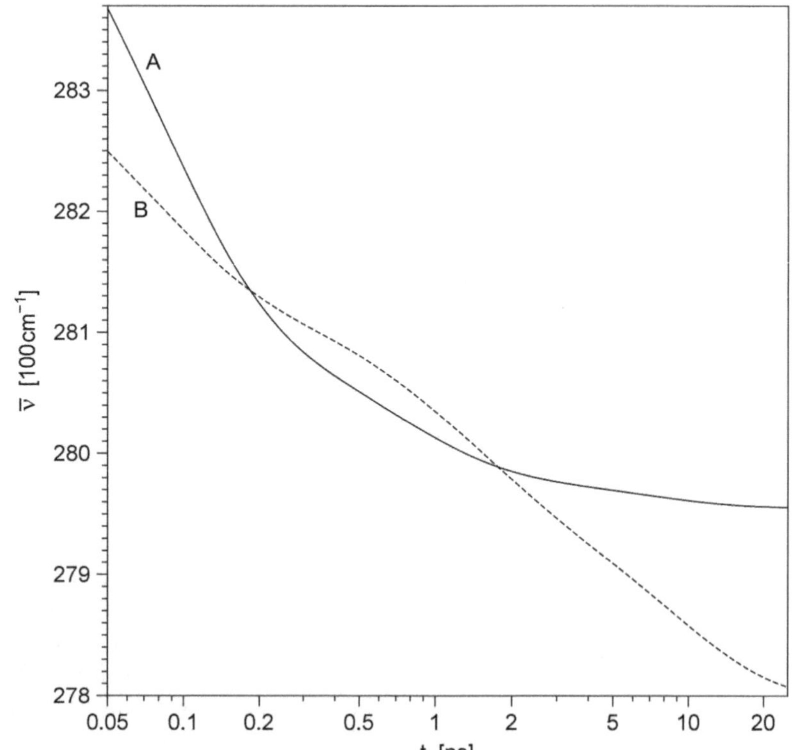

Fig. 12 Center of gravity of $\nu^{-3}\cdot F_\nu(\nu,t)$ spectra as a function of time, plotted on a logarithmic time scale. *A*: the $\nu^{-3}\cdot F_\nu(\nu,t)$ spectra from the *bottom panel* of Fig. 10, adapted with permission from [3]. Copyright (2006) American Chemical Society. *B*: adapted with permission from [2]. Copyright (2001) American Chemical Society

and saves the result in numerical format. Any plotting program can be used to plot the data. In Fig. 12 line A depicts the TDRS in the fluorescence emission from the single Trp residue of the B1 domain of streptococcal protein G (GB1) at +5 °C and 296-nm excitation; the data are from [3]. MD simulations [13] helped to identify all processes contributing the TDRS in this protein. Line B depicts the TDRS in the fluorescence emission from the single Trp residue of the E21W variant of IIA^{Glc} protein at +5 °C and 288.5-nm excitation; the data are from [2]. Since the emission from both of these proteins appears to be homogeneous, the TDRS describes the relaxation of the protein matrix in both cases. The curves A and B have little in common; furthermore, the relaxation dynamics of each protein has unique features that cannot be found in the TDRS of other proteins.

4 Notes

1. Preliminary results of a tryptophan photobleaching study were presented in the form of a poster [14]. The chemical structures of the products of tryptophan photobleaching have not been identified; however, it was shown that some or all of the products are fluorescent, that there are two or more fluorescent products with different excitation and emission spectra, and that the composition of the fluorescent product mixture depends on the local environment at the time of photobleaching. All fluorescent products of tryptophan photobleaching have red-shifted emission spectra and red-shifted excitation spectra as compared to normal tryptophan. The latter makes it possible to observe the emission of the photobleached products by exciting at wavelengths above 310 nm, where normal tryptophan residues do not absorb. Substantially different emission spectra were observed when a photobleached single-tryptophan protein sample was excited at 315 nm and at 325 nm, proving that the emission is from a mixture of two or more fluorescent molecules (no emission was observed with 315-nm or 325-nm excitation prior to photobleaching). Furthermore, the emission spectra of the photobleached product mixture changed depending on whether the protein was in the native state or in the denatured state at the time of photobleaching, proving that the outcome of photobleaching depends on the local environment of the tryptophan residue. Since all the fluorescent products of tryptophan photobleaching have red-shifted emission spectra and short lifetimes, the contamination of time-resolved tryptophan fluorescence by the emission from the photobleached products usually masks the negative amplitudes on the red side of the emission maximum [14].

2. Time-correlated single-photon counting data are characterized by the real time, the dead time, and the live time. The real time is the wall clock time elapsed between the instances when the photon counting was turned on and turned off. The live time equals the difference between the real time and the dead time. The dead time represents the total time during which the TCSPS equipment was busy (unable to process new photon events). ORTEC model 457 time-to-amplitude converter (TAC in Fig. 1) has the dead time of about 5 μs per photon event; the exact value of the TAC dead time depends on the setting of the "output delay" potentiometer on its front panel. ORTEC model 918 ADCAM (MCA in Fig. 1) has a fixed dead time of 10 μs per photon event. If neither the TAC nor the MCA is busy when a pulse arrives from the pico-Timing discriminator (*see* Fig. 1), then the dead times of the TAC and MCA start almost simultaneously, the dead time of the TAC ends first, and the total dead time is close to 10 μs per photon count, unless the TAC accepts a new input pulse from the discriminator after the end if its dead time but before the end of the MCA's dead time. To make sure that this does not happen, the busy logical output on the rear panel of the MCA is connected by a properly terminated RG-62A cable to the gate input on the front panel of the TAC and the gate switch is set to anticoincidence. This MCA → TAC logical feedback (not shown in Fig. 1) ensures that the TAC is not busy at the end of the MCA's dead time. The MCA keeps track of its dead time, real time, and live time. When the computer saves TCSPC data to a file, the values of the real time and live time are saved with the data.
3. The total memory of ORTEC model 918 ADCAM MCA is divided in four segments, each containing 2,048 channels. The segments are used to store different kinds of data. For example, in a time-resolved anisotropy measurement, one segment is used to store the IRF (*see* Subheading 2.1, **item 5**), the other segment is used to store the vertically polarized fluorescence data, and the third segment is used to store the horizontally polarized fluorescence data. Alternatively, one segment can be used to store the IRF, the other segment can be used to store data from the first fluorescence sample, and the third segment can be used to store data from the second fluorescence sample (two fluorescence cells and one scatterer cell can be measured quasi-simultaneously using the instrument; *see* Fig. 1). If the data from two fluorescent samples were obtained quasi-simultaneously and for equal live times (*see* **Note 2**), then not only the shapes of the decay curves but also the relative intensities constitute accurate data. It is worth to mention, however, that this method should not be used to subtract the

background fluorescence of the solvent from the total fluorescence of the protein in that solvent, since the direct subtraction of photon counts results in non-Poissonian statistics, whereas all data analysis programs require that the input data strictly follow Poissonian statistics (*see* Subheading 3.3, **step 3**). It is always better to use nonfluorescent solvents, e.g., instead of organic pH buffers, such as Tris or Hepes, which always contain traces of fluorescent contaminants; one can switch to inorganic phosphate buffer, which has no background fluorescence. In the case of organic solvents (glycerol, acetonitrile, etc.), one can either buy those without added preservatives and antioxidants or distill the solvent using a glass distiller. If, however, it is not possible to get rid of the background fluorescence, then the TCSPC data from the sample and the blank should not be subtracted; instead, they should be globally fit by the linear combination of a sufficient number of exponentials with the time constants τ_n playing the role of the global parameters (the same set of time constants apply to the sample and to the blank), after which the corresponding intensities f_{mn} can be subtracted.

4. The presence of two identical fluorescence emission wings in the TCSPC instrument (*see* Fig. 1) serves two purposes. First, using both wings, the data of the same quality can be acquired in half the time as compared to the case where just one wing is used (*see* Subheading 2.1, **item 6**). Second, in a time-resolved anisotropy measurement, while the right wing measures vertically polarized fluorescence, the left wing measures horizontally polarized fluorescence and vice versa. Combining polarized fluorescence data from both wings as described in Subheading 3.2 cancels the effect of laser intensity fluctuations (since the vertical and the horizontal polarization are measured truly simultaneously) and also cancels the subtle differences between the left and the right wing (since each wing is used to measure each polarization exactly one half of the time). The two wings can be also used for a truly simultaneous collection of the data at two different wavelengths; however, since a truly simultaneous collection at 31 emission wavelengths is not possible, it is necessary to use the renormalization described in Subheading 3.4, **step 1**, Eqs. 23 and 24. This renormalization takes care not only of the random fluctuations in laser intensity but also of the spectral sensitivity variation and of the differences in the live times. It is not a good idea to use one wing for the even wavelengths (450, 440, 430 nm, etc.) and the other wing for the odd wavelengths (445, 435, 425 nm, etc.), since this does not decrease the data collection time, but the amplitude spectra $\alpha_{\lambda n}(\lambda_m)$ corresponding to short τ_n may acquire a subtle periodic structure with a 10-nm period. Thus, in a typical

multiwavelength data acquisition, both wings collect data at the same wavelength, which starts from 450 nm and is then decremented by 5 nm between each data set and the next one.

5. Each segment of ORTEC model 918 ADCAM MCA contains 2,048 channels (*see* **Note 3**); however, not all these channels can be used. Channels 0 through 53 always contain zero counts, since the amplitudes of the corresponding input pulses are lower than the threshold that must be exceeded to trigger analog-to-digital conversion. The threshold is set to about 260 mV; a threshold value below 210 mV allows self-triggering of the MCA by internal electronic noise and results in a dead time equal to the real time (*see* **Note 2**). Channels 2,000 through 2,047 contain excessive counts, since input pulses with amplitudes higher than 10.000 V, which result in ADC overflow, are not discarded, as one would expect, but are counted in the channels 2,000 through 2,047. To determine the range of usable MCA channels, the system shown in Fig. 1 is modified in the following ways: a scatterer solution is used in place of the fluorescent sample, the exciting laser beam is blocked by a black metal plate where it normally passes through the ultraviolet filter (UVF), a dimmable miniature 12 V incandescent lamp is placed after the black metal plate, and the brightness of the lamp is adjusted to achieve about 20,000 counts per second in each wing. Both wings are set to count photons for 5 h; at the end a histogram of the photon counts versus the channel number is generated. The histogram is inspected at different magnifications and a range of channels is selected where the histogram is completely flat (within random Poissonian noise). In our case the flat part of the histogram comprises channels 74 through 1,999, making a total of 1,926 usable MCA channels. The range of usable MCA channels is saved together with the data; only the data from these channels are involved in the data analysis. We never limit the range of the data in order to improve the fit at the front or at the tail of a decay curve; the range is defined objectively as described above, and then it is changed only when an MCA is replaced or its threshold is readjusted.

6. Normally a cell with a protein solution is placed in the "sample" beam, and an identical cell filled with just the solvent (e.g., a pH buffer) is placed in the "reference" beam of a spectrophotometer. The resulting absorption spectrum represents the amount of light absorbed by the protein, but not that absorbed by the solvent or lost to reflections at the interfaces between the cell walls and air. Using such an absorption spectrum for the inner-filter correction (*see* Subheading 3.1, **step 1**) would result in incomplete correction, since it would not correct for the light absorbed by the solvent or lost to reflections at the

interfaces. Thus, for the purpose of the inner-filter correction, it is necessary to measure the absorption of the protein sample with nothing in the "reference" beam.

7. Not only the model of discrete exponentials is based on the function in Eq. 20. The model of exponential series and the-method of maximum entropy also use this function; however, there is a significant difference between the three models. Only in the case of the discrete exponential model the time constants τ_n play the roles of the fitting parameters, whereas in the cases of the exponential series model and the maximum entropy method, the values of τ_n are equally spaced on the logarithmic scale and fixed; thus, only the values of b_m, s_m, and f_{mn} play the roles of the fitting parameters. The exponential series model uses well-separated time constants $\tau_{n+1} = 1.75{\cdot}\tau_n$, which ensures that the Hessian matrix is not singular. Since the function in Eq. 20 is linear in the parameters b_m, s_m, and f_{mn}, the model fitting is done by the method of linear least squares, which converges in just one iteration. The model of exponential series gives a perfect fit to any TCSPC data at very low computational cost, although it uses more fitting parameters than the model of discrete exponentials. Programs *tcphoton* and *l_global* use the model of exponential series for calculating the weights $W_m(i)$; *see* Subheading 3.3, **step 3**. The maximum entropy method uses closely separated time constants (e.g., $\tau_{n+1} = 1.044274{\cdot}\tau_n$, $\tau_{n+16} = 2{\cdot}\tau_n$) that result in a singular Hessian matrix; therefore an infinite variety of different f_{mn} combinations produce indistinguishable fits to the data (as judged by the χ^2 goodness-of-fit test or by the weighted residuals or by the autocorrelation of residuals). To select one out of the infinite variety of equally good solutions, an additional requirement of maximum entropy is imposed. The program *tcphoton* has the capability of fitting TCSPC data using either the model of discrete exponentials or the method of maximum entropy. All three models (discrete exponentials, exponential series, and maximum entropy) have been tested as prospective tools for reconstructing time-resolved emission spectra of Trp residues in proteins. The discrete exponential model worked reliably, and the present chapter is entirely devoted to the application of this model. The model of exponential series resulted in about 13 discrete amplitude spectra $\alpha_{\lambda n}(\lambda_m)$, which were overwhelmed by random noise. The maximum entropy method resulted in about 145 discrete amplitude spectra $\alpha_{\lambda n}(\lambda_m)$, which looked like discontinuous functions, i.e., had nonzero values over a range of wavelengths λ_m and then abruptly plunged to zero at the next wavelength. Both models required polynomials of power $L = 30$ in Eq. 26 for an adequate fit (since a polynomial of power 30 can pass through any 31 points randomly scattered on

a plane), and the resulting time-resolved emission spectra contained a significant contribution of high-frequency noise, which made it impossible to apply the shape conservation criterion to these spectra or to calculate the time variation of their FWHM as described in Subheading 3.5, **step 1**. Thus, the exponential series model and the maximum entropy method were found unsuitable for reconstructing the shapes of the time-resolved emission spectra.

8. The confidence intervals for the values of the model parameters are determined from the functional dependence of S defined in Eq. 21 on the values of all model parameters (including local and global parameters). The quantity S is also known as the χ^2, which should not be confused with the reduced χ^2 defined in Eq. 22. To remove the grounds for the ambiguity, the symbol S is used here to denote the nonreduced χ^2. As it follows from Eqs. 18–21, S is a function of the model parameters b_m, s_m, f_{mn}, and τ_n, which here will be all referred to as p_k for notation simplicity; thus, $S = S(p_1, p_2, \ldots, p_{N_P})$. A TCSPC data analysis program first finds the minimum of this function with respect to all p_k. The value of S at the minimum will be denoted S_0, and the corresponding parameter values will be denoted p_k^0. The confidence interval for a parameter p_k is defined as such interval that for any p_k value from this interval, it is possible to find such values of the remaining model parameters that $S(p_1, p_2, \ldots, p_{N_P}) < S_T$. The threshold value S_T is strictly defined in **Note 9**. An efficient method for calculating confidence intervals relies on the following Taylor expansion of $S(p_1, p_2, \ldots, p_{NP})$ in the vicinity of the minimum:

$$S = S_0 + \sum_{j=1}^{N_P} \sum_{k=1}^{N_P} H_{jk} \left(p_j - p_j^0\right) \left(p_k - p_k^0\right) + O\left[\left(p_n - p_n^0\right)^3\right] \quad (31)$$

In Eq. 31 H_{jk} are the elements of the Hessian matrix $\mathbf{H}$:

$$H_{jk} = \sum_{m=1}^{N_W} \sum_{i=1}^{N_C} W_m(i) \frac{\partial G_m(i)}{\partial p_j} \frac{\partial G_m(i)}{\partial p_k} \quad (32)$$

and $O[(p_n - p_n^0)^3]$ denotes a combination of terms of the third and higher powers in $p_n - p_n^0$; this is known as the big O notation. For a model function that is linear in all parameters, the term $O[(p_n - p_n^0)^3]$ is strictly equal to zero. For nonlinear models, $O[(p_n - p_n^0)^3]$ is not equal to zero, but in the vicinity of the minimum, this term is small as compared to the second-power term in Eq. 31; therefore, as an approximation, one can neglect $O[(p_k - p_k^0)^3]$ in the calculation of the confidence intervals. This results in a symmetrical confidence interval for every model parameter, which can be denoted as

$p_k^0 \pm \delta p_k$ and includes the values from $p_k^0 - \delta p_k$ to $p_k^0 + \delta p_k$. It is tempting to calculate the half-width δp_k using Eq. 33:

$$\delta p_k = \sqrt{(S_T - S_0)/H_{kk}} \tag{33}$$

The half-width calculated in this way does not take into account the correlation between parameters, and this often results in a gross underestimation of δp_k. Indeed, the information about the correlation between parameters is contained in the off-diagonal elements of the Hessian matrix **H**, which are not used when δp_k is calculated using Eq. 33. The inverted Hessian matrix **H** will be denoted **C** rather than $\mathbf{H}^{-1}$, since when the notation $\mathbf{H}^{-1}$ is used, $H_{jk}{}^{-1}$ can be understood either as the inverse of the element H_{jk} or as an element of the inverted matrix $\mathbf{H}^{-1}$, thus, creating ambiguity. The matrix **C** also represents the variance–covariance matrix for the model parameters [11]. Some algorithms that search for the minimum of S, such as the Gauss–Newton algorithm, calculate both matrices **H** and **C** at every iteration; therefore, once the minimum of S is found, the elements of the matrix **C** are already present in the computer memory, and they do not have to be calculated again. Note that the Marquardt algorithm and the Levenberg–Marquardt algorithm add λH_{kk} to each diagonal element of the Hessian matrix defined in Eq. 32; the inverse of the Hessian matrix modified in this manner does not represent the variance–covariance matrix, and it should not be used to obtain confidence intervals. The correct method of calculating δp_k uses the diagonal elements of **C**:

$$\delta p_k = \sqrt{(S_T - S_0)C_{kk}} \tag{34}$$

The half-width calculated in this way does take into account the correlation between parameters, since the diagonal elements of the inverted Hessian matrix **C** depend on the off-diagonal elements of the Hessian matrix **H**. The confidence intervals calculated using Eq. 34 are considerably wider and closer to reality than those obtained using Eq. 33. For a model function that is linear in all parameters, the confidence intervals obtained using Eq. 34 are precisely correct; however, for nonlinear models they are approximate. The exact confidence intervals for the parameters of nonlinear models are not symmetrical and can be found using the exhaustive search method described in [15]. According to this method, one parameter at a time, e.g., p_k, is assigned a value different from p_k^0, after which the process of S minimization is repeated with the value of p_k fixed and the values of all remaining parameters subject to optimization. If the new minimum S does not exceed the S_T value (*see* **Note 9**), then the fixed p_k value is within the confidence interval. Systematic repetition of the above with many different p_k values makes it

Table 2
95 % confidence intervals for the global parameters τ_n involved in the model function in Eq. 20

τ_n^{0}	0.07333	0.57859	3.09200	7.14611
$\pm\delta\tau_n$	±0.00286	±0.02335	±0.05399	±0.00931
$UL_n - \tau_n^{0}$	+0.00299	+0.02492	+0.05638	+0.00980
$LL_n - \tau_n^{0}$	−0.00292	−0.02371	−0.05477	−0.00930

Fluorescence emission is from the single Trp residue of the B1 domain of streptococcal protein G (GB1) at +5 °C and 296 nm excitation; the data include 31 TCSPC data sets obtained at emission wavelengths from 300 to 450 nm at 5-nm wavelength intervals; these 31 TCSPC data sets were used to generate the time-resolved emission spectra published in [3]; however, no confidence intervals was included in [3]. The confidence intervals calculated using Eq. 34 are presented in the form $\tau_n^{0} \pm \delta\tau_n$; the τ_n^{0} values and $\pm\delta\tau_n$ values are shown in the top two rows. The confidence intervals $[LL_n, UL_n]$ calculated using the exhaustive search method [15] are represented by the numbers $UL_n - \tau_n^{0}$ and $LL_n - \tau_n^{0}$ in the bottom two rows. It is not difficult to see that the absolute values of $UL_n - \tau_n^{0}$ and $LL_n - \tau_n^{0}$ differ from each other and form $\delta\tau_n$. All values are in nanoseconds

possible to define the confidence interval. A comparison of the confidence intervals calculated using the exhaustive search method [15] with those obtained using Eq. 34 allowed us to make the following four qualitative conclusions. First, if a model function is linear in some of its parameters and nonlinear in other parameters, then for the linear parameters the confidence intervals calculated using both methods agree better than for the nonlinear parameters. Second, if a model function is linear in all of its local parameters, and the number of localities is large, then the confidence intervals for the local parameters calculated using both methods agree with a great deal of accuracy. For example, the model function in Eq. 20 is linear in all local parameters b_m, s_m, and f_{mn}, and if the TCSPC data are collected at 31 emission wavelengths, then the confidence intervals for b_m, s_m, and f_{mn} calculated using both methods agree to at least four significant digits; in this case the confidence intervals calculated using the exhaustive search method are practically symmetrical. Third, for the nonlinear model parameters, such as τ_n in the model function in Eq. 20, the confidence intervals calculated using the two methods may differ; *see* Table 2 for a specific example. Fourth, in terms of the CPU time, the exhaustive search method takes about 1–10 million times longer than the method based on Eq. 34. By default, the programs *tcphoton* and *l_global* use the fast method; but if necessary, they can be tricked into doing the exhaustive search method by means of a simple script, as it was done to generate the numbers for Table 2.

9. The threshold value S_T is defined in terms of the *F*-distribution [11]:

$$S_T = S_0 + \frac{N_S \cdot S_0}{N_D - N_P} F(\alpha; N_S, N_D - N_P) \tag{35}$$

Here S_0 represents the absolute minimum of the sum S defined in Eq. 21, N_D and N_P are the number of data points and the number of model parameters (*see* Subheading 3.3, **step 4**), and N_S is the dimensionality of the confidence subspace in the parameter space. For confidence intervals, which can be expressed as two limits between which the value of the parameter is found with a certain degree of probability, $N_S = 1$. For confidence areas in the space of any two model parameters, $N_S = 2$. Note that the projections of a confidence area on the axes representing the two parameters do not coincide with the confidence intervals for these parameters. For confidence volumes in the space of any three model parameters, $N_S = 3$. Note that for $N_S > 2$ it is not possible to map the confidence subspace on plain paper. All the discussion in **Note 8** is limited to the case $N_S = 1$. The function $F(\alpha;m,n)$ is the cumulative distribution function for the F-distribution. The first argument of this function, α, represents the probability that the parameter value (values) falls (fall) in the specified confidence interval (area, volume, etc.). For example, for a 95 % confidence interval, $\alpha = 0.95$. Mathematically the function $F(\alpha;m,n)$ is calculated by first solving Eq. 36 with respect to x and then substituting the value of x into Eq. 37:

$$\frac{B(x;\, m/2, n/2)}{B(m/2, n/2)} = \alpha \tag{36}$$

$$F(\alpha;\, m, n) = \frac{nx}{m(1-x)} \tag{37}$$

In Eq. 36 $B(x;a,b)$ denotes the incomplete beta-function and $B(a,b)$ denotes the complete beta-function:

$$B(x;\, a, b) = \int_0^x t^{a-1}(1-t)^{b-1}\mathrm{d}t \tag{38}$$

$$B(a, b) = B(1;\, a, b) = \frac{\Gamma(a)\Gamma(b)}{\Gamma(a+b)} \tag{39}$$

For practical applications it is convenient to use the fact that for $m = 1$ and $n > 1{,}000$, the value of $F(\alpha;m,n)$ is very close to the limit of this function at $n \to \infty$; this limit will be denoted as $F(\alpha;1,\infty)$. Here are the most commonly used values: $F(0.683;1,\infty) = 1.00$, $F(0.90;1,\infty) = 2.706$, $F(0.95;1,\infty) = 3.841$, $F(0.99;1,\infty) = 6.635$, $F(0.995;1,\infty) = 7.879$, and $F(0.999;1,\infty) = 10.828$.

Acknowledgment

This work was supported by NSF grants 0416965 and 0719248.

References

1. Toptygin D, Brand L (2000) Spectrally- and time-resolved fluorescence emission of indole during solvent relaxation: a quantitative model. Chem Phys Lett 322:496–502
2. Toptygin D, Savtchenko RS, Meadow ND, Brand L (2001) Homogeneous spectrally- and time-resolved fluorescence emission from single-tryptophan mutants of IIAGlc protein. J Phys Chem B 105:2043–2055
3. Toptygin D, Gronenborn AM, Brand L (2006) Nanosecond relaxation dynamics of protein GB1 identified by the time-dependent red shift in the fluorescence of tryptophan and 5-fluorotryptophan. J Phys Chem B 110:26292–26302
4. DeToma RP, Brand L (1977) Excited state solvation dynamics of 2-anilinonaphthalene. Chem Phys Lett 47:231–236
5. Gafni A, DeToma RP, Manrow RE, Brand L (1977) Nanosecond decay studies of a fluorescence probe bound to apomyoglobin. Biophys J 17:155–168
6. Easter JH, DeToma RP, Brand L (1978) Fluorescence measurements of environmental relaxation at the lipid-water interface region of bilayer membranes. Biochim Biophys Acta 508:27–38
7. Badea MG, Brand L (1979) Time-resolved fluorescence measurements. Methods Enzymol 61:378–425
8. Gakamsky DM, Petrov EP, Rubinov AN (1991) Decay time distribution of fluorescence kinetics in systems with inhomogeneous broadening of electronic spectra. Inst Phys Conf Ser 126:523–526
9. Gakamsky DM, Goldin AA, Petrov EP, Rubinov AN (1992) Fluorescence decay time distribution for polar dye solutions with time-dependent fluorescence shift. Biophys Chem 44:47–60
10. Velapoldi RA, Mielenz KD (1980) Standard reference materials: a fluorescence standard reference material: quinine sulfate dihydrate. National Bureau of Standards Special Publication. National Bureau of Standards, Washington, DC, pp 260–264
11. Hamilton WC (1964) Statistics in physical science. Ronald, New York
12. Bertragna A, Toptygin D, Brand L, Barrick D (2008) The effects of conformational heterogeneity on the binding of the Notch intracellular domain to effector proteins: a case of biologically tuned disorder. Biochem Soc Trans 36:157–166
13. Toptygin D, Woolf TB, Brand L (2010) Picosecond protein dynamics: the origin of the time-dependent spectral shift in the fluorescence of the single Trp in the protein GB1. J Phys Chem B 114:11323–11337
14. Toptygin D, Rodgers ME, Brand L (1999) Non-exponential fluorescence decay of photobleached indole and tryptophan. Biophys J 76: A361
15. Beechem JM, Gratton E, Ameloot M et al (1991) The global analysis of fluorescence intensity and anisotropy decay data: second-generation theory and programs. In: Lakowicz JR (ed) Topics in fluorescence spectroscopy, vol 2. Plenum, New York, pp 241–305

Chapter 10

Global Analysis of Time-Resolved Fluorescence Data

Anatoli V. Digris, Eugene G. Novikov, Victor V. Skakun, and Vladimir V. Apanasovich

Abstract

In this chapter, we describe the global analysis approach for processing time-resolved fluorescence spectroscopy data of molecules in the condensed phase. Combining simultaneous analysis of data measured under different experimental conditions (spatial coordinates, temperature, concentration, emission wavelength, excitation intensity, etc.) with the fitting strategy, enabling parameter linkage and thus decreasing the total amount of estimated variables, makes global analysis more robust and more consistent compared to a sequential fit of single experimental data. We consider the main stages of the global analysis approach and provide some details that are important for its practical implementation. The application of the global approach to the analysis of time-resolved fluorescence anisotropy is demonstrated on experimental data of (enhanced) green fluorescent protein in aqueous solution.

Key words Global analysis, TCSPC, Time domain, Frequency domain, Deconvolution

1 Introduction

Structural and dynamical properties of complex, fluorescent biological and chemical systems can be successfully studied using time- and frequency-domain methods of time-resolved fluorescence spectroscopy [1]. The most commonly used time-domain experimental method is time-correlated single-photon counting (TCSPC). In this method the fluorescent sample is excited by a series of short laser pulses of fixed repetition rate and the fluorescence decay is measured and stored in computer memory [1]. Quantitative information about the explored molecular system is obtained by fitting the model-generated decay curve to the experimental sample fluorescence decay. This gives the values of model parameters that characterize the sample under study. In the time-resolved frequency-domain approach [1] the sample is excited by intensity-modulated light at different modulation frequencies. The emission signal of the molecular system is then demodulated and

Yves Engelborghs and Antonie J.W.G. Visser (eds.), *Fluorescence Spectroscopy and Microscopy: Methods and Protocols*, Methods in Molecular Biology, vol. 1076, DOI 10.1007/978-1-62703-649-8_10,

phase-shifted with respect to the modulated excitation light. The dependencies of degree of modulation and phase shift on the modulation frequency are measured and analyzed by the appropriate model to estimate the values of fit parameters characterizing the sample under study.

The experimental methods of time-resolved fluorescence spectroscopy are frequently applied to measure fluorescent molecules in the condensed phase under different experimental conditions such as different temperatures, emission wavelengths, and concentrations. Since some parameters characterizing the investigated molecular system may be independent on the experimental conditions, these sets of data can be simultaneously analyzed using a global analysis approach [2, 3]. The fit parameters that are the same in different experiments are linked and thus kept equal for the corresponding measured curves participating in the fit. The other parameters that change with experimental conditions are fitted independently. As the overall number of fitted parameters decreases, whereas the amount of experimental data remains the same, the global analysis approach has proved to be more robust and more consistent compared to the separate analysis of single curves.

The abilities of the time-resolved fluorescence spectroscopy are significantly extended with introduction of the fluorescence lifetime imaging microscopy (FLIM) technique [1, 4, 5]. This technique uses a wide-field or a confocal microscope to perform the time- or frequency-domain measurements in various spatial points of the sample, thus resolving the immediate molecular environment and the state of a fluorescent molecule in a microscopic image. Another example is the use of genetically encoded fluorescent proteins such as green fluorescent protein (GFP), which can be attached to the protein of interest and expressed in living cells [6]. When two different proteins are labeled with two different variants of GFP, for instance cyan fluorescent protein (CFP) and yellow fluorescent protein (YFP) in living cells, the distance between these proteins when interacting can be mapped in a living cell by using FLIM combined with Förster resonance energy transfer (FRET) [7]. In the following we use the term donor for a protein that is tagged with CFP and the term acceptor for the other protein that is tagged with YFP. In other biosensors donor and acceptor are both linked to a common protein that changes conformation when a ligand is bound and thus alters the donor–acceptor distance. When FRET occurs (i.e., when donor and acceptor are a few nanometers apart) the reaction between excited donor and acceptor in the ground state will decrease the fluorescence lifetime of the donor in various points of the image. The appearance of double-exponential fluorescence intensity decays is frequently encountered because of the presence of a mixture of FRET-active and FRET-inactive donor molecules [8, 9]. To detect the spatially dependent relative contribution of donors participating in FRET global analysis may be applied. The fluorescence intensity decays measured in different points of the image are analyzed globally using double-exponential models. The corresponding fluorescence lifetimes for all space points are linked whereas the pre-exponential factors for each exponent are fitted independently (for an example *see* ref. 10).

In this chapter we present the main stages required to perform the global analysis of time-resolved fluorescence spectroscopy data. We focus on some practical details important for its implementation. As illustration we show the application of global analysis for fitting the time-resolved fluorescence anisotropy of enhanced GFP from *Aequorea victoria* (abbreviated as EGFP) in aqueous, buffered solution.

2 Materials

2.1 EGFP

The purification of EGFP has been described previously [11, 12]. Protein solutions were prepared in 20 mM Tris buffer (pH 7.0).

2.2 Time-Resolved Polarized Fluorescence Measurements

Time-resolved fluorescence measurements (20 °C) were carried out using a mode-locked laser pumped by a continuous-wave laser for excitation and TCSPC as the detection technique, as extensively described previously [8, 11, 13]. The samples were excited with plane-polarized light pulses (0.2 ps FWHM) at an excitation frequency of 3.8 MHz and both parallel- and perpendicular-polarized fluorescence intensities were detected.

Experiments with EGFP were conducted at 470-nm excitation and fluorescence detection via the combination of 515-nm cutoff and 512-nm interference filters.

The reference compound for time-resolved fluorescence decay experiments with 470-nm excitation and 512-nm detection was erythrosine B in aqueous solution having a fluorescence lifetime of 85 ps. In all experiments the binning time was 5 ps and 4,096 time channels were used.

3 Methods

3.1 General Global Analysis Algorithm

The global analysis method is based on the simultaneous processing of more than one separately measured data set of a fluorescent species in the condensed phase (*see* Fig. 1). Such data sets are usually obtained by measuring samples under different experimental conditions (temperature, emission wavelength, excitation intensity, etc.). The application of global analysis becomes meaningful if at least one of the unknown parameters can be assumed to be the same for different measurements. Therefore these parameters, even

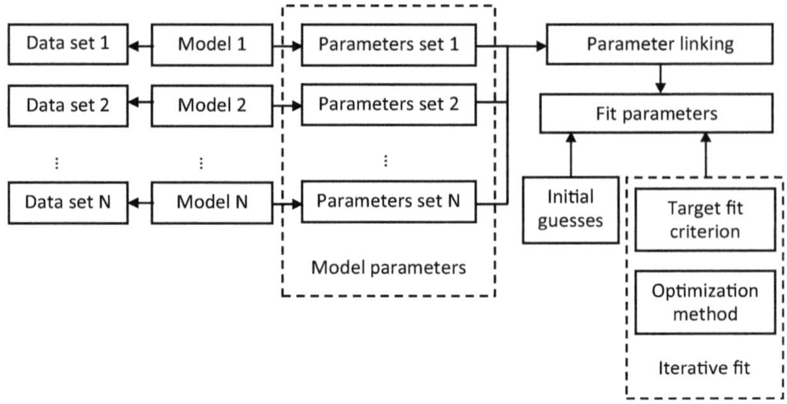

Fig. 1 General scheme of global analysis

though being assigned to different experimental data sets, are united into one parameter in the fitting procedure. The global analysis is more robust and consistent than the separate analysis of individual data sets, because it exploits measured information available from several data sets in a more efficient way by reducing the total number of fit parameters.

The global analysis method can be applied to fit the time-resolved fluorescence data obtained from both time-domain and frequency-domain measurements and includes several consecutive steps.

1. Collect the data to be analyzed and prepare the data sets. The term "data set" refers to the set of information that can be analyzed by a single mathematical model with a well-defined set of parameters. The data set contains the measured, possibly multidimensional, data from the sample (e.g., fluorescence intensity decay, phase shift, demodulation). In addition, either prior known or experimentally obtained information required for the analysis can also be included.

   For time-domain measurements, each data set contains the temporal information (either as time points or as time step and the number of time points), sample fluorescence intensity decay, and either the instrumental response function (IRF, obtained with a scattering solution) or the fluorescence intensity decay of a reference compound with known fluorescence lifetime. The latter is used for deconvolution of measured data (*see* **step 2**). A background sample such as solvent alone, obtained at the same conditions as the investigated sample, accounting for impurities or Raman scattering in the sample may also be added to the data set.

   For frequency-domain measurements the data set includes the sequence of modulation frequencies and values for phase shift and degree of modulation obtained for each frequency.

Optionally, the data set can include the list of external parameters reflecting the conditions of the measurement (e.g., temperature, excitation/emission wavelength, polarization angle). As these parameters are set by the user during the measurement and are typically known precisely, they can be used by some models as preset (fixed) values to calculate the theoretical curve.

2. Select the model for each data set created in **step 1**. Depending on the model, an algorithm should be developed and implemented generating the theoretical curve that is fitted to the measured one.

   The analytical models that are applied for fitting the fluorescence intensity decays are based on the following general equations.

   If an IRF is used for deconvolution, the basic model for calculating the theoretical decay $f^{\mathrm{th}}(t)$ takes the form

$$f^{\mathrm{th}}(t) = (g(t+\delta) - b) \otimes I(t, \vec{A}) + \gamma B(t) + c + n \times (g(t+\delta) - b), \quad (1)$$

   where $g(t+\delta)$ denotes the instrument response function with a time shift of δ relative to the emission response; $B(t)$ is the measured intensity of time-dependent background emission contributing with a relative weight of γ; b and c are constant values for dark noise and time-uncorrelated background photons in g and f, respectively; n is the scattered light coefficient and $I(t, \vec{A})$ represents a model function with the vector of fit parameters $\vec{A}$.

   For deconvolution using a single-lifetime reference compound the basic model takes the form

$$f^{\mathrm{th}}(t) = (f_{\mathrm{ref}}(t) - b) \otimes \left(\delta(t) I(0, \vec{A}) + \frac{I(t, \vec{A})}{\tau_{\mathrm{ref}}} + \frac{\mathrm{d}I(t, \vec{A})}{\mathrm{d}t} \right) + \gamma B(t) + c, \quad (2)$$

   where $f_{\mathrm{ref}}(t)$ denotes the measured one-exponential reference compound fluorescence intensity decay with decay time τ_{ref} and $\delta(t)$—Dirac delta-function.

   To complete the model, the function $I(t, \vec{A})$ should be defined. It reflects typical physical processes taking place in the sample under study. A number of models for $I(t, \vec{A})$ have been published [1, 14, 15]. These include simple models, such as a sum of exponentials or stretched exponential, as well as more complicated ones used in the compartmental formalism. One of the most widely used equations for $I(t, \vec{A})$ is the sum of exponential terms:

$$I(t, \vec{A}) = \sum\nolimits_{j=1}^{M} p_j \mathrm{e}^{-t/\tau_j}, \quad (3)$$

where p_j and $\tau_j (j = 1, \ldots, M)$ are, respectively, the amplitudes and decay times of corresponding exponents and M is the number of exponents in the sum. The multi-exponential model can be used to fit the fluorescence intensity decays obtained from the mixture of noninteracting species, donor molecules when resonance energy transfer process takes place, etc. [1].

The analytical models for fitting the data obtained from frequency-domain measurements are based on the sine $N(\omega, \vec{A})$ and cosine $D(\omega, \vec{A})$ Fourier transforms of the function $I(t, \vec{A})$:

$$N(\omega, \vec{A}) = \frac{\int_0^\infty I(t, \vec{A}) \sin(\omega t) \mathrm{d}t}{\int_0^\infty I(t, \vec{A}) \mathrm{d}t},$$

$$D(\omega, \vec{A}) = \frac{\int_0^\infty I(t, \vec{A}) \cos(\omega t) \mathrm{d}t}{\int_0^\infty I(t, \vec{A}) \mathrm{d}t},$$

where ω is the angular modulation frequency of the excitation light. The general model equations for phase shift $\varphi(\omega, \vec{A})$ and degree of modulation $m(\omega, \vec{A})$ are

$$\begin{aligned} \varphi(\omega, \vec{A}) &= \operatorname{arctg}\left(N(\omega, \vec{A}) / D(\omega, \vec{A}) \right), \\ m(\omega, \vec{A}) &= \sqrt{N^2(\omega, \vec{A}) + D^2(\omega, \vec{A})}. \end{aligned} \tag{4}$$

If $I(t, \vec{A})$ is a sum of exponents and defined by Eq. 3, the models for $\varphi(\omega, \vec{A})$ and $m(\omega, \vec{A})$ become

$$\varphi(\omega, \vec{A}) = \operatorname{arctg}\left(\frac{\sum_{j=1}^{M} \frac{\omega p_j \tau_j^2}{1+\omega^2 \tau_j^2}}{\sum_{j=1}^{M} \frac{p_j \tau_j}{1+\omega^2 \tau_j^2}} \right), \tag{5}$$

$$m(\omega, \vec{A}) = \frac{1}{\sum_{j=1}^{M} p_j \tau_j} \sqrt{\left(\sum_{j=1}^{M} \frac{\omega p_j \tau_j^2}{1 + \omega^2 \tau_j^2} \right)^2 + \left(\sum_{j=1}^{M} \frac{p_j \tau_j}{1 + \omega^2 \tau_j^2} \right)^2}. \tag{6}$$

Note that parameters $p_j (j = 1, \ldots, M)$ in Eqs. 5 and 6 define the contribution of corresponding components in the sum, whereas amplitudes in Eq. 3 are also responsible for fitting the amplitude of the fluorescence decay.

3. Link parameters: Every model selected in **step 2** contains the set of fit parameters. The term "fit parameter" refers to the value of interest that should be estimated during the analysis. For each fit parameter the range of admissible values can be set by defining the minimum and maximum constraints. The number of fit parameters and their physical interpretation depend

on the selected model. The models based on Eqs. 1 and 2 have experiment-specific fit parameters (δ, b, γ, $c\,n$ for Eq. 1 and b, γ, c, τ_{ref} for Eq. 2) and sample-specific parameters defined by $I(t, \vec{A})$. For the multi-exponential model defined by Eq. 3 for time-domain measurements and by Eqs. 5 and 6 for frequency-domain measurements the model-specific parameters are p_j and $\tau_j (j = 1, \ldots, M)$. The fit parameters from all models make up the common list of fit parameters. If for certain parameters the value is known or has been already determined from other, independent experiments, its value can be fixed so that it will remain unchanged during the fit.

The key feature of the global analysis is the parameter linkage. The linking procedure can be introduced as creating a group of fit parameters that must be kept equal during the fit. The settings (value, minimum, maximum, etc.) of fit parameters included in such group always match the appropriate settings of the group itself. In most cases one parameter group contains the fit parameters that are supposed to have the same value. The linked fit parameters are excluded from the common fit parameter list, whereas the created fit parameter group is added to this common list as a new fit parameter. Since the link procedure replaces several fit parameters in the common list by a group, the total number of parameters to be estimated is reduced. For the analysis to be global, at least one fit parameter group should be created. Generally the number of such global groups is more than one.

4. Initial guess generation: Initial guesses (IG) are required to make the forthcoming iterative fitting procedure fast and reliable. Therefore the IG should be reasonably close to the optimal values of fit parameters that correspond to the best fit of experimental data. If IG are chosen randomly and are located relatively far from the optimal values of fit parameters, the iterative algorithm may converge rather slowly and in some cases may stop untimely, thus leading to biased parameter estimates. To automatically generate the initial guesses a fast non-iterative algorithm should be applied. Such algorithms always take into account the particularity of the models selected for the fit and therefore are not available for any combination of models within global analysis. If such algorithms exist, it is always recommended to use them in combination with the iterative fit routine.

   In many practical applications such as FLIM the multi-exponential model can be applied for the global analysis of fluorescence intensity decays [2, 3]. The model is defined by Eqs. 1 and 3, for which the corresponding decay times are assumed to be equal for different data sets and therefore they should be linked, whereas the amplitudes are allowed to be different. If instrumental distortions in Eq. 1 are negligibly small (parameters δ, b, γ, cn equal to 0), the phase plane method can be used to generate initial guesses [16, 17]. It is based on the equivalence of the model given by Eq. 1 with $I(t, \vec{A})$ defined by Eq. 3 to the following integral equation:

$$\sum_{j=0}^{M} c_j F_{ij}(t) = \sum_{j=1}^{M} \rho_{ij} G_{ij}(t), \tag{7}$$

where

$$F_{i0}(t) = f_i(t),$$

$$F_{ij}(t) = f_i(t) \otimes \varphi_j(t) = \int_0^t \frac{f_i(x)(t-x)^{j-1}}{(j-1)!} \mathrm{d}x,$$

$$G_{ij}(x) = g_i(t) \otimes \varphi_j(t) = \int_0^t \frac{g_i(x)(t-x)^{j-1}}{(j-1)!} \mathrm{d}x,$$

$$\varphi_j(t) = \frac{t^{j-1}}{(j-1)!},$$

$i = 1, \ldots, m$, where m is the number of data sets. The coefficients c_k are functions of the decay times only. As the decay times are linked, c_j do not depend on the data set index i. Coefficients ρ_{ij} depend on both decay times and amplitudes. These coefficients are different for each data set because exponential amplitudes are not linked. Taking into account Eq. 7 the algorithm of the phase plane method can be described as follows.

(a) Calculate integral functions $F_{ij}(t)$ and $G_{ij}(t)$. Since in practice $f_i(t)$ and $g_i(t)$ are obtained on discrete set of time points $t_k (k = 1, \ldots, N)$ the integration is replaced by summation and discrete convolution of $f_i(t_k)$ and $g_i(t_k)$ with $\varphi_j(t_k)$ is calculated giving $F_{ij}(t_k)$ and $G_{ij}(t_k)$.

(b) For each data set construct the following matrices:

$$\overline{F_i} = \begin{pmatrix} F_{i0}(t_1) & \cdots & F_{i0}(t_N) \\ F_{i1}(t_1) & \cdots & F_{i1}(t_N) \\ \vdots & \ddots & \vdots \\ F_{iM}(t_1) & \cdots & F_{iM}(t_N) \end{pmatrix}; \quad \overline{G_i} = \begin{pmatrix} G_{i1}(t_1) & \cdots & G_{i1}(t_N) \\ \vdots & \ddots & \vdots \\ G_{iM}(t_1) & \cdots & G_{iM}(t_N) \end{pmatrix};$$

$$\overline{W_i} = \begin{pmatrix} 1/f_i(t_1) & 0 & \cdots & 0 \\ 0 & 1/f_i(t_2) & \cdots & 0 \\ \vdots & \vdots & \ddots & \vdots \\ 0 & 0 & \cdots & 1/f_i(t_N) \end{pmatrix}.$$

(c) Obtain the vector of coefficients c_j by the linear least-squares method. To do this prepare for each data set the matrices:

$$\overline{R_i} = \overline{F}_i \overline{W}_i \overline{G_i}' \left(\overline{G}\ \overline{W}\ \overline{G_i}' \right)^{-1}; \quad \overline{D_i} = \overline{F_i} - \overline{R_i}\ \overline{G_i}; \qquad (8)$$
$$\bar{S} = \sum_{i=0}^{m} \overline{D_i}\ \overline{D_i}'.$$

Find the coefficients c_j by solving the system of linear equations. This solution in matrix form can be represented as follows:

$$\bar{C} = \bar{A}/\bar{B}, \qquad (9)$$

where $\bar{C}$ is a vector of coefficients $c_j(j = 1, \ldots, M)$; matrices $\bar{A}$ and $\bar{B}$ are obtained from matrix $\bar{S}$:

$$\bar{A} = \begin{pmatrix} & S_{22} \cdots \\ & \vdots \ddots \\ S_{(M+1)2} \cdots & \end{pmatrix}; \quad \bar{B} = -\left(S_{21} \vdots S_{(M+1)1} \right).$$

(d) Calculate estimations for decay times as roots of the following polynomial:

$$\sum_{j=0}^{M} (-1)^j c_j \tau^{j-M} = 0, \qquad (10)$$

where $c_0 = 1$; $c_j(j = 1, \ldots, M)$ are taken from vector $\bar{C}$ obtained above from Eq. 9. In most cases the roots of polynomial in Eq. 10 are calculated numerically [18].

(e) For each data set calculate the amplitudes of exponents $p_{ij}(i = 1, \ldots, m;\ j = 1, \ldots, M)$ from the following equation:

$$p_{ij} = \frac{\sum_{k=0}^{M-1} (-1)^k \rho_{i(k+1)} \tau_j^k}{\prod_{m=1, m \neq j}^{M} \left(1 - \frac{\tau_j}{\tau_m} \right)}, \qquad (11)$$

where decay times τ_j are obtained by solution of Eq. 10; coefficients ρ_{ij} are elements of vector $\overline{r_i}$ calculated for each data set as follows:

$$\overline{r_i} = \left(1\overline{C} \right) R_i, \qquad (12)$$

where $\overline{C}$ is calculated from Eq. 9 and $\overline{R_i}$ obtained from Eq. 8.

5. Iterative fit: The iterative fit procedure is used for getting final estimations of fit parameters that provide the best match of

theoretical curves generated by the particular model and the measured data. To organize the iterative fit the target fit criterion (TFC) and optimization method are required.

The TFC is used to quantify the distance between the theoretical curves generated by the model and the corresponding measured data. The choice of the appropriate TFC is based on the type of statistical noise in the experimental data. For global analysis of time-domain data three TFCs are often used. All these TFCs are based on the maximum likelihood approach [19–22].

When the number of detected photons in each time channel is relatively high, the noise distribution can be well approximated by Gaussian statistics [19] and the TFC is defined by the chi-square equation:

$$\chi_G^2 = \frac{1}{n-s-1} \sum_{i=1}^{m} \sum_{k=1}^{N_i} \frac{\left(x_{ik} - F_{ik}^{\mathrm{th}}(\vec{A})\right)^2}{\sigma_{ik}^2}, \tag{13}$$

where $v = n - s - 1$ is the number of degrees of freedom; n is the total number of data points from all data sets that participate in the calculation of sums in Eq. 13; s is the total number of fit parameters (non-linked parameters from all models and all parameter groups) available for optimization; x_{ik}, $F_{ik}^{\mathrm{th}}(\vec{A})$, and σ_{ik}^2 are, respectively, experimental value, theoretical value, and variance in the k-th point of the i-th data set. The application of TFC based on Eq. 13 leads to the iterative fit procedure, which is well known as the weighted least-squares method [1, 19, 21]. Some practical remarks related to application of Eq. 13 are given in **Note 1**.

For some time-domain measurements, such as single-molecule fluorescence detection and FLIM, the number of photons registered in each time channel of fluorescence decay is relatively small. At the same time the number of time channels in each decay is large, so the probability to detect a photon within a separate time channel is small. Application of the TFC defined by Eq. 13 for such data leads to getting biased estimates of fit parameters [20]. For such cases the statistical noise is better approximated by the Poisson distribution. Under this assumption the TFC for global analysis takes the form

$$\chi_P^2 = \frac{2}{n-s-1} \sum_{i=1}^{m} \sum_{k=1}^{N_i} x_{ik} \ln\left[\frac{x_{ik}}{F_{ik}^{\mathrm{th}}(\vec{A})}\right] - x_{ik} + F_{ik}^{\mathrm{th}}(\vec{A}). \tag{14}$$

The scheme of the single-photon counting technique also allows applying the multinomial distribution for description of statistical noise in the measured fluorescence decays. At each registration cycle of the fluorescence decay measurement no more than one photon is detected in one of the N channels. Registering the photon in a certain time channel is one of N possible outcomes occurring each time when the photon is detected. Such experimental scheme leads to the multinomial distribution of statistical noise in the measured data. In this case the TFC for global analysis is defined by the following equation:

$$\chi_{\mathrm{M}}^{2}=\frac{2}{n-s-1}\sum_{i=1}^{m}\sum_{k=1}^{N_i}x_{ik}\ln\left[\frac{x_{ik}}{F_{ik}^{\mathrm{th}}(\vec{A})}\right]. \quad (15)$$

While using Eq. 15 in practice the following condition should be ensured during each iteration of the fit:

$$\sum_{k=1}^{N_i}x_{ik}=\sum_{k=1}^{N_i}F_{ik}^{\mathrm{th}}(\vec{A}),\quad i=1,\ldots,m. \quad (16)$$

The iterative fit procedure based on either Eq. 14 or Eq. 15 is known as the maximum likelihood estimation method [20–22].

The global analysis of frequency-domain data is performed by least-squared method with the TFC defined by Eq. 13 [3]. The variances σ_{ik}^2 in this case are obtained experimentally.

The optimization method finds the estimations of fit parameters that match the minimum of the TFC, thus ensuring the best correspondence between measured and model-generated data. Each optimization method is based on the unique mathematical algorithm that finds the minimum of the TFC by calculating and processing its values for different points in the fit parameter space. One of the most popular classes of the optimization algorithms is based on gradient optimization. It requires calculating the derivatives of the TFC with respect to fit parameters. Gradient optimization methods demonstrate better performance in comparison with non-gradient methods, when the required derivatives are available. Although the derivatives of model functions can often be calculated numerically with sufficient accuracy, analytical derivatives are preferable ensuring higher numerical accuracy and higher processing speed. Gradient methods are not suitable when the theoretical function is obtained by the model that is based on Monte-Carlo simulations. In this case, the model-generated data are obtained by a stochastic simulation algorithm instead of an analytical function and thus contain statistical noise. Therefore, the derivatives required by the gradient method cannot be calculated analytically, whereas the application of numerical algorithms leads to significant errors.

One of the most frequently used gradient algorithms is the Marquardt–Levenberg optimization algorithm [19, 23]. This algorithm consists of the following steps.

(a) Set the initial guesses for fit parameters as a current point in parameter space $a_j^{\text{curr}} = a_j^{\text{in}}, \quad j = 1, \ldots, s$. Initialize the regularization value λ (for example $\lambda = 0.001$).

(b) Calculate the TFC χ^2_{curr} for the current point.

(c) Calculate matrices of the system of linear equations A and B. The elements of these matrices depend on the derivatives of the TFC with respect to the fit parameters and are defined as follows:

$$A_{ij} = -\frac{\partial^2 \chi^2}{\partial a_i \partial a_j}; \; B_j = -\frac{\partial \chi^2}{\partial a_j} \quad i = 1, \ldots, s; j = 1, \ldots, s. \tag{17}$$

All derivatives are calculated at the current point in the parameter space (at the beginning this is the point defined by initial guesses).

(d) Modify the diagonal elements of matrix A:

$$A_{ii} = (1 + \lambda) A_{ii}, \quad i = 1, \ldots, s. \tag{18}$$

(e) Solve the system of linear equations:

$$AX = B, \tag{19}$$

where X is a column vector that contains the fit parameter increments Δa_j for the current iteration.

(f) Calculate a new trial point in the fit parameter space:

$$a_j^{\text{tr}} = a_j^{\text{curr}} + \Delta a_j, \quad j = 1, \ldots, s. \tag{20}$$

For the new point calculate the value of the TFC, χ^2_{tr}. For some practical recommendations on this step *see* **Note 2**.

(g) If $\chi^2_{\text{tr}}\ \chi^2_{\text{curr}}$ the new trial point in fit parameter space is accepted and is set as the current point, i.e., $a_j^{\text{curr}} = a_j^{tr}, \quad \chi^2_{\text{curr}} = \chi^2_{\text{tr}}$. The value of the method parameter λ is decreased: $\lambda = 0.1\lambda$. Alternatively, if $\chi^2_{\text{tr}} \geq \chi^2_{\text{curr}}$ the trial point is ignored and the value of method parameter λ is increased: $\lambda = 10\lambda$.

(h) The stop optimization conditions are checked. The first one is defined as follows:

$$\lambda > \lambda_{\max}. \tag{21}$$

This condition is used to stop iterations when the method cannot find the direction to the minimum of the TFC. In practice $\lambda_{\max} = 10^{10}$.

The second stop condition is used when the optimal point in fit parameter space has been successfully found with the required accuracy:

$$\frac{\left|\chi^2_{\text{tr}} - \chi^2_{\text{curr}}\right|}{\chi^2_{\text{curr}}} \leq \epsilon, \tag{22}$$

where ε is the small accuracy value that is set beforehand. In practice it can be equal to 10^{-5}.

When at least one of the stop conditions is satisfied, the iterations are interrupted and the current point defined by $a_j^{\text{curr}}, \quad j = 1, \ldots, s$ is returned as the result. Otherwise the method moves to **step c** and a new iteration starts.

6. Judging the quality of the fit: The final stage of the global analysis procedure consists in evaluation of the quality of the obtained results. For this purpose several fit quality criteria are used. The first is the final value of the TFC calculated by Eqs. 13–15. This value should be close to 1 for a good fit. Besides the TFC two graphical criteria, weighted residuals and autocorrelation function of weighted residuals, are widely used for judging the goodness of the fit. Both these criteria are built individually for each data set and are defined on the same argument range as the measured data.

The weighted residuals are used for visual inspection of the agreement between measured and model-generated data. The weighted residuals correspond to the type of the TFC used for the fit. In the case of chi-square TFC (Eq. 13) the residuals are calculated according to the following equation:

$$R_{ik} = \frac{x_{ik} - F_{ik}^{\text{th}}(\vec{A})}{\sigma_{ik}}, \quad i = 1, \ldots, m; \; k = 1, \ldots, N_i. \tag{23}$$

In the case of Poisson statistics (Eq. 14) the residuals are calculated as

$$R_{ik} = \operatorname{sign}\left(x_{ik} - F_{ik}^{\text{th}}(\vec{A})\right) \times \sqrt{2\left(x_{ik} \ln\left(x_{ik}/F_{ik}^{\text{th}}(\vec{A})\right) + F_{ik}^{\text{th}}(\vec{A}) - x_{ik}\right)}, \tag{24}$$

where $i = 1, \ldots, m; \; k = 1, \ldots, N_i$ and function $\operatorname{sign}(x)$ is equal to -1 in the case of negative x and 1 otherwise. For the multinomial statistics (TFC is defined by Eq. 15) the equation for the weighted residuals takes the form

$$R_{ik} = \operatorname{sign}\left(x_{ik} - F_{ik}^{\text{th}}(\vec{A})\right) \left|2x_{ik} \ln\left(x_{ik}/F_{ik}^{\text{th}}(\vec{A})\right)\right|, \quad i = 1, \ldots, m; \; k = 1, \ldots, N_i. \tag{25}$$

For the successful fits the weighted residuals should be randomly distributed around 0.

The plot of the autocorrelation function of weighted residuals [1, 24] provides another valuable visual check of the quality of the fit. The autocorrelation function of the weighted residuals for i-th data set in the k-th point of argument $C_{ik}(i = 1, \ldots, m; \; k = 1, \ldots, N_i)$ is calculated by the following equation:

$$C_{ik} = \frac{1}{N_i - k + 1} \sum_{j=1}^{N_i - k + 1} R_{ij} R_{i(j+k-1)} \Bigg/ \left(\frac{1}{N_i} \sum_{j=1}^{N_i} R_{ij}^2 \right), \tag{26}$$

where R_{ij} is the value of weighted residuals calculated for i-th data set in the j-th point of argument and N_i is the number of points for which weighted residuals are calculated. The plots of autocorrelation are usually made across half of the data points ($N_i/2$). In many cases the visual inspection of the autocorrelation function is more sensitive for the goodness of the fit than a plot of weighted residuals. For successful fits the autocorrelation function of weighted residuals is randomly distributed around zero. Bad fits provide low-frequency periodicity in an autocorrelation plot that can easily be detected by eye. For additional details related to calculation of autocorrelation plots *see* **Note 3**.

Besides fit quality criteria described above additional fit quality criteria (standard deviation of χ^2, Durbin–Watson parameter, etc.) can be considered [24].

As the parameters are estimated from the measured, statistically distorted data, they are known with some error. To quantify the error in the estimated parameters, we use confidence intervals. The confidence interval is the range where the true value of the fit parameter is located with some confidential probability.

The confidence intervals for fit parameters can be found by the exhaustive search method [1, 25]. The advantage of this approach is that it takes into account possible correlations between parameters. The search procedure starts from the optimal value of examined parameter obtained after a successful global fit. The value of the selected fit parameter is shifted from the optimal value to the left or to the right (depending on the bound of the confidence interval that should be found) with the predefined step. After each move the examined parameter is fixed to the shifted value and other parameters are adjusted to find the minimum of the TFC. The new TFC is compared with the χ^2_{lim} level calculated from the statistical F-test as follows:

$$\chi^2_{\text{lim}} = \chi^2_{\text{min}} \left(1 + \frac{s}{n - s - 1} F(s,\, n - s - 1,\, 1 - \beta) \right), \tag{27}$$

where χ^2_{min} is the minimum value of TFC obtained from the global analysis before calculation of confidence intervals, s is the total number of fit parameters available for optimization, $v = n - s - 1$ is the number of degrees of freedom, β is the confidential probability (0.67, 0.95, etc.), and $F(s,\, n - s - 1,\, 1 - \beta)$ is the F-statistic. The search procedure stops when the TFC value becomes equal (with some predefined accuracy) to the χ^2_{lim} and the corresponding shifted value of the examined fit parameter is taken as a bound of confidence interval.

3.2 Application of Global Analysis for Fitting Time-Resolved Fluorescence Anisotropy

The global analysis approach described in Subheading 3.1 can be applied for fitting time-resolved anisotropy. The algorithm of global analysis (*see* Subheading 3.1) adopted for this application consists of the following steps.

1. Prepare the data sets. The data sets for anisotropy analysis contain the fluorescence decays measured for different polarization angles Θ. In practice two data sets for parallel ($\Theta = 0$ degrees) and perpendicular ($\Theta = 90$ degrees) polarization components are used. Some practical recommendations related to preparing the data sets with the IRF for anisotropy analysis are given in **Note 4**.
2. Select the models. The model for global analysis of sample polarization components is based on Eqs. 1 or 2 where the function $I(t, \vec{A})$ is defined by the formula

$$I(t, \vec{A}) = H\left(\sum_{j=1}^{M} \alpha_j e^{-t/\tau_j}\right) \times \left[1 + \left(3\cos^2(\Theta) - 1\right)\left(r_\infty + \sum_{K=1}^{L} T_{jk}\, \beta_k e^{-t/\phi_k}\right)\right], \quad (28)$$

   where H is an amplitude parameter; α_j and τ_j are, respectively, the contributions and decay times of j-th fluorescence exponent $\left(\sum_{j=1}^{M} \alpha_j = 1\right)$; r_∞ is the residual polarization; β_k and ϕ_k are, respectively, the amplitudes and rotational correlation times of corresponding anisotropy exponents; T_{jk} controls the associations $T_{jk} = 1$, if fluorescence and anisotropy exponents are associated, and $T_{jk} = 0$, if not; and M and L are, respectively, the number of fluorescence and anisotropy exponential components.
3. Link the parameters. Since each data set for anisotropy analysis is fitted by the model based on Eq. 16, the set of fit parameters includes H, α_j, τ_j, r_∞, β_k and ϕ_k, and $j = 1, \ldots, M$; $k = 1, \ldots, L$. The value α_1 is not fitted directly because the sum of contributions α_j should be kept equal to 1. Therefore the value α_1 is simply calculated as $\alpha_1 = 1 - \sum_{j=2}^{M} \alpha_j$. The fit parameters with the same names should be linked for all data sets participating in the global fit. The only exception is parameter H, as it takes into account the absolute amplitude of the sample polarization components and therefore it is different for different data sets.
4. Generate initial guesses. Although no special initial guesses algorithm exists for Eq. 28, the start values of some parameters can nevertheless be obtained automatically. For example, if possible ranges for decay times and rotational correlation times are known and a number of fluorescence and anisotropy exponents is selected, the initial guesses for parameters τ_j and ϕ_k can be uniformly distributed in corresponding ranges.

The initial guesses for all contributions α_j can be selected equal to $1/M$. In most cases the initial values for β_k can be set equal to $0.4/L$.

5. Perform the iterative fit. The instructions for the iterative fit are given in Subheading 3.1.
6. Judge the goodness of the fit. To judge the quality of the fit of sample polarization components the fit quality criteria presented in Subheading 3.1 can be applied. Additionally, the measured and model-generated anisotropy decays can be built from polarization components and presented graphically. The anisotropy decays are calculated as follows:

$$r(t) = \frac{I_1(t)/H_1 - I_2(t)/H_2}{I_1(t)(1 - 3\cos^2(\Theta_2))/H_1 - I_2(t)(1 - 3\cos^2(\Theta_1))/H_2}, \quad (29)$$

where $I_1(t)$ and $I_2(t)$ are sample polarization components (measured or model-generated) that correspond to the polarization angles Θ_1 and Θ_2, respectively, and H_1 and H_2 are estimated values of amplitude parameters (Eq. 28). When a shift δ_s exists between measured polarization components $I_1(t)$ and $I_2(t)$, then before calculating the anisotropy decays this shift should be compensated by moving, for example, $I_2(t)$ to $I_1(t)$ over δ_s time units.

When the total IRF is used for fitting polarization data (*see* **Note 4**) the sample *G*-factor G_s can be estimated using amplitude parameters that correspond to the parallel $H_{||}$ and perpendicular $H_{\perp}$ polarization components:

$$G_s = \frac{H_{||}}{H_{\perp}}. \quad (30)$$

At the same time it is recommended to fix *G*-factor to the value obtained from the additional measurements, since the relation between parameters $H_{||}$ and $H_{\perp}$ is correlated with anisotropy parameters β_k and ϕ_k. This can be done by fixing $H_{||}$ and $H_{\perp}$ to some values that being substituted to the Eq. 30 give the required value of G_s. In this case the condition $\left(\sum_{j=1}^{M} \alpha_j = 1\right)$ cannot be kept anymore. Therefore all parameters α_j with the same indexes should be linked and the obtained parameter groups should be estimated to fit the amplitudes of the sample decays.

3.3 Example of Global Anisotropy Analysis of Enhanced EGFP

To demonstrate the application of global approach for time-resolved anisotropy analysis the parallel and perpendicular fluorescent components of enhanced green fluorescence protein (EGFP) were measured and fitted globally according to the algorithm described in Subheading 3.2. Global analysis was performed with the TRFA Data Processor Advanced software (SSTC, http://www.sstcenter.com). Since one-exponential reference compound was used for deconvolution, the general model Eqs. 2 and 28 were selected to obtain model-generated curves. The total decays for reference

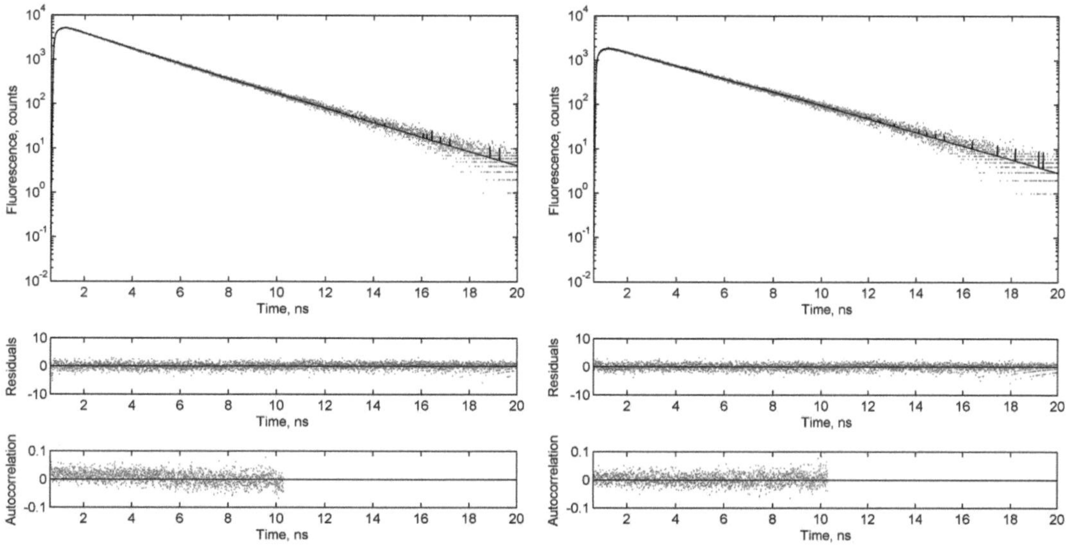

Fig. 2 Experimental (points) and fitted (*solid line*) curves of parallel (*left panel*) and perpendicular (*right panel*) polarization components of EGFP. The results are obtained by the global analysis approach. The quality of the fit can be judged by graphs of weighted residuals and their autocorrelation shown at the *bottom* of both panels. The target fit criterion $\chi^2_G = 1.089$. The estimated values of fluorescence lifetimes are $\tau_1 = 0.771$ ns (with contribution of 8.4 %) and $\tau_2 = 2.735$ ns (with contribution of 91.6 %). The rotational correlation time was estimated at $\phi_1 = 15.6$ ns with initial anisotropy $\beta_1 = 0.383$

compound were prepared from its polarization components according to **Note 4** and added to the corresponding data sets. For both data sets the models defined by Eq. 28 were configured to include two fluorescence exponents ($M = 2$) and one anisotropy exponent ($L = 1$). The values of parameters $H_{||}$ and $H_{\perp}$ were fixed to the same values to keep the $G_s = 1$. Parameters α_1 were linked for both data sets and obtained parameter group was fitted as a separate fit parameter. Reference lifetime parameters τ_{ref} for both data sets were fixed to the prior known value. Since constant background was not observed in the measured data the parameters b and c from Eq. 2 were fixed to 0. The intensity of time-dependent background emission $B(t)$ was available from the measurements for both parallel and perpendicular components with known relative weight equal to 5 (parameters γ for both data sets were fixed to this value). The number of registered photons was high enough to accept the Gaussian approximation of statistical noise in both polarization components of the sample. Therefore the TFC defined by the chi-square Eq. 13 was applied. The Marquardt–Levenberg optimization method was used in the iterative fit. The confidence intervals for fit parameters at 95 % level were estimated by the exhaustive search method.

The results of the global analysis of both EGFP polarization components are presented in Fig. 2. The graphs for anisotropy are shown in Fig. 3. The final value of the TFC obtained after the fit $\chi^2_G = 1.089$. This value as well as the presented graphs for weighted residuals and their autocorrelation functions for both polarization

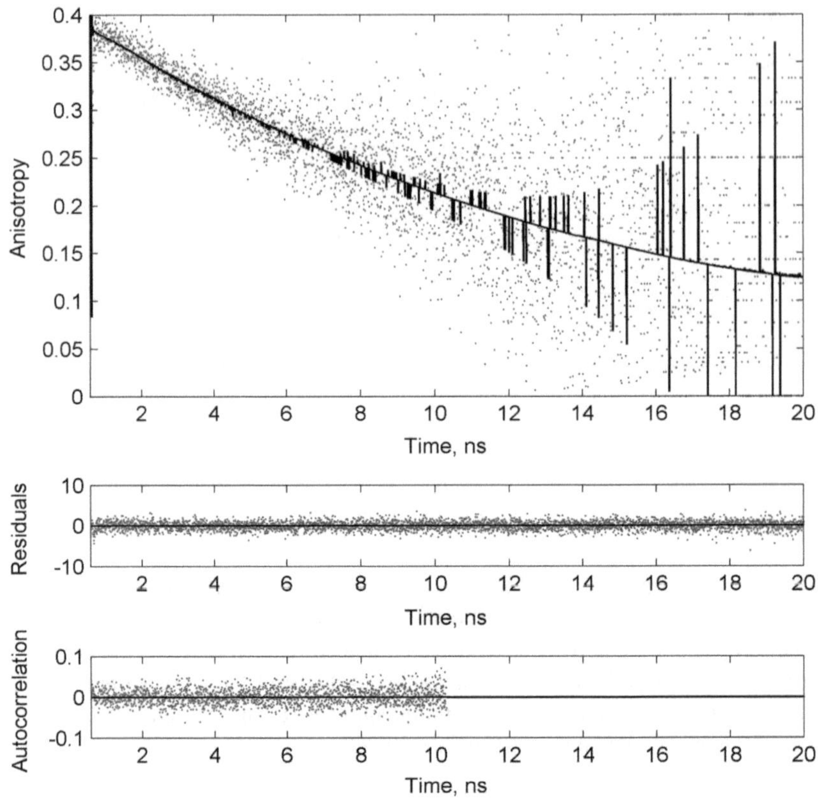

Fig. 3 Experimental (points) and fitted (*solid line*) curves of EGFP anisotropy. The quality of the fit can be judged by presented graphs of weighted residuals and their autocorrelation. Parameter values are given in the legend of Fig. 2

components prove that the fit was successful. The estimated values of EGFP lifetimes are $\tau_1 = 0.771$ [0.714; 0.816] ns (with contribution of 8.4 [7.9; 8.8] %) and $\tau_2 = 2.735$ [2.711; 2.742] ns (with contribution of 91.6 [91.5; 92.3] %). The rotational correlation time was estimated at $\phi_1 = 15.6$ [15.2; 16.0] ns with initial anisotropy $\beta_1 = 0.383$ [0.382; 0.386]. The values shown in square brackets represent the estimated confidence intervals at the 95 % level. The obtained parameter values for EGFP are in good agreement with the results reported previously [11, 26, 27]. This result also demonstrates that the global analysis scheme presented in this section can be successfully applied in practice for fitting the experimental data.

4 Notes

1. If Eq. 13 is applied to fit time-domain data and analytical models are selected for the fit, then the variances in each time point are defined as $\sigma_{ik}^2 = 1/x_{ik}$. The measured fluorescence intensity decays can contain time channels where no photons were detected and therefore $x_{ik} = 0$. Since σ_{ik}^2 for such points

cannot be calculated these points are excluded from the sum in Eq. 13 and total number of points n decreases.

2. In most cases the constraints are defined for all fit parameters. The simplest type of the constraint is setting minimum and maximum values but more complex functional constraints are also possible. In the presence of constraints the new values of fit parameters obtained by the optimization method (Eq. 20) should be checked to satisfy them. If the values of some parameters do not satisfy their constraints, the new values of these parameters are replaced by values from the current point in the fit parameter space. Then the new set of fit parameter values will contain the mixture of new values a_j^{tr} obtained by Eq. 20 and current values a_j^{curr}, $j = 1, \ldots, s$. Forming such mixed vector of fit parameter values is more effective than completely decline the new set of values obtained by Eq. 20, because it makes the behavior of the optimization method more flexible in the area close to the boundaries of possible values of fit parameters.
3. As it follows from the Eq. 26, $C_{i1} = 1$. This means that each value of weighted residual R_{ik} is fully correlated to itself. In order to make autocorrelation plots more convenient for visual inspection the value C_{i1} is replaced by 0 on the graphs.
4. For the measured IRF/reference data an additional preprocessing is required before they can be included to the data sets for anisotropy analysis. Let us consider this procedure for the case of an IRF (for a reference compound the same steps should be performed). In order to accurately estimate the value of the sample G-factor after the fit the data sets for anisotropy analysis should include the total IRF that corresponds to the magic angle, although the IRF curves are usually measured on the same polarization angle as sample fluorescence intensity decay. To calculate these total curves the polarization components of IRF $g_{\|}(t)$ and $g_{\perp}(t)$ can be used and possible time shift between these components should be taken into account. The time shift δ_{IRF} of perpendicular IRF polarization component in respect to the parallel one is detected by a superposition of their rising edges, thus ensuring better fit of sample decays at the beginning. The IRF for parallel component data set is prepared by the formula

$$g_0(t) = g_{\|}(t) + 2 \times G_{\mathrm{IRF}} g_{\perp}(t + \delta_{\mathrm{IRF}}), \tag{31}$$

where G_{IRF} is the G-factor at the wavelength used for IRF measurement. The IRF for perpendicular component data set is calculated as

$$g_{90}(t) = g_{\|}(t - \delta_{\mathrm{IRF}}) + 2 \times G_{\mathrm{IRF}} g_{\perp}(t). \tag{32}$$

Application of total IRFs to fit the sample polarization component data potentially allows to restore the sample G-factor after the fit.

References

1. Lakowicz JR (2006) Principles of fluorescence spectroscopy, 3rd edn. Springer, New York
2. Beechem JM, Gratton E, Ameloot M et al (2002) The global analysis of fluorescence intensity and anisotropy decay data: second-generation theory and programs. In: Lakowicz JR (ed) Topics in fluorescence spectroscopy, vol 2. Springer, USA, pp 241–305
3. Verveer PJ, Squire A, Bastiaens PI (2000) Global analysis of fluorescence lifetime imaging microscopy data. Biophys J 78(4):2127–2137
4. Elson D, Requejo-Isidro J, Munro I et al (2004) Time-domain fluorescence lifetime imaging applied to biological tissue. Photochem Photobiol Sci 3(8):795–801
5. Borst JW, Visser AJWG (2010) Fluorescence lifetime imaging microscopy in life sciences. Meas Sci Technol 21:1–21
6. Tsien RY (1998) The green fluorescent protein. Annu Rev Biochem 67:509–544
7. Bastiaens PIH, Squire A (1999) Fluorescence lifetime imaging microscopy: spatial resolution of biochemical processes in the cell. Trends Cell Biol 9(2):48–52
8. Borst JW, Laptenok SP, Westphal AH et al (2008) Structural changes of yellow Cameleon domains observed by quantitative FRET analysis and polarized fluorescence correlation spectroscopy. Biophys J 95(11):5399–5411
9. Laptenok SP, van Stokkum IH, Borst JW et al (2012) Disentangling picosecond events that complicate the quantitative use of the calcium sensor YC3.60. J Phys Chem B 116 (9):3013–3020
10. Laptenok SP, Borst JW, Mullen KM et al (2010) Global analysis of Förster resonance energy transfer in live cells measured by fluorescence lifetime imaging microscopy exploiting the rise time of acceptor fluorescence. Phys Chem Chem Phys 12(27):7593–7602
11. Visser AJWG, Laptenok SP, Visser NV et al (2010) Time-resolved FRET fluorescence spectroscopy of visible fluorescent protein pairs. Eur Biophys J 39(2):241–253
12. Malikova NP, Visser NV, van Hoek A et al (2011) Green-fluorescent protein from the bioluminescent jellyfish *Clytia gregaria* is an obligate dimer and does not form a stable complex with the Ca(2+)-discharged photoprotein clytin. Biochemistry 50(20):4232–4241
13. Borst JW, Hink MA, van Hoek A et al (2005) Effects of refractive index and viscosity on fluorescence and anisotropy decays of enhanced cyan and yellow fluorescent proteins. J Fluoresc 15(2):153–160
14. Van den Bergh V, Kowalczyk A, Boens N et al (1994) Experimental design in the global compartmental analysis of intermolecular two-state excited-state processes. J Phys Chem 98: 9503–9508
15. Boens N, Kowalczyk A (1996) Identifiability of competitive intermolecular three-state excited-state processes. Chem Phys Lett 260: 326–330
16. Apanasovich VV, Novikov EG (1995) Method for simultaneous reconstruction of parameters for a set of fluorescence intensity decay curves. J Appl Spectrosc 62(6):1069–1074
17. Apanasovich VV, Novikov EG (1996) The method of fluorescence decays simultaneous analysis. Rev Sci Instrum 67(1):48–54
18. Press WH, Teukolsky SA, Vetterling WT et al (2007) Numerical recipes: the art of scientific computing, 3rd edn. Cambridge University Press, New York
19. Bevington PR, Robinson DK (2003) Data reduction and error analysis for the physical sciences, 3rd edn. McGraw-Hill, New York
20. Laurence TA, Chromy BA (2010) Efficient maximum likelihood estimator fitting of histograms. Nat Methods 7(5):338–339
21. Maus M, Cotlet M, Hofkens J et al (2001) An experimental comparison of the maximum likelihood estimation and nonlinear least-squares fluorescence lifetime analysis of single molecules. Anal Chem 73(9):2078–2086
22. Hauschild T, Jentschel M (2001) Comparison of maximum likelihood estimation and chi-square statistics applied to counting experiments. Nucl Instr Meth A 457:384–401
23. Marquardt DW (1963) An algorithm for least-squares estimation of non-linear parameters. J Soc Ind Appl Math 11:431–441
24. Eaton DF (1990) Recommended methods for fluorescence decay analysis. Pure Appl Chem 62(8):1631–1648
25. Johnson ML, Faunt L (1992) M. Parameter estimation by least-squares methods. Methods Enzymol 210:1–37

26. Heikal AA, Hess ST, Webb WW (2001) Multiphoton molecular spectroscopy and excited-state dynamics of enhanced green fluorescent protein (EGFP): acid–base specificity. Chem Phys 274:37–55

27. Hess ST, Sheets ED, Wagenknecht-Wiesner A et al (2003) Quantitative analysis of the fluorescence properties of intrinsically fluorescent proteins in living cells. Biophys J 85 (4):2566–2580

Chapter 11

Nanometrology

David J.S. Birch and Philip Yip

Abstract

Methods and protocols are described when using fluorescence metrology to determine the average nanoparticle (np) size in colloids in the range of 1–10 nm. The technique is based on determining the rotational correlation time of the np from the decay of fluorescence anisotropy of a dye that is electrostatically or covalently attached to the np as it undergoes Brownian rotation. The np size is then calculated from the Stokes–Einstein equation. The exemplar of silica nps is presented, but the approach can also be applied to other types of nps.

Key words Fluorescence anisotropy decay, Fluorescence lifetime, Nanoparticle, Silica, Nanometrology, LUDOX®, Sol–gel, TCSPC

1 Introduction

Measurement science on the nanoscale (nanometrology) is widely seen as a key to progress in the many burgeoning applications of nanotechnology. These applications span not only science and technology but also engineering and medicine. Global interest in the metrology of nanoparticles (nps) in particular has undergone quite a renaissance in recent times. This is because of the development of novel manifestations of nps such as carbon nanotubes, noble metal, diamond, and semiconductor quantum dots. Irrespective of physical and chemical differences between nps, there is a certain common concern in respect of the potential toxicological and environmental issues surrounding not only new types of nps but also older structures. Carbon nps from combustion and metal oxide nps such as silica and titania are some of the latter that immediately spring to mind. As a paradigm here we consider only silica. Also, we confine ourselves to the 1–10 nm np range as this is where there is a particularly urgent need for globally accepted standards and precise measurement techniques. Such small nps

Yves Engelborghs and Antonie J.W.G. Visser (eds.), *Fluorescence Spectroscopy and Microscopy: Methods and Protocols*, Methods in Molecular Biology, vol. 1076, DOI 10.1007/978-1-62703-649-8_11,

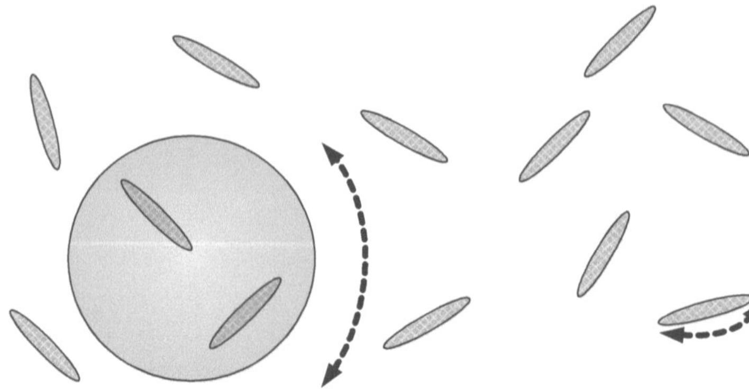

Fig. 1 Depicts a dye rotating free ($\phi_d \sim$ 100 ps) and rotating bound to a nanoparticle ($\phi_p \sim$ 10–100 ns) in a colloid

are known to traverse cellular membranes with ease, are at the limit of present-day measurement techniques, and yet are most appropriate for study in a colloid using fluorescence nanometrology. Traditional methods for nanoparticle metrology, such as small angle X-ray and neutron scattering, are much more expensive, complex, and difficult to use when compared to fluorescence nanometrology, as we will describe here.

Fluorescence nanometrology can be defined as the use of fluorescence to determine nanoscale lengths. When it comes to tracking rapidly changing phenomena involving nps on the scale of supermolecular dimensions, fluorescence decay time techniques have few equals [1]. This is because they can be performed in situ, have an appropriate timescale (ps to ns), offer high sensitivity (down to the single-molecule, single-photon level), and benefit from the wide range of well-characterized fluorescence probes with which to customize a study. The associated equipment is also now extremely easy to use, reliable, and relatively inexpensive. However, the interpretation of data still benefits from skill and experience. Here, we describe the use of the decay of fluorescence anisotropy to determine np size.

The measurement approach is depicted in Fig. 1 for the case of dye molecules partitioned between dual Brownian motions relating to when bound to a np and when rotating freely.

The rotational correlation time of the free dye of known radius gives the local viscosity of the solution using the Stokes–Einstein equation. When the viscosity is combined in the same equation with the rotational correlation time of the dye bound to nps this leads to determination of the np radius. Both stable [2] and changing [3] np sizes can in principle be measured (*see* **Note 1**).

2 Materials: Chemical Preparation

As fluorescence is an extremely sensitive technique, we stress that all glassware must be sufficiently cleaned before use (*see* **Notes 2** and **3**). Prepare and store all reagents at room temperature (unless indicated otherwise). Use distilled water (unless indicated otherwise). Cap all chemicals and create an additional seal with parafilm (unless indicated otherwise). Label each chemical/sample being used (*see* **Note 4**) and preferably store in the dark. Diligently follow all waste disposal regulations when disposing of waste chemicals.

2.1 Stöber Synthesis of Silica Nanoparticles

The Stöber process is the ammonia-catalyzed reaction of an orthosilicate, e.g., tetraethyl orthosilicate (TEOS: $Si(OR)_4$; $R{=}C_2H_5$) with water in low molecular weight alcohols to produce monodispersed, spherical silica nanoparticles (*see* Fig. 2). This method, as specified by Green et. al. [4], produces a narrow size distribution of spherical particles as in the case of LUDOX®.

1. 10 ml sols will be produced using a conical flask.
2. The solutions required for their synthesis are 33 % ammonia in water solution (*see* **Note 5**), water, tetraethyl orthosilicate (TEOS), and methanol (*see* **Note 6**).
3. We recommend making a tertiary diagram like our Fig. 3. From this, calculate the ratio of ammonia, water, and TEOS for the desired sample. For accuracy, we recommend calculating the mass of each component you will require for your sample (*see* **Note 7**).
4. Make a water–ammonia solution of desired concentration. This solution should be made in a separate glass container which is placed on a balance and zeroed. The water should be first added by using a pipette. The 33 % ammonia solution should be quickly added to the water using a Pasteur pipette. The container should be immediately capped and shaken to ensure that it has completely mixed (*see* **Note 8**).

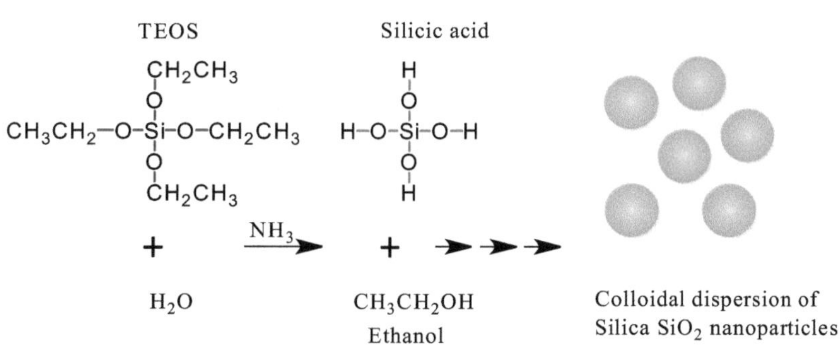

Fig. 2 Schematic of the Stöber synthesis of nps

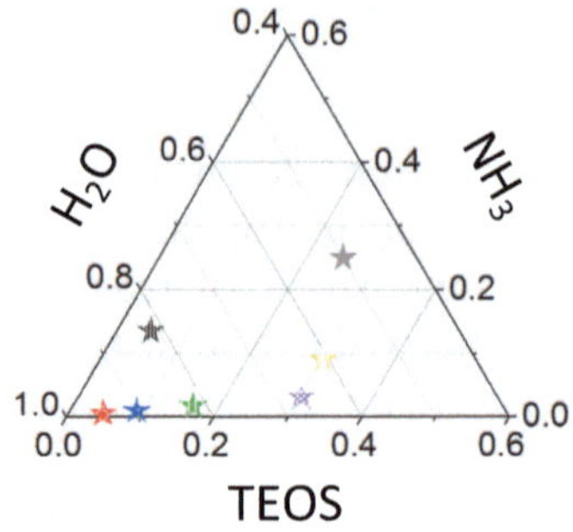

Sample	[TEOS] M l^{-1}	[NH_3] M l^{-1}	[H_2O] M l^{-1}
★PY1	0.50	0.05	10.00
★PY2	0.50	0.05	5.00
★PY3	0.50	0.05	2.50
★M4	0.50	0.05	1.10
★S1	0.17	0.50	3.00
★S2	0.50	0.50	1.00
★S3	0.17	0.05	0.34

Fig. 3 Ternary diagram of colloidal silica compositions

5. Place the 10 ml conical flask on the balance and zero it. Add the desired volume of TEOS using a Pasteur pipette.
6. Fill the conical flask up to the halfway mark with methanol and re-zero it again (*see* **Note 9**).
7. Next, add the water–ammonia solution and fill up to the 10 ml mark using methanol.
8. Add a 5 mm stirring bar and cap (*see* **Note 10**).
9. Place the conical flask on a magnetic stirrer for 24 h at a gentle setting.
10. Decant the solution into a glass container and store at room temperature (*see* **Note 11**).

2.2 Fluorescent Dye Preparation

Like all fluorescence measurements, an appropriate dye must be selected. The dye must meet two criteria: first, it must attach to the np under investigation and second have appropriate photophysical properties (e.g., fluorescence decay) for tracking the np rotation. For anisotropy measurements of nanoparticles, the dye may attach to the system electrostatically or covalently. Before working with a new dye, it is important to record the steady-state absorption and emission spectra so one has a basic understanding of the photophysical properties of the dye. We will take rhodamine 6G as an example here.

1. Prepare a stock solution of dye by dissolving the dye in methanol or some other common solvent at a concentration of ~10–50 μM (*see* **Notes 12–15**).
2. Pipette 3 ml of solvent into a cuvette. Pipette 50 μl of stock solution (*see* **Notes 16** and **17**).
3. Record a steady-state absorption spectrum like that of Fig. 4 using a spectrophotometer (*see* **Note 18**). In our case, we used a Jasco V660 spectrophotometer. A second identical, and preferably matched, cuvette filled with only the pure solvent should be used in the reference channel.

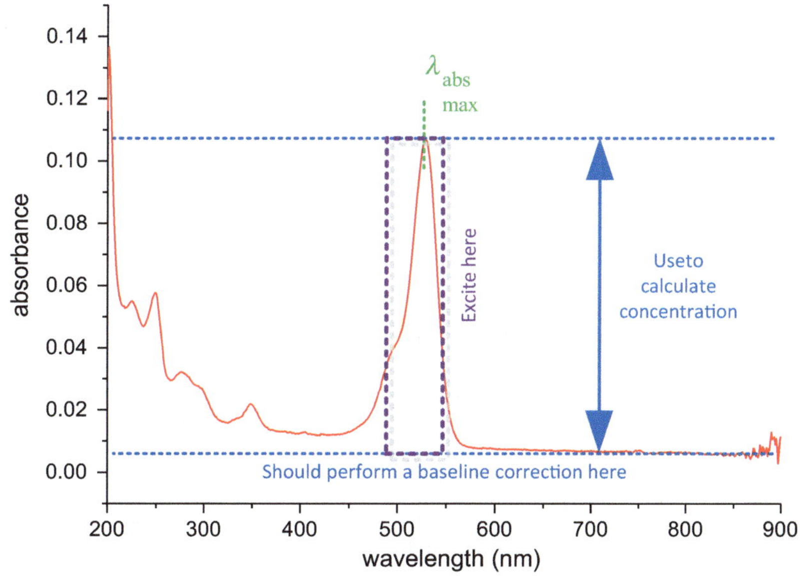

Fig. 4 Absorption spectra of rhodamine 6G in methanol. In this example the path length was 10 cm and therefore the concentration in the cuvette is ~ 0.1 μM. One can clearly see the most suitable wavelength region for excitation

4. From the absorption spectra you can verify the dye concentration in the cuvette using the Beer–Lambert law Eq. 1 (*see* **Note 19**)

$$A = \log\left(\frac{I_0}{I}\right) = \varepsilon c l \tag{1}$$

where A is the absorbance, I_0 and I are the intensity of the incident and transmitted light, respectively, c is the concentration in Mol l^{-1}, l is the path length in cm, and ε is the molar extinction coefficient in Mol^{-1} l cm^{-1}.

5. From the concentration in the cuvette and the dilution factor, you can calculate the concentration of the stock solution (*see* **Notes 20** and **21**).
6. From the absorption spectra, select a suitable pulsed light source for excitation. This should have a peak wavelength in a region where the dye has a high absorbance. In our case, we selected an IBH 503 nm DeltaDiode.
7. Record a steady-state emission spectrum using a spectrofluorometer and select the excitation wavelength to match that of your selected light source (*see* **Note 22**). In our case, we used a HORIBA Fluorolog 3 spectrophotometer.
8. Normalize and overlay both absorption and emission spectra as our Fig. 5.
9. Measure the Stokes shift and select an appropriate long-pass filter or monochromator wavelength for later use with lifetime measurements (*see* **Note 23**).

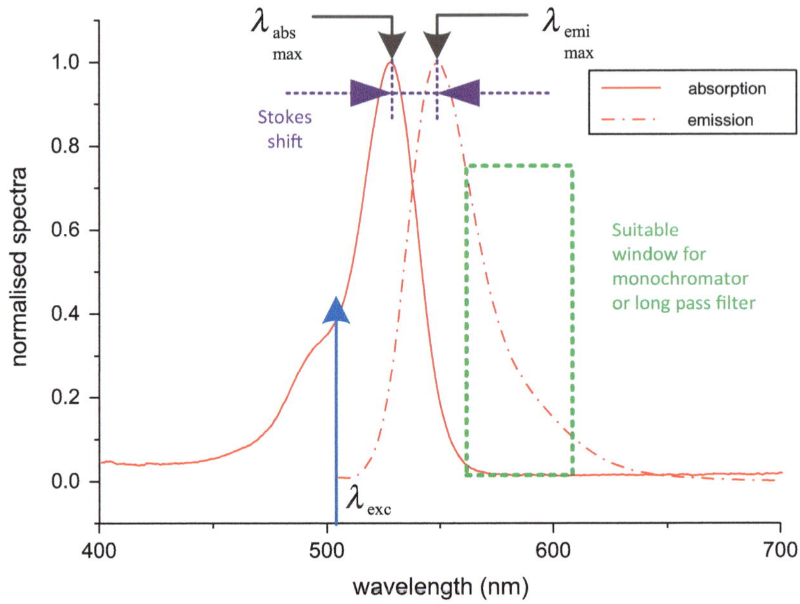

Fig. 5 Normalized absorption and emission spectra of rhodamine 6G, excited at 503 nm. From this one can calculate the Stokes shift and see what wavelength range is the best for measuring the fluorescence

2.2.1 Electrostatic Labelling

Due to electrostatic interactions, cationic dyes will readily bind to negatively charged silica colloids in alkaline solution (*see* **Note 24**) and vice versa.

1. Measure the pH of the colloid.
2. Prepare an oppositely charged dye that will affix to the np by electrostatic interaction.
3. In this case, rhodamine 6G as prepared above will be used as an example.
4. Add a small quantity of the dye to the nanoparticles using a pipette, mix the solutions using the pipette, and place a lid on the cuvette (again *see* **Notes 16–19**).
5. Measure the absorbance spectra to have a peak at ~0.1 (again *see* **Note 20**).
6. Prepare a similar dye that does not bind to the np, for example, Rhodamine B, in exactly the same manner as rhodamine 6G.
7. A dye such as Rhodamine B that does not bind to the np can be used to determine the microviscosity. We recommend preparing both dyes for a comparison. *See* Fig. 6 for the chemical structures of both dyes.

2.2.2 Covalent Labelling

For covalent labelling, a dye needs to not only be selected with appropriate properties as listed previously but must also possess chemical linkers for binding to the np [5]. In this example,

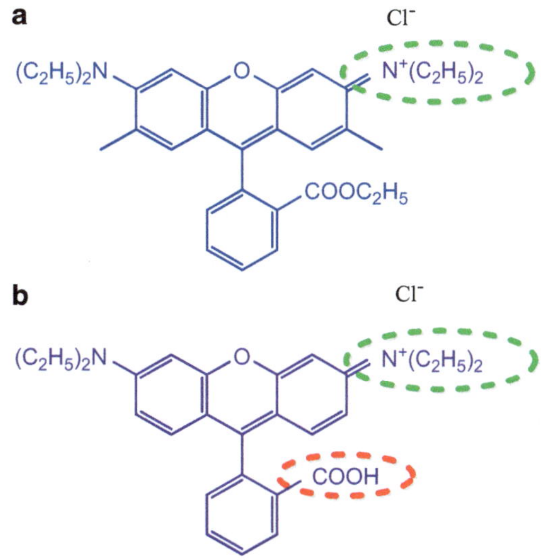

Fig. 6 Chemical structure of (**a**) rhodamine 6G and (**b**) rhodamine B. Note the electrostatic attraction *circled green* for both dyes and the electrostatic repulsion from the carboxylic acid group *circled red* for rhodamine B

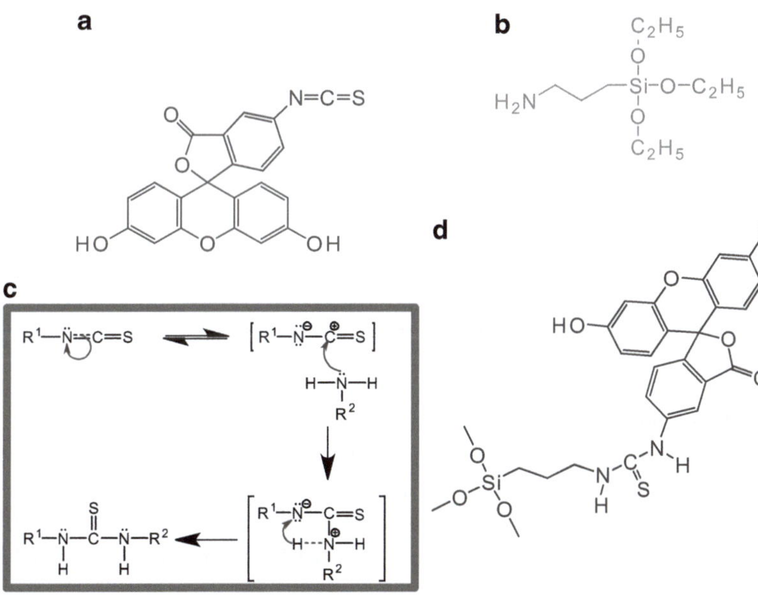

Fig. 7 (**a**) Fluorescein 5(6)-isothiocyanate (FITC) is bound to (**b**) 3-aminopropyl trimethoxysilane (APS) via the mechanism (**c**) to create (**d**) FITC–APS

Fluorescein 5(6)-isothiocyanate (FITC) will be bound to 3-aminopropyl trimethoxysilane (APS). Such a labelling approach has previously been used with silica nanoparticles [6–8]. *See* Fig. 7 for chemical structures and the reaction mechanism.

1. Dry a glass vial under nitrogen (*see* **Note 25**).
2. Measure out 110 mg of FITC and add to this to a 10 ml glass vial (*see* **Note 14**).
3. Add 10 ml of anhydrous ethanol to the glass vial and a magnetic stirrer (*see* **Note 25**).
4. Seal and stir at room temperature at a high speed for 30 min.
5. Add 50 μl of APS (*see* **Note 26**).
6. Dry the top of the vial with Nitrogen.
7. Seal and stir at a high speed at room temperature for ~24–72 h.

2.3 Adding the Dye to the Nanoparticle

1. Pipette 3 ml of the nps into a glass or quartz cuvette (*see* **Notes 17** and **18**).
2. Add a small quantity of your covalent dye or electrostatic dye (*see* **Notes 16** and **19**).
3. Measure the absorbance spectrum, aim for a peak at ~0.1 (*see* **Note 20**).
4. Measure the emission spectra, excited by the appropriate wavelength from the light source.

2.4 Preparation of a Scattering Colloid for Recording the Excitation Pulse

1. Add 17 ml of water to a 30 ml glass vial.
2. Add 2 ml of 0.1 M sodium hydroxide.
3. Add 1 ml of LUDOX SM30.
4. Pipette 3 ml of your scatter colloid into a plastic, glass, or quartz cuvette (*see* **Notes 17**, **18** and **27**).
5. The scatter colloid is quite stable if left sealed, but we recommend replacing it every month (*see* **Note 28**).

3 Methods: Decay Time Techniques

3.1 Acquisition of Data

It is generally accepted that anisotropy decay time measurements are best performed in the time domain using time-correlated single-photon counting techniques (TCSPC) [9, 10]. A schematic of such a fluorometer is shown in Fig. 8. In our case, we have used a FluoroCube from Horiba Jobin Yvon IBH Ltd, Glasgow.

3.1.1 Fluorescence Decay Time Measurement

The fluorescence lifetime is the average time a fluorescent dye remains in an excited state before emission of a photon (*see* **Notes 29** and **30**).

1. Choose an excitation source with a compatible wavelength for exciting the dye (*see* earlier notes about absorption spectra). The most appropriate light sources are pulsed laser diodes sources or pulsed light-emitting diodes (LEDs) [11–13] which operate ~1–10 MHz repetition rate (*see* **Note 31**).

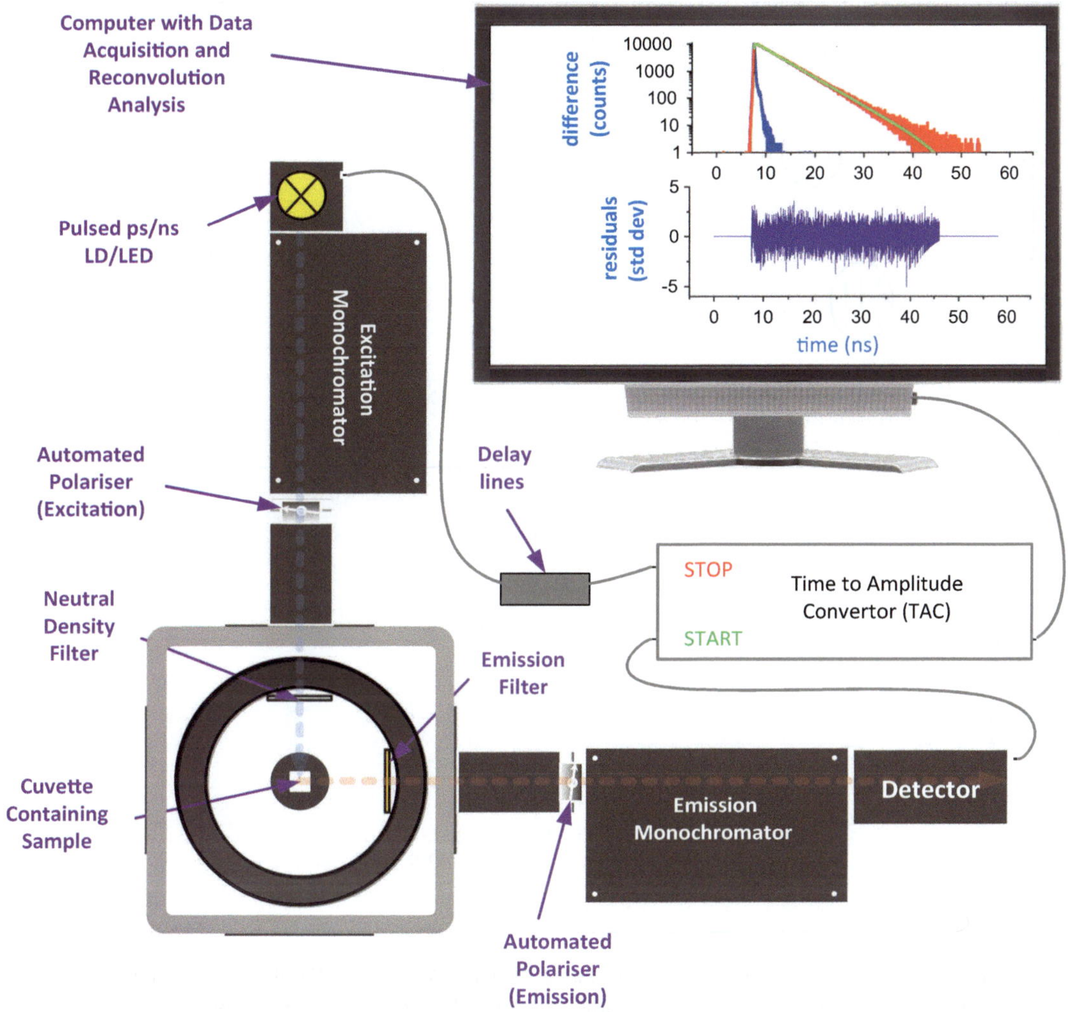

Fig. 8 Schematic of the fluorometer used for fluorescence anisotropy decay measurements. Fluorescence photons are detected using a single-photon timing detection system. Fluorescence decay data are accumulated using the FluoroHub electronics which incorporate a time-to-amplitude converter (TAC), operated in reverse mode to minimize dead time, and multichannel analyzer. The operation of all functions is fully controlled under Microsoft Windows software

2. Choose a suitable detector with a compatible wavelength response and single-photon timing sensitivity for your sample. In our case, a compact photomultiplier in a T0-8 package (Hamamatsu Model R7400) (*see* **Note 32**).
3. Ensure the light source and the detector are at right angles to each other in L-format orientation (*see* Fig. 8 and **Note 33**).
4. Choose a time-to-amplitude range which suitably includes all the fluorescence decay of the dye as in (*see* **Notes 34** and **35**). A logarithmic view helps to achieve this.
5. Use a scattering colloid to simulate the optical geometry of the sample (*see* **Note 28**).

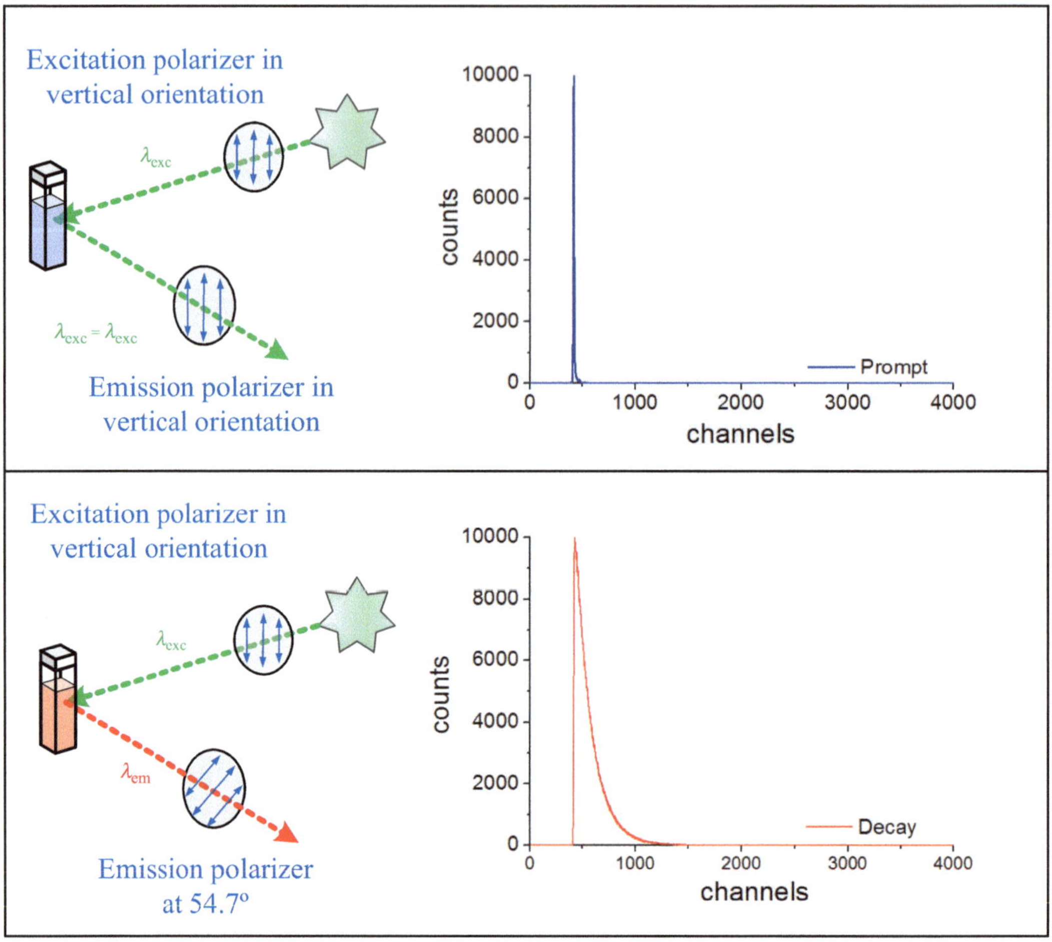

Fig. 9 Polarizer orientations for a fluorescence decay measurement. Typically the axes would be displayed using semilogarithmic scales with *x* being the time axis and *y* the counts (log); however, in this case a linear scale is used to emphasize the differences between the decay curves

6. Set the excitation monochromator to the required excitation wavelength (*see* **Note 36** and Fig. 9).
7. Set the emission monochromator to the excitation wavelength (*see* **Note 37** and Fig. 9).
8. Set the excitation polarizer to the vertical position (*see* **Note 38**).
9. Set the emission polarizer to the vertical position.
10. Change the delay lines as appropriate (*see* **Note 39** and Fig. 10) to find a suitable time range for measurement.
11. Add a neutral density filter if necessary before the sample (on the excitation arm) or iris diaphragm to ensure that the photon count rate is kept ≤ 2 % of the source repetition rate (*see* **Notes 40, 41** and Fig. 8).

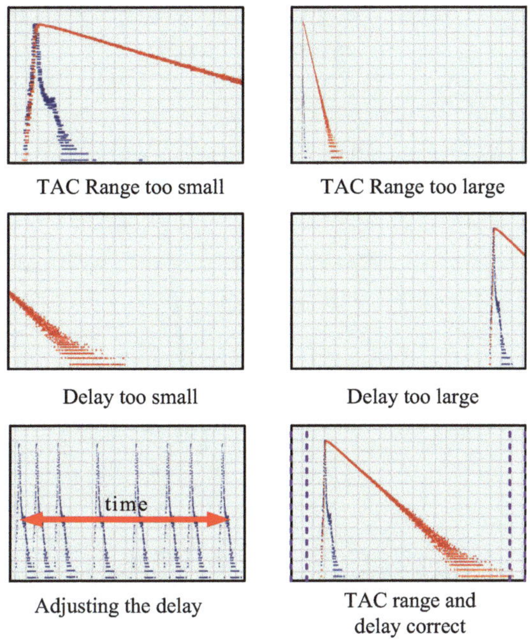

Fig. 10 Adjustment of the TAC range and delay lines so the measurement window covers the area of interest. The axes are semilogarithmic with *x* being the time *axis* and *y* the counts (log). Examples illustrated are too small a TAC range, too large a TAC range, too small a delay, too large a decay, adjusting the delay, and a correct measurement. *Purple lines* indicate likely nonlinear regions of the TAC at the beginning and end of the time range

12. Acquire the instrumental pulse, typically to 10^4 counts in the peak at a channel width of ~100 ps.
13. Replace the scattering colloid with the colloid which contains the labelled nps of interest and tune the emission monochromator to the peak wavelength of fluorescence. Given the different optical densities of the scattering colloid and the fluorescent sample, it is usually necessary to adjust the focusing of the collection optics in order not to waste fluorescence counts (*see* **Note 42**).
14. Set the emission polarizer to the magic angle (54.7 °) (*see* **Note 43** and Fig. 9).
15. Acquire the fluorescence decay typically to 10^4 counts in the peak at a channel width of ~100 ps (*see* **Note 44**).
16. Fit to the fluorescence decay to obtain the fluorescence decay times using reconvolution analysis in order to correct for the finite duration of the excitation pulse.
17. Figure 11 illustrates the principle of reconvolution whereby each time channel of the instrumental pulse (i.e., the measured excitation pulse) is treated as a δ function and the sum of the fluorescence impulse responses is then fitted to the measured fluorescence decay using nonlinear least squares (NLLS).

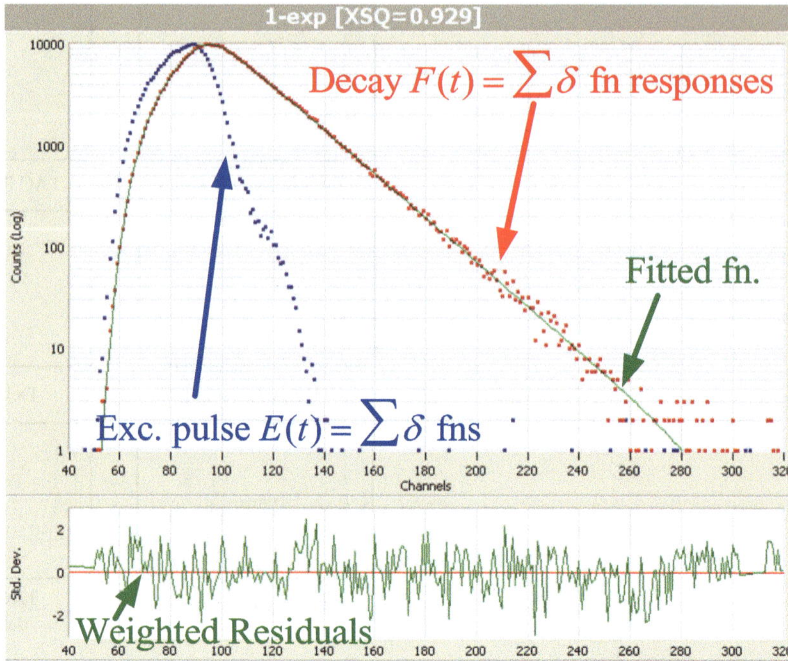

$$\chi^2 = \sum_{\text{Data pts}} \left\{ \frac{F(t) - E(t) \otimes i(t)}{[F(t)]^{1/2}} \right\}^2$$

$i(t)$ is the fluorescence impulse response

Fig. 11 Reconvolution analysis of fluorescence decay. The axes are semilogarithmic with *x* being the time axis and *y* the counts (log)

In NLLS analysis, the sum of the squares of the weighted residuals is minimized and the fit to the data assessed using statistical criteria. In the first instance, these are usually the trend in the residuals (random for a good fit) and the reduced chi-squared ($\chi^2 \sim 1$ for good fit) (*see* **Note 45**).

3.1.2 Anisotropy Decay Time Measurements

1. Your experimental apparatus should be set up exactly in the same manner as for the lifetime measurement.
2. Set the excitation polarizer to the vertical position.
3. Insert the scattering colloid and record a prompt as instructed for the lifetime measurement (*see* Fig. 9) with the emission polarizer also vertical.
4. Replace the scattering sample with the colloid which contains the labelled nps of interest and again tune the emission monochromator to the peak wavelength of fluorescence.
5. Record the polarized fluorescence decay curves by alternating the emission polarizer between the vertical and horizontal positions for equal dwell times (e.g., every 30 s) using automated polarizers (*see* Fig. 12 and **Note 46**).

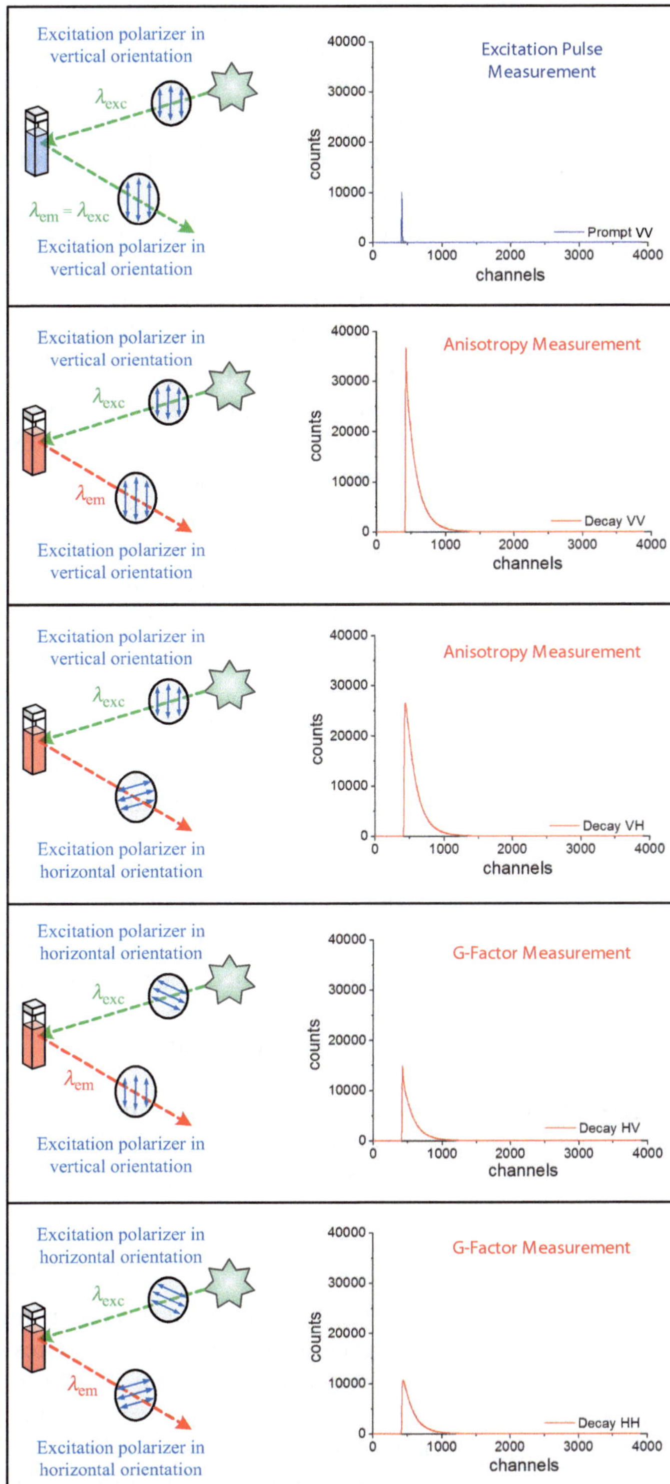

Fig. 12 Polarizer orientations for an anisotropy measurement. Typically the axes would be displayed using semilogarithmic scales with *x* being the time axis and *y* the counts (log); however, here a linear scale is used to emphasize the differences between the decay curves

6. The key statistic to take under consideration is the peak number of counts in the difference between these curves which may vary over 10^3–10^5 counts depending on the precision required.
7. Record the relative optical throughput of the vertical and horizontal emission polarizations, the so-called G-factor (*see* Fig. 12).
8. To do this place the excitation polarizer in the horizontal orientation and record the decay curves for both vertical and horizontal emission polarizer orientations. 10 cycles of 30 s is usually sufficient (*see* **Notes 31** and **47**).
9. When taking anisotropy measurements again ensure that the photon count rate is kept ≤ 2 % of the source repetition rate, use neutral density filters on the excitation arm as required (*see* **Note 40** and Fig. 8).
10. Data analysis is performed using nonlinear least squares (NLLS) iterative reconvolution using a library of fluorescence and anisotropy decay functions as described in Subheading 3.2. In our case, we use IBH DAS 6.6 software.
11. This involves first fitting to the fluorescence decay $I_{VV}(t) + 2GI_{VH}(t)$, i.e., the denominator of Eq. 2 as outlined in Subheading 3.1.1.
12. Next the difference curve $I_{VV}(t) - GI_{VH}(t)$, i.e., the numerator in Eq. 2, is fitted using different models for the anisotropy decay $R(t)$, in order to obtain the best-fitting rotational parameters. Selection of the most appropriate model is aided using the χ^2 goodness of fit criterion.
13. The anisotropy decay curves are recorded for typically up to a few hours as necessary to acquire 10,000 counts in the peak, at 28 ps per channel [5] (*see* **Note 48**). The errors are best quoted as 3 standard deviations.

3.2 Analysis of Data

By recording vertically and horizontally polarized fluorescence decay curves, $I_{VV}(t)$ and $I_{VH}(t)$, orthogonal to vertically polarized excitation, generate an anisotropy function $r(t)$ [14], i.e.,

$$r(t) = \frac{I_{VV}(t) - GI_{VH}(t)}{I_{VV}(t) + 2GI_{VH}(t)} \quad (2)$$

where $G = I_{HV}(t)/I_{HH}(t)$ is determined for horizontal orientation of the excitation polarizer (*see* **Notes 49** and **50**).

In order to analyze Eq. 2, first fit the denominator, which is in fact the fluorescence decay, to the minimum number of decay components needed using NLLS reconvolution of the instrumental pulse [9, 10] with a chi-sq (χ^2) goodness of fit criterion, i.e., as a rule of thumb $\chi^2 < 1.2$ for a good enough fit (*see* **Note 51**). This yields the best-fit fluorescence decay parameters τ_{f1}, τ_{f2}, τ_{f3}, etc.

associated with the fluorescence impulse response had we been using a δ function for excitation.

Again using NLLS reconvolution of the instrumental pulse fit the product $r(t) \times [I_{VV}(t) + 2GI_{VH}(t)]$ to the difference data $I_{VV}(t) - GI_{VH}(t)$ where $I_{VV}(t) + 2GI_{VH}(t)$ is now described by the fluorescence impulse response parameters and $r(t)$ describes the rotational kinetics in terms of the minimum number of rotational correlation times needed to describe the difference data.

The simplest case occurs for a spherical rigid rotor in an isotropic medium such as a solvent [14]. If all the dye is rigidly attached to the np, the decay of $r(t)$ describes fluorescence depolarization due to Brownian rotation of the np according to

$$r(t) = r(0) \exp\left(-\frac{t}{\phi_p}\right) \tag{3}$$

where $r(0)$ is the initial anisotropy with a maximum value of 0.4 for one photon excitation (*see* **Note 52**).

In this case, ϕ_p is the np rotational correlation time described by the Stokes–Einstein equation:

$$\phi_p = \frac{\eta V}{kT} \tag{4}$$

where η is the microviscosity, V the hydrodynamic volume $= 4\pi R_p^{\,3}/3$ prescribed by the np, T the temperature, and k the Boltzmann constant.

The radius (*see* **Note 53**) of the particle is then given by

$$R_p = \left(\frac{3kT\phi_p}{4\pi\eta}\right)^{1/3} \tag{5}$$

In the case where the dye partitions between being bound to the np and freely rotating we have

$$r(t) = (1 - f)r(0) \exp\left(-\frac{t}{\phi_d}\right) + fr(0) \exp\left(-\frac{t}{\phi_p}\right) \tag{6}$$

where f is interpreted as the fraction of fluorescence due to dye bound silica nps rotating with a correlation time ϕ_p and $1 - f$ to dye molecules unbound in the colloid and rotating faster with a correlation time ϕ_d [3]. Substitution of ϕ_d and the dye hydrodynamic radius R_d in Eq. 4 yields the microviscosity η. This, when combined with ϕ_p, gives the particle radius R_p again through use of Eqs. 4 and 5 (*see* **Note 54**).

If there are probe molecules also bound to aggregates of the "fundamental nanoparticles" and of similar composition, but a much larger size, then we can similarly approximate this by

Table 1
Typical results for nanoparticle metrology via fluorescence anisotropy measurements on silica colloids compared with other methods

Sample	Measured particle radius via other methods (nm)	Measured particle radius by fluorescence anisotropy (nm)
$M4_{R6G}$	4.2 ± 0.2 [4]	3.5 ± 0.1 [5]
$M4_{FITC\text{-}APS}$	4.2 ± 0.2 [4]	3.8 ± 0.1 [5]
LUDOX® SM30	3.5 [16]	4.0 ± 0.4 [2]
LUDOX® AM30	6.0 [16]	6.4 ± 0.5 [2]
LUDOX® AS40	11.0 [16]	11.0 ± 1.6 [2]

$$r(t) = (1 - f - g)\, r(0)\, \exp\left(-\frac{t}{\phi_d}\right) + fr(0)\, \exp\left(-\frac{t}{\phi_p}\right) + gr(0) \tag{7}$$

where $gr(0) = r(\infty)$ can also effectively describe a very slow rotation, i.e., g is the fraction of fluorescence derived from dye attached to the aggregate and $1 - f$ describes free dye or bound dye wobbling on a nanoparticle or a compromise between the two. When dealing with more complex kinetics, *see* **Notes 55–57**.

3.3 Typical Results

Table 1 shows typical silica np radii measurements using electrostatic and covalent labelling.

4 Notes

1. The fluorescence lifetime of the dye τ_f should be (1) monoexponential as this simplifies the decay analysis, although since the dye fluorescence tracks the rotation irrespective of the decay profile this is not too critical; (2) of a value with respect to the np rotational correlation time ϕ_p given by $\tau_f \leq \phi_p \leq 10\tau_f$, although in fact ϕ_p/τ_f ratios up to 50 have been used successfully [2]; and (3) stable during the course of the measurement.
2. Fluorescence is an extremely sensitive technique and all glassware should be cleaned thoroughly. Our cleaning procedure involves rinsing all glassware out with hot tap water and scrubbing with a bristled brush. All glassware is then steeped overnight in a 5 l polypropylene beaker containing 3 % DECON 90 solution. After the glassware has steeped, it is then rinsed out once again with hot water (ensuring all detergent bubbles are

removed). Finally the glassware is rinsed with a wash flask of cleaning alcohol (methanol or ethanol) and a wash flask of distilled water and allowed to dry.

3. For cuvettes (and smaller items), a similar cleaning procedure is used. To scrub the inside and outside of cuvettes, cotton buds are used (pick cotton buds which don't fragment heavily when wet). The cuvettes are rinsed with a wash flask of water and steeped overnight in a 100 ml polypropylene beaker containing 3 % DECON 90 solution. Polypropylene beakers are preferentially used over glass beakers to prevent glass-to-glass scratching. Cuvettes are then rinsed as with a wash flask of cleaning alcohol (methanol or ethanol) and a wash flask of distilled water (again ensuring all detergent bubbles are removed). Cuvettes are generally dried using nitrogen gas.
4. Marker pens may be used to scribble on temporary glassware, but for glassware kept in use for more than 1 day, it is recommended to use a label printer. If a splash of alcohol runs on the outside of a label with a marker pen, the ink will run and the label will become illegible. Brother P-Touch labels are chemical resistant and don't present such a problem.
5. Ammonia at 33 % concentration tends to bubble away rapidly; we recommend checking the concentration of ammonia by titration before nanoparticle synthesis.
6. Other low molecular weight alcohols such as ethanol and propanol may be used as solvents; in addition, other orthosilicates may be used. The nanoparticle sizes will be different than those produced with methanol, water, ammonia, and TEOS [15].
7. When calculating the ammonia and water content, recall that the ammonia solution is 33 % ammonia and 67 % water.
8. 33 % ammonia solution is very volatile; it is extremely difficult to measure small amounts. To achieve greater accuracy, we upscale it so the water–ammonia solution is ~30 ml.
9. Filling the container up to the halfway mark before addition of the water–ammonia solution prevents an uncontrolled initial reaction from taking place.
10. Ensure that the stirring bar is added after the conical flask is filled up to the 10 ml mark.
11. We have found the Stöber silica nanoparticles produced via this method to be very stable and have observed no changes in them for up to a month when sealed and left at room temperature. Samples left for over a year were found to form gels.
12. Fluorescent dyes are generally shipped in powder form; you will need to dissolve some dye in solvent to make a stock solution. Skip this step if you purchased the dye as a solution.

13. Since there is usually only a small quantity of fluorescent dye used per measurement (~50 μl of dye stock solution), it is not necessary to make huge volumes. In general, we make the stock solution in 3 ml volumes.
14. We recommend using a 4 ml amber-glass container for storing and making dye stock solutions as fluorescent dyes can be extremely photosensitive. If you don't have amber-glass vials, wrap each vial up with dye in aluminum foil.
15. Some dyes will not readily dissolve in solvent, and we recommended trying to sonicate these at ~37 Hz for 30 min at room temperature to induce them to dissolve. It should be noted that some dyes will be insoluble in particular solvents and hence a range of solvents needs to be available for making the stock solution.
16. Pipette liquid up and down to mix the dye with the sample. To ensure thorough mixing, cap the cuvette and gently tip the cuvette up and down a few times. The cuvettes should remain sealed as repeated exposure to atmospheric CO_2 has the effect of lowering the pH, eventually resulting in auto-gelation of silica sol.
17. Plastic cuvettes may not be suitable for methanol or ethanol solvents as they will become dissolved by such solvents if left exposed for a prolonged time. Plastic cuvettes may also have strain-induced birefringence that will affect the polarization.
18. When taking absorption spectra, be wary of the wavelength range your cuvette can transmit. Glass and plastic cuvettes, for example, will transmit at ~350–800 nm, while quartz ones will work from 200 to 800 nm. In addition, some solvents will transmit at certain wavelength ranges. The baseline measurement is a transmission measurement, and from this you can directly observe the range your cuvette and cuvette/solvent will work at. Although in general you will be given the specification of all the optical components (e.g., neutral density filters, long-pass filters, and cuvettes) that you order, it is worthwhile verifying them experimentally. Absorption measurements recorded outside the working transmission ranges are meaningless.
19. For the majority of dyes, the molar extinction coefficient ε can be found in their supplied datasheet in addition to normalized spectra.
20. We often just use the absorption spectra to verify the dye concentration as this is often easier than attempting to measure out minute quantities of dye powder.
21. We generally calculate the volume of dye that has to be added to 3 ml of solvent to generate an absorbance of 0.1 and label

the vial accordingly. An absorption of ~0.1 is sufficiently strong enough to generate a good fluorescence signal but not strong enough for dye–dye interactions such as self-quenching to become a major issue.

22. Record the steady-state emission spectra just after the excitation wavelength; if you have narrow slits in your spectrofluorometer, then you may begin 1–2 nm after your excitation wavelength. If however you have broader slits, then you should start 5–10 nm after the excitation pulse. Otherwise, you will observe a peak from your excitation wavelength which will be much larger in magnitude than your fluorescence emission. There is no point in measuring the wavelengths before the excitation wavelength.

23. A suitable wavelength is one where there is strong emission and weak self-absorption.

24. An example of those already demonstrated would be negatively charged colloidal silica nps such as those in Dupont LUDOX® [16] at pH 10, and a cationic range of methoxyquinolinium dyes or acidic hydrogels and certain xanthene dyes such as rhodamines can be bound to both anionic and cationic nps and provide free dye for determining the microviscosity.

25. APS is very reactive with water and once it reacts with water it will lose its ability to bind to the np. The mechanism is the same as the Stöber reaction for np preparation. APS does not readily dissolve in methanol or ethanol which is why the reaction needs to be left mixing for such a long period of time.

26. A 1 mol FITC–1 mol APS ratio should be made. In practice, we find lumps of APS at the bottom, leaving a significant proportion of free dye (~40 % bound and ~60 % free). To bind all the dye, we have attempted to add APS in 10 M excess; however, in this case adding the dye to the nps becomes similar to adding more TEOS and results in the formation of silica flakes. These can be removed from the sample via centrifuge; however, the absorption peak becomes much weaker going from ~0.1 to ~0.012. Similar anisotropy results were found via both methods; however, it should be noted that there is a trade-off between optimizing the synthesis of FITC–APS and disturbing the np by adding too much APS. Ideally the reaction to create FITC–APS should be carried out with a huge excess of APS; later a high molecular weight molecule should be added in even larger excess and with appropriate chemistry to bind to the remaining free APS in order to isolate the FITC–APS by centrifuge.

27. We recommend you to match the cuvette type with that used for the sample.

28. LUDOX SM30 [16] when purchased is generally too concentrated to use directly as a light scattering colloid. It will aggregate under acidic conditions, and the sodium hydroxide is necessary to ensure the pH is kept between 10 and 12 when diluting. Other colloids can also be used, e.g., when exciting at visible wavelengths, diluted milk may be used. Chalk is also commonly used right down to the UV measurements.
29. The term "lifetime" is strictly only applicable to a one-component fluorescence decay, and the term "decay times" rather than "lifetimes" is better when describing multicomponent behavior.
30. The decay time measurement is an inherent part of the anisotropy measurement described in Subheading 3.1.2, but the separate description given here serves to check out the instrument in a simpler mode of operation before proceeding to the np measurement.
31. Note most suitable pulsed light sources display some level of polarization.
32. Avalanche photodiodes with single-photon timing capabilities are an alternative to photomultipliers, but they have a stronger temporal dependence on the wavelength which can adversely affect reconvolution analysis and less active area of response than a photomultiplier.
33. L-format orientation reduces the transmission of scattered excitation light by the emission arm.
34. In order to optimize data collection, it is often necessary to alter the range of the time-to-amplitude convertor (TAC). The TAC range should be altered to closely match the lifetime of the fluorophore; if the TAC range is much smaller than the lifetime of the fluorophore under investigation, then much of the fluorescence decay will be missed; if, on the other hand, one has a much larger TAC time than the dyes lifetime, the resolution will be significantly reduced (*see* Fig. 10).
35. Ensure that the light source repetition rate isn't too high for the TAC range being used; otherwise, the dye will be excited twice in the measurement window and a second lifetime decay may overlap with the end of the first.
36. LEDs emit light over a broader spectrum in comparison to laser diodes and can sometimes generate weaker emissions at longer wavelengths than their fundamental wavelength band [11–13]. For these reasons, the excitation was selected using an optical monochromator (in our case a Seya–Namioka geometry).
37. In the case where a long-pass filter is being used to separate the fluorescence emission from the scattered light, insert no emission filter during the measurement of the excitation pulse.

38. Both dichroic and prism optical polarizers can be used successfully to select polarization; however, dichroic polarizers are unsuitable for measurements in the UV.
39. The time delay at the stop signal should be varied, so that the prompt (excitation pulse) is measured near the start of the TAC. This can be done by running a prompt measurement with a scatter sample and altering the delay lines accordingly. Care should be taken to ensure that the prompt is not at the very edge due to nonlinear effects which occur at the edge of the TAC range (*see* Fig. 10). Linearity can be checked using uncorrelated pulses, e.g., photomultiplier noise.
40. In all the measurements, keep the photon count rate to ≤2 % of the source repetition rate to eliminate data pileup [9, 10]. A neutral density filter or iris diaphragm may be inserted before the sample to ensure that this requirement is met. Use the pulsed source to generate "stop" signals and the detector to generate "start" signals in order to minimize dead time. This is known as running the TAC in reverse mode. Although the measurement is ran in reverse mode, the data can be reversed on the computer and later analyzed in the conventional way.
41. Alternatively use a long-pass filter on the emission arm. The choice between optical monochromator and filter is determined by the count rate available, the monochromator always being preferable if a fluorescence count rate of ~2 % can be achieved. To reduce the detection of scattered excitation light, both components are sometimes used.
42. Motor-driven lenses are preferable to manual adjustment as they enable the excitation and emission focusing to be automatically computer optimized in order to maximize the data collection rates for samples of differing optical density.
43. Fluorescence lifetimes are measured at the magic angle (54.7 °) to eliminate polarization bias in the instrument [9, 10], e.g., due to the diffraction grating response in the monochromator.
44. We would recommend testing your kit with a few reference samples. For example, fluorescein in methanol, rhodamine 6G in water. The lifetimes of these (and of other common dyes) are well covered in the literature. Compare your results to the literature values to verify your electronics (linearity and calibration) are working correctly. Time calibration can be checked using a crystal oscillator to generate start and stop signals or using calibrated delay lines for the same signal input to the start and stop channels. A series of calibration points of known time difference can then be analyzed using least squares analysis.
45. Nonlinear least squares (NLLS) is an iterative process with the fluorescence impulse response adjusted until a good fit to the data is obtained. A good fit is taken to be when the reduced

χ^2 is ~1.0 (reduced means χ^2 divided by the number of channels of data minus the number of fitted parameters). Because the data is statistically imperfect due to the effects of systematic errors (such as scattered excitation light), in practice this figure is often taken to be ≤ 1.2. In the first instance, iteration usually means increasing the number of decay components. However, care should be taken not to over-parameterize as this can give an anomalously low χ^2 value, i.e., use the minimum number of decay parameters and do not increase their number when doing so produces little change in the χ^2. An extra fitted parameter, which needs to be included in reconvolution analysis, is that of a time shift. This corrects for the quantization of the time channels and provides a first-order approximation to wavelength-dependent timing response in the photomultiplier. The latter arises from the effect on the photoelectric effect caused by differences in wavelength between excitation and fluorescence.

46. Continuous alternation of the emission polarizers between horizontal and vertical orientation will correct for any drift in light intensity. Motorized control is recommended for optimal accuracy and ease. If motor control is unavailable, manual alternation can be used.
47. This procedure can be used as for the fluorescence decay time measurement described in Subheading 3.1.1.
48. The measurements in our early work [2] were taken using the IBH NanoLED which had a repetition rate of 1 MHz. Measurements with the newer more intense IBH DeltaDiodes have a repetition rate of up to 100 MHz, significantly reducing measurement acquisition time.
49. G is just a number and the decay curves $I_{HV}(t)$ and $I_{HH}(t)$ can simply be integrated to obtain numbers of counts to calculate their ratio.
50. $I_{VH}(t)$ denotes the excitation polarizer to be orientated in the vertical (V) direction and the emission polarizer to orientated in the horizontal (H) direction. $I_{VV}(t)$ relates to both polarizers in the vertical orientation.
51. As a rule of thumb χ^2 for a good enough fit. It is essential to obtain a good fit to the fluorescence decay as any errors here will translate into the anisotropy determination.
52. Higher $r(0)$ values and hence higher dynamic range can be achieved using two-photon excitation [17], though at a significant increase in measurement complexity.
53. The cube root reduces the effect of errors in ϕ_p and η on R_p.
54. A different and nonbinding dye can also be added to determine the microviscosity [18], *see* Subheading 2.2.1.

55. If $\phi_p \gg \tau_f$ then the second term in Eq. 6 becomes a residual anisotropy and ϕ_p is immeasurable. If the dye is not free but bound and wobbling on the np, this can also be approximated by Eq. 6 or if both free and wobbling on the np, then $1 - f$ and ϕ_d describes an average of free and wobbling dye and the effect of the associated different viscosities. It should be noted that only if the fluorescence quantum yield of the probe is the same when unbound and bound to silica particles, does the fraction f exactly describe the probe partition ratio between the nanoparticle and fluid. Also, τ_f for unbound and bound probe needs to be similar for Eqs. 6 and 7 to be valid. Moreover, a single rotational correlation time description is only strictly valid for a spherical rotor of fixed radius of rotation. Hence, in the case of a nonspherical or a distribution of nanoparticle sizes, the analysis becomes more complicated and the above treatment becomes more approximate [17].
56. The determination of size distributions is generally not possible as such samples don't yield unique solutions to reconvolution analysis.
57. The presence of other depolarization mechanisms can be sources of error, e.g., diffusion of the dye on the surface of the np, homo-fluorescence resonance energy transfer between dye molecules. These can however be corrected for to an extent by incorporation of additional terms in Eq. 6.

Acknowledgements

The authors wish to thank the EPSRC and SFC for research support and SUPA, NPL, and Horiba Jobin Yvon IBH Ltd for a studentship held by PY.

References

1. Birch DJS (2011) Fluorescence detections and directions. Meas Sci Technol 22:8. doi:10.1088/0957-0233/22/5/052002
2. Apperson K, Karolin J, Martin RW, Birch DJS (2009) Nanoparticle metrology standards based on the time-resolved fluorescence anisotropy of silica colloids. Meas Sci Technol 20:11. doi:10.1088/0957-0233/20/2/025310
3. Birch DJS, Geddes CD (2000) Sol-gel particle growth studied using fluorescence anisotropy: an alternative to scattering techniques. Phys Rev E Stat Phys Plasmas Fluids Relat Interdiscip Topics 62:2977–2980. doi:10.1103/PhysRevE.62.2977
4. Green DL, Lin JS, Lam YF, Hu MZ, Schaefer DW, Harris MT (2003) Size, volume fraction, and nucleation of Stober silica nanoparticles. J Colloid Interface Sci 266:346–358. doi:10.1016/S0021-9797(03)00610-6
5. Yip P, Karolin J, Birch DJS (2012) Fluorescence anisotropy metrology of electrostatically and covalently labelled silica nanoparticles. Meas Sci Technol 23:8. doi:10.1088/0957-0233/23/8/084003
6. Kačenka M, Kaman O, Kotek J et al (2011) Dual imaging probes for magnetic resonance imaging and fluorescence microscopy based on perovskite manganite nanoparticles. J Mater Chem 21:157–164. doi:10.1039/C0JM01258K
7. Van Blaaderen A, Vrij A (1992) Synthesis and characterization of colloidal dispersions of fluorescent, monodisperse silica spheres. Langmuir 8:2921–2931. doi:10.1021/la00048a013

8. Imhof A, Megens M, Engelberts JJ et al (1999) Spectroscopy of fluorescein (FITC) dyed colloidal silica spheres. J Phys Chem B 9:1408–1415. doi:10.1021/jp983241q
9. Birch DJS, Imhof RE (2002) Time-domain fluorescence spectroscopy using time-correlated single-photon counting. In: Lakowicz JR (ed) Topics of fluorescence spectroscopy, vol 1. Springer, New York, pp 1–96. doi:10.1007/0-306-47057-8_1
10. O'Connor DV, Phillips D (1984) Time-correlated single photon counting. Academic, London
11. O'Hagan WJ, McKenna M, Sherrington DC et al (2002) MHz LED source for nanosecond fluorescence sensing. Meas Sci Technol 13:84–91. doi:10.1088/0957-0233/13/1/311
12. McGuiness CD, Sagoo K, McLoskey D, Birch DJS (2004) A new sub-nanosecond LED at 280 nm: application to protein fluorescence. Meas Sci Technol 15:L19–L22. doi:10.1088/0957-0233/15/11/L02
13. McGuinness CD, Sagoo K, McLoskey D, Birch DJS (2005) Selective excitation of tryptophan fluorescence decay in proteins using a subnanosecond 295 nm light-emitting diode and time-correlated single-photon counting. Appl Phys Lett 86:261911. doi:10.1063/1.1984088
14. Steiner R (1991) Fluorescence anisotropy: theory and applications. In: Lakowicz JR (ed) Topics of fluorescence spectroscopy, vol 2. Springer, New York, pp 1–51. doi:10.1007/0-306-47058-6_1
15. Ibrahim IAM, Zikry AAF, Sharaf MA (2010) Preparation of spherical silica nanoparticles: stöber silica. J Amer Sci 6:985–989
16. LUDOX Colloidal Silica. Grace Davison Data Sheet September 2008. Reference ludox_broch_E_08
17. Karolin J, Geddes CD, Wynee K, Birch DJS (2002) Nanoparticle metrology in sol-gels using multiphoton excited fluorescence. Meas Sci Technol 13:21–27. doi:10.1088/0957-0233/13/1/303
18. Tleugabulova D, Duft AM, Brook MA, Brennan JD (2004) Monitoring solute interactions with poly(ethylene oxide)-modified colloidal silica nanoparticles via fluorescence anisotropy decay. Langmuir 20:101–108. doi:10.1021/la035333z

Chapter 12

Upconversion Spectrophotofluorometry

Arianna Biesso, Jianhua Xu, and Jay R. Knutson

Abstract

As the other chapters attest, sensitivity of fluorescent molecules to their local environment has created powerful tools in the study of molecular biology, particularly in the study of protein, DNA, and lipid dynamics. Surprisingly, even events faster than the nanosecond lifetimes of fluorophores are important in protein function, and in particular, events lasting just a few ps reflect on water motion and the coupled dynamics of proteins. These ultrafast phenomena can best be studied by using the same laser that excites fluorescence to also "strobe" the emission, providing sub-picosecond time slices of the action. We explain the strobing "upconversion" technique and some limits on its execution.

Key words Time-resolved fluorescence, Upconversion, Tryptophan, Protein dynamics and photodamage

1 Introduction

1.1 A Brief History of Upconversion Spectroscopy

The newly discovered red ruby laser was focused into quartz in 1961 by Franken et al. [1], and the output beam was passed through a *prism* spectrograph, allowing them to observe a much weaker 347 nm (doubled frequency) spot. While frequency mixing of many types was known in microwave circuits, RF cavities, and (of course) low-frequency electronics (esp. household radio heterodyne circuit stages), *optical* mixing had not previously been seen.

The other nonlinear optical phenomena followed, including higher harmonic generation and generation of both optical difference frequency (e.g., for novel monochromatic IR beams) and sum frequency generation. The latter, conceptually identical to second harmonic generation, gained favor as a method to convert weak infrared signals into visible signals amenable to detection with more conventional cameras and detectors. The redder signal was said to have been frequency "upconverted" into the visible.

To accomplish upconversion, a stronger (and usually redder) optical field was needed to mix with the weak signal. Just as in electronics, mixing demands the voltages be high enough to elicit

Yves Engelborghs and Antonie J.W.G. Visser (eds.), *Fluorescence Spectroscopy and Microscopy: Methods and Protocols*, Methods in Molecular Biology, vol. 1076, DOI 10.1007/978-1-62703-649-8_12,

a distorted (nonlinear) response of the mixer material to the sinusoidal waves. Otherwise, the beams simply superimpose in the crystal, and the electrons simply track the superposition of electric fields. The strong electric fields of short laser pulses were just what was needed to reveal second-order crystal nonlinearity and mixing. Of course, to create mixing, the electric fields had to superimpose not only in space, but also in time. This created the opportunity to auto or cross-correlate laser pulses, since the mixed signal output envelope is a temporal convolution of the input pulses.

Autocorrelation and cross-correlation of pulses set the stage for not only transforming wavelengths, but also quantifying ultrafast kinetics. The photocathodes all found at the heart of photomultipliers, vacuum diodes, and streak cameras invariably produce a spread of photoelectron ejection energies in response to a laser pulse, meaning those electrons take slightly different paths (and times) to their eventual detection. As a result, phenomena faster than several ps were intractable for even the fastest photodetectors. Optical mixing was the best solution.

In chemistry, the first published use of upconversion to trace a luminescent process was by Halliday and Topp [2]. The method spread rapidly in photophysics, as briefly reviewed by Barbara and colleagues [3].

Ultrafast spectroscopy (particularly time-resolved fluorescence) has been a fruitful area of biological research. While it is, at first, counterintuitive to think that nanosecond (or worse, picosecond and faster) events bear relevance to enzymatic and structural changes taking a billion times longer, the facts have been clear: changes in the fluorescent signals on or in proteins, membranes, and DNA give intimate details about the size, flexibility, fluidity, and interrelationships (including proximity and motion) of these various macromolecules. Several other chapters in this book will provide examples of this windfall.

Among the various biophysical targets of interest, upconversion spectroscopy quickly became most popular in a field that, unsurprisingly, is hungriest for time-resolved data tracing the flow of photonic energy: Photosynthesis. The study of photosynthetic pigments and their photochemical and functional physical arrangements includes relevant kinetics on everything from femtosecond through minute timescales [4–6].

1.2 Use of Upconversion to Study Proteins

Tryptophan is both the blessing and curse of protein dynamics spectroscopy. Present in most proteins, yet sparse enough that most have only a few Trp (W) in their sequence, it represents a local reporter that is intricately linked to its surroundings as it performs its native structural and/or electronic duties [7]. Trp has been studied for years with time-resolved spectroscopy on the nanosecond scale, and the best software and genetic manipulations have been applied to the task of isolating the signals belonging to single Trps within proteins [8, 9]. Such signals tell us about local mobility and folding, overall tumbling of the protein, proximity to FRET acceptors such as NADH [10] or heme [11], and the availability of the Trp ring to solvent quenchers [12].

Single Trp in proteins often provides multiexponential decay behavior, a fact that had led to a long-standing controversy about why the protein environment is not homogeneous on that timescale. Two camps evolved, one (observing comparable features of indole in glycerol) and seeing Trp in proteins also tended to have longer decays at the red side of the spectrum, invoked ns solvent relaxation [13, 14]. A second camp invoked Occam's razor to point out the decay surface was compatible with simple heterogeneity—perhaps by mechanisms as simple as side-chain rotamers (or more broadly, conformers) [15, 16]. The nanosecond fluorometry "heterogeneity vs. relaxation" debate went on throughout the last three decades, and the combination of studies on crystals (where orientation modified preexponential weights, a result incompatible with relaxation) [17] and studies on constrained Trp derivatives [18] pushed toward a consensus that on the nanosecond scale, Trp decay in proteins was often dominated by heterogeneity.

This was not to say relaxation was absent; instead, it was suggested that it was mostly complete within ~100 ps (the lower time resolution typical of nanosecond instruments). Most telling was the absence of a negative preexponential term on the red side of the Trp emission, a signal known to flag relaxation (or any other excited state reaction) [19]. Advocates of relaxation models, while agreeing this rising term should be present, rightly argued that the negative preexponential could be quite small and therefore easily lost in either noise or the contributions of short-lived contaminants (perhaps photoproducts) [20].

A progression of nanosecond lifetimes associated with a progression to the red in spectral position (redder "DAS" correlating with longer lifetime) was also suggested as reflective of relaxation, and the paucity of proteins breaking that pattern was noted. Only recently, the intrinsic nature of the nanoheterogeneous quenching mechanism has been found to favor this correlation [21].

Early picosecond studies of Trp, meanwhile, were carried out laser-based chemical physics community, and they were generally remote from the nanosecond Trp controversy. Treating Trp like other dyes (e.g., coumarins) in solvents, they obtained time-dependent fluorescence Stokes shifts (TDFSS) (or fluorescence dynamic Stokes shift, FDSS) and deduced correlation functions, concerned primarily with the spectral movement rather than deactivation [22, 23]. Exceptions to this focus were studies of Trp in heme-containing proteins, where efficient FRET was anticipated from Trp to Soret bands, leading to sub -30 ps overall lifetimes [24, 25]. The early behavior of Trp was also probed for emission anisotropy [26], where, unfortunately, some Raman contamination to curves (apparently) made Trp appear to be a complex, vibrational-cooling dominated system (likely too complex to be useful in ps motion studies). In the late 1990s, however, Shen et al. [27] showed that Trp not only produced strong 1.6 ps solvent relaxation spectra in water, but also had simple anisotropy behavior; oscillators within Trp analogs were seen to perform internal conversion (reaching stable orientation) well within 100 fs. This opened the door to exploiting Trp without the complexity concerns previously mentioned.

A recurrent theme in protein dynamics is the protein–water interface. Data available from X-ray to neutron diffraction point to an ordered water layer in direct contact with the protein. The dynamic properties of this ordered layer of water are not clear, and if they are different it poses a question: Are these water molecules simply slaved to the protein charge distribution and dynamics or does the ordered water also affect protein dynamics? The nuclear overhauser effect (NOE) allows the investigator to probe water relaxation and compare it to bulk water with sub-nanosecond time resolution; further, nuclear magnetic relaxation dispersion (NMRD) has shorter effective time resolution. FDSS, using fluorescence upconversion on Trp residues of Monellin, showed spectral downshifting that was slower compared to those observed for single indole molecules in solution. It was then proposed that such a slow (8–20× slower) shift was the signature of a layer of organized water (also called "biological water"). This new "slow" water on protein would have solvation dynamics in the range of 30 to hundreds of picoseconds, in contrast to the few picoseconds measured for bulk water. In protein dynamics, X-ray diffraction can provide initial structures that, along with MD simulation, can predict stepwise information on this layer of solvation water. Recent results have also been gleaned from ^{17}O spin relaxation, providing values for the dynamics of water at the interface that are consistently faster that those inferred using FDSS.

When the laboratory of laureate A. Zewail published relaxation profiles for Trp within the sweet protein Monellin, finding not only the 1.6 ps "bulk water" relaxation term but also a redshift that required 10s of ps, they (as mentioned above) attributed this slower process to exchanging water between bulk and bound sites with off-rates near 20 ps [28]. Importantly, they also made emission anisotropy measurements in a variety of protein states, confirming the viability of Trp as a probe of segmental motion [29, 30].

Noting that the raw TRES (Time Resolved Emission Spectra) of Monellin did not rise after a few ps, Xu et al. [45] showed that this 20 ps behavior of Monellin was actually dominated by an ultrafast quenching process (the blue side was simply being consumed more rapidly than red) rather than a conservative steady redshift like that seen in bulk water. *In other words, no negative preexponential was found for the 20 ps term.* This essentially caused the two model controversy from ns Trp studies to reemerge at 100× faster rates.

Other proteins [31, 32] clearly did show the negative amplitude feature. Was relaxation always present and was it simply "masked" in some proteins by superposed quenching? Was it unique to proteins with particular folds, partial unfolding, or denaturation? Monellin in denaturants was certainly a strong exhibitor of ~50 ps relaxation, but the presence of a viscous denaturant complicates any dynamics gleaned from that picture [28].

One approach to the "masking" question was mutagenesis—by removing likely quenchers in the neighborhood of Trp (and replacing them with benign side chains), a quencher-free relaxation was seen [33].

Meanwhile, the rapid quenching, naturally unseen in nanosecond instrumentation, was termed QSSQ (Quasi-Static Self Quenching) in accordance with Chen et al. and it was found to be effective even in simple dipeptides [34, 35]. More importantly, it was seen that there were not only < 100 ps quenching rates, but also losses faster than the instrument response of 0.2 ps (200 fs). The "initial brightness" (taken typically at ~1 ps) could be used to sense this, and the sum of these ultrafast terms reconciled yield defects in a variety of peptides [36]. An actual *biological use* for QSSQ was found in the eye lens proteins termed crystallins; apparently, animals have evolved to keep these proteins from oxidizing in ultraviolet light by creating very strongly quenched Trp environments [37]. Thus, the ultrafast QSSQ is a useful probe of proteins independent of the relaxation phenomenon it masks.

1.3 Molecular Dynamics (MD) as a Tool to Decode Protein Dynamics

Molecular dynamics (MD) simulations have contributed in shaping today's awareness of proteins acting as dynamic complex systems, and they also provide a unique window on the behavior of protein and solvation water in the femto- to picosecond timescale.

Attempts to follow protein dynamics involving the use of covalently bound extrinsic fluorescence probes produced mixed results. In some cases it was difficult to deconvolute the protein internal dynamics from those of the dye [38]. The protein dynamics community pursued many different routes. Alongside fluorescence dynamic Stokes shift (FDSS), a number of different techniques were employed to probe protein dynamics, esp. NMR and nuclear overhauser effect (NOE) with MD simulations performed in parallel [39–42].

Decoding of the Trp fluorescence signal and its correlation to short- and long-range molecular dynamics in proteins is not an easy task. MD simulations proved to be essential in helping with the data interpretation; hybrid QM-MM simulation methods have indeed been successful in predicting fluorescence emission maxima and quantum yields for Trp residues in a variety of proteins [43, 44]. Importantly, the study done on γS and γD-crystallin revealed multiexponential decays in the ps time domain along with a very rapidly quenched population, and QM-MM simulations done on such residues clearly disclosed that nearby charges were the cause for the fast quenching: A Trp charge transfer (CT) state was produced with the amide backbone, stabilized by nearby water and charged residues; thus, the *predicted* Trp fluorescence lifetime was drastically reduced [37]. As mentioned above, nature might have selected such "enhanced" quenching mechanisms in crystallin to insure longevity in an eye lens proteins heavily exposed to UV light.

The most recent work in this area by our group concludes that relaxation is ubiquitous but can be masked by the equally valuable QSSQ phenomenon. We find relaxation in GB1 is remarkably refractory to pH-induced changes in protein structure that have profound effects on ns emission [45]. Further, substitution of Trp analogs via auxotrophic biosynthesis can eliminate the QSSQ term. For example, relaxation within the aforementioned Monellin protein (dominated by 20 ps quenching) is "unmasked" when 5FTrp replaces Trp [46].

Upconversion of Trp then gives us access to two different valuable indicators of protein dynamics: relaxation, with wider extent, and QSSQ, with precise dependence on Trp nearest neighbors.

2 Materials

2.1 Upconversion Spectrophotofluorometer and Experimental Instructions

Fluorescence upconversion is a derivative of the nonlinear pump probe technique, and a spectrophotofluorometer was developed especially for studying tryptophan fluorescence dynamics in proteins, which is shown in Fig. 1. A ~400 mW pulse train with a typical pulse duration of 120 fs and a repetition rate of 82 MHz was generated from a mode-locked Ti: sapphire laser (Tsunami, Spectra-Physics) pumped by a diode-pumped solid state (DPSS) laser with the power of 5 W (Millennia, Spectra-Physics). It seeds a Ti: sapphire regenerative amplifier (Spitfire, Spectra Physics) pumped by a diode-pumped nanosecond green laser with average power of 15 W at 5 kHz (Empower, Spectra-Physics). The output amplified pulses at 885 nm typically had a pulse energy of 0.2 mJ and an autocorrelation pulse width of 350 fs (*see* Fig. 2) at a repetition rate of 5 kHz (i.e., ~1 W output power). Ultraviolet excitation pulses at 295 nm with an average power up to 10 mW were obtained from nonlinear harmonic generation (a 1 mm BBO crystal and 0.5 mm BBO crystal for doubling and tripling, respectively). This UV beam was separated from the infrared beam (fundamental) and visible (doubled) by two dichroic mirrors and by spatial filtering (using a pinhole). Power should be carefully attenuated to below 1 mW before excitation of the sample to avoid photodegradation, hole burning, and other undesirable effects. A zero-order half wavelength plate (HWP) was put on a computer-controlled servo stage

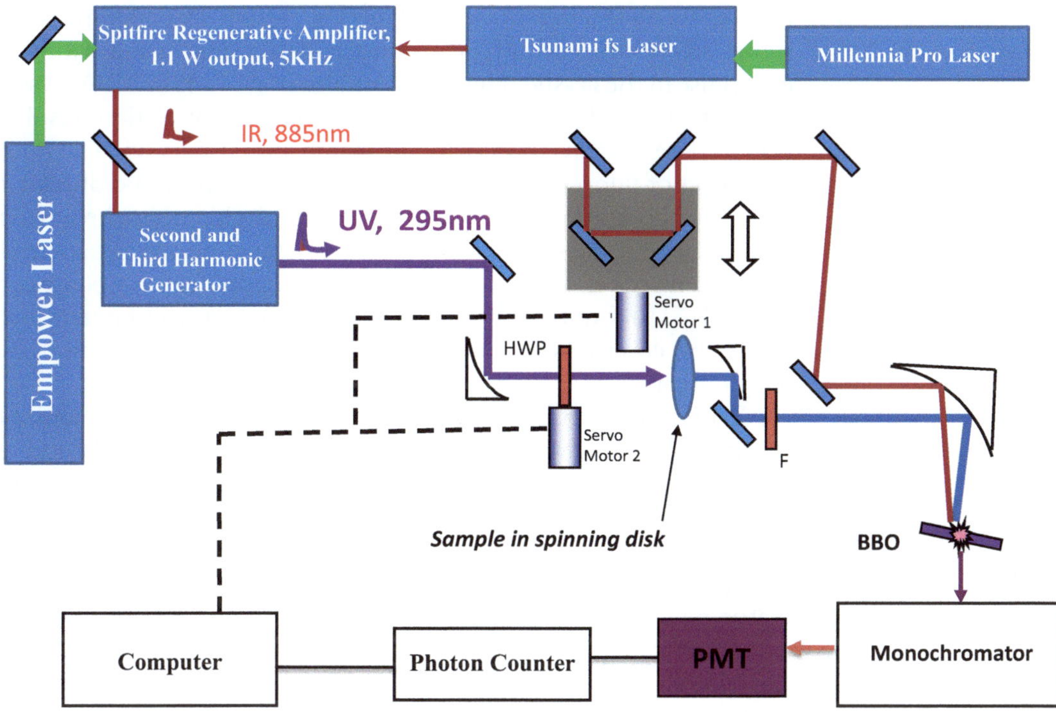

Fig. 1 A layout of the upconversion spectrometer setup

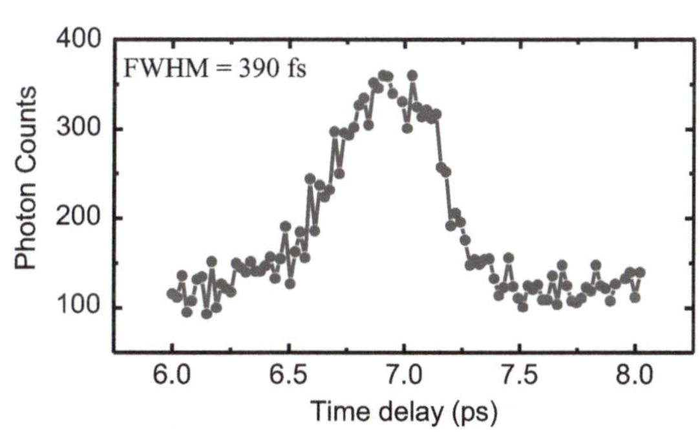

Fig. 2 A typical snapshot of the response time profile of upconversion spectrophotofluorometer

and used to change the polarization of the UV beam. A circular array of thin cells (T-20, NSG Precision Cells) with a path length of 1 mm in a delrin stacked slotted disk was used to hold samples, and the total volume was about 2 ml. The disk was spun continuously (up to 5 m/s circumferential motion) to make each UV pulse excite a slightly different sample spot. The remaining infrared fundamental pulse was reflected from a broadband hollow retroreflector on a computer-controlled precision stage and used as a gate pulse for the upconversion process. The fluorescence emission was collected by parabolic mirrors, passed through a long pass filter (F), and focused into a 0.5 mm thick BBO mixing crystal. The upconversion signal was produced via type I sum frequency generation with the gate pulse in the crystal. Here the combination of a thin BBO crystal and mirrors is selected for reducing the limitations of time resolution imposed by group velocity mismatch and dispersion. The time evolution of fluorescence at any given wavelength can be obtained by using a computer to set the BBO crystal angle, monochromator wavelength (at $\omega_F + \omega_P$), and scan delay stage. The delay stage is comprised of a hollow cube corner retroreflector (Newport UBBR1-5I) mounted on a cross roller slide (Parker) linear stage driven by compumotor motor, combined with appropriate NIR relay mirrors.

To reject the strong background signals (infrared laser and its second harmonic generation, remnant UV, and unconverted fluorescence) accompanying the upconverted signal, a non-configuration was arranged between infrared probe and fluorescence. The use of a monochromator helps in reducing these unwanted signals at the PMT, also improving the spectral bandwidth. A double monochromator may be necessary if this problem is severe. Polarizations of gated fluorescence were determined by the orientation of nonlinear crystals, so no extra linear polarizer was needed (for anisotropy calculation, the grating factor G ~ 1). By angle tuning the mixing crystal, the upconverted fluorescence signal, with a wavelength in the range 230–280 nm, always polarized in the same direction, was directed into a monochromator (Triax 320, Jobin Yvon Inc. with a bandwidth of 0.5 nm) and a solar blind photomultiplier tube (R2078, Hamamatsu, dark rate <1cps). Amplified SBPMT signals were discriminated and then recorded by a gated single photon counter (994, EG&G Ortec). Photon arrival events were held to less than 5 % of the repetition rate to minimize "pileup" (which might flatten transient). One should mention that there are two main sources of noise for the detection. If the signal is sufficiently greater than the PMT dark current, the power fluctuations of the infrared laser beam may be the primary noise source. If the signal is very low, the shot noise from the dark current of the PMT will be the limiting factor, and in this case, cooling the PMT is advisable.

A precise determination of the bandwidth and zero time delay is essential in all ultrafast experiments. The best way of determining the zero in upconversion system is to get a cross correlation trace between the scattered laser light from the sample and a delayed laser beam in the crystal. Such a trace provides not only an accurate zero but also an accurate measurement of the system response time. In our system, the "lamp" (AKA "apparatus" or "instrument response") function was determined by measuring the cross-correlation trace either between the UV and infrared pulses or UV-generated spontaneous Raman scattering in water and the infrared. In both ways, the lamp function was found to be around 400 fs (FWHM) shown in Fig. 2, with a timing jitter of less than 30 fs. The use of unamplified pulses and a more collinear geometry can potentially reduce this FWHM to <150 fs, with at much lower selectivity.

When collecting on longer timescales, we often found focal "detuning" to ~1 ps FWHM optimized other geometric parameters. Instrument calibration was further verified with indoles and, e.g., the linear fluorophores like *p*-terphenyl, which yielded $r_0 = 0.40 \pm 0.01$ and a single rotational correlation time φ of 41 ps in cyclohexane. A schematic of the upconversion setup is shown below.

3 Methods

After tripling our 885 nm pulses, we can obtain an average energy of 5 μJ at 295 nm, but we only use 0.2 μJ.

The upconversion process itself, while linearly dependent on both IR power and fluorescence, is quite inefficient (~10^{-4}) and therefore, lasers with higher repetition rates and IR "gate" pulse powers are certainly to be desired.

3.1 Upconversion Sample Preparation and Data Collection

1. Degas the buffer of choice beforehand. Although this step is not strictly necessary, it will improve the sample photostability upon UV exposure.
2. Dissolve the biomolecule of interest in the buffer to a concentration of about 0.5–0.8 mM depending on the quantum yield (QY) of the Trp under analysis. Concentrations as low as 150 μM have been used for single Trp containing molecules with QYs around 0.3.
3. Absorption spectra must be taken before and after upconversion measurement to check for photodamage; in particular, one looks for loss of native Trp absorbance using the band at 280 nm.
4. The solution is loaded on the sample holder (*see* Fig. 3) minimizing the amount of air bubbles. One of the six embedded cuvettes is usually filled with buffer only for stopping to selectively collect the system response function (AKA lamp function) from spontaneous Raman in situ.
5. When starting "cold," the first step to get an upconversion signal is to directly mix (in the BBO crystal) the 295 nm excitation with the 885 nm output, as these are powerful beams (~100× more than the divergent fluorescence emission from the sample) and it will be easier to use them to test the alignment of beams inside the crystal as well as benchmark the crystal angle.
6. Set the monochromator to the upconversion (quadrupling, in this case) wavelength of 221 nm, attenuate the UV tenfold, and open the monochromator slits and look for the signal by scanning delays to match pathlenghts. To make sure that the source

Fig. 3 Sample holder for fluorescence upconversion measurements. Slotted delrin disk flat cuvette holder for sample spinning. Each slot is about 1 cm wide × 4 cm long

of your signal is in fact from 295 to 221 upconversion, move the delay line a few ps, and the impulse signal should disappear.

7. Once the signal is optimized, close the entrance slit so that you still have ~80 % signal and repeat the alignment optimization one more time.
8. Put the fluorescence standard sample in place, add about 10 ps delay, spin the disk array, and rotate the BBO to again look for an upconversion signal; here it is easier to start with the sample emission maximum ~350 nm creating a signal near 251 nm. Also remember when retracing the "lamp" temporal profile with disk stopped that the zero delay line will have to be moved; now that the 295 nm beam is going through the sample, it will be delayed a few picoseconds just by the difference in transit time between air and the silica-water sandwich now present. The same will be true whether observing 295 or the spontaneous Raman scatter "lamp" at ~328 nm. This makes the latter very useful.
9. After each run, to ascertain the system has time-invariant sensitivity, place a known long-lived monoexponential standard material (e.g., melatonin in H_2O) in the disk and collect repeated time traces until ca. 1000 photon counts are present at each interval (of several tens of points), yielding overall slope accuracy better than 1 %. Over short (<40 ps) intervals even ns decays appear level. If unexpected declines are seen, it is likely due to walk of the beam in the cube-corner delay stage, and one must realign using fine pinholes to detect any misalignment during stage movement.
10. Each trace should either be collected randomly in delay time or with averaging of both increasing and decreasing time traces, to reduce the influence of any bleaching or long-term laser intensity

drift. Alternatively, one may recollect a few selected delay points to ascertain no drift/bleach reductions in intensity have occurred.

11. Since the traces provide photon counts vs. time, ordinary fitting routines for TCSPC data may be employed for fitting, or global analysis (preferred) can be applied to the entire data surface (several traces vs. wavelength).
12. The delay line requires calibration, which could be done by placing a slab of fused silica of known thickness (d) and index of refraction (n) on the path of the IR beam being careful not to disturb its positioning. The slab will create a delay (D) equal to $d(n-1)$. Find the new zero delay. The ratio between D and the number of steps needed to find the new zero will provide the delay imposed by each step
13. The number of counts per second obtained out of the sample should be much less than 10 % of the repetition rate of the system. By doing so we reduce the probability of 2 photons reaching the detector simultaneously, a phenomenon called "pileup". This produces a measured count rate (*see* Graphs 1 and 2) significantly lower than the true count rate, according to Poisson statistics. The deviation from the true count rate is more severe in a system with lower repetition rate; this is one reason we upgraded our system to 5 kHz.

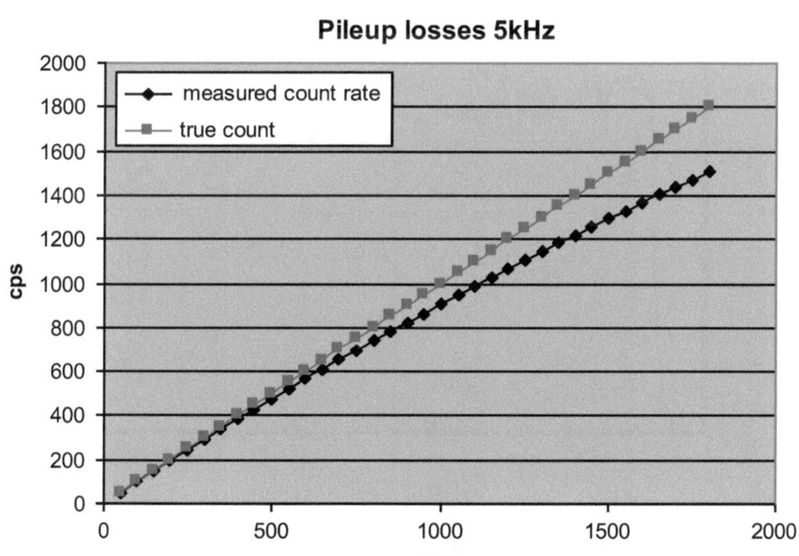

Graph 1 Poisson statistics of pileup losses for a 5 kHz system as function of the sample photon count. At low photon counts (up to 10 % the rep. rate) the probability of pileup (two photons reaching the detector simultaneously) is low. Such probability increases at higher photon count (30–40 % of the rep rate) especially for a low rep rate system

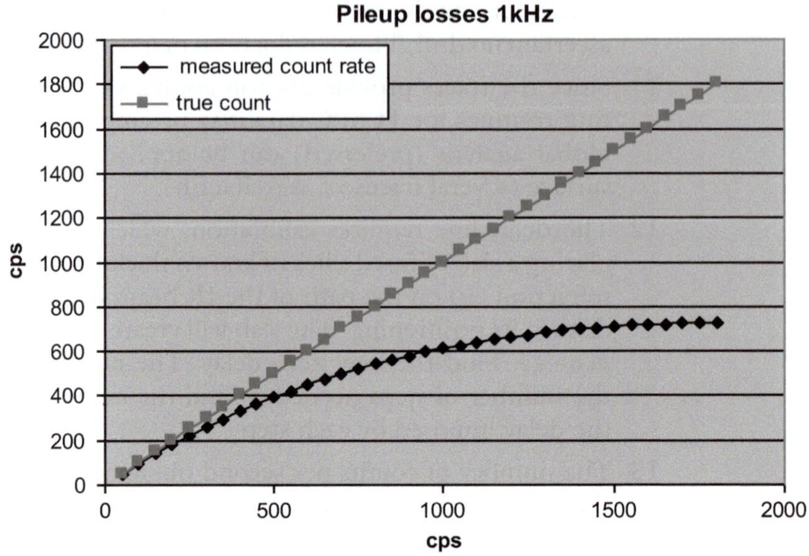

Graph 2 Poisson statistics of pileup losses for a 1 kHz system as function of the sample photon count. At low photon counts (up to 10 % the rep. rate) the probability of pileup (two photons reaching the detector simultaneously) is low. Such probability increases at higher photon count (30–40 % of the rep rate) especially for a low rep rate system

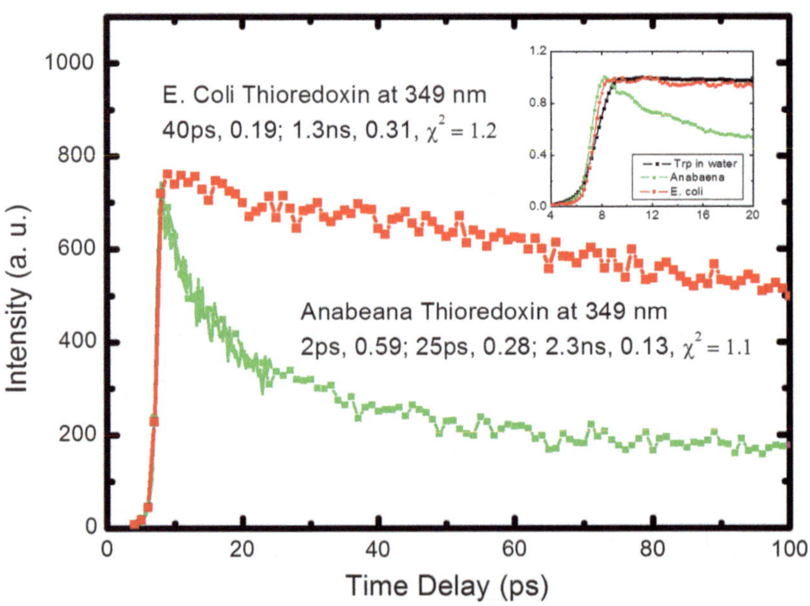

Fig. 4 Example of a protein decay transient near the emission peak. Thioredoxin proteins in the oxidized state are known to have very low quantum yield, and the mechanism of such quenching has been ascribed to both disulfide bond quenching and to proximate carboxyl groups. Here two different species' thioredoxins are compared at $\lambda_{em} = 349$ nm

An example of a typical upconversion transient is shown in Fig. 4 after averaging five datasets; the number of laser flashes used exceeds the peak count by over 20×.

4 Notes

4.1 Tryptophan Photodamage

As reviewed above Trp is a widely studied fluorescence probe, scientifically relevant for being the brightest fluorophore among amino acids, and largely used as a natural probe to report on a growing number of biological systems [47–50].

Photooxidation is the most common damage mechanism both in Trp and Tyr when irradiated with UV photons in an aqueous environment [51]. The oxidation mechanism is initiated by the formation of reactive oxygen species (ROS) along with hydroxyl radicals and singlet oxygen which ultimately will modify these residues into a variety of photoproducts [52]. There is an ongoing effort in understanding the photodegradation mechanism of Trp in aqueous solution, partly driven by the role played by photodegradation in the progression of a variety of diseases like yellowing of the human lens [53].

Fluorescence upconversion is a powerful technique that allows reporting on changes in the Trp microenvironment in the picosecond timescale, and as described in this chapter, the setup involves using short laser pulses (few hundred femtoseconds) in the UV range (280–295 nm) in order to excite Trp to its first excited state (1La). In order to generate a detectable upconversion signal, less than 1 mW energy is sent into the sample. The energy is packed, however, in a few very short pulses, resulting in relatively high peak power being applied to the sample. In comparison to TCSPC measurements where the average energy is about 1 order of magnitude lower, for an upconversion type of measurement, the peak power applied to the sample is on average 4 orders of magnitude higher. These figures are a reminder that upconversion data are also more at risk for photodamage than other "linear" spectroscopies; as a result, more precautions need to be taken in preparing the sample like degassing the buffer, using oxygen scavengers, and avoiding reexcitation of the same volume.

We have found that some samples (particularly flexible peptides of 4–40 residues) are quite sensitive to the UV damage process during upconversion.

The generation of photoproducts in Trp-containing peptides was tested by taking multiple upconversion datasets over a 2 h period, after repeatedly exposing the sample to 295 nm excitation light. As shown in Fig. 5, remaining Trp lifetime at 375 nm is clearly shortened upon sample exposure to large doses of pulsed UV light. Electrospray ionization tandem mass spectroscopy (ESI-MS/MS) confirmed that oxidation took place almost exclusively at the Trp residue level. The take-home message is that photodamage can compromise fluorescence upconversion data resulting in an alteration of detected Trp fluorescence dynamics at the picosecond level, especially at the low energy side of the Trp emission spectrum.

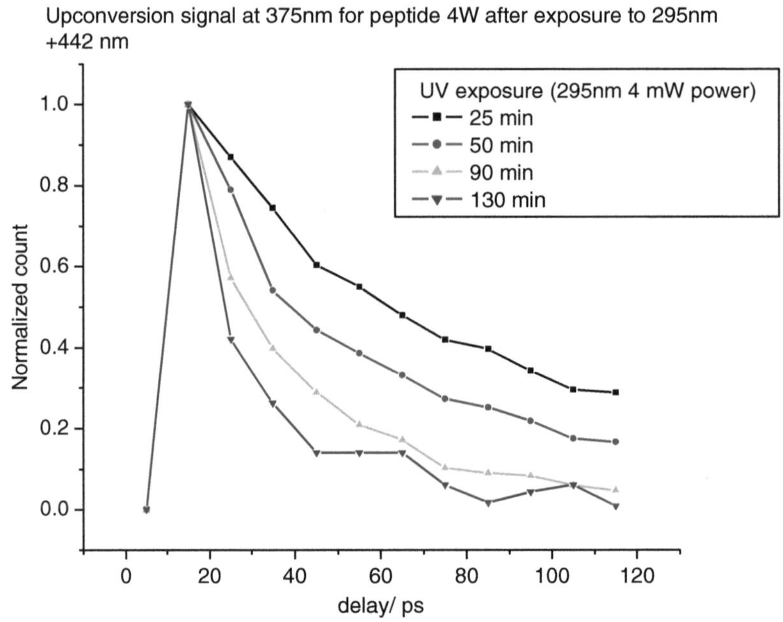

Fig. 5 Low resolution upconversion traces for a single tryptophan amyloid-related peptide (Ac-MDVWM) upon exposure to UV pulses. The apparent decay rate at 375 nm is drastically reduced over time, owing to photodamage

A number of studies have been conducted on the mechanism and conditions that favor Trp photoproducts from UV light absorption, but the photophysics are not fully understood yet. Many of the proposed mechanisms point at photoproducts being mainly generated in the S_1 excited state over a nanosecond timescale. Some of these mechanisms invoke a proton transfer; others suggest that an electron transfer is involved.

Two of the best known products of Trp photodegradation are Kynurenine and N-Formylkynurenine [54, 55]. Their absorption and emission spectra show a clear redshift compared to Trp; they also have intrinsically low fluorescence quantum yield, with N-Formylkynurenine being a slightly better emitter between the two [56]. At the excitation wavelength of Trp, usually in the range of 285 to 295 nm, a smaller amount of the photoproducts get excited. Whether they are directly observed in the milieu below/at 442 nm on the red edge, or whether they act as general fluorescence quenchers, a sub-nanosecond Trp lifetime definitely gets reduced (as shown in Fig. 5).

Recent investigations on Trp damage via UV irradiation uncovered that a large variety of photoproducts coexist with multiple oxidations taking place at the Trp residue (up to 4), providing a cascade of modifications [57]. Photodamage is an important artifact to carfully monitor in any laser study.

References

1. Franken PA, Weinreich G, Peters CW et al (1961) Generation of optical harmonics. Phys Rev Lett 7:118–119
2. Hallidy LA, Topp MR (1977) Picosecond luminescence detection using type-2 phase-matched frequency-conversion. Chem Phys Lett 46:8–14
3. Kahlow MA, Jarzeba W, Dubruil TP et al (1988) Ultrafast emission-spectroscopy in the ultraviolet by time-gated upconversion. Rev Sci Instrum 59:1098–1109
4. Bradforth SE, Jimenez R, Vanmourik F et al (1995) Excitation transfer in the core light-harvesting complex (LH-1) of rhodobacter-sphaeroides-an ultrafast fluorescence depolarization and annihilation study. J Phys Chem 99:16179–16191
5. Kennis JTM, Gobets B, van Stokkum IHM et al (2001) Light harvesting by chlorophylls and carotenoids in the photosystem I core complex of Synechococcus elongatus: a fluorescence upconversion study. J Phys Chem B 105:4485–4494
6. Macpherson AN, Gillbro T (1998) Solvent dependence of the ultrafast S-2-S-1 internal conversion rate of beta-carotene. J Phys Chem A 102:5049–5058
7. Longworth JW (1983) Time-resolved fluorescence spectroscopy in biochemistry and biology. Plenum Press, New York
8. Beechem JM, Brand L (1985) Time-resolved fluorescence of proteins. Annu Rev Biochem 54:43–71
9. Christiaens B, Symoens S, Vanderheyden S et al (2002) Tryptophan fluorescence study of the interaction of penetratin peptides with model membranes. Eur J Biochem 269:2918–2926
10. Schauerte JA, Schlyer BD, Steel DG et al (1995) Nanosecond time-resolved circular-polarization of fluorescence-study of NADH bound to horse liver alcohol-dehydrogenase. Proc Natl Acad Sci USA 92:569–573
11. Gryczynski Z, Lubkowski J, Bucci E (1995) Heme-protein interactions in horse heart myoglobin at neutral pH and exposed to acid investigated by time-resolved fluorescence in the picosecond to nanosecond time range. J Biol Chem 270:19232–19237
12. Eftink MR, Ghiron CA (1981) Fluorescence quenching studies with proteins. Anal Biochem 114:199–227
13. Lakowicz JR, Keating-Nakamoto S (1984) Red-edge excitation of fluorescence and dynamic properties of protein and membranes. Biochemistry 23:3013–3021
14. Alcala JR, Gratton E, Prendergast FG (1987) Interpretation of fluorescence decays in proteins using continuous lifetime distribuition. Biophys J 51:925–936
15. Szabo AG, Rayner DM (1980) Fluorescence decay of tryptophan conformers in aqueous solution. J Am Chem Soc 102:554–563
16. Knutson JR, Beechem JM, Brand L (1983) Simultaneous analysis of multiple fluorescence decay curves-a global approach. Chem Phys Lett 102:501–507
17. Dahms TES, Willis KJ, Szabo AG (1995) Conformational heterogeneity of tryptophan in a protein crystal. J Am Chem Soc 117:2321–2326
18. Adams PD, Chen Y, Ma K et al (2002) Intramolecular quenching of tryptophan fluorescence by the peptide bond in cyclic hexapeptides. J Am Chem Soc 124:9278–9286
19. Davenport L, Knutson JR, Brand L (1986) Excited-state proton-transfer of equilenin and dihydroequilenin-interaction with bilayer vesicles. Biochemistry 25:1186–1195
20. Toptygin D, Brand L (2000) Spectrally- and time-resolved fluorescence emission of indole during solvent relaxation: a quantitative model. Chem Phys Lett 322:496–502
21. Pan CP, Muino PL, Barkley MD et al (2011) Correlation of tryptophan fluorescence spectral shifts and lifetimes arising directly from heterogeneous environment. J Phys Chem B 115:3245–3253
22. Jimenez R, Fleming GR, Kumar PV et al (1994) Femtosecond solvation dynamics of water. Nature 369:471–473
23. Maroncelli M, Fleming GR (1987) Picosecond solvation dynamics of Coumarin-153-the importance of molecular aspects of solvation. J Chem Phys 86:6221–6239
24. Janes SM, Holtom G, Ascenzi P et al (1987) Fluorescence and energy-transfer of tryptophans in aplysia mioglobin. Biophys J 51:653–660
25. Gryczynski Z, Lubkowski J, Bucci E (1997) Intrinsic fluorescence of hemoglobins and myoglobins. In: Fluorescence spectroscopy, vol 278. (L. Brand ed) Elsevier Academic Press Inc, San Diego, pp 538–569
26. Ruggiero AJ, Todd DC, Fleming GR (1990) Subpicosecond fluorescence anisotropy studies of tryptophan in water. J Am Chem Soc 112:1003–1014
27. Shen XH, Knutson JR (2001) Subpicosecond fluorescence spectra of tryptophan in water. J Phys Chem B 105:6260–6265

28. Peon J, Pal SK, Zewail AH (2002) Hydration at the surface of the protein Monellin: dynamics with femtosecond resolution. Proc Natl Acad Sci USA 99:10964–10969
29. Pal SK, Peon J, Zewail AH (2002) Biological water at the protein surface: dynamical solvation probed directly with femtosecond resolution. Proc Natl Acad Sci USA 99:1763–1768
30. Zhong DP, Pal SK, Zhang DQ et al (2002) Femtosecond dynamics of rubredoxin: tryptophan solvation and resonance energy transfer in the protein. Proc Natl Acad Sci USA 99:13–18
31. Zhang LY, Wang LJ, Kao YT et al (2007) Mapping hydration dynamics around a protein surface. Proc Natl Acad Sci USA 104: 18461–18466
32. Qiu WH, Zhang LY, Okobiah O et al (2006) Ultrafast solvation dynamics of human serum albumin: correlations with conformational transitions and site-selected recognition. J Phys Chem B 110:10540–10549
33. Zhang LY, Kao YT, Qiu WH et al (2006) Femtosecond studies of tryptophan fluorescence dynamics in proteins: local solvation and electronic quenching. J Phys Chem B 110:18097–18103
34. Chen RF, Knutson JR, Ziffer H et al (1991) Fluorescence of tryptophan dipeptides—correlation with the rotamer model. Biochemistry 30:5184–5195
35. Chen Y, Liu B, Yu HT et al (1996) The peptide bond quenches indole fluorescence. J Am Chem Soc 118:9271–9278
36. Xu JH, Knutson JR (2009) Quasi-static self-quenching of Trp-X and X-Trp dipeptides in water: ultrafast fluorescence decay. J Phys Chem B 113:12084–12089
37. Xu JH, Chen JJ, Toptygin D, Tcherkasskaya O et al (2009) Femtosecond fluorescence spectra of tryptophan in human gamma-crystallin mutants: site-dependent ultrafast quenching. J Am Chem Soc 131:16751–16757
38. Halder M, Mukherjee P, Bose S et al (2007) Solvation dynamics in protein environments: comparison of fluorescence upconversion measurements of coumarin 153 in monomeric hemeproteins with molecular dynamics simulations. J Chem Phys 127:055101
39. Wang W, Donini O, Reyes CM, Kollman PA (2001) Biomolecular simulations: recent developments in force fields, simulations of enzyme catalysis, protein–ligand, protein–protein, and protein–nucleic acid noncovalent interactions. Annu Rev Biophys Biomol Struct 30:211–243
40. Halle B, Andersson T, Forsen S et al (1981) Protein hydration from water O-17 magnetic-relaxation. J Am Chem Soc 103:500–508
41. Halle B, Wennerstrom H (1981) Interpretation of magnetic-resonance data from water nuclei in heterogeneous systems. J Chem Phys 75:1928–1943
42. Roccatano D, Amadei A, Di Nola A et al (1999) A molecular dynamics study of the 41–56 beta-hairpin from B1 domain of protein G. Protein Sci 8:2130–2143
43. Callis PR, Liu TQ (2004) Quantitative prediction of fluorescence quantum yields for tryptophan in proteins. J Phys Chem B 108:4248–4259
44. Callis PR, Vivian JT (2003) Understanding the variable fluorescence quantum yield of tryptophan in proteins using QM-MM simulations. Quenching by charge transfer to the peptide backbone. Chem Phys Lett 369:409–414
45. Biesso A, Xu J, Muino PL, Callis P et al Tryptophan dynamics in GB1 sense very different phenomena on the picosecond and nanosecond timescales (submitted)
46. Xu J, Chen B,Callis P, Rozeboom H et al (2012) Femtosecond fluorescence dynamics of tryptophan and 5-fluorotryptophan in monellin: slow water relaxation unmasked Biophys J 104(2):681a
47. Vivian JT, Callis PR (2001) Mechanisms of tryptophan fluorescence shifts in proteins. Biophys J 80:2093–2109
48. Elove GA, Chaffotte AF, Roder H et al (1992) Early steps in cytochrome-c-folding probed by time-resolved circular-dichroism and fluorescence spectroscopy. Biochemistry 31:6876–6883
49. Pal SK, Peon J, Bagchi B et al (2002) Biological water: femtosecond dynamics of macromolecular hydration. J Phys Chem B 106:12376–12395
50. Chen Y, Barkley MD (1998) Toward understanding tryptophan fluorescence in proteins. Biochemistry 37:9976–9982
51. Berlett BS, Stadtman ER (1997) Protein oxidation in aging, disease, and oxidative stress. J Biol Chem 272:20313–20316
52. Davidson RS (1996) The photodegradation of some naturally occurring polymers. J Photochem Photobiol B 33:3–25
53. Paik DC, Dillon J (2000) The nitrite/alpha crystallin reaction: a possible mechanism in lens matrix damage. Exp Eye Res 70:73–80
54. Grossweiner LI (1984) Photochemistry of proteins-a review. Curr Eye Res 3:137–144

55. Pirie A (1971) Formation of N'-Formylkynurenine in proteins from lens and other sources by exposure to sunlight. Biochem J 125:203–208
56. Fukunaga Y, Katsuragi Y, Izumi T et al (1982) Fluorescence characteristics of Kynurenine and N'-Formylkynurenine-their use as reporters of the environment of tryptophan-62 in hen egg-white lysozyme. J Biochem 92:129–141
57. Grosvenor AJ, Morton JD, Dyer JM (2010) Profiling of residue-level photo-oxidative damage in peptides. Amino Acids 39:285–296

Chapter 13

Subpicosecond Kerr-Gate Spectrofluorometry

Sergey P. Laptenok, Patrick Nuernberger, Andras Lukacs, and Marten H. Vos

Abstract

This chapter describes an experimental layout for time and spectrally resolved fluorescence measurements with femtosecond time resolution based on Kerr gating. The combination of data recorded using different Kerr media allows a temporal dynamic range from ~100 fs to several nanoseconds. Simultaneous analysis of multiple datasets is described.

Key words Kerr shutter, Time-resolved fluorescence, Ultrafast spectroscopy, NK88, Decay-associated spectrum, Global analysis

1 Introduction

Time-resolved fluorescence spectroscopy techniques monitoring the intrinsic decay and rise of emission from fluorophores in the condensed phase provide powerful tools that allow studying the structure and dynamics of molecular systems. These methods use the sensitivity of fluorescence properties to the environment of the fluorophore(s) to obtain important information about the investigated system. The diversity of the physical parameters that can be studied by time-resolved fluorescence is remarkable, and include inter- and intramolecular distances, polarity, viscosity, structural ordering, molecular mobility, and electric potential [1]. In biophysical studies, using natural or nonnatural fluorophores, time-resolved fluorescence can be used to study a wealth of phenomena including photo-induced redox reactions, Förster resonance energy transfer (FRET) between fluorophores, and rotational dynamics of macromolecules. For many applications measurements at single wavelengths are used, employing time-correlated single-photon counting (tens of picosecond resolution) or classical fluorescence

Yves Engelborghs and Antonie J.W.G. Visser (eds.), *Fluorescence Spectroscopy and Microscopy: Methods and Protocols*, Methods in Molecular Biology, vol. 1076, DOI 10.1007/978-1-62703-649-8_13,

upconversion (femtosecond resolution) techniques. However, the determination of the complete time- and spectrally resolved fluorescence matrix is often desired, especially for multi-fluorophore systems, fluorophore distributions, and at short timescale where relaxation processes play a significant role. There are several types of experimental methods that provide combined time and spectral resolution: streak camera [2–4], broadband upconversion [5–7], and Kerr gating [8–10], all of them having advantages and disadvantages. The streak camera has high sensitivity but limited time resolution. Broadband upconversion setups have very good time resolution but are quite demanding in alignment and operation. Kerr-gate setups have intrinsic high time resolution but require special attention in the case of large long-lived background components. In the present work, we focus on the latter technique based on the optical Kerr shutter and give a detailed description of the measurement procedure and data analysis.

A Kerr shutter for time-resolved fluorescence detection was first used by Duguay and Hansen in 1969 [11, 12] with picosecond laser technology. The optical Kerr effect is due to light-induced anisotropy changes in the medium. In Kerr-based gating setups, the fluorescence light pathway contains crossed polarizers and therefore light does not reach the detector unless it is "gated." To do so, the Kerr medium is placed between the two polarizers. A strong, linearly polarized gate pulse induces changes in the refractive index of the medium so that the medium effectively becomes birefringent. The polarization of the (fluorescence) light, which passes the Kerr medium while this induced birefringence persists, will be changed from linear to elliptical. Thus the shutter is "opened" and part of the fluorescence light can pass through the second polarizer and reach the detector.

2 Materials

The Kerr media CS_2 and benzene were purchased from Sigma-Aldrich. Suprasil (1 mm) was purchased from Hellma. The dye NK88 was purchased from Exciton, Inc.

3 Methods

3.1 Principles of the Setup

A schematic presentation of the setup is shown in Fig. 1.

Excitation light pulses are generated with a femtosecond amplifier system. A mode-locked home-built Ti:sapphire laser is used to generate short light pulses at a central wavelength of 800 nm with a duration of about 50 fs and a pulse energy of ~2 nJ (200 mW) at a repetition rate of 100 MHz. The pulses are then passed through a regenerative amplification system (Spitfire, Spectra-Physics),

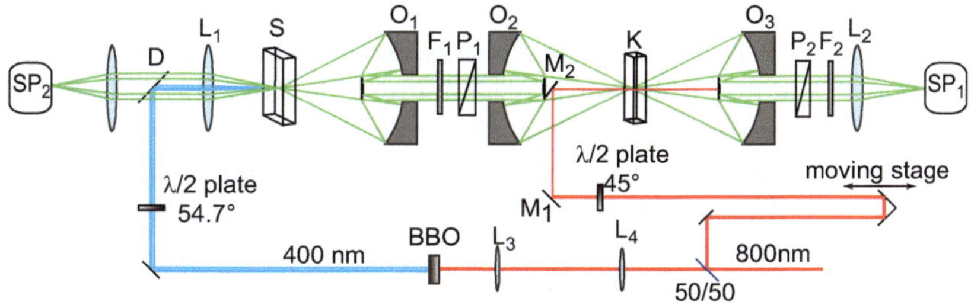

Fig. 1 Schematic representation of the setup. *S* sample; *K* Kerr cell; P_1, P_2 crossed polarizers; O_1, O_2, O_3 reflective objectives; SP_1 spectrometer connected to cooled CCD for gated fluorescence detection; SP_2 reference spectrometer; *D* dichroic mirror; F_1 filter to block pump beam; F_2 filter to block gate pulse; M_1 mirror to direct gate pulse; M_2 pickup mirror

increasing the pulse energy to ~400 μJ (400 mW) at a repetition rate of 1 kHz, with a pulse duration of about 100 fs. The output of the amplifier is split into two. One fraction is directed into the optical delay line and then used as a gate pulse. The second fraction is used for the generation of the pump pulse, by frequency doubling in a ß-barium-borate (BBO) crystal yielding pulses centered at 400 nm, or alternatively by frequency conversion in a noncollinear optical parametric amplifier (NOPA) which yields pulses with adjustable central wavelengths in the visible spectral range [13]. The intensity of the pump and gate beams is adjusted with neutral density filters. The polarization of the pump pulses is set to "magic angle" (54.7°) with respect to the polarization direction of the first polarizer (P_1) to avoid polarization effects from anisotropy in the excited sample (*see* **Note 1**). The pump pulse is directed into the sample via a dichroic mirror and focused to ~150 μm with a lens with a focal length of 5 cm (Fig. 1, L_1). This lens is also used to collect fluorescence emitted in the backward direction and that is transmitted through the dichroic mirror (Fig. 1, D, reflecting at the excitation wavelength and transmitting in the emission spectral range, see below) and focused into a spectrometer (Fig. 1, SP_2) (Ocean Optics USB2000 + VIS-NIR) in order to monitor fluctuations in total fluorescence (*see* **Note 2**). The sample is contained in a moving, 1 mm internal thickness, quartz cell (Fig. 1, S) (*see* **Note 3**).

To avoid strong dispersion that cannot be compensated and would deteriorate the time resolution, reflective optics are employed to collect the emitted light. Fluorescence is collimated in the forward direction by a Cassegrain-type reflective objective (RO) (Newport 50105-02, effective focal length 13 mm, working distance 24 mm) (Fig. 1, O_1) (*see* **Note 4**). A vertically oriented thin polarizer (Moxtek PFU04C, thickness 0.7 mm) is placed (Fig. 1, P_1) in the collimated beam. The linearly polarized fluorescence is then refocused by a second RO (Fig. 1, O_2) and overlapped spatially within the gate beam inside the Kerr medium (Fig. 1, K) (*see* **Note 5**). After the Kerr medium, a third RO (Fig. 1, O_3) is used to recollimate the fluorescence beam. A second, horizontally oriented polarizer is placed after this RO (Fig. 1, P_2) so that fluorescence undergoing the gate-induced polarization rotation may partially pass. This "gated" fluorescence is directed into the slit of the spectrograph (Acton, SP2150) and detected by a nitrogen-cooled CCD (Fig. 1, SP_1) (Princeton Instruments, SPEC-10) configured as an array-detector.

3.2 Calibration of the Setup

Temporal dispersion: The remaining unavoidable transmissive optics between the sample and the Kerr medium (i.e., cells, filter, and polarizer) will give rise to temporal spreading of different spectral components due to dispersion. This dispersion can be taken into account during data processing and analysis by using a dispersion curve. This curve can be estimated by gating white light (super continuum) generated on the position of the sample and assuming that the signal resembles the instrument response function (IRF) of the system at each wavelength.

Color sensitivity of the detector: The wavelength-dependent sensitivity of the system determined by the transmission/reflection curves of the optical components and the CCD sensitivity can be taken into account using an incoherent broadband ("white") light source with known spectrum (calibrated lamp). The spectrum of the lamp placed at the sample position is measured with high S/N, and intrinsic fluorescence spectra are calculated by weighting the recorded spectra by the ratio of "true" lamp spectrum divided by measured lamp spectrum. The wavelength dependence of the gate efficiency is taken into account by comparing the time-integrated gated fluorescence spectrum (or, for long-lived dyes, long-time gated fluorescence) with the total fluorescence spectrum measured with a calibrated fluorimeter; several dyes can be used to cover a large spectral range [10].

Determination of the temporal IRF: The most convenient way to determine the temporal IRF, i.e., the convolution of the pump pulse and the gate response, is to measure the gated scatter from the pump pulse (Fig. 2). The Kerr effect has electronic and nuclear components. For Suprasil only the electronic component is significant and the gate response is quasi-instantaneous, following the gate pulse. The drawback for the ultrafast response is the low efficiency (defined as the fraction of the fluorescence at the maximum of the gate passing through the crossed polarizers) of the gate, less than 1 %. In liquids (benzene and CS_2) the Kerr effect is predominantly due to the nuclear response that has inertial components. Therefore the total response slows down and becomes asymmetric (Fig. 2). However, the gate efficiency in such cases is higher (on the order of few percent). As the Kerr response in liquids may depend on the power of the gate pulse [14], the IRF should be

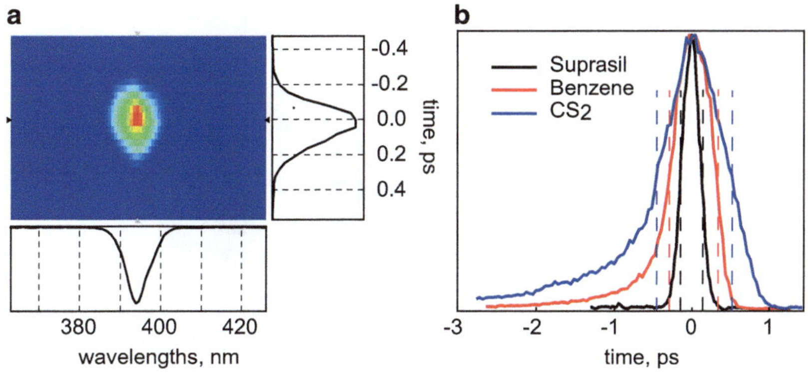

Fig. 2 (**a**) Spectro-temporal response of scattered pump pulse with Suprasil as Kerr medium. Cross section in time and spectral domains. (**b**) Gate profiles obtained with different Kerr media; *dashed lines* display FWHM: Suprasil ~200 fs, benzene ~ 550 fs, CS_2 ~ 1 ps. Note the asymmetric tail observed for negative delay times for benzene and CS_2 due to the dynamics induced by the gate pulse in the Kerr medium

determined under gate conditions of the experiment. In practice, the power of the gate beam is limited by continuum generation in the Kerr medium, to ~10, ~4, and ~2.5 mW for Suprasil, benzene and CS_2, respectively, under our experimental conditions. By using different Kerr media it is possible to vary width and efficiency of the gate. Combination of these experiments (together with simultaneous analysis) then optimizes the dynamic range and data accumulation time.

3.3 Alignment of the Setup

The aim of the following alignment procedure is to optimize the amount of gated fluorescence delivered to the spectrometer while minimizing possible aberrations leading to polarization distortions between the two polarizers and hence enhance background fluorescence and noise. In our setup an extinction ratio of 1,500 for the non-gated fluorescence is routinely achieved.

In a first assembly the axis of the setup is defined by directing the pump beam directly to the slit of the spectrometer. Subsequently, step by step all elements starting from lens L_1 (*see* Fig. 1 for a sketch of the optical alignment) are added in such a way that the fluorescence always propagates along the axis defined in the first step. A complication in the alignment arises from the layout of the Cassegrain objectives, in which the central part of the aperture on the focal side is obscured by the convex interior mirror (*see* Fig. 3a). However, in our setup this feature is also exploited by using it as a beam stop for the excitation (O_1) and gate (O_3) pulses which are much more intense than the fluorescence (*see* **Note 6**). Also the blocked central part of O_2 is exploited as a support for the mount of the pickup mirror M_2 that directs the gate beam into the Kerr medium (O_2) along the axis of the setup (Fig. 3b) [9] (*see* **Note 5**).

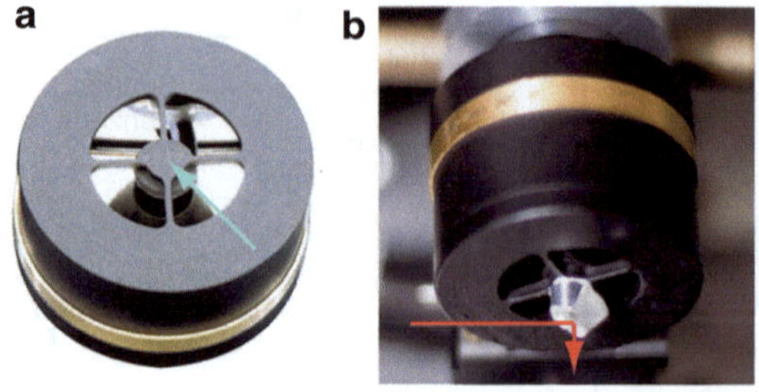

Fig. 3 (**a**) Cassegrain-type reflective objective from Newport. (**b**) Pickup mirror mounted on the output of O_2 for directing of the gate pulse. The *red arrow* represents the path direction of the gate beam

Because of the blocked central part of the reflective objectives, the alignment is done using scattered pump light, fluorescence from a sample, or white light generated at the position of the sample. For the initial alignment, it is convenient to use a photostable high-yield dye with a long fluorescence lifetime (coumarin, Alexa, rhodamine etc., depending on the central wavelength of the pump pulse), because it can be placed in the same cell as the sample. The design features of the RO can be used as a reference in aligning O_1 and O_3 (Fig. 1) (*see* **Note 7**). In the course of optimizing the path of the fluorescence from the sample through the three ROs, the polarizers (initially set parallel ||), and the lens to the spectrometer, the Kerr medium should be placed as well.

Subsequently the gate pulse is introduced into the setup. The spatial overlap of the focused fluorescence with the focused gate pulse can initially be obtained by placing a pinhole that is somewhat smaller than the fluorescence focal spot at the position of the Kerr cell such that fluorescence transmission is optimized, and steering the gate pulse through. The delay line defines the temporal delay between excitation and gating; therefore the maximum time window of the experiment is defined by the optical length of the delay line. Care must be taken to ensure optimal spatial overlap along the entire delay line (*see* **Note 8**).

Subsequently, the temporal overlap of the fluorescence and gate is determined. For ease of detecting gated fluorescence, changing to a short-lived fluorescent dye (with an associated low fluorescence yield) is preferred. In this case the ratio of the gated signal and the background fluorescence (that leaks through the crossed polarizers) is the highest (*see* **Note 9**). Among the molecular species with favorable properties are the short-chain cyanine dyes (as, e.g., listed in [15]) such as NK88 (see below) for which internal conversion through a conical intersection occurs on the picosecond timescale. To find the signal, the gate pulse must coincide with the fluorescence; for dyes with a lifetime in the order of a few picoseconds this corresponds to a delay line range on the order of 1 mm. The optical delay line is therefore slowly scanned. When the gate pulse and the fluorescence temporally overlap, the detector will measure a signal of gated fluorescence light and the delay then can be adjusted for optimal signal (*see* **Note 10**).

3.4 Performing Experiment

The delay between the pump pulse and the gate pulse defines the relative moment in time for which a spectral snapshot is taken (Kerr gate open); therefore the set of delay line positions defines the time axis of the measurement. Often, it is practical to use a multiscale or a nonlinear time axis so that the step size is adapted to the timescale of the changes. The step size in the time around $t = 0$ (abbreviated "t_0") should be constant, should be adapted to the IRF, and in particular should extend long enough to negative times to fully encompass the IRF (*see* Fig. 2). As explained above, the highest dynamic range may be obtained by using different active media, thus varying the width and efficiency of the Kerr gate and combining the data from these experiments (*see* **Note 11**).

As the gated fluorescence is often weak with respect to the background, signal averaging is generally required. There are two possible strategies to do that: (a) average over a long time for each delay position and performing only one scan and (b) average over a shorter time for each delay position and do several scans, with one extra averaging step during data analysis. The first possibility is most time economic, but when working with fragile biological samples, the second strategy is more favorable, because it provides a possibility to inspect the data for slow changes during measurements, which might be due to degradation, precipitation, or generation of photoproducts with different properties.

3.5 Data Processing

The fluorescence data are accumulated as matrices $F(\lambda,t)$. Global analysis procedures, often in terms of exponential transfer functions, take into account such full datasets and in many cases combine multiple datasets. The aim of the analysis is to obtain a model-based description of the time-resolved data in terms of a number of rate constants and fluorescence spectra. The basis for global analysis is the superposition principle, which assumes that the measured data result from a linear combination of n_{comp} components, each with a distinct time profile and spectrum. Mathematically this principle reads

$$F(\lambda, t) = \sum_{i=1}^{n_{\text{comp}}} c_i(t) a_i(\lambda) \quad (1)$$

where $c_i(t)$ is the time profile of the component i and $a_i(\lambda)$ spectrum of the component i.

In the simplest case the data is fitted to a parallel model, in which n_{comp} excited species decay mono-exponentially in parallel. In this procedure, spectral shapes are left unconstrained yielding decay-associated spectra (DAS). In this case Eq. 1 becomes

$$F(\lambda, t) = \sum_{i=1}^{n_{\text{comp}}} c_i^{DAS}(t)\text{DAS}_i(\lambda) \quad (2)$$

where $c_i^{\text{DAS}}(t)$ is the exponentially decaying concentration of component i convolved with the IRF. More complex models yield linear combinations of these DAS as species-associated spectra (SAS). For more information about global analysis and how to implement alternative models for the dynamics *see* [16].

Several such methods have been described and implemented in freely available software [17, 18]. In the following example we apply global analysis by using the open-source and freely available software Glotaran (http://glotaran.org).

3.6 Exemplary Dataset

As an example, the fluorescence of the dye NK88 in ethanol (EtOH) upon 400 nm excitation is presented. The sample had a thickness of 1 mm and an optical density of 0.3 at 400 nm. To illustrate the possibility of combining multiple time resolutions, two Kerr media were used: a 1 mm Suprasil window and benzene in a 1 mm quartz cell. The gate pulse had a central wavelength of 800 nm.

As mentioned in Subheading 3.3, this short excited-state lifetime dye can be very helpful to find the gated signal. NK88 (3,3′-diethyl-2,2′-thiacyanine iodide, also called DTI or THIA) is a dye with many remarkable properties. Its fluorescence lifetime in methanol was predicted in 1974 to be as short as 1.7 ps [19], and it could be confirmed in ultrafast experiments that the fastest component decays with a time constant on the order of 2 ps [20, 21]. The fluorescence yield is strongly enhanced when the dye is bound for instance to DNA [22], and because of its affinity for bacterial cells it can be used as an agent in fluorescence staining experiments [23].

The time-resolved fluorescence spectra of NK88 corrected for the background and the spectral response of the instrument are shown in Fig. 4 (*see* **Note 12**). Along the x-axis, the wavelength of the fluorescence is shown, whereas the relative time delay between pump and gate pulse is displayed on the y-axis. The color represents the amplitude of the signal. It is clear from the images that the maximum of the fluorescence spectrum evolves on the sub-picosecond timescale. The fluorescence intensity then decays on the picosecond timescale. Figure 5 shows measured spectra at different times (a) and detected decay traces at different wavelengths (b). Initially, the fluorescence maximum is around 460 nm, but then shifts to 470 nm within a few hundred femtoseconds and then further to 480 nm on the picosecond scale. This behavior is typical of a Stokes shift and can be attributed to solvation dynamics and the torsional motion of the molecule in the excited state; the molecule will actually perform an internal conversion to the ground state via a conical intersection at a twisted intermediate configuration; it then can either relax back to the

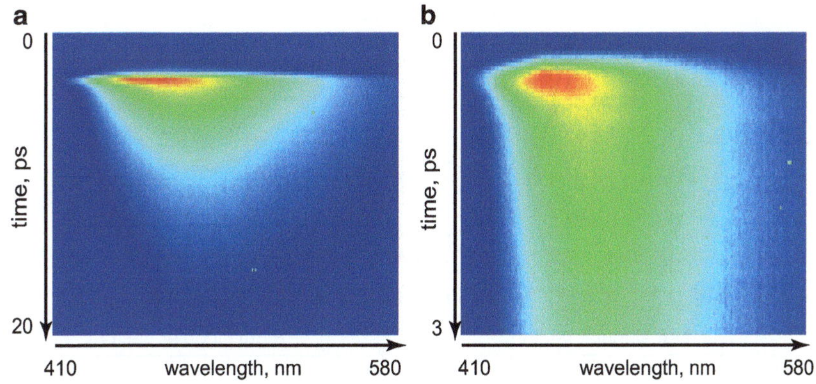

Fig. 4 Time-resolved fluorescence of NK88 in EtOH upon 400 nm excitation. (**a**) Measured with benzene as Kerr medium (time window 20 ps, time step 0.2 ps); (**b**) measured with Suprasil as a Kerr medium (time window 3 ps, time step 0.04 ps); The color coding of the signal ranges from *blue* (low signal) to *red* (high signal)

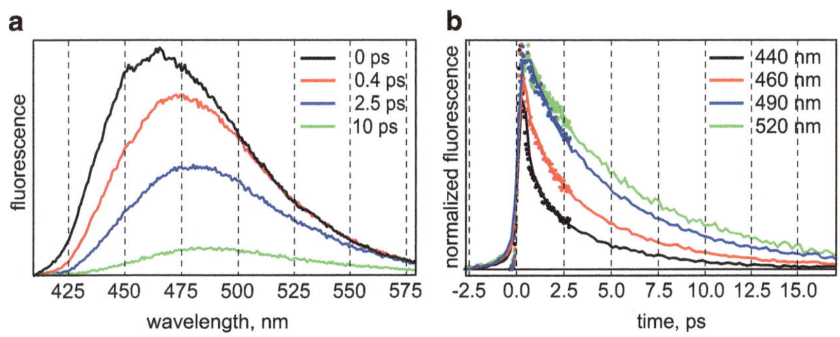

Fig. 5 (**a**) Measured spectra at different times, using benzene as Kerr medium. (**b**) Measured time traces at different wavelengths measured with two different Kerr media: *solid lines*—benzene, dots—Suprasil. Note that the different temporal widths of the Kerr gate are evident from the rise in (**b**)

initial *trans* configuration or complete the ultrafast photoisomerization to end up in the *cis* ground-state configuration [20, 21].

The most basic form of analysis available in Glotaran is singular value decomposition (SVD) of the data matrix (*see* **Note 13**). SVD analysis allows one to make a rough estimate of the number of different processes required to describe the data in a global analysis. Results of the SVD analysis of the dataset measured with benzene as Kerr medium are displayed in Fig. 6.

At least three linearly independent vectors significantly differ from the noise, as observed from the presence of significant structure in both singular vectors. The fourth vectors (green lines in Fig. 6) have almost no structure in the left singular vector (LSV) (time domain), only around t_0, and some structure around 450 nm in the right singular vector (RSV) (spectral domain). Most likely it contains a Raman scattering contribution that is expected at

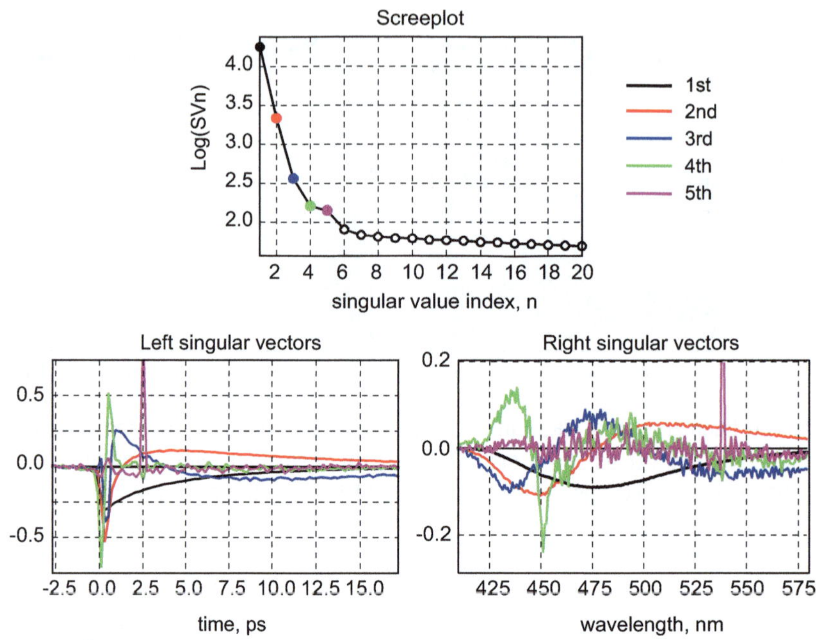

Fig. 6 The results of the SVD decomposition of the dataset measured with benzene as Kerr medium. The upper "Screeplot" displays singular values on a logarithmic scale. The two lower panels display the first five *left* (temporal) and *right* (spectral) singular vectors, respectively. The analysis indicates the presence of at least three significant components (see text)

~454 nm upon 400 nm excitation [24] and a dispersion-related contribution. The fifth singular vectors (pink lines in Fig. 6) are clearly lacking structure; the spike at 2.5 ps/540 nm presumably is due to a remaining cosmic ray detected by the CCD camera. Therefore a global analysis, with three rate constants, that takes into account dispersion and also allows for an instantaneous Raman scatter component (*see* **Note 14**), is initially performed.

A global analysis of both datasets using a three-exponential model (*see* **Note 15**) results in the time constants of 0.2, 2.3, and 6.3 ps. The corresponding DAS (*see* **Note 16**) are shown in Fig. 7. Inspection of the residuals highlights that most features are essentially described by the fit. Indeed, adding a fourth component somewhat improves the fit (*see* **Note 17**), but yields time constants in the same range. The DAS associated with the fastest component (Fig. 7, black) exhibits a positive signal (corresponding to decay) peaking at 450 nm and a negative signal (corresponding to a rise) around 525 nm and reflects a red shift. It can therefore be assigned to the Stokes shift dynamics. The DAS is not conservative (the positive area is larger than the negative) indicating that a decay process is also present on the same timescale. The other two DAS indicate that the fluorescence basically decays in a multiphasic way on the timescale of a few picoseconds. The DAS corresponding to the intermediate time constant also has a slight negative

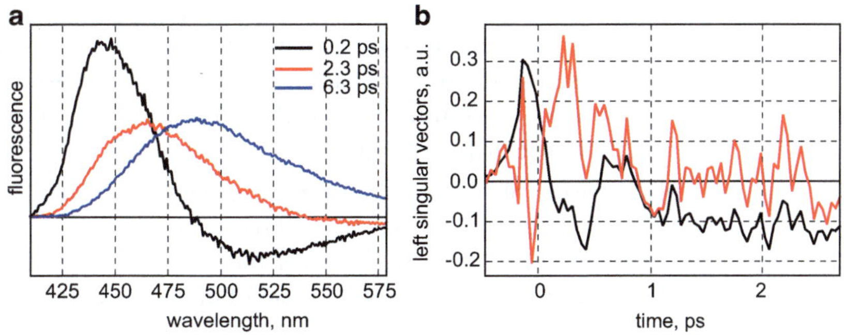

Fig. 7 (**a**) DAS of NK88, obtained from simultaneous global analysis of the datasets measured with two time resolutions using a three-exponential model. The corresponding lifetimes are given in the figure. (**b**) The two first *left* singular vectors from the SVD of the residual matrix from the dataset obtained with Suprasil, as a measure of the fit quality in the time domain

contribution above 550 nm, which indicates further yet slower Stokes shift dynamics; therefore this DAS represents two roughly concomitant processes: decay to the ground-state and excited-state relaxation. These results are in agreement with transient absorption experiments on this system and with the proposed presence of multiple excited-state decay pathways [21].

Further analysis in terms of an exponential reaction scheme using target analysis (*see* **Note 18**) resulting in "SAS" (which represent spectra of the intermediate species) is possible if additional physical constraints are known. For further discussion on the underlying molecular processes associated with the determined time constants we refer to the literature (*see* refs. 20, 21).

4 Notes

1. The combination of a half-wave plate (for rotating the polarization of the pump pulse) and a polarizer (to determine the polarization of the fluorescence entering the Kerr cell) allows setting the polarization direction of the pump light to the desired value. Instead of using the "magic angle" condition, one can also use parallel and perpendicular settings between the directions of pump pulse polarization and the fluorescence polarization direction after the first polarizer. This procedure allows measuring the anisotropy dynamics.
2. Simultaneous collection of the total fluorescence (by the additional spectrometer) with the signal of the gated detection using the same integration time allows *a posteriori* correction for fluctuations.
3. The alignment of the moving cell, with both the cell and the plane of motion perpendicular to the optical axis, is important

in order to avoid noise due to sample movements. To avoid inhomogeneity in the residence time of the sample in the beam we used a cell holder that performs a circular movement with a diameter of ~8 mm. Flow cells with peristaltic pumps or micro-annular gear pumps can also be used.

4. In principle it is possible to use parabolic mirrors instead of Cassegrain-type objectives. However, they have depolarizing effects [25], which increases the amount of non-gated background fluorescence when used between the polarizers P_1 and P_2 of Fig. 1.
5. Usually a lens with a rather long focal length is used to focus the gate beam (in our case $f = 150$ mm). To prevent damage of the pickup mirror (Fig. 1 M_2, Fig. 3b) and to ensure a gate area larger than the focused fluorescence, the gate beam is focused behind the Kerr cell. The efficiency of the gating process is very sensitive to the angle between polarizations of the fluorescence and the gate pulse. The maximum efficiency is reached when this angle is equal to 45° [10].
6. Substantial amounts of scattered pump and gate light can still reach the spectrometer; therefore filters are also necessary.
7. Upon proper alignment, after the first RO (Fig. 1, O_1) the image of the transmitted light should be symmetrical, here a circle of light with a dark cross and dark spot in the center. For precise positional control, the ROs are mounted on x-, y-, z-, and φ-adjustable stages. After O_3 in Fig. 1 one should see identical overlapping symmetrical circles (the orientation of the crosses can be used to discriminate them). A separate camera, or the CCD with the spectrometer set to zero order, can be used for monitoring the position of the beams.
8. The direction of the incoming and outgoing collimated beams in the delay line should both be strictly parallel to the direction of the stage motion; a retroreflector as reflecting element on the stage strongly facilitates this procedure. In order to increase the accessible time delays, in principle multiple pass configurations are also possible (see, e.g., the double-pass delay line in [26]).
9. It is also possible to use scattered pump light, a Raman scatter peak, or a white light super continuum generated at the position of the sample. However using a solution of a very short-lived fluorescence dye mimics the sample better. The dye can be placed in the same cell as the sample and can be selected in a way that its fluorescence spectrum overlaps with that of the sample. Furthermore, it is much easier to find gated light first using CS_2 as the Kerr medium, because it has a much higher gate efficiency. Once "t_0" is known, fine-tuning with Suprasil glass can be done, which provides the highest time resolution.

10. After "t_0"is found, the spatial overlap of the fluorescence and the gating pulse can be further optimized on the amplitude of the gated light by tweaking mirror M_1.
11. Scattered light from the pump beam may have additional effects distorting the measured signal. If the Kerr medium can itself be excited by the scattered pump light (e.g., CS_2 in the near-UV spectral range [27]), a fluorescence signal may arise from the Kerr medium. Such a signal may be eliminated by a background correction (*see* **Note 15**). However, as the pump scattering overlaps with the gate pulse in the Kerr medium as well, in addition a real "pump-probe" contribution may occur [28] which depends on the time delay but is unrelated to the fluorescence of the sample. Therefore, the amount of scattered pump light reaching the Kerr medium should be minimized.
12. Possible preprocessing steps of the data:
    (a) Averaging of multiple images.
    (b) Background (BG) correction:
    There are at least two types of BG that can be detected during measurements on the Kerr-based spectrometers, (1) ambient light detected by the CCD in the absence of fluorescence and (2) fluorescence leaking through the crossed polarizers. The first, in principle, can be measured by performing the experiment with identical settings as the actual measurement but with the pump beam blocked. The second type will also be present at negative time delays (when the gate pulse precedes the pump pulse) and measurements at such delay conveniently provide a BG to be subtracted before further analysis.
    (c) Shading correction (correction for spectral sensitivity of the instrument):
    This correction does not affect kinetic parameters; however it can significantly change spectral shapes.
    (d) Outlier correction:
    Often cosmic rays can be detected by the CCD during measurements. Such signals in most cases are localized on one pixel of the CCD and have an amplitude value which is much higher than a real signal. They can be corrected for in the raw data by standard CCD software or in Glotaran using a scanning median filter.
    (e) Averaging within the image:
    Averaging the dataset along the wavelength dimension (spectral binning) while maintaining a reasonable resolution (typically 1–3 nm for fluorescence) can increase S/N without losing information.

13. SVD [29, 30] is a model-free mathematical operation that rewrites the data as a sum of orthonormal vector products scaled by singular values. Here, the LSV represent time traces and the RSV represent spectra. The singular vectors are ordered by decreasing information content. It is possible to use a limited number of components for filtering; however, as any data manipulation technique this must be used with caution.
14. The temporal profile of the Raman scatter is instantaneous. Therefore it can be taken into account by an additional component which follows the time profile of the IRF.
15. In the present study we use two datasets measured under the same conditions but using different Kerr media (benzene and Suprasil). Therefore, we can analyze such datasets simultaneously, allowing better determination of the different components by imposing the same ("linking") kinetic parameters and spectral shapes between datasets. Under other conditions multiple dataset analysis by linking either only kinetic or only spectral components can also be useful.

    The two datasets differ in IRF. The IRF can often be satisfactorily modeled by a Gaussian function with two parameters μ and Δ for temporal position of the maximum and full width at half maximum (FWHM): $i(t) = \frac{1}{\tilde{\Delta}} \exp\left(-\log(2)\left(\frac{2(t-\mu)}{\Delta}\right)^2\right)$, where $\tilde{\Delta} = \Delta/\left(2\sqrt{2\log(2)}\right)$. Parameters μ and Δ can be fit parameters. Here, Δ was ~200 fs for Suprasil and ~550 fs for benzene. The wavelength dependence of the position $\mu(\lambda)$ (temporal dispersion) often can be modeled by a polynomial function; in the present study a linear function was sufficient. For more details on the use of Glotaran see http://glotaran.org/wiki.html.
16. DAS represent the pre-exponential factors of the exponential evolutions (Eq. 2). In many cases DAS do not represent real spectra of the fluorescent species, unless they correspond to independently decaying species, but they can be used to visualize possible relaxation processes and conversions between different species. Positive DAS represent fluorescence decay while negative DAS are a signature of the rise of the fluorescence.
17. Statistical characteristics of the residuals are usually used to check the quality of the fit. Glotaran has three available tools for analysis of the fit quality. The root mean square of the residuals can be used to compare analysis of the same datasets with different fit models. Assessment whether SVD of the residual matrix deviates from noise provides information about the overall quality of the fit in time domain and the spectral domain. Finally, individual residual traces in the time and spectral domain can also be analyzed.

18. The target model, fuelled by scientific hypotheses and assumptions, includes all possible branching routes and equilibria between intermediate species. This allows estimating the spectra of each of the excited species (SAS). In this case Eq. 1 is rewritten as

$$\Psi(\lambda, t) = \sum_{i=1}^{n_{\text{comp}}} c_i^{\text{SAS}}(t, \theta)\text{SAS}_i(\lambda) \tag{3}$$

in which $c_i^{\text{SAS}}(t, \theta)$ are linear combinations of all $c_i^{\text{DAS}}(t, \theta)$. For more details about the use of compartmental models for global and target analysis see reviews by van Stokkum and coworkers [4, 16].

References

1. Valeur B (2002) Molecular fluorescence: principles and applications. Wiley-VCH, Weinheim
2. Laptenok SP, van Stokkum IHM, Borst JW et al (2012) Disentangling picosecond events that complicate the quantitative use of the calcium sensor YC3.60. J Phys Chem B 116:3013–3020
3. van Oort B, Eremeeva EV, Koehorst RBM et al (2009) Picosecond fluorescence relaxation spectroscopy of the calcium-discharged photoproteins aequorin and obelin. Biochemistry 48:10486–10491
4. van Stokkum IHM, van Oort B, van Mourik F et al (2008) (Sub)-picosecond spectral evolution of fluorescence studied with a synchroscan streak-camera system and target analysis. In: Aartsma TJ, Matysik J (eds) Biophysical techniques in photosynthesis, vol II. Springer, Dordrecht, pp 223–240
5. Cannizzo A, Bräm O, Zgrablic G et al (2007) Femtosecond fluorescence upconversion setup with broadband detection in the ultraviolet. Opt Lett 32:3555–3557
6. Zhang X-X, Würth C, Zhao L et al (2011) Femtosecond broadband fluorescence upconversion spectroscopy: improved setup and photometric correction. Rev Sci Instrum 82:063108
7. Schanz R, Kovalenko SA, Kharlanov V et al (2001) Broad-band fluorescence upconversion for femtosecond spectroscopy. Appl Phys Lett 79:566–568
8. Nakamura R, Kanematsu Y (2004) Femtosecond spectral snapshots based on electronic optical Kerr effect. Rev Sci Instrum 75:636–645
9. Schmidt B, Laimgruber S, Zinth W et al (2003) A broadband Kerr shutter for femtosecond fluorescence spectroscopy. Appl Phys B 76:809–814
10. Arzhantsev S, Maroncelli M (2005) Design and characterization of a femtosecond fluorescence spectrometer based on optical Kerr gating. Appl Spectrosc 59:206–226
11. Duguay MA (1969) An ultrafast light gate. Appl Phys Lett 15:192–194
12. Duguay MA, Hansen JW (1969) Direct measurement of picosecond lifetimes. Opt Commun 1:254–256
13. Cerullo G, de Silvestri S (2003) Ultrafast optical parametric amplifiers. Rev Sci Instrum 74:1–18
14. Yan L, Si J, Yan Y et al (2011) Pump power dependence of femtosecond two-color optical Kerr shutter measurements. Opt Express 19:11196–11201
15. Weigel A, Pfaffe M, Sajadi M et al (2012) Barrierless photoisomerisation of the "simplest cyanine": joining computational and femtosecond optical spectroscopies to trace the full reaction path. Phys Chem Chem Phys 14:13350–13364
16. van Stokkum IHM, Larsen DS, van Grondelle R (2004) Global and target analysis of time-resolved spectra. Biochim Biophys Acta 1657:82–104
17. van Wilderen LJGW, Lincoln CN, van Thor JJ (2011) Modelling multi-pulse population dynamics from ultrafast spectroscopy. PLoS One 6:e17373
18. Snellenburg JJ, Laptenok SP, Seger R et al (2012) Glotaran: a Java—based graphical user interface for the R-package TIMP. J Stat Soft 49:1–23
19. Roth NJL, Craig AC (1974) Predicted observable fluorescent lifetimes of several cyanines. J Phys Chem 78:1154–1155
20. Nuernberger P, Vogt G, Gerber G et al (2006) Femtosecond study on the isomerization

dynamics of NK88. I. Ground-state dynamics after photoexcitation. J Chem Phys 125:44512

21. Vogt G, Nuernberger P, Gerber G et al (2006) Femtosecond study on the isomerization dynamics of NK88. II. Excited-state dynamics. J Chem Phys 125:44513
22. Pal MK, Ghosh JK (1994) Energy transfer and formation of exciplex between thiacyanine and acridine orange facilitated by anionic biopolymers and synthetic polymers. J Photochem Photobiol A 78:31–37
23. Thomas MS, Nuñez V, Upadhyayula S et al (2010) Kinetics of bacterial fluorescence staining with 3,3′-diethylthiacyanine. Langmuir 26:9756–9765
24. Mammone JF, Sharma SK, Nicol M (1980) Raman spectra of methanol and ethanol at pressures up to 100 kbar. J Phys Chem 84:3130–3134
25. Brückner C, Notni G, Tünnermann A (2010) Optimal arrangement of 90° off-axis parabolic mirrors in THz setups. Optik - Int J Light Electron Optics 121:113–119
26. Schmidt A, Chiesa M, Chen X et al (2008) An optical pump-probe technique for measuring the thermal conductivity of liquids. Rev Sci Instrum 79:064902
27. Ganeev RA, Ryasnyanskiĭ AI, Kuroda H (2006) Nonlinear optical characteristics of carbon disulfide. Opt Spectrosc 100: 108–118
28. Thomsen CL, Madsen D, Thøgersen J et al (1999) Femtosecond spectroscopy of the dissociation and geminate recombination of aqueous CS_2. J Chem Phys 111:703
29. Press WH, Teukolsky SA, Vetterling WT et al (2007) Numerical recipes 3rd Edition: the art of scientific computing. Cambridge University Press, Cambridge
30. Golub GH, Van Loan CF (1996) Matrix computations, 3rd edn. The Johns Hopkins University Press, Baltimore

Chapter 14

Photoinduced Electron Transfer Modeling to Simulate Flavoprotein Fluorescence Decay

Nadtanet Nunthaboot, Kiattisak Lugsanangarm, Arthit Nueangaudom, Somsak Pianwanit, Sirirat Kokpol, and Fumio Tanaka

Abstract

A method of analysis is described on the photoinduced electron transfer (PET) from aromatic amino acids as tryptophans (Trp) and tyrosines (Tyr) to the excited isoalloxazine (Iso*) in FMN-binding proteins (FBP) from *Desulfovibrio vulgaris* (strain, Miyazaki F). Time-dependent geometrical factors as the donor–acceptor distances are determined by means of a molecular dynamics simulation (MDS) of the proteins. Fluorescence decays of the single mutated isoforms of FBP are used as experimental data. The electrostatic (ES) energy between the photoproducts and ionic groups in the proteins is introduced into the Kakitani and Mataga (KM) model, which is modeled for an electron transfer process in solution. The PET parameters contained in the KM rate are determined by means of a nonlinear least square method, according to the Marquardt algorithm. The agreement between the observed and calculated decays is quite good, but not optimal. Characteristics on PET in flavoproteins, obtained by the present method, are described. Possible improvements of the method are discussed.

Key words Flavin, Flavoprotein, Photoinduced electron transfer, Excited isoalloxazine as acceptor, Tryptophan and tyrosine as donor, Marcus–Hush theory, Kakitani–Mataga theory, Ultrafast fluorescence dynamics, Molecular dynamics simulation, Calculated fluorescence decay, Electrostatic energy between photoproducts and ionic groups inside protein

1 Introduction

Flavoprotein was first discovered as a yellow pigment in cow milk in 1879 according to Massey [1]. By the early 1930s, this pigment had been recognized as a component of vitamin B2. The chemical structure of the pigment was identified, and it was named as riboflavin in 1934 [1] (*see* Fig. 1). The chromophore of flavins is the isoalloxazine ring (Iso). Theorell first isolated a yellow-colored yeast protein and found that the pigment is a cofactor of the protein [2]. Today, it is known that flavoproteins function as electron transport carrier, oxidase, oxygenase, dehydrogenase, etc. [3].

Yves Engelborghs and Antonie J.W.G. Visser (eds.), *Fluorescence Spectroscopy and Microscopy: Methods and Protocols*, Methods in Molecular Biology, vol. 1076, DOI 10.1007/978-1-62703-649-8_14,

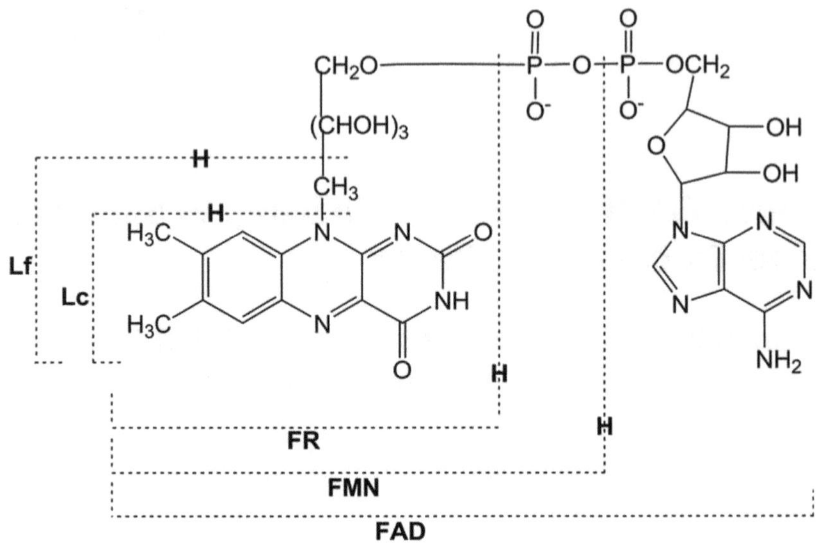

Fig. 1 Chemical structures of flavins. FR, riboflavin; FMN, flavin mononucleotide; FAD, flavin adenine dinucleotide; Lf, lumiflavin; Lc, lumichrome

Characterization of the photochemistry of flavins and flavoproteins had been pioneered by Weber, who first quantitatively studied the fluorescence of riboflavin for his Ph.D. work in 1947 [4]. Weber also studied the fluorescence quenching of flavins by various substances [5, 6]. The quenching mechanism of Iso fluorescence upon complex formation with adenine was first characterized by means of fluorescence lifetime measurements [7, 8]. McCormick studied the fluorescence quenching of Iso by indole with Iso-$(CH_2)_n$-indole diads [9]. Time-resolved fluorescence of flavins and flavoproteins has been reviewed by van den Berg and Visser [10].

In the last decade a number of new flavoproteins has been found, which function as photoreceptors [11]. Among six families of the photoreceptors, phototropins [12], cryptochromes [13], and BLUF (blue-light sensing using flavin) contain flavins as the reaction center [14]. Cryptochromes are blue-light receptors for development processes such as hypocotyl elongation in plants and circadian rhythm in plants and insects [13, 15]. The BLUF unit of AppA is involved in the repression of photosynthesis genes (PpsR) located at N-terminal region of the AppA protein from *Rhodobacter sphaeroides* [16]. The BLUF in TePixD from the thermophilic unicellular cyanobacterium *Thermosynechococcus elongatus* BP-1 regulates pili-dependent cell mobility [17]. In these BLUFs, the initial step of the photoreceptor functions is considered to be photoinduced electron transfer (PET) from the essential tyrosine (Tyr) to the excited isoalloxazine (Iso*) through hydrogen bonding (H bond) with glutamine (Gln). Thus, photochemistry and photophysics of flavins is becoming an important field of science [18].

The fluorescence quenching mechanism was first studied in systems of riboflavin tetrabutyrate with indole or *N,N′*-dimethylaniline in organic solvents [19] and in flavodoxin from *Desulfovibrio vulgaris* (strain, Miyazaki F) [20] by means of picosecond transient-absorption spectroscopy. It is concluded that the remarkable quenching of Iso fluorescence can be ascribed to photoinduced electron transfer (PET) from indole, *N,N′*-dimethylaniline, and Trp to the Iso*. The PET mechanism in riboflavin-binding protein from egg white was also reported in the femtosecond time domain by means of transient-absorption spectroscopy [21]. A number of flavoproteins display apparently no fluorescence. However, they exhibit fluorescence decays with ultrashort lifetimes upon excitation with a sub-picosecond pulse laser [22–29].

Since the seminal works on electron transfer theory by Marcus [30–32], several research workers further developed the theory [33–38]. These theories were modeled for electron transfer processes in solution. Hush introduced a donor–acceptor distance dependence into the electronic coupling term of the electron transfer rate [39]. The resulting equation is called Marcus–Hush theory [40]. The electron transfer theory in proteins has been worked out by Warshel et al. [41–43] and Beratan et al. [44]. Experimental studies on the electron transfer processes in proteins have been reviewed by Bendall [45] and by Gray and Winkler [46].

Transient fluorescence spectroscopy is a most useful experimental method to study the PET mechanism in flavoproteins. The ultrafast fluorescence of flavoproteins often decays with non-exponential functions. However, it has been difficult to obtain the PET rates from the fluorescence decay function, because any PET theory contains several unknown parameters, which cannot be determined by experiments. We have been developing a new method to evaluate the PET parameters contained in the analytical PET theories and, consequently, the PET rate [47–53]. In this method the Kakitani and Mataga (KM) model [36–38] was used, because it could satisfactorily reproduce the experimental fluorescence decays of flavoproteins.

In this chapter we describe a method to obtain the calculated fluorescence decays of FMN-binding proteins (FBP) from *Desulfovibrio vulgaris* (strain, Miyazaki F). FBP is a small flavoprotein, which contains FMN as cofactor and which is considered to play an important role in the electron transport process among bacterial proteins [54]. The three-dimensional structures of wild-type (WT) FBP were determined by NMR spectroscopy [55] and X-ray crystallography [56]. According to these structures, tryptophan 32 (Trp32) is closest to Iso and then tyrosine 35 (Tyr35) and Trp106 among the aromatic amino acids. Ultrafast fluorescence dynamics of WT FBP [27] and E13T (Glu13 was replaced by Thr), E13Q (Glu13 replaced by Gln), W32Y (Trp32 replaced by Tyr), and W32A (Trp32 replaced by Ala) FBP isoforms [28, 29] were measured by means of a fluorescence up-conversion method. These decays were simultaneously analyzed to obtain PET rate from individual donor and related physical quantities [49]. The limitations and possible improvements of the method are discussed in the Subheading 4.

2 Method of Analyses

2.1 Outline of the Method

The principle of the method to calculate the fluorescence decay, PET rate, and related physical quantities are illustrated in Fig. 2. More precise procedures are described below.

Procedure A: The time-dependent atomic coordinates in the flavoproteins are determined by molecular dynamics simulation (MDS).

Procedure B: Any PET rates contain several unknown parameters (PET parameters). The PET rate by the KM model is given by Eq. 1.

Procedure C: Initial values of all PET parameters are set, which are dependent on the model used. Sometimes plausible values obtained by earlier works are chosen as the initial values.

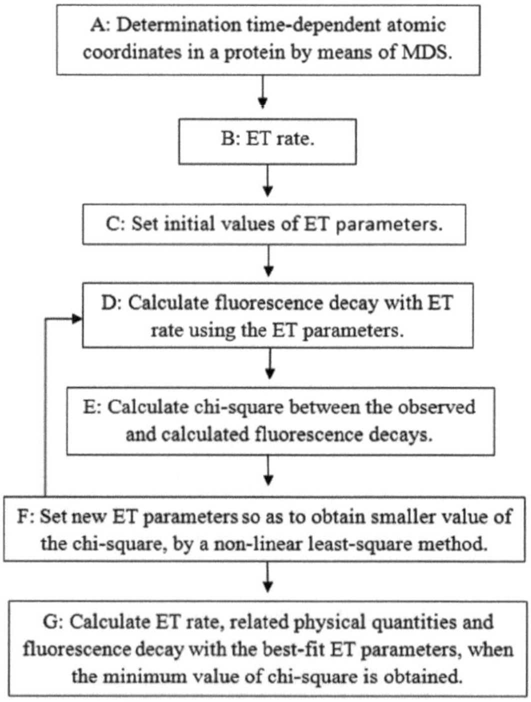

Fig. 2 Schematic diagram for the determination of PET rate and related physical quantities. Details of Procedures A–G are described in the text

Procedure D: The fluorescence decay is calculated with Eq. 7. In the primary process before taking the average with the MDS-snapshot ensemble, the probability of Iso* should decrease with a simple exponential function of the sum of the PET rates from potential PET donors. The PET rate itself is time independent. However, it becomes time dependent on the MDS time (t'), because the rate is dependent on the atomic coordinates, which vary with the MDS time. The calculated decay function is obtained by taking the average over the MDS time.

Procedure E: The chi-square is calculated between the calculated and experimental decay functions by Eq. 8.

Procedure F: A new set of PET parameters is obtained by a nonlinear least square method, which gives a lower value for the chi-square than the previous value. Procedures D to F are repeated until the minimum value of the chi-square is obtained.

Procedure G: The calculated decay, PET rates, and related physical quantities are obtained when the value of the chi-square becomes the minimum. These procedures are repeated with several models to obtain better fitting of the calculated decay with the observed decay.

All calculations of these procedures are performed with a personal computational program written with a BASIC compiler.

2.2 Determination of Atomic Coordinates in Five FBP Isoforms by Means of MDS

The starting structures of the WT and its mutated FBPs are obtained by using the X-ray structure of the *D. vulgaris* (strain Miyazaki F) WT FBP (PDB code 1FLM) [56]. In the E13T and E13Q single substitution isomers, the glutamate (E) at position 13 is replaced by threonine (T) and glutamine (Q), respectively, and the replacements of amino acids from the WT are conducted using the LEap module implemented in AMBER 8 [57]. All calculations are carried out using the AMBER 8 suite of programs. The parm99 force field [58] is used to describe the protein atoms, whereas force field parameters for Iso are obtained from the work by Schneider and Suhnel [59]. All missing hydrogen atoms of the protein are added using the LEap module of AMBER 8. The simulated systems are subsequently solvated with a cubic box of 6,336 TIP3P water molecules in WT, E13T, and E13Q, 6,345 in W32Y, and 6,347 in W32A. One chloride counterion is added to electronically neutralize the system. The total energy of the added water molecules is first minimized, while the protein and Iso coordinates are kept fixed. The entire system is then optimized using 2,000 steps of steepest descent and conjugated gradient minimizations. Afterwards, the whole system was heated from 10 K to 298 K over 50 ps and is further equilibrated under periodic boundary conditions at 298 K. The systems are set up under the isobaric–isothermal ensemble

(NPT) with a constant pressure of 1 atm and constant temperature of 298 K. The electrostatic interaction is corrected by the Particle Mesh Ewald method [60]. The SHAKE algorithm [61] is employed to constrain all bonds involving hydrogen atoms. A cutoff distance of 1 nm is employed for nonbonded pair interactions. MDS based calculations are performed with time steps of 2 fs and the coordinates of the MDS snapshots are collected every 0.01 ps. The simulations are performed over 5 ns. Data collected during the last 2–3 ns are used for the analysis.

2.3 PET Rate

PET rate by the KM model is used and expressed by Eq. 1:

$$k_{\mathrm{ET}}^{jk} = \frac{\nu_0^q}{1+\exp\{\beta^q(R_{jk}-R_0^q)\}}\sqrt{\frac{k_B T}{4\pi\lambda_S^{jk}}} \times \exp\left[-\frac{\left\{\Delta G_q^0 - e^2/\varepsilon_0^{\mathrm{DA}}R_{jk} + \lambda_S^{jk} + \mathrm{ES}_j(k)\right\}^2}{4\lambda_S^{jk}k_B T}\right] \quad (1)$$

Here k_{ET}^{jk} is the PET rate constant of donor k in jth FBP isoform. $k = 1$ for Trp32 (Tyr32 in W32Y), $k = 2$ for Tyr35, and $k = 3$ for Trp106. $j = 1$ for WT, $j = 2$ for E13T, $j = 3$ for E13Q, $j = 4$ for W32Y, and $j = 5$ for W32A FBP. ν_0^q is an adiabatic frequency, β^q is the PET process coefficient, and R_0^q is a critical distance between the adiabatic and nonadiabatic PET processes. It was assumed that these quantities depend only on q (Trp or Tyr), not on j and k. When $R_{jk} < R_0^q$, the PET process is adiabatic, whereas when $R_{jk} > R_0^q$, the PET process is nonadiabatic. R_{jk} is the center-to-center distance (often Rc is used for it) between Iso and the PET donor k in the jth FBP system. The physical constants k_B, T, and e are Boltzmann constant, temperature (298 K), and electron charge, respectively. $\varepsilon_0^{\mathrm{DA}}$ is a static dielectric constant of medium between PET donors and acceptor. $\mathrm{ES}_j(k)$ is a net electrostatic (ES) energy between the kth aromatic ionic species and all other ionic groups in the jth FBP isoform, as described below. λ_S^{jk} is the solvent reorganization energy [30, 31] of the Iso* and the kth donor in the jth FBP isoform, as shown by Eq. 2:

$$\lambda_S^{jk} = e^2\left(\frac{1}{2a_{\mathrm{Iso}}}+\frac{1}{2a_q}-\frac{1}{R_{jk}}\right)\left(\frac{1}{\varepsilon_\infty}-\frac{1}{\varepsilon_0^{\mathrm{DA}}}\right) \quad (2)$$

Here, a_{Iso} and a_q are the radii of Iso and the donor q (Trp or Tyr), assuming these reactants are spherical, and ε_∞ is the optical dielectric constant (a value of 2 being used). The radii of Iso, Trp, and Tyr were determined in the following way: (1) Three-dimensional sizes of lumiflavin for Iso, 3-methylindole for Trp, and p-methylphenol for Tyr were obtained by a semiempirical molecular orbital method (PM3). (2) The volumes of these molecules were determined as asymmetric rotors. (3) Radii of the spheres having the same volumes as the asymmetric rotors were obtained. a_{Iso} is 0.224 nm, a_{Trp} 0.196 nm, and a_{Tyr} 0.173 nm.

The standard free energy gap (ΔG_q^0) is expressed with the ionization potential (E_{IP}^q) of the PET donor q (Trp or Tyr), as in Eq. 3:

$$\Delta G_q^0 = E_{\mathrm{IP}}^q - G_{\mathrm{Iso}}^0 \tag{3}$$

where G_{Iso}^0 is the standard Gibbs energy related to the electron affinity of Iso*. The values of E_{IP}^q for Trp and Tyr are 7.2 eV and 8.0 eV, respectively [62].

2.4 Electrostatic Energy in the Proteins

Proteins, including FBPs, contain many ionic groups, which may influence the PET rate. The cofactor in all these FBP systems is FMN, which has two negative charges at the phosphate group. The WT, W32Y, and W32A contain eight Glu, five Asp, four Lys, and nine Arg residues, whereas E13T and E13Q contain seven Glu, five Asp, four Lys, and nine Arg residues, since the Glu13 is replaced by the neutral Thr13 and Gln13, respectively. The ES energy between the Iso anion or kth donor cation and all other ionic groups in jth isoform is expressed by Eq. 4:

$$E_j(k) = \sum_{i=1}^{l} \frac{C_k \cdot C_{\mathrm{Glu}}}{\varepsilon_0^j R_k(\mathrm{Glu}-i)} + \sum_{i=1}^{5} \frac{C_k \cdot C_{\mathrm{Asp}}}{\varepsilon_0^j R_k(\mathrm{Asp}-i)} + \sum_{i=1}^{4} \frac{C_j \cdot C_{\mathrm{Lys}}}{\varepsilon_0^j R_k(\mathrm{Lys}-i)} + \sum_{i=1}^{9} \frac{C_j \cdot C_{\mathrm{Arg}}}{\varepsilon_0^j R_k(\mathrm{Arg}-i)} + \sum_{i=1}^{2} \frac{C_j \cdot C_P}{\varepsilon_0^j R_k(P-i)} \tag{4}$$

where $n = 8$ for the WT, W32Y, and W32A, while $l = 7$ for E13T and E13Q. Here, $k = 0$ for the Iso$^-$, $k = 1$ for Trp32$^+$, $k = 2$ for Tyr35$^+$, and $k = 3$ for Trp106$^+$. ε_0^j is the static dielectric constant inside the jth isoform. C_k is the charge of the aromatic ionic species k and is $-e$ for $k = 0$ (Iso anion) and $+e$ for $k = 1$–3. $C_{\mathrm{Glu}}(= -e)$, $C_{\mathrm{Asp}}(= -e)$, $C_{\mathrm{Lys}}(= +e)$, $C_{\mathrm{Arg}}(= +e)$, and $C_P(= -e)$ are the charges of Glu, Asp, Lys, Arg, and phosphate anions, respectively. We assume that these groups are all in an ionic state in solution. The values of pK_a of ionic amino acids in water are 4.3 in Glu, 3.9 in Asp, 10.5 in Lys, and 12.5 in Arg. In the proteins these values of pK_a may be modified in the range of ±0.3. Histidine (His) displays a pK_a of 6.0 in water. All measurements were performed in a 10 mM Tris–HCl buffer (pH 8.0) containing 200 mM NaCl, in which His should be neutral. Distances between the aromatic ionic species k and the ith Glu ($i = 1$–8 or 7) are denoted as $R_k(\mathrm{Glu}-i)$, those between kth ion and the ith Asp ($i = 1$–5) are denoted as $R_k(\mathrm{Asp}-i)$, and so on.

Table 1
Experimental decay parameters in five FBP isoforms

FBP	τ_1 (ps)	(α_1)	τ_2 (ps)	(α_2)	τ_3 (ps)	(α_3)	τ_{AV}[a] (ps)
WT[b]	0.167	(0.96)	1.5	(0.04)	–	–	0.220
E13T[c]	0.107	(0.86)	1.5	(0.12)	30	(0.02)	0.872
E13Q[c]	0.134	(0.85)	0.746	(0.12)	30	(0.03)	1.10
W32Y[d]	3.4	(0.23)	18.2	(0.74)	96	(0.03)	17.1
W32A[d]	30.1	(1.00)	–	–	–	–	30.1

[a]Average lifetimes were obtained as $\tau_{AV} = \sum_{i=1}^{n} \alpha_i \tau_i$ where $i = 1–3$
[b]Data taken from ref. 27
[c]Data taken from ref. 29
[d]Data taken from ref. 28

$ES_j(k)$ (k =1–3) in Eqs. 1 and 4 is expressed as follows:

$$ES_j(k) = E_j(0) + E_j(k) \tag{5}$$

Here, j varies from 1 to 5.

2.5 Fluorescence Decays

The ultrafast fluorescence dynamics of the WT FBP [27], W32Y and W32A [28], and E13T and E13Q FBP isoforms [29] were obtained by Chosrowjan and coworkers by means of the fluorescence up-conversion method. The fluorescence decay functions of these flavoproteins are expressed by multiple-exponential function as Eq. 6. The pre-exponential factor (α_i^j) of the ith lifetime component (τ_i^j) in the jth isoform is given in Table 1. The lifetimes are expressed in units of picoseconds:

$$F_{\text{obs}}^j = \sum_{i=1}^{n} \alpha_i^j \exp(-t/\tau_i^j) \quad (n = 1\text{-}3) \tag{6}$$

The calculated decay function of the jth FBP system is expressed by Eq. 7:

$$F_{\text{calc}}^j(t) = \left\langle \exp\left[-\left\{\sum_{k=1}^{3} k_{\text{ET}}^k(t')\right\}t\right]\right\rangle_{\text{AV}} \tag{7}$$

The fluorescence decays of FBP isoforms are calculated up to 3 ps with 0.003 ps time intervals. The fluorescence decays of the WT, E13T, and E13Q FBP isoforms are calculated up to 3 ps with 0.003 ps time intervals, while that for the W32Y and W32A isoforms are calculated up to 20 ps with 0.02 ps intervals. Note that $\langle \ldots \rangle_{\text{AV}}$ denotes the averaging procedure of the exponential function in Eq. 7, which is performed over t' up to 2 ns with 0.1 ps time intervals for the WT, W32Y, and W32A isoforms and up to

3 ns with 0.1 ps intervals for the E13T and E13Q isoforms. In Eq. 7 we assume that the decay functions at every instant of time (t') during the MDS time range can always be expressed by an exponential function. The present method is equivalent to the reported one [63] when the time range ($t' = 2$–3 ns) of MDS data is much longer than that for the fluorescence data (t is at most 20 ps) [48].

2.6 Determination of Unknown PET Parameters

In the KM model the unknown parameters are G^0_{Iso}, $\varepsilon_0^{\mathrm{WT}}$, $\varepsilon_0^{\mathrm{E13T}}$, $\varepsilon_0^{\mathrm{E13Q}}$, $\varepsilon_0^{\mathrm{DA}}$, $\varepsilon_0^{\mathrm{W32Y}}$, and $\varepsilon_0^{\mathrm{W32A}}$. The previously reported values are used for ν_0^q, β^q, and R_0^q ($q =$ Trp or Tyr) [48]. The chi-squared value of the jth FBP system between the observed and calculated fluorescence intensities (χ_j^2) is defined by Eq. 8:

$$\chi_j^2 = \frac{1}{N}\sum_{i=1}^{N}\frac{\left\{F^j_{\mathrm{calc}}(t_i) - F^j_{\mathrm{obs}}(t_i)\right\}^2}{F^j_{\mathrm{calc}}(t_i)} \tag{8}$$

Here, N denotes the number of time intervals of the fluorescence decay and was 1,000 for all evaluations. The deviation between the observed and calculated intensities is expressed by Eq. 9:

$$\mathrm{Dev}_j(t_i) = \frac{\left\{F^j_{\mathrm{calc}}(t_i) - F^j_{\mathrm{obs}}(t_i)\right\}}{\sqrt{F^j_{\mathrm{calc}}(t_i)}} \tag{9}$$

The PET parameters are determined by the KM model so as to obtain the minimum values of χ_5^2 given by Eq. 10, by means of a nonlinear least square method, according to the Marquardt algorithm:

$$\chi_5^2 = \chi^2_{\mathrm{WT}} + \chi^2_{\mathrm{E13T}} + \chi^2_{\mathrm{E13Q}} + \chi^2_{\mathrm{W32Y}} + \chi^2_{\mathrm{W32A}} \tag{10}$$

3 Global Analyses Among the Five FBP Isoforms

3.1 Donor–Acceptor Distance

The three-dimensional structures of WT, E13T, E13Q, W32Y, and W32A FBP isoforms are shown in Fig. 3. E13 in WT, W32Y, and W32A have negative charges. T13 in E13T and Q13 E13Q are neutral. The charge states of these amino acids may influence the PET rates from Trp32 (Tyr32), Tyr35, and Trp106 to Iso*. The mean distances of Glu13 from Iso, Trp32, Tyr35, and Trp106 in WT over the MDS time range are 1.5, 1.0, 1.0, and 1.7 nm, respectively. The mean distances of Thr13 from Iso, Trp32, Tyr35, and Trp106 in E13T are 1.5, 0.9, 1.2, and 2.0 nm, respectively. The distances of Gln13 from Iso, Trp32, Tyr35, and Trp106 in E13Q are 1.6, 1.0, 1.1, and 1.8 nm, respectively.

The time evolutions of the donor–acceptor distances between Iso and Trp32 are shown in Fig. 4. The distance between Iso and

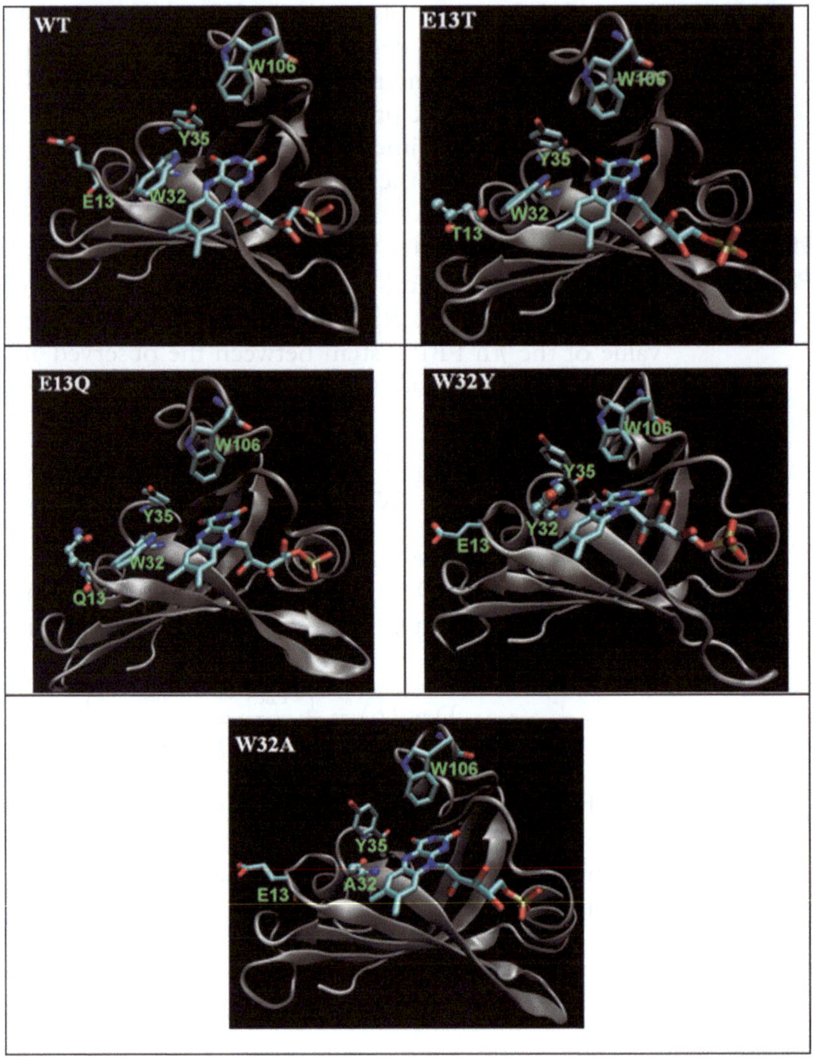

Fig. 3 MDS snapshots at FMN-binding sites in five FBP isoforms. W and Y denote Trp and Tyr. E, T, Q, and A denote Glu, Thr, Gln, and Ala, respectively. Data are taken from the work by Nadtanet et al. [49] under permission of the PCCP Owner Societies

Tyr32 is shown in W32Y. The distance is expressed with a center-to-center distance (Rc is used for R_{jk} in Eq. 1). Table 2 lists mean values of Rc over MDS time range. The distance to Iso is shortest in Trp32 among the donors in WT, E13T, and E13Q. The distance of Tyr32 in W32Y is shortest among all donors.

3.2 Calculation of Fluorescence Decay

Fluorescence decays of the five FBP isoforms are shown in Fig. 5. Agreements between the calculated and observed decays are quite good, except for W32A FBP. In W32A the observed intensity decays exponentially in the time domain up to 20 ps, while the

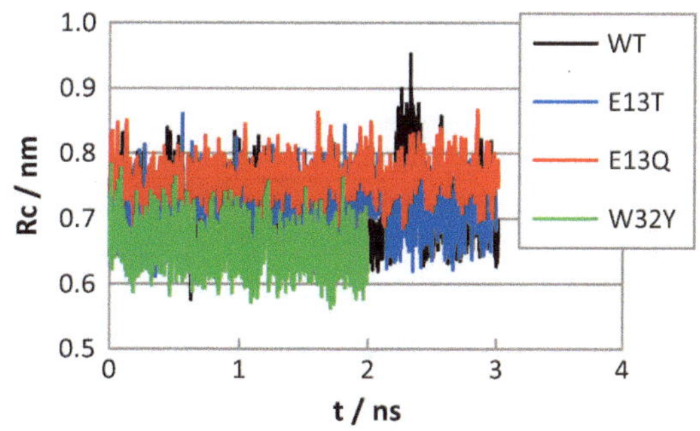

Fig. 4 Time evolution of the donor–acceptor distance in five FBP isoforms. *Rc* denotes center-to-center distance between Iso and Trp32. In W32Y *Rc* between Iso and Tyr32 is shown. Data are taken from the work by Nadtanet et al. [49] under permission of the PCCP Owner Societies

Table 2
The donor–acceptor distance in FBP isoforms[a]

Protein	Trp32	Tyr32	Tyr35	Trp106
WT	0.73	–	1.02	1.05
E13T	0.72	–	0.87	0.91
E13Q	0.75	–	0.85	0.94
W32Y	–	0.65	0.83	0.91
W32A	–	–	0.77	0.90

[a]Mean values of center-to-center distances over MD time domain are listed in nm unit. Data are taken from Nadtanet et al. [49] (Reproduced by permission of the PCCP Owner Societies)

calculated decay is non-exponential. It is apparent that the calculated decay functions in WT, E13T, and E13Q well reproduce the observed ones at the shorter time domain, but rather poor in the longer time domain.

3.3 Best-Fit PET Parameters

The best-fit PET parameters contained in the KM model are listed in Table 3. Static dielectric constants inside the proteins are 15 in WT, 6.0 in E13T, 6.7 in E13Q, 5.9 in W32Y, and 6.3 in W32A. The dielectric constant (ε_0^{DA}) in the domain between the donors and acceptor is 2.2 (*see* footnote in Table 3). It is reasonable to have less polarity in the domain between the donors and Iso, because no amino acid exists in this domain, which is thus like a kind of vacuum, while many amino acids are present between the photoproducts and ionic amino acids in the proteins.

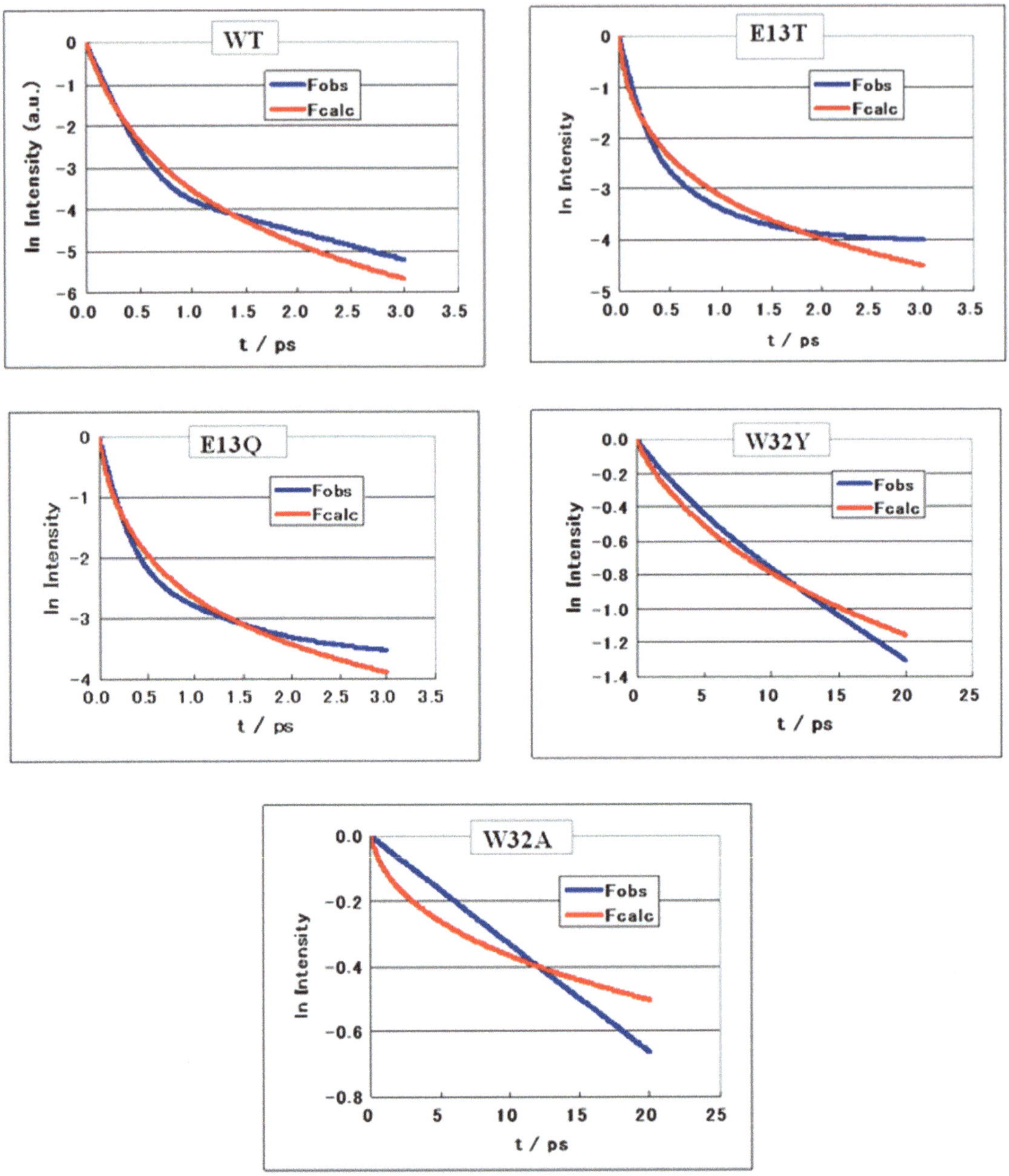

Fig. 5 Calculated fluorescence decay of the five FBP isoforms. F_{obs} and F_{calc} in inserts denote the observed and calculated fluorescence decays. Data are taken from the work by Nadtanet et al. [49] under permission of the PCCP Owner Societies

3.4 PET Rates and Related Physical Quantities

The PET rate from the individual donor is obtained with the best-fit PET parameters. Figure 6 shows time evolutions of the PET rates from potential donors to Iso* in the five FBP isoforms. The PET rate from Trp32 is fastest among the donors in each FBP isoform except for W32Y and W32A. PET rates from Tyr35 are

Table 3
Static dielectric constant inside the proteins[a]

Dielectric constant	WT	E13T	E13Q	W32Y	W32A
ε_0^j	14.8	5.99	6.69	5.89	6.29

[a]The dielectric constants appear in Eq. 4. The other ET parameters common among the five FBP isoforms are $\nu_0^{\mathrm{Trp}} = 1016$ (ps^{-1}), $\nu_0^{\mathrm{Tyr}} = 197$ (ps^{-1}), $\beta^{\mathrm{Trp}} = 21.0$ (nm^{-1}), $\beta^{\mathrm{Tyr}} = 6.25$ (nm^{-1}), $R_0^{\mathrm{Trp}} = 0.663$ (nm), $R_0^{\mathrm{Tyr}} = 0.499$ (nm), $G_{\mathrm{Iso}}^0 = 6.71$ (eV), $\varepsilon_0^{\mathrm{DA}} = 2.19$. Data are taken from Nadtanet et al. [49] (Reproduced under permission of the PCCP Owner Societies)

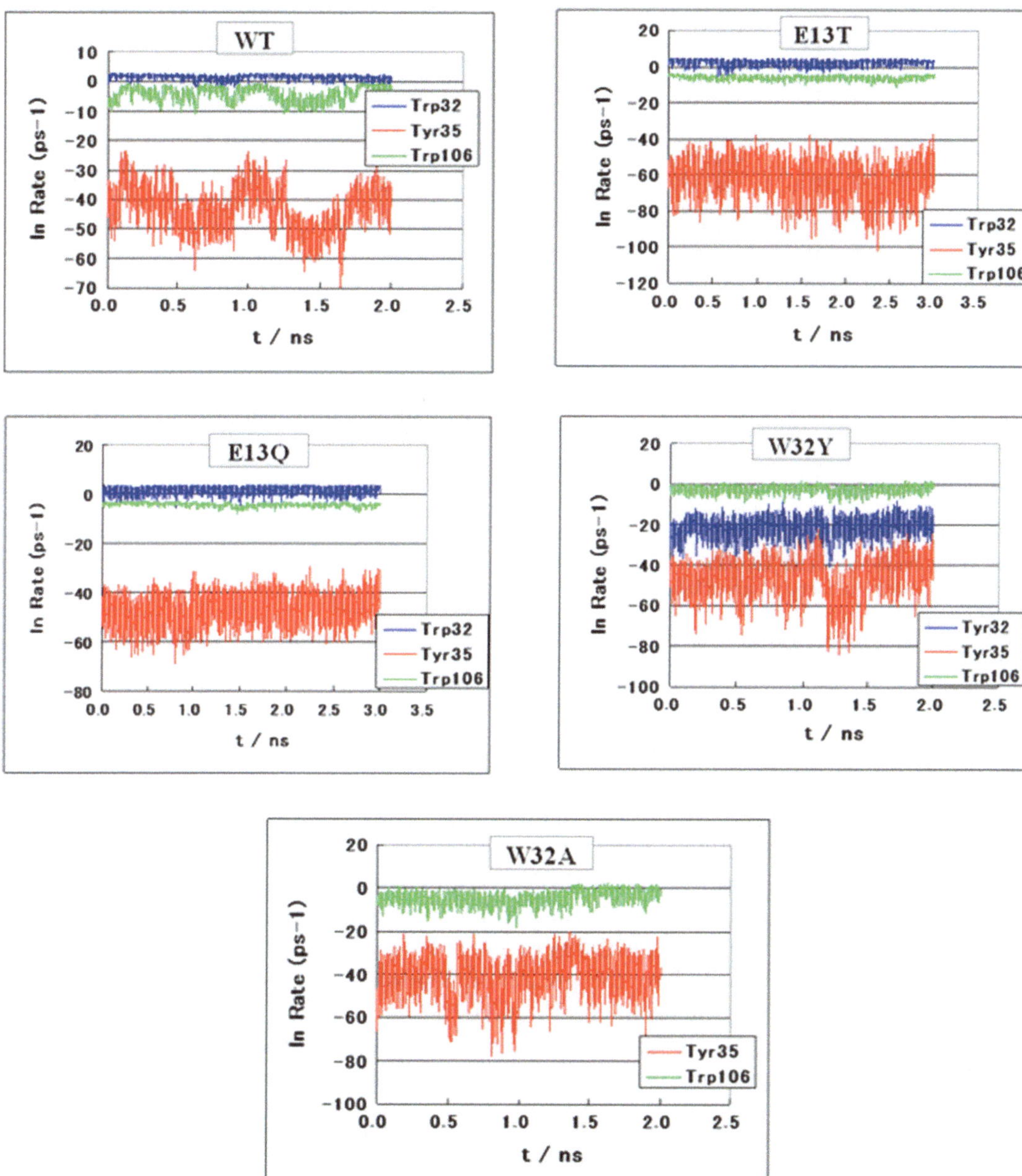

Fig. 6 Time evolution of PET rate from the individual donor to Iso* in the five FBP isoforms. Data are taken from the work by Nadtanet et al. [49] under permission of the PCCP Owner Societies

Table 4
The ET rate and related physical quantities[a]

Quantity	Donor	WT	E13T	E13Q	W32Y	W32A
k_{ET}^{jk} [b] (ps^{-1})	Trp32	7.10	17.22	10.81	–	–
	Tyr32	–	–	–	1.6×10^{-7}	–
	Tyr35	4×10^{-14}	6.4×10^{-21}	7×10^{-17}	3.7×10^{-14}	5×10^{-13}
	Trp106	0.082	0.003	0.018	0.192	0.176
λ_S^{jk} [c] (eV)	Trp32	0.202	0.206	0.208	–	–
	Tyr32	–	–	–	0.217	–
	Tyr35	0.249	0.240	0.229	0.236	0.231
	Trp106	0.232	0.223	0.225	0.223	0.222
$ES_j(k)$ [d] (eV)	Trp32	0.076	0.312	0.290	–	–
	Tyr32	–	–	–	0.197	–
	Tyr35	0.150	0.449	0.412	0.330	0.321
	Trp106	−0.069	−0.349	−0.276	0.215	0.309
$-e^2/\varepsilon_0^{DA}R_{jk}$ [e] (eV)	Trp32	−0.949	−0.912	−0.883	–	–
	Tyr32	–	–	–	−1.009	–
	Tyr35	−0.660	−0.759	−0.883	−0.803	−0.863
	Trp106	−0.627	−0.722	−0.703	−0.728	−0.728
ΔG_T^0 [f] (eV)	Trp32	0.371	0.098	0.090	–	–
	Tyr32	–	–	–	−0.491	–
	Tyr35	−0.792	−0.992	−0.832	−0.829	−0.760
	Trp106	0.194	0.569	0.477	0.011	−0.073

[a]Mean values are obtained by taking an average over the entire MDS time range. Data are taken from the work by Nadtanet et al. [49] (Reproduced by permission of the PCCP Owner Societies)
[b]k_{ET}^{jk} is given by Eq. 1. $j = 1$ for WT, 2 for E13T, 3 for E13Q, 4 for W32Y, and 5 for W32A. $k = 1$ for Trp32 (Tyr32 in W32Y), 2 for Tyr35, and 3 for Trp106
[c]Solvent reorganization energy is given by Eq. 2
[d]Net ES energy is given by Eq. 5
[e]ES energy between a donor and acceptor. ε_0^{DA} is static dielectric constant in the domain between the donors and acceptor
[f]Total free energy gap is given by Eq. 11

much slower than Trp32 and Trp106. The mean PET rates over the MDS time range from the individual donor to Iso* are listed in Table 4. The PET rate from Trp32 is fastest in E13T, followed by E13Q and WT. The PET rate of Tyr32 in W32Y is much slower than the one of Trp32, despite the fact that the Tyr32–Iso distance is shorter by 0.1 nm than the Trp32–Iso distances. The electron donating capability of Tyr is much weaker than that of Trp, which is ascribed to the greater ionization potential of Tyr (8.0 eV) than that of Trp (7.2 eV) [62].

The solvent reorganization energy does not vary appreciably among the donors of the five FBP isoforms. This is because ε_0^{DA} (*see* Eq. 2) is common among the donors of the proteins. The ES energies between Iso^- and $Trp32^+$ and between Iso^- and $Trp106^+$ are around −0.9 eV and −0.7 eV, respectively. However, net ES energies [$ES_j(k)$] obtained by Eq. 5 vary markedly among the

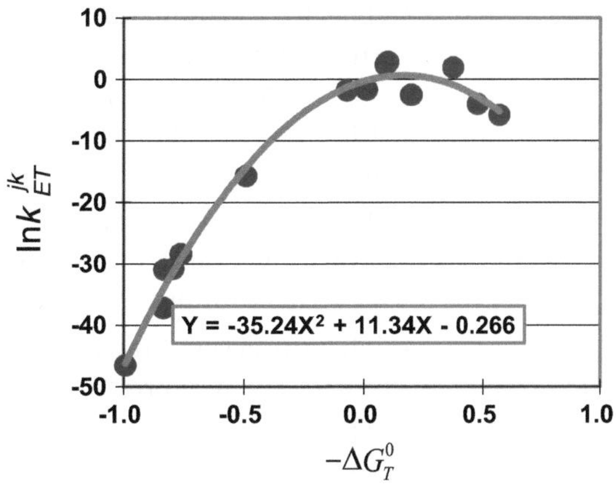

Fig. 7 Energy gap law in the five FBP isoforms. ET rate is expressed in unit of ps^{-1} and $-\Delta G_T^0$ in eV unit. Insert denotes an approximate parabolic function, $Y(\ln k_{ET}^{jk})$ vs. $X(-\Delta G_T^0)$. Data are taken from the work by Nadtanet et al. [49] under permission of the PCCP Owner Societies

donors and the FBP isoforms. The net ES energy of Trp32 is obtained to be 0.076 eV in WT, while 0.31 eV in E13T. The net ES energies of Trp106 are negative, −0.07 eV in WT and −0.35 eV in E13T, while those in W32Y and W32A are positive, 0.22 eV and 0.31 eV, respectively.

3.5 Energy Gap Law

The total free energy gap ($-\Delta G_T^0$) is expressed by Eq. 11:

$$-\Delta G_T^0 = -\Delta G_q^0 + e^2/\varepsilon_0^{DA} R_{jk} - ES_j(k) \qquad (11)$$

Figure 7 shows the energy gap laws obtained with all five FBP isomers. All required numerical data are listed in Table 4. The logarithmic PET rates with the KM model along $-\Delta G_T^0$ are expressed well by a parabolic function, being $Y = -35.24 X^2 + 11.34X - 0.266$, where Y is $\ln k_{KM}^{jk}$ (in ps^{-1} unit) and X is $-\Delta G_T^0$ (in eV unit). The values of $-\Delta G_T^0$ are dependent much on $ES_j(k)$, but not much on the other energies in the right-hand side of Eq. 11, as stated above. The PET in these FBP isoforms mostly takes place in the normal region. The value of $-\Delta G_T^0$ at the maximum rate is ca. 0.2 eV.

4 Notes

The PET mechanism in Flavoproteins obtained from Ultrafast Fluorescence Dynamics and MDS [26, 40, 47–53, 64] may be summarized as follows:

1. The donor–acceptor distance is expressed with the center-to-center distances [26, 64], rather than edge-to-edge distances [40].

2. The PET parameters (ν_0^{Trp}, ν_0^{Tyr}, β^{Trp}, β^{Tyr}, R_0^{Trp}, and R_0^{Tyr}) in the electronic coupling term are dependent on the donor [47, 48]. Effects of variations in the electronic parameters on PET rate are rather small compared to those in the nuclear term.
3. The ultrafast fluorescence of flavoproteins normally decays with a non-exponential function. This observation can be explained in terms of rapid fluctuations of the protein structure [47, 48].
4. The observed fluorescence dynamics of flavoproteins can be satisfactorily reproduced by the KM model.
5. The PET theories in solution are modified by introducing the ES energy between the photoproducts and ionic groups inside the proteins into the nuclear term (exponential term) of the theories [47, 48].
6. The ES energy turns out to be very important in PET processes in flavoproteins [47–53].
7. The ES energy depends on the static dielectric constants inside the flavoproteins. The dielectric constant between the donors and acceptor is much less than those of entire proteins [49].
8. Hydrogen bonding between Iso and surrounding amino acids may modify the PET rate through the free energy-related electron affinity of Iso (G_{Iso}^{0}) in Eq. 3 [53].

Sometimes it is quite difficult to completely reproduce the observed decay by the present method. As shown in Fig. 5, the calculated decays hardly reproduce the longer tails of the observed decays. This demonstrates the limitation of the present method. However, it is expected that the accuracy of the method is improved upon the following modifications.

9. The reliability of the PET parameters determined may be limited, when the number of unknown PET parameters increases with the PET analysis. When a crystal structure is used, the average fluorescence lifetime is used as an experimental data point [26, 29, 65]. The experimental fluorescence decay may be a more critical factor than the fluorescence lifetime, because a decay has many experimental points, while the lifetime has only one value. In this regard the results may be more reliable when several experimental decays are simultaneously analyzed.
10. The ES energies between the photoproducts and ionic groups in the proteins and between Iso^- and donor^+ are evaluated with point charges of the respective ionic species. These may be replaced by partial charges as determined by a molecular orbital theory.
11. The dielectric constant introduced in the present method is time independent. It should be time dependent, though it

is difficult to evaluate this at the moment. According to Onsager theory [65] based upon a continuum model, the dielectric constant depends on dipole–dipole interaction among solvent molecules. However, this model is for a uniform solution with an infinite molecular assembly. It is not clear whether the Onsager model can be applied for such heterogeneous protein system with limited molecular size.

12. Experimental values of ionization potentials for Trp and Tyr are used as in Eq. 3. These ionization potentials were determined in a vacuum. The ionization potential should be dependent on the polarity of the medium. Evidently, PET processes in a protein occur in a medium with some polarity expressed with the static dielectric constant. The dependence of the ionization potential on the dielectric constant is not clear yet. The dielectric constant-dependent ionization potentials should be introduced in the future.
13. Iso contains four proton acceptors and one donor. Hydrogen bondings between Iso and surrounding amino acids may directly relate to electron affinity of Iso* (G^0_{Iso}) as stated in **Note 8**. If an analytical relationship between the hydrogen bonding pair and G^0_{Iso} is known, then G^0_{Iso} may not be an unknown parameter anymore. It is also noted that G^0_{Iso} may be dependent on the polarity of the medium.

References

1. Massey V (2000) The chemical and biological versatility of riboflavin. Biochem Soc Trans 28:283–296
2. Theorell H (1935) Preparation in pure state of the effect group of yellow enzymes. Biochem Zeitsch 275:344–346
3. Frago S, Gomez-Moreno C, Medina M (eds) (2009) Flavins and flavoproteins 2008. Prensas Universitarias de Zaragoza, Zaragoza
4. Weber G (1947) Fluorescence of riboflavin, diaphorase and related substances. PhD Thesis in Biochemistry, St John's College, University of Cambridge
5. Weber G (1950) Fluorescence of riboflavin and flavin-adenine dinucleotide. Biochem J 47:114–121
6. Weber G (1948) The quenching of fluorescence in liquids by complex formation: determination of the mean life of the complex. Trans Faraday Soc 44:185–189
7. Spencer RD, Weber G (1972) Thermodynamics and kinetics of the intramolecular complex in flavin-adenine dinucleotide. In: Åkeson A, Ehrenberg A (eds) Structure and function of oxidation-reduction enzymes. Pergamon, Oxford–New York, pp 393–399
8. Weber G, Tanaka F, Okamoto BY et al (1974) The effect of pressure on the molecular complex of isoalloxazine and adenine. Proc Natl Acad Sci USA 71:1264–1266
9. McCormick DB (1977) Interactions of flavins with amino acid residues: assessments from spectral and photochemical studies. Photochem Photobiol 26:169–182
10. van den Berg PA, Visser AJ (2001) Tracking molecular dynamics of flavoproteins with time-resolved fluorescence spectroscopy. In: Valeur B, Brochon JC (eds) New trends in fluorescence spectroscopy applications to chemical and life sciences. Springer, Berlin, pp 457–485
11. Crosson SK, Moffat K (2001) Structure of a flavin-binding plant photoreceptor domain: insights into light-mediated signal transduction. Proc Natl Acad Sci USA 98:2995–3000
12. Kinoshita T, Doi M, Suetsugu N et al (2001) phot1 and phot2 mediate blue light regulation of stomatal opening. Nature 414:656–660
13. Bouly PJ, Schleicher E, Dionisio-Sese M et al (2007) Cryptochrome blue light photoreceptors are activated through interconversion of flavin redox states. J Biol Chem 282:9383–9391

14. Masuda S, Bauer CE (2002) AppA is a blue light photoreceptor that antirepresses photosynthesis gene expression in *Rhodobacter sphaeroides*. Cell 110:613–623
15. Millar AJ (2004) Input signals to the plant circadian clock. J Exp Bot 55:277–283
16. Moskvin OV, Kaplan S, Gilles-Gonzalez M-A et al (2007) Novel heme-based oxygen sensor with a revealing evolutionary history. J Biol Chem 282:28740–28748
17. Kita A, Okajima K, Morimoto Y et al (2005) Structure of a cyanobacterial BLUF protein, Tll0078, containing a novel. FAD-binding blue light sensor domain. J Mol Biol 349:1–9
18. Silva E, Edwards AM (eds) (2006) Flavins: photochemistry and photobiology. RCS Publishing, Cambridge, UK
19. Karen A, Ikeda N, Mataga N et al (1983) Picosecond laser photolysis studies of fluorescence quenching mechanisms of flavin: a direct observation of indole-flavin singlet charge transfer state formation in solutions and flavoenzymes. Photochem Photobiol 37:495–502
20. Karen A, Sawada MT, Tanaka F et al (1987) Dynamics of excited flavoproteins-picosecond laser photolysis studies. Photochem Photobiol 45:49–53
21. Zhong DP, Zewail AH (2001) Femtosecond dynamics of flavoproteins: charge separation and recombination in riboflavin (vitamin B2)-binding protein and in glucose oxidase enzyme. Proc Natl Acad Sci USA 98:11867–11872
22. van den Berg PA, van Hoek A, Walentas CD et al (1998) Flavin fluorescence dynamics and photoinduced electron transfer in *Escherichia coli* glutathione reductase. Biophys J 74:2046–2058
23. Mataga N, Chosrowjan H, Shibata Y et al (1998) Ultrafast fluorescence quenching dynamics of flavin chromophores in protein nanospace. J Phys Chem B 102:7081–7084
24. Mataga N, Chosrowjan H, Shibata Y et al (2000) Dynamics and mechanisms of ultrafast fluorescence quenching reactions of flavin chromophores in protein nanospace. J Phys Chem B 104:10667–10677
25. Mataga N, Chosrowjan H, Taniguchi S et al (2002) Femtosecond fluorescence dynamics of flavoproteins: comparative studies on flavodoxin, its site-directed mutants, and riboflavin binding protein regarding ultrafast electron transfer in protein nanospaces. J Phys Chem B 106:8917–8920
26. Tanaka F, Chosrowjan H, Taniguchi S et al (2007) Donor-acceptor distance-dependence of photoinduced electron-transfer rate in flavoproteins. J Phys Chem B 111:5694–5699
27. Chosrowjan H, Taniguchi S, Mataga N et al (2007) Comparison between ultrafast fluorescence dynamics of FMN binding protein from *Desulfovibrio vulgaris*, strain miyazaki, in solution vs crystal phases. J Phys Chem B 111:8695–8697
28. Chosrowjan H, Taniguchi S, Mataga N et al (2008) Ultrafast fluorescence dynamics of FMN-binding protein from *Desulfovibrio vulgaris* (Miyazaki F) and its site-directed mutated proteins. Chem Phys Lett 462:121–124
29. Chosrowjan H, Taniguchi S, Mataga N et al (2010) Effects of the disappearance of one charge on ultrafast fluorescence dynamics of the FMN binding protein. J Phys Chem B 114:6175–6182
30. Marcus RA (1956) On the theory of oxidation-reduction reactions involving electron transfer. J Chem Phys 24:966–978
31. Marcus RA (1956) Electrostatic free energy and other properties of states having nonequilibrium polarization. J Chem Phys 24:979–989
32. Marcus RA (1964) Chemical and electrochemical electron-transfer theory. Annu Rev Phys Chem 15:155–196
33. Bixon M, Jortner J (1991) Non-Arrhenius temperature dependence of electron-transfer rates. J Phys Chem 95:1941–1944
34. Bixon M, Jortner J (1993) Charge separation and recombination in isolated supermolecules. J Phys Chem 97:13061–13066
35. Bixon M, Jortner J, Cortes J, Heitele H, Michel-Beyerle ME (1994) Energy-gap law for nonradiative and radiative charge-transfer in isolated and in solvated supermolecules. J Phys Chem 98:7289–7299
36. Kakitani T, Mataga N (1985) New energy gap laws for the charge separation process in the fluorescence quenching reaction and the charge recombination process of ion pairs produced in polar solvents. J Phys Chem 89:8–10
37. Kakitani T, Yoshimori A, Mataga N (1991) Theoretical analysis of energy-gap laws of electron-transfer reactions. In: Bolton JR, Mataga N, McLendon G (eds) Electron transfer in inorganic, organic, and biological systems. American Chemical Society, Washington, DC, pp 45–69
38. Kakitani T, Yoshimori A, Mataga N (1992) Effects of the donor-acceptor distance distribution on the energy gap laws of charge separation and charge recombination reactions in polar solutions. J Phys Chem 96:5385–5392
39. Hush NS (1961) Adiabatic theory of outer sphere electron-transfer reactions in solution. Trans Faraday Soc 57:557–580

40. Moser CC, Keske JM, Warncke K et al (1992) Nature of biological electron transfer. Nature 355:796–802
41. Warshel A, Chu TZ, Parson WW (1989) Dispersed polaron simulations of electron transfer in photosynthetic reaction center. Science 246:112–116
42. Warshel A, Parson WW (1991) Computer simulations of electron transfer reactions in solution and photosynthetic reaction centers. Annu Rev Phys Chem 42:279–309
43. Warshel A, Parson WW (2001) Dynamics of biochemical and biophysical reactions: insight from computer simulation. Quart Rev Biophys 34:563–679
44. Beratan DN, Balabin IA (2008) Heme-copper oxidases use tunneling pathways. Proc Nat Acad Sci USA 105:403–404
45. Bendall DS (1996) Protein electron transfer. BIOS Scientific Publishers Ltd., Oxford, UK
46. Gray HB, Winkler JR (1996) Electron transfer in proteins. Annu Rev Biochem 65:537–561
47. Nunthaboot N, Tanaka F, Kokpol S et al (2009) Simulation of ultrafast non-exponential fluorescence decay induced by electron transfer in FMN binding protein. J Photochem Photobiol A 201:191–196
48. Nunthaboot N, Tanaka F, Kokpol S et al (2008) Simultaneous analysis of ultrafast fluorescence decays of FMN binding protein and its mutated proteins by molecular dynamic simulation and electron transfer theory. J Phys Chem B 112:13121–13127
49. Nunthaboot N, Pianwanit S, Kokpol S et al (2011) Simultaneous analyses of photoinduced electron transfer in the wild type and four single substitution isomers of the FMN binding protein from *Desulfovibrio vulgaris*, Miyazaki F. Phys Chem Chem Phys 13:6085–6097
50. Lugsanangarm K, Pianwanit S, Kokpol S et al (2011) Analysis of photoinduced electron transfer in flavodoxin. J Photochem Photobiol A 217:333–340
51. Lugsanangarm K, Pianwanit S, Kokpol S et al (2011) Photoinduced electron transfer in wild type and mutated flavodoxin from *Desulfovibrio vulgaris*, strain Miyazaki F: energy gap law. J Photochem Photobiol A 219:32–41
52. Lugsanangarm K, Pianwanita S, Kokpol S et al (2011) Theoretical analyses of photoinduced electron transfer in medium chain acyl-CoA dehydrogenase: electron transfer in the normal region. J Photochem Photobiol A 224:80–90
53. Nueangaudom A, Lugsanangarm K, Pianwanit S et al (2012) Structural basis for the temperature-induced transition of D-amino acid oxidase from pig kidney revealed by molecular dynamic simulation and photoinduced electron transfer. Phys Chem Chem Phys 14:2567–2578
54. Kitamura M, Kojima S, Ogasawara K et al (1994) Novel FMN-binding protein from *Desulfovibrio vulgaris* (Miyazaki F). Cloning and expression of its gene in Escherichia coli. J Biol Chem 269:5566–5573
55. Liepinsh E, Kitamura M, Murakami T et al (1997) Pathway of chymotrypsin evolution suggested by the structure of the FMN-binding protein from *Desulfovibrio vulgaris* (Miyazaki F). Nat Struct Mol Biol 4:975–979
56. Suto K, Kawagoe K, Shibata N et al (2000) How do the X-ray structure and the NMR structure of FMN-binding protein differ? Acta Crystallogr D 56:368–371
57. Case D, Darden T, Cheatham T et al (2004) AMBER, ver 8. University of California, San Francisco
58. Wang JM, Cieplak P, Kollman PA (2000) How well does a restrained electrostatic potential (RESP) model perform in calculating conformational energies of organic and biological molecules? J Comput Chem 21:1049–1074
59. Schneider C, Suhnel J (1999) The kinetic deuterium isotope effect of proton transfer reactions in solution. Biopolymers 50:287–302
60. Essmann U, Perera L, Berkowitz ML et al (1995) A smooth particle mesh Ewald method. J Chem Phys 103:8577–8593
61. Ryckaert JP, Ciccotti G, Berendsen HJC (1977) Numerical integration of the Cartesian equations of motion of a system with constraints: molecular dynamics of n-alkanes. J Comp Phys 23:327–341
62. Vorsa V, Kono T, Willey KF et al (1999) Femtosecond photoionization of ion beam desorbed aliphatic and aromatic amino acids: fragmentation via α-cleavage reactions. J Phys Chem B 103:7889–7895
63. Henry ER, Hochstrasser RM (1987) Molecular dynamics simulations of fluorescence polarization of tryptophans in myoglobin. Proc Natl Acad Sci USA 84:6142–6146
64. Tanaka F, Rujkorakarn R, Chosrowjan H et al (2008) Analyses of donor–acceptor distance-dependent rates of photo-induced electron transfer in flavoproteins with three kinds of electron transfer theories. Chem Phys 348:237–241
65. Böttcher CJF (1973) Theory of electric polarization, 2nd edn. Elsevier, Amsterdam

Part III

Fluorescent Probe Development (FPD)

Chapter 15

Biosynthetic Incorporation of Tryptophan Analogs in Proteins

Jaap Broos

Abstract

Biosynthetic incorporation of Trp analogs in a protein can help in its characterization using fluorescence spectroscopy and other methodologies like NMR and phosphorescence. Here a protocol is presented resulting in the efficient incorporation of Trp analogs in a recombinant protein, using an *Escherichia coli* Trp auxotroph. An overview of recent developments in the Trp analog incorporation field is also presented.

Key words Trp auxotroph, Fluorescence, Alloprotein

1 Introduction

Replacing Trp in a protein by a Trp analog can be advantageous in its characterization using various spectroscopy techniques including fluorescence spectroscopy. Trp is a popular fluorescence probe in protein science because it is an intrinsic probe and its quantum yield (Q) and emission energy are very sensitive for changes in its microenvironment [1]. Under certain experimental conditions, one cannot exploit these advantages of Trp photophysics, for example, when there are other Trp-containing proteins in the sample or when the absorbance of sample components, like DNA at high concentration, overlaps with Trp absorbance. This can result in a strong inner filter effect, limiting efficient excitation of Trp. Replacing Trp by a Trp analog absorbing at higher wavelengths may help in overcoming this. The absorbance and fluorescence characteristics of Trp can be engineered via a subtle modification in the indole moiety, and many Trp analogs can be efficiently biosynthetically incorporated. *Escherichia coli* is the most popular expression host for the incorporation of Trp analogs in recombinant proteins, and the following Trp analogs with an indole moiety are translated: 4-, 5-, and 6-fluoroTrp [2–4]; 2-azaTrp [5], 4-azaTrp [6], 5-azaTrp [6], 6-azaTrp [7], and 7-azaTrp [5];

Yves Engelborghs and Antonie J.W.G. Visser (eds.), *Fluorescence Spectroscopy and Microscopy: Methods and Protocols*, Methods in Molecular Biology, vol. 1076, DOI 10.1007/978-1-62703-649-8_15,

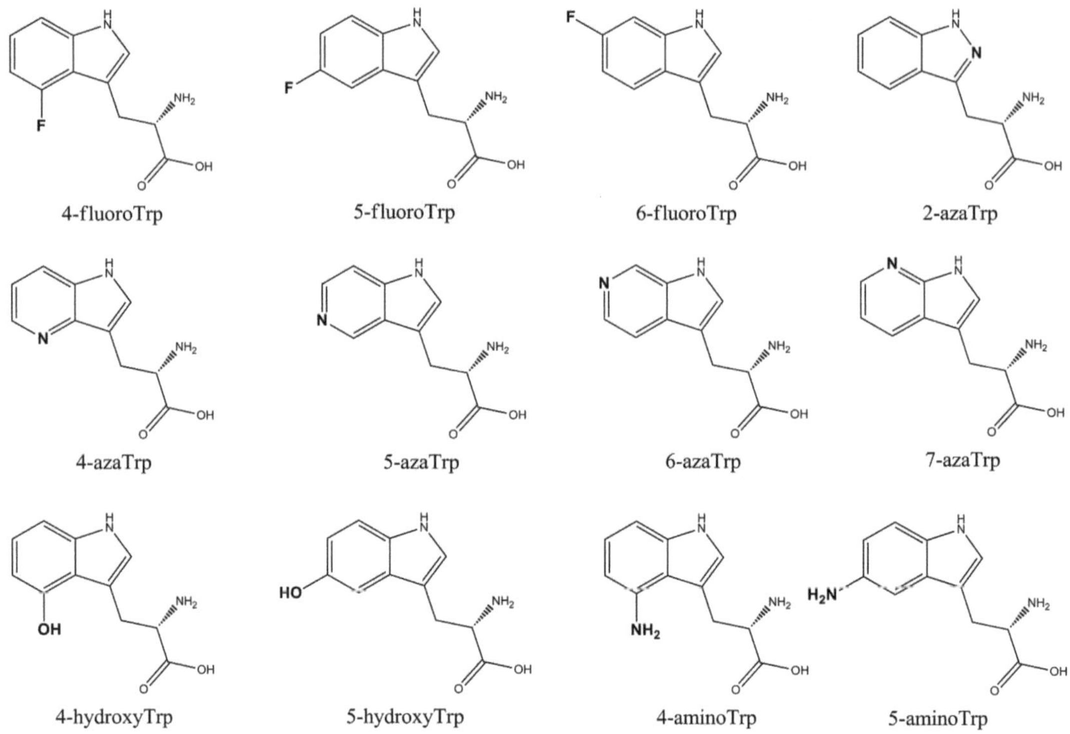

Fig. 1 Structures of Trp analogs biosynthetically incorporated in proteins using an *Escherichia coli* Trp auxotroph

4- and 5-hydroxyTrp [8, 9]; and 4- and 5-aminoTrp [10] (Fig. 1). The absorbance spectra of Trp and three popular Trp analogs, 7-azaTrp, 5-hydroxyTrp, and 5-fluoroTrp, are shown in Fig. 2.

For the efficient replacement of Trp in the recombinant protein by a Trp analog, a Trp auxotroph expression host is needed and the gene of the target protein should be cloned behind an inducible non-leaky promotor. Most groups use the so-called "two-step" incorporation protocol. In brief, during the first step the auxotroph, harboring the plasmid containing the gene of the target protein, is grown in a synthetic medium, like M9 medium, supplemented with Trp. Cells are pelleted at mid-exponential phase and washed several times to remove Trp. Culturing of the cells is continued in the same medium without Trp for ~30 min to reach full Trp depletion. Trp analog is added and after ~15 min, allowing uptake of the analog by the cells, expression of the target protein is induced and incubation continued during a period optimal for the expression of the target protein.

In this chapter, a "two-step" protocol is presented in Subheadings 2 and 3. Below a brief description of those analogs of interest for fluorescence spectroscopy is presented; a more comprehensive description of these analogs can be found in several excellent reviews [11–14] and in the cited references. An overview of recent developments in the Trp analog incorporation field, not included in the reviews mentioned above, is also presented.

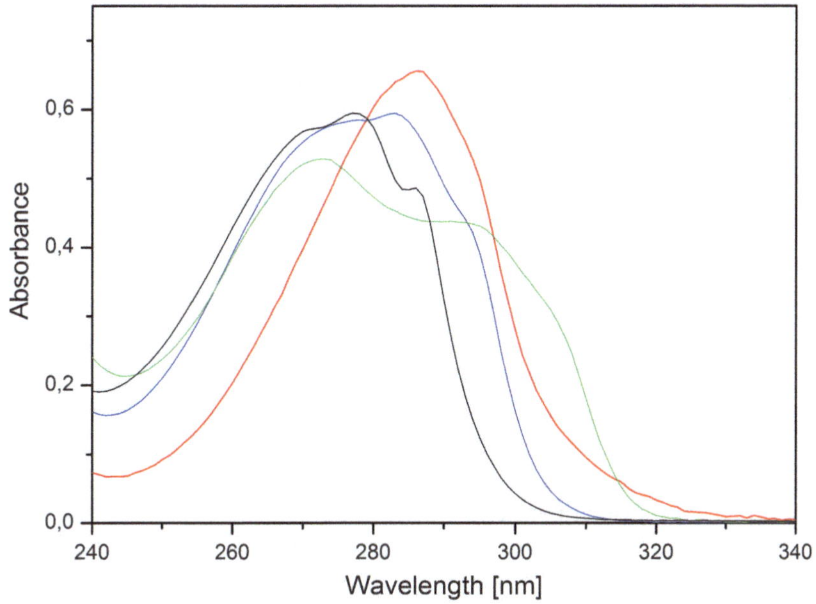

Fig. 2 Absorbance spectra of equimolar concentrations of Trp (*black*), 7-azaTrp (*red*), 5-hydroxyTrp (*green*), and 5-fluoroTrp (*blue*) in neutral buffer

1.1 Trp Analogs Translated by E. coli

1.1.1 Fluorotryptophan

FluoroTrp analogs have been biosynthetically incorporated in many proteins, mainly because of their application as intrinsic probe in ^{19}F NMR experiments. Replacing Trp by fluoroTrp can also be attractive when the protein is analyzed using fluorescence spectroscopy. 4-FluoroTrp is not fluorescent at ambient temperatures [4]. Replacing Trp by this analog in a protein results in a fluorescently silent protein, which can be an advantage if a Trp residue in another protein, also present in the sample, is the target of the investigation.

The absorbance spectrum of 5-fluoroTrp is 7 nm red shifted compared to Trp (Fig. 2). 5-FluoroTrp shows a ~50 % higher Q than Trp [15] and this has been related to the increase of the ionization potential of indole upon 5-fluoro substitution [16]. The higher ionization potential suppresses electron transfer of the excited 5-fluoroTrp to a nearby peptide bond, acting as electron acceptor. This electron transfer process is an efficient quenching route of Trp in a protein, and its rate can vary between ground-state subpopulations of the protein, resulting in different Trp lifetimes. Efficient suppression of this electron transfer process in 5-fluoroTrp explains its higher Q and the homogeneous fluorescence decay, which is often monoexponential, in particular if the 5-fluoroTrp is at a buried position and features a blue emission [17–20]. The homogeneous decay of 5-fluoroTrp makes it an attractive donor in time-resolved FRET experiments. The high sensitivity of the Q of Trp for a change in microenvironment is mainly due to changes in the efficiency of the electron transfer to the peptide bond [1]. Therefore, the Q of 5-fluoroTrp embedded in a protein is less sensitive for a change in microenvironment than Trp at that position [16].

The ionization potential of 6-fluoroTrp is in between that of Trp and 5-fluoroTrp, and its decay was found more homogeneous than for Trp, but less than when it is replaced by 5-fluoroTrp [16].

1.1.2 7-AzaTrp

Replacing Trp by 7-azaTrp in a protein introduces two attractive properties: (1) its absorbance is red shifted compared to Trp, allowing its specific excitation at 310–315 nm where Trp does not absorb (Fig. 2), and (2) the Q is very sensitive for changes in the microenvironment. Larger than tenfold changes in Q for a 7-azaTrp containing protein have been observed [11]. Compared to Trp, the emission spectrum is also 30–60 nm red shifted. Free 7-azaTrp is almost nonfluorescent ($Q = 0.01$–0.02), while free Trp shows a Q of 0.14. The large variation in Q makes 7-azaTrp an attractive reporter probe in titration experiments. For monitoring conformational changes during protein (un)folding, including conformations where the 7-azaTrp is fully solvent exposed, loss of the emission signal limits its use as reporter group. 4-AzaTrp and 5-azaTrp have been proposed as alternatives to overcome this limitation, as they show a modest Q when free in solution [6]. These analogs can also be selectively excited >310 nm and show larger Stokes shifts than 7-azaTrp. Only few proteins have been labeled with these analogs, and the reported data indicate the λ_{em}^{max} and Q of 4-azaTrp and 5-azaTrp are not very sensitive for changes in microenvironment.

1.1.3 5-HydroxyTrp

Like 7-azaTrp, the absorbance spectrum of 5-hydroxyTrp allows its selective excitation at 315 nm (Fig. 2). The Q of free 5-hydroxyTrp, ~0.25, is relatively high for a Trp analog. 5-HydroxyTrp is one of the few Trp analogs known to emit from the ^1L_b state. Only a small change in permanent dipole is induced when 5-hydroxyTrp is excited from ground to the ^1L_b state, and this explains why the emission energy of 5-hydroxyTrp is not very sensitive for changes in microenvironment. The red shoulder of 5-hydroxyTrp is due to ^1L_b absorption, and red-edge excitation gives a high intrinsic anisotropy because no depolarization from ^1L_a to ^1L_b takes place. This makes 5-hydroxyTrp an attractive probe for anisotropy studies [8, 9].

1.2 Recent Developments in the Trp Analog Incorporation Field

1.2.1 E. coli Trp Auxotroph Expression Hosts

Most Trp analog-containing proteins reported in the literature have been produced using an *E. coli* Trp auxotroph. We often use the M5219 Trp auxotroph, a strain engineered with the cI_{857} repressor gene in its chromosome. This makes it an attractive strain if the gene of the target protein is cloned behind the heat-inducible λP_R or λP_L promoters [21].

Employing a Trp auxotrophic strain results in the global replacement of all Trp positions in the target protein by a Trp analog. When site-specific information is needed of a multi-Trp-containing protein via Trp analog spectroscopy, first a single Trp construct needs to be created, and replacing many Trp positions can compromise structure and function of the protein. In such a case, the amber codon suppression technology [22] is attractive as it has the potential to label a multi-Trp-containing protein site selectively with a Trp analog. In this approach, the expression host is outfitted with an orthogonal tRNA/aminoacyl-tRNA synthetase pair, able to label a protein with a Trp analog in response to an engineered stop codon in the gene of the target protein.

One example is the orthogonal yeast tRNA/ aminoacyl-tRNA synthetase pair reported by Kwon and Tirrell [23]. A single mutation, T415G, was introduced in phenylalanyl-tRNA synthetase making it able to aminoacylate the tRNA with 6-chloroTrp, 6-bromoTrp, or benzothienylalanine. This orthogonal pair was introduced in a Phe/Trp double auxotrophic *E. coli* strain, and the three analogs could be incorporated in a protein with high fidelity in response to an amber codon in the gene of the target protein. The engineered synthetase also recognizes Trp and Phe, and to prevent incorporation of Trp at the position corresponding with the amber codon, the synthetic medium was supplemented during expression of the target protein with a very low concentration of Phe and Trp and a high concentration of Trp analog. In this way, the endogenous *E. coli* translation system, showing a higher affinity for Trp than the engineered synthetase, translated the Trp codon positions in the target protein, while the position corresponding to the amber codon was labeled for >98 % with Trp analog. Unfortunately, the three analogs used in this study are not or minimally fluorescent. Because attractive fluorescent Trp analogs like 7-azaTrp and 5-hydroxyTrp are translated by *E. coli*, this system cannot be used to label a protein, in response to an engineered amber codon, with one of these analogs.

1.2.2 Other Trp Auxotroph Expression Hosts

Besides *E. coli*, several other expression systems were recently reported for the incorporation of Trp analogs.

The gram-positive bacterium *Lactococcus lactis* is known as an attractive expression host because many proteins, including membrane proteins, can be functionally expressed at high levels [24]. The organism grows under aerobic and anaerobic conditions, shows a relative low proteolytic activity, and a tightly regulated nisin-inducible promoter system [25] is available. A *L. lactis* Trp auxotroph was created by El Khattabi et al. via deletion of the Trp synthetase gene *trpBA* [26]. A single Trp-containing target protein, cloned behind the nisin-inducible promoter P_{nisA}, was expressed in the presence of Trp analogs using this Trp auxotroph. 7-AzaTrp and 5-fluoroTrp could be efficiently incorporated (>97 %) and also 5-hydroxyTrp, although its incorporation yield (89 %) is somewhat lower than typically found for an *E. coli* Trp auxotroph system. Interestingly, this system can also efficiently incorporate 5-methylTrp (92 %), an analog not translated by

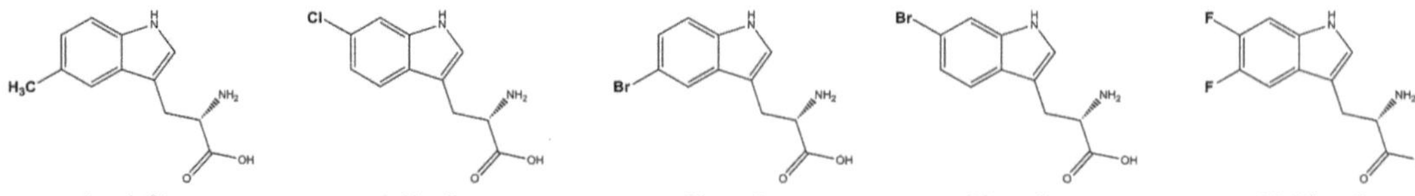

Fig. 3 Structures of Trp analogs biosynthetically incorporated in a protein by a *Lactococcus lactis* Trp auxotroph, coexpressing the *L. lactis* tryptophanyl-tRNA synthetase enzyme [27]

E. coli. This latter result suggests that the substrate specificity of *L. lactis* tryptophanyl-tRNA synthetase is more relaxed than *E. coli* tryptophanyl-tRNA synthetase. The number of Trp analogs translated by *L. lactis* could be increased when the *L. lactis* tryptophanyl-tRNA synthetase was coexpressed together with the target protein [27]. Chloro and bromo atoms containing Trp analogs and also 5,6 difluoroTrp can now be efficiently incorporated (>89 %) (Fig. 3). Coexpression of the *L. lactis* tryptophanyl-tRNA synthetase also increased the incorporation efficiency of 5 hydroxyTrp and 5-methylTrp to 95 %. This new expression system features the most relaxed specificity for Trp analog structures reported to date and gives a high alloprotein yield.

Zhang et al. reported the development of an orthogonal tryptophanyl-tRNA/ tryptophanyl-tRNA synthetase (TrpRS) pair for use in mammalian 293 T cells and able to incorporate 5-hydroxyTrp in response to an opal (TGA) codon in a target protein [28]. A high fidelity was reported for this orthogonal pair, based on tryptophanyl-tRNA and TrpRS of *Bacillus subtilis*. One mutation, V144P, was needed to alter its specificity from Trp to 5-hydroxyTrp. The high fidelity of this system has recently been questioned by Antonczak et al. [29]. The developed TrpRS protein by Zhang et al. was characterized in more detail using a complementation assay, X-ray crystallography, molecular modeling, and calorimetry. Their analysis indicates this TrpRS prefers Trp over 5-hydroxyTrp as substrate, similar to wt TrpRS from *Bacillus subtilis*. Labeling a Trp-containing protein with 5-hydroxyTrp in response to an opal codon position is therefore not possible [29].

Baby hamster kidney (BHK) cells are unable to synthesize Trp, and these cells have been used to incorporate 5-hydroxyTrp in two mammalian proteins, human serum transferrin and a soluble portion of the transferrin receptor, containing 8 and 11 Trp residues, respectively. 5-HydroxyTrp incorporation efficiencies of 20 and 26 %, respectively, were reported [30]. The relative low incorporation efficiencies may be the result of protein turnover as the expression of the target proteins took almost a week.

1.2.3 Promoter Systems

Only tightly controlled promoters will result in the efficient replacement of Trp by a Trp analog, as basal expression of the target protein during growing the cells in the first Trp-containing medium is minimal. The *tac*, T5, λP_R, and λP_L promoters, among others, have been used successfully [11, 12]. In many cases, Trp was replaced by >95 % and yield of target protein was similar as when expressed in a rich medium. For recombinant protein expression in *E. coli*, many laboratories use pET vectors which contain the T7 promoter. As discussed before [11, 12], the T7 promoter is not very suitable for labeling a target protein with a Trp analog. Upon induction, T7 RNA polymerase is expressed and labeling with Trp analogs compromises its transcription activity.

The p_{BAD} expression system is another popular system to express proteins in *E. coli*, and this promoter can be employed to label proteins with Trp analogs. 4-FluoroTrp, 5-fluoroTrp, 5-hydroxyTrp, and 7-azaTrp were incorporated in exotoxin A using this system and an *E. coli* Trp auxotroph strain [31]. The protein was expressed in a good yield, and incorporation efficiencies >82 % were reported for these analogs.

Below a protocol is presented for the biosynthetic incorporation of Trp analogs in a protein using the M5219 *E. coli* Trp auxotroph (*see* **Note 1**) and a plasmid containing the gene of the target protein cloned behind the temperature-inducible λP_R promoter. We used this protocol to incorporate with very high efficiency (≥95 %) 7-azaTrp, 5-hydroxyTrp, or 5-fluoroTrp in many single Trp mutants of the mannitol transporter from *E. coli* and in apoflavodoxin mutants from *Azotobacter vinelandii* [17, 18, 21]. There is no expression of the target protein when the cells are cultured at 30 °C, and induction starts upon increasing the temperature to 42 °C. With only small changes, the same protocol can be used for other promoter systems and/or *E. coli* Trp auxotroph strains (*see* **Note 2**).

2 Materials

2.1 Growth Media

LB medium (1 L).

- 5 g Bacto yeast extract.
- 10 g Bacto tryptone.
- 10 g NaCl.
- Make up to 1 L with deionized water and autoclave.

For M9 minimal medium, the following stock solutions are needed:

M9 mineral salts stock solution

- 17.5 g $Na_2HPO_4 \cdot 12\ H_2O$.
- 3 g KH_2PO_4.
- 0.5 g NaCl.
- 1 g NH_4Cl.
- Make up to 1 L with deionized water and autoclave.

20 % glucose

- 5 g glucose.
- Make up to 25 ml and autoclave.

*20 % thiamine*HCl stock solution*

- 5 g thiamine*HCl.
- Make up to 25 ml with deionized water and filter sterile.

1 M $MgSO_4$

- 6.2 g $MgSO_4 \cdot 7\ H_2O$.
- Make up to 25 ml with deionized water and autoclave.

0.01 M $FeSO_4$

- 0.07 g $FeSO_4 \cdot 7\ H_2O$.
- Make up to 25 ml with 0.01 N HCl and autoclave.

0.1 M $CaCl_2$

- 0.35 g $CaCl_2.2H_20$.
- make up to 25 ml and autoclave..

Trace elements

- 1.0 g EDTA.
- 0.147 g $CaCl_2 \cdot 2H_2O$.
- 0.198 g $MnCl_2 \cdot 4H_2O$.
- 0.029 g $ZnSO_4 \cdot 7H_2O$.
- 0.254 g $CoSO_4 \cdot 6H_2O$.
- 0.013 g $CuCl_2$.
- Make up to 100 ml with deionized water, adjust pH at 4.0 and autoclave.

0.25 M Tryptophan

- 1.3 g L-tryptophan.
- Add 20 ml deionized water.
- While stirring solution with magnetic bar, add a conc. NaOH solution drop by drop till all Trp is dissolved.
- Make up to 25 ml and filter sterile.

The Trp and thiamine stocks are stored in a freezer; the other solutions can be stored at room temperature.

1 L M9 medium, without Trp or Trp analog, is made by mixing the above stock solutions as follows:

- 965 ml M9 mineral salts.
- 25 ml 20 % glucose.
- 5 ml 20 % thiamine*HCl.
- 1 ml 1 M $MgSO_4$.

- 0.1 M $CaCl_2$.
- 1 ml 0.01 M $FeSO_4$.
- 1 ml trace elements.
- Antibiotic marker(s).

2.2 Washing Buffer

- Add to 487 ml sterile M9 mineral salts stock solution 12.5 ml sterile 20 % glucose solution.

3 Methods

Day 1

1. 25 ml of sterile LB medium is transferred to a sterile 100 ml Erlenmeyer, an antibiotic marker(s) is added, and the content inoculated with M5219, transformed with a plasmid containing the gene of the target protein behind the λP_R promoter. A colony of a fresh plate can be used for the inoculating or a few microliters of a frozen stock solution, stored at −80 °C.
2. Incubate overnight at 30 °C and 200 rpm in an incubator.
3. Prepare 1 L M9 medium, including 1 mM L-tryptophan (4 ml 0.25 M L-Trp stock per liter) and transfer it to a sterile 5 L Erlenmeyer.
4. Add 25 ml overnight culture and incubate at 30 °C and 200 rpm in a shaker.
5. Follow growth by taking a sample using a sterile pipette and measure the optical density at 600 nm.
6. When the optical density at 600 nm has reached 0.7–0.9 (1 cm optical patch length), the culture is transferred to centrifuge tubes and pelleted (7 min 5,000 × *g*) at room temperature (*see* **Note 3**).
7. Decant the supernatant and place the centrifuge tube upside down on a piece of paper during several minutes, so most of the remaining medium in the tube is absorbed by the tissue. Rinse the tube lids with water or wipe them dry with a tissue.
8. Resuspend the pellets with washing buffer and collect them in one centrifuge tube. Use in total ~0.5 L of washing buffer and repeat the centrifugation step and **step** 7.
9. Resuspend the pellets in 1 L of M9 medium (total volume), without Trp or Trp analog, and transfer the culture to a clean 5 L culture flask. Add antibiotic marker(s).
10. Start the Trp starvation phase by placing the flask in an incubator at 30 °C, 200 rpm during 30 min.

11. Weigh 1 mmol Trp analog in a 10 ml beaker glass and add a magnetic stirrer and ~5 ml of deionized water. Add slowly a concentrated NaOH solution drop by drop till all Trp analog is dissolved; immersion of the beaker in a sonification batch for a short moment can speed up the solubilization process.
12. At the end of the 30 min starvation time, pour the Trp analog solution into the medium and continue incubation at 30 °C and 200 rpm for 15 more minutes.
13. Start induction by increasing the set temperature to 42 °C and continue shaking the culture till enough protein is produced (4 h in the case of the mannitol transporter) (*see* **Note 2**).
14. Harvest cells via centrifugation and store the pellet at –20 °C.
15. After purification of the protein (*see* **Note 4**), establish the incorporation efficiency using mass spectrometry (*see* **Note 5**).

4 Notes

1. This strain can be ordered at BBCM/LMBP, Belgium, for a small fee (*see* http://bccm.belspo.be/db/lmbp_host_strains/). Many other *E. coli* Trp auxotrophs used for Trp analog incorporation [12] are available at the Genetic Stock Center at Yale University.
2. An available *E. coli* expression protocol for a target protein using a promoter system compatible for Trp analog incorporation (*see* Subheading 1.2.3) can be adapted for Trp analog work by just changing the strain for a Trp auxotroph strain and using M9 medium instead of rich medium. Parameters like growth temperature, inducer concentration, and induction time often do result in a protein expression level similar as with the non-Trp auxotroph strain cultured in rich medium.
3. Centrifuge tubes, the Erlenmeyer used for the second medium with Trp analog, and the Trp analog solution do not need to be sterile.
4. The isolation procedure developed for the Trp-containing protein can be followed as the alloprotein is not expected to behave (significant) differently during the chromatography steps.
5. Mass spectrometry is nowadays the most straightforward approach to analyze the analog incorporation efficiency as the sensitivity of the instruments has improved considerably in the last decade. For example, the mass difference between Trp and 7-azaTrp is just 1Dalton, but its incorporation efficiency in a protein could be accurately established without the need of using a very advanced instrument [26].

References

1. Callis PR (2009) Exploring the electrostatic landscape of proteins with tryptophan fluorescence. In: Geddes CD (ed) Reviews in fluorescence 2007. Springer, pp 199–248
2. Sykes BD, Weingarten HI, Schlesinger MJ (1974) Fluorotyrosine alkaline phosphatase from *Escherichia coli*: preparation, properties, and fluorine-19 nuclear magnetic resonance spectrum. Proc Natl Acad Sci USA 71: 469–473
3. Pratt EA, Ho C (1975) Incorporation of fluorotryptophans into proteins of *Escherichia coli*. Biochemistry 14:3035–3040
4. Hott JL, Borkman RF (1989) The nonfluorescence of 4-fluorotryptophan. Biochem J 264:297–299
5. Schlesinger S (1968) The effect of amino acid analogues on alkaline phosphatase formation in *Escherichia coli* K-12: II. replacement of tryptophan by azatryptophan and by tryptazan. J Biol Chem 243:3877–3883
6. Lepthien S, Hoesl MG, Merkel L, Budisa N (2008) Azatryptophans endow proteins with intrinsic blue fluorescence. Proc Natl Acad Sci USA 105:16095–16100
7. Twine SM, Murphy L, Phillips RS et al (2003) The photophysical properties of 6-azaindole. J Phys Chem B 107:637–645
8. Ross JBA, Senear DF, Waxman E et al (1992) Spectral enhancement of proteins: biological incorporation and fluorescence characterization of 5-hydroxytryptophan in bacteriophage lambda cI repressor. Proc Natl Acad Sci USA 89:12023–12027
9. Hogue CW, Rasquinha I, Szabo AG, MacManus JP (1992) A new intrinsic fluorescent probe for proteins. Biosynthetic incorporation of 5-hydroxytryptophan into oncomodulin. FEBS Lett 310:269–272
10. Budisa N, Rubini M, Bae JH et al (2002) Global replacement of tryptophan with aminotryptophans generates non-invasive protein-based optical pH sensors. Angew Chem Int Ed 41:4066–4069
11. Ross JBA, Szabo AG, Hogue CWV (1997) Enhancement of protein spectra with tryptophan analogs: fluorescence spectroscopy of protein-protein and protein-nucleic acid interactions. Methods Enzymol 278:151–190
12. Ross JBA, Rusinova E, Luck LA, Rousslang KW (2000) Spectral enhancement of proteins by in vivo incorporation of tryptophan analogues. Protein fluorescence. Kluwer Academic/ Plenum, New York, pp 17–42
13. Twine SM, Szabo AG (2003) Fluorescent amino acid analogs. Methods Enzymol 360:104–127
14. Budisa N, Pal PP (2004) Designing novel spectral classes of proteins with a tryptophan-expanded genetic code. Biol Chem 385:893–904
15. Broos J, Maddalena F, Hesp BH (2004) In vivo synthesized proteins with monoexponential fluorescence decay kinetics. J Am Chem Soc 126:22–23
16. Liu TQ, Callis PR, Hesp BH et al (2005) Ionization potentials of fluoroindoles and the origin of nonexponential tryptophan fluorescence decay in proteins. J Am Chem Soc 127:4104–4113
17. Visser NV, Westphal AH, Nabuurs SM et al (2009) 5-Fluorotryptophan as dual probe for ground-state heterogeneity and excited-state dynamics in apoflavodoxin. FEBS Lett 583:2785–2788
18. Opacic M, Vos EP, Hesp BH, Broos J (2010) Localization of the substrate binding site in the homodimeric mannitol transporter, EIImtl, of *Escherichia coli*. J Biol Chem 285: 25324–25331
19. Sarkar SS, Udgaonkar JB, Krishnamoorthy G (2011) Reduced fluorescence lifetime heterogeneity of 5-fluorotryptophan in comparison to tryptophan in proteins: implication for resonance energy transfer experiments. J Phys Chem B 115:7479–7486
20. Opacic M, Hesp BH, Fusetti F et al (2012) Structural investigation of the transmembrane C domain of the mannitol permease from *Escherichia coli* using 5-FTrp fluorescence spectroscopy. Biochim Biophys Acta 1818:861–868
21. Broos J, Gabellieri E, Biemans-Oldehinkel E, Strambini GB (2003) Efficient biosynthetic incorporation of tryptophan and indole analogs in an integral membrane protein. Protein Sci 12:1991–2000
22. Xie JM, Schultz PG (2006) Innovation: a chemical toolkit for proteins–an expanded genetic code. Nat Rev Mol Cell Biol 7:775–782
23. Kwon I, Tirrell DA (2007) Site-specific incorporation of tryptophan analogues into recombinant proteins in bacterial cells. J Am Chem Soc 129:10431–10437
24. Kunji ERS, Slotboom DJ, Poolman B (2003) *Lactococcus lactis* as host for overproduction of functional membrane proteins. Biochim Biophys Acta 1610:97–108

25. Kuipers OP, Kok J (2007) Nisin- and subtilin-controlled gene expression systems for gram-positive bacteria. In: Dyson MR, Durocher Y (eds) Expression systems. Scion, Bloxham, pp 225–240
26. El Khattabi M, van Roosmalen ML, Jager D et al (2008) *Lactococcus lactis* as expression host for the biosynthetic incorporation of tryptophan analogues into recombinant proteins. Biochem J 409:193–198
27. Petrovic DM, Leenhouts K, van Roosmalen ML, Broos J (2013) Amino acids 44:1329–1336
28. Zhang ZW, Alfonta L, Tian F et al (2004) Selective incorporation of 5-hydroxytryptophan into proteins in mammalian cells. Proc Natl Acad Sci USA 101:8882–8887
29. Antonczak AK, Simova Z, Yonemoto IT et al (2011) Importance of single molecular determinants in the fidelity of expanded genetic codes. Proc Natl Acad Sci USA 108:1320–1325
30. James NG, Byrne SL, Mason AB (2009) Incorporation of 5-hydroxytryptophan into transferrin and its receptor allows assignment of the pH induced changes in intrinsic fluorescence when iron is released. Biochim Biophys Acta 1794:532–540
31. Mohammadi F, Prentice GA, Merrill AR (2001) Protein-protein interaction using tryptophan analogues: novel spectroscopic probes for toxin-elongation factor-2 interactions. Biochemistry 40:10273–10283

Chapter 16

Optimization of Fluorescent Proteins

Daphne S. Bindels, Joachim Goedhart, Mark A. Hink, Laura van Weeren, Linda Joosen, and Theodorus W.J. Gadella, Jr.

Abstract

Nowadays, fluorescent protein (FP) variants have been engineered to fluoresce in all different colors; to display photoswitchable, or photochromic, behavior; or to show yet other beneficial properties that enable or enhance a still growing set of new fluorescence spectroscopy and microcopy techniques. This has allowed the (in situ) study of biomolecules with unprecedented resolution, specificity, sensitivity, and ease of labeling. However, brighter FPs, more photostable FPs, and FPs that display an even better compatibility with biophysical microspectroscopic techniques are still highly desired. The key characteristics of FPs—absorption spectrum, emission spectrum, brightness, fluorescence lifetime, maturation rate, oligomeric state, photostability, pH sensitivity, and functionality in protein fusions—determine their application. This chapter will describe these key features and present several experimental protocols to optimize them.

The optimization procedure contains three steps. First the amino acid sequence of a template FP is changed via random or site-directed mutagenesis. A primary screening based on fluorescence intensity, fluorescence lifetime, and emission spectrum is applied on the FP libraries expressed in bacteria. The most promising mutants are isolated, purified, and characterized in vitro. In this step all key characteristics are determined experimentally. Finally the new FPs are evaluated for use in vivo. The protein production and maturation is monitored in bacteria, while transfected mammalian cells report on the photostability, relative brightness, and correct localization to various subcellular compartments.

Key words Green fluorescent protein, Fluorescence lifetime imaging microscopy, Extinction coefficient, Quantum yield, Fluorescence fluctuation spectroscopy, Site-directed mutagenesis, Brightness, mTurquoise2

1 Introduction

Fluorescent proteins (FPs) have become indispensable for visualizing structures and processes in molecular cell biological research [1]. FPs can be used for many purposes, for example, structural studies (i.e., cell organelle and protein labeling), but also for functional studies (i.e., biosensors and drug screening). Because an FP is genetically encoded, protein molecules or cellular structures of interest can be FP labeled within living cells. This minimizes perturbation of living systems. FPs have greatly accelerated

Yves Engelborghs and Antonie J.W.G. Visser (eds.), *Fluorescence Spectroscopy and Microscopy: Methods and Protocols*, Methods in Molecular Biology, vol. 1076, DOI 10.1007/978-1-62703-649-8_16,

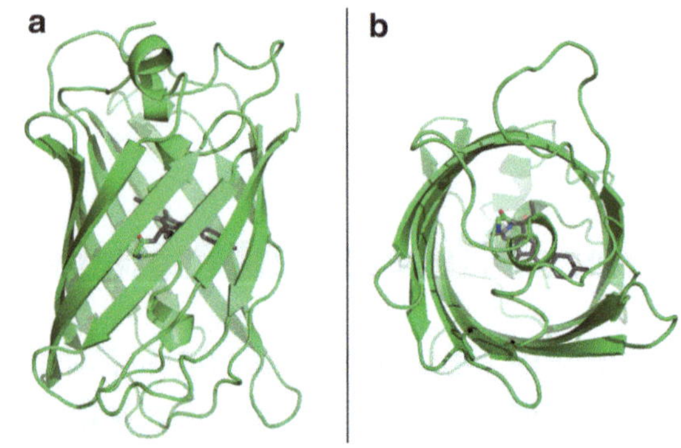

Fig. 1 Side view of crystal structure of avGFP (**a**) and bottom view (**b**) (pdb entry 1ema). The 11 β-sheets form a protective environment for the chromophore, which is located on a distorted helix in the center of the barrel

developments in fluorescence spectroscopy and microscopy. Nowadays, FPs are available in multiple colors and have different photochromic behavior and optimized fluorescence properties. In conjunction, a growing set fluorescence spectroscopy and imaging techniques have emerged. This combination has enabled the study of biomolecules with unprecedented resolution, sophistication, sensitivity, and ease of labeling.

An FP consists of 220–240 amino acid residues. The residues are arranged in 11 β-sheets that form the β-barrel (Fig. 1a). A distorted helix on which the chromophore is situated is located in the center of the barrel (Fig. 1b), and thereby, the β-barrel provides a protected environment for the chromophore. The chromophore is formed after a series of autocatalytic reactions from amino residues 65–67 (numbering in accordance of the *Aequorea victoria* green fluorescent protein (avGFP)). These reactions are displayed in Fig. 2. Although the exact sequence of reactions and rearrangements are still under debate [2], it is evident that molecular oxygen is the only required external factor.

Most engineered FPs that are used nowadays have been derived from intensely mutating avGFP or DsRed (a red FP from the *Discosoma* species). Substitution of amino acid residues located in the chromophore and/or its direct proximity can lead to altered excitation and emission spectra. Also, mutations near the chromophore have resulted in increased brightness and faster maturation speed. Amino acid substitutions in the barrel, especially on the outside, have caused a lowered tendency of FPs to dimerize or oligomerize. These characteristics will be discussed in more detail in the next section.

For example, the Tyr66Trp mutation of avGFP resulted in cyan fluorescent protein (CFP) [3]. The chromophore structures of avGFP and CFP are presented in Fig. 3. The structural difference caused a blue shift in the absorption and emission spectrum of avGFP. The bulkier chromophore has caused spatial shifts within

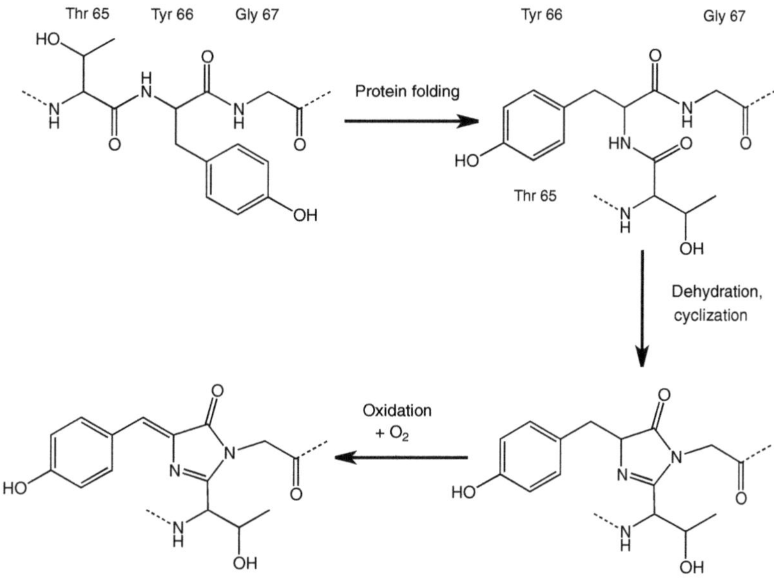

Fig. 2 avGFP chromophore formation. First a rearrangement takes place to facilitate the cyclization. Second, dehydration takes place, resulting in cyclization. In the last step one oxidation reaction occurs with molecular oxygen, resulting in a completely conjugated chromophore

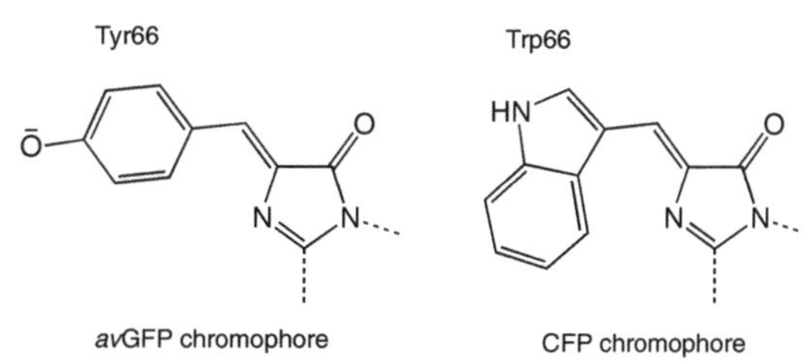

Fig. 3 Structure of the avGFP chromophore (*left*) and CFP chromophore (*right*), only amino acid residue 66 in the chromophore differs

the β-barrel, resulting in a reduced brightness. Further engineering resulted in enhanced cyan fluorescent protein (ECFP) [4], mCerulean [5], SCFP3A [6], mTurquoise [7], and recently mTurquoise2 [8]. The brightness and photostability are substantially increased from CFP to mTurquoise2.

1.1 Fluorescent Protein Characteristics

The key characteristics of FPs include absorption spectrum, emission spectrum, brightness, fluorescence lifetime, maturation rate, oligomeric state, photostability, pH sensitivity, and functionality in protein fusions. By mutating the DNA sequence, key FP characteristics can be optimized. However, optimizing one characteristic may lead to a downgrade of another one. During optimization all characteristics need to be monitored.

1.1.1 Brightness and Fluorescence Lifetime

The brightness is dependent on the molar extinction coefficient and the quantum yield. The molar extinction coefficient is a measurement for the efficiency of the chromophore to absorb photons and can be calculated using the law of Lambert–Beer. The molecular extinction coefficient is usually constant within an FP spectral class, due to the given physical size of the chromophore. Still largely varying extinction coefficients have been published even for the same FP by different laboratories. While measuring the absorbance of purified FPs in solution at the absorbance maximum is straightforward, the determination of the actual protein concentration easily introduces errors up to 10 %. Also the extent of fully matured protein within a purified sample or presence of impurities can seriously interfere with the extinction coefficient assay. In some cases indirect measurements of denatured protein and comparison to FP standards are used [9] in order to circumvent low extinction coefficient estimation, due to poor maturation. Another factor that can interfere when determining the extinction coefficient is the pH sensitivity of the chromophore. Especially in GFP and yellow fluorescent protein (YFP) spectral classes, protonation of the chromophore causes a dramatic (blue) shift in the absorbance spectrum at the cost of the absorbance of the deprotonated species [10, 11]. Hence, for these spectral classes absorbance measurements at relatively high pH (8.0) are advised. In this chapter we restrict ourselves to determine the extinction coefficient by measuring the absorbance of the chromophore and determine the protein concentration (Subheading 3.2.2). In addition, the relative intrinsic molecular brightness of an FP can be estimated with single-molecule fluctuation microscopy techniques (Subheading 3.2.5). The advantage of this method is that only fully maturated FPs contribute to the signal and nonfluorescent (non-folded) FPs do not contribute to the signal.

The fluorescence quantum yield is the ratio of the number of emitted photons over the number of absorbed photons. Hence, the quantum yield is minimally 0 and its theoretical upper limit is 1 (or 100 %). Ideally, a bright FP has a large extinction coefficient and a high quantum yield. These desired characteristics give the advantage that enough fluorescence is produced by the FP even under conditions of low levels of excitation light. Within cells, this reduces phototoxicity and undesired photobleaching of the FP (*see* Subheading 1.1.4). Furthermore, bright FPs characterized by high signal-to-noise ratios allow working at low, physiologically relevant concentrations of FP-labeled proteins in the cell.

In contrast to the extinction coefficient of a fully maturated protein, the fluorescence quantum yield can vary a lot within one spectral class of FPs. Usually decreased fluorescence quantum yields are paralleled by reduced excited state duration, hence shortened fluorescence lifetimes. This can be understood by realizing that the fluorescence quantum yield is the quotient of the rate of radiative decay over the rate of total decay, whereas the fluorescence lifetime is the inverse of the total decay rate. In addition, usually the radiative decay rate is constant for a fluorophore. Accordingly, decreased quantum yields or fluorescence lifetimes result from additional or enhanced non-radiative decay from the excited state. While proper determination of the quantum yield of a fluorophore generally requires purified protein, fluorescence lifetimes can be accurately measured even within living cells by fluorescence lifetime imaging microscopy (FLIM). Therefore, FLIM (Subheading 3.1.2) is an excellent technique for screening FP libraries for FPs with increased quantum yields [7]. Such screens will select FP variants with increased fluorescent lifetimes and quantum yields. Consequently, the selected FPs will show reduced non-radiative decay, which are believed to be related to minimized fluctuations of side chains or chromophore movements within the β-barrel. The presence of other chromophores within a close proximity of an FP can reduce its lifetime, resulting from Förster resonance energy transfer (FRET). A mono-exponential fluorescence lifetime decay can be greatly beneficial for the analysis of time-resolved lifetime measurements, for example, in FRET microscopy.

1.1.2 Maturation Rate

After synthesis at the ribosome, the folding of the protein structure and the formation of the chromophore by autocatalytic reactions determine the so-called maturation speed of the FP. When an FP is used as reporter gene, the maturation rate should be faster than the gene-induction process itself. In addition, maturation should occur at the incubation temperature of the cells or organism used. As shown in Fig. 2, the formation of the GFP chromophore only requires one oxidation step. The chromophore of a red fluorescent protein (RFP) requires an additional oxidation step, and therefore, the maturation rate of a RFP will be lower. Still, the two-step oxidation process for RFPs can be exploited for gene-induction studies: so-called timer variants were developed [12, 13] that showed a relatively fast first oxidation step leading to a green fluorescent intermediate and a slower second oxidation step yielding the final RFP. Here, the degree of red-to-green coloration seen in cells can be used to time the onset of gene expression.

1.1.3 Oligomeric State

Naturally occurring FPs have a tendency to dimerize or tetramerize (the latter especially occurs in the coral-derived FPs). For many applications, it is of great importance that FPs do not form oligomers. For instance, if the fusion protein to which the FP is attached can also oligomerize, FP oligomerization can lead to clusters and even larger protein aggregates. For FRET experiments where FPs are used to probe molecular proximity, FP oligomerization can be detrimental. Instead of observing interactions/proximity of the targeted objects, interactions between FPs themselves can cause "false-positive" FRET signals. Therefore, FPs ideally must be inert to each other and monomeric. This can be accomplished by altering certain amino acid residues located on the outside of the barrel. Note, when FPs are expressed at high (local) levels, e.g., when fused to membrane proteins, the possibility of very close proximity of even monomeric FPs remains. This causes (acceptor) concentration-dependent FRET also called "bystander FRET." For other situations, oligomerization can even be desired, as for certain FRET-based biosensors.

1.1.4 Photostability

For quantitative measurements it is important that the fluorescence intensity of the sample is not declining over time due to photobleaching or photochromic changes [14, 15]. The photostability will vary under different illumination conditions, i.e., the selected light source (scanning laser, wide-field lamp illumination, spinning disk, or total internal reflection excitation), intensity, wavelength, and the frequency of pulsed light sources. The boundary conditions for setting the threshold for minimal desired FP photostability are dictated by the fluorescence microscopy technique of choice, the imaging time that is applied, and the desired signal-to-noise ratio. Preferably, newly developed FPs are photostable under all conditions.

1.1.5 pH Sensitivity

The pH within the cytoplasm and within cell organelles usually differs, and during a studied process pH fluctuations can occur. Therefore, it is important that FPs are stable under a wide pH range. For many FPs in the cyan and red spectral class, the pH sensitivity is rather low, in contrast to FPs in the green and yellow spectral classes. Especially for quantitative experiments, the difference in fluorescence intensity should not be a consequence of merely a pH shift. However, some highly pH-sensitive FPs are being exploited as intracellular pH sensors [16].

1.1.6 Functionality in Protein Fusions

As mentioned before, FPs are used for a wide variety of applications, and researchers want to express FPs in all sorts of cells and cellular compartments, either alone or as fusion protein. The justification for an FP that is properly produced in one compartment but not in another compartment is not always evident. However, FPs that are correctly produced in all compartments have great advantages for scientists. Also, FPs with low phototoxicity, which perform well in fusion proteins (like membrane or cytoskeletal fusion proteins) and which are insensitive to aggregation, are highly desired.

1.2 Design and Mutagenesis

Genetic engineering of FPs is usually performed by a combination of site-directed and random mutagenesis. Random mutagenesis has been very successful for optimization of FPs and usually identifies mutations that improve folding/maturation. Because multiple mutations can be introduced simultaneously, it is often uncertain what the contribution is of each of the individual mutations.

However, it should be noted that with random mutagenesis the position at which a mutation will be introduced may be randomly selected, but mostly one base pair within one codon will be changed at a time. This severely limits or even prevents variations in the corresponding amino acids in view of the degenerate genetic code. Hence, random mutagenesis is very biased towards the chosen template. Site-directed mutations on the other hand allow encoding of all possible amino acids but only at a preselected position. The positional selection is usually based on structural information and targets specific areas within the protein, e.g., residues in proximity of the chromophore. Methods to introduce targeted mutations at multiple positions have been developed [17, 18]. Site-directed mutagenesis can also be used to introduce step-by-step mutations that have been previously identified by random mutagenesis. Site-directed mutagenesis can either be performed with a predefined mutation codon encoding a specific set of amino acids or with a degenerate codon. The degenerate codon encodes all 20 amino acid residues and will be inserted at a given position in the DNA sequence. Besides random and site-direct mutagenesis, DNA shuffling [19] and in vivo evolution approaches have been described. It is expected that, due to decreasing costs of gene synthesis, the use of synthetic gene libraries will become increasingly popular.

1.3 General Approach for Optimizing FPs

The general approach to optimize FPs is to start with a rational design on the chromophore structure and spatial structure of the FP. A schematic overview is shown in Fig. 4. The corresponding DNA modifications can be obtained either by site-directed mutagenesis or by random mutagenesis to a carefully chosen template. This results in a library of mutated FP DNA sequences, which can be expressed in bacteria. An example of a site-directed mutagenesis protocol is worked out below in Subheading 3.1.1. The advantage of using bacteria is that one bacterium will contain one mutated FP variant. After growing the bacteria on agar plates, a primary screen can be performed on bacterial colonies. In this round bacterial colonies producing improved FPs are selected and DNA can be isolated for sequence confirmation. The primary screening (Subheading 3.1) entails a screening based on fluorescence lifetime (Subheading 3.1.2) and on the emission spectrum (Subheading 3.1.3). By screening on basis of fluorescence lifetime and thereby indirect on intrinsic brightness, factors such as expression level, colony size, and maturation rate do not interfere with the screening results.

In Subheading 3.2 spectroscopic characteristics will be measured that are most accurately determined from purified FPs, which are easily produced and isolated from bacteria (Subheading 3.2.1). In this round absorption spectrum and molar extinction coefficient are measured (Subheading 3.2.2), as well as the corrected emission spectrum and quantum yield (Subheading 3.2.3). We also provide protocols for quantifying the fluorescence lifetime

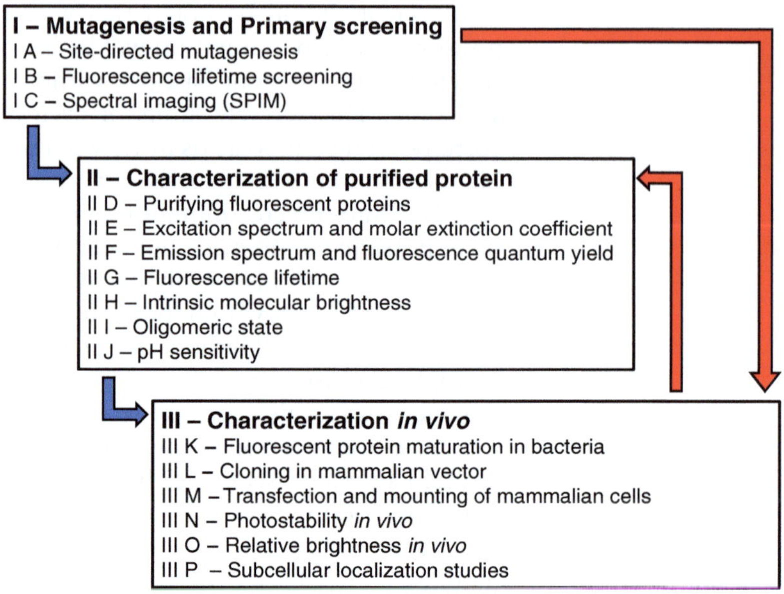

Fig. 4 General approach for optimizing FPs

(Subheading 3.2.4), relative intrinsic molecular brightness (Subheading 3.2.5), oligomeric state (Subheading 3.2.6), and pH sensitivity (Subheading 3.2.7).

Subheading 3.3 includes characterization in different expression systems. The maturation rate of the FP variants in bacteria is evaluated (Subheading 3.3.1). In addition, the photostability using different image techniques is determined in mammalian cells (Subheading 3.3.4). Furthermore, the relative brightness is determined in mammalian cells (Subheading 3.3.5). And finally, the functionality in protein fusions is assessed in different cell types and cellular compartments (Subheading 3.3.6).

The experiments in Subheadings 3.2 and 3.3 can be performed consecutively (blue arrows in Fig. 4), in opposite order (red arrows in Fig. 4), or simultaneously. There is no absolute order for the screening experiments. If the rational design of the new FP was targeted to one of the key characteristics, this screening could be performed as one of the first screenings. It is not required to complete all assays from Fig. 4 for every intermediate FP. As soon as the new characteristics of the newly synthesized FP are found as non-desired, one can return to design a new FP. However, by performing all experiments mentioned above, one might acquire more valuable information for the new design. In this chapter we describe the spectroscopic aspects of the screening methods. All methods are worked out for the optimization of CFP variants with which we have considerable experience. For characterization of FPs of other spectral classes and/or origin, the chosen FP libraries, the excitation and detection specifications, and fluorescence references given in these methods should be adapted accordingly.

2 Materials

2.1 Mutagenesis and Primary Screening

2.1.1 Site-Directed Mutagenesis

1. Bacterial expression vector containing a leaky bacterial expression promotor and an FP template for mutagenesis (*see* **Note 1**).
2. The selected bacterial expression vector containing a reference FP for the primary screening (*see* **Note 2**).
3. Electro-competent bacterial cells which are suitable for protein production in combination with the selected vector, such as BL21(DE3) (NEB, C2527H), for expression with pRSET.
4. Sequence primers for the FP, for RSET vector: T7 promotor (TAATACGACTCACTATAGGG) and T7 terminator (GCTAGTTATTGCTCAGCGG).
5. Primers for site-directed mutagenesis, for design; *see* Box 1 (Eurofins).
6. Pfu polymerase (includes 10× Pfu buffer with $MgSO_4$, Fermentas, EP0501).
7. dNTPs (Fermentas, R1121).
8. DpnI (Fermentas, ER1701).
9. GeneRuler™ DNA Ladder Mix (Fermentas, SM0333).
10. 10× Orange G DNA loading buffer (100 mg of Orange G, 30 % (w/v) glycerol and ddH_2O, total volume 50 ml).
11. 1 % (w/v) agarose gel (Roche 11388991001) in TAE (40 mM Tris, 1 mM ethylenediaminetetraacetic acid (pH 8.0), 2.2 ml glacial and ddH_2O, total volume 50 ml).
12. Growth medium, Luria-Bertani medium (LB): 10 g Bacto Tryptone, 5 g Bacto Yeast extract, 10 g NaCl and ddH_2O, total volume of 1 l. Autoclaved.
13. Agar plates, 1 l of LB supplemented with 15 g agar and 100 μg/ml ampicillin, 10 ml per Petri dish (diameter 90 mm) (*see* **Note 3**).
14. Electroporation cuvettes, gap width 0.2 cm (Sigma Z706086).
15. PCR apparatus.
16. DNA electrophoresis equipment (cell, power source, UV visualization setup).
17. Horizontal shaker for growing bacterial suspensions (37 °C and 250 rpm).
18. Incubator at 37 °C.
19. Electroporation equipment (Eppendorf electroporator 2510).

Box 1
Design of Primers for Site-Directed Saturation Mutagenesis

Primer Design: We typically design our mutagenic primers by using 15 nucleotides before and 15 nucleotides after the codon that is targeted for annealing to the plasmid. For saturated random mutagenesis, we employ the degenerate codon NNK in the forward primer, while the reverse primer has the complementary codon MNN (N = A,C,G,T and K = G,T and M = A,C). These degenerate codons encode all 20 possible amino acids.

The GC content should be above 40 % and the nucleotide at the 3′-end is preferably a G or a C. For site-directed mutagenesis the melting temperature (T_m) should be close to 78 °C or above. The T_m can be calculated according to this equation:

$$T_m = 81.5 + 0.41(\%GC) - \frac{675}{n} - \%\,\text{mismatch}$$

where n = total number of bases.

Since the codon is degenerate, the %GC and % mismatch will vary, depending on the exact codon that is present in the primer. Consequently, the T_m is a range for a primer with a degenerate codon, rather than a single value. We do not calculate the T_m and our simple design rules have worked so far for every primer tested (>10 different primers with degenerate codons). Example primers can be found in the following table.

Name	Sequence[a]	%GC (>40 %)	Anneals on
fw-mTq-165X	GGCATCAAGGCCAACNNKAAGATCCGCCACAAC	52–61	mTurquoise
rv-mTq-165X	GTTGTGGCGGATCTTMNNGTTGGCCTTGATGCC		

[a]N = A,C,G,T and K = G,T and M = A,C

2.1.2 Fluorescence Lifetime Screening

1. Bacteria producing the newly synthesized FP(s), bacteria producing a reference FP, and bacteria producing the template FP growing on agar Petri dishes (*see* Subheading 3.1.1). The colonies should be at least 1 day old. Depending on the fluorescent protein maturation rate, the agar plates with colonies can be older.
2. Microscopy equipment: home-build frequency-domain FLIM setup. The custom-build apparatus comprises an inverted Zeiss fluorescence microscope (Axiovert-200 M), a II18MD modulated image intensifier (Lambert Instruments), a CoolSNAP HQ CCD camera (Roper Scientific), and two computer-controlled

HF-frequency synthesizers (Rohde & Schwartz model SML 01), one driving the intensifier and the other driving a 440 nm modulated laser diode (PicoQuant, LDH-M-C-440) through an MDL-300 driver unit. For CFP variants the laser is intensity modulated at a frequency of 75.1 MHz. Excitation light is reflected by a 455DCLP dichroic mirror (Chroma) onto the sample, and the CFP emission is filtered with a D480/40 band-pass emission filter (Chroma). For reference phase and modulation measurements, a modified filter cube is used in which a mirror is mounted 90° tilted as compared to the normal orientation of a dichroic mirror that directly reflects the laser and passes it directly onto the image intensifier light, attenuated by an ND 5 neutral density filter [20].

3. Software for control, acquisition, processing, and analysis of the data was written in C++, using MATLAB 6.1 (The Mathworks, USA) and the image processing library DIPlib (Pattern Recognition Group, TU Delft, the Netherlands, http://www.diplib.org/).
4. Cardboard postal tubes (for A0 format poster) with plastic end caps, diameter 100 mm, length 1,000 mm (thickness of cardboard 2 mm): http://www.verzendkokers.nl.
5. Low-magnification glass plano-convex lenses (Melles Griot, LPX-25.0-415.0-C-SLMF-400-700).
6. A computer running ImageJ software (free download for different platforms: http://rsbweb.nih.gov/ij/).
7. Alternatively, we use a commercially available time-correlated single-photon counting (TCSPC) detection module available from PicoQuant, implemented on an inverted Olympus FluoView 1000 confocal microscope. The excitation light of a 440 nm 20 MHz pulsing laser diode (PicoQuant) (*see* **Note 4**), as controlled by a Sepia II laser driver unit (PicoQuant), was attenuated ten times by a neutral density filter. The light was guided via a D440 primary dichroic mirror (Chroma) into the tube. The emission light was guided via a size-adjustable pinhole, set at 170 μm, through the Olympus confocal detection box to the fiber output channel. The optical fiber was coupled to a custom-made detection box (PicoQuant) containing four PDM avalanche photodiodes (MPD). The light is filtered by a 475/45 nm emission filter (Chroma) and directed into one of the MPDs. The photon arrival times were recorded by a PicoHarp 300 time-correlated single-photon counting system (PicoQuant) and controlled and analyzed using the SymPhoTime 5.2.3 (PicoQuant) software.

2.1.3 Spectral Imaging (SPIM)

1. Bacteria growing in Petri dishes as outlined in Subheading 3.1.1. Spectral imaging only has to be performed on the colonies that are concluded to be interesting from Subheading 3.1.2.

2. Microscopy equipment: Axiovert 200 M motorized microscope (Zeiss, Germany), 10×/NA 0.3 air objective. AttoArc 100 W mercury lamp (Zeiss, Germany). CFP spectra are recorded using a 436/10 nm excitation and an 80/20 (transmission/reflection) dichroic mirror and a LP460 longpass filter, using a slit spectrograph (ImSpector, SPECIM, Finland), mounted in front of a CCD camera (ORCA II, Hamamatsu, Japan). The instrument is described in Subheading 2.1.2 and is a multimode platform for FLIM and SPIM [20, 21].
3. Software for control, acquisition, processing, and analysis of the data was written in C++, using MATLAB 6.1 (The Mathworks, USA) and the image processing library DIPlib (Pattern Recognition Group, TU Delft, The Netherlands, http://www.diplib.org/).

2.2 Characterization of Purified Protein

2.2.1 Purifying Fluorescent Proteins

Ni^{2+} Affinity Columns

1. 0.1 M $NiSO_4$ in ddH_2O.
2. 50 mM EDTA pH 7.5–8.0 in ddH_2O.
3. 20 % (w/v) ethanol in ddH_2O.
4. His-Bind resin (Novagen, 69670-2).
5. 15 ml tubes with screw cap.
6. Table top swing-out centrifuge (Eppendorf 5810R).

Protein Isolation

7. Bacterial strain optimized for protein production, e.g., BL21 DE3 (electro-competent).
8. Bacterial expression vector with selected FP fused to a 6xHis tag and pRSET including 6xHis tag.
9. Growth medium and medium with selection marker suited for protein production of the strain in combination with the vector. For BL21 DE3 in combination with pRSET vectors, LB supplemented with 100 μg/ml ampicillin can be used for protein production (*see* Subheading 2.1.1).
10. Lysozyme (L7651, Sigma-Aldrich).
11. Benzonase nuclease 25 U/μl (previously Novagen, now Merck/Millipore, 71205-3).
12. 100 mM phenylmethanesulfonyl fluoride (PMSF) in isopropanol.
13. 10 % NP40 (v/v) in ddH_2O.
14. Imidazole.
15. Buffer 1: 20 mM Tris and 200 mM NaCl in ddH_2O pH 8.0 (HCl).
16. Buffer A (20 mM Tris–HCl in ddH_2O, pH 8.5, filter sterilized 0.22 μm).

17. Buffer B (20 mM Tris–HCl and 1 M NaCl in ddH_2O, pH 8.5, filter sterilized 0.22 μm).
18. Electroporator (Eppendorf 2510) and electroporation cuvettes (gap size 0.2 cm).
19. Incubator at 37 °C, shaking at 250 rpm, for 250 ml Erlenmeyer flasks and 25 ml cultivation flasks.
20. Table top swing-out centrifuge (Eppendorf 5810R).
21. Centrifuge (Sorvall RC 5B plus, rotor Sorvall SS-34 with compatible buckets).
22. Rotator at 4 °C.
23. Sonicator (Branson Sonic Power Company, sonifier B12, with tip size ½″ diameter).
24. 50 ml tubes with screw cap.
25. 0.22 μm filter (Millipore) and syringe.
26. Dialysis tube (molecular porous membrane tubing #3), dialysis clamps, beaker glass ~2 L, and large stirring bean.
27. FPLC machine (ÄKTA FPLC, Amersham Bioscience).
28. MonoQ 5/50 GL (GE Healthcare 17-5166-01).

Protein Concentration Determination

29. BCA protein assay kit including bovine serum albumin reference (Thermo Scientific).

Protein Purity Assay: SDS-PAGE

30. 2× sample buffer: 2.5 ml 0.5 M Tris–HCl (pH 6.8), 2.0 ml 20 % (w/v) sodium dodecyl sulfate in ddH_2O, 5 ml 100 % glycerol, 2 mg bromophenol blue, and 0.5 ml ddH_2O.
31. PageRuler™ protein ladder (Fermentas, SM0669).
32. Coomassie staining (0.2 % Coomassie BB R-250, 2 % acetic acid, 50 % methanol).
33. Protein cell and casting system (Biorad, Mini-PROTEAN II).
34. Platform shaker.

2.2.2 Excitation Spectrum and Molar Extinction Coefficient

1. Purified FP solution with known concentration (*see* Subheading 3.2.1).
2. Buffer for diluting the selected FPs (*see* **Note 5**).
3. Two quartz absorption cuvettes (200 μl).
4. Biochrom Libra S70 spectrophotometer.

2.2.3 Emission Spectrum and Fluorescence Quantum Yield

1. Selected purified FP solution (*see* Subheading 3.2.1).
2. Buffer for diluting the selected FPs, CFP variants PBS or 10 mM Tris–HCl (pH 8.0).

3. Reference fluorescence dye with known quantum yield and emission and excitation spectra comparable to the selected FPs with unknown quantum yield. For CFP variants fluorescein (borate buffer pH 9.1) can be used [22]. It is also possible to use a well-characterized FP with known quantum yield and comparable emission and excitation spectra.
4. Spectral Fluorescence Standard Kit (certified by BAM) (Sigma 97003).
5. Two quartz absorption cuvettes (200 μl).
6. One quartz fluorescence cuvette (1 ml).
7. Biochrom Libra S70 spectrophotometer.
8. PTI QuantaMaster 2000-4 fluorescence spectrofluorometer (Photon Technology International, Lawrenceville, NJ).
9. Acquisition software: FelixGX software.

2.2.4 Fluorescence Lifetime

1. Selected purified FP solution (*see* Subheading 3.2.1).
2. Purified FP solution of the same spectral class with known fluorescence lifetime, which can be used as reference.
3. Phosphate buffered saline (PBS: mix 8 g of NaCl, 0.2 g of KCl, 1.44 g of Na_2HPO_4, 0.24 g of KH_2PO_4, and 800 mL of ddH_2O. Set to pH 7.4, fill up to 1 l with ddH_2O, and autoclave)
4. 96-well plate (black polystyrene, glass bottom, Whatman 7706-2370)
5. Microscope equipment as described in Subheading 2.1.2, TD-FLIM, but no adaptations are performed on the inverted microscope. Here a standard microscope objective is used.

2.2.5 Intrinsic Molecular Brightness

1. Selected purified FP solution (*see* Subheading 3.2.1).
2. Purified FP solution of the same spectral class for molecular brightness comparison, for CFP variants ECFP was used.
3. Calibration dye with known diffusion time [23] and roughly similar spectral characteristics as the selected FPs, e.g., Alexa 488 for CFP variants.
4. 10 mM Tris (pH 7.4).
5. Diluted samples may be stabilized by adding bovine albumin serum (BSA) or TritonX-100 (Tx-100) [24].
6. Microscope setup as described in Subheading 2.1.2, TD-FLIM.
7. High-quality objectives are required, matching the refractive index of the sample like the 60× water immersed UPLS Apochromat, N.A. 1.2 (Olympus), and UPlanSApo 60× silicone, 1.30 N.A. (Olympus) for in vivo measurements.

8. A computer running fluorescence fluctuation spectroscopy analysis software, e.g., FFS data processor 2.3 (www.sstcenter.com).

2.2.6 Oligomeric State

1. Selected purified FP solution (*see* Subheading 3.2.1).
2. Dimer FP reference (FP with a high tendency to dimerize, but is not fused as a oligomer, e.g., dTomato and dsRed [25]), and a monomer FP reference (e.g., mEGFP [26]).
3. *See* Subheading 2.2.1 (Protein purity assay—SDS-PAGE) but a different sample buffer and additional FP references.
4. 2× sample buffer: 2.5 ml 0.5 M Tris–HCl (pH 6.8), 5 ml 100 % glycerol, 2 mg bromophenol blue, and 2.5 ml ddH_2O.

2.2.7 pH Sensitivity

1. Selected purified FP solution (*see* Subheading 3.2.1).
2. pH-titration buffers: 50 mM citric acid/Na^{2+}citrate (pH 3.0, 3.5, 4.0, 4.5, 5.0, 5.5), KH_2PO_4/Na_2HPO_4 (pH 6.0, 6.5, 7.0, 7.5, 8.0), and/or glycine/NaOH (pH 8.5, 9.0, 9.5, 10.0).
3. 96-well plate (black plate clear bottom, Costar 3603).
4. Microtiter plate reader (BioTek Synergy Mx plate reader).
5. Software: BioTek Gen5 Data analysis software.
6. Plate reader settings: excitation 440 nm and emission filter 480/20 nm, measure from the bottom.
7. A computer running KaleidaGraph or alternative software package in order to fit the data to a sigmoid equation.

2.3 Characterization In Vivo

2.3.1 Fluorescent Protein Maturation in Bacteria

1. Vector containing selected FP sequence, e.g., pRSET (*see* **Note 6**).
2. Vector containing a reference FP sequence (template, other related FPs for comparison, or optional dark FP) (*see* **Note 6**).
3. Strain bacteria optimized for protein production in combination with the selected vector, e.g., BL21 (DE3) (electro-competent).
4. Growth medium and medium with selection marker suited for protein production of the strain in combination with the vector (*see* Subheading 2.1.1). For BL21 (DE3) in combination with pRSET vectors, LB supplemented with 100 μg/ml ampicillin can be used for protein production.
5. 96-well plate (black plate clear bottom, Costar 3603) including lid.
6. Small bacteria-cultivating flasks (volume 25 ml).
7. Incubator at 37 °C and vigorous shaking (~250 rpm).
8. Absorbance spectrophotometer, to measure absorbance at 600 nm (*see* **Note 7**).

9. Plastic cuvettes.
10. Microtiter plate reader: BioTek Synergy Mx plate reader.
11. Software: BioTek Gen5 data analysis software.

2.3.2 Cloning in Mammalian Vector

1. Bacteria expressing the selected FPs.
2. Equipment and kits for mammalian vector cloning.

2.3.3 Transfection and Mounting of Mammalian Cells

1. FP mammalian expression plasmid (e.g., mTurqoise2-C1).
2. 1–3 μl of plasmid (~250 ng/μl; from a column purified mini-prep (GeneJET PCR purification kit, #k0702, Fermentas)).
3. Confluent T25 flask with HeLa cells.
4. Growth medium (DMEM-GlutaMAX (#61965-059), 10 % FBS, penicillin (100 units/ml) and streptomycin (100 μg/ml) all from Invitrogen).
5. Lipofectamine 2000 (Invitrogen).
6. Opti-MEM (Invitrogen).
7. Microscopy medium (140 mM NaCl, 5 mM KCl, 1 mM $MgCl_2$, 1 mM $CaCl_2$, 10 mM glucose, 20 mM HEPES (Sigma), pH 7.4) (*see* **Note 8**).
8. 70 % EtOH.
9. 6-well plate.
10. Cover glasses (diameter 24 mm, thickness #1, Thermo Scientific).

2.3.4 Photostability In Vivo

Wide-Field Conditions

1. Transient transfected cells with the selected FPs produced (*see* Subheading 3.3.3).
2. Transient transfected cells with a reference FP (from the same spectral class) produced (*see* Subheading 3.3.3).
3. Zeiss Axiovert 200 M.
4. 63× (Plan Apochromat NA 1.4 oil).
5. 10× objective (10× Plan Neofluar NA 0.3 air).
6. 100 W HBO Hg lamp.
7. Princeton Instruments CCD camera with Kodak KAF1300 chip and ST-133 controller interfaced with a PCI board to a PC.
8. For CFP variants: excitation filter 436/20, dichroic mirror 80/20, and emission filter LP460 nm.
9. Melles Griot Broadband Powermeter (13PEM001).
10. Stage micrometer (Agar Scientific, L4078).
11. Grey filter (ND1.3).
12. A computer running ImageJ software (free download for different platforms: http://rsbweb.nih.gov/ij/).

Laser Scanning Conditions

1. Mammalian cells with the selected immobilized FP produced, e.g., HeLa cells with mTurquoise2 fused to histone2A to immobilize most FP protein to the DNA in the nucleus. This can be accomplished by transfecting HeLa cells with an mTurquoise-histone2A-C1 construct (*see* Subheading 3.3.3).
2. Mammalian cells with a reference FP (from the same spectral class) immobilized in the same manner as the selected FP. It is of great importance to keep all settings equal for the different FP variant (*see* Subheading 3.3.3).
3. Microscopy equipment: Nikon A1 laser scanning microscope, Nikon Plan Apo VC 60×, NA1.4 oil immersed objective. CFP was excited by a 443 nm diode laser and the fluorescence was detected using a 482/35 nm band-pass filter. The pinhole was opened completely.
4. A computer running ImageJ software (free download for different platforms: http://rsbweb.nih.gov/ij/).

2.3.5 Relative Brightness In Vivo

1. Mammalian cells that produced the selected FP, which is quantitatively coproduced with a reference FP. The selected FP and the reference FP must have separated spectra. The coproduction is realized by linking the DNA sequences of the selected FP and reference FP with a viral 2A peptide sequence [7]. Here we used CFP variants as selected FPs and SYFP2 as reference.
2. Microscopy equipment: Zeiss 200 M inverted microscope, 40× Plan Neofluar (1.3 NA) oil-immersion lens (Zeiss). Illumination by a Cairn Xenon Arc lamp was selected by a monochromator (Cairn Research). For the CFP channel, excitation at 420 nm (slit 30 nm), reflection by a 455DCLP dichroic mirror, and fluorescence passing through a 470/30 nm band-pass filter. For the YFP channel, excitation at 500 nm (slit 30 nm), reflection by a 515DCXR dichroic mirror, and the emission passing through a 535/30 nm band-pass filter. Images are acquired with a Photometrics CoolSNAP HQ CCD camera (4 × 4 binning).
3. Acquisition software: MetaMorph software (Molecular Devices).
4. A computer running ImageJ software (free download for different platforms: http://rsbweb.nih.gov/ij/) and ObjectJ plug-in (http://simon.bio.uva.nl/objectj/) for ImageJ.

2.3.6 Subcellular Localization Studies

1. Isolated DNA of the selected FPs.
2. Mammalian vectors containing a target sequence. Plasmids encoding mTurquoise2 organelle markers are available (http://www.addgene.org/browse/article/5268/).
3. Mammalian vector cloning kit.

4. Commercial available organelle stains (e.g., MitoTracker, Life Technologies).
5. Microscopy equipment: Nikon A1 laser scanning microscope, Nikon Plan Apo VC 60× oil objective N.A. 1.4. Excitation by a 443 nm diode laser. The fluorescence is directed by a 440/514/593 dichroic mirror, through a 482/35 nm band-pass filter.

3 Methods

3.1 Mutagenesis and Primary Screening

3.1.1 Site-Directed Mutagenesis

1. Set up the reaction mix for site-directed mutagenesis for a mutagenesis and control reaction according to the following scheme:

Reagent	Volume per 50 μl	Final concentration
Pfu polymerase (2.5 U/μl)[a]	1 μl	2.5 U
10× Pfu buffer (with $MgSO_4$)	5 μl	1×
2.5 mM dNTPs	4 μl	0.2 mM
100 ng/μl forward mutagenesis primers[a]	2 μl	4 ng/μl
100 ng/μl reverse mutagenesis primers[a]	2 μl	4 ng/μl
Plasmid template (100–400 ng/μl)	1 μl	2–8 ng/μl
ddH_2O	35 μl	–

[a]For the control reaction, replace the primers with ddH_2O or replace the polymerase

2. Run the PCR according to the following program:

Cycle number	Denature	Anneal	Extend
1	5 min at 95 °C		
2–17	1 min at 95 °C	1 min at 54 °C	14 min at 68 °C
18			10 min at 68 °C

3. Add 1 μl of *Dpn*I to each individual sample and incubate at least 1 h at 37 °C to digest the (methylated) template (*see* **Note 9**).
4. Check the formation of DNA. Mix 2–5 μl of the PCR reaction with 10× DNA loading dye (9:1 PCR reaction: loading dye) and load this on a 1 % agarose gel using GeneRuler as reference markers. A clear band should be visible with the same size as the original linear vector which should be absent in the control (*see* **Note 10**).

5. Thaw 4 × 50 μl electro-competent BL21 (DE3) cells on ice and cool the electroporation cuvettes on ice.
6. For mutagenesis, control the template and reference sample, add 1 μl of the mutagenesis PCR reaction sample, control PCR reaction, and reference DNA, respectively.
7. Transfer the solution to the electroporation cuvette and apply 2,500 V (time constant should be >4 ms).
8. Immediately add 950 μl of growth medium to the cuvette and transfer the suspension to an Eppendorf vial and incubate for 1 h at 37 °C (growth culture).
9. Plate 10, 100 and 890 μl of the sample cell suspension on different agar plates. Plate 1,000 μl of the control sample. Plate 50 μl of the reference and template cell suspension (*see* **Note 11**).
10. Incubate at 37 °C overnight.
11. The control plate should contain zero to a few colonies, and the plates from transformation of the mutagenesis reaction should contain about 200 individual colonies (*see* **Notes 12** and **13**).

3.1.2 Fluorescence Lifetime Screening

Fluorescence lifetime screening can be performed in the frequency domain (FD-FLIM) as well as in the time domain (TD-FLIM). The following method is written for FD-FLIM. Protocol adaptations for TD-FLIM will follow directly.

Adapting an Inverted Microscope

1. Remove the standard objective from the inverted microscope and put the low-magnification lens on the revolver (Fig. 5a).
2. Adjust the length of the postal tube in such a manner that when the tube is placed on the microscopy stage, the top will be around 80 cm above the lens (assuming the Melles Griot LPX-25.0-415.0-C-SLMF-400-700 lens is used). For higher-magnification lenses the length of the poster tube must be smaller.
3. Cut a 85 mm diameter hole in one of the plastics end caps to create a holder for a Petri dish (Fig. 5b, c). (Outer diameter of the bottom of the Petri dish is 87 mm; outer diameter of the lid of the Petri dish is 90 mm.)
4. Place the Petri dish holder on one of the sides of the postal tube.
5. Remove the other end cap and place the postal tube on the microscope stage (adapter side up), such that it is centered above the lens (Fig. 5d).

Calibration of Frequency-Domain Fluorescence Lifetimes

Absolute Lifetime

In general, measurements with the screening setup can be verified by measuring the fluorescent lifetime of FP producing colonies, which a known lifetime from the same spectral class (e.g., SCFP3A,

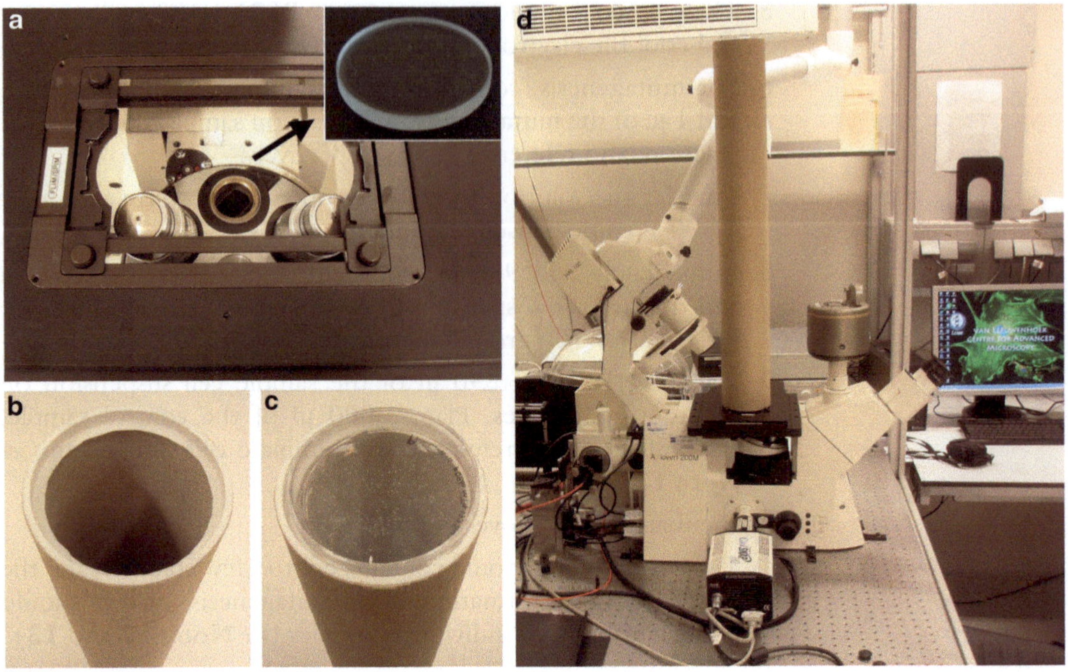

Fig. 5 Microscope setup for imaging Petri dishes. The objective is replaced by a low-magnification lens (**a**). A Petri dish with colonies can be mounted on a custom-made adapter that fits the poster tube (**b** and **c**). The poster tube of appropriate length with respect to the focal distance of the lens on the objective turret is placed on the microscope stage (**d**)

mTurquoise, mTurquoise2 for a CFP screening [8]). Alternatively, a chemical dye with a known lifetime can be used [27, 28].

Phase Correction

Each FD-FLIM measurement is calibrated by a reference measurement of the reflected laser light using a modified filter cube for correcting the phase and modulation drift of the excitation light [20]. Due to the increased path length, the travel time of the emission light that originates from the sample has increased. This increases the phase of the emission light that originates from the sample. The modulation depth of the signal is not affected. To account for the phase delay, we calculate the extra travel time by dividing the extra travel space (assuming a poster tube of 75 cm, the sample is 80 cm above the lens) by the speed of light: 75×10^{-2} m $\times\ 2/3.0 \times 10^{8}$ m/s = 5.0 ns

The according phase difference is $5.0 \times 10^{-9} \times$ frequency $\times\ 360°$. Thus, it can be calculated that at the modulation frequency (f) of 75.1 MHz, that we typically use, the phase has to be corrected by 135°. If the phase (φ_{exc}) and modulation (M_{exc}) of the excitation light are estimated by mounting a reference sample with known lifetime (τ_{ref}) on top of the poster tube, the light paths of reference and sample are identical and no 135° phase correction is necessary. Instead, the fluorescence lifetime of the reference compound will require corrections in order to estimate φ_{exc} and M_{exc} from the reference phase (φ_{ref}) and modulation (M_{ref}) measured at the top of the poster tube. These corrections are $\varphi_{exc} = \varphi_{ref} = \arctan(2\pi f \tau_{ref})$ and $M_{exc} = M_{ref}\sqrt{1 + (2\pi f \tau_{ref})^2}$.

Frequency-Domain Fluorescence Lifetime Image Acquisition

1. At least 30 min before imaging, take the plates with colonies from the refrigerator and allow reaching room temperature to avoid condensation.
2. Place the Petri dish in the Petri dish holder (Fig. 5c). To reduce background, the dish can be covered with black paper or a black metal plate.
3. Select the appropriate excitation and emission settings. For CFP, the sample is excited at 440 nm and fluorescence is detected using 480/40 nm or 475/45 nm emission filters.
4. Focus on the plate by visual inspection.
5. Acquire a fluorescence lifetime image ("sample"), using 12–36 phase steps with binning 2 × 2 pixels and an exposure time that depends on the intensity of the colonies.
6. Determine the average fluorescence (DC) intensity from acquired phase images, the phase lifetime image, and the modulation lifetime images. Put these three images into one FD-FLIM output image stack [20].

Quantitative Analysis of Frequency-Domain Fluorescence Lifetime

First a mask is created in ImageJ based on the intensity image in the FD-FLIM output stack:

1. Analyze > Set Measurements > activate checkbox "Mean gray value."
2. Image > Type > 8-bit.
3. Process > Binary > Make Binary.
4. Process > Binary > Watershed.
5. Process > Binary > Erode.
6. Analyze > Analyze Particles > Size, "0–1,000"; Circularity, "0.8–1.00"; Show, "Outlines"; activate checkbox; and "Add to Manager."

   Subsequently, the image of interest is analyzed:
7. Open a lifetime image.
8. Analyze > Tools > ROI manager: select all ROIs (click on the first ROI in the list, shift–click last ROI in the list).
9. Click on the button "Measure."

This procedure provides a results list with the lifetime per colony. This list can be used to identify the top highest lifetime colonies. The ROI manager can be used to retrieve the position of these colonies (select the colony number and the corresponding colony will be highlighted by an ROI on the fluorescent image). Colonies with high lifetime, single exponential lifetime (*see* **Note 14**), or low lifetime can be of interest (*see* **Note 15**). The DNA of colonies of interest can be isolated from bacteria (alkaline lysis method or a plasmid purification kit). The DNA sequence can be determined by sequencing.

Time-Domain Fluorescence Lifetime Image Acquisition

1. Adapt the microscope, switch on the system for stabilization, and implement the tube as was written for FD-FLIM (*see* **Note 16**).
2. Measure the sample but check if the detected fluorescence intensity is below 500 kHz. Typically the laser power is set not higher than 2.1 kW/cm^2 for the 440 nm laser line to prevent photobleaching, cellular damage, photophysical effects of the dyes, and detector artifacts (*see* **Notes 17** and **18**). This excitation intensity should still be sufficient to achieve a reasonable signal-to-noise ratio (5–10) within measurement times of 2–5 min.
3. The raw fluorescence intensity files are saved for processing. Datasets containing large intensity spikes (>10 times the intensity standard deviation around the mean intensity), significant signal drift, or photobleaching (>20 % intensity loss per minute) are discarded.

Analysis of Time-Domain FLIM

1. After loading the raw data into the SymPhoTime software, a region that contains at least 50,000 photons is selected out of the complete raw data trace. The data is summed and plotted to create the lifetime histogram. When plotted in semilogarithmic space, a linear decaying curve is expected for FPs having a single fluorescence lifetime. Only datasets are analyzed that have more than 10^3 counts in the peak and have a ratio of peak intensity over background that is at least two orders of magnitude.
2. The lifetime histogram can be fitted over the whole time range or will only include the last part of the decaying histogram, called tail fit, when excluding the initial rise. This latter option should be used only when no short-lived lifetime components, smaller than 1 ns, are being present.
3. When the whole histogram is fitted, the data is convolved with the instrumental response function (IRF), describing the peak broadening effect of the detector, electronics, optics, and especially the laser pulse profile and duration. The IRF can be calculated by the fitting algorithm from the dataset itself or will be determined from a calibration measurement using a

dye with a fluorescence lifetime smaller than the laser pulse duration, like erythrosin B in KCl solution.

4. The lifetime histogram $F(t)$ is fitted to a mono- or multi-exponential model (Eq. 1) including a background offset, b, resulting in values for the amplitude, A_i, and lifetime, τ_i, of all (n) fluorescence lifetime components (*see* **Note 19**):

$$F(t) = b + \sum_{i=1}^{n} A_i \exp\left(\frac{-t}{\tau_i}\right). \tag{1}$$

5. The quality of the fit is judged by the calculated reduced χ^2, which should approximate 1.0, and a visual inspection of the residuals. Residuals should have a low amplitude and be scattered around zero without showing a correlated pattern.
6. Also here, colonies with high lifetime, single exponential lifetime, or low lifetime can be of interest (*see* **Note 14**).

Figure 6a illustrates a typical fit of a TD fluorescence lifetime measurement. The signal, fit, and IRF are shown in grey, black, and red, respectively. In Fig. 6b the residuals are shown for a fit that has been executed from 0 to 40 ns (black). Visual inspection of the residuals concludes a pattern from 0 to 3 ns, due to the IRF. When only the tail is fitted (4.5–35 ns, shown in stripped borders in Fig. 6a), the residuals (Fig. 6b, black curve) show no correlated pattern.

Frequency-Domain and Time-Domain FLIM Results

In Fig. 7 a typical fluorescence lifetime result is depicted in false color-coding. Bacteria were transformed with a library in which amino acid 165 of mTurquoise was mutated (*see* Box 1). The plate with colonies was subsequently analyzed using both FD-FLIM and TD-FLIM, giving similar results.

Clearly, the acquired fluorescence intensity and excited state lifetime varies between individual colonies. The lifetime analysis shows a prominent peak around 3.7 ns, in agreement with the mTurquoise fluorescence lifetime. The corresponding colonies express either the mTurquoise starting material or mutants that do not have an altered lifetime. In addition, the lifetime map shows colonies with lifetimes as low as 3 ns. Thus, a clear effect of the mutagenesis of position 165 is thus observed. Moreover, comparison of the FD-FLIM and TD-FLIM data shows an excellent correlation between the lifetimes that were obtained by the two measurements.

3.1.3 Spectral Imaging (SPIM)

1. Place the Petri dish directly on the microscope stage and align the colony to the center of the objective.
2. Take a spectral image of a selected colony. The wavelength is displayed on the x-axis and the spatial information on the y-axis (*see* **Note 20**).

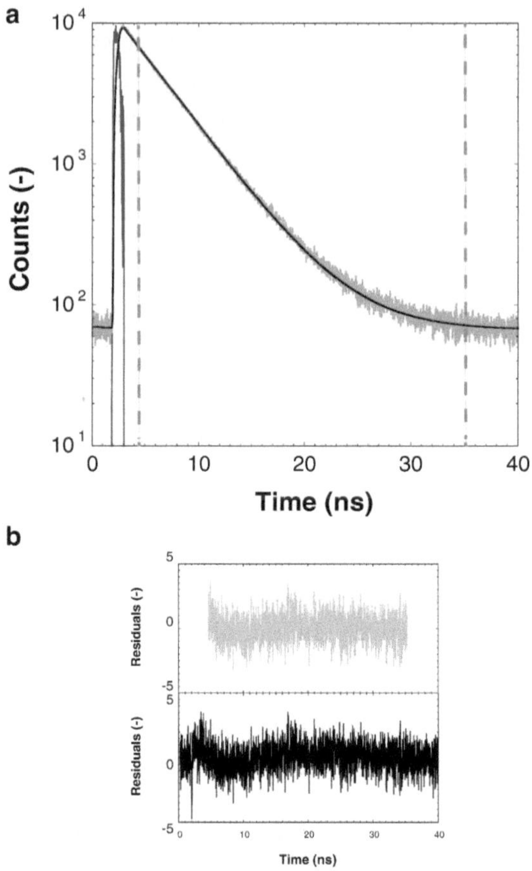

Fig. 6 A typical fit of a TD fluorescence lifetime decay of mTurquoise, resulting in a fluorescence lifetime of 4.00 ± 0.02 ns for mTurquoise. (**a**). The data, fit, and IRF are shown in *grey*, *black*, and *red*, respectively. The IRF is obtained through the fit algorithm from SymPhoTime software (PicoQuant). The residuals are shown in graph (**b**). The residuals for a fit from 0 to 40 ns (complete fit) and 4.5–35 ns (tail fit) are shown in *black* and *grey*, respectively. The residuals of the tail fit have a lower amplitude and show no correlated behavior, concluding that the fit is more accurate when the IRF is not included

3. Subtract the background spectrum.
4. Extract the emission spectrum from the spectral image (*see* **Note 21**).

3.2 Characterization of Purified Protein

The assays in this section require purified FPs.

3.2.1 Purifying Fluorescent Proteins

Preparation Ni^{2+} Affinity Columns

1. Add 1 ml His-Bind resin to a 15 ml tube (*see* **Note 22**)
2. Wash 3× with ~14 ml ddH_2O, centrifuge after each wash (3 min, 550 g, 4 °C, Eppendorf swing-out). Remove supernatant carefully by suction after every wash.

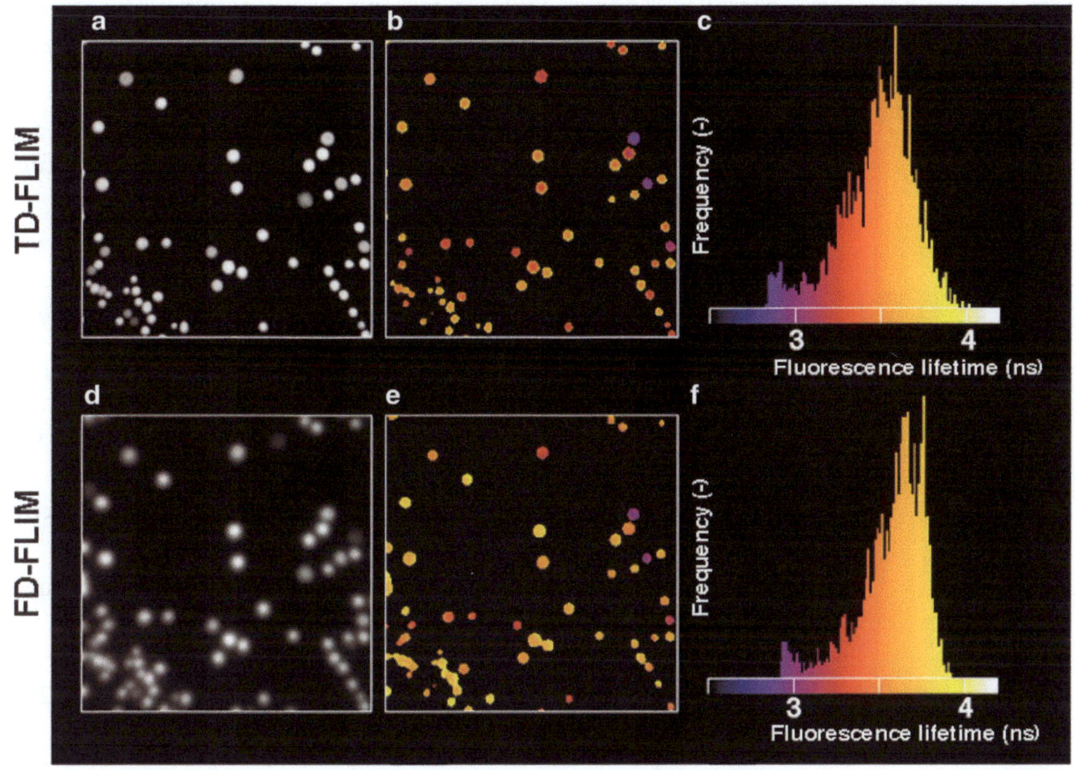

Fig. 7 Site-directed mutation on amino acid 165 of mTurquoise (Box 1) was analyzed using TD-FLIM (*top*) and FD-FLIM (*bottom*). The *left* and *middle panels* show an intensity-based image (*left*) and false-colored fluorescence lifetime image (*middle*) of bacterial colonies each expressing an mTurquoise variant. A fluorescence lifetime histogram of the bacterial colonies is shown in the *right panel*. Both methods give similar results and reveal a group of FPs with a fluorescent lifetime around 3.7 ns and a variety of lower fluorescent lifetimes. Both methods conclude a clear effect of the mutagenesis of position 165

3. Resuspend in ~14 ml 0.1 M $NiSO_4$ in ddH_2O and incubate on roller for 5 min.
4. Wash 3× ~14 ml with buffer 1, centrifuge after each wash (3 min, 550 g, 4 °C, Eppendorf swing-out), and remove supernatant. The Ni^{2+} affinity column is now ready for usage.

Protein Isolation

5. Transform BL21 (DE3) with 1 μl of plasmid (concentration plasmid ~250 ng/μl) as described in **steps 5–8** of Subheading 3.1.1.
6. Add 100 μl of the growth culture to 50 ml medium for protein production.
7. Grow culture at 37 °C, 300 rpm, overnight.

8. Incubate for an additional 6 h at RT, 300 rpm to allow maturation.
9. Transfer culture to a 50 ml Falcon tube and centrifuge (3,000 × *g*, 30 min, Eppendorf swing-out).
10. Remove supernatant, resuspend in 1 ml buffer 1, and add buffer 1 to a final volume of 20 ml and centrifuge (3,000 × *g*, 30 min, Eppendorf swing-out).
11. Remove supernatant and resuspend in 5 ml buffer 1 [the cell suspension can be stored at −20 °C].
12. From now on everything is kept ice cold. Add 5 mg lysozyme and 1 μl 25 U/μl benzonase nuclease.
13. Incubate at least 30 min on ice.
14. Add 100 μl 100 mM PMSF (in isopropanol) and 100 μl 10 % (w/v) NP40 to the bacterial suspension.
15. Sonicate directly for 5 min at 60 W continuously on ice-cold water.
16. Transfer to centrifuge bucket and centrifuge solution 30 min at 40,000 × *g* at 4 °C.
17. Transfer clear (and colored) solution directly onto the Ni^{2+} affinity column. Fill tube maximal with buffer 1 (*see* **Note 23**).
18. Carefully resuspend and rotate at 4 °C for 1 h.
19. Centrifuge (3 min, 550 g, 4 °C, Eppendorf swing-out), remove supernatant, and add 14 ml buffer 1. Repeat two more times.
20. After centrifugation (3 min, 550 g, 4 °C, Eppendorf swing-out) remove supernatant and add 1 ml 0.3 M imidazole in buffer 1. Gently tap to mix and incubate 10 min and centrifuge (3 min, 550 rcf, 4 °C, Eppendorf swing-out) (*see* **Note 24**).
21. Collect supernatant, be careful not to transfer any beads (containing FP).
22. Add 1 ml 0.2 M imidazole in buffer 1 to the Ni^{2+} affinity column, incubate 10 min, and centrifuge (3 min, 550 rcf, 4 °C, Eppendorf swing-out).
23. Collect supernatant.
24. Filter pooled supernatant using a 0.22 μm filter and syringe.
25. Boil dialysis tube about 5 cm in length in ddH_2O (for 2 min using a microwave).
26. Rinse with ddH_2O.
27. Close one side of the dialysis tube with a dialysis clamp, transfer protein mixture to the tube, close other side, and dialyze in 1 l 20 mM Tris (pH 8.0) at 4 °C overnight with gentle stirring (*see* **Note 25**).
28. Pass FP solution through a 0.22 μm filter.
29. Load FP solution onto MonoQ column and start program (flow rate 1 ml/min, max pressure 5 mPa, loading range

20–50 mg, linear NaCl gradient: 0–8 min 100 % buffer A, 8–58 min 0 to >100 % buffer B, 58–63 min 100 % buffer B, 63–70 min 100 % buffer A).

30. The FP fraction will elute around 20 % buffer B (200 mM NaCl).

Protein Concentration Determination

31. Use BCA protein kit and included BSA reference according to manufacturer's instructions to determine protein concentration.

Protein Purity Assay: SDS-PAGE

32. Prepare the gel (10 % running gel and 4 % stacking buffer) and set up the electrophoreses cell according to manufacturer's instructions (*see* **Note 26**).
33. Add loading buffer to the samples and boil for 3 min, include a monomeric pure FP as reference (e.g., mTurquoise2).
34. Load the samples and protein ladder.
35. Run the gel at 150 V for 45 min.
36. Stain the gel 30 min in Coomassie staining on a platform shaker (75 rpm).
37. Destain in ddH_2O on a platform shaker (75 rpm) overnight.
38. An FP will result in a band around 27 kDa; the gel is screened for additional bands.

Regeneration Ni^{2+} Affinity Columns

39. The resin can be regenerated at least five times by adding ~14 ml 50 mM EDTA pH 7.5–8.0 (in ddH_2O) to resin and incubate on roller for 10 min.
40. Perform **steps 2** and **3**.
41. For storage, wash 3× with ~14 ml ddH_2O and centrifuge after each wash (3 min, 550 g, 4 °C, Eppendorf swing-out). Remove supernatant carefully by suction after every wash and resuspend in ~14 ml 20 % (v/v) ethanol (store at 4 °C). For direct usage, go to **step 4**.

3.2.2 Excitation Spectrum and Molar Extinction Coefficient

1. Add 200 μl buffer to both cuvettes (sample and reference cuvette, *see* **Note 27**) and place them in the spectrophotometer.
2. Acquire an absorbance spectrum (350–500 nm for CFP variants, 1 nm step size).
3. Inspect the spectrum for deviations other then instrumental response. If not, auto-zero the baseline.
4. Add 10 μl FP solution to the sample cuvette (*see* **Note 28**).

5. Acquire an absorbance spectrum (350–500 nm, 1 nm step size, integration time 0.1 nm/s).
6. Determine the wavelength of the absorption maximum (λ_{max}), for CFP variants 434 nm is used.
7. Repeat **steps 4** and **5** until the absorbance at λ_{max} is between 0.1 and 0.2 (*see* **Note 29**).
8. Acquire an absorbance spectrum (350–500 nm, 0.1 nm step size).
9. Calculate the FP concentration (in M) in the sample cuvette and the molar extinction coefficient (*see* **Note 30**) at λ_{max} using the law of Beer–Lambert (Eq. 2):

$$A = \varepsilon c l \tag{2}$$

In Eq. 2, A is the actual absorbance of chromophore, l is light path length (in cm), c is the protein concentration (M), and ε is extinction coefficient (M^{-1} cm^{-1}) at the absorption maximum.

3.2.3 Emission Spectrum and Fluorescence Quantum Yield

1. Select an excitation wavelength that is suited for the reference dye and the selected FP. For measuring fluorescein and CFP variants, 455 nm is used.
2. Switch on the lamp of the fluorescence spectrophotometer and wait 15 min for the lamp to stabilize (75 W).
3. **Steps 1–4** of Subheading 3.2.2 using an FP with unknown quantum yield.
4. Acquire an absorbance spectrum (350–500 nm, 1 nm step size).
5. Determine the wavelength at the absorption maxima (λ_{max}) at 455 nm.
6. Repeat **steps 3–5** until $A_{455} = 0.2$. Note the exact value.
7. Add 750 μl buffer to a fluorescence cuvette and add 250 μl of the FP solution with absorbance at $A_{455} = 0.2$ (*see* **Note 31**).
8. Place the cuvette in the fluorescence spectrophotometer (*see* **Note 32**).
9. Acquire an emission spectrum: emission slits at 1 nm precision, excitation slits at 4 nm precision, excitation 455 nm, emission 460–650 nm, and 0.1 s integration time. Repeat **steps 3–8** for the reference dye.
10. If different buffers are used for the reference dye and the selected FP, a correction for refractive index should be performed (*see* **Note 33**). Add 1 ml of buffer to the cuvette and repeat **steps 8** and **9** for each buffer.
11. Obtain an instrumental correction factor using Spectral Fluorescence Standard Kit according to the manufacturer's instruction and the same experimental settings.

Analysis Fluorescence Quantum Yield

1. Correct the spectra of the reference dye and FP for buffer and instrumental settings using Eq. 3 [22]. In Eq. 3 $E_{cor}(\lambda)$ is the corrected signal value, $E_s(\lambda)$ is the fluorescent value of the sample, $E_b(\lambda)$ is the fluorescent value of buffer only, and $CF(\lambda)$ is the correction factor, all in arbitrary units:

$$E_{cor}(\lambda) = [E_s(\lambda) - E_b(\lambda)]CF(\lambda) \quad (3)$$

2. Integrate the corrected spectra over λ.
3. Calculate the quantum yield of the selected FP, using Eq. 4 [22]:

$$QY_U = QY_R \frac{A_R \eta_U^2 F_U}{A_U \eta_R^2 F_R} \quad (4)$$

where QY_U = unknown quantum yield of selected FP,
QY_R = quantum yield of reference,
A_R = absorbance of reference at excitation wavelength,
A_U = absorbance of selected FP at excitation wavelength,
η_U = refractive index of selected FP in buffer,
η_R = refractive index of reference in buffer,
F_U = integrated corrected spectrum of selected FP, and
F_R = integrated corrected spectrum of reference.

3.2.4 Fluorescence Lifetime

For this method purified FPs (HPLC purified or directly after dialysis), a cell lysate containing FPs or cells producing FPs, can be used. The following method is written for purified FPs. A method to prepare a cell lysate is followed directly. In the following section (Subheading 3.3), a general protocol (Subheading 3.3.3) for mammalian cell preparation producing FPs can be found.

1. Dilute the selected purified FP with PBS (*see* **Note 34**).
2. Place the 96-well plate at the microscope stage.
3. Optimize the detected fluorescence to its highest value by moving the *X*-, *Y*-, and if available the *Z*-position of the pinhole(s) in the confocal microscope. Optimize the lens position in front of the APD, if possible.
4. Measure the sample as described in Subheading 3.1.2. Typically the laser power is set not higher than 2.1 kW/cm^2 for FPs in solution (*see* **Notes 17** and **18**). This excitation intensity should still be sufficient to achieve a reasonable signal-to-noise ratio (5–10) within measurement times of 60–180 s.
5. The data is saved and analyzed as in Subheading 3.1.2, TD-FLIM (*see* **Note 35**).

Lysing Mammalian Cells

1. Wash transient transfected HeLa cells (*see* Subheading 3.3.3) grown in a 6-well plate twice with ice-cold PBS buffer (pH 7.4).
2. Remove buffer completely.

3. Add 300 μl 1 % Triton X-100 in PBS.
4. Incubate for a few minutes, swirl now and then.
5. Transfer solution to an Eppendorf tube spin at 16,000 × g speed for 5 min at 4 °C.
6. Transfer 200 μl supernatant to a fresh tube.
7. The supernatant can be used for in vitro measurements TD-FLIM, FCS, and PCH (*see* also Subheading 3.2.5).
8. Typically, for untagged FPs dilute 5–10 μl with 90 μl 1 % Triton X-100 in PBS [24].

3.2.5 Intrinsic Molecular Brightness

1. Experimental setup and optimization as in Subheading 3.2.4.
2. Optimize the objective correction collar and the confocal pinhole position such that the highest fluorescence intensity is detected from the calibration solution.
3. Perform a FCS measurement of Alexa488 in 10 mM Tris and adjust the measurement time such that the resulting correlation curve is smooth in the decaying part of the curve.
4. The intensity traces are imported into the FFS data processor 2.3 software, autocorrelated, and fitted to a model including terms for triplet state kinetics and three-dimensional Brownian diffusion (Eq. 5) [29]:

$$G(\tau) = 1 + \frac{\gamma}{N} \frac{1 + \frac{T}{1-T} \exp\left(\frac{-\tau}{\tau_{\text{dark}}}\right)}{\left(1 + \frac{\tau}{\tau_{\text{dif}}}\right)\sqrt{1 + \frac{\tau}{\tau_{\text{dif}}}\left(\frac{\omega_{xy}}{\omega_z}\right)^2}} \quad (5)$$

where $G(\tau)$ = autocorrelation function,
γ = the shape factor of the observation volume,
N = average number of fluorescent particles in the detection volume,
T = fraction of fluorophores present in the dark state (triplet or longer duration dark states),
τ_{dark} = the average time a fluorophore resides in the dark state (s),
τ_{dif} = average diffusion time of the fluorophores (s),
ω_{xy} = equatorial radius of the observation volume (m), and
ω_z = axial radius of the observation volume (m).

5. Validate if the obtained fitting parameters are within the expected range (*see* **Note 36**). For our system the value for a ($=\omega_z/\omega_{xy}$) ranges from 4 to 15 and the value of τ_{dif} for Alexa488 around 35 μs (*see* **Note 37**). The observed brightness values vary between 8 and 12 kHz per molecule. Note that these values will be different for other microscope systems. In general, the higher the brightness per particle, the better

the quality of the obtained data will be. The system has to be calibrated until the highest possible brightness is reached without having photobleaching and/or saturation artifacts [30].

6. Perform a background measurement, measure 0.2 % (w/v) BSA in 10 mM Tris.
7. Measure the sample as described in Subheading 3.2.4.
8. For FCS measurements, if a clear sigmoid-shaped autocorrelation curve is not being built up during acquisition, check the calibration and/or dilute the experimental sample when count rates exceed 100 kHz.

Brightness Analysis

1. After loading the raw data into the FFS data processor software, a region that contains at least 1,000,000 photons is selected out of the complete raw data trace. The selected intensity trace should be stable, so without an intensity drift, and the presence of (small) intensity spikes should be minimal. The data was binned to time windows of 0.2 μs (for FCS) or 5, 10, 50, and 100 μs (for photon counting histogram (PCH)) and autocorrelated or plotted as a photon counting histogram.
2. The FCS curve was fitted using Eq. 5 and the PCH curves were analyzed with a two-component three-dimensional PCH model containing corrections for dark-state kinetics, diffusion, and the non-Gaussian detection volume [31] (*see* Chapter 33 by Skakun et al. in this volume).
3. The molecular brightness of the first (background) component was fixed to the value obtained in the control measurement consisting of 10 mM Tris buffer with 0.2 % BSA, using a one-component fitting model. FCS and PCH curves were fitted simultaneously in which parameters as dark-state relaxation time, diffusion time, volume structure parameter, particle number, and molecular brightness were globally linked.

In Fig. 8 the PCH data (symbols) and fits (solid lines) are shown for mTurquoise. The curves represent the data binned using 5, 10, 50, and 100 μs and are shown in squares, diamonds, triangles, and circles, respectively. Fitting the data points to a PCH model containing dynamics corrections, a molecular brightness for mTurquoise was found. By applying the same protocol to a related FP of the same spectral class, a relative molecular brightness can be calculated (*see* **Note 38**).

3.2.6 Oligomeric State

1. Prepare the gel (10 % running gel and 4 % stacking buffer) and set up the electrophoreses cell according to manufacturer's instructions.
2. Add sample buffer to the purified FP samples, dimer FP reference, and monomer FP reference (do not boil).

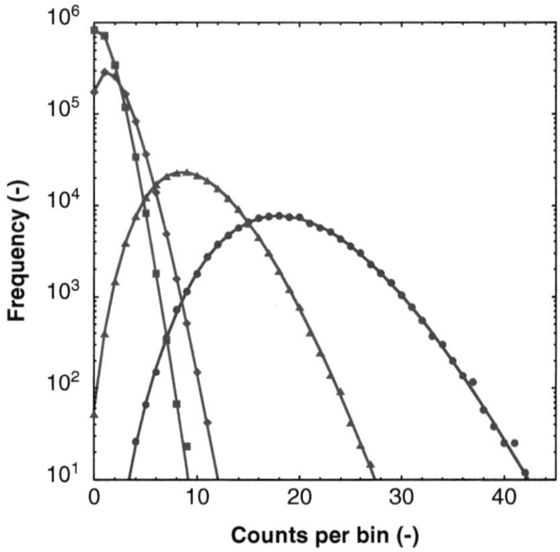

Fig. 8 PCH data of mTurquoise. The four datasets show the same data binned with 5 μs (*squares*), 10 μs (*diamonds*), 50 μs (*triangles*), and 100 μs (*circles*). By fitting the data points to a PCH model containing optical and mobility corrections, the molecular brightness of 171 ± 26 kHz/molecule for mTurquoise was obtained using our setup

3. Run the gel at 150 V for 45 min.
4. Stain with Coomassie staining and destain using ddH_2O; *see* Subheading 3.2.1 (Protein purity assay—SDS-PAGE).
5. The gel is screened for the presence of oligomers. A monomer will result in a band around the height of the monomer reference; a dimer will show a higher band (*see* **Notes 39** and **40**).

3.2.7 pH Sensitivity

1. Pipet the various pH solutions in the 96-well plate (*see* **Note 41**), equal volume in all wells.
2. Add FP solution to each well, equal concentration in all wells.
3. Measure the fluorescence from the bottom.
4. Fit the data in a software program to Eq. 6 and determine the pK_a. Using this equation one assumes a single (average) protonation effect in the sample. For more complicated behaviors, a multicomponent analysis can be performed. In Eq. 6 F is the fluorescence signal (arbitrary units), B is signal baseline (arbitrary units), and G is the gain (maximal fluorescence minus the base line, in arbitrary units):

$$F = B + \frac{G}{1 + 10^{(pK_a - pH)}} \tag{6}$$

The pH sensitivity of mTurquoise2 is shown in Fig. 9, resulting in a pK_a of 3.1. mTurquoise2 is not pH sensitive from pH 5 to 9.

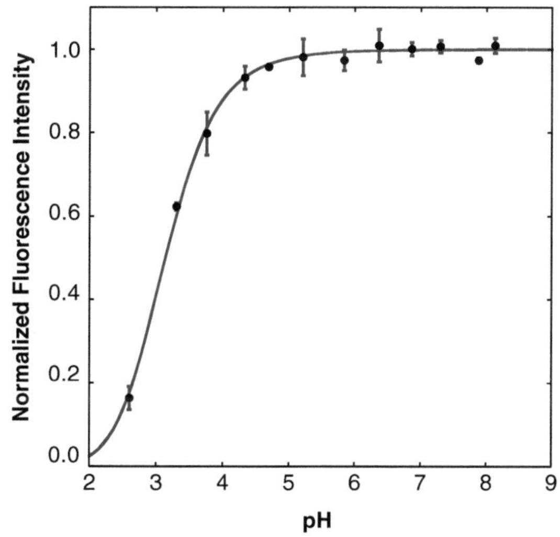

Fig. 9 pH sensitivity of mTurquoise2 [8]. The fluorescence of mTurquoise2 is not affected within the pH range 5–9

3.3 Characterization In Vivo

3.3.1 Fluorescent Protein Maturation in Bacteria

1. Transform the bacteria with the vectors (*see* Subheading 3.1.1, **steps 5–8**).
2. Transfer the growth culture to a cultivation flask with 4 ml protein production medium (LB supplemented with ampicillin).
3. Incubate at 37 °C under vigorous shaking for an additional ~4 h.
4. During incubation, measure the absorbance at 600 nm in plastic cuvettes, use LB as reference. Continue if the $0.1 < A_{600} < 1.0$ (*see* **Note 42**).
5. Preheat the incubator of the microtiter plate reader to 37 °C.
6. Dilute all cultures to $A_{600} = 0.1$ with protein production medium (LB supplemented with ampicillin).
7. Fill the microtiter plate with 180 μl protein production medium (LB supplemented with ampicillin) and add 20 μl bacteria culture ($A_{600} = 0.1$) to the wells (each FP variant in triplicate).
8. Start the plate reader experiment. Plate reader settings: temperature 37 °C and moderate continuous shaking (*see* **Note 43**). Measure every 10 min during 16 h, the absorbance at 600 nm and the fluorescence (excitation 440, emission 480/20 nm for CFP), measure from the bottom (*see* **Note 44**).
9. Construct the growth curves for each well using the absorbance at 600 nm. If all wells show similar growth curves, the rise of the fluorescence of the FPs can directly be compared without correction. The time point where the fluorescence starts to increase is indicative for the FP production and maturation rate in *E. coli*. By comparing the upcoming fluorescence among the FP variants, a relative maturation rate in bacteria can be determined.

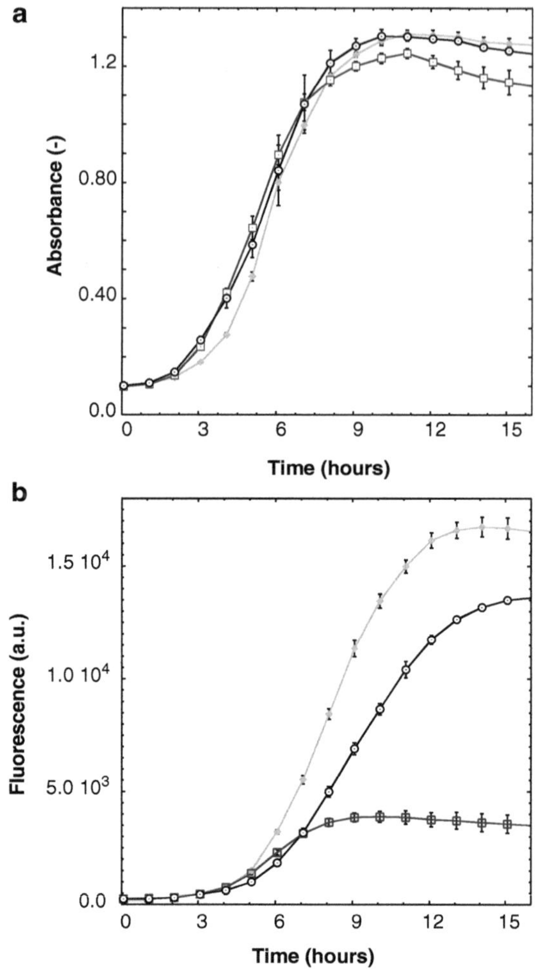

Fig. 10 Maturation experiment of CFP variants in *E. coli*. The variants mCerulean2 (*squares*), mCerulean3 (*circles*), and mTurquoise (*diamonds*) were used. The growth curves measured at 600 nm show equal growth rate (**a**). The rise of the fluorescence (**b**). mCerulean2 and mTurquoise2 show initially (2–5 h) equal production and maturation rate. However, after 6 h, the production and maturation rate of mCerulean2 is reduced, with respect to mTurquoise2. mCerulean3 has an decreased production and maturation rate in *E. coli* compared to mCerulean2 and mTurquoise2

Figure 10 shows the result of a maturation experiment of the following CFP variants: mCerulean2, mCerulean3, and mTurquoise. In Fig. 10a the growth curves, measured at 600 nm, are shown. The curves are the average of 3 wells and the standard deviations are given by the bars. All three variants show equal growth. Figure 10b shows the fluorescence rise of the same experiment. From Fig. 10b it can be concluded that initial maturation of mCerulean2 is equal to mTurquoise2 (t = 3–5 h). However, the production and maturation rate of mCerulean2 is reduced after

$t = 5$ h. While the production and maturation rate of mTurquoise2 increases, mCerulean3 has a decreased maturation and production rate compared to mCerulean2 and mTurquoise2.

3.3.2 Cloning in Mammalian Vector

Isolate the DNA from bacteria (alkaline lysis method or a plasmid purification kit) and place the DNA sequence in a mammalian vector. We use a modified pRSET vector that allows simply exchange of an FP DNA sequence to pC1 vector containing an FP (Clontech) using the restriction sites *Age*I/*BsrG*I. Plasmids and sequence information are available on request.

3.3.3 Transfection and Mounting of Mammalian Cells

1. Place 24 mm cover slips 70 % (v/v) EtOH solution for 10 min.
2. Transfer 24 mm cover slips from the 70 % EtOH solution to a 6-well plate and dry for >10 min in the flow cabinet.
3. Seed HeLa cells in the 6-well plates. Typically 1/5th of the cell suspension from a confluent T25 flask is diluted in a total volume of 12 ml growth medium. 2 ml of this suspension is transferred to each well.
4. Perform transfection the next day, using 1 μl of plasmid DNA (from a purified miniprep), 1–2 μl of lipofectamine, and 50 μl Opti-MEM, according to manufacturer's instructions.
5. 1 or 2 days after transfection, mount the cover slip in an Attofluor cell chamber (Invitrogen) and add 1 ml of microscopy medium (room temperature).

3.3.4 Photostability In Vivo

Wide-Field Conditions

Preparing the Microscope

1. Switch on the mercury lamp and wait 15 min for the lamp to stabilize.
2. Measure the illumination power using the 10× objective in combination with and without a 1.3 ND grey filter, for "normal" illumination condition and intense illumination, respectively.
3. Place the stage micrometer on the stage and measure the diameter of the illumination field, observed through the ocular.
4. Calculate the illumination intensity under the two circumstances by dividing the illumination power by the area illuminated at 63× magnification.

For example, the illumination power on our setup using 10× objective and grey filter is 1.2 mW. The diameter of the illumination field of our setup is 310 μm. Dividing the total illumination intensity by the area at 63× magnification gives 1.2 mW/75,477 $\mu m^2 = 1.6\ W/cm^2$ for "normal" illumination intensity. For intense illumination the light intensity is 23 W/cm^2 on our setup.

Photostability Measurements

1. Focus cells in the center of the field of view (*see* **Note 45**).

Photobleaching Under "Normal" Illumination Conditions

2a. Continuously illuminate the cells at a magnification of 63× using a 1.3 ND filter. Acquire images every 3 s for a total of 150 s.

Photobleaching Under Intense Illumination Conditions

2b. Continuously illuminate the cells at a magnification of 63× using no ND filter. Acquire images every 3 s for a total of 150 s.

Discontinuous Illumination to Measure Reversible Bleaching

2c. Illuminate the cells at a magnification of 63× using (with 1.3 ND filter and without ND filter) for some time and interrupt illumination and continue afterwards. Acquire every 3 s an image during illumination for a total of 150 s.

Image Processing Using ImageJ

1. Measure the average intensity of an ROI that represents background.
2. Measure the CFP fluorescence from an ROI that represents a single cell.
3. Subtract the background data from the CFP intensity data (per frame).
4. Normalize the fluorescence intensity to 1 at $t = 0$ s.
5. Average the background-corrected CFP intensity from at least six cells, obtained from at least three independent measurements.
6. Plot the CFP intensity over time.

Figure 11 shows the decay of fluorescence over time for CFP variants using continuous illumination intensity of 1.6 W/cm^2. mTurquoise2 shows almost no bleaching over time under these conditions.

From the obtained data the average number of photons a molecule emits before bleaching (N_{em}) can be calculated, using Eq. 7 (justification can be found in [8]):

$$N_{\text{em}} = \frac{k_{\text{f}}}{k_{\text{bl}}} = \frac{1{,}000 \cdot \ln 10 \cdot \varepsilon \cdot W \cdot \text{QY} \cdot \lambda}{h \cdot c \cdot N_{\text{A}} \cdot k_{\text{bl}}} \tag{7}$$

k_{f} = radiative decay (photons/molecules · s)
k_{bl} = rate of photobleaching, fraction of molecules that are photobleached per unit of time (s^{-1})
ε = extinction coefficient at the employed excitation wavelength ($\text{M}^{-1}\ \text{cm}^{-1}$)
W = excitation power density (W/cm^2)
QY = quantum yield (–)
λ = excitation wavelength (m)
h = Planck's constant (6.6×10^{-34} J s)
c = speed of light (3×10^8 m/s)
N_{A} = Avogadro's number

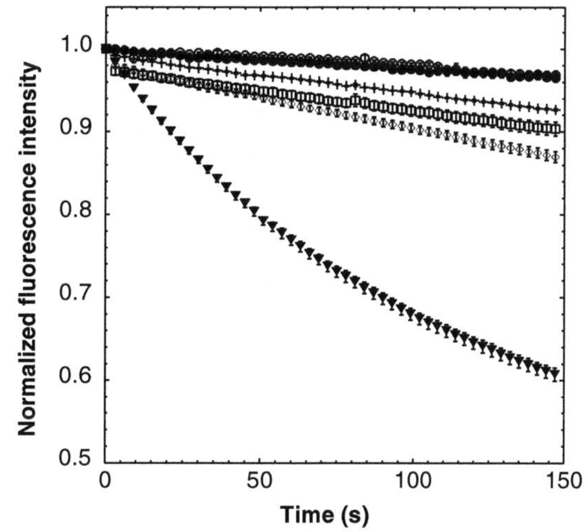

Fig. 11 Photostability of CFP variants was measured in living HeLa cells ($n = 7$, three independent experiments) using wide-field illumination at 436 nm light at an intensity of 1.6 W/cm^2. The CFP variants ECFP (*open diamonds*), SCFP3A (*crosses*), mCerulean (*open squares*), mCerulean3 (*triangles*), and mTurquoise (*closed circles*) were measured. mTurquoise and mTurquoise2 are the most photostable CFP variants in vivo [8]

For CFP variants, Eq. 7 can be approximated by substituting $\varepsilon = 30{,}000\ \text{M}^{-1}\text{cm}^{-1}$ and $\lambda = 435 \times 10^{-9}$ m, yielding Eq. 8:

$$N_{\text{em}} = \frac{251 \cdot W \cdot \text{QY}}{k_{\text{bl}}} \tag{8}$$

By approximating k_{bl} from the curves of Fig. 11, N_{em} can be calculated. At the illumination intensity of 1.6 W/cm^2, k_{bl} is approximated for mTurquoise2 (QY = 0.93) and SCFP3A (QY = 0.56) at $0.23 \times 10^{-3}\ \text{s}^{-1}$ and $0.50 \times 10^{-3}\ \text{s}^{-1}$, respectively. N_{em} is calculated to be 1.62×10^6 and 0.45×10^6 photons for mTurquoise2 and SCFP3A, respectively. So it is noted that although the bleach rate of SCFP3A is only half compared to mTurquoise2, N_{em} of SCFP3A is 25 % of N_{em} of mTurquoise2, due to the influence of the quantum yield.

Laser Scanning Conditions

1. Start with the FP that is thought to be the most photostable; with this sample the settings can be explored. For CFP variants the following settings were used: 512 × 512 pixels, dwell time 1.9 μs, pixel size 0.07, laser power 80 %, pinhole open, and 1.03 frames/s.
2. Measure all samples with the settings that were selected in **step 2** (*see* **Notes 45** and **46**).
3. Image processing is similar to photostability measurements under wide-field illumination.

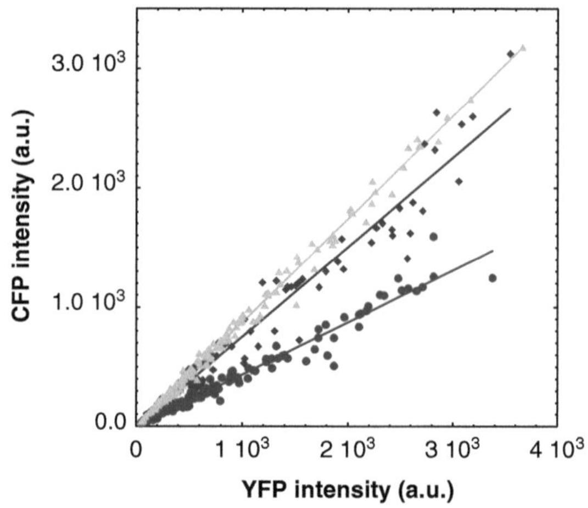

Fig. 12 Relative brightness of CFP variants compared to sYFP2. Transient transfected HeLa cells producing an equivalent number of sYFP2 and CFP proteins, accomplished by a viral T2A construct. From the CFP variants mTurquoise-146G (*circles*), mTurquoise (*diamonds*), and mTurquoise2 (*triangles*), mTurquoise2 is clearly the most bright variant

3.3.5 Relative Brightness In Vivo

1. Preheat the cell chamber and medium at 37 °C (*see* **Note 47**).
2. Acquire first a YFP fluorescence image and second a CFP fluorescence image to prevent YFP bleaching. Keep acquisition parameters (exposure time and binning) constant during the entire experiment (*see* **Note 48**).
3. Acquire the fluorescence from at least 50 cells.
4. Measure a non-illuminated image for background correction
5. Measure a homogeneous sample for flat-field correction

Image Processing

1. Perform background correction and flat-field correction.
2. Measure CFP fluorescence using the average intensity from an ROI that represents a single cell in the CFP channel. Measure YFP fluorescence using the average intensity from an ROI that represents a single cell in the YFP channel. The CFP ROI and YFP ROI should be identical.
3. Plot the CFP intensity versus YFP intensity ($n > 50$) and calculate the relative brightness by fitting the data to a linear equation without offset (ax + b). The slope (a) is a relative measurement for the brightness of the CFP variants compared to SYFP2.

Figure 12 represents the relative brightness of CFP variants compared to SYFP2. The YFP and CFP fluorescence are plotted on the *x*-axis and *y*-axis, respectively. The slope differs for the three CFP variants. Clearly, the relative brightness of mTurquoise2 is higher than of mTurquoise-146G.

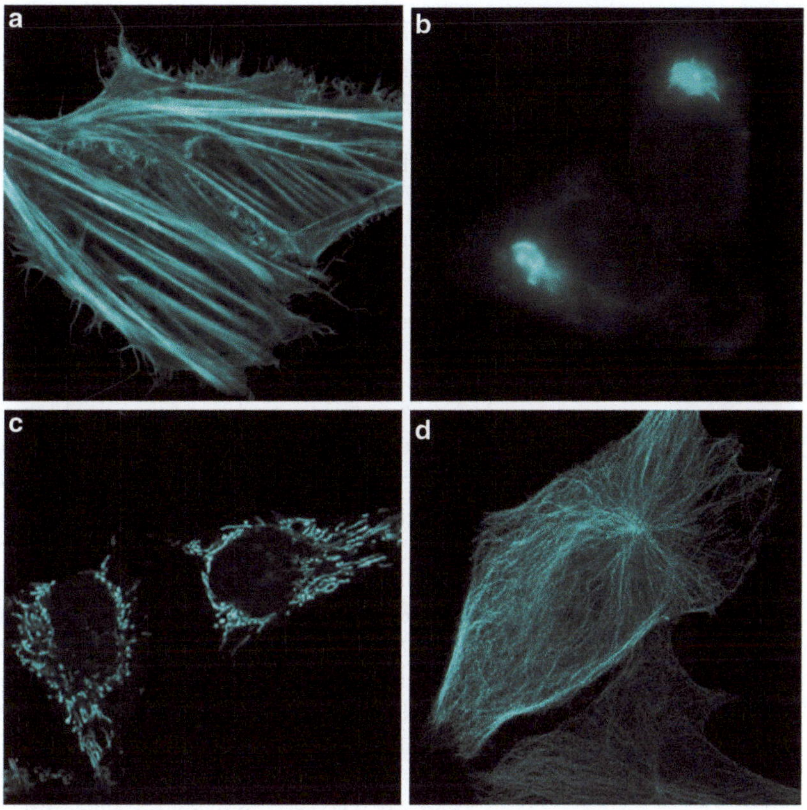

Fig. 13 Localization of mTurquoise2 targeted to the actin skeleton (**a**), Golgi apparatus (**b**), mitochondria (**c**), and tubulin skeleton (**d**) (markers are available from http:/www.addgene.org/browse/article/5268/)

3.3.6 Subcellular Localization Studies

In general, the production and localization of the selected FPs in HeLa or other mammalian cells can be evaluated using a similar protocol for cell preparation and microscopy methods as in Subheading 3.3.3. For the evaluation of correct localization in different cell compartments, the DNA sequence of the selected FP has to be fused to a signal motif or target proteins; several examples can be found in [32]. Confocal microscopy is adequate in most cases to evaluate the localization and production of the selected FPs in a variety of cell types and cell organelles. A method to evaluate co-localization of a selected FP is shown below as an example.

1. Clone mammalian vectors with a selected FP fused to a target sequence. Here we used a target sequence for actin, the Golgi apparatus, mitochondria, and tubulin marker.
2. Transient transfected mammalian cells with the expression vectors (*see* Subheading 3.3.3).
3. Image the cells. The following settings were used for Fig. 13: pinhole size 27.06 μm, scan size is 512 × 512 pixels, 2 frames/s, and laser power 7.8 %.

4. Confirm the location of the selected FP using commercially available organelle stains. If the selected FP and the commercial organelle stain are from different spectral classes, one can stain the organelles and produce the selected FP in the same cell to confirm the overlay of the two fluorescent probes.

mTurquoise2 targeted to the actin skeleton, Golgi apparatus, mitochondria, and tubulin skeleton is shown in Fig. 13.

4 Notes

1. For screening and protein production, we employ bacterial expression vectors based on pRSET (Invitrogen). The RSET is modified in such a way that the cDNA encoding the fluorescent protein can be exchanged between pRSET and Clontech-like vectors (e.g., pEGFP-C1 and pEGFP-N1), which are widely used for mammalian cell expression using the *Age*I/*BsrG*I restriction sites. The pRSET vectors are used since they show leaky expression when transformed into BL21(DE3) cells which is helpful for screening. In addition, these bacterial expression vectors can be used for protein production and purification by immobilized metal affinity chromatography (IMAC). We used a modified version of pRSET (Invitrogen). Several bacterial expression plasmids with an FP are available from our laboratory upon request under material transfer agreement.
2. An FP with a well-characterized fluorescence lifetime from the same spectral class. In same cases the template FP and the reference FP can be identical. Bacteria producing a reference and template FP are required for Subheading 3.1.2.
3. Optimal growth conditions can provide colonies of a moderate size, which makes screening on plates easier.
4. Since the time-correlated single-photon counting principle is being used for detection [33], one relates the arrival time of the detected photon to the time of the previous laser pulse. However, if the laser repetition frequency is set too high with respect to the fluorescence lifetime of the fluorophore, the photon may originate not from the previous but from earlier excitation pulses. To prevent this artifact a minimal laser pulse delay is chosen that is at least five times the fluorescence lifetime.
5. Deprotonation effects could occur when buffers are used at low pH. In addition, halogens could also affect the chromophore, especially for YFP variants. Use for YFP variants a buffer with high pH and low halogens and do not set the pH with HCl. For CFP variants PBS or 10 mM Tris (pH 8.0) can be used.

6. It is important to use the same strain of bacteria and the same vector for all FP variants, for example, the *E. coli* strain BL21 (DE3) carrying a pRSET vector with a FP variant.
7. Measure the absorbance for the growth rate at a wavelength that does not interfere with the absorbance of the chromophore itself. CFP variants are excited using 380–480 nm light; therefore, the growth rate of CFP variants is monitored at 600 nm.
8. HEPES is not stable during autoclaving, filter sterilize HEPES.
9. The procedure is based on amplification of methylated DNA, which can subsequently be selectively cleaved by *Dpn*I.
10. When no product is formed after mutagenesis PCR reaction, try a different annealing temperature. Most modern PCR apparatuses are able to perform a temperature range, to check for the most suited annealing temperature. The band size should be 3.6 kbp for pRSET vector with an FP.
11. For plating <100 μl, centrifuge the cell suspension (13,000 × *g*, 1 min) and resuspend in 100 μl.
12. A high background of colonies in the negative control indicates that the *Dpn*I digestion is incomplete. Prolong incubation time with *Dpn*I or try another batch of *Dpn*I.
13. When no individual colonies are formed, dilute the mutagenesis PCR reaction or the growth culture after recovery.
14. Resemblance of the modulation (τ_m) and phase shift lifetime (τ_φ) indicates a mono-exponential lifetime, as was the case for mTurquoise ($\tau_m = 3.7$ ns and $\tau_\varphi = 3.8$ ns), but not for SCFP3A ($\tau_m = 2.7$ ns and $\tau_\varphi = 3.2$ ns) and Cerulean ($\tau_m = 2.3$ ns and $\tau_\varphi = 3.1$ ns) [7]. A more accurate analysis of the decay components of the fluorescence lifetime can be found in Subheading 3.2.4.
15. Absorption of fluorescence and reemission can occur when FPs concentration is too high and when the Stokes shift is small. Due to this effect a higher fluorescence lifetime is observed than the actual fluorescence lifetime. The fluorescence lifetime is determined again in Subheading 3.2.4, there the FP concentration can be adapted.
16. The increased path length of the excitation/emission light increases the traveling time of light by several nanoseconds. In case of TD-FLIM the delay in photon arrival times is not affecting the lifetime analysis. The fluorescence lifetime decay curve is composed of the accumulated arrival times which all have the same delay. Therefore, the acquired decay will have a shift, but its slope, which determines the lifetime, remains unaltered.

17. Many fluorescent probes like the FPs are susceptible to photobleaching and show blinking behavior, due to continuous conversions between fluorescent and dark states [34–36]. Both processes become more pronounced at higher laser powers. These effects, although interesting for the development of dyes for some specific spectroscopic applications, are not included in the screening of fluorescent proteins as discussed here. In addition to these effects, at higher laser intensities, probe saturation might occur in the center of the focus [30]. This will change the effective shape of the detection volume, resulting in a deviation from the 3D Gaussian-shaped detection volume that is assumed in the fitting models. Therefore, a trade-off should be found using a laser power where the brightness of the molecule (expressed as detected counts s^{-1} molecule^{-1}) is as high as possible but where artifacts caused by the processes described above are minimized. To find this saturation level, one should plot the fluorescence intensity of the fluorophore versus laser power.
18. For TCSPC detection too high count rates (>500 kHz) should be prevented. If more than one photon arrives after the excitation pulse, only the first one will be detected, resulting in a bias of the retrieved fluorescence lifetime histogram. This effect is called pileup.
19. Without a priori knowledge the fitting model is kept as simple as possible, always starting with a mono-exponential model ($n = 1$, Eq. 1). Only when the fit quality is considered to be mediocre (Subheading 3.1.2, Analysis of time-domain FLIM, **step 5**), the model is extended with additional exponential components ($n > 1$).
20. The software automatically scales to 1 arbitrary fluorescent unit. Most LP filters are not 100 % closed at lower wavelengths. If the fluorescence intensity of the sample is very low, the software will auto-scale the excitation light peak to 1, which could be used as an internal standard. The wavelength axis of the SPIM device can be calibrated by measuring (attenuated) laser or lamp lines.
21. A corrected emission spectrum can be obtained with purified FPs using a fluorescence spectrophotometer and applying corrections for the instrument response (*see* Subheading 3.2.3).
22. Carefully remove supernatant from the resin by suction; do not decant. The pellet is only loosely attached to the bottom.
23. During incubation of Ni^{2+} affinity column and the protein solution, it is very important to prevent foam formation (foams are denatured proteins).
24. To obtain a slightly more pure fraction after eluting with imidazole, a pre-elution with a lower concentration of imidazole

(0.02 M) can be performed to remove unspecific binding to the Ni^{2+} resin.

25. Place the dialysis tubes and the stirring bean in such a position that the clamps will not open by the stirring motion, stir at a modest speed.
26. The final purity assay can be combined with the oligomeric state assay (*see* Subheading 3.2.6).
27. Assign one cuvette as reference and one cuvette as sample. Always place them in the same orientation in the absorbance spectrophotometer.
28. Mix by carefully pipetting up and down and do not remove the cuvettes after the machine has been set to auto-zero.
29. The contribution of fluorescence will result in a decrease of the measured absorption. To minimize the fluorescence contribution, absorption measurements should be performed with low concentration but in the linear range of the detector. This effect can be circumvented by placing band-pass filters between the cuvettes and the detectors that block the fluorescence.
30. The molar extinction coefficient can easily be underestimated if the FP solution contains a small fraction of protein impurity or immature FP.
31. To prevent inner filtering during fluorescence measurements, the absorbance at excitation wavelength should be <0.1, preferably 0.05.
32. Use the same cuvette during one experiment and place this cuvette always in the same orientation in the fluorescence spectrophotometer.
33. The refractive index of the buffers could be measured using an Abbe refractometer [37].
34. The detection of too high intensities can be prevented by diluting the in vitro samples with buffer. When measuring transfected cells, one could change promotor sequence, reduce the amount of DNA in the transfection procedure, or shorten the time between transfection and measurement.
35. The fluorescence lifetime of FPs can differ in vitro and in vivo, due to the different environment of the FPs and the refractive index of the samples [38, 39]. Absorption of fluorescence and reemission will not occur in these fluorescence lifetime measurements as a consequence of low concentrations (as was described in **Note 34**).
36. Take care that τ_{dark} is fitted correctly, typically 1–200 μs depending on dye and laser power, and is not mixing with τ_{dif}. In order to distinguish the two processes, one can change the size of the detection volume by slightly altering the pinhole diameter. Only τ_{dif} will be affected.

37. The shape factor of the observation volume (γ) equals 0.3535 for a 3D Gaussian or 1.0 for a cylindrical-shaped observation volume. From the autocorrelation fits the shape factor, a, describing the ratio between the axial and lateral e^{-2} radii of the observation volumes ($a = \omega_z/\omega_{xy}$), and the τ_{dif} for Alexa488 are obtained. Although the shape of the detection volumes is close to a 3D Gaussian in "ideal conditions," the optical aberrations introduced by measuring inside the living cell make it possible to approximate the detection volumes by cylinders. Then the amplitude of the autocorrelation function, $G(0)$, equals to γ/N. Here cylindrical detection volumes are assumed with $\gamma = 1$.

38. The molecular brightness depends on the excitation and the detection efficiency; in order to compare results of different days, one can include a reference dye. For CFP variants Alexa488 can be used.

39. This method does not only rely on the size of the protein, also on the dielectric point and the interaction with the SDS from the gel and running buffer. This method will yield an initial idea of the oligomeric state of the selected FP. Quantitative methods to determine the dimerization/affinity based on ultra centrifuge and fluorescence polarization have been published [26, 40].

40. The oligomeric state can be tested spectroscopically by the photon counting histogram analysis as described in Subheading 3.2.5. The retrieved average brightness is expected to be roughly a multitude of the brightness of the monomeric control; thus, a solution of dimers should result in the double brightness. However, for mixtures of monomers with oligomers, a more complicated brightness-weighted average will come out [41].

41. It is not recommended to use buffers or set the pH value with solutions that contain chloride or other halides, since some FP variants can be sensitive to halides, especially the YFP variants.

42. A_{600} should not exceed 1.0. If one culture will exceed $A_{600} = 1.0$, while another culture has not reached $A_{600} = 0.1$ yet, dilute the first culture before reaching $A_{600} = 1.0$. After a dilution incubate the cultures for minimal 30 min before continuing with **step 6**.

43. During the plate reader experiment, shake as fast as possible, but without spilling or cross contaminating the medium between the wells.

44. It is also possible to record the emission spectrum. This is especially useful if the emission spectrum of the new FP is still unknown. The data has to be acquired quickly to minimize the growth disturbance due to the absence of shaking during the spectral measurement.

45. When searching and focusing on the cells, make sure that excitation light is dim and observation time is minimized to avoid bleaching.
46. It is of great importance to keep all settings equal for the different FP variant.
47. For this experiment it is not necessary to set the temperature at 37 °C. However, if the temperature is set at 37 °C, the relative brightness can directly be extrapolated to other biological experiments for which 37 °C is demanded.
48. The excitation and emission settings should be chosen in such a way that cross-talk free images are obtained.

Acknowledgments

We acknowledge Richard N. Day (Department of Cellular and Integrative Physiology, Indiana University School of Medicine, Indianapolis, USA) for the kind donation of the plasmid with cDNA encoding mCerulean3. This work was enabled through "Middelgroot" investment grants (834.07.003 and 834.09.003) and Echo grant (711.01.01812) from the Netherlands Organization for Scientific Research (NWO).

References

1. Chudakov DM, Lukyanov S, Lukyanov KA (2005) Fluorescent proteins as a toolkit for in vivo imaging. Trends Biotechnol 23:605–613
2. Chudakov DM, Matz MV, Lukyanov S, Lukyanov KA (2010) Fluorescent proteins and their applications in imaging living cells and tissues. Physiol Rev 90:1103–1163
3. Heim R, Prasher DC, Tsien RY (1994) Wavelength mutations and posttranslational autoxidation of green fluorescent protein. Proc Natl Acad Sci USA 91:12501–12504
4. Heim R, Tsien RY (1996) Engineering green fluorescent protein for improved brightness, longer wavelengths and fluorescence resonance energy transfer. Curr Biol 6:178–182
5. Rizzo M, Piston D (2005) Optimization of cyan fluorescent protein fluorescence for Förster resonance energy transfer. Microsc Microanal 11:4–5
6. Kremers G-J, Goedhart J, van Munster EB, Gadella TWJ (2006) Cyan and yellow super fluorescent proteins with improved brightness, protein folding, and FRET Förster radius. Biochemistry 45:6570–6580
7. Goedhart J, van Weeren L, Hink MA, Vischer NOE, Jalink K, Gadella TWJ (2010) Bright cyan fluorescent protein variants identified by fluorescence lifetime screening. Nat Methods 7:137–139
8. Goedhart J, Stetten von D, Noirclerc-Savoye M, Lelimousin M, Joosen L, Hink MA, van Weeren L, Gadella TWJ, Royant A (2012) Structure-guided evolution of cyan fluorescent proteins towards a quantum yield of 93%. Nat Commun 3:751–759
9. Lelimousin M, Noirclerc-Savoye M, Lazareno-Saez C, Paetzold B, Le Vot S, Chazal R, Macheboeuf P, Field MJ, Bourgeois D, Royant A (2009) Intrinsic dynamics in ECFP and Cerulean control fluorescence quantum yield. Biochemistry 48:10038–10046
10. Elsliger MA, Wachter RM, Hanson GT, Kallio K, Remington SJ (1999) Structural and spectral response of green fluorescent protein variants to changes in pH. Biochemistry 38:5296–5301
11. McAnaney TB, Zeng W, Doe CFE, Bhanji N, Wakelin S, Pearson DS, Abbyad P, Shi X, Boxer SG, Bagshaw CR (2005) Protonation,

photobleaching, and photoactivation of yellow fluorescent protein (YFP 10C): a unifying mechanism. Biochemistry 44:5510–5524

12. Terskikh A, Fradkov A, Ermakova G, Zaraisky A, Tan P, Kajava AV, Zhao X, Lukyanov S, Matz M, Kim S, Weissman I, Siebert P (2000) "Fluorescent timer": protein that changes color with time. Science 290:1585–1588
13. Bevis BJ, Glick BS (2002) Rapidly maturing variants of the Discosoma red fluorescent protein (DsRed). Nat Biotechnol 20:83–87
14. Sinnecker D, Voigt P, Hellwig N, Schaefer M (2005) Reversible photobleaching of enhanced green fluorescent proteins. Biochemistry 44:7085–7094
15. Shaner NC, Lin MZ, McKeown MR, Steinbach PA, Hazelwood KL, Davidson MW, Tsien RY (2008) Improving the photostability of bright monomeric orange and red fluorescent proteins. Nat Methods 5:545–551
16. Hanson GT, McAnaney TB, Park ES, Rendell MEP, Yarbrough DK, Chu S, Xi L, Boxer SG, Montrose MH, Remington SJ (2002) Green fluorescent protein variants as ratiometric dual emission pH sensors. 1. Structural characterization and preliminary application. Biochemistry 41:15477–15488
17. Dennig A, Shivange AV, Marienhagen J, Schwaneberg U (2011) OmniChange: the sequence independent method for simultaneous site-saturation of five codons. PLoS One 6:e26222
18. Sawano A, Miyawaki A (2000) Directed evolution of green fluorescent protein by a new versatile PCR strategy for site-directed and semi-random mutagenesis. Nucleic Acids Res 28:E78
19. Crameri A, Whitehorn EA, Tate E, Stemmer WP (1996) Improved green fluorescent protein by molecular evolution using DNA shuffling. Nat Biotechnol 14:315–319
20. Van Munster EB, Gadella TWJ (2004) phi-FLIM: a new method to avoid aliasing in frequency-domain fluorescence lifetime imaging microscopy. J Microsc 213(29–38)
21. van Munster EB, Gadella TWJ (2004) Suppression of photobleaching-induced artifacts in frequency-domain FLIM by permutation of the recording order. Cytometry A 58:185–194
22. Velapoldi RA, Tønnesen HH (2004) Corrected emission spectra and quantum yields for a series of fluorescent compounds in the visible spectral region. J Fluoresc 14:465–472
23. Kapusta P (2010) Absolute diffusion coefficients: compilation of reference data for FCS calibration. http://www.picoquant.com/technotes/appnote_diffusion_coefficients.pdf
24. Zhang D, Lans H, Vermeulen W, Lenferink A, Otto C (2008) Quantitative fluorescence correlation spectroscopy reveals a 1000-fold increase in lifetime of protein functionality. Biophys J 95:3439–3446
25. Shaner NC, Campbell RE, Steinbach PA, Giepmans BNG, Palmer AE, Tsien RY (2004) Improved monomeric red, orange and yellow fluorescent proteins derived from *Discosoma* sp. red fluorescent protein. Nat Biotechnol 22:1567–1572
26. Zacharias DA, Violin JD, Newton AC, Tsien RY (2002) Partitioning of lipid-modified monomeric GFPs into membrane microdomains of live cells. Science 296:913–916
27. Sun Y, Day RN, Periasamy A (2011) Investigating protein-protein interactions in living cells using fluorescence lifetime imaging microscopy. Nat Protoc 6:1324–1340
28. Boens N, Qin W, Basarić N, Hofkens J, Ameloot M, Pouget J, Lefèvre J-P, Valeur B, Gratton E, vandeVen M, Silva ND, Engelborghs Y, Willaert K, Sillen A, Rumbles G, Phillips D, Visser AJWG, van Hoek A, Lakowicz JR, Malak H, Gryczynski I, Szabo AG, Krajcarski DT, Tamai N, Miura A (2007) Fluorescence lifetime standards for time and frequency domain fluorescence spectroscopy. Anal Chem 79:2137–2149
29. Skakun VV, Hink MA, Digris AV, Engel R, Novikov EG, Apanasovich VV, Visser AJWG (2005) Global analysis of fluorescence fluctuation data. Eur Biophys J 34:323–334
30. Enderlein J, Gregor I, Patra D, Fitter J (2004) Art and artefacts of fluorescence correlation spectroscopy. Curr Pharm Biotechnol 5:155–161
31. Skakun VV, Engel R, Digris AV, Borst JW, Visser AJWG (2011) Global analysis of autocorrelation functions and photon counting distributions. Front Biosci (Elite Ed) 3:489–505
32. VanEngelenburg SB, Palmer AE (2008) Fluorescent biosensors of protein function. Curr Opin Chem Biol 12:60–65
33. Becker W (2010) The bh TCSPC handbook. Berlin. http://www.becker-hickl.de/handbook.htm
34. Haupts U, Maiti S, Schwille P, Webb WW (1998) Dynamics of fluorescence fluctuations in green fluorescent protein observed by fluorescence correlation spectroscopy. Proc Natl Acad Sci USA 95:13573–13578
35. Schwille P, Kummer S, Heikal AA, Moerner WE, Webb WW (2000) Fluorescence correlation spectroscopy reveals fast optical excitation-driven intramolecular dynamics of yellow fluorescent proteins. Proc Natl Acad Sci USA 97:151–156

36. Hendrix J, Flors C, Dedecker P, Hofkens J, Engelborghs Y (2008) Dark states in monomeric red fluorescent proteins studied by fluorescence correlation and single molecule spectroscopy. Biophys J 94:4103–4113
37. Pitot C (2012) Copie de ABBE Zeiss refractometer1. Zeiss. http://macro.lsu.edu/howto/Abbe_refractometer.pdf
38. Borst JW, Hink MA, van Hoek A, Visser AJWG (2005) Effects of refractive index and viscosity on fluorescence and anisotropy decays of enhanced cyan and yellow fluorescent proteins. J Fluoresc 15:153–160
39. Suhling K, Siegel J, Phillips D, French P (2002) Imaging the environment of green fluorescent protein. Biophys J 83:3589–3595
40. Espagne A, Erard M, Madiona K, Derrien V, Jonasson G, Lévy B, Pasquier H, Melki R, Mérola F (2011) Cyan fluorescent protein carries a constitutive mutation that prevents its dimerization. Biochemistry 50:437–439
41. Chen Y, Wei L-N, Müller JD (2003) Probing protein oligomerization in living cells with fluorescence fluctuation spectroscopy. Proc Natl Acad Sci USA 100:15492–15497

Chapter 17

Monitoring Membrane Properties and Apoptosis Using Membrane Probes of the 3-Hydroxyflavone Family

Zeinab Darwich, Andrey S. Klymchenko, and Yves Mély

Abstract

Environment-sensitive fluorescent membrane probes are attractive tools for investigating the membrane properties and their changes under perturbing conditions. Membrane probes of the 3-hydroxyflavone family are of particular interest due to their excited-state intramolecular proton transfer (ESIPT) reaction, which confers a dual emission highly sensitive to the polarity and hydration of the environment. In the present work, we will describe the protocols used with these probes in order to monitor the physicochemical properties of lipid membrane models and cell plasma membranes and to detect apoptosis.

Key words Fluorescent membrane probe, Environment-sensitive dye, Excited-state intramolecular proton transfer, F2N12S, Ratiometric response, Cell membrane, Lipid order, Lipid domains, Apoptosis

1 Introduction

Fluorescent membrane probes are important tools to investigate the physicochemical properties of model and cellular lipid membranes. The most common properties addressed by these probes are the membrane fluidity, hydration, and electrostatics, such as surface, dipole, and transmembrane potential [1]. These properties correlate with other membrane properties, which generally cannot be directly measured by these probes, such as lipid order, interlipid distance, phase transitions, and phase separations. Among the existing membrane probes, the 3-hydroxyflavone derivatives are of particular interest [2–4]. These probes exhibit the remarkable property to emit two bands, due to an excited-state intramolecular proton transfer (ESIPT). This dual emission provides an additional channel of spectroscopic information and through the ratio of the two bands offers a signal that is independent of the probe concentration. Particularly attracting is the probe F2N12S, which due to its apolar chain and zwitterionic group stains specifically the outer

Yves Engelborghs and Antonie J.W.G. Visser (eds.), *Fluorescence Spectroscopy and Microscopy: Methods and Protocols*, Methods in Molecular Biology, vol. 1076, DOI 10.1007/978-1-62703-649-8_17,

leaflet of plasma membranes, without rapid internalization inside the cells. In contrast to fluorescent lipids, this probe binds spontaneously lipid bilayers and thus does not require any helper agent, such as surfactants or cyclodextrins. Being bound to lipid membranes, the dual fluorescence spectrum of F2N12S is characterized by a short-wavelength band, originating from the emission of the normal N* (H-bonded and H-bond free) emissive species and a long-wavelength band assigned to the emission of the ESIPT tautomer T* [5, 6]. The dual emission of F2N12S is highly sensitive to several membrane properties. By deconvoluting the emission spectrum of F2N12S, both the polarity and hydration, two different properties of lipid membranes can be monitored simultaneously [4, 7]. The polarity parameter correlates well with the membrane electrostatics, i.e., surface and dipole potentials, while the hydration parameter shows strong variation in response to changes in the membrane phase state [1]. This bi-parametric sensitivity of F2N12S allowed its applications for detecting apoptosis and probing lipid domains.

Apoptosis, the programmed cell death, is a fundamental process controlling cell homeostasis. In normal cells, the plasma membrane is asymmetric, so that the outer leaflet presents a large quantity of sphingomyelin (SM), while phosphatidylserine (PS) and phosphatidylethanolamine (PE) are mainly concentrated in the inner leaflet [8]. In apoptotic cells, due to inactivation of flippases, this asymmetry is lost, so that PS is exposed at the outer leaflet. Moreover, PS exposure is associated with a decrease in SM at the outer leaflet [9], which may decrease the fraction of the liquid-ordered (Lo) phase at this leaflet. Both exposure of PS and loss of Lo phase induce a significant change in the dual emission of F2N12S, providing an increase in the relative intensity of the short-wavelength band [4].

Another application of probe F2N12S is to monitor lipid domains. In cell membranes, lipid domains attracted large attention since they are thought to be involved in the regulation of various cell functions such as signal transduction, lipid trafficking, and membrane protein activity [10]. Lipid domains composed of saturated lipids (SM in particular) and cholesterol form a liquid-ordered (Lo) phase. These domains are thought to float in the liquid-disordered (Ld) phase, formed by unsaturated lipids and cholesterol. As the F2N12S probe is sensitive to the phase state, it was applied for imaging lipid domains in giant vesicles, showing a good contrast between Lo and Ld phases in ratiometric images [3]. In cell membranes of astrocytoma, we further evidenced the presence of a high fraction of Lo phase, which could be substantially decreased by cholesterol extraction [4]. Remarkably, cholesterol extraction gave the same spectroscopic response as apoptosis, suggesting that both phenomena lead to similar loss in the Lo phase at the outer membrane leaflet. Thus, the F2N12S probe appears as a promising tool for monitoring the biophysical properties of model and cell membranes and for probing lipid domains and apoptosis. In the present work, we review the sensing methodology based on F2N12S for these two applications. In particular, we will present protocols for (a) spectroscopic studies of phase state in model lipid membranes, (b) imaging membrane domains in giant unilamellar vesicles, (c) spectroscopic studies of membrane modifications due to apoptosis and cholesterol extraction, (d) detection of apoptosis by flow cytometry, and (e) imaging apoptosis by fluorescence microscopy.

2 Materials

Prepare stock solution of 0.1 mM (0.64 g/L) F2N12S in DMSO and store it at −20 °C. Dioleoylphosphatidylcholine (DOPC) and cholesterol were purchased from Sigma-Aldrich (Lyon, France). Bovine brain sphingomyelin was from Avanti Polar Lipids (Alabaster, USA).

All lipids were dissolved in chloroform.

2.1 Large Unilamellar Vesicles

1. 20 mM of phosphate buffer, pH 7.4 (10 mL).
2. 1 mM of lipids (DOPC, sphingomyelin, cholesterol) in chloroform.
3. Lipex Biomembranes extruder (Vancouver, Canada) with two sizes of filters, 0.2 and 0.1 μm, respectively.

2.2 Giant Unilamellar Vesicles

1. Homebuilt liquid cell (University of Odense, Denmark) with platinum wires.
2. 0.1 mM of lipids (DOPC, sphingomyelin, cholesterol) in chloroform.
3. 300 mM of sucrose solution in mQ water (500 μL/sample).
4. 300 mM of glucose solution in mQ water (200 μL/sample).
5. Thermostat.
6. Power supply: 2 V, 10 Hz.

2.3 Cell Culture

1. X-VIVO 15 serum-free medium (Cambrex, France) for human lymphoid CEM T cells.
2. EMEM (Eagle's minimal essential medium, Lonza): with 10 % heat-inactivated fetal bovine serum (PAN Biotech GmbH), 0.6 mg/mL glutamine (Biowhittaker) for U87MG human glioblastoma cell line.
3. DPBS (Dulbecco's Phosphate Buffered Saline-Lonza): without Ca^{2+} and Mg^{2+}.

4. HBSS (Hank's Buffered Salt Solution-Sigma): with $NaHCO_3$ (0.35 g/L), without phenol red, calcium chloride and magnesium sulfate.
5. 1× of trypsin diluted in phosphate buffer.
6. Opti-MEM: Reduced Serum Medium from Invitrogen.
7. Actinomycin D (Sigma-Aldrich): dissolved in DMSO, stored at −20 °C.
8. Methyl-β-cyclodextrin (Sigma-Aldrich): dissolved in DPBS, stored at +4 °C.
9. 35 mm imaging dishes (μ-Dish$^{35mm,\ low}$, IBiDi) to prepare cells for microscopy experiments.
10. Cell lines: human lymphoid CEM T cells and U87MG human glioblastoma cell line.

2.4 Fluorescence Spectroscopy and Microscopy

1. Jobin-Yvon Fluoromax 3 spectrofluorometer equipped with a thermostated cuvette holder was used to record fluorescence spectra in vesicles and cells.
2. Quartz cuvettes of 1 mL were used to measure fluorescence spectra.
3. For fluorescence microscopy experiments, a homebuilt two-photon laser scanning setup based on Olympus IX70 inverted microscope with an Olympus 60× 1.2 NA water immersion objective was used. Two-photon excitation was provided by a titanium-sapphire laser (Tsunami, Spectra Physics), and photons were detected with avalanche photodiodes (ADP SPCM-AQR-14-FC, Perkin Elmer) connected to a counter/timer PCI board (PCI6602, National Instrument). Imaging was carried out using two fast galvo mirrors in the descanned fluorescence collection mode. Typical acquisition time was 20 ms with an excitation power around 20 mW ($\lambda = 830$ nm) at the sample level. Images corresponding to the short-wavelength and the long-wavelength emission bands were recorded simultaneously using a dichroic mirror (Beamsplitter 550DCXR) and two band-pass filters (Brightline HC 520/20 and HQ 585/40) in front of the APDs.

2.5 Flow Cytometry

1. Annexin V-FITC Kit: from Immunotech.
2. Propidium iodide (PI): from Sigma.
3. Binding buffer provided with the Annexin V-FITC kit: 10 mM Hepes/NaOH, pH 7.4, 140 mM NaCl, 2.5 mM $CaCl_2$.
4. F2N12S.
5. FACSAria Cell Sorter (BD Biosciences, CA) equipped with an argon laser (488 nm) and a diode laser (405 nm) and emission

filters DF30, DF42, and DF20 centered at 530, 585, and 610 nm, respectively.

6. FACSDiva Software (flow cytometer analysis software from BD Biosciences).

3 Methods

Carry out all procedures at room temperature unless otherwise specified.

3.1 Spectroscopic Studies of Phase State in Model Lipid Membranes

3.1.1 Large Unilamellar Vesicle Preparation and Staining

1. Dissolve 5 mM of lipids: DOPC (19.65 mg), DOPC/35 % Chol (12.78 mg DOPC + 3.38 mg Chol), or SM/35 % Chol (11.88 mg of SM + 3.38 mg Chol) in 5 mL chloroform to ensure their homogeneous distribution.
2. Evaporate chloroform by rotary evaporation to get a dry thin lipid film on the wall of a round-bottom flask with gentle heating (about 35 °C) during 30 min.
3. Hydrate the lipid film by adding 10 mL phosphate buffer to the flask.
4. Leave the obtained solution 1 h at RT for unsaturated lipids or 55 ° C for lipids with saturated alkyl chains.
5. Vortex (70 × *g*) the solution for 2 min to obtain a suspension of multilamellar vesicles with a heterogeneous size distribution.
6. Use the lipid extruder to downsize the vesicles. Lipid suspension is forced through a polycarbonate filter, firstly, by passing through a large pore size (0.2 μm) filter seven times. Then the suspension is passed ten times through a filter with 0.1 μm pore size. Finally, LUVs obtained have a size about 110–120 nm.
7. Label LUVs by adding aliquots (~2 μL) of F2N12S probe stock solution to 1 mL of vesicle suspension (200 μM lipid concentration) in phosphate buffer.
8. Incubate LUVs with the probe (1 μM) for 3–5 min at RT.

3.1.2 Fluorescence Spectroscopy Measurements

Transfer stained LUVs to 1 mL quartz cuvettes. Record the emission fluorescence spectrum of the LUV solution at 20 °C, using 400 nm excitation wavelength on the spectrofluorometer described in Subheading 2.4. All the spectra should be corrected from the fluorescence of the corresponding blank solution (vesicles without the probe). The ratio of their two bands (N*/T* ratio) is obtained by dividing the emission intensity at the maximum of the N* band by the intensity at the maximum of emission of the T* band. To better analyze and compare the spectra, a normalization to the T* band maximum could be applied.

The fluorescence spectrum of F2N12S in DOPC vesicles is characterized by a high N*/T* ratio close to 1 (Fig. 1, solid

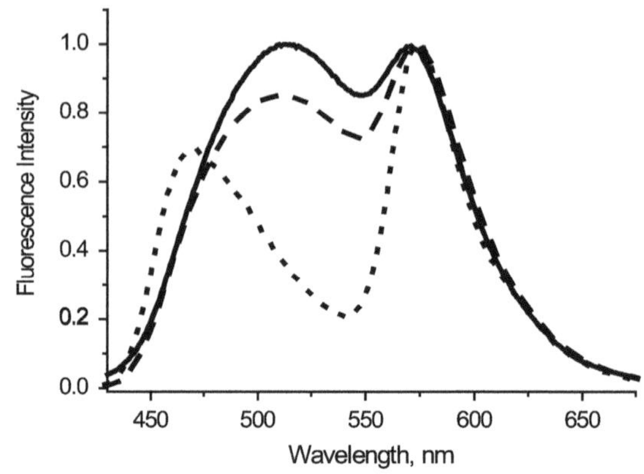

Fig. 1 Fluorescence spectra of F2N12S in LUVs composed of DOPC (*solid curve*), DOPC + Chol (*dashed curve*), and SM + Chol (*dotted curve*). Data from [3]

curve). When cholesterol is added to DOPC, this ratio decreases to 0.8 (Fig. 1, dashed curve). In SM/Chol vesicles, the fluorescence spectrum of F2N12S is characterized by a blue-shifted N* band with a decrease of the N*/T* ratio to 0.7 (Fig. 1, dotted curve). This decrease is related to the low polarity and hydration in SM + cholesterol domains corresponding to Lo phase.

3.2 Imaging Membrane Domains in Giant Unilamellar Vesicles

3.2.1 Giant Unilamellar Vesicle Preparation and Staining

1. Dissolve lipids (DOPC, SM/Chol or DOPC/SM/Chol) in chloroform to get 1 M final concentration.
2. Deposit 3 μL of lipid solution on the platinum wires of the chamber with a micro-syringe.
3. Evaporate the solvent under vacuum for 20 min.
4. Fill the chamber with 500 μL solution of 300 mM sucrose solution.
5. Connect the chamber to an alternative power supply 2-V, 10-Hz for ca. 2 h.
6. Store obtained solutions of GUVs at RT in the darkness.
7. For GUV staining, add the probe aliquot to 200 μL of 300 mM glucose solution.
8. Quickly, add 50 μL of obtained solution of GUVs to the solution of dye in glucose and mix gently (*see* **Note 1**).
9. Transfer the obtained stained GUV suspension to the Lab-Tek^{TM} 8-well chamber and incubate it for 3–5 min to allow GUVs to sediment at the bottom of the chamber.

3.2.2 Microscopic Measurements

Visualize stained GUVs with a two-photon microscope using 830 nm excitation wavelength. Alternatively, a single-photon laser excitation source at 405 nm can be used in a confocal microscope.

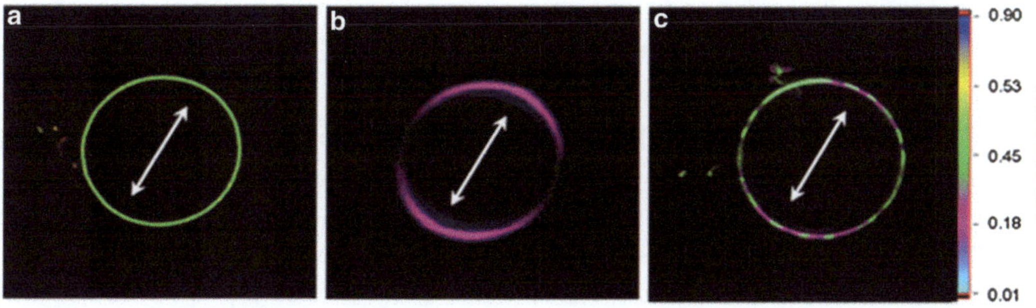

Fig. 2 Fluorescence ratiometric images (520/580 nm) of GUVs composed of (**a**) DOPC, (**b**) SM + cholesterol, and (**c**) the ternary mixture (DOPC/SM/cholesterol 1/1/0.7) (*see* **Note 4**). Adapted from [3]

Record images corresponding to blue and red channels simultaneously, using a 550 DCXR dichroic mirror and two band-pass filters (Brightline HC 520/20 and HQ 585/40) in front of the two APDs. Process these images with a program that generates a ratiometric image by dividing the image of the short-wavelength channel by the long-wavelength channel. In our case, a homemade program under LabView was used to calculate for each pixel the ratio of the integral intensity of the N* band (below 550 nm) to the integral intensity of the T* band (above 550 nm) and code it in a pseudo-color scale. To avoid artifacts, the N*/T* ratio at a given pixel was only calculated when the integral intensity of both bands exceeds a threshold value.

In DOPC GUVs presenting a Ld phase, the fluorescence intensity of F2N12S is homogeneously distributed all over the vesicles, with a blue/red intensity ratio of 0.45 (Fig. 2a). For SM/Chol GUVs presenting a Lo phase, this ratio decreases to 0.2 (Fig. 2b). In vesicles composed of a ternary mixture of DOPC/SM/Chol, a heterogeneous color distribution was observed, showing different domains with different blue/red ratios, corresponding to the separated Lo and Ld phases (Fig. 2c).

3.3 Spectroscopic Studies of Cell Membrane Modifications due to Apoptosis or Cholesterol Extraction

Cell culture: Maintain cell concentration at 5–10 × 10^4 cells/mL by removing a portion of the culture and replacing it with a fresh EMEM medium three times per week.

For cellular spectroscopic measurements, prepare cells in Nunc Flasks Nunclon (80 cm^2) 2 days in advance.

In apoptosis and cholesterol depletion studies, respectively:

1. Add the apoptosis agent, actinomycin D (0.5 μg/mL) in D-MEM medium to cells 18 h before the experiment, and incubate at 37 °C in a humidified 5 % CO_2 atmosphere.
2. For cholesterol extraction, incubate cells with 5 mM methyl-β-cyclodextrin for 2 h at 37 °C.

For adherent cells (U87MG astrocytes, in our case), the day of the measurement:

1. Remove old medium.
2. Wash cells two times with 10 mL DPBS to remove all traces of medium and dead cells.
3. Add 3 mL of 1× diluted trypsin to detach cells, then incubate them at 37 °C for 4 min.
4. Dilute trypsin with 6 mL DPBS to stop its action. Then, resuspend cells and transfer them to a Falcon tube.
5. Centrifuge cell suspension at 190 × *g* for 5 min at RT.
6. Resuspend the pellet with 3 mL HBSS.
7. Recentrifuge one more time at 190 × *g* for 5 min at RT to remove all trypsin and media traces.
8. Resuspend the pellet with 1–2 mL HBSS.
9. Count cells and dilute them to obtain a final concentration of 2×10^6 cells/mL.

For cells in suspension (human lymphoid CEM T cells in our case), just centrifuge cells at 190× *g* RPM for 5 min to remove old medium, then apply the same protocol as for adherent cells, centrifuge twice, and resuspend to obtain a final concentration of 2×10^6 cells/mL.

The protocol of staining with F2N12S is the same for both types of cells and for all treatments (control cells, cholesterol-depleted cells, and apoptotic cells).

1. Mix 500 μL of the cell suspension with 500 μL of dye diluted in HBSS (*see* **Note 2**) to obtain a final probe concentration of 0.1 μM (*see* **Note 3**).
2. Incubate cells with probe for 7 min at 20 °C in the dark.
3. Transfer the cell suspension to a 1 mL quartz cuvette.

Record the emission spectrum of intact, apoptotic, or cholesterol-depleted cells at 20 °C, using 400 nm excitation wavelength on the spectrofluorometer described in Subheading 2.4. All the spectra should be corrected from the fluorescence of the corresponding blank solution (cell suspension without the probe). The ratio of their two bands (N*/T* ratio) is obtained by dividing the intensity at the maximum of emission of the N* band by the intensity at the maximum of emission of the T* band. The emission spectrum of F2N12S in cells could be normalized to the T* band maximum to facilitate the comparison of the spectra. For intact cells, a low N*/T* ratio was recorded (Fig. 3, solid curve). After cholesterol depletion, the N*/T* ratio increases (Fig. 3, dotted

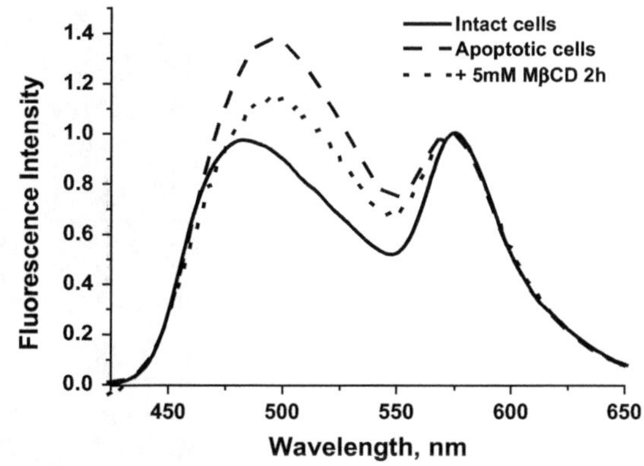

Fig. 3 Effect of apoptosis (*dashed curve*) and cholesterol depletion (*dotted curve*) on the emission spectrum of F2N12S in cell membranes. Excitation wavelength was 400 nm. Final probe concentration was 0.1 μM. Data from [4]

curve). This increase of the ratio is also observed with apoptotic cells (Fig. 3, dashed curve). These data suggest that during apoptosis, a loss of the Lo phase in the outer leaflet occurred in addition to PS exposure to the cell surface.

3.4 Detection of Apoptosis by Flow Cytometry

1. Treat cells with actinomycin D (0.5 μg/mL), 18 h before the measurements.
2. For labeling, cells were washed twice with HBSS and then resuspended in the binding buffer.
3. Label cells with PI (2.5 μg/mL) by adding stock solution of PI to the cell suspension and mix gently.
4. Keep tubes in ice for 15 min in the dark.
5. Add 400 μL of ice-cold 1× binding buffer and mix gently.
6. Label cells with F2N12S (0.2 μM) for 5 min in the dark at room temperature (*see* **Note 8**).
7. Sort cells treated with actinomycin D with the FACSAria Cell Sorter equipped with an argon laser (488 nm) and a diode laser (405 nm) and emission filters DF30, DF42, and DF20 centered at 530, 585, and 610 nm, respectively. The argon laser (488 nm) was used for PI selective excitation, where F2N12S does not absorb and the diode laser 405 nm was used for F2N12S selective excitation.
8. Analyze obtained data with FACSDiva Software.

F2N12S-PI labeling allows discriminating 3 populations of cells: apoptotic cells (low T*/N* ratio and PI negative), living cells (high T*/N* ratio and PI negative), and dead cells (low T*/ N* ratio and PI positive) (Fig. 4).

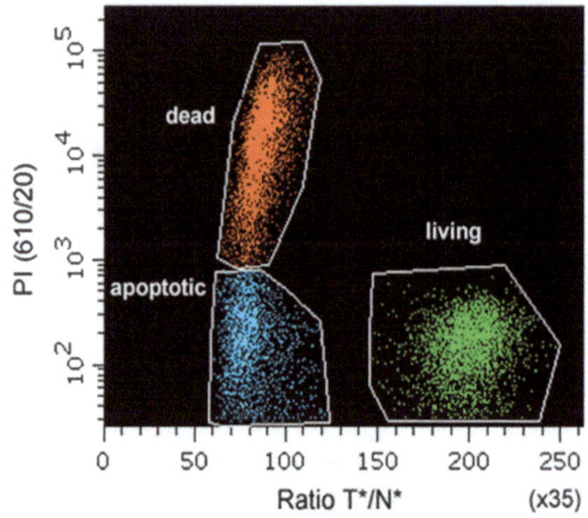

Fig. 4 Flow cytometry of T lymphoblastoid cells treated by actinomycin D. Cells were stained by PI and F2N12S. F2N12S is selectively excited at 405 nm, and its fluorescence was collected simultaneously at 530 and 585 nm. PI was selectively excited at 488 nm, and its fluorescence was detected at 610 nm. The PI fluorescence intensities are plotted versus the T*/N* ratio (585/530 nm) of F2N12S. Three distinct populations of cells were detected, namely, living cells in *green*, apoptotic cells in *blue*, and dead cells in *orange*. Data from [2]

3.5 Imaging Membrane Modifications in Cells by Two Photon Microscopy: Apoptosis and Cholesterol Extraction

1. Seed cells in a chambered coverglass (IBiDi) at a density of 5×10^4 cells/IBiDi.
2. Next day, wash gently cells twice with 1 mL HBSS.
3. Add freshly diluted F2N12S in 400 μL HBSS to 400 μL of HBSS to a final concentration of 0.3 μM.
4. Add this fresh solution to the cells.
5. Incubate 7 min at RT in the darkness.
6. Wash once gently with HBSS to remove excess of the probe.
7. Visualize cells with a two-photon microscope using 830 nm excitation wavelength. Alternatively, a single-photon laser excitation source at 405 nm can be used in a confocal microscope. Perform the image recording and processing as in Subheading 3.2.2 (*see* **Notes 6** and 7).

Intact cells stained with F2N12S show a low blue/red ratio around 0.2 (Fig. 5a). After cholesterol depletion and apoptosis induction, an increase of this ratio is noticed (Fig. 5b, c, respectively). These data are consistent with the fluorescence spectroscopy data described above.

In conclusion, the 3-HF probe F2N12S constitutes a good tool for studying model lipid membranes and cell plasma membranes. It is able to bind to both Lo and Ld phases and to discriminate them by the ratio of its two emission bands. In cellular applications,

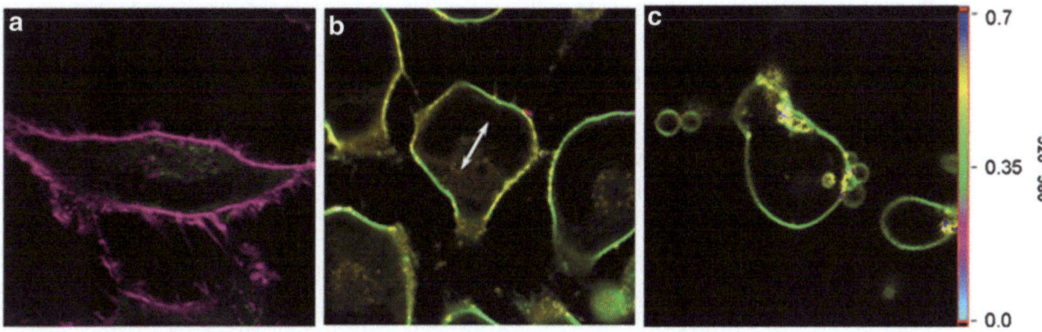

Fig. 5 Fluorescence ratiometric images (520/580 nm) of U87MG cells stained with F2N12S. (**a**) Intact, (**b**) cholesterol-depleted, and (**c**) apoptotic cells. F2N12S concentration was 0.3 μM. Data from [4]

F2N12S stains selectively the plasma membrane and shows different intensity ratios for intact cells and treated cells (cholesterol depleted and apoptotic). Thus, F2N12S can be used to sense changes in lipid composition and order and to detect apoptosis.

4 Notes

1. During the staining of GUVs with the probe, respect the order of mixing: first add the probe aliquot to the glucose solution, mix gently, then immediately add GUVs, to get a homogenous staining of GUVs.
2. To stain cells for spectroscopic measurements, prepare each time a fresh solution of F2N12S in HBSS. Diluting the dye in the buffer should be very fast since the solubility of F2N12S is limited in water, leading to probe precipitation. Therefore, the probe solution should be used immediately after preparation.
3. During all spectroscopic measurements, avoid using a high concentration of F2N12S. An excess of dye can lead to an unspecific binding, which modifies the spectrum of the probe. In the optimal concentration range (0.1–1 μM), the shape of the fluorescence spectrum of F2N12S should be independent of the probe concentration.
4. In SM/Chol vesicles, the polarization of the excitation light induces some effect on the fluorescence intensity distribution. The fluorescence intensity of F2N12S depends on the orientation of the lipid bilayer relatively to the polarization plane of the excitation light.
5. For flow cytometry studies, vortex cells immediately before measurement to avoid sedimentation of the cells on the bottom of the tube.

6. For two-photon imaging, the number of photons collected in the green and red channels of a given pixel should be high enough to get a reliable blue/red ratio value.
7. For two-photon imaging, a threshold was applied with the aim to remove all pixels with too low signal, which would provide erratic blue/red values.
8. F2N12S is available as a product from Molecular Probes, Invitrogen, Life Sciences, under the name *Violet Ratiometric Membrane Asymmetry Probe/Dead Cell Apoptosis Kit* (http://tools.invitrogen.com/content/sfs/manuals/mp35137.pdf).

Acknowledgements

This work was supported by ANR grant (fluo-aa-nt) and ANR JCJC (ANR-11-JS07-014-01). Z.D. received a fellowship from CNRS and Region Alsace.

References

1. Demchenko AP, Mely Y et al (2009) Monitoring biophysical properties of lipid membranes by environment-sensitive fluorescent probes. Biophys J 96(9):3461–3470
2. Shynkar VV, Klymchenko AS et al (2007) Fluorescent biomembrane probe for ratiometric detection of apoptosis. J Am Chem Soc 129 (7):2187–2193
3. Klymchenko AS, Oncul S et al (2009) Visualization of lipid domains in giant unilamellar vesicles using an environment-sensitive membrane probe based on 3-hydroxyflavone. Biochim Biophys Acta 1788(2):495–499
4. Oncul S, Klymchenko AS et al (2010) Liquid ordered phase in cell membranes evidenced by a hydration-sensitive probe: effects of cholesterol depletion and apoptosis. Biochim Biophys Acta 1798(7):1436–1443
5. Klymchenko AS, Duportail G et al (2002) Novel two-band ratiometric fluorescence probes with different location and orientation in phospholipid membranes. Chem Biol 9 (11):1199–208
6. Klymchenko AS, Duportail G et al (2003) Ultrasensitive two-color fluorescence probes for dipole potential in phospholipid membranes. Proc Natl Acad Sci U S A 100 (20):11219–11224
7. Klymchenko AS, Mely Y et al (2004) Simultaneous probing of hydration and polarity of lipid bilayers with 3-hydroxyflavone fluorescent dyes. Biochim Biophys Acta 1665 (1–2):6–19
8. Zwaal RF, Schroit AJ (1997) Pathophysiologic implications of membrane phospholipid asymmetry in blood cells. Blood 89 (4):1121–1132
9. Tepper AD, Ruurs P, Wiedmer T et al (2000) Sphingomyelin hydrolysis to ceramide during the execution phase of apoptosis results from phospholipid scrambling and alters cell-surface morphology. J Cell Biol 150(1):155–64
10. Brown DA, London E (1998) Structure and origin of ordered lipid domains in biological membranes. J Membr Biol 164(2):103–14

Part IV

Fluorescence Microscopy: Fluorescence Recovery After Photobleaching (FRAP)

Chapter 18

Rectangle FRAP for Measuring Diffusion with a Laser Scanning Microscope

Ranhua Xiong, Hendrik Deschout, Jo Demeester, Stefaan C. De Smedt, and Kevin Braeckmans

Abstract

Fluorescence recovery after photobleaching (FRAP) is one of the most useful microscopy techniques for studying the mobility of molecules in terms of a diffusion coefficient. Here, we describe a FRAP method that allows such measurements, relying on the photobleaching of a rectangular region of any size and aspect ratio. We start with a brief overview of the rectangle FRAP theory, and next we provide guidelines for performing FRAP measurements, including a discussion of the experimental setup and the data analysis. Finally, we discuss how to verify correct use of the rectangle FRAP method using test solutions.

Key words Fluorescence recovery after photobleaching, Confocal laser scanning microscopy, Fluorescence microscopy, Diffusion coefficient

1 Introduction

Fluorescence recovery after photobleaching (FRAP) has been used extensively to study molecular mobility on a micrometer scale in terms of a diffusion coefficient in a variety of systems like cell membranes [1, 2], polymer gel systems [3–8], and inside living cells [9–12]. FRAP is based on photobleaching of fluorescently labeled molecules, a photochemical process through which fluorescent molecules lose their fluorescence properties after being excited by incoming photons. By illuminating an area in the sample with high-intensity excitation light, the fluorescent molecules within that area are photobleached, leading to a local reduction of fluorescence intensity. A gradual recovery of the fluorescence inside the area occurs due to diffusion of the photobleached molecules outside that area and diffusion of intact fluorescent molecules into that area. The rate of fluorescence recovery is proportional to the rate of diffusion of the fluorescently labeled molecules. Fitting a FRAP

Yves Engelborghs and Antonie J.W.G. Visser (eds.), *Fluorescence Spectroscopy and Microscopy: Methods and Protocols*, Methods in Molecular Biology, vol. 1076, DOI 10.1007/978-1-62703-649-8_18,

model to the observed fluorescence recovery can yield the physical quantities describing the local diffusion in the sample, such as the diffusion coefficient.

By means of a confocal laser scanning microscope (CLSM), it is possible to define a photobleached area of any size and shape and to obtain both spatial and temporal information from the recovery images. However, due to the mathematical complexity, FRAP data analysis remains mostly limited to temporal analysis of the average fluorescence in the photobleached area. In this case, accurate results are only achievable if the initial concentration of bleached molecules after photobleaching is correctly characterized. However, because of nonlinear saturation effects during the highly intense photobleaching, it is very difficult if not impossible to estimate or calibrate the initial bleaching profile exactly [13–16]. One approach to avoid this extra calibration and to increase the accuracy is to take the full spatial and temporal information into account. Recently, we introduced a rectangle FRAP (rFRAP) model that follows this approach [17]. In this model, a rectangular area is assumed instead of the more typical circular region, because this allows to obtain a closed-form expression for the full spatial and temporal recovery process. Moreover, by taking the microscope's effective photobleaching and imaging resolution into account, the rectangle can have any size and aspect ratio, thus providing for maximum flexibility.

2 Theory

When a rectangle is bleached by a CLSM, it is done sequentially, pixel by pixel and line by line, as is illustrated in Fig. 1.

Similar to any analytical FRAP model, in its theoretical derivation, a number of assumptions are made:

1. The diffusion process is isotropic and takes place in an infinite medium (*see* **Note 1**) in the absence of flow (*see* **Note 2**).
2. Excitation is a single-photon process for imaging and bleaching (*see* **Note 3**).
3. After photobleaching, 2-D diffusion takes place in (or parallel to) the focal plane (*see* **Note 4**).
4. The bleaching phase is sufficiently short to avoid recovery during bleaching (*see* **Note 5**).
5. The photobleaching process causes a linear decrease of the fluorescence (*see* **Note 6**).

For those assumptions, the single-photon 2-D rFRAP model of the fluorescence recovery $F(x,y,t)$ is given by the following formula:

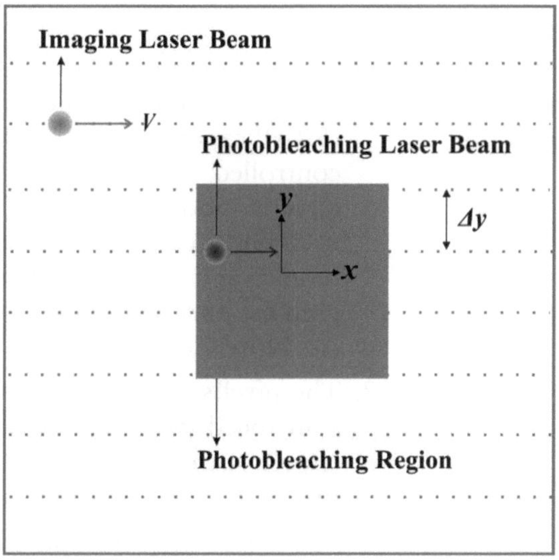

Fig. 1 On a CLSM, it is possible to photobleach a user-defined rectangle with any size and aspect ratio. While scanning in a raster pattern, line by line, the intensity of the laser beam is modulated between a low and high intensity to obtain the user-defined photobleached rectangle

$$\frac{F(x,\ y,t)}{F_0} = 1 - \frac{K_0}{4} \times \left[\operatorname{erf}\left(\frac{x + l_x/2}{\sqrt{r^2 + 4Dt}}\right) - \operatorname{erf}\left(\frac{x - l_x/2}{\sqrt{r^2 + 4Dt}}\right)\right] \times \left[\operatorname{erf}\left(\frac{y + l_y/2}{\sqrt{r^2 + 4Dt}}\right) - \operatorname{erf}\left(\frac{y - l_y/2}{\sqrt{r^2 + 4Dt}}\right)\right] \quad (1)$$

with F_0 being fluorescence before photobleaching, K_0 the amount of photobleaching, l_x the length of the bleached rectangle in x-direction, l_y the width of the bleached rectangle in y-direction, r^2 the average of the squared lateral imaging and photobleaching resolution, and D the diffusion coefficient. The parameter r takes the finite imaging and effective bleaching resolution into account. Thanks to making use of the full spatial and temporal information in Eq. 2, in practice the parameter r can be determined from the recovery analysis directly without any prior calibration.

An interesting situation is the combination of a mobile and immobile fraction. Let k be the fraction of mobile molecules, then the fluorescence recovery $F_k(x,y,t)$ is given by

$$F_k(x,y,t) = F(x,y,0) + k[F(x,y,t) - F(x,y,0)] \quad (2)$$

where $F(x,y,t)$ is defined by Eq. 1. For more details about the derivation of the general rFRAP expression, we refer to our previous work [17].

3 Methods

3.1 Experimental Setup and Parameter Settings

1. A regular (single-photon) CLSM can be used to perform rFRAP experiments. Usually, the intensity of laser scanning is controlled by an AOTF (acousto-optic tunable filter) or a directly modulated laser. The CLSM software typically allows to define a rectangular area in which the intensity of the laser beam is increased to obtain photobleaching. The laser power used for photobleaching also typically is a user-defined setting (*see* **Note 7**).
2. The pixel size (controlled through the CLSM zoom setting), scan speed, and size of the rectangle (*see* **Note 8**) should be chosen such that the total bleach time is sufficiently short relative to the expected characteristic diffusion time (*see* **Note 5**).
3. The FRAP sequence should include a number of pre-bleach and post-bleach images (*see* **Note 9**). The pre-bleach images will be used to normalize the recovery images in the subsequent data analysis.
4. The time interval between subsequent frames needs to be determined. The characteristic recovery time as defined in **Note 5** is a good choice (*see* **Note 10**).

A typical sequence of recovery images of a rFRAP measurement is shown in Fig. 2.

3.2 Data Analysis

The data analysis is done by fitting the rFRAP model (Eqs. 1 and 2) to the pixel values of the recovery images, with the diffusion coefficient (D), the extent of photobleaching (K_0), the fraction of mobile molecules (k), and the average squared resolution (r^2) as free fitting parameters. The fitting procedure can be a simple non-weighted least-squares procedure, or a maximum-likelihood approach for maximum precision [17]. The following steps should be performed:

1. Before fitting of the data to the rFRAP model, the recovery data should be normalized to the fluorescence before bleaching (*see* **Note 11**).
2. Correction for laser fluctuations and bleaching during imaging is performed by dividing the pixels of each recovery image by the average value from one or more reference background regions in the same image, as shown in Fig. 2 (*see* **Note 12**).
3. To limit the computation time, the fitting is performed on a limited region centered around the bleach area, defined as the region of interest (ROI), as shown in Fig. 2 (*see* **Note 13**).

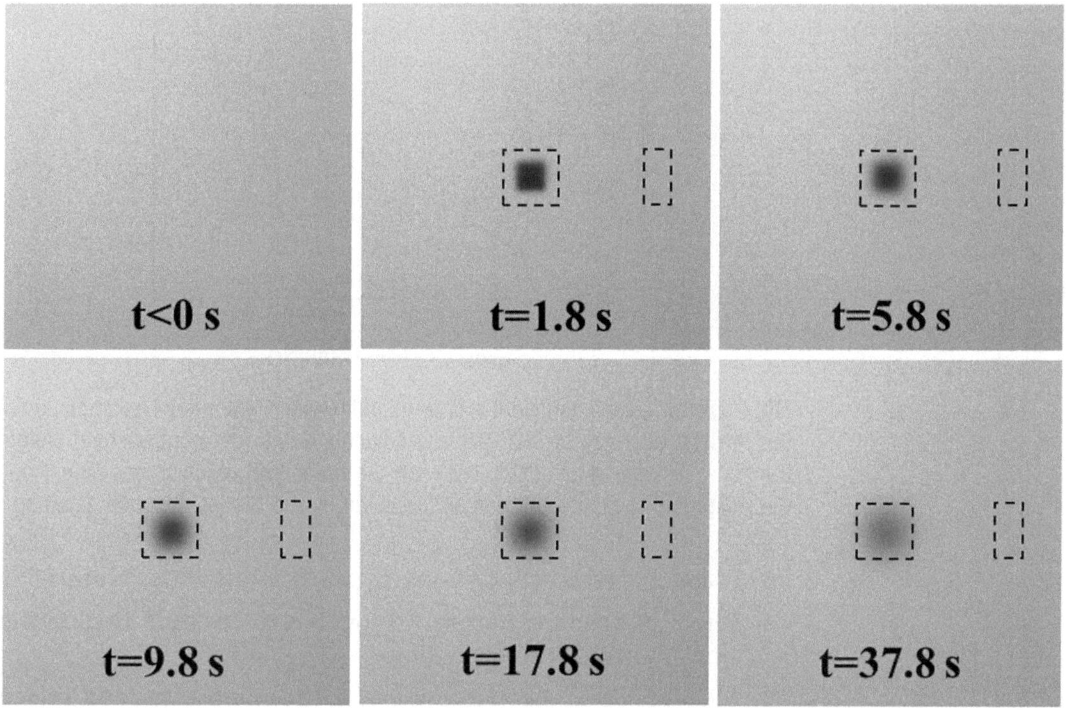

Fig. 2 A typical sequence of recovery images from a rFRAP measurement on a 150 kDa FITC-dextran solution in HEPES buffer at pH = 7.4 with 40 % sucrose. The *top left image* ($t < 0$) shows the pre-bleach situation. The next five images show the fluorescence recovery at different times after photobleaching as indicated. The bleach region is 20 μm by 20 μm with the center position at center of the image (256 by 256 pixels). The black dashed square around the bleached region is the region of interest that will be analyzed. The black dashed rectangle to the right of the bleached square shows the region that will be used for background correction (e.g., because of bleaching during imaging or laser fluctuations) in the data analysis

4. The fitting of the model is performed to the pixel values within the ROI of all frames in the recovery time-lapse movie yielding the parameters D , K_0, k, and r^2 (*see* **Note 14**).

3.3 Application: Diffusion Measurement in a Test Solution

To validate the rFRAP method and to become experienced with the method, it can be applied to a series of solutions of FITC-dextrans in HEPES buffer at pH 7.4 with different viscosities by adding different amounts of sucrose. According the Stokes–Einstein relationship, a linear relationship should be found between the measured diffusion coefficients and the reciprocal viscosity of the solutions. These experiments can be performed on a CLSM equipped with a blue laser, such as an Ar ion or solid-state laser.

1. Determine if the FITC-dextran concentration is within the linear fluorescence range (*see* **Note 15**).
2. Prepare FITC-dextran solutions with different amounts of sucrose to obtain different diffusion coefficients (*see* **Note 16**).

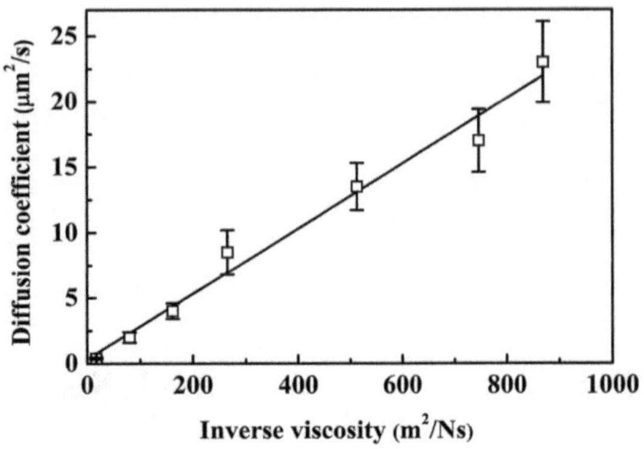

Fig. 3 The diffusion coefficient is determined by the rFRAP model as a function of the inverse viscosity of FITC-dectran soluctions with different sucrose weight fractions. A linear relationship between the measured diffusion coefficient and the reciprocal viscosity is found as predicted by the Stokes–Einstein equation

3. Prepare a microscope sample for each solution (*see* **Note 17**).
4. Select an objective lens with a low NA to ensure 2-D diffusion after photobleaching (*see* **Note 4**).
5. Make sure that the objective lens of the microscope is calibrated to yield the correct pixel size.
6. Perform the FRAP experiments according to the guidelines in Subheading 3.1.
7. For each sample, 5–10 experiments can be performed. For each measurement, make sure to select a new region in the sample sufficiently far from the previous one.
8. Perform the analysis as described in Subheading 3.2. Plot the measured diffusion coefficients in function of inverse viscosity as shown in Fig. 3.

4 Notes

1. In practice, this condition of an infinite medium means that, during the time period of observing the recovery process, the diffusion front should not have reached any boundaries that can influence the free diffusion process.
2. This can be verified by examining the position of the photobleached rectangle in the recovery images. If there is drift in the sample, this will be noticeable from a translation of the center of the bleach area over time in the FRAP time-lapse movie.
3. Actually, a general formula is available for an n-photon illumination and m-photon imaging process [17]. However, the formula given here (Eq. 1) is for single-photon bleaching and imaging, as it is of most practical relevance.

4. Diffusion measurements in 3-D extended samples are possible when using an objective lens of sufficiently low NA (typically <0.5) which produces a cylindrical laser beam in a substantial area above and below the focal plane. In that case, the bleaching will be quite uniform over an extended region along the optical axis, so that only 2-D radial diffusion (i.e., perpendicular to the optical axis) effectively takes place.
5. Conveniently, the effect of diffusion during bleaching, which causes broadening of the bleach profile, can be taken care of by the parameter r in Eq. 1. It was shown experimentally that a total bleach time of at least 15 % of the characteristic recovery time $\tau = \frac{(L/2)^2}{4D}$ is allowed when r is included as a free fitting parameter.
6. Typically an exponential decrease is assumed in FRAP modeling. It was shown that the linear approximation holds for up to 50 % photobleaching [17] and also *see* **Note 7**.
7. The photobleaching laser intensity should not be too high nor too low. Due to the assumption of the linear photobleaching process, the amount of bleaching should be less than 50 %. On the other hand, insufficient laser power will result in little bleaching, leading to a low SNR and low precision of the measurement. In practice, 25–50 % of photobleaching is recommended.
8. For example, for a diffusion coefficient of 10 $\mu m^2/s$, a line scan time of 2 ms and a pixel size of 0.2 μm, the bleached area should be no less than 10.6 μm by 10.6 μm for the bleaching time to be less than 15 % of the characteristic recovery time as defined in **Note 5**.
9. At least one frame should be recorded before photobleaching. Enough frames should be recorded for imaging the fluorescence recovery (30–50 typically). More images do not necessarily increase the precision of the measurement but increase photobleaching during imaging and the calculation time.
10. If variable timing between the images is allowed by the software, it is useful to have fast sampling shortly after bleaching and less sampling at longer times after bleaching. This maximizes information while minimizing the number of images and thus photobleaching during imaging.
11. Normalization to the pre-bleach intensity can be performed by dividing every pixel in the recovery images by the corresponding pixel in the pre-bleach image. To limit the corresponding amplification of noise, the pre-bleach image can be smoothed first with an average or median filter with a kernel of, e.g., 3 × 3 or 5 × 5 pixels.

Table 1
The viscosity vs. different weight fraction of sucrose at room temperature of 20 °C

Weight fraction of sucrose	5 %	10 %	20 %	32 %	40 %	48 %	60 %
Viscosity (mN/m^2 s)	1.15	1.34	1.95	3.77	6.18	12.55	58.66

12. The reference background region (BG) should be placed sufficiently far from the bleach region so as to remain unaffected by the diffusion front during the observation time. Normalization to the BG region can correct to some extent for bleaching during imaging. However, it is recommended that bleaching during imaging should not be more than 10 %.
13. Usually the region of interested consists of the bleached area extended with a certain amount pixels in each direction. According to our experience, more than 30 pixels did not substantially improve the analysis precision.
14. The pixel values of all regions of interest in the entire stack of images of the time-lapse movie are simultaneously fitted (least-squares fit) to the rFRAP model (Eqs. 1 and 2). The 3-D stack of images should be converted to a 1-D matrix by the MATLAB (The MathWorks Inc., Natick, MA) function "reshape." The least-squares fit can be performed in MATLAB by using the function "lsqcurvefit" in the Optimization Toolbox.
15. The concentration range in which there exists a linear relation with the observed fluorescence can be determined by imaging FITC-dextran solutions of different concentrations with the same instrument settings (identical laser power and detector gain). Typically the highest concentration for which there is still a linear relation with the image intensity is used for further experiments.
16. The viscosity at room temperature (20 °C) of solutions with different weight fractions of sucrose is shown in Table 1.
17. A good way to prepare the sample is by making use of a double-sided adhesive sticker with a perforation in the middle. First attach the sticker to a microscope slide. Then place a drop of the solution in the sample space in the middle and close it off with a cover slip. We have good experience using the 120 μm thick Secure-Seal Spacers (Molecular Probes, Leiden, The Netherlands). In this way, a 3-D extended sample is obtained while avoiding flow in the solution.

Acknowledgements

Financial support by the Ghent University Special Research Fund (Centre for Nano- and Biophotonics) is acknowledged with gratitude. Ranhua Xiong gratefully acknowledges the financial support from China Scholarship Council (CSC). Hendrik Deschout is a doctoral fellow of the Institute for the Promotion of Innovation through Science and Technology in Flanders (IWT), Belgium.

References

1. Ishihara A, Jacobson K (1993) A closer look at how membrane-proteins move. Biophys J 65 (5):1754–1755
2. Umenishi F, Verbavatz JM, Verkman AS (2000) cAMP regulated membrane diffusion of a green fluorescent protein-aquaporin 2 chimera. Biophys J 78(2):1024–1035
3. Alvarez-Mancenido F, Braeckmans K, De Smedt SC et al (2006) Characterization of diffusion of macromolecules in konjac glucomannan solutions and gels by fluorescence recovery after photobleaching technique. Int J Pharm 316(1–2):37–46. doi:10.1016/j.ijpharm.2006.02.029
4. Burke MD, Park JO, Srinivasarao M, Khan SA (2000) Diffusion of macromolecules in polymer solutions and gels: a laser scanning confocal microscopy study. Macromolecules 33 (20):7500–7507. doi:10.1021/Ma000786l
5. Censi R, Vermonden T, van Steenbergen MJ et al (2009) Photopolymerized thermosensitive hydrogels for tailorable diffusion-controlled protein delivery. J Control Release 140(3):230–236. doi:10.1016/j.jconrel.2009.06.003
6. DeSmedt SC, Meyvis TKL, Demeester J et al (1997) Diffusion of macromolecules in dextran methacrylate solutions and gels as studied by confocal scanning laser microscopy. Macromolecules 30(17):4863–4870. doi:10.1021/Ma970100y
7. Van de Manakker F, Braeckmans K, el Morabit N et al (2009) Protein-release behavior of self-assembled PEG-beta-cyclodextrin/PEG-cholesterol hydrogels. Adv Funct Mater 19 (18):2992–3001. doi:10.1002/adfm.200900603
8. Van Tomme SR, De Geest BG, Braeckmans K et al (2005) Mobility of model proteins in hydrogels composed of oppositely charged dextran microspheres studied by protein release and fluorescence recovery after photobleaching. J Control Release 110(1):67–78. doi:10.1016/j.jconrel.2005.09.005
9. Braga J, Desterro JMP, Carmo-Fonseca M (2004) Intracellular macromolecular mobility measured by fluorescence recovery after photobleaching with confocal laser scanning microscopes. Mol Biol Cell 15(10):4749–4760. doi:10.1091/mbc.E04-06-0496
10. Seksek O, Biwersi J, Verkman AS (1997) Translational diffusion of macromolecule-sized solutes in cytoplasm and nucleus. J Cell Biol 138(1):131–142. doi:10.1083/jcb.138.1.131
11. Verkman AS (2003) Diffusion in cells measured by fluorescence recovery after photobleaching. Methods Enzymol 360:635–648
12. Axelrod D, Koppel DE, Schlessinger J et al (1976) Mobility measurement by analysis of fluorescence photobleaching recovery kinetics. Biophys J 16(9):1055–1069
13. Braeckmans K, Stubbe BG, Remaut K et al (2006) Anomalous photobleaching in fluorescence recovery after photobleaching measurements due to excitation saturation—a case study for fluorescein. J Biomed Opt 11(4)
14. Braeckmans K, Remaut K, Vandenbroucke RE et al (2007) Line FRAP with the confocal laser scanning microscope for diffusion measurements in small regions of 3-D samples. Biophys J 92(6):2172–2183. doi:10.1529/biophysj.106.099838
15. Mazza D, Braeckmans K, Cella F et al (2008) A new FRAP/FRAPa method for three-dimensional diffusion measurements based on multiphoton excitation microscopy. Biophys J 95(7):3457–3469. doi:10.1529/biophysj.108.133637
16. Braeckmans K, Peeters L, Sanders NN et al (2003) Three-dimensional fluorescence recovery after photobleaching with the confocal scanning laser microscope. Biophys J 85 (4):2240–2252. doi:10.1016/S0006-3495(03)74649-9
17. Deschout H, Hagman J, Fransson S et al (2010) Straightforward FRAP for quantitative diffusion measurements with a laser scanning microscope. Opt Express 18(22):22886–22905

Part V

Fluorescence Microscopy: Förster Resonance Energy Transfer Imaging (FRETim)

Chapter 19

A Quantitative Protocol for Intensity-Based Live Cell FRET Imaging

Clemens F. Kaminski, Eric J. Rees, and Gabriele S. Kaminski Schierle

Abstract

Förster resonance energy transfer (FRET) has become one of the most ubiquitous and powerful methods to quantify protein interactions in molecular biology. FRET refers to the sensitization of an acceptor molecule through transfer of energy from a nearby donor, and it can occur if the emission band of the donor exhibits spectral overlap with the absorption band of the acceptor molecule. Numerous methods exist to quantify FRET levels from interacting protein labels including fluorescence lifetime, acceptor photobleaching, and polarization-resolved imaging (Lakowicz, Principles of fluorescence spectroscopy, 2006; Jares-Erijman and Jovin, Curr Opin Chem Biol **10**(5):409–416, 2006; van Munster and Gadella, Microscopy techniques, 2005). For live cell imaging, however, sensitized emission FRET (seFRET) is the most powerful and robust method of FRET signal quantification (Chakrabortee et al., Proc Natl Acad Sci USA **107**(37): 16084–16089, 2010). It is fast, can be applied using straight forward microscopy equipment, and offers information not only on strength of interaction but, uniquely, also on the relative changes between interacting and noninteracting moieties in the reaction, referred to as FRET stoichiometry. A rigorous and quantitative application of seFRET is far from trivial, however, and requires appropriate calibration experiments and constructs, control over hardware settings, and appropriate image processing steps.

This protocol presents a rigorous method to perform quantitative seFRET measurements in live cells, providing the maximum possible information content from the measurement. The theoretical development and validation of the method is described in detail in Elder et al. (J R Soc Interface **6**:S59–S81, 2009) where it is also demonstrated in the kinetic ("time-lapse") analysis of protein interactions governing mitosis. The present protocol gives a detailed recipe for application of seFRET. It is written specifically for use with CFP (cyan fluorescence protein) as donor fluorophore and YFP (yellow fluorescent protein) as acceptor fluorophore, a popular choice for many experiments. The protocol is however valid for any other FRET fluorophore pair, and we indicate how to adapt the protocol in such situations. We also provide a software program that automates the calibration tasks outlined in this protocol and which is available for free to download (http://wiki.laser.ceb.cam.ac.uk/wiki/index.php/Resources).

Key words Fluorescence, FRET, Förster resonance energy transfer, Live cell imaging, Microscopy, Sensitized emission

Yves Engelborghs and Antonie J.W.G. Visser (eds.), *Fluorescence Spectroscopy and Microscopy: Methods and Protocols*, Methods in Molecular Biology, vol. 1076, DOI 10.1007/978-1-62703-649-8_19,

1 Introduction to seFRET Methodology

Sensitized emission FRET (seFRET) is one of the key microscopy techniques for measuring FRET efficiency: it is particularly well suited to measurements in which both acceptor- and donor-normalised FRET efficiencies need to be determined, and in this role it can be superior to alternative FRET imaging methods such as fluorescence lifetime, acceptor photobleaching and polarisation resolved measurements [1–4].

The principal features of seFRET measurements are shown in Fig. 1. Two fluorescence filter sets are required: one which is optimized to detect fluorescence from the donor ("donor emission channel," here CFP) and similarly one optimized for emission from the acceptor ("acceptor emission channel," here YFP).

Excitation of the donor molecule, D (solid blue arrow), leads to fluorescence emission into the donor emission channel (light blue). In the absence of acceptors and FRET, this is the only signal emitted. If FRET occurs (red arrow, labeled FRET), the acceptor molecule, A, becomes sensitized, and the seFRET signal emitted from A is detected in the acceptor channel. This is indicated as pathway 3 (solid red arrow). The efficiency of energy transfer from D to A is denoted by E, and the so-called Förster radius is defined to be the distance at which $E = 50$ %.

However, in practice the seFRET signal is contaminated by *signal cross talk*. There are two contributions to cross talk in seFRET based on the CFP/YFP—and most other fluorophore—pairings. *Cross excitation* refers to the direct excitation of the acceptor at the donor excitation wavelength (path 1, dashed blue arrow). For example, YFP has a finite probability of being excited at the CFP excitation wavelength. Contamination also arises from *donor bleed through* (path 2, red dashed red arrow), e.g., CFP fluorescence

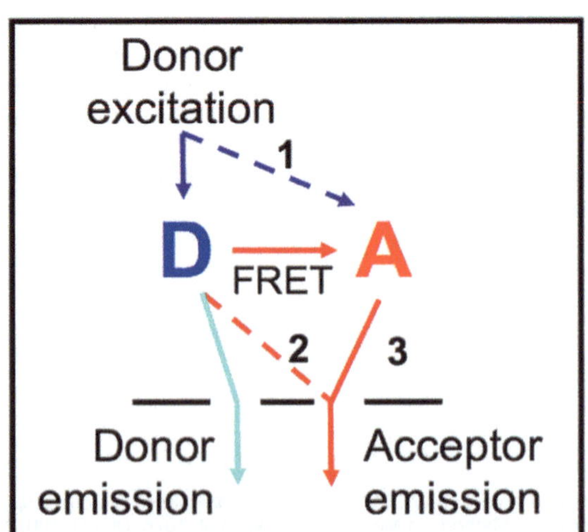

Fig. 1 Principle of seFRET

Table 1
Quantities required to correct for signal cross talk in seFRET

I^{DA}	Uncorrected "FRET" signal: signal in acceptor channel upon donor excitation
AER	Acceptor excitation ratio: corrects for direct excitation of acceptor (path 1 in Fig. 1)
DER	Donor emission ratio: determines bleed through of donor fluorescence into the acceptor channel (path 2 in Fig. 1)
I^{AA}	Signal obtained in acceptor channel, upon excitation at acceptor excitation wavelength
I^{DD}	Signal obtained in donor channel, upon excitation at the donor excitation wavelength

will leak into the YFP fluorescence band. Signal cross talk is indistinguishable from the true seFRET signals and, if unaccounted for, can generate false positives.

To correct for these effects, we calculate a corrected seFRET efficiency, which we refer to as cFRET [5], to yield

$$\mathrm{cFRET} = I^{\mathrm{DA}} - \mathrm{AER} \times I^{\mathrm{AA}} - \mathrm{DER} \times I^{\mathrm{DD}}$$

Here the first term on the right-hand side denotes the uncorrected signal, the second term accounts for the acceptor cross excitation, and the last term for the donor bleed through. Table 1 denotes the nomenclature in detail.

The AER and DER are dye-specific parameters that are ideally constant for a given sample and can be determined from control samples containing, respectively, donor (CFP)-only and acceptor-only (YFP).

A measureable value for cFRET signifies that FRET takes place between donor and acceptor; however, its value also depends on the *stoichiometries* of free donors and acceptors and interacting FRET pairs. To obtain information on the relative number of free and interacting fluorophores, we further normalize cFRET by either the acceptor or the donor concentration. This leads to the finally sought quantities:

$$\mathrm{dFRET} = \frac{\mathrm{cFRET}\frac{\alpha}{\mathrm{DER}}}{I^{\mathrm{DD}} + \mathrm{cFRET}\frac{\alpha}{\mathrm{DER}}} = \chi_{\mathrm{D}} E$$

and

$$\mathrm{aFRET} = \frac{\mathrm{cFRET}}{\mathrm{AER}\ I^{\mathrm{AA}}\beta} = \chi_{\mathrm{A}} E$$

The nomenclatures for the terms in these equations are given in Table 2. dFRET is the true FRET efficiency E normalized by χ_{D} which is the fraction of the donor molecules that undergo FRET.

Table 2
Nomenclature of seFRET calibration constants

$\chi_D E$	Donor-normalized FRET efficiency (i.e., fraction of interacting donor χ_D times FRET efficiency E)
$\chi_A E$	Acceptor-normalized FRET efficiency (i.e., fraction of interacting acceptor χ_A times FRET efficiency E)
α	Relates to relative quantum yields and signal detection efficiencies between donor and acceptors
β	Relates to line shapes and excitation efficiencies of donor and acceptor molecules

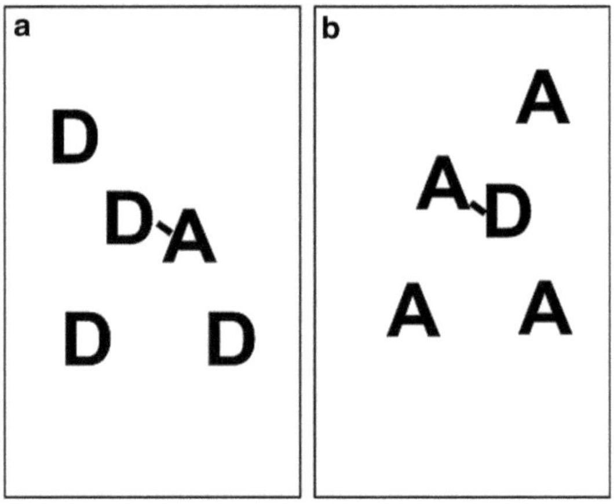

Fig. 2 Illustration to exemplify the difference between donor-normalized and acceptor-normalized FRET values. In (**a**) there is an abundance of noninteracting donors, whereas all acceptors are interacting. Here dFRET is less than aFRET. In (**b**) there is an overabundance of acceptor, and the opposite holds true. A simultaneous determination of aFRET and dFRET thus establishes both whether energy transfer is taking place and also whether there is an overabundance of noninteracting donors/acceptors

Similarly aFRET is the acceptor-normalized FRET efficiency. The two may be significantly different as Fig. 2a, b illustrate. For the situation depicted in Fig. 2a, we have aFRET >> dFRET signifying that the number of donor molecules participating in FRET interactions is much smaller than the number of acceptor molecules. Similarly in Fig. 2b, the opposite is true. The evolution of aFRET and dFRET and their interdependence thus offer important information in the stoichiometry of biological interactions, information that is usually hidden in other modalities of performing FRET (e.g., lifetime measurements). Examples of how this can be used to follow kinase activity in living cells are given in [5].

In what follows we give a recipe on how to quantify aFRET and dFRET in biological samples. Example MATLAB scripts are provided together with example data as a guide to performing the image processing required for seFRET.

2 Materials

2.1 Biological Sample Preparation

In addition to the biological FRET sample of interest, three control samples are required to determine α, β, DER, and AER (*see* **Note 1**):

1. Donor-only control sample (used as negative control and for determination of DER): Transfect cells (the protocol is verified for mammalian cells) using the CFP-tagged protein of interest (*see* **Note 2**).
2. Acceptor-only control sample (used as negative control and for determination of AER): Transfect cells with YFP-tagged protein of interest in analogy to **item 1**.
3. Positive control (required for determination of α, β): Transfect cells with CFP-YFP linker construct exhibiting a known amount of FRET. The protocol as follows is valid for a CFP-YFP tandem construct obtained by excising the EYFP gene (enhanced yellow fluorescent protein) from the pEYFP-N1 vector and ligating it into the multiple cloning site (MCS) of the pECFP-C1 vector (enhanced cyan fluorescent protein; both originally obtained from Clontech, Palo Alto, CA, USA) such that the two fluorophores in the construct are separated by a sequence of 18 amino acids (GLRSRAQASNSAVEGSAM). The plasmid vector maps are available from http://www.addgene.org/vector-database/2689/, and http://www.addgene.org/vector-database/2445/, respectively (*see* **Note 3**). Propagate the plasmid in *E. coli* (recommended host strains are DH5-alpha or HB101) and use kanamycin (30 μg/ml) to select positive strains. Isolate DNA and transfect cells according to standard protocols. If required, stable transformants can be selected by G418 (neomycin). *See* also **Note 4**.
4. Samples for FRET analysis: Transfect cells with pECFP-fusion and pEYFP-fusion proteins. Aim to obtain similar expression levels for both proteins.

2.2 Microscope Setup

The protocol described here permits system-independent determination of FRET efficiency, and system-specific attributes (filter transmission, detector gains, etc.) are taken into account through control experiments. However, for optimal sensitivity such filter sets and excitation, wavelengths should be used as to maximize FRET signals while minimizing donor bleed through and acceptor direct excitation. For CFP-/YFP-based FRET, the following settings are representative examples for either wide-field or confocal imaging setups.

Table 3
Wide-field microscope filter setup

Excitation wavelength	
Donor excitation filter	431–441 nm
Acceptor excitation filter	490–510 nm
Donor emission filter	455–485 nm
Acceptor emission filter	520–550 nm

Table 4
Confocal microscope filter setup

Donor excitation wavelength	458 nm
Acceptor excitation wavelength	514 nm
Donor emission filter	470–494 nm
Acceptor emission filter	530–545 nm

2.3 Wide-Field Microscope Settings (See Table 3)

Sensitive CCD cameras with good dynamic range (12 bit or greater) should be chosen as detectors, and a linear response ensured. Note: Camera/detector settings must remain constant for control and sample measurements. If acceptor and donor channels are measured sequentially, ensure that no sample movement occurs between images (*see* **Note 5**).

2.4 Confocal Microscope Settings (See Table 4)

Choose PMT voltage settings to ensure a linear signal response. Set laser excitation powers to obtain similar signal levels for cfp and yfp. Check for photobleaching and adjust settings to minimize this effect. Minimize sample movement artifacts (e.g., use line averaging rather than frame averaging if available and minimize time delays between consecutive measurements of donor and acceptor channels).

3 Methods

3.1 Measurement of DER

Use the donor-only sample. Keep magnification and gain settings constant throughout and subtract any d.c. background signal from all obtained images before further processing (*see* **Note 6**). Set up the microscope in sequential acquisition mode (if available) and obtain the following set of images, from which the DER can be calculated:

1. Set excitation light source to donor excitation setting and the emission channel to the donor emission bandpass to obtain signal I^{DD} (e.g., 431–441 nm excitation/455–485 nm emission filter for wide-field system).

2. Set excitation light source to donor excitation setting and the emission channel to acceptor emission bandpass to obtain signal I^{DA} (e.g. 431–441 nm excitation/520–550 nm emission filter for wide-field system).
3. Calculate the DER for every image pixel according to:

$$\mathrm{DER} = \frac{I^{\mathrm{DA}}}{I^{\mathrm{DD}}}$$

4. Repeat this calibration measurement on at least five different areas of the donor-only sample.

3.2 Measurement of AER

Use the acceptor-only sample. Keep magnification and gain settings same as above and constant throughout and subtract d.c. background signal from all obtained images before further processing (*see* **Note 6**). Set up the microscope in sequential acquisition mode (if available) and obtain the following set of images, from which the AER can be calculated:

1. Set excitation light source to donor excitation setting and the emission channel to acceptor emission bandpass to obtain signal I^{DA} (e.g., 431–441 nm excitation/520–550 nm emission filter for wide-field system).
2. Set excitation light source to acceptor excitation setting and the emission channel to acceptor emission bandpass to obtain signal I^{AA} (e.g., 490–510 nm excitation/520–550 nm emission filter for wide-field system).
3. Calculate the AER for every image pixel according to

$$\mathrm{AER} = \frac{I^{\mathrm{DA}}}{I^{\mathrm{AA}}}$$

4. Repeat this calibration measurement on at least five different areas of the acceptor-only sample.

3.3 System Calibration with Tandem FRET Construct

Use the sample containing the CFP-YFP linker construct. Keep magnification and gain settings same as above and constant throughout and subtract any d.c. background signal from all obtained images before further processing (*see* **Note 6**). Set up the microscope in sequential acquisition mode (if available) and obtain the following set of images, from which the cross talk corrected FRET signal, cFRET, can be calculated:

1. Set excitation light source to donor excitation setting and the emission channel to the donor emission bandpass to obtain signal I^{DD} (e.g., 431–441 nm excitation/455–485 nm emission filter for wide-field system).
2. Set excitation light source to donor excitation setting and the emission channel to acceptor emission bandpass to obtain

signal I^{DA} (e.g., 431–441 nm excitation/520–550 nm emission filter for wide-field system).

3. Set excitation light source to acceptor excitation setting and the emission channel to acceptor emission bandpass to obtain signal I^{AA} (e.g., 490–510 nm excitation/520–550 nm emission filter for wide-field system).
4. Calculate the cFRET for every image pixel according to

$$\text{cFRET} = I^{\text{DA}} - \text{AER} \times I^{\text{AA}} - \text{DER} \times I^{\text{DD}}$$

5. Select regions (pixels) in your image where FRET occurs and determine α through calculation of (*see* **Notes 7** and **8**)

$$\alpha = \frac{E}{1 - E} \frac{\text{DER} \times I^{\text{DD}}}{\text{cFRET}}$$

6. Select regions (pixels) in your image where FRET occurs and determine β through calculation of (*see* **Note 1**)

$$\beta = \frac{\text{cFRET}}{\text{AER} \times I^{\text{AA}} \times E}$$

7. Repeat measurements on five areas of linker-only sample.

3.4 Measurement of dFRET and aFRET

Use the biological sample with CFP- and YFP-fusion proteins of interest. Keep magnification and gain settings same as above and constant throughout and subtract any d.c. background signal from all obtained images before further processing (*see* **Note 6**). Set up the microscope in sequential acquisition mode (if available) and obtain the following set of images, from which dFRET and aFRET can be determined:

1. Set excitation light source to donor excitation setting and the emission channel to the donor emission bandpass to obtain signal I^{DD} (e.g., 431–441 nm excitation/455–485 nm emission filter for wide-field system).
2. Set excitation light source to donor excitation setting and the emission channel to acceptor emission bandpass to obtain signal I^{DA} (e.g., 431–441 nm excitation/520–550 nm emission filter for wide-field system).
3. Set excitation light source to acceptor excitation setting and the emission channel to acceptor emission bandpass to obtain signal I^{AA} (e.g., 490–510 nm excitation/520–550 nm emission filter for wide-field system).
4. Calculate the cFRET for every image pixel according to

$$\text{cFRET} = I^{\text{DA}} - \text{AER} \times I^{\text{AA}} - \text{DER} \times I^{\text{DD}}$$

5. Calculate dFRET for every image pixel according to

$$\text{dFRET} = \frac{\text{cFRET}\frac{\alpha}{\text{DER}}}{I^{\text{DD}} + \text{cFRET}\frac{\alpha}{\text{DER}}} = \chi_{\text{D}} E$$

See **Notes 9** and **10**.

6. Calculate aFRET for every image pixel according to

$$\text{aFRET} = \frac{\text{cFRET}}{\text{AER}\ I^{\text{AA}}\beta} = \chi_{\text{A}} E$$

See **Notes 9** and **10**.

4 Notes

1. The protocol here is specifically written for use with CFP- and YFP-fusion proteins and for the linker construct presented in Subheading 2, **item 3**. It has furthermore been validated for use in live HeLa cells. The protocol is however valid for other FRET fluorophore pairs, cell types, and linker constructs. In this case, appropriate filter sets and excitation lines have to be chosen, and, crucially, the FRET efficiency of the positive FRET tandem control construct needs to be determined independently, as described in **Note 3**.
2. To ensure strong FRET signal levels, it is advisable to devise constructs labeled near the binding sites of the proteins. For example, if the interaction is likely to occur in the C-terminal region of the protein of interest, it is advisable to construct a C-terminal fusion protein, e.g., via use of the pECFP-N1 vector (see http://www.addgene.org/vector-database/2445/). Vice versa if the interaction is more likely to happen in the N-terminal region, use the pECFP-C1 plasmid, instead. Follow similar procedure for the yfp-fusion proteins.
3. The plasmid for the positive control construct is available from the authors. In live HeLa cells, this construct was determined to yield a FRET efficiency of $E = 38$ %. Other linker constructs may also be used but require an independent determination of FRET efficiency (*see* ref. 5 for details).
4. Optionally, a further negative control sample containing empty pECFP and pEYFP plasmids can be used. These are not strictly necessary but provide a useful means to validate the calibration procedure described in the Subheadings 3.1, 3.2, and 3.3 and in **Note 9**.
5. Use other filter sets/laser settings as appropriate if FRET pairs other than CFP/YFP are used in the experiments.
6. For background subtraction, identify dark regions in measured images. The average brightness of these regions allows for a d.c. background subtraction from the images. Incorrect background subtraction would yield erroneous FRET results.

Perform background subtraction for all measurements in this protocol. The MATLAB script provided in [6] as an example of seFRET image processing already includes a background correction which is suitable for the sample data.

7. For the 18 aa CFP-YFP linker construct expressed in live HeLa cells (*see* Subheading 2), substitute $E = 0.38$ in the equations for α and β. If fixed cells are used or a different calibration construct is used (e.g., different FRET fluorophores or different linker system), then E needs to be determined independently for this construct and the appropriate value for E substituted. E can be determined from a measurement of the donor fluorescence lifetime as

$$E = 1 - \frac{\tau}{\tau_0}$$

where τ is the donor fluorescence lifetime in the presence of FRET and τ_0 is the donor lifetime in the absence of FRET (i.e., the natural lifetime of the donor). For details, refer to [5].

8. Measure in at least five different image areas to determine α and β to check for consistency of and enable averaging.

9. A control experiment to check that the calibration has been successful and the correct data for AER and DER have been obtained is to measure a negative control sample, which contains noninteracting pECFP and pEYFP plasmids (or corresponding plasmids if other fluorophores are used in the experiment). For this sample, both aFRET and dFRET should be 0 if correct values for DER, AER, and α and β were used.

10. The consistency of obtained datasets can be verified through a repeat determination of the AER from a corresponding measurement of the acceptor-only sample at the end of the experiment. If this value differs from that determined at the start of the experiment, this is indicative of changes in the experimental conditions (e.g., laser power fluctuations, microscope settings) which may invalidate results.

References

1. Lakowicz JR (2006) Principles of fluorescence spectroscopy. Springer, New York. ISBN 978-0-387-46312-4
2. Jares-Erijman EA, Jovin TM (2006) Imaging molecular interactions in living cells by FRET microscopy. Curr Opin Chem Biol 10(5):409–416
3. van Munster EB, Gadella TWJ (2005) Microscopy techniques. doi: 10.1007/b14097
4. Chakrabortee S, Meersman F, Kaminski Schierle GS et al (2010) Catalytic and chaperone-like functions in an intrinsically disordered protein associated with desiccation tolerance. Proc Natl Acad Sci USA 107(37):16084–16089
5. Elder AD, Domin A, Kaminski Schierle GS et al (2009) A quantitative protocol for dynamic measurements of protein interactions by Förster resonance energy transfer-sensitized fluorescence emission. J R Soc Interface 6:S59–S81
6. http://wiki.laser.ceb.cam.ac.uk/wiki/index.php/Resources

Part VI

Fluorescence Microscopy: Fluorescence Lifetime Imaging (FLIM)

Chapter 20

Wide-Field Fluorescence Lifetime Imaging with Multi-anode Detectors

Roland Hartig, Yury Prokazov, Evgeny Turbin, and Werner Zuschratter

Abstract

Fluorescence lifetime imaging microscopy (FLIM) has become a powerful and widely used tool to monitor inter- and intramolecular dynamics of fluorophore-labeled proteins inside living cells.

Here, we present recent achievements in the construction of a positional sensitive wide-field single-photon counting detector system to measure fluorescence lifetimes in the time domain and demonstrate its usage in FRET applications.

The setup is based on a conventional fluorescence microscope equipped with synchronized short-pulse lasers that illuminate the entire field of view at minimal invasive intensities, thereby enabling long-term experiments of living cells. The system is capable to acquire single-photon counting images and measures directly the transfer rate of fast photophysical processes as, for instance, FRET, in which it can resolve complex fluorescence decay kinetics.

Key words Time-correlated single-photon counting (TCSPC), Fluorescence lifetime imaging (FLIM), Förster resonance energy transfer (FRET), Positional sensitive photon multiplier, Multichannel plate

1 Introduction

1.1 High-Resolution Light Microscopy Beyond the Diffraction Limit

During the past two decades, the traditional fluorescence microscopy has developed from a discipline interested in the description of static microstructures to a tool of molecular cell biology aiming to analyze physiologically relevant molecular processes on the cellular and systems network level.

Exploring the dynamics and function of macromolecular networks requires high-resolution microscope techniques below the diffraction limit of visible light [1]. With the introduction of new microscopes and optical techniques, as, for instance, the introduction of multiphoton microscopy [2–5, 4pi-microscopy [6–10],

Yves Engelborghs and Antonie J.W.G. Visser (eds.), *Fluorescence Spectroscopy and Microscopy: Methods and Protocols*, Methods in Molecular Biology, vol. 1076, DOI 10.1007/978-1-62703-649-8_20,

stimulated-emission-depletion microscopy (STED) [11, 12], structured illumination (SIM) [13, 14], localization microscopy (ground-state-depletion microscopy (GSD)) [15], spatial precision-distance microscopy (SPDM) [16, 17], photoactivated localization microscopy (PALM/STORM) [18, 19], and light sheet illumination [20, 21], a significant improvement of spatial resolution beyond the diffraction limit of classical microscopy was recently achieved. Most of these concepts take advantage of the fact that fluorescence can be switched between a dark state and a bright state, either in a stochastic (PALM/STORM/GSD) or in a deterministic way (STED).

Unfortunately, most fluorescence microscopes work at high illumination intensities. The strive to collect as many photons as possible necessarily results in high average and peak powers [22] (typically >1 kW/cm^2), which leads to irreversible photobleaching of dyes and significant photodamage of living cells [23]. Therefore, the abovementioned visualization techniques (1) do not allow the direct observation of molecular interactions over a longer period of time, (2) induce photodynamic reactions of the molecular probes, and (3) distort the living state. Prerequisite for non-distorted visualization of dynamic processes within living cells (e.g., during synapse formation and signal transduction) by fluorescence microscopy is a minimal invasive approach [24] to excite and detect the molecular constituents at conditions that preserve the undisturbed living state. Consequently, the development of minimal invasive observation systems that yield information about protein–protein or protein–target structure interactions on the cellular, subcellular, and single molecule level, without disturbing the intracellular metabolism, is an outstanding goal in modern cell microscopy.

1.2 Fluorescence Lifetime Imaging

The improvements of the spatial resolution in modern superresolution microscopes were an important step forward to visualize the molecular organization of cells but do not exploit the full informativeness of the fluorescence process. A fundamental technique capable to grant highest resolution, sensitivity, and comprehensive information about a fluorophore's environment is fluorescence lifetime imaging microscopy (FLIM) (Fig. 1). Measuring the lifetime of the fluorophore's excited state reveals incidents occurring in nanometer range of the molecular environment like changes in pH, temperature, ion concentration, viscosity, or polarity among others.

At present, two methods have been evolved to measure fluorescence lifetimes: frequency domain methods [25–27] and time domain-based methods [28–30] (Fig. 2a, b).

Whereas in the frequency domain, the fluorescence lifetime is extracted from the phase shift and demodulation of the fluorescent light with respect to the phase and the modulation depth of a modulated excitation source, in time domain the decay information

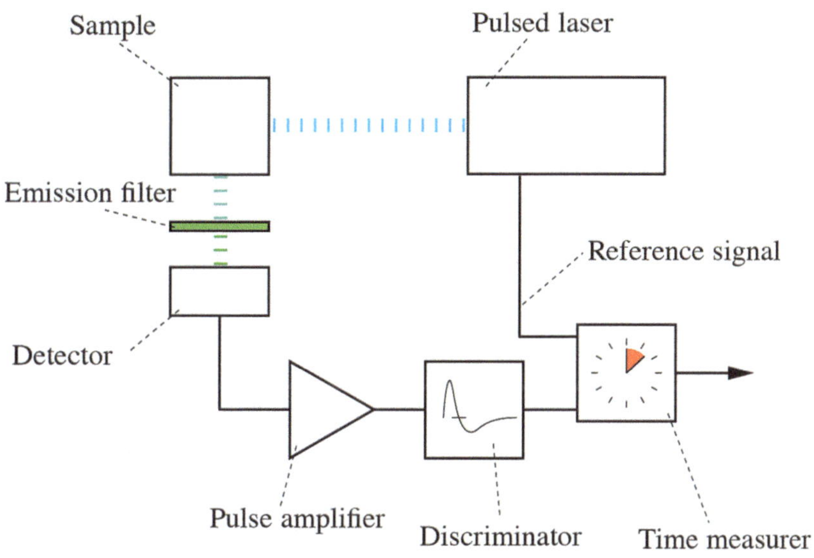

Fig. 1 Principle of FLIM. A basic single-photon counting setup comprises of a pulsed laser, a set of emission filters, a sensitive detector, and electronics for measuring time. The sample is excited with ultrashort laser pulses. Emission filters block the excitation wavelength of the laser and let the fluorescence pass. The detector generates an electrical pulse for each detected photon. This pulse is further electronically amplified and fed into a discriminator. The time difference between the moment of emitting the laser excitation pulse and the detection of the photon is measured

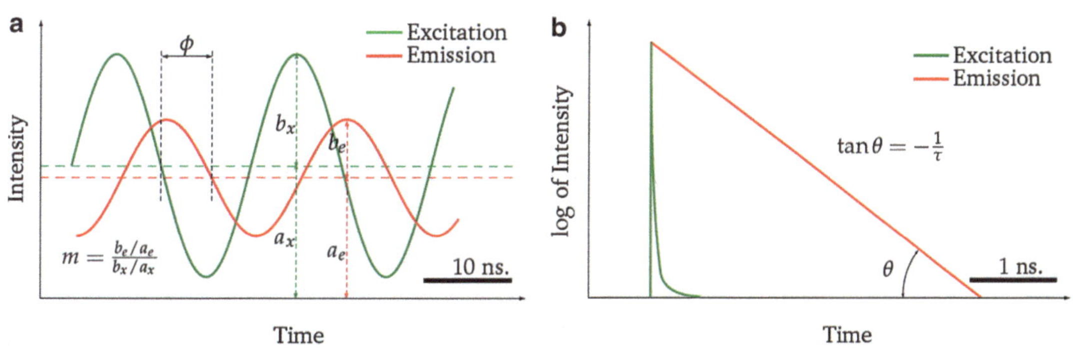

Fig. 2 Comparison between the principles of frequency-domain and time-domain FLIM measurement. (**a**) In frequency domain the fluorescence emission light (*red line*) is emitted with a phase shift ϕ after excitation by a modulated light source (*green line*). The amplitude of the emission light is characterized by the demodulation factor *m*. (**b**) In time-domain lifetime measurements, the sample is excited with a narrow pulse of light (*green line*). The emission intensity is detected (*red line*). The decay is well described by the exponential law and, therefore, looks like a straight line in log scale in case of a mono-exponential decay

is acquired directly by recording the decrease of the fluorescent light intensity over time after exciting the specimen in a single-photon counting mode (Figs. 2b and 3). That means, a sample is illuminated with short laser pulses (fs to ps range), and the time difference between the emitted laser pulse and the appropriate

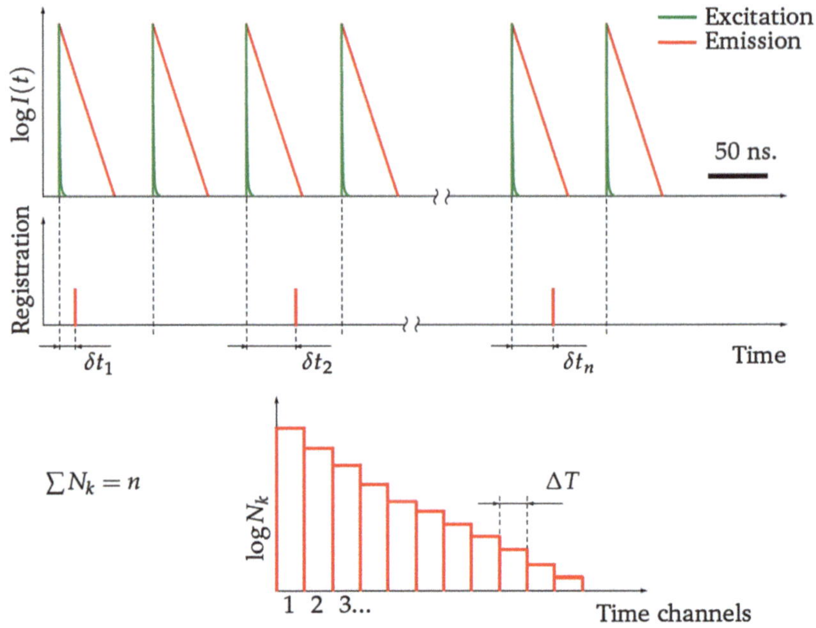

Fig. 3 Fluorescence decay acquisition using single-photon counting method. The sample is excited periodically with ultrashort pulses of light. The time difference δt_i between the moment of the photon detection and the moment of the pulse emission is measured and stored for each registered photon. A histogram of the arrival times δt is build. The values of the histogram relate to the intensity integral inside the interval ΔT

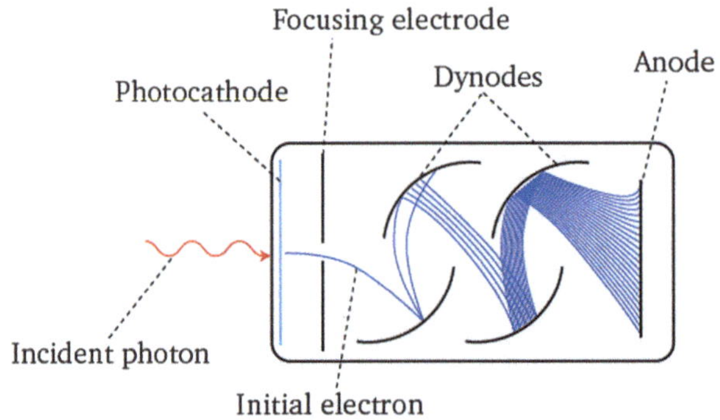

Fig. 4 Principle of a conventional photomultiplier tube (PMT). An incident photon strikes the surface of a photocathode and releases a photoelectron, which follows an applied electrical field towards a dynode system thorough the focusing electrode. Dynodes have a voltage gradient to accelerate secondary electrons and deliver more and more electrons. The anode electrode acquires all the electrons, resulting in a sharp current pulse

fluorescent photon at the detector is measured resulting in a histogram of arrival times (Fig. 3). Among the other techniques, only single-photon counting is capable to utilize the detected photons as efficiently as it is physically possible [31] with short excitation pulses [32].

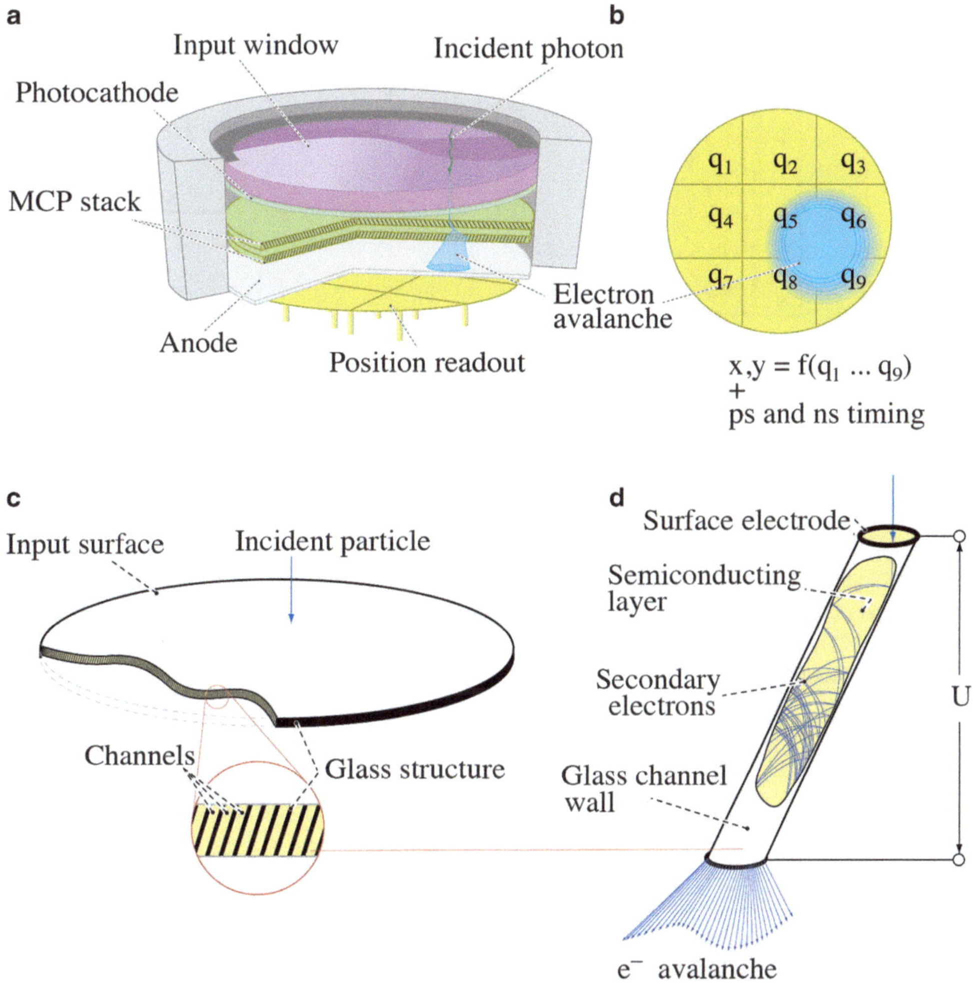

Fig. 5 Detector head. (**a**) Scheme of the microchannel plate (MCP)-based photomultiplier tube. The photocathode converts an incident photon to a photoelectron. A stack of two MCPs multiplies the photoelectron, and the resulting electron avalanche (indicated by a *boxed area* in **a**, **b**) is collected by a position-sensitive planar anode (**b**), which consists in the depicted case of a 3 $\times$ 3 readout pattern. (**c**) A microchannel plate is a glass plate of diameter varying in range of 18–80 mm that holds millions of individual channels 3–12 μm diameter. Input and output surfaces of a MPC are covered by a thin layer of metal for applying a voltage to grand electron multiplication. (**d**) Each individual channel of the plate acts like a miniature electron amplifier, because the inner walls of the channels are covered with a semiconducting layer (showing good secondary emission coefficient)

1.3 Detectors for Measuring FLIM

A large number of detection techniques are employed for FLIM applications nowadays. The most widespread detectors, implementing single-photon counting-based approach, are photomultiplier tubes (PMTs) (Fig. 4, *see* **Note 1**), microchannel plate-based PMTs (MCP-PMTs) (Fig. 5a, c, *see* **Notes 2** and **3**), and single-photon avalanche diodes (SPADs) (*see* **Note 4**). In addition there are a number of integrating (opposite to counting) detectors: gated

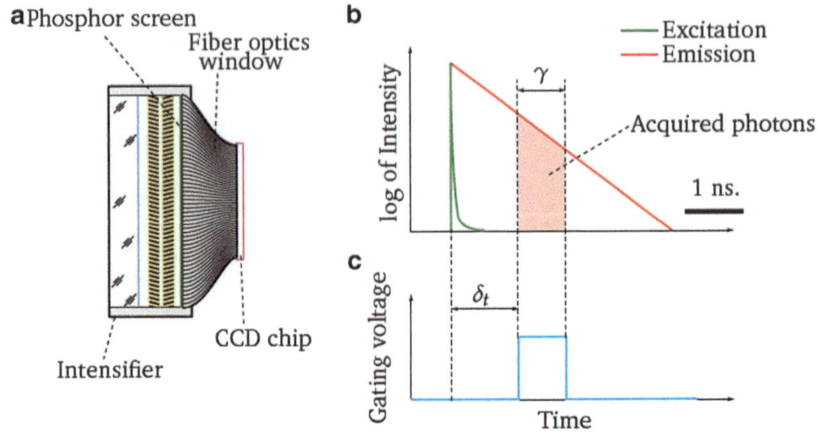

Fig. 6 Measurement principle of gated CCD. (**a**) A gated camera is a combination of a MCP-based light typically intensifier coupled to a CCD sensor by a fiber-optic window with a layer of phosphor as scintillator for back conversion of the electrons into photons. If the diameter of the intensifier is larger than the CCD-sensitive area, a fiber light guide is required to project the image to the sensor. (**b**) The fluorescence decay curve profile is acquired in several time windows ("gates") along the time axis. (**c**) An interval of time within the photons are registered is formed by applying a voltage pulse in the photocathode–MCP gap. The duration γ and offset δt of the pulse define the period of the photons acquisition

CCD cameras (Fig. 6, *see* **Note 5**), streak cameras (*see* **Note 6**), and conventional photodiodes. Advantages and drawbacks of the individual systems are listed in Table 1.

1.4 Time and Position-Resolved Single-Photon Counting

In contrast to time only single-photon counting techniques that acquire positional information by scanning methods in a wide-field FLIM-system, a vacuum-sealed microchannel plate-based photomultiplier tube (MCP-PMT) with a positional sensitive multianode is used as detector.

A position-sensitive photomultiplier (*see* **Note 3**) combines a photocathode (*see* **Note 7**), a microchannel plate (MCP) stack (*see* **Notes 2** and **3**), and a position-sensitive anode system (*see* **Note 8**) (Fig. 5).

To register single photons, the detector needs an amplification element to convert single quanta of light into an electrical signal that can be processed and measured. This conversion process is performed by the photocathode. The photoelectron following an applied electrical field is subsequently multiplied by a microchannel plate stack (Fig. 5c). The resulting avalanche, depending on the number of amplifying elements, consists of 10^5 to 10^7 electrons. The avalanche is detected by an anode (Fig. 5a). A conventional disk or cone-like anode enables only time-resolved registration. Alternatively, the electron cloud is converted back to the visible light by a phosphor screen for CCD-based acquisition (e.g., gated CCD).

Table 1
Advantages and disadvantages of various photon counting methods used for FLIM

Method	Advantage	Disadvantage
Scanning FLIM	• High signal-to-noise ratio due to sensitive point detectors (SPADs) • Optical sectioning • Lifetimes in combination with spectral detection	• Count rate must be kept below excitation rate • Accumulation of TCSPC signals necessary • Sequential data acquisition
Gated CCD	• Simultaneous lifetime information in whole image • High-speed acquisition	• Loss of sensitivity due to discarding of photons • Limited lifetime accuracy • High speed at cost of higher excitation power
Wide-field FLIM	• Acquisition of all parameters describing the physical properties of individual photons • Minimal invasive excitation allows long-term observation	• Limited count rate • Accumulation of photons

However, the latter approach cannot be used in applications where single particle counting is required because of the long readout time of a CCD. Therefore, it is necessary to employ a position-sensitive anode that provides spatial information for each individual detected event in combination with timing information.

2 Materials

2.1 Cell Culture and Transfection Reagents

1. DMEM → Dulbecco's Modified Eagle Medium.
2. DMEM(+) → DMEM, 10 % fetal calf serum (FCS), 2 mM L-glutamine, 100 U/mL penicillin, 100 μg/mL streptomycin.
3. Lipofectamine™ 2000 Transfection Reagent.
4. PolyFect Transfection Reagent.
5. RPMI 1640 medium without phenol red, 10 %, FCS, streptomycin.

2.2 Cell Culture Material

6. Cell culture flasks, 40 mL.
7. Glass bottom dishes, 35 mm.

2.3 Buffers Used in FLIM Measurements

8. Krebs–Ringer solution: 10 mM HEPES, pH 7.0, 140 mM NaCl, 4 mM KCl, 1 mM $MgCl_2$, 10 mM Glucose, and 1 mM $CaCl_2$.
9. Phosphate-buffered saline (PBS): 10 mM phosphate, pH 7.4.

2.4 Staining Material

10. Auramine in H_2O freshly prepared.
11. Erythrosin B in H_2O freshly prepared.
12. Light-proofed, humidified chamber.

3 Methods

3.1 Transfection of HeLa Cell Cultures with PolyFect

Transfect cells about 24 h after splitting.

Add 20 μg of DNA of the construct (in total 3 μg/well) to 100 μL of cell culture.

Keep mixture for 10 min incubation time at RT.

Add 100 μL of 37 °C DMEM + media.

Keep the mixture for 48 h in the incubator.

3.2 Transfection of Jurkat T-Cell Cultures by Electroporation

Exchange of media: Centrifuge cells (300 × *g*, 8–10 min) remove supernatant.

Collect 20 mL of old media and mix with fresh media in cell culture flask (conditioned media).

Resuspend pellet in 1 mL PBS add 35 mL PBS.

Centrifuge and remove supernatant.

Resuspend pellet in 350 mL PBS.

Transfer 350 μL of cell suspension into the electroporation cuvette.

Add 20 μg of DNA, mix, and perform electroporation 230 V, 950 μF, 1 pulse (26–28 ms).

Remove DNA and transfer cell suspension into cell culture flask containing conditioned media, cultivate over night at 37 °C, 5 % CO_2.

3.3 Camera for Wide-Field FLIM

The FLIM camera presented here can be employed in combination with existing conventional fluorescence microscopes eliminating the need of complex and expensive scanning microscopic equipment. Its detection system operates as easy as a conventional camera with approximately 1,000 × 1,000 pixels in 25 mm diameter but with a temporal resolution below 50 ps. The core of the camera consists of a vacuum-sealed position-sensitive photomultiplier tube (PMT) produced by ProxiVision (Bensheim, Germany) with a multialkali photocathode (S20), a stack of MCPs with a pore diameter of 10 μm, and a resistive area anode [33]. The tube operates in a single-photon counting mode enabling acquisition of time information simultaneously with the position. The extraordinary timing characteristic of the MCPs allows to acquire fluorescent decay information with highest precision.

The position readout relies on measuring the electron cloud footprint emerging from the MCP stack that falls on the resistive

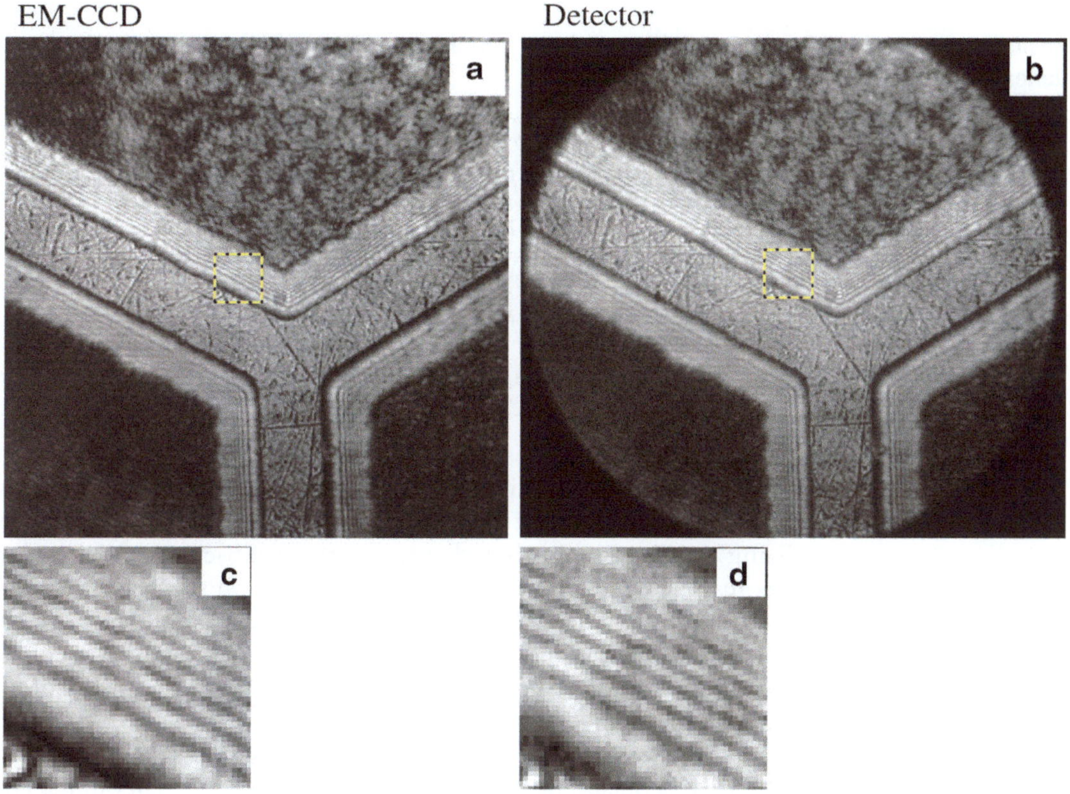

Fig. 7 Comparison of the spatial resolution between an EMCCD (Andor Ixon 897) and the FLIM camera by imaging a test sample (EM grid) used for beam and focus calibration. (**a** and **c**) Image and detail taken by EMCCD; (**b** and **d**) image and detail taken by FLIM camera. Notably, both systems show typical diffraction patterns indicating a resolution close to the diffraction limit of visible light

area anode where it is subsequently read out by an external 3 × 3 anode scheme (Fig. 5b). The total charge is shared between these electrically isolated electrodes; thus the coordinate of an incident photon can be precisely calculated using the charge parts (*see* **Note 8**). This readout method grants superior positional resolution compared to timing-based approaches, e.g. delay line or cross delay line.

Figure 7 compares the spatial resolution of a highly sensitive EMCCD (Andor Ixon 897) with the FLIM camera described here. Whereas the spatial resolution of both cameras is nearly identical, the additional time information in the FLIM camera allows wide-field multiparameter acquisition of the photon characteristics. The acquired parameters including position, absolute arrival and travel time, wavelength, and polarization are recorded for each detected photon and stored in list mode (*see* **Note 9**).

Simultaneous acquisition and list mode storage of all parameters of individual photons, in combination with minimal invasive conditions, significantly increases the power of fluorescence microspectroscopy and opens up a wide range of new opportunities for

Table 2
Simultaneous 4-channel image acquisition of donor and acceptor fluorescence after pulsed interleaved excitation by 473 and 532 nm lasers

Channel	Excitation (nm)	Detection/ bandwidth (nm)	Application
473 GFP	473	520/35	Fluorescence signal of the donor molecules
473 RFP	473	605/55	FRET-mediated acceptor signal
532 GFP	532	520/35	May contain small quantity of donor fluorescence excited by 532 nm
532 RFP	532	605/55	Natural fluorescence signal of the acceptor

sensitive analytical applications in biophysics. Possible applications might be the study of intra- or intermolecular dynamics within living cells by monitoring changes in the multi-exponential ps/ns fluorescence dynamics, anisotropy, and emission spectra. As example of multiparameter FLIM (*see* Table 2), a simultaneous acquisition of the donor and acceptor fluorescence inside a living cell is shown in Fig. 10 and quantified in Table 3 following pulsed interleaved excitation [34].

3.4 Wide-Field FLIM Setup

1. Microscopes: The wide-field FLIM setups (Fig. 8) in our lab are either based on a Nikon TI Eclipse (Nikon GmbH, Düsseldorf, Germany) or an Olympus IX81 (Olympus GmbH, Hamburg, Germany) wide-field fluorescence microscope equipped with 100×/N.A. 1.4 Plan Apo objectives.
2. Illumination source: As illumination sources, the following ultrashort-pulse lasers are integrated within the setups. For the Olympus setup, a 80 MHz Millennia/Tsunami tunable fs-laser with a frequency-doubled/pulse selector is adjusted to 840 nm resulting in 420 nm, 8 MHz after pulse selection. The Nikon setup consists of two 8 MHz frequency-doubled Nd:Vanadate pulsed lasers (473 nm and 532 nm, 10 ps, High Q Laser, Hohenems, Austria) as shown in Fig. 8(1). The pulse repetition rate is synchronized by a phase-lock module (High Q Laser) resulting in two interleaved trains of pulses (PIE: pulsed interleaved excitation) with an interval of about 62.5 ns. This delay time is sufficient to separate the fluorescence emission of most organic dyes, meaning that the fluorescence emission generated from one laser pulse is completely declined before the next excitation pulse arrives.
3. Laser power: Average laser powers are measured every 10 s by power meters (PD 300, Spiricon GmbH, Ahrensburg, Germany, Fig. 8(2)) to monitor variations in the illumination intensity during the experiment.

Table 3
The table summarizes results of the lifetime analysis, which revealed six different lifetimes in the sample

Channel	t_1 0.39 ± 0.01 ns	t_2 1.42 ± 0.01 ns	t_3 2.53 ± 0.01 ns	t_4 2.44 ± 0.01 ns	t_5 0.23 ± 0.01 ns	t_6 1.23 ± 0.01 ns
473GFP	28 % ± 10	42 % ± 10	28 % ± 4			
473RFP	−30 % ± 5			70 % ± 5		
532RFP				67 % ± 4	8 % ± 4	29 % ± 4

The lifetime of each component is given in ns and shows normalized relative contributions of each component α_i in %

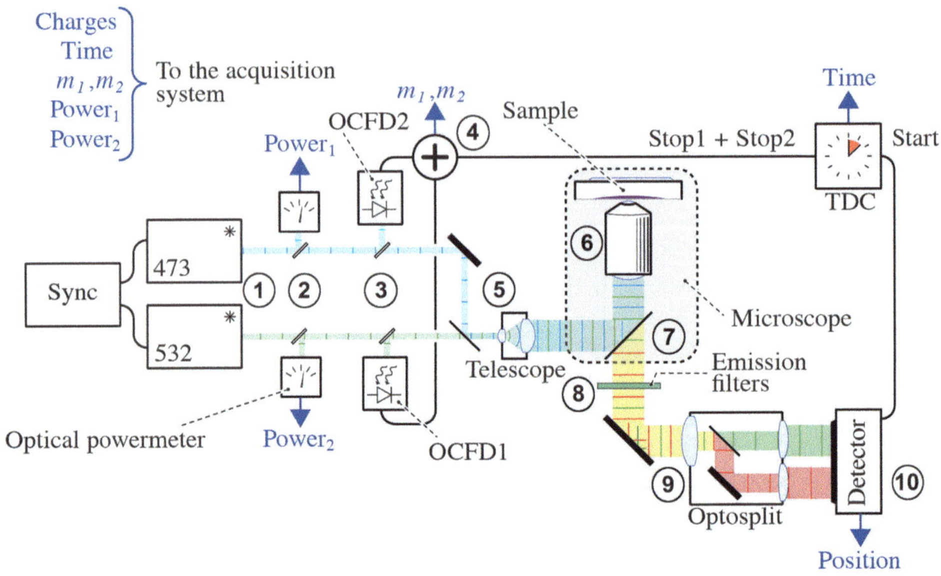

Fig. 8 Setup configuration. Scheme of the four-channel wide-field FLIM setup with two synchronized short-pulse lasers (picoTrain, HQ Laser, Hohenems, Austria) building the light sources (1). Fractions of the light from each laser were directed into optical power meters (2) (PD 300, Spiricon GmbH, Ahrensburg, Germany) to monitor the stability of operation lasers and to optical CFDs (3) (OCF-401, Becker & Hickl GmbH, Berlin, Germany) to form timing reference signals "Stop 1" and "Stop 2" (4). The collimated beams (5) are focused into the back focal plane of the objective (6) using a telescope (Thorlabs GmbH, Dachau, Germany) to form an uniform wide-field illumination. The sample is placed in a temperature and CO_2-controlled incubation chamber (PeCon, Erbach, Germany). A dual-band dichroic mirror (7) (470-532-rdc, AHF, Tübingen, Germany) is used as a beam splitter for excitation/emission signals. Two emission filters (8) are inserted to block remaining excitation light of 532 nm laser and to pass the fluorescence signal to the detector (532 nm Notch and 480LP AHF, Tübingen, Germany). The image is then split into two spectral bands using an Optosplit module (9) (Optosplit II, Cairn Research Limited, Faversham, UK). A dichroic cube (565DXCR, Cairn Research Limited, Faversham, UK) and two band-pass filters (520/35 and 605/55 AHF, Tübingen, Germany) are used to select the spectral channels. The resulting image is projected on the surface of the positional sensitive detector (10)

4. A fraction of each laser beam is focused on Optical Constant-Fraction discriminators (OCF-401, Becker & Hickl GmbH, Berlin, Germany) to define the synchronization pulses (Fig. 8(3)).
5. A photon-marking module (Fig. 8(4)) generates markers (m_1, m_2) and combined stop signals ($stop_1$ + $stop_2$). The result of separation of photons on the basis of m_1, m_2 is presented in Fig. 9. A test sample consisting of Rhodamine G, dissolved in oil (Fig. 9a), and Calcein, dissolved in water (Fig. 9b), was illuminated via PIE, and emission of both fluorophores was separated just by time correlation of the detected photons with the appropriated laser pulses without any further spectral filtering. A sharp segregation of the two dyes with zero cross-talk (Fig. 9c) and individual decays (Fig. 9d) was achieved using alternating 473 and 532 nm laser pulses.

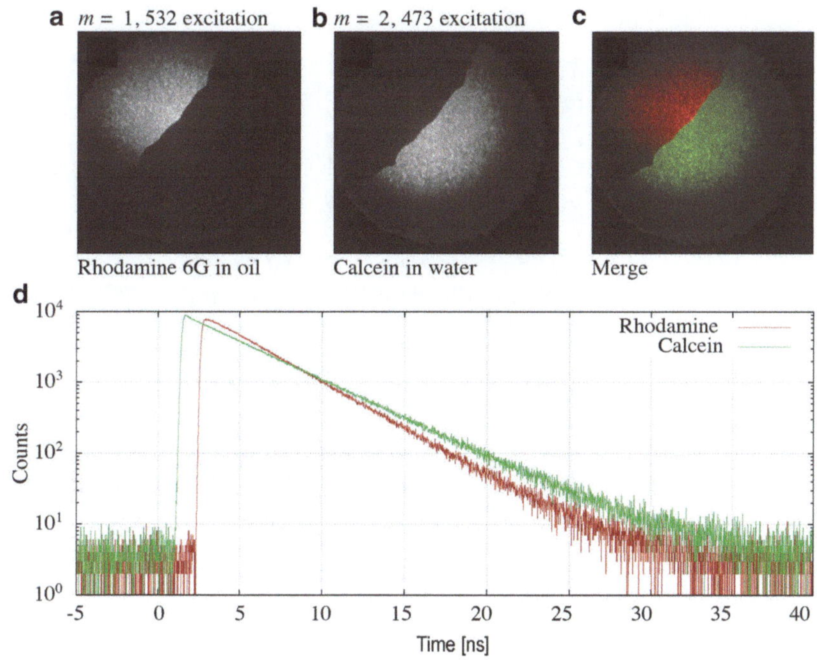

Fig. 9 Rhodamine–calcein measurement. Multiparameter single-photon counting of a test sample consisting of two separated fluorescent dyes: rhodamine G dissolved in oil (**a**) and calcein in water (**b**). The solvents are selected to achieve a sharp separation of the dyes with zero crosstalk during pulsed interleaved excitation with 473 and 532 nm lasers. (**c**) Merged fluorescence channels. (**d**) Two individual fluorescent decay signals are depicted biased due to their time shifted starting points relative to each other

6. The collimated interleaved trains of the laser pulses are merged (Fig. 8(5)) and focused into the back focal plane of the objective (Fig. 8(6)) to produce a uniformly illuminated field of view.

7. Detection light path: A dual-band dichroic mirror (470-532-rdc, AHF, Tübingen, Germany) is used to discriminate between the excitation and the fluorescence signal (Fig. 8(7)). Emission filters (HQ 460 ALP (in case of 420 nm excitation) or 532 nm Notch and 480LP (in case of 473 and 532 nm PIE excitation), AHF, Tübingen, Germany) remove remaining laser signals (Fig. 8(8)), which are transmitted through the dichroic mirror.

8. Multiparameter data acquisition: Fluorescence signals collected via the objective are split into two spectral bands by a two-channel simultaneous imaging device (Fig. 8(9)) (Optosplit II, Cairn Research Limited, Faversham, UK). A dichroic cube (565DXCR, Cairn Research Limited, Faversham, UK) and two band-pass filters (520/35 and 605/55, AHF) are used to select the spectral channels.

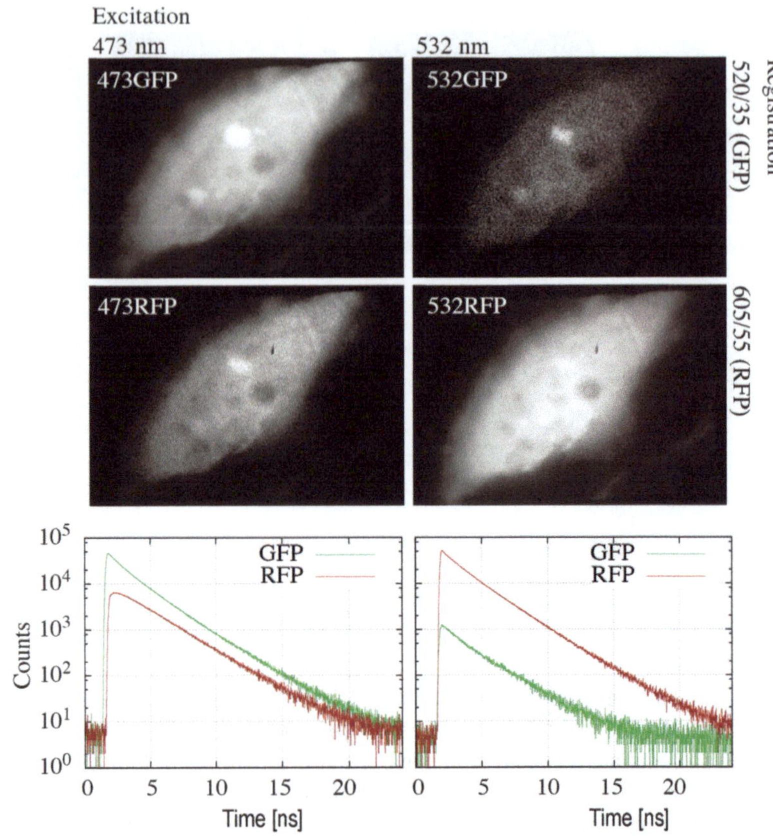

Fig. 10 4-Channel FLIM–FRET. Four FLIM images from donor and acceptor channels simultaneously acquired from a single HeLa cell transfected with an EGFP–TagRFP construct. The separation of the two excitation sources (473 and 532 nm) was performed by the photon-marking method in combination with pulsed interleaved excitation as described in subheading 3.4 and summarized in Table 2. Two band-pass spectral channels are used for registration: 520/35 nm and 605/55 nm. Decay histograms are built for every channel. Notably, the fluorescence lifetime curve acquired by the 473RFP channel demonstrated a slower dynamics at the rise compared to the 532RFP (direct excitation of the acceptor). Analysis of the data is shown in Table 3. This phenomenon is known from the literature [53] as a visual confirmation of FRET presence and relates to the contribution of negative components due to the FRET-mediated fluorescence

9. The photon stream is projected on the FLIM detector (Fig. 8 (10)) where the electronic registers a set of parameters for each collected photon (*see* **Note 9**) in addition to the spatial coordinates. In case of PIE illumination, four spectral channels are simultaneously detected as summarized in Table 2 and depicted in Fig. 10.
10. Spectral FLIM (sFLIM): Spectral measurements are widely used for sensitized emission detection in FRET experiments.

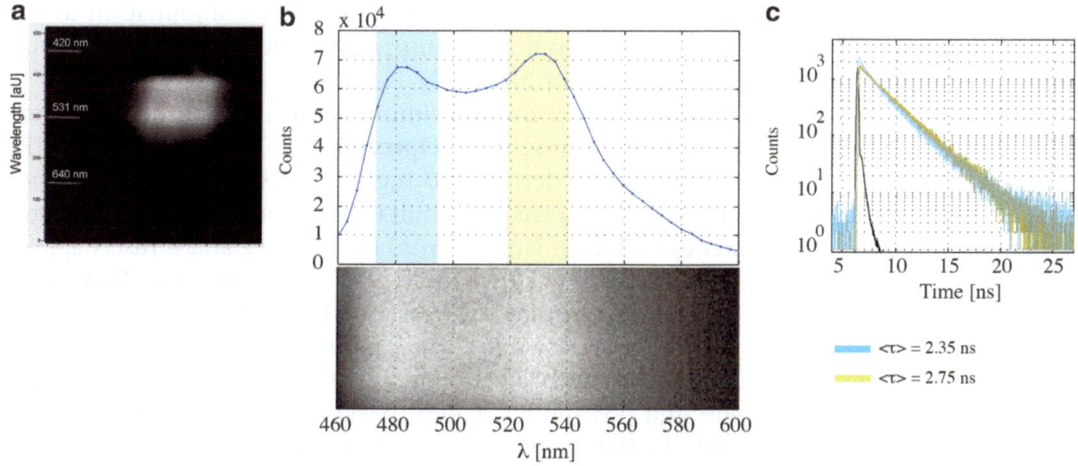

Fig. 11 sFLIM. Spectrally resolved fluorescence lifetime imaging microscopy (sFLIM) of Jurkat T cells transfected with a FRET-pair construct consisting of ECFP and Topaz spaced by a 16AA linker. S-FLIM was performed using the FLIM detector combined with a polychromator. (**a**) Wavelength calibration with three different laser lines (420, 531, and 640 nm) and spectrally splitted intensity signals of a cell expressing the FRET construct. Analysis of this measurement resulted in the spectral histogram (**b**) and emission-wavelength resolved decay curves (**c**) *blue* = 475–495 nm, *yellow* = 520–540 nm)

To record time-resolved spectra, a polychromator (Imspector V8, Spectral Imaging Ltd., Oulu, Finland) can be inserted in front of the wide-field FLIM camera instead of the simultaneous imaging device. The polychromator translates the spectral emission of the fluorophores into spatially resolved information, which is then recorded with respect to its lifetime. An example is presented in Fig. 11.

11. Postprocessing of the photon stream is performed by time and space binning of the stored raw data according to the requirements of the analysis as well as deconvolving the records with the instrument response function (IRF, *see* **Note 10**).

3.5 FLIM Measurement

1. After switching on the microscope system, computer, main laser oscillator (MLO), pulse picker, and detector electronics, let the system warm up for at least 1 h to stabilize the system and let the detector head cool down to −14 °C.

   Tune MLO to appropriate wavelength, e.g., 840 nm, if ECPF is used as donor molecule. Set pulse picker to 1/10 resulting in 8 MHz pulse rate at 420 nm. Measure laser pulses acquired by the stop diode (*see* **Note 11**) by using an oscilloscope and check whether the laser pulses are homogeneous and stable. If necessary, optimize multiphoton laser.

2. To measure the instrument response function (IRF) (*see* **Note 10**), use a dye like auramine or erythrosin B in ETOH with a known fast decay time instead of a mirror. Use these

measurements to optimize homogeneous illumination in the field of view. Limit the count rate to 500,000 cps to minimize pileup effects (*see* **Note 12**).

Insert a two-channel simultaneous imaging device (Dual-View, Optical Insights or Optosplit II, Cairn, Faversham, UK) in front of the detector to register emission signals of donor and acceptor molecules simultaneously and optimize the alignment of the dual view to avoid overlapping areas. Insert appropriate filters to perform multi color acquisition (*see* **Note 13**).

Collect photons (at least 1×10^6 in the peak) and check for pre- and post-pulses within the IRF signal; optimize the illumination and detection light path in the system to minimize pre- and post-pulses. Check background counts; a ratio of at least 1:10,000 between background and peak should be achieved.

3. To calibrate the time-channel width of the detector, add a delay time to the stop signal (*see* **Note 11**) of 2, 4, and 8 ns and calculate how many channels correspond to the different delay times.
4. Measure IRF ($2–4 \times 10^6$ counts in the peak) and calculate the full width half maximum (FWHM) resolution of the whole system.
5. In case of PIE illumination with two lasers, check repetition rate and synchronization of laser pulses with an oscilloscope.
6. Measurement of living cells: The buffer must not contain autofluorescent molecules or dyes like phenol red. Wash cells two times with the buffer (e.g., Krebs–Ringer solution: 10 mM HEPES, pH 7.0, 140 mM NaCl, 4 mM KCl, 1 mM $MgCl_2$, 10 mM glucose, 1 mM $CaCl_2$), and plate cells on glass bottom dishes. Let the cells settle down for 5 min and choose cells showing moderate brightness.
7. Recording: Switch to laser illumination light path and detector light path. Switch on high voltage at the detector control software. Open illumination shutter and adjust laser intensity to acquire sufficient counts per second (typical values are 20,000–200,000 cps/cell) using variable neutral density filters. Store data by typing a file name and record cells until you have enough counts for good statistics (typically 1×10^6 counts in the peak). During the measurement, you can check the quality of the data by loading the file into a FLIM analysis software. This online preview helps to identify FRET signals, or other excited states reactions as well as autofluorescence signals before the detailed post-recording analysis of the decay curves of donor and acceptor fluorescence as shown in Fig. 10.
8. Finishing: Stop recording by pressing the stop button in the acquisition software, switch off high voltage of the detector,

and close laser shutter. Close software, shut down computer, and turn off lasers, other illumination sources, filter wheels, and microscope.

3.6 Spectral Resolved FLIM Measurements (sFLIM)

1. Insert a polychromator (Imspector V8, 50 μm slit width, 380–780 nm visible range, Spectral Imaging Ltd. Oulu, Finland), instead of the two-channel simultaneous imaging device in front of the detector.
2. To calibrate wavelength channels, collect different laser wavelengths, e.g., 420, 531, and 640 nm, as shown in Fig. 11a.
3. Use a dye with a known fast decay time to measure IRF [41–49].
4. Plate cells on a glass bottom dish, choose cells, and reduce illumination to the area of the region of interest. Switch to laser illumination and detector light path, switch on high voltage of the detector, open shutter, and adjust laser intensity to appropriate count rate.
5. Record fluorescence signals of the cell, check for the numbers of photons in the wavelength band of interest (FLIM analysis software), and stop measurement when sufficient counts are recorded. Results of a sFLIM measurement are shown in Fig. 11.

4 Notes

1. *PMT*
   A conventional PMT (Fig. 4) is a vacuum-sealed device comprising a photocathode, a dynode system, and an anode. The photocathode converts an incident photon into an electron via the photoelectric effect (*see* **Note** 7). The dynode system is an electron multiplier comprising a series of electrodes enabling an electron multiplication in the range of 10^5 to 10^7 via secondary emission. An output electrical pulse is acquired from the anode. The fastest commercially available conventional PMTs have a response time down to 500 ps.
2. *Microchannel plate operational principle*
   A microchannel plate (MCP) is an electron-multiplying device featuring subnanosecond reaction time and position sensitivity (Fig. 5). Its planar structure consists of 10^4 to 10^7 tiny channels oriented parallel to each other. Depending on the technology of fabrication and the purpose, channel diameters vary in range between 3 and 25 μm. The axis of the channel is normal to the surface of the plate or has a slight angle of 4–13°. The plate is usually fabricated from glass, and the outer surfaces of the plate are covered by metal to provide charge shipment. The inner channel walls are coated by semiconducting material optimized

for secondary electron emission [35]. Each individual channel of the plate acts like a miniature electron amplifier (Fig. 5d). An incident particle hitting the channel wall releases one or several electrons. These electrons are accelerated along the channel following an applied electrical field. Some of the electrons interacting with the wall free more electrons via secondary emission. These electrons are further accelerated and release more and more electrons from the semiconducting layer. Eventually, this avalanche process yields a cloud of thousands of electrons, which emerge out of the channel. The diameter of an individual channel defines the minimal resolvable position step and therefore limits the spatial resolution. Commonly, plates with 6 or 10 μm channel diameter are used in various imaging devices nowadays.

3. *MCP-based PMT*

   Combining the MCP assembly with a photocathode results in a sensing element for photons within the visible part of the spectra. The MCP-based PMT is a photomultiplier where an MCP stack acts as an electron multiplication element. Like a conventional PMT, this vacuum-sealed device comprises a photocathode, a MCP electron amplification assembly, and an anode (Fig. 5a). An incident photon interacting with the photocathode results in an initial photoelectron. The electron following an applied electrical field is driven towards to the MCP stack where the amplification occurs. An electron cloud emerging from the MCP stack is acquired by the anode (Fig. 5). In the simplest case, the anode is a disk or cone-shaped electrode providing a timing electrical signal. In practically used MCP-PMTs, the gain varies in a range of 10^6 to10^7. An achievable timing resolution of such tube is $s = 6.2$ ps resulting in 14.6 ps FWHM [36].

4. *SPAD*

   Single-photon avalanche diodes (SPADs) are the most popular type of detectors employed in TCSPC available nowadays. High quantum efficiency, short dead time, and low-cost made the SPADs the most attractive point detectors used in time-resolved spectroscopy. SPADs are photodiodes operating in a reverse bias mode over the breakdown voltage [37]. Under such conditions of operation, a single photo-generated carrier results in the development of a self-sustaining avalanche current, thus making them capable to register single optical photons. An external circuit is required to stop an avalanche and return the detector to its initial state. The best achievable time resolution of SPAD is 20 ps [37].

5. *Gated CCD camera*

   Basically, an intensifier module has the same construction as a MCP-PMT (Fig. 6a) comprising a photocathode, a MCP stack, and an anode. First, a photon hitting the photocathode is converted to the photoelectron. This electron is multiplied by the MCP stack, and the resulting avalanche is fed onto the anode. A layer of luminophore (typically phosphor) plays the role of the anode, i.e., charge delivered by the MCP will be converted back to the visible light by a phosphor screen. The resulting light intensity is high enough to be detected by a charge-coupled device (CCD). The luminescent screen is coupled to the CCD chip by a fiber optic window.

To enable gated operational mode, the photocathode of an intensified CCD is triggered by a short subnanosecond voltage pulse (Fig. 6b, c). In this mode, the voltage difference between the MCP stack and the photocathode is applied only for the short time γ forming a so-called time gate when the photoelectrons are driven to the MCP stack. Consequently, only the photons arriving within the period of the gate are registered by the CCD. In order to acquire the fluorescence decay information, series of frames are acquired. The gate of duration γ is formed in synchrony with the excitation pulse after a certain delay δt. This delay defines a part of the decay being acquired. The decay is scanned by variations of the delay and gate width in synchrony with the excitation source. The best achievable time resolution of gated CCD is currently 80 ps (PicoStar UF, La Vision GmbH, Bielefeld, Germany).

6. *Streak camera*

   Streak cameras employ another approach to obtain timing information. Basically the device construction is similar to the intensified CCD camera with a single extension. In the simplest case, two electrodes are placed between the photocathode and the first MCP enabling to change the trajectories of the photoelectrons by applying a voltage difference between the electrodes. Modulating the voltage in synchrony with the pulsed irradiation source converts one of the positional axes of the detector into a time axis. The best achievable time resolution of a streak camera for line scanning is currently 1 ps (C 10910, Hamamatsu Corp., Japan).

7. *Photocathodes*

   The efficiency of photon detection and spectral sensitivity are defined by the photocathode. In the simplest case, the photocathode is a layer of metal with a typical thickness varying from 10 to 500 nm that covers an input window of the PMT [38]. All the photocathodes share the same operational principle employing a phenomenon of photoelectricity. As a reaction to an incident photon, an electron is released. The quantum efficiency (QE) of the photocathode is the probability to emit an electron out of the surface as a reaction for an incident photon. The most conventional photocathode types suitable for visible light registration are: bialkali, multialkali (S20, S25), and GaAs. Spectral response and dark noise are the major parameters discriminating the photocathodes. For example, the QE of a biakali photocathode has a peak of 20 % of QE around 400 nm becoming blind at 700 nm. Multialkali photocathodes sensitivity curve lasts up to 800–900 nm with the drawback of higher dark noise 30–50 cps/cm^2 comparing to 5–10 cps/cm^2 for biakali.

8. *Position-sensitive anode system*

   In order to acquire the position of the electron avalanche centroid, it is required to employ some kind of sensing element capable to resolve the position of the cloud emerging from the stack. This resolves the position of an incident particle hitting an MCP assembly because the centroid position and the incident particle coordinates are related. There are two commonly used methods for the position readout: charge dividing and time-based approach.

   In contrast to the time-based position readout method where the position of the electron avalanche center is measured by acquiring the signal travel time, the charge division anode readout schemes rely on measuring charge distribution acquired from the elements of an electrode pattern (Fig. 5b). In order to evaluate the position of the electron avalanche center, its parts are measured independently. In its simplest form, a positional sensitive anode consists of a quadrant (QA), i.e., a planar charge division readout pattern comprising four electrically isolated electrodes. Consequently, an electron cloud induced by an incident particle is shared between four electrodes. The combination of the measured charges uniquely determinate the position of the incident photon by simple ratio equations:

   $$x = (q_1 + q_4 - q_2 - q_3))/\Sigma q$$
   $$y = (q_1 + q_2 - q_3 - q_4)/\Sigma q$$
   where $\Sigma q = q_1 + q_2 + q_3 + q_4$.

   Since the vacuum outputs of an anode system are the main source of leakage and contamination, there are good reasons for constructing the detector with as few vacuum-air interconnections as possible. Therefore, in order to simplify the production of the vacuum-sealed devices, a method of indirect detection of a MCP avalanche has been developed [39, 40]. In this method, the charge of the electron avalanche falls on the surface of a highly resistive material such as germanium instead of directly to the anode electrode. The charge held by the film polarizes dielectric material of the detector bottom part. The image of the initial charge appears on the outer surface of the detector where it can be measured by an external anode.

9. *List mode storage*
List mode storage of all available parameters of the recorded photons (e.g., *x*, *y*-coordinate, absolute arrival time, travel time) offers full flexibility in subsequent handling, processing, and analysis of the data. Particularly, the access to each parameter allows the generation of high-quality images or fast time-lapse series from the same data set by varying the binning of time channels and thereby the number of photons per frame. Moreover, time gates can be applied for filtering certain lifetime ranges to improve contrast by cutting off autofluorescence. In other words, recorded photons can be sorted differentially according to the requirements of the specific analysis.

10. *Instrument response function (IRF)*
In FLIM measurements, the instrument response function (IRF) is usually recorded by detecting the reflected pulses of the excitation light using a mirror. The IRF is used in the analysis of time domain FLIM data. A theoretical model decay is convolved with the measured IRF, and the result is compared with the recorded fluorescence decay. A new, more precise, theoretical decay of the sample can be generated.

The wavelength dependency of many detectors requires IRF measurements at the same wavelength range as the fluorescence emission of the sample. Changing the detection light path like adding optical filters and beam splitters can additionally change the IRF. This will cause a less precise analysis of the FLIM data.

Reference dyes with known short lifetimes (<100 ps) solve these problems. The IRF measurements can be performed exactly with the same optical elements (beam splitters and filters) in front of the detector like in the final measurements. Appropriate fluorophores have been already described [41–49].

11. *Time measurement*
Fluorescence lifetime imaging is based on the recording of the arrival time of each individual photon in combination with positional information. In order to resolve the nanosecond-scale lifetime of the chromophore's excited state, the sample is excited by ultrashort pulses of light. The time difference between the moment of excitation and photon detection is recorded. Registering fluorescent photons one after another allows to build a histogram of arrival times and to extract the lifetime information (Fig. 3) analyzing the distribution of the arrival times [29].

Two most commonly used types of time measuring devices operating in a picosecond time range are time-to-amplitude converter (TAC) followed by analog-to-digital converter (ADC) and direct time-to-digital converter (TDC). An operational principle of a TAC is based on the conversion of the time interval into a voltage signal by charging a capacity by a current source. The current source is connected to the capacity through the electronic key that is opened by a "start" signal and closed by a "stop" signal. The charge delivered to the capacitor is proportional to the time interval between the "start" and the "stop" signals. Such method provides a resolution down to 6.5 ps but has a relatively long dead time typically of more than 100 ns.

12. *Pileup rejection*

    To avoid pileup effects [32] due to the finite dead time of the detecting system (t_{dead}), the count rate of the detection should not exceed $3/t_{dead}$. Modern high repetition rate lasers allow to avoid pileup by working at count rates less than 1 % of the repetition rate of the laser. A pileup effect relates to the higher probability registrating the first event rather than the second coming within the dead time period of the detector. In other words, if the detector is processing the photon and after a short delay (shorter than t_{dead}), the following arrived photons will be ignored. This results in a distortion of the decay histogram because photons would have non-equal probability of registration. The distortion looks like a sharp peak in the beginning of the decay corresponding to the earlier photons.

    Several methods are known to avoid measurement distortions by effects pileup [50]. One of the methods is based on generating a frame gate covering the dead time of the system. Upon an arrival of the event, a time frame signal is generated of duration roughly equal or exceeding t_{dead}. If there is any further signal appearing during this time period, a time frame is extended and all the events including the initial one are ignored.

13. *Multi-color data acquisition*

    The advantage of the PIE method is the ability to monitor the dynamics of several spectrally separated chromophores. This feature is especially useful to perform FRET-based experiments. A conventional method of FRET measurement relies on the observation of the donor and acceptor kinetics by excitation of the donor molecules only [51, 52]. Here, the kinetic information of the acceptor is acquired in two channels. The first channel is dedicated to register the fluorescence resulting from the FRET process, and the second channel enables the observation of the acceptor molecules alone after being directly excited by the laser (Fig. 10 bottom). This additional channel is acquired simultaneously with two others, thus maximizing the information, which can be obtained from a dual-molecule FRET system.

Acknowledgments

This work was supported by the following grants: DFG FOR 521 HA3498/1-3; BMBF "Quantum" VDI 13N10077 and DFG SFB 854 TPZ (W.Z.). The authors thank Thomas Kuner, Institute of Anatomy and Cell Biology, University of Heidelberg, Germany, Fernando Picazo and Juan Llopis, University of Castilla-La Mancha, Albacete, Spain for providing FRET constructs.

References

1. Hell SW (2007) Far-field optical nanoscopy. Science 316:1153–1158
2. Straub M, Hell SW (1998) Multifocal multiphoton microscopy: a fast and efficient tool for 3-d fluorescence imaging. Bioimaging 6:177–185
3. Straub M, Lodemann P, Holroyd P et al (2000) Live cell imaging by multifocal multiphoton microscopy. Eur J Cell Biol 79:726–734
4. Bewersdorf J, Egner A, Hell SW (2006) Multifocal multi-photon microscopy. Springer, USA, pp 550–560
5. Andresen V, Egner A, Hell SW (2001) Time-multiplexed multifocal multiphoton microscope. Opt Lett 26:75–77
6. Hell SW, Nagorni M (1998) 4pi-confocal microscopy with alternate interference. Opt Lett 23:1567–1569
7. Nagorni M, Hell SW (1998) 4pi-confocal microscopy provides three dimensional images of the microtubule network with 100- to 150-nm resolution. J Struct Biol 123:236–247
8. Bahlmann K, Jakobs S, Hell SW (2001) 4pi-confocal microscopy of live cells. Ultramicroscopy 87:55–64
9. Kano H, Jakobs S, Nagorni M et al (2001) Dual-color 4pi-confocal microscopy with 3d-resolution in the 100 nm range. Ultramicroscopy 90:207–213
10. Hell SW, Stelzer EHK, Lindek S et al (1994) Confocal microscopy with an increased detection aperture: type-B 4Pi confocal microscopy. Opt Lett 19:222–224
11. Klar TA, Dyba M, Hell SW (2001) Stimulated emission depletion microscopy with an offset depleting beam. Appl Phys Lett 78:393–395
12. Hell SW (2003) Toward fluorescence nanoscopy. Nat Biotechnol 21:1347–1355
13. Gustafsson MG (2000) Surpassing the lateral resolution limit by a factor of two using structured illumination microscopy. J Microsc 198:82–87
14. Gustafsson MG (2005) Nonlinear structured-illumination microscopy: Wide-field fluorescence imaging with theoretically unlimited resolution. Proc Natl Acad Sci USA 102:13081–13086
15. Hell SW, Kroug M (1995) Ground-state-depletion: a concept for breaking the diffraction resolution limit. Appl Phys B 60:495–497
16. Esa A, Edelmann P, Kreth G et al (2000) Three-dimensional spectral precision distance microscopy of chromatin nanostructures after triple-colour DNA labelling: a study of the BCR region on chromosome 22 and the Philadelphia chromosome. J Microsc 199:96–105
17. Lemmer P, Gunkel M, Baddeley D et al (2008) SPDM: light microscopy with single- molecule resolution at the nanoscale. Appl Phys B 93:1–12
18. Betzig E, Patterson GH, Sougrat R et al (2006) Imaging intracellular fluorescent proteins at nanometer resolution. Science 313:1642–1645
19. Rust MJ, Bates M, Zhuang X (2006) Sub-diffraction-limit imaging by stochastic optical reconstruction microscopy (STORM). Nat Methods 3:793–795
20. Huisken J, Swoger J, Bene D et al (2004) Optical sectioning deep inside live embryos by selective plane illumination microscopy. Science 305:1007–1009
21. Vois AH, Burns DH, Spelman FA (1993) Orthogonal-plane fluorescence optical sectioning: three dimensional imaging of macroscopic biological specimens. J Microsc 170:229–236
22. Schönle A, Glatz M, Hell SW (2000) Four-dimensional multiphoton microscopy with time-correlated single-photon counting. Appl Opt 39:6306–6311
23. Koester HJ, Baur D, Uhl R et al (1999) Ca2+ fluorescence imaging with pico- and femtosecond two-photon excitation: signal and photodamage. Biophys J 77:2226–2236
24. Delic J, Coppey J, Magdelenat H et al (1991) Impossibility of acridine orange intercalation in

nuclear dna of the living cell. Exp Cell Res 194:147–153
25. Fantini S, Franceschini M-A, Maier JS et al (1995) Frequency-domain multichannel optical detector for noninvasive tissue spectroscopy and oximetry. Opt Eng 34:32–42
26. Gratton E, Breusegem S, Sutin J et al (2003) Fluorescence lifetime imaging for the two-photon microscope: time-domain and frequency-domain methods. J Biomed Opt 8:381–390
27. Kumar ATN, Raymond SB, Bacskai BJ et al (2008) Comparison of frequency-domain and time-domain fluorescence lifetime tomography. Opt Lett 33:470–472
28. Gadella TWJ (ed) (2008) FRET and FLIM Techniques, vol 33. Elsevier, Amsterdam
29. Bastiaens PI, Squire A (1999) Fluorescence lifetime imaging microscopy: spatial resolution of biochemical processes in the cell. Trends Cell Biol 9:48–52
30. Jares-Erijman EA, Jovin TM (2006) Imaging molecular interactions in living cells by FRET microscopy. Curr Opin Chem Biol 10:409–416
31. Philip J, Carlsson K (2003) Theoretical investigation of the signal-to noise ratio in fluorescence lifetime imaging. J Opt Soc Am A 20:368–379
32. Becker W (2005) Advanced time-correlated single photon counting techniques. Springer, Berlin
33. Schmidt-Bocking H (1997) United States Patent 5,686,721
34. Vitali M, Picazo F, Prokazov Y et al (2011) Long term wide-field FLIM: minimal invasive long term observation of proteins in living cells. PLoS One 6:e15820
35. Wiza JL (1979) Microchannel plate detectors. Nucl Instrum Methods 162:587–601
36. Inami K, Kishimoto N, Enari Y (2006) A 5 ps tof-counter with an mcp-pmt. Nucl Instrum Methods Phys Res A 560:303–308
37. Cova S, Lacaita A, Ghioni M et al (1989) 20-ps timing resolution with single-photon avalanche diodes. Rev Sci Instrum 60:1104–1110
38. Hallensleben S, Harmer SW, Townsend PD (2000) Optical constants for the s20 photocathode, and their application to increasing photomultiplier quantum efficiency. Opt Commun 180:89–102
39. Battistoni G, Campana P, Chiarella V et al (1982) Resistive cathode transparency. Nucl Instrum Methods Phys 202:459–464
40. Jagutzki O, Lapington JS, Worth LBC et al (2002) Position sensitive anodes for mcp readout using induced charge measurement. Nucl Instrum Methods Phys Res A 477:256–261
41. Bastiaens PI, van Hoek A, Wolkers WF et al (1992) Comparison of the dynamical structures of lipoamide dehydrogenase and glutathione reductase by time-resolved polarized flavin fluorescence. Biochemistry 31:7050–7060
42. van den Berg PA, Mulrooney SB, Gobets B et al (2001) Exploring the conformational equilibrium of *E. coli* thioredoxin reductase: characterization of two catalytically important states by ultrafast flavin fluorescence spectroscopy. Protein Sci 10:2037–2049
43. Szabelski M, Ilijev D, Sarkar P (2009) Collisional quenching of erythrosine B as a potential reference dye for impulse response function evaluation. Appl Spectrosc 63:363–368
44. Harris JM, Lytle FE (1977) Measurement of subnanosecond fluorescence decays by sampled single. Rev Sci Instrum 48:1469–1476
45. Hanley QS, Subramaniam V, Arndt-Jovin DJ et al (2001) Fluorescence lifetime imaging: multi-point calibration, minimum resolvable differences, and artifact suppression. Cytometry 43:248–260
46. Zuker M, Szabo AG, Bramall L et al (1985) Delta function convolution method (DFCM) for fluorescence decay experiments. Rev Sci Instrum 56:14–22
47. Van Den Zegel M, Boens N, Daems D et al (1986) Possibilities and limitations of the time-correlated single photon counting technique: a comparative study of correction methods for the wavelength dependence of the instrument response function. Chem Phys 101:311–335
48. Boens N, Qin W, Basarić N et al (2007) Fluorescence lifetime standards for time and frequency domain fluorescence spectroscopy. Anal Chem 79:2137–2149
49. Luchowski R, Gryczynski Z, Sarkar P (2009) Instrument response standard in time-resolved fluorescence. Rev Sci Instrum 80:033109-1–033109-6
50. Davis CC, King TA (1970) Correction methods for photon pile-up in lifetime determination by single-photon counting. J Phys A Gen Phys 3:101–109
51. Nair DK, Jose M, Kuner T et al (2006) Fret-Flim at nanometer spectral resolution from living cells. Opt Express 14:12217–12229
52. Jose M, Nair DK, Reissner C et al (2007) Photophysics of clomeleon by film: discriminating excited state reactions along neuronal development. Biophys J 2:2237–2254
53. van der Meer BW, Coker G, Chen SY (1994) Resonance energy transfer theory and data. VCH, New York

Chapter 21

Global Analysis of FRET–FLIM Data in Live Plant Cells

Sergey P. Laptenok, Joris J. Snellenburg, Christoph A. Bücherl, Kai R. Konrad, and Jan Willem Borst

Abstract

This chapter describes the procedure for globally analyzing *fl*uorescence *li*fetime *im*aging (FLIM) data for the observation and quantification of Förster resonance energy transfer (FRET) in live plant cells. The procedure is illustrated by means of a case study, for which plant protoplasts were transfected with different visible fluorescent proteins and subsequently imaged using two-photon excitation FLIM. Spatially resolved fluorescence lifetime images were obtained by application of global analysis using the program Glotaran, which is open-source and freely available software. Using this procedure it is possible to extract the fraction and distance of interacting species between, or conformational changes within proteins, from complex experimental FRET–FLIM datasets, even at low signal-to-noise ratios. In addition, the software allows excluding inherently present autofluorescence from the plant cells, which improves the accuracy of the FRET analysis. The results from the case study are presented and interpreted in the context of the current scientific understanding of these biological systems.

Key words FLIM, FRET, mTurquoise1, Global analysis, Time-resolved fluorescence, Microscopy, Glotaran

1 Introduction

Fluorescence microscopy in combination with Förster resonance energy transfer (FRET) is nowadays a widely used method for visualizing protein interactions in living cells. FRET is a process in which the excited-state energy from a fluorescent donor molecule is transferred non-radiatively to an acceptor molecule. FRET is based on very weak dipole–dipole coupling and therefore can only occur at very short intermolecular distances, typically less than 10 nm. However, by carefully selecting the FRET pair one can detect interactions over longer distances. There are several methods to visualize FRET, of which donor fluorescence lifetime imaging (FLIM) is the most straightforward one for estimating FRET efficiencies. The fluorescence lifetime is an intrinsic property of a fluorophore giving the advantage that measurements are independent of

Yves Engelborghs and Antonie J.W.G. Visser (eds.), *Fluorescence Spectroscopy and Microscopy: Methods and Protocols*, Methods in Molecular Biology, vol. 1076, DOI 10.1007/978-1-62703-649-8_21,

fluorophore concentration. Though, the fluorescence lifetime is sensitive to the immediate environment of the fluorophore, which is the basis for FRET–FLIM measurements. Typically, FRET–FLIM experiments consist of measuring donor fluorescence lifetimes in the absence (τ_D) and presence (τ_{DA}) of acceptor molecules resulting in spatially resolved, color-coded fluorescence lifetime images. Observation of a decreased donor fluorescence lifetime due to FRET is used as an indicator for molecular interactions [1].

Well-designed data analysis protocols are indispensable for accurate and quantitative analysis of FRET–FLIM data. Global analysis is a technique which offers significant advantages in terms of improved accuracy, decreased sensitivity to noise, and the possibility of combining fluorescence decay data from different measurements [2]. A fluorescence decay can often be described by single or multiple kinetic processes depending on the electronic configuration of the fluorophore and its immediate environment. With global analysis, the kinetic rate constants can be determined simultaneously for all pixels of an image, even though the relative intensity values may vary from pixel to pixel [3]. Global analysis can be used to accurately detect FRET phenomena by estimating the fluorescence lifetime of donor molecules in the absence and presence of acceptor molecules. A complicating issue is the fact that FRET systems in cell biology are not always purely homogeneous, since they can contain a population of donor molecules that cannot transfer their excitation energy, because of either unwanted photobleaching (photodestruction) of the acceptor molecules or incomplete maturation of the acceptor proteins (this is especially the case of genetically encoded visible fluorescent proteins) [4]. Therefore, average donor lifetimes do not reflect true FRET efficiencies, because they originate from a pool of interacting and noninteracting species [5, 6].

In this chapter, a step-by-step protocol is described for the procedure of analyzing FRET–FLIM data on live plant cells by means of global analysis. A complete procedure is described ranging from sample preparation to data acquisition and subsequent analysis of the acquired data. Data analysis is carried out using the open-source and freely available software program Glotaran [7]. The data analysis protocol is illustrated using a case study, for which plant protoplasts were transiently transfected with two different plasmids: one expressing the mTurquoise1 fluorescent protein only and the second expressing a mTurquoise1–mVenus fusion protein connected by a two amino acid linker. The latter is a FRET pair, where mTurquoise1 is serving as a donor and mVenus as an acceptor with a Förster radius (see below) of 5.4 nm. Descriptive screenshots at different steps of the global analysis procedure are included as a visual guide. Finally, essential parameters resulting from the analysis, as well as the final donor fluorescence lifetime images of mTurquoise1 in presence and absence of acceptor molecules, are presented.

2 Materials and Imaging Setup

2.1 Expression Vectors, Plant Protoplasts, Microscopy Features

1. mTurquoise1 (kindly provided by Prof. Th. J. W. Gadella, University of Amsterdam) and mTurquoise1–Venus (two amino acid (LN) linker) constructs were cloned into pSAT-based vector harboring the Ubiquitin 10 promoter.
2. Plant material: Rosette leaves of *Arabidopsis thaliana* plants (ecotype Columbia) grown for 4–5 weeks under long-day conditions (16 h light/8 h dark) at 22 °C.
3. Protoplast isolation and transfection protocol can be found in Bücherl et al. [8] (*see* **Note 1**).
4. Coherent Ti:Sapphire laser (Coherent Mira D900) (Coherent Inc., Santa Clara, CA).
5. Biorad Radiance 2100MP confocal system.
6. Nikon TE 300 inverted microscope (Tokyo, Japan).
7. B&H HPM-100-40 Hybrid detectors (Becker & Hickl, Germany) (*see* **Note 2**).
8. B&H SPC 730 module (Becker & Hickl, Germany).

2.2 FRET–FLIM Data Acquisition

Spatially resolved fluorescence lifetime imaging is performed on a Biorad Radiance 2100 MP system in combination with a Nikon TE 300 inverted microscope as described in Russinova et al. [9]. Two-photon excitation pulses at 860 nm are generated by a Ti-Sapphire Laser model Mira 900 (Coherent Inc.), pumped by a 5 W continuous wave solid state laser (Coherent Verdi V5), resulting in excitation pulses of 200 fs at a repetition of 76 MHz. A 60×/1.2 water immersion objective is used (*see* **Note 3**). mTurquoise1 fluorescence emission is selected by a 480DF30 band pass filter and detected by a Becker & Hickl HPM-100-40 Hybrid detector with a time resolution of 120 ps. Fluorescence images of 64 × 64 pixel size are acquired using the B&H SPC 730 module. The resolution for the analog-to-digital (ADC) converter was set to 256 channels, resulting in a channel spacing of 48 ps/channel. The average photon count rate is around 10^4 photons per second for an acquisition time of 90 s for one measurement. Measurements of single transfected protoplasts expressing mTurquoise1 represent the reference. To elucidate FRET, donor fluorescence lifetimes of the mTurquoise1–Venus fusion are determined and compared to the mTurquoise1-only samples (*see* **Note 4**).

3 Methods

3.1 FLIM Data Analysis Summary

For each pixel in a FLIM image, the fluorescence intensity is measured as a function of time. Therefore, each pixel contains a recorded fluorescence-time trace that can be considered as a separate experiment. The time traces of all pixels are stored as a single matrix, in which each column represents the fluorescence intensity decay associated with a single pixel.

The fluorescence decay is typically well described by a sum of exponential decay components convolved with the instrumental response function (IRF), so that the time dependent fluorescence for a single pixel reads as

$$F(t) = \sum\nolimits_{I}^{n_{\text{comp}}} \alpha_i \exp\left(-\frac{t}{\tau_i}\right) \otimes g(t), \qquad (1)$$

where the summation is over the number of exponential fluorescence decay components n_{comp}, τ_i is the lifetime of component i, α_i is the fractional contribution of component i to the fluorescence decay, $g(t)$ is the IRF, and $\otimes$ is the convolution operator.

The basis for global analysis is the superposition principle stating that the experimental decay is composed of a linear combination of n_{comp} components. The fact that the fluorescence lifetime components are spatially invariant is crucial for global analysis of a FLIM image (*see* **Note 9**), which has been justified by Verveer and coworkers [10]. More details about global analysis on FLIM data can be found in [11].

3.2 Summary of FRET Theory

This section summarizes the concepts of fluorescence lifetimes and FRET, which can be found in textbooks on fluorescence spectroscopy [12, 13]. The fluorescence lifetime of a fluorophore τ is the average time that the molecule stays in the excited state after response to a very short excitation pulse and is given as a function of the radiative (k_r) and non-radiative (k_{nr}) rate constants: $\tau = 1/(k_r + k_{nr})$. Since the fluorescence lifetime of a fluorophore is sensitive to the local environment (e.g., pH, charge, presence of quenchers, refractive index, temperature), lifetime measurements under a microscope offer the important advantage of contrast by spatial variations of lifetimes [1]. FRET is a bimolecular process in which the excited-state energy of a donor fluorophore is non-radiatively transferred with rate constant (k_T) to a ground-state acceptor molecule by very weak dipole–dipole coupling. The fluorescence lifetime of donor molecules that involves FRET becomes shorter compared to donor molecules in the absence of acceptor and can be calculated using $\tau_{da} = 1/(k_r + k_{nr} + k_T)$. The transfer rate constant can be calculated knowing the fluorescence lifetime of the donor in presence and absence of acceptor as

$$k_T = k_{da} - k_d = \frac{1}{\tau_{da}} - \frac{1}{\tau_d}. \qquad (2)$$

The transfer rate (k_T) is proportional to the inverse 6th power of the distance r between donor and acceptor molecules, which makes this technique an extremely sensitive parameter for obtaining distances as

$$k_{\mathrm{T}} = k_{\mathrm{d}}\left(\frac{R_0}{r}\right)^6, \tag{3}$$

where r is the actual distance between donor, kd is equal to $1/\tau$d and acceptor and R_0 is the so-called critical or Förster radius [14], defining the distance between donor and acceptor at which 50 % of the donor energy is transferred to the acceptor.

The efficiency of energy transfer (E) can be obtained as

$$E = \frac{k_{\mathrm{T}}}{k_{\mathrm{d}} + k_{\mathrm{T}}} = \frac{R_0^6}{R_0^6 + r^6}. \tag{4}$$

3.3 Software Installation and Compatibility

All experimental data described in this chapter are visualized and analyzed using the Glotaran software. Glotaran was designed to be cross-platform, so that it will run on a wide variety of different operating systems (Windows, Linux, Mac OS). Detailed system requirements, installation instructions, and installers for Glotaran can be obtained from http://www.glotaran.org (*see* **Note 5**). FLIM data need to be in the correct format before visualization or analysis. Glotaran accepts either raw data files obtained from B&H and PicoQuant photon counting cards or plain text data files that are formatted as described in [15].

3.4 Data Preprocessing

1. Once Glotaran is installed and set up as described in Subheading 3.3, the analysis can be started by first creating a *New Project* (File—New Project), where project data such as raw data, models, and analysis results are stored (*see* Fig. 1).
2. Right click on the *Datasets* folder and select "Open dataset file" to create a new data reference file. A dialog window will appear from which data files can be selected and imported (*see* **Note 6**). The data reference file will appear in the project explorer under the *Datasets* folder (*see* Fig. 1).
3. After opening a data reference file or a dataset (double click on file in project explorer), the corresponding data displayer will show the content of the opened file. The first tab of the FLIM data displayer, labeled Data (Fig. 2), displays a fluorescence intensity image (*see* **Note 7**) (Fig. 2). The pixels that are selected for analysis are shown in green, and the corresponding fluorescence decay profile of the last selected pixel is displayed in the *Selected trace* panel (Fig. 2). The sum of the fluorescence decay profiles of all selected pixels is plotted in the *Sum Trace* panel (Fig. 2). The time window can be adjusted by changing the position of the sliders (indicated below the decay profile). The selected time window appears highlighted in the *Sum*

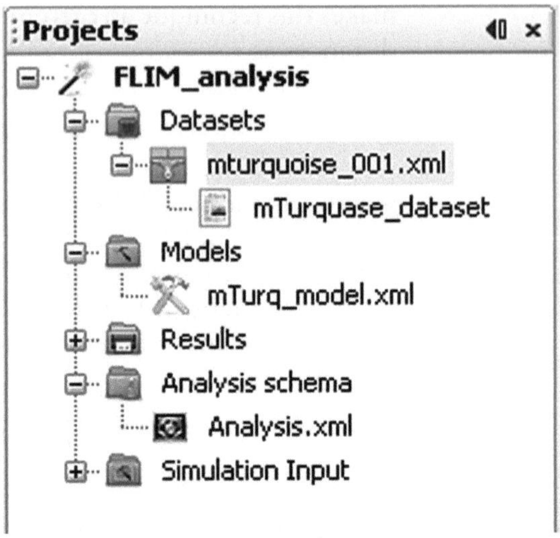

Fig. 1 Screenshot of the project manager in Glotaran. Here, "FLIM_analysis" is the opened project representing a Glotaran project folder on disk. *Datasets*, *Models*, *Results*, *Analysis schema*, and *Simulation Input* are the standard subfolders of any Glotaran project folder. The *Datasets* folder contains references to data files (here *mturquoise_001.xml*), and every data reference file can store datasets that can be used for analysis (here *mTurquoise_dataset*). The *Models* folder includes model files (*mTurq_model.xml*) that are used in the Analysis scheme (*Analysis.xml*), stored under the *Analysis schema* folder

Trace plot. Changes to the selected time window can be applied by clicking the "Select time" button in the toolbar at the top of the Data tab.

4. For Glotaran data analysis, one should first create a new dataset from a data reference file (*see* **Note 8**). This can be done by clicking on the "create dataset" button in the toolbar of the data displayer and specify a name for the new dataset. The dataset will be created under the data reference file node (*see* Fig. 1). In principle it is possible to select all pixels and analyze the whole image, but in most cases, it is more appropriate to just select pixels of interest (*see* **Note 9**). Pixels can be selected by either clicking on the pixels in the fluorescence intensity image or by specifying a threshold value. If the photon counts per pixel are low, there is an option to perform pixel binning by pressing the button "Binning" (*see* **Note 10**).

3.5 Setting Up an Analysis Scheme

All information necessary for analysis (which datasets are selected, which model to use, etc.) is stored in the analysis scheme file, which is enclosed within the project's "Analysis schema" folder (*see* Fig. 1). When a new project is created using the Empty Project template, it will already contain one empty analysis scheme in the "Analysis schema" folder as well as an empty model file in the folder "Models."

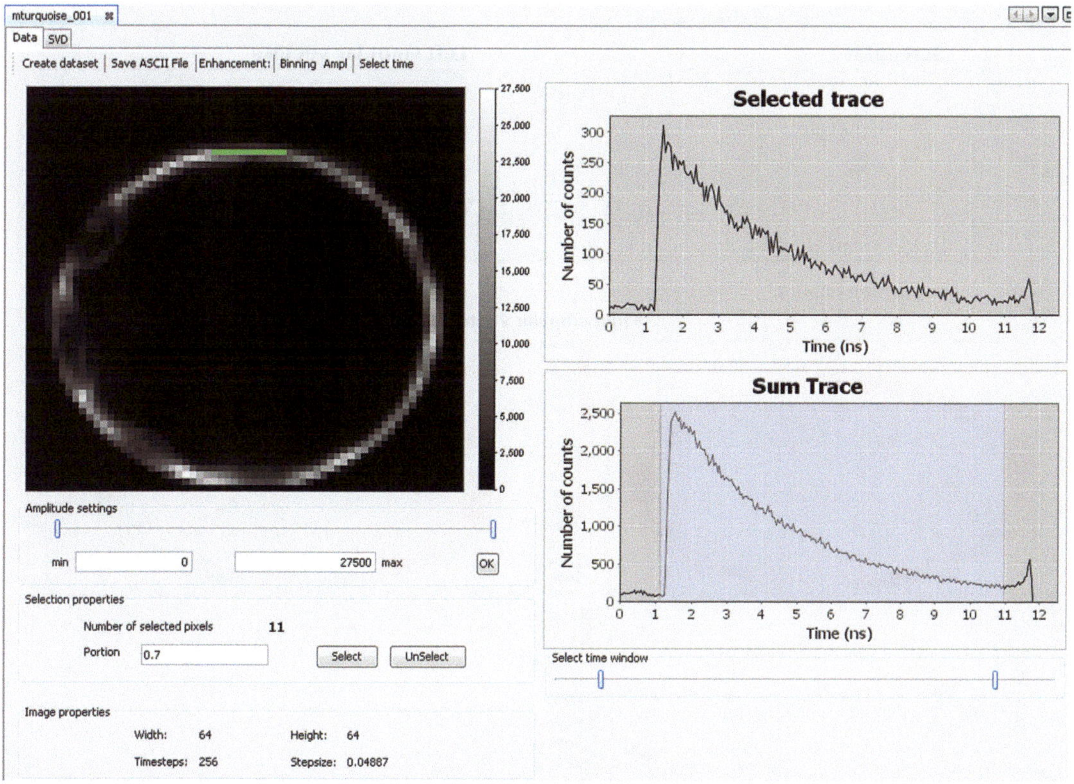

Fig. 2 Screenshot showing the Data tab of the FLIM data displayer displaying a data reference file or a dataset. Shown here is a fluorescence intensity image of a protoplast expressing mTurquoise1 acquired using the B&H 730 module. A description of the various components of the FLIM data displayer is given in Subheading 3.4, **step 3**

The first step in performing global analysis on a new dataset is typically to make an estimate of the number n_{comp} of distinct kinetic components underlying the data. In FLIM analysis this is estimating the number of unique fluorescence components present in the data. The singular value decomposition of the data matrix (SVD) can be a powerful tool to make the initial estimate (*see* **Note 11**) and is available directly from the data displayer window.

1. By clicking on "SVD" tab in the data displayer window (Fig. 2), the SVD of the corresponding dataset will be automatically calculated. Figure 3 shows the SVD of the *mTurquoise1_001* dataset for the first three SVD components. In this example, we can observe two distinct fluorescence components: one relatively slow component with a high contribution (first singular vector) and one very fast component with a lower contribution (second singular vector). Both processes should be taken into account when defining the starting model for analysis. Note that the third singular vector cannot be distinguished from the noise.

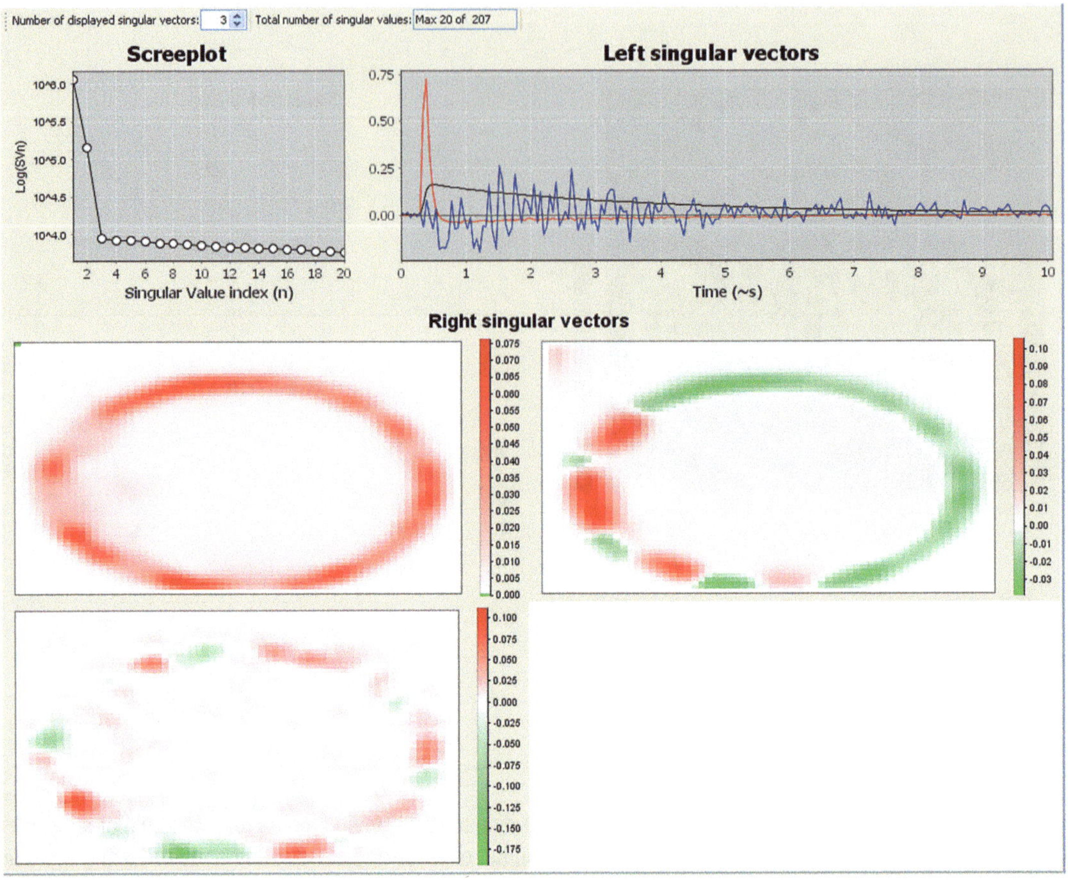

Fig. 3 SVD decomposition of a fluorescence intensity image of a protoplast expressing mTurquoise1. The *top right panel* shows the first three left singular vectors (LSV) (in time), the *bottom panel* shows the corresponding right singular vectors (RSV) (spatial), and, finally, the *top left panel* shows the screeplot of the first 20 singular values. Looking at the LSV suggests the presence of at least two linearly independent components (one slow, *black line*, and one fast, *red line*) significantly different from noise, whereas the third component (*blue line*) is likely due to noise. The RSV depicts the spatial distribution of independent components, showing a few regions with more pronounced fast component

2. *Double click* on the analysis scheme file (Analysis.xml in Fig. 1) to open the analysis scheme editor window, which allows setting up all relevant parameters necessary for analysis (*see* **Note 12**). Figure 4 shows the analysis scheme that was used to analyze a single dataset *mTurquoise_dataset* (*see* **Note 13**).
3. As observed from the SVD analysis, there are at least two distinct kinetic components underlying the data from this experiment. One of these components can likely be attributed to autofluorescence from the chloroplasts and the other to mTurquoise1 fluorescence. The fluorescence decay of mTurquoise1 is modeled using a kinetic parameter shown as “KinPar” in the model with one rate parameter (*see* **Note 14**).

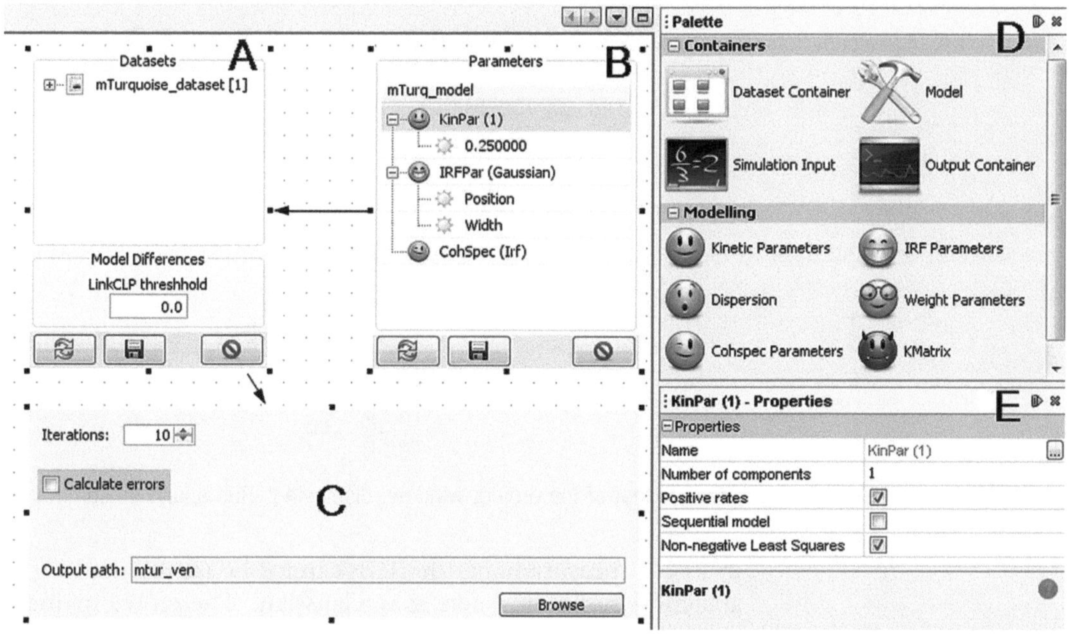

Fig. 4 Screenshot of the analysis scheme editor windows and associated palette and properties windows. On the analysis scheme editor canvas, the following containers are shown: (**a**) *Dataset container*, which contains a list of datasets that will be analyzed (*see* **Note 16**); (**b**) *Model container*, which includes relevant model parameters used in analysis; and (**c**) *Output container*, which is used to specify the name of the folder where the result files will be written (*see* **Note 17**) and specifies the maximum number of iterations used in the optimization algorithm by the underlying software (*see* **Note 18**). The *right side* of the screenshot contains (**d**) the *Palette*, which contains various containers that can be dragged onto the canvas and modeling nodes can be dragged into the *Model container*, and (**e**) the *Properties* window allows for editing of the properties of the selected node (in this example, the kinetic parameters node (KinPar) from the *Model Container* is selected)

The starting value for the decay rate ($1/\tau$) is given in the name of the subnode and can be adjusted using the properties window (Fig. 4e). In this example, the autofluorescence component is modeled by a parameter "CohSpec." This option can be used to account for all processes that occur in parallel to any normal decay process and which rise and decay within the time duration of the instrument response function (IRF) of the measurement setup, such as autofluorescence or scattering. This process, which is typically modeled as having the time profile of the IRF, can be subtracted from the data prior to estimating the real kinetic components of donor and/or acceptor. In the present study, the FLIM setup has a time resolution of around 120 ps, which makes it impossible to resolve the autofluorescence component, and therefore the "CohSpec" option is used to account for this effect (*see* **Note 15**).

4. The final element to add to the model container in the analysis scheme is the "IRFPar" option. This node contains all relevant parameters to model the IRF of the experiment. Typically,

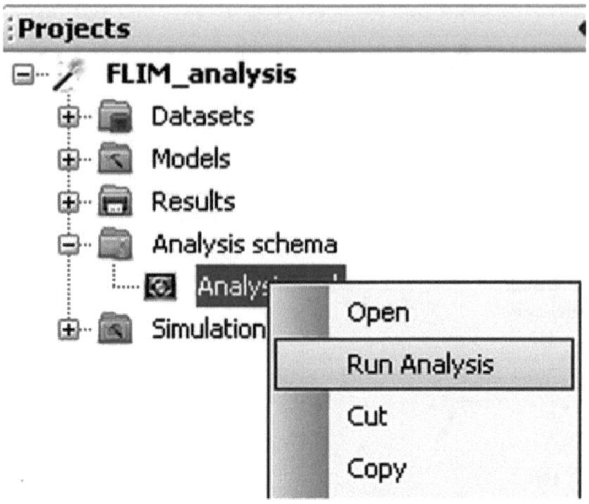

Fig. 5 Screenshot of the project window, displaying "Run analysis" command

in FLIM measurements the IRF cannot be represented by an analytical function, such as a Gaussian. Therefore, in many cases the IRF is measured and numerically convolved with the exponential decay model (*see* **Note 16**). Nevertheless, the time resolution of this particular experiment (*see* **Note 15**) is such that it is possible to approximate the IRF with a Gaussian to obtain reliable fits. For this analysis, the IRF is approximated using a Gaussian function depending on the parameters μ and Δ, being the temporal position of the maximum and the full width at half maximum (FWHM) respectively:

$$\mathrm{IRF}(t) = \frac{1}{\tilde{\Delta}\sqrt{2\pi}} \exp\left(-\log(2)\left(\frac{2(t-\mu)}{\Delta}\right)^2 \right), \tag{5}$$

where $\tilde{\Delta} = \Delta / \left(2\sqrt{2\log(2)}\right)$.

Reasonable starting values for the parameters μ and Δ should be provided, which are then optimized during the analysis. In principle, μ and Δ are free parameters, although they can be fixed, if they can be determined more accurately by other means.

5. In the output container, a specific folder name is created in which the analysis results are saved to. The analysis results will be stored in this folder under the "Results" folder in the project (*see* Fig. 5). If a results folder already exists (for instance, when the analysis is rerun a second time, without specifying a different name), a number will be appended to the name of the folder.

6. Also in the output container the maximum number of iterations can be specified, for which the software should optimize the nonlinear parameters of the model (such as the rate constants and IRF parameters). The conditionally linear parameters

(the amplitudes of every component per pixel) are always optimized, even at 0 iterations. Usually one starts any modeling process with a sanity test by specifying 0 iterations to test the validity of the initial starting parameters. If the starting values are found to be reasonable, the number of iterations can be increased to, for instance, 5 or 10 (*see* **Note 17**).

7. After setting up the complete analysis scheme correctly, the file can be executed to start global analysis of the data. To do this, select the "Run Analysis" option from the context menu of the analysis scheme file by right clicking on the file (*see* Fig. 5). After this, a progress indicator in the bottom right part of the software window will show that the analysis is running.

3.6 Validating Analysis Results

1. When the analysis is complete, the results will appear in the specified folder under the "Results" folder in the project folder (*see* **Note 18**). The results of the analysis of the mTurquoise1 dataset are shown in Fig. 6.
2. Figure 6 shows the estimated fluorescence lifetimes, corresponding error, the fluorescence intensity map, and corresponding color-coded average fluorescence lifetime image as well as a distribution of the average fluorescence lifetime (*see* **Note 19**). Here, a single number for the average fluorescence lifetime (3.6 ns) is shown, because a mono-exponential model is used for analysis. The lack of structure in the SVD of the residual matrix is a good indicator for the quality of the fit.

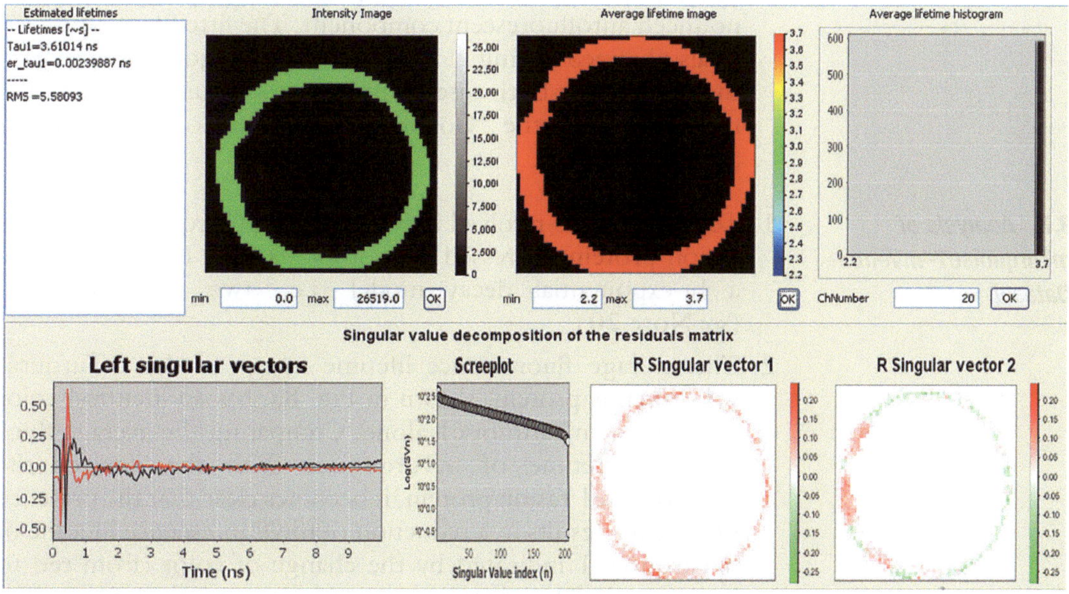

Fig. 6 FLIM results displayer giving an overview of the results of the analysis

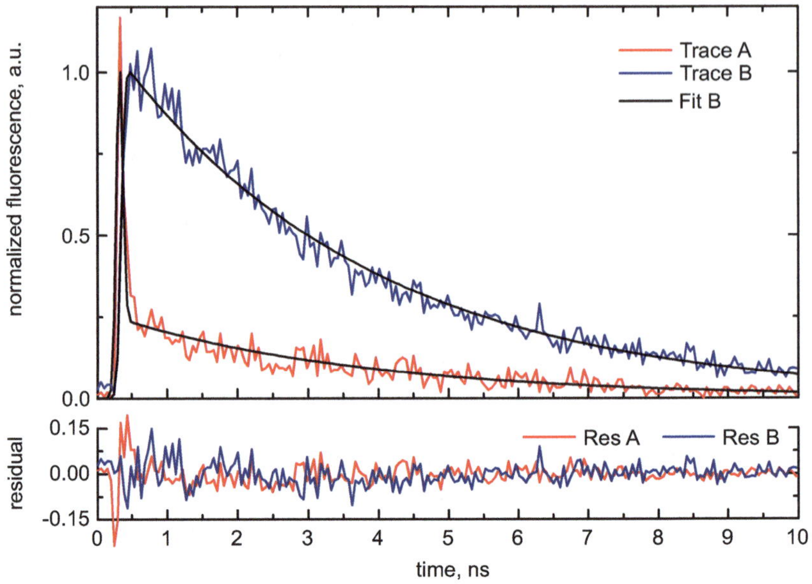

Fig. 7 Two fluorescence decay traces and fits from pixels with (Trace A, *red*) and without (Trace B, *blue*) autofluorescence. Weighted residuals are presented in the *bottom panel*

3. The results in Fig. 6 show that fluorescence of the protoplasts transfected with the mTurquoise1 displays a mono-exponential decay with a lifetime of 3.6 ns in plant cells, which is in agreement with studies performed in mammalian cells as well as in vitro experiments [16, 17]. Figure 7 shows fluorescence decay traces from two selected pixels with and without a pronounced autofluorescent component. The fitted trace and residuals obtained during global analysis shows good overlap with the raw data in both cases, which serves as justification of the relevance using the "CohSpec" parameter for compensating autofluorescence.

3.7 Analysis of mTurquoise1–mVenus Dataset

1. To analyze time-resolved fluorescence of mTurquoise1–mVenus fusion protein expressed in protoplasts, it is necessary to use a bi-exponential decay model (i.e., two rate constants) (*see* **Note 20**).
2. The average fluorescence lifetime image of the mTurquoise1–mVenus protein, shown in Fig. 8a, has an identical color scale as for mTurquoise1 alone. Comparing the average fluorescence lifetime of mTurquoise1 with the one of the mTurquoise1 fusion protein, it becomes clear that the presence of mVenus results in a reduction of the fluorescence lifetime of mTurquoise1 indicated by the change of color (from red to blue/cyan). This decrease can also be observed from time traces of selected pixels in Fig. 9.

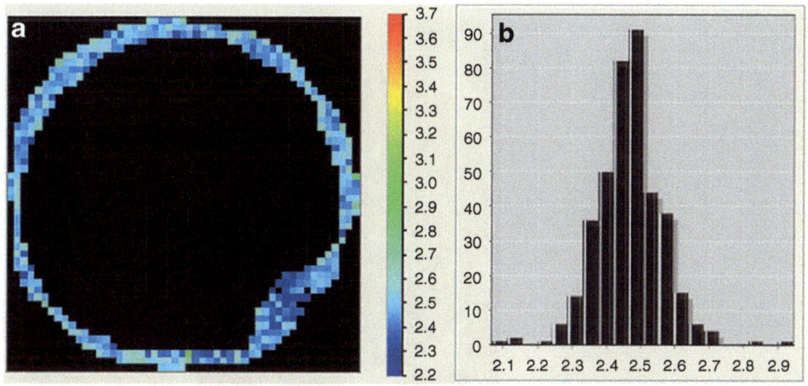

Fig. 8 Results of the analysis of the protoplasts transfected with mTurquoise1–mVenus using a bi-exponential decay model. (**a**) Average lifetime image. (**b**) Average lifetime histogram

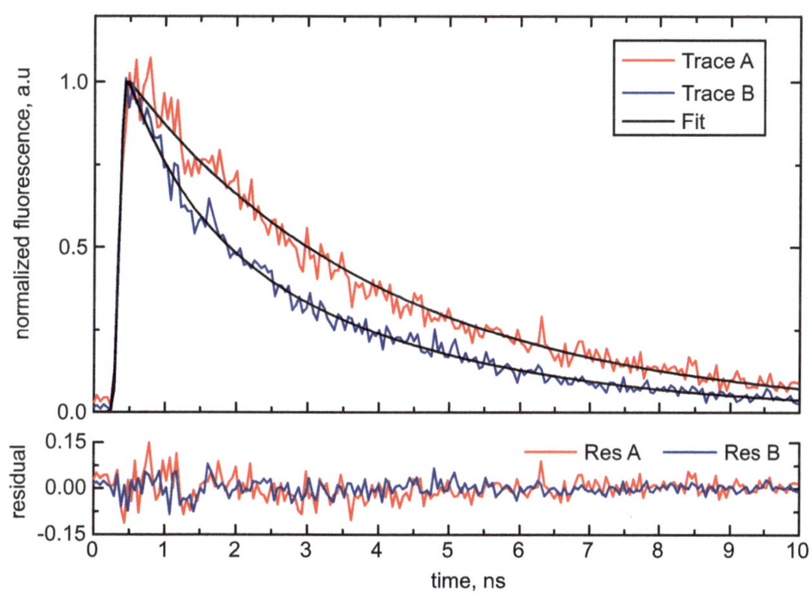

Fig. 9 Comparison of experimental and fitted fluorescence decay traces of mTurquoise1 alone (Trace A, *red*) and mTurquoise1–mVenus (Trace B, *blue*). Weighted residuals belonging to each fitted curve are presented in the *bottom panel*

3. Despite the clear decrease of the donor average fluorescence lifetime of the mTurquoise1–mVenus sample due to energy transfer, it remains difficult to quantify the FRET process. To improve analysis, one needs to analyze the estimated lifetime components and distributions of the corresponding

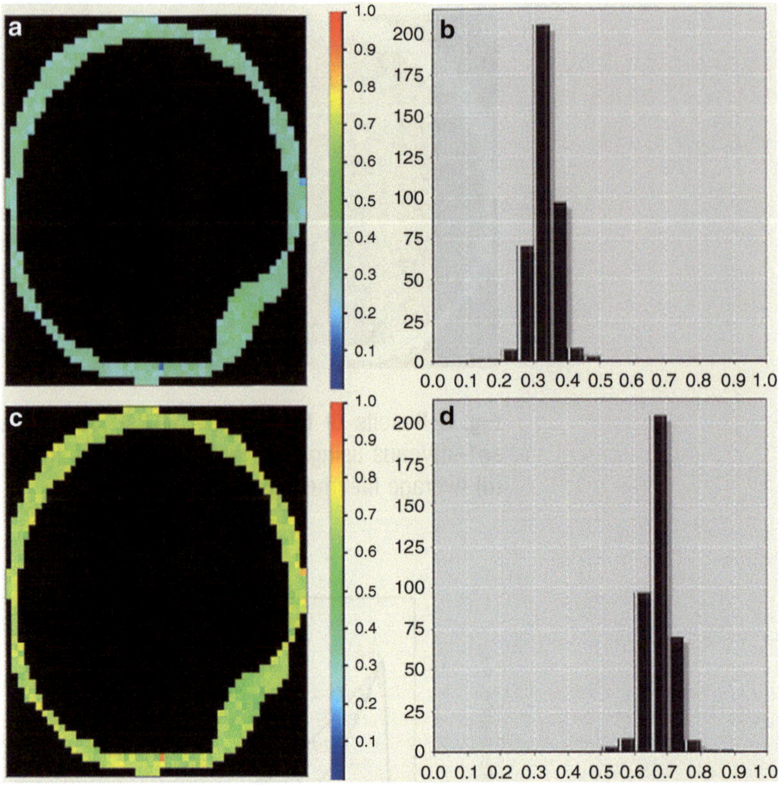

Fig. 10 Estimated distributions of the two estimated components. Here, (**a**) is the spatial distribution and (**b**) is the histogram of fractional distribution of component with fluorescence lifetime of 0.83 ns. And (**c**) is the spatial distribution and (**d**) is the histogram of fractional distribution of component with fluorescence lifetime of 3.3 ns

amplitudes. In the presented case, the estimated lifetimes are 0.83 ns with average amplitude around 35 % and 3.3 ns with average amplitude of 65 % (Fig. 10).

The long lifetime component (3.3 ns) is almost identical to the fluorescence lifetime of mTurquoise1 alone (3.6 ns) and can be attributed to "non-FRETting" mTurquoise1 molecules of the FRET sample (*see* **Note 21**). Because mTurquoise1 follows mono-exponential decay kinetics, the fast component of 0.83 ns of the FRET sample can be attributed to the donor molecules undergoing energy transfer from mTurquoise1 to mVenus. It is important to realize that, if one is interested in FRET efficiency quantification or distance calculation, the short component instead of the average fluorescence lifetime must be used. However, describing a whole population, one should take into account the relative contributions of the different components (*see* **Note 22**).

This type of analysis would be very difficult in case of using bi-exponential decaying donor molecules, since it will require more complicated models using three or even four exponential terms [11, 18]. The use of donor molecules that exhibit a mono-exponential fluorescence decay significantly simplifies quantitative analysis of FRET data.

4 Notes

1. The isolation of the protoplast from the leaves is carried out according to the Tape sandwich method described in ref. 19.
2. The B&H card only operates in combination with the Detector Controller Card (DCC) card.
3. The image size in both case studies was 15 × 15 μm. In order to have sufficient photon counts per pixel, the pixel size was set to 64 × 64 pixels resulting in a pixel size of ca 350 nm.
4. The fusion proteins could be observed both in the cytoplasm and the nucleus of the plant cell.
5. Installers for Linux, Mac OS X, and Windows operating systems are available and can be downloaded from http://glotaran.org/downloads.html. Documentation about the installation protocol can be found at: http://glotaran.org/wiki.html. A video demonstration can be found here: http://glotaran.org/demonstration.html.
6. To add a new dataset to the project, *right mouse click* on the "Datasets" folder or a subfolder and select "Open dataset file." In the resulting Open file dialog, select the respective file to be added to the project. Using the CTRL + *left mouse click* combination in the Open file dialog, more data files can be added simultaneously. After opening the data file(s), a data reference file is created and added to the project. The data reference file in the project folder only contains a reference or link to the actual data file. To store the actual data file inside the project folder, *right mouse click* the data reference file and select "Cache dataset to project folder." In this way the entire project can be copied onto a different computer for analysis. To organize the data reference files, more folders can be created in the Datasets folder.
7. The grey scale of the fluorescence intensity can be adjusted using the slider or by filling in the numbers in the amplitude settings box.
8. When a data file is opened in Glotaran, a data reference file is created in the Datasets folder within the project. The data reference file only contains a link to the actual data file to protect the user accidentally modifying the original data during

preprocessing of the data. Therefore, before starting the analysis on any kind of data file, first, a new dataset should be created from the data file.

9. Selection of pixels for analysis is performed for several reasons. First, analysis of pixels containing only fluorescence of donor molecules or molecules of interest significantly decreases computation time. Second, it allows excluding those pixels with pronounced, undesirable (auto)fluorescent components from the analysis. In this way the number of required parameters for the analysis can be minimized, and the accuracy by which these parameters can be estimated improved.
10. In case the number of photons is too low for quantitative analysis, a binning factor can be used. Binning is a procedure where the selected pixel is analyzed, but the neighboring pixels are included for calculation of the fluorescence lifetime. The binning factor can be calculated according to the following formula: binning factor = $(2n + 1)^2$, where n is number of pixels.
11. The singular value decomposition (SVD) is a model-free matrix factorization technique, which decomposes the data into a sum of orthonormal vector products scaled by singular values. Here, the left singular vectors (LSV) represent time dimension and the right singular vector (RSV) represents spatial dimension. The contribution to the data is the product of the nth left singular vector and right singular vector scaled by the nth singular value. The singular vectors are ordered based on their contribution to the data as represented by the magnitude of the corresponding singular values as shown in the screeplot [20, 21]. The technique can be used to explore the number of spectrally and spatially independent components in the data matrix, which is an important aspect of defining an initial model. However, it should be noted that an independent component in the SVD analysis does not necessarily represent a single lifetime component [22].
12. A functional analysis scheme should have three connected containers, one *model container* linked to one *dataset container*, linked to one *output container*. To create the *dataset container* and *output container*, drag and drop the corresponding icons from the palette (Menu: Window—Palette), which is available on the right of the analysis scheme editor (Fig. 4d), to the canvas. The *model container* can be created either from an existing model file, by dragging the model file from the *Models* folder in the project manager to the canvas, or from the palette, by dragging the Model icon from the palette to the canvas, thus creating a new model file which will be added to the Models folder in the project. Finally, the model container should be

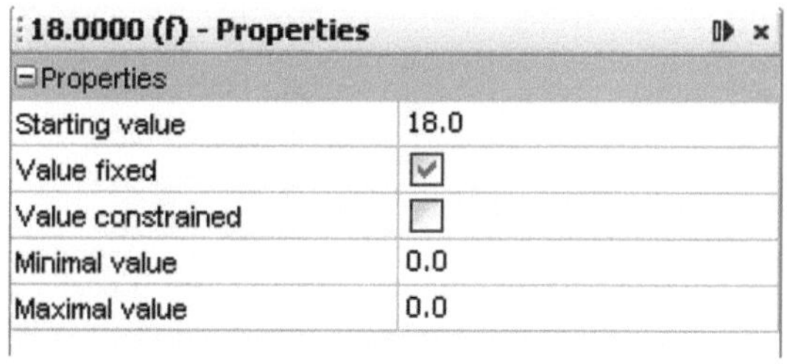

Fig. 11 Example property window of a parameter node

connected to the dataset container, and the dataset container should be connected to the output container. To connect two containers, press and hold the Ctrl (Windows, Linux) or Cmd+CTRL (Mac)key, click the top of a container, and drag the connector arrow that appears from one container to another.

13. To add datasets to the dataset container, drag and drop them from the project manager. By default, all added datasets would be placed in the same group (number in the square brackets next to the dataset name). For all datasets from the same group, the amplitudes will be linked. For simultaneous analysis of multiple FLIM images, all datasets have to be in the separate groups. It is possible to change the group assigned to the dataset in the properties of the dataset (right click menu or Properties window) (Fig. 5e).

14. To add different parameters to the model, drag the corresponding icon from the palette (Fig. 5d). Each parameter has its own properties that can be edited from the property window (Fig. 5e). The property window in Fig. 5 represents the property of the kinetic parameters. The number of components property defines the number of rates (or reciprocal lifetimes) that will be fitted. The positive rates property will constrain rates to be positive (for FLIM analysis, it is always true), and the nonnegative least-squares property will force amplitudes of the component to be positive (in most cases of FLIM analysis, this is also true; however, in case of analyzing rise times of the fluorescence [11], that property should be set to false). When the number of kinetic parameters is set, the node for each kinetic parameter will be created as a subnode for "KinPar" node. Figure 11 represents a property window corresponding to the parameter node. In this property, the starting value for the parameter that will be used for analysis should be set. The parameter can be fixed or constrained by setting minimal and maximal values.

15. The time channel spacing in the present study is 48 ps/channel. The excitation source, the photodetector, and associated electronics define the FWHM of the IRF. A femtosecond laser,

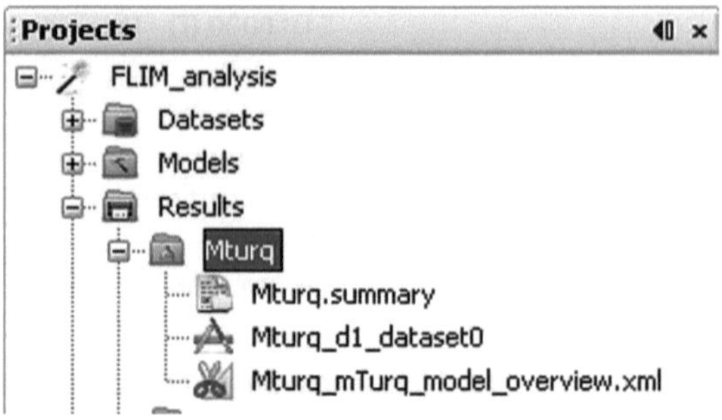

Fig. 12 Screenshot of the project manager window showing the generated result files

together with the used single-photon counting detection system, makes the FWHM of the IRF in the present experiments in the order of 120 ps. With the given signal-to-noise ratio (SNR), processes that are faster than 120 ps are not possible to estimate during analysis. Therefore, we can use the "CohSpec" option to compensate for an autofluorescence component. In experiments with higher time resolution [23], such modeling of the autofluorescence would be impossible, and one would need to use additional kinetic parameter(s) to account for the contribution of autofluorescence.

16. Various algorithms for numerical convolution of an exponential decay with a measured IRF can be applied. The iterative reference and scatter methods described in [24–26] perform well in practice and are implemented in Glotaran. The reference convolution method uses a measured fluorescence decay of a reference compound with a well-defined mono-exponential decay in a convolution procedure. This allows measuring the IRF of the setup exactly under the same conditions as the experiment, as well as estimate an IRF in multi-photon excitation systems. This is in sharp contrast when using a scattering solution to obtain the IRF, because one would need to change the detection wavelength.

17. Valid starting values are found if the analysis returns results at 0 iterations (i.e., the analysis does not crash and returns no valid results at all). However, this does not imply that the starting values are reasonable. The analysis is especially sensitive to the IRF parameters starting values.

18. An analysis progress indicator in the lower right corner of the window will indicate the progress of the data analysis. If the analysis is finished, three files will be created in the results folder (*see* Fig. 12). The file with extension "summary" provides a textual summary of the data analysis. Given are the final residual standard error and values of the fitted parameters with their standard errors, the number of iterations for the analysis,

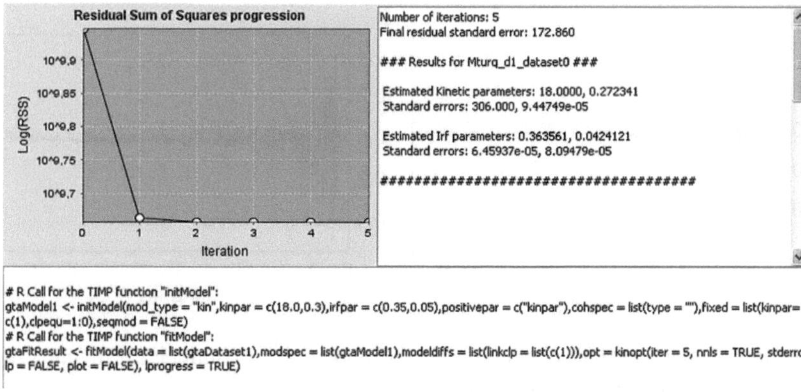

Fig. 13 Screenshot of the "Overview" file windows showing the residual sum of squares progression plot, the estimated parameters of the analysis with corresponding errors, and the calls made to the R package TIMP [15, 27] to perform the analysis

and the function calls that were made to the R package TIMP [15, 27] in order to perform that analysis. If the analysis failed, because of a wrong model or the initial guess for the parameters is too far off, then this file will be the only generated file, and it will contain a diagnostic error message rather than the parameters. The file with the name ending with "overview" shows a graphical view of the progress of the residual sum of squared errors. This graph will appear, if the fit was converged. Here, there was no progress in the last iterations from 2 to 5 (Fig. 13). This is an important criterion, because before convergence it is not possible to judge the quality of the model. The last generated file contains the results of the fit for a dataset. In case of simultaneous analysis of multiple datasets, there will be a file for each analyzed dataset.

19. The average fluorescence lifetime is calculated with the formula $\langle\tau\rangle = \sum_i \alpha_i\tau_i$ where τ_i is the lifetime of *i*th component and α_i is its amplitude such that $\sum_i \alpha_i = 1$.

20. The SVD analysis of the mTurquoise1–mVenus dataset indicates the presence of at least two independent components; one of them can be attributed to autofluorescence and the other to fluorescence from mTurquoise1. However, using a mono-exponential model including a "CohSpec" parameter (as it was used in donor only case) does not describe the data correctly. Figure 14 shows the left singular vector of the residuals after single- and bi-exponential analysis. The single-exponential analysis has clear systematic bias in the shape, which is an indication for a misfit of the data.

21. In principle, the "non-FRETting" donor molecules should have a similar fluorescence lifetime as the donor only sample. However, in the fusion protein the local environment of the

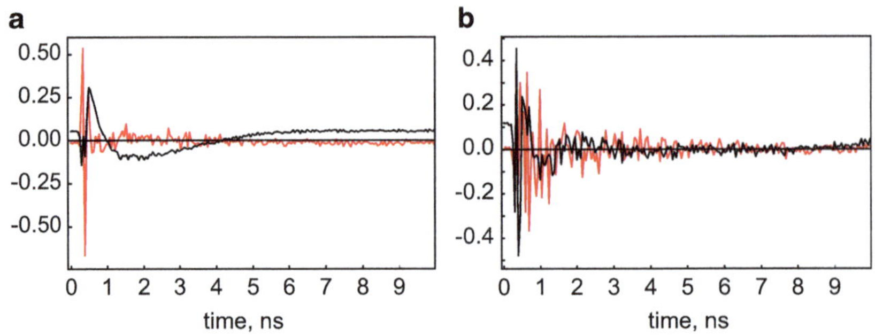

Fig. 14 First two left singular vectors obtained after the analysis of mTurquoise1–mVenus dataset using a mono-exponential (**a**) and a bi-exponential model (**b**)

Table 1
Summarized estimates of FRET efficiency and donor–acceptor distance using the average fluorescence lifetime or the shorter lifetime of the estimated FRET component

	Using lifetime of the estimated component involved in FRET	Using the average fluorescence lifetime
Estimated efficiency (%)	75	25
Estimated distance (nm)	4.2	6.2

The fluorescence lifetime of donor only is 3.6 ns and used for calculation of FRET efficiency (*see* Eq. 4). Using the lifetime of 3.3 ns estimated as "non-FRETting" population of the donor in the FRET sample does not change the values significantly. A critical distance $R_0 = 5.4$ nm is used for distance calculations

donor might be slightly different because of the presence of the (acceptor) protein. Another reason for observing differences of donor lifetimes could be crosstalk of acceptor fluorescence in the donor detection channel. The fluorescence lifetime of mVenus is in the order of 3 ns, and therefore with the given time resolution and SNR, this component would be indistinguishable from the donor fluorescence lifetime. Thus, a mixture of acceptor fluorescence in the donor detection channel can potentially give rise to a different estimate of the donor fluorescence lifetime.

22. Using the average fluorescence lifetime to calculate FRET efficiencies and distances will result in an underestimation of the expected values. In the present study, the calculation of the FRET efficiency was compared using the average fluorescence lifetime and the estimated short lifetime (*see* Table 1). This underestimation of the parameters will depend on the amount of

"non-FRETting" donor molecules. There are several reasons for this population to be present in the data:

(a) Direct excitation of the acceptor molecules—if acceptor molecules are in the excited state, energy transfer from excited donor to acceptor molecules cannot take place.

(b) Photobleaching of the acceptor—irreversibly destroyed acceptor molecules cannot accept energy from excited donor molecules. Photobleaching can occur by direct excitation of the acceptor or by the energy transfer process.

(c) Photoconversion of acceptor molecules—many fluorescent proteins (FPs) that are used as acceptors can be photoconverted to a species of which the fluorescence properties are different from the original species [28].

(d) Protonation of fluorescent proteins may also change spectroscopic properties of the particular FP as has been observed for YFP (mVenus) [29].

(e) Differences in maturation time of the fluorescent proteins within the FRET pair can give a rise to a large population of "non-FRETting" donor molecules [11].

When studying protein–protein interactions with the FRET method in which FPs are fused to the proteins of interest, one should be careful with assignment of a noninteracting population. The presence of donor molecules not participating in FRET does not necessarily imply that the proteins of interest are not interacting, but interactions cannot be detected by FRET.

References

1. Borst JW, Visser AJWG (2010) Fluorescence lifetime imaging microscopy in life sciences. Meas Sci Technol 21:102002
2. Beechem JM (1992) Global analysis of biochemical and biophysical data. Methods Enzymol 210:37–54
3. Bednarkiewicz A, Whelan MP (2008) Global analysis of microscopic fluorescence lifetime images using spectral segmentation and a digital micromirror spatial illuminator. J Biomed Opt 13:041316
4. Barber P, Ameer-Beg S, Gilbey J et al (2009) Multiphoton time-domain fluorescence lifetime imaging microscopy: practical application to protein-protein interactions using global analysis. J R Soc Interface 6:S93–S105
5. Wlodarczyk J, Woehler A, Kobe F et al (2008) Analysis of FRET signals in the presence of free donors and acceptors. Biophys J 94: 986–1000
6. Laptenok SP, van Stokkum IHM, Borst JW et al (2012) Disentangling picosecond events that complicate the quantitative use of the calcium sensor YC3.60. J Phys Chem B 116: 3013–3020
7. Snellenburg JJ, Laptenok SP, Seger R et al (2012) Glotaran: a Java-based graphical user interface for the R-package TIMP. J Stat Softw 49(3):1–23
8. Bücherl C, Aker J, de Vries S et al (2010) Probing protein-protein Interactions with FRET-FLIM. Methods Mol Biol 655:389–399
9. Russinova E, Borst JW, Kwaaitaal M et al (2004) Heterodimerization and endocytosis of Arabidopsis brassinosteroid receptors BRI1 and AtSERK3 (BAK1). Plant Cell 16: 3216–3229
10. Verveer PJ, Squire A, Bastiaens PIH (2000) Global analysis of fluorescence lifetime imaging microscopy data. Biophys J 78:2127–2137
11. Laptenok SP, Borst JW, Mullen KM et al (2010) Global analysis of Förster resonance energy transfer in live cells measured by fluorescence lifetime imaging microscopy

exploiting the rise time of acceptor fluorescence. Phys Chem Chem Phys 12:7593–7602
12. Valeur B (2002) Molecular fluorescence: principles and applications. Wiley-VCH, New York
13. Lakowicz JR (2006) Principles of fluorescence spectroscopy, 3rd edn. Springer, New York
14. Förster T (1949) Experimentelle und theoretische Untersuchung des zwischenmolekularen Ubergangs von Elektronenanregungsenergie. Z Naturforsch 4a:321–327
15. Laptenok SP, Mullen KM, Borst JW et al (2007) Fluorescence lifetime imaging microscopy (FLIM) data analysis with TIMP. J Stat Softw 18(8):1–20
16. Klarenbeek JB, Goedhart J, Hink MA et al (2011) A mTurquoise-based cAMP sensor for both FLIM and ratiometric read-out has improved dynamic range. PLoS One 6: e19170
17. Goedhart J, van Weeren L, Hink MA et al (2010) Bright cyan fluorescent protein variants identified by fluorescence lifetime screening. Nat Methods 7:137–139
18. Borst JW, Laptenok SP, Westphal AH et al (2008) Structural changes of yellow Cameleon domains observed by quantitative FRET analysis and polarized fluorescence correlation spectroscopy. Biophys J 95:5399–5411
19. Wu F-H, Shen S-C, Lee L-Y et al (2009) Tape-Arabidopsis sandwich—a simpler Arabidopsis protoplast isolation method. Plant Methods 5:16
20. Press WH, Teukolsky SA, Vetterling WT et al (2007) Numerical recipes: the art of scientific computing, 3rd edn. Cambridge University Press, Cambridge, MA
21. Golub GH, Van Loan CF (1996) Matrix computations, 3rd edn. The Johns Hopkins University Press, Baltimore, MD
22. Laptenok SP, Visser NV, Engel R et al (2011) A general approach for detecting folding intermediates from steady-state and time-resolved fluorescence of single-tryptophan-containing proteins. Biochemistry 50:3441–3450
23. Krumova SB, Laptenok SP, Borst JW et al (2010) Monitoring photosynthesis in individual cells of Synechocystis sp. PCC 6803 on a picosecond timescale. Biophys J 99:2006–2015
24. Grinvald A, Steinberg IZ (1974) Analysis of fluorescence decay kinetics by method of least-squares. Anal Biochem 59:583–598
25. Vos K, van Hoek A, Visser AJWG (1987) Application of a reference convolution method to tryptophan fluorescence in proteins. A refined description of rotational dynamics. Eur J Biochem 165:55–63
26. Boens N, Ameloot M, Yamazaki I et al (1988) On the use and the performance of the delta-function convolution method for the estimation of fluorescence decay parameters. Chem Phys 121:73–86
27. Mullen KM, van Stokkum IHM (2007) TIMP: an R package for modeling multi-way spectroscopic measurements. J Stat Softw 18(3):1–46
28. Kremers G-J, Hazelwood KL, Murphy CS et al (2009) Photoconversion in orange and red fluorescent proteins. Nat Methods 6:355–358
29. McAnaney TB, Zeng W, Doe CFE et al (2005) Protonation, photobleaching, and photoactivation of yellow fluorescent protein (YFP 10C): a unifying mechanism. Biochemistry 44: 5510–5524

Chapter 22

Time-Resolved Fluorescence Anisotropy Imaging

Klaus Suhling, James Levitt, and Pei-Hua Chung

Abstract

Fluorescence can be characterized by its intensity, position, wavelength, lifetime, and polarization. The more of these features are acquired in a single measurement, the more can be learned about the sample, i.e., the microenvironment of the fluorescence probe. Polarization-resolved fluorescence lifetime imaging—time-resolved fluorescence anisotropy imaging, TR-FAIM—allows mapping of viscosity or binding or of homo-FRET which can indicate dimerization or generally oligomerization.

Key words FAIM, TR-FAIM, Polarization-resolved FLIM, TCSPC, Fluorescence spectroscopy, Fluorescence microscopy, Homo-FRET

1 Introduction

Fluorescence imaging techniques are powerful and versatile tools in the life sciences, because they use nonionizing radiation, are minimally invasive, and can be applied to living cells and tissues. Measurements can be made in situ, thus allowing access to biological function within a true physiological context [1–4].

In order to maximize the information available from a limited fluorescence photon budget, it is advantageous to record multiple fluorescence parameters—such as lifetime, spectrum, and polarization—in a single imaging experiment [5]. Fluorescence is polarized due to the existence of a transition dipole moment of the fluorophore and thus the electric dipole characteristics of the emission. Indeed, polarization-resolved fluorescence measurements have been performed since the 1920s [6]. When using fluorescence as a probe, polarization-resolved measurements can yield information on the properties of a sample that cannot be extracted by intensity and lifetime methods alone [7, 8].

In a polarization-resolved fluorescence microscopy experiment, a fluorescently labeled sample is excited using linearly polarized light, and the (time-resolved) fluorescence intensity is measured at

Yves Engelborghs and Antonie J.W.G. Visser (eds.), *Fluorescence Spectroscopy and Microscopy: Methods and Protocols*, Methods in Molecular Biology, vol. 1076, DOI 10.1007/978-1-62703-649-8_22,

polarizations parallel and perpendicular to that of the exciting light. The difference between the parallel and perpendicular fluorescence signal is due to depolarization of the fluorescence. The time-dependent fluorescence anisotropy r is then defined as

$$r(t) = \frac{I_{\parallel}(t) - GI_{\perp}(t)}{I_{\parallel}(t) + zGI_{\perp}(t)} \tag{1}$$

where $I_{\parallel}(t)$ and $I_{\perp}(t)$ are the fluorescence intensity decays parallel and perpendicular to the polarization of the exciting light. The value of z depends on the NA of the microscope objective, where $1 \leq z \leq 2$ ($z \approx 1$ for a high NA objective, $z = 2$ for a collimated beam) [9–16]. Although a rigorous treatment of the effect of high NA objectives to "see around" the fluorophore and therefore collect all three emission components I_x, I_y, I_z leads to a slightly more complex description than Eq. 1 [10, 15, 17], this empirical approach is attractive due to its simplicity and similarity with that of a collimated beam and has worked well in our laboratory and others [18]. The empirical constant z is a function of the NA of the microscope objective and is chosen such that (1) a time-resolved fluorescence anisotropy decay starts at the correct initial anisotropy r_0 (as determined by spectroscopic measurements using collimated excitation light) and (2) the total fluorescence intensity decay $I_{\parallel}(t) + zI_{\perp}(t)$ is the same as a decay collected using collimated beams with magic angle detection such that polarization contributions are removed [18]. The denominator is proportional to the total fluorescence emission, and G accounts for differences in the transmission and detection efficiencies of the imaging system at parallel and perpendicular polarization. If necessary, an appropriate background intensity has to be subtracted from each polarization channel [19]. Due to the nature of the photoselection for absorption and emission transition dipoles, multiphoton excitation provides a greater dynamic range for anisotropy measurements than single photon excitation [20].

The depolarization of the fluorescence, i.e., the decay of the anisotropy r as a function of time, can either be due to the rotational diffusion of the fluorophore in its excited state before emission of a fluorescence photon or due to energy migration or homo-FRET as described in section 1.2 below.

1.1 Rotational Diffusion

The rotational diffusion of the fluorophore in its excited state before emission of a fluorescence photon depends on its volume and the viscosity and temperature of its environment. For a spherical molecule, $r(t)$ decays as a single exponential and is related to the rotational correlation time θ according to

$$r(t) = (r_0 - r_{\infty})e^{-t/\theta} + r_{\infty} \tag{2}$$

where r_0 is the initial anisotropy (maximum value is 0.4 for single photon excitation) and r_{∞} accounts for a restricted rotational mobility. For a freely rotating spherical molecule in an isotropic medium, $r_\infty = 0$ and θ is directly proportional to the viscosity η of the solvent and the hydrodynamic volume V of the rotating molecule:

$$\theta = \frac{\eta V}{kT} \tag{3}$$

where k is the Boltzmann constant and T the absolute temperature. Therefore, if the volume of the fluorophore is known, the rotational correlation time can report on the viscosity of the fluorophore's immediate environment. Alternatively, as the rotational diffusion can be slowed down by binding or sped up by cleavage, θ can yield information about the size of the tumbling unit. In addition, evidence of a hindered rotation of the fluorophore due to geometrical restrictions, e.g., in the cell membrane, can be gleaned from r_∞.

Note that the steady-state anisotropy is calculated from the fluorescence intensities, i.e., integrated fluorescence decays, from Eq. 1 and obeys the Perrin equation

$$r = \frac{r_0 - r_\infty}{1 + \frac{\tau}{\theta}} + r_\infty \tag{4}$$

where τ is the fluorescence lifetime [21]. While the steady-state anisotropy is relatively easy to measure, and in particular to image, it may not be unambiguous to interpret in the absence of time-resolved measurements.

Steady-state fluorescence anisotropy imaging has, for example, been used to study viscosity, enzyme activity or binding in cells [22–29], and DNA digestion [30] or to identify FRET between fluorescent proteins [31–35]. However, it is difficult to obtain information about a hindered rotational mobility as indicated by a nonzero r_∞, and time-resolved measurements are needed to determine this parameter.

Time-resolved fluorescence anisotropy has been used on cells for single-point measurements [23, 36–38] and for mapping solvent interactions in microfluidic devices [39], as well as the viscosity in the cell cytoplasm [19, 21, 40] and membrane [41].

In the absence of rotational diffusion, polarization-resolved measurements can be used to elucidate the orientation of fluorophores, e.g., in the membrane [42, 43], muscle fibers [44], or in DNA [45]. In these cases, neither the depolarization due to Brownian rotational motion nor homo-FRET is measured but rather the angle between the electric vector of the light exciting the sample and the transition dipole moment of the static fluorophore, thus yielding its orientation.

1.2 Energy Migration or Homo-FRET

FRET can occur if the absorption spectrum of the acceptor overlaps with the emission spectrum of the donor, the fluorophores are in close proximity, and their orientation is favorable (i.e., orientation factor $\kappa^2 \neq 0$ [46–48]), as extensively discussed in reference [49]. These conditions can apply to fluorophores with a small Stokes shift and hence lead to the donor and acceptor being the same type of fluorophore. Thus resonance energy transfer between the same type of fluorophore can take place, known as energy migration or homo-FRET. This phenomenon depolarizes the fluorescence emission [6] and has been exploited in single-point measurements and imaging, e.g., to monitor the proximity of isoforms of the glycosylphosphatidylinositol (GPI)-anchored folate receptor bound to a fluorescent analogue of folic acid to study lipid rafts [50, 51], to monitor actin polymerization [52], or to image the aggregation of protein α-synuclein, relevant for Parkinson's disease [53].

Time-resolved fluorescence anisotropy measurements to identify homo-FRET have been carried out to study conformational changes in G-protein-coupled receptors [54], dimerization [55], and quantification of protein cluster sizes [56–59]. It has also been employed to show that a neuronal isoform of Venus-tagged calcium-calmodulin-dependent protein kinase-II (CaMKIIa) holoenzyme forms catalytic domain pairs, and to show that glutamate receptor activation in neurons triggers an increase in anisotropy consistent with a structural transition from a paired to unpaired conformation [60, 61] and to study the homodimerization of amyloid precursor protein at the plasma membrane, relevant for Alzheimer's disease [62]. In these cases, it is advantageous to have negligible rotational diffusion (τ/θ is small), so that homo-FRET can be identified, *see* Fig. 1.

For homo-FRET involving two fluorophores, and in the absence of any rotational diffusion, $r(t)$ decays as a single exponential and is related to the FRET efficiency E_{TR} according to [63]

$$r(t) = r_0 e^{-2E_{TR}t} \tag{5}$$

While hetero-FRET between different donors and acceptors to identify protein interaction can routinely be imaged with FLIM, mapping energy migration or homo-FRET to identify protein dimerization requires polarization-resolved FLIM, i.e., TR-FAIM. The only way to detect homo-FRET is by polarization measurements, because homo-FRET does not affect spectra or fluorescence lifetime—as long as the fluorescence lifetime of both fluorophores is the same—and thus cannot be identified by intensity and lifetime methods. If the fluorescence lifetimes of the two fluorophores are different, however, then FRET can be identified by fluorescence lifetime measurements [64, 65]. This has, for example, been done in the case of tryptophan to tryptophan homo-FRET in barnase, where the tryptophans are located in different environments yielding different fluorescent lifetimes [66].

Homo-FRET between fluorescent proteins can be extremely fast, 2 ps has been quoted for yellow fluorescent protein [67], and is

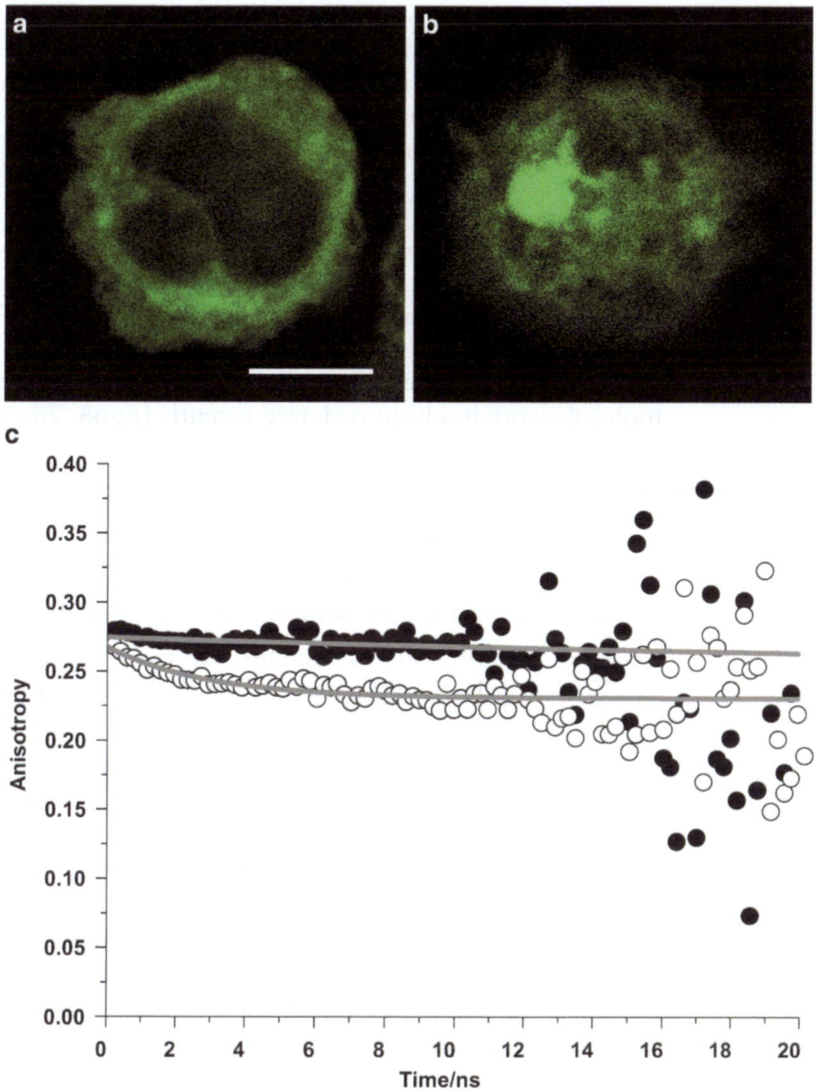

Fig. 1 Green fluorescent protein-labeled proteins in fixed leukocytes. (**a**) Monomeric proteins. (**b**) Clustered proteins. (**c**) Time-resolved fluorescence anisotropy decays. In the upper curve (filled black circles, corresponding to scan over confocal fluorescence image (**a**)), there is almost no depolarization of the fluorescence over time due to the restricted motion of the monomeric GFP in the fixed cell. In the lower curve (open circles, corresponding to scan over confocal fluorescence image (**b**)), a fast decay component in the GFP fluorescence anisotropy decay appears, which is the characteristic signature of GFP undergoing homo-FRET. Scale bar 5 μm

an indicator of protein dimerization or oligomerization, and TR-FAIM is the only technique which can image it.[1]

Anisotropy imaging can be performed as steady-state or time-resolved measurements in the time domain or frequency domain

[1]Only in the specific case of the cerulean fluorescent protein (CFP) the fluorescence lifetime has been reported to change due to homo-FRET [92].

[68] using scanning or wide-field methods [5, 69, 70] and has been combined with spectral imaging [71]. Photon counting approaches are particularly attractive because of their excellent signal-to-noise ratio and single photon sensitivity [72–74].

The combination of total internal reflection fluorescence (TIRF) microscopy with time-resolved fluorescence anisotropy measurements allows excitation with *s*- and *p*-polarized evanescent waves and provides spatial information on the fluorescence depolarization processes near an interface. This has, for example, facilitated the observation of membrane dyes in and out of plane [75, 76] and has indeed been combined with imaging [77].

The use of fluorescence anisotropy in imaging and for single molecule work has been reviewed recently [5, 63, 78–82].

2 Materials

2.1 Hardware

1. Inverted Leica TCS SP2 confocal scanning microscope with internal analogue photomultiplier detectors.
2. Coherent Mira 900 Ti:Sapphire femtosecond laser with a Verdi V6 pump laser or Hamamatsu PLP-10,470 ps pulsed diode laser excitation sources.
3. Becker & Hickl SPC 830 board for time-correlated single photon counting (TCSPC) in 3 GHz, Pentium IV, 1 GB RAM computer with Windows. (The minimum requirements recommended by the manufacturer are PC Pentium, multi-core CPU.)
4. Cooled Becker & Hickl PMC100-01 detector head based on Hamamatsu H5773P-01 photomultipliers, mounted on microscope's X1 port, or hybrid detectors (external detectors).
5. DCC 100 detector control module.

2.2 Software

TRI-2 [83] or SPCImage 2.8 by Becker & Hickl.

3 Methods

3.1 Align the Optical System

A schematic representation of the time-resolved fluorescence anisotropy experiment is shown in Fig. 2.

Ensure that the incident laser (excitation beam) is linearly polarized and that this linear polarization state is retained to the highest degree possible at the sample plane. Accurate alignment through the microscope scan head along with a half-wave plate and linear polarizer combination in the incident beam path will ensure that this is the case and allow for rotation of the polarization of the beam.

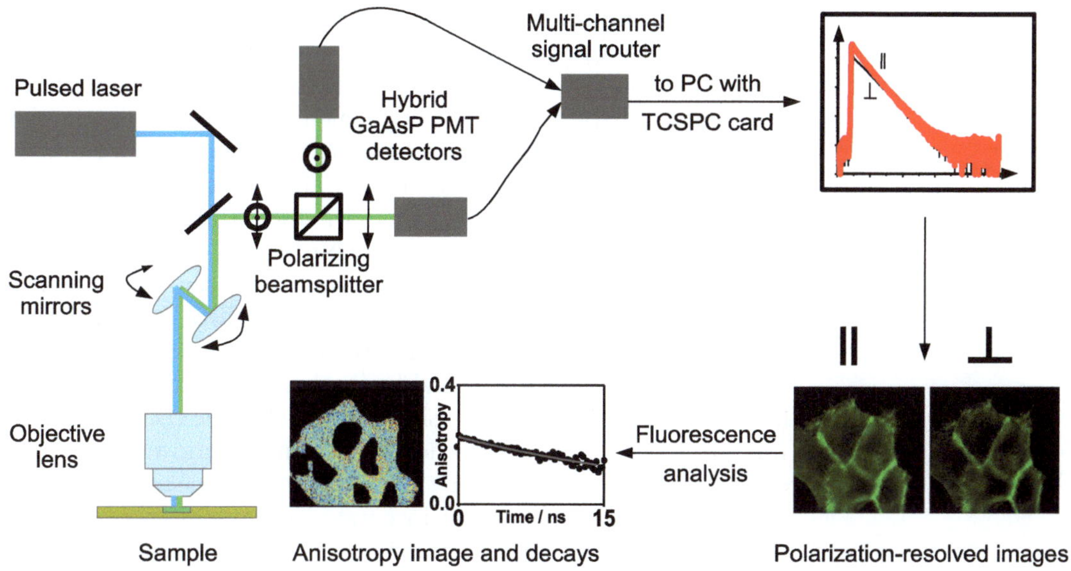

Fig. 2 Experimental arrangement for confocal microscope-based time-resolved anisotropy measurements. The sample is excited by linearly polarized pulsed laser light. The fluorescence polarized parallel and perpendicular to the incident light is separated using a polarizing beam splitter and detected using multichannel time-correlated single photon counting (TCSPC) [18]. The time- and polarization-resolved images are processed to give anisotropy images

Ensure that the polarization components of the detected light, parallel and perpendicular, are split adequately by the polarizing beam splitter. Using unpolarized light (e.g., from a lamp) to test the detection, the intensity of each polarization component after the beam splitter should be equal. Circularly polarized light, created by placing a quarter waveplate in the beam, can also be used. Ensure correct alignment of cleanup polarizers in each arm of the detection.

Select a suitable microscope objective for the sample of interest you wish to study. Phase objectives should be avoided due to the phase plate, and note that high NA objectives partially depolarize the light (an effect that can be corrected for, as described in the context of Eq. 1).

Select the correct emission filter/wavelength for the sample of interest.

Check polarized fluorescence decays of a test sample. Select time window size, adjust delay to place peak at the beginning of the time window, and select the number of time channels, typically 4,096 for recording single decays and 256 for imaging.

3.2 Measure the G-Factor

Place a suitable test sample of a small fluorophore in solution on the microscope stage. An example of a suitable test sample with peak emission wavelength ~530 nm is rhodamine 123 in homogeneous solution (methanol) with micromolar concentration. The sample used to measure the G-factor should be freely rotating in a single, uniform environment, ideally with monoexponential fluorescence decay kinetics. This will avoid artifacts arising from hindered rotations and associated anisotropy which can be found in complex samples [84]. This sample should also have similar spectral emission characteristics to the sample of interest. This is because the G-factor may vary as a function of wavelength due to differences in the spectral response of the detectors (*see* **Note 1**).

Select the objective you wish to use for the sample of interest and select/note the confocal pinhole size (*see* **Note 2**). Record the fluorescence decays with polarization parallel and perpendicular to the excitation beam with a suitable number of time channels in the decay (*see* **Note 3**). Where possible, ensure that the complete decay (into the noise floor) is recorded in the time window. This allows for the steady-state anisotropy to be correctly measured from the time-resolved data using the integrated decays in each detector (*see* **Note 4**). The photon count rate should be kept at around 1 % (or below) of the laser repetition rate to avoid photon pileup, both conventional pileup [85] and cross-detector pileup [86]. Ideally, these decays should be recorded simultaneously (using two detectors) to ensure that there are no artifacts due to sample or laser drift [87]. If the parallel and perpendicular decays are recorded sequentially using a single detector, then repeat measurements should be performed to ensure reproducibility.

Save the decays, and import into Origin (or similar spreadsheet/analysis software).

Subtract the background from each of the decays using the intensity values from the end of each decay trace, after the fluorescence has decayed completely, where the intensity value in each time channel should be only noise (*see* **Note 5**).

Perform any timing corrections necessary to ensure correct temporal overlap of the decays measured in each detector (*see* **Note 6**).

Divide the parallel signal by the perpendicular signal in each time channel. The limiting value of the resulting decay at long times is the G-factor—this is the tail-matching method. For an optimized system, this value will be close to 1.

Representative fluorescence decays of 3 μM rhodamine 123 in a 25/75 (V/V) methanol/glycerol mixture, yielding a G-factor of 1.15, are shown in Fig. 3a along with b the resulting fluorescence anisotropy decay curve.

3.3 Preparation and Measurement of Cell Samples

3.3.1 Living Cell Samples

Cells labeled with fluorescent proteins, quantum dots/nanoparticles, or stained with fluorescent dyes should be transferred to a suitable multi-well plate/sample chamber with a coverslide underside for imaging. This may be a micro-incubation system with a resistively heated multi-well plate with CO_2 input (and medium exchange, if necessary) or dish or multi-well plate placed inside

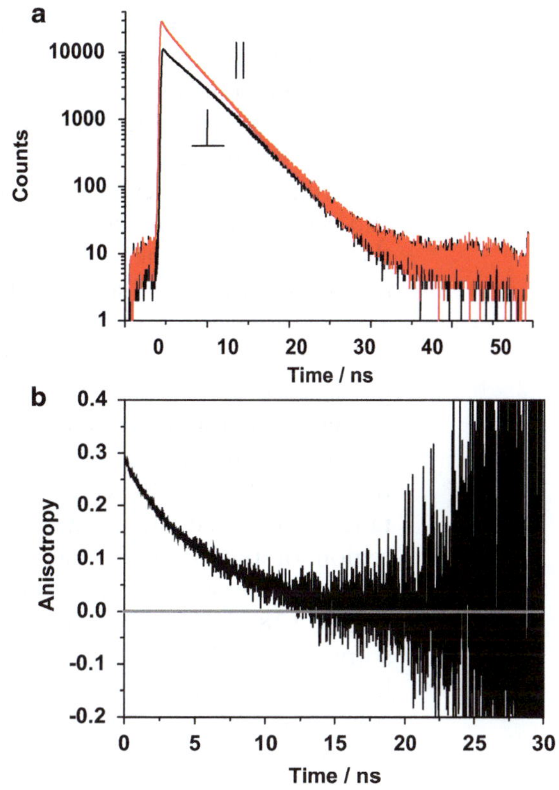

Fig. 3 Time and polarization-resolved fluorescence data from 3 μM rhodamine 123 in a 25/75 (V/V) methanol/glycerol mixture recorded using an inverted confocal laser scanning microscope with a 63 NA 1.2 objective and TCSPC detection. (**a**) Fluorescence decays recorded with polarization components parallel and perpendicular to that of the excitation light and (**b**) the time-resolved anisotropy decay created using Eq. 1 and the decays in (**a**)

a specialized temperature-controlled housing around the microscope. The samples should be kept at 37 °C and 5 % CO_2. The imaging medium should be free of fluorescent indicators (i.e., OptiMEM or equivalent) to eliminate measurement artifacts arising from the medium itself.

3.3.2 Fixed Cell Samples

Cells which have been fixed on microscope coverslides can be imaged at room temperature. Note that particularly thick samples such as tissue scatter and thus depolarize light.

3.3.3 Imaging the Cell Samples

Place samples on microscope stage and optimize CO_2 levels/temperature as necessary.

Select an appropriate microscope objective for the measurements, making sure that the G-factor for the system is known for measurements using this objective. We typically use a 63× NA 1.2

water immersion objective (Leica) for living cell samples and a 63× NA 1.4 oil immersion objective (Leica) for fixed cell samples.

Obtain transmitted light and fluorescence images of the fluorescent cells, verifying that the fluorescence originates from the regions/organelles expected. Measure a fluorescence emission spectrum, from a series of wavelength-resolved images in order and verify that it corresponds to that of the dye or protein used in the cell labeling. These measurements, along with those from an unstained sample, will serve to check for and assess possible artifacts arising from autofluorescence. This procedure is not specific to polarization-resolved fluorescence experiments but is generally good practice.

Focus on the sample and select the region of interest (ROI) for imaging. If the ROI is found using software/hardware other than the B&H SPCM/SPC card (e.g., Leica software), then it is important to note that the image settings in SPCM will determine the region of the original image that will be imaged in the time-resolved measurements. Specifically, we typically image two 256 × 128 pixel images (one from each detector) using the B&H software with 256 time channels—a 256 × 256 pixel image would limit the maximum number of time channels to <256 available using the SPC-830 TCSPC card (*see* **Note 7**). (The measured 256 × 128 pixel image corresponds to the top half of the image recorded using the Leica confocal software).

Switch to time-resolved detection by removing the internal mirror which directs the fluorescence to the internal Leica photomultiplier detectors, such that the fluorescence is incident on the external detector(s). Image the sample for as long as necessary—typically 5–10 min is sufficient for photon count rates of ~10^5 s^{-1} and 256 × 128 pixel images with 256 time channels.

Save the resulting data in the SPCM software.

Open the *.sdt files using TRI2 or OptiSpec (Becker & Hickl) software. In TRI2, the user is prompted to select each channel (corresponding to each polarization component of the emission) individually [83].

Multiply the perpendicular image by the G-factor.

Threshold the G-factor-corrected perpendicular image. We typically set threshold intensity levels to ~75 counts per pixel (*see* **Note 8**).

Fit the entire image or ROI using Eq. 2 or 5. Simultaneous fitting of the two orthogonal fluorescence decay curves

$$I_{\parallel}(t) = \frac{1}{3} I_0 \exp\left(-\frac{t}{\tau}\right) \cdot \left[1 + z r_0 \exp\left(-\frac{t}{\theta}\right)\right] \quad (6a)$$

and

$$I_{\perp}(t) = \frac{1}{3} I_0 \exp\left(-\frac{t}{\tau}\right) \cdot \left[1 - r_0 \exp\left(-\frac{t}{\theta}\right)\right] \quad (6b)$$

is better than creating the anisotropy from Eq. 1 directly. The latter method requires the decays to be measured under exactly the same conditions, so that they can be manipulated according in Eq. 1 [88].

4 Notes

1. *Spectral dependence of the G-factor*
   Depending on the type of detection used, the G-factor can vary markedly as a function of the emission wavelength. This is usually the case if there is a grating in the detection arm. Calibration of the G-factor as a function of wavelength is particularly useful and can help avoid multiple laborious measurements each time a new fluorophore is employed.
2. *Effect of microscope objective and pinhole size on anisotropy*
   The absolute measured anisotropy values in imaging experiments are critically dependent on the NA of the microscope objective used. Increasing the NA increases the out-of-plane (axial) polarization components in the excitation and detection light paths. This has the effect of depolarizing the optical system and leads to a decrease in the measured anisotropy values. This effect is well known and documented [9, 12–14, 40], and there have also been attempts to correct for this effect [10, 11, 15]. Increasing the size of the confocal detection pinhole increases the contribution from emission above and below the focal plane of the sample, reducing the confocality of the image. In a high NA imaging system, this also has the effect of altering the contribution from out-of-plane polarization components which arise from the large cone angle of the microscope objective.
3. *Selecting the appropriate number of time channels*
   As with fluorescence lifetime imaging (FLIM) experiments, careful consideration of the number of time channels and the time window used for detection is imperative for high quality data collection and analysis. The parameters which characterize the resultant fluorescence anisotropy decay curve in the time-resolved experiments are the initial anisotropy, r_0, and r_∞, and the rotational correlation time, θ. A small number (<256) of time channels in a relatively large time window (e.g., 50 ns) may be adequate to characterize monoexponential fluorescence decays. Although the signal-to-noise ratio depends on the total number of photons collected, the same sample may have a very fast θ. This may mean that after processing using Eq. 1, the number of time channels used is inadequate for measurement of a fluorescence anisotropy decay, as the decay is complete within a very short amount of time corresponding to a small

number of time channels. If it is known that the value of θ is much greater than that of τ_f, e.g., in the case of monomeric green fluorescent protein (GFP) where $\tau_f \sim 2.5$ ns and $\theta \sim 15$ ns in aqueous solution, then this discrepancy becomes less of an issue, but accurate determination of r_∞ may then become problematic as the fluorescence decay is complete before the molecules have rotated.

4. *Steady-state anisotropy from time-resolved measurements*
   Provided complete fluorescence decays (into the noise floor) are recorded in each polarization detection arm, it is possible to extract steady-state anisotropy values from the time-resolved data. In TCSPC experiments, an accurate measurement of both the initial anisotropy and r_∞ values requires the collection of several thousand photon counts in the peaks of the polarization-resolved fluorescence decays. Steady-state anisotropy values are dependent on the total number of photons detected in each polarization channel. Therefore, they typically require shorter acquisition times than those required for accurate time-resolved anisotropy decay measurements.

5. *Subtraction of the background*
   The data in Fig. 3a show that after ~35 ns the intensity in each detector channel becomes "flat" indicating that the signal is dominated by noise at acquisition times longer than this rather than the intensity of the fluorescence. For longer acquisition times, this noise is higher, and it is also not necessarily equal in each detector for multi-detector arrangements. It is important to remove this noise when generating fluorescence anisotropy decays as it can lead to artifacts in the value of r_∞ and the possibility of incorrect identification of binding events. For optimized detection electronics and optical alignment, the signal-to-noise ratio should be high, and the background value should have little or no influence on the anisotropy values at short times.

6. *Timing and signal overlap*
   It is necessary to have good overlap in the timing for each (parallel and perpendicular) detector in order to produce an accurate fluorescence anisotropy decay curve. The start of the fluorescence decay transient in each detector must occur at the same time; otherwise there will be an artifact peak in the anisotropy decay, and an incorrect value for r_0 will be measured.

   The signal overlap can be optimized using the optical path length (light travels ~3 mm in 10 ps), the timing electronics of the TCSPC card, and/or the voltage applied to the detector via the DCC-100 software. Alternatively, we correct for any timing mismatch between the detectors in the data processing stage. In this case, the same voltage is applied to each detector giving the benefit of comparable instrument response functions from each detector [89, 90]. If vastly different arrival times are measured in each detector, it is worth checking that the optical path lengths of the fluorescence to each detector are matched and that the length of the cables linking each detector to the signal router is equal.

7. *Image acquisition parameters using the SPC-830 board*
For a single detector, the maximum image size with 256 time channels in each pixel is 256 × 256 pixels. For two detectors, the maximum image size for the same number of time channels per pixel is 256 × 128 pixels, due to the limited memory on the SPC830 board in this mode of operation. If the value of θ is long and it is unnecessary to have as many time channels to record useful fluorescence intensity and anisotropy decays, then it is feasible to use 64 (or fewer, though not advisable in the interests of good decay fitting) time channels per pixel and still measure a 256 × 256 pixel image in each detector. We find that the trade-off in spatial and time-resolution offered by recording rectangular 256 × 128 images with 256 time channel in each pixel typically serves our samples adequately. Note that limited on-board memory is not an issue if the FIFO buffer is used for imaging, and the data transferred to the memory of the computer [18].

8. *Setting the threshold intensity level for anisotropy image processing*
The setting of the threshold level is largely dependent on the needs of the application and the sample under study. The error in the calculated anisotropy value as a function of number of detected photons has been reported [91]. For samples with significant spatial heterogeneity in their anisotropy yet only a small range of anisotropy values, it is necessary to measure very high numbers of photons to reveal this heterogeneity with any degree of certainty. Conversely, if the range of anisotropy values is high, fewer photons are necessary to see the heterogeneity. It should be noted also that pixel binning in the image processing stage also means that it may not be necessary to threshold individual pixel intensities at such high values as summing over several pixels increases the intensity in the binned regions.

Acknowledgements

We would like to thank Dr Simon Ameer-Beg in the Randall Division for Cell and Molecular Biophysics at King's College London for fruitful discussions and Dr Aleksandar Ivetic from the Membrane/Cytoskeleton Signalling Group in the Cardiovascular Division at King's College London for the samples in Fig. 1.

We would also like to thank Dr Steven Vogel from the National Institute on Alcohol Abuse and Alcoholism in Bethesda, USA, and editor Prof Yves Engelborghs for valuable comments on the manuscript. Finally, we would like to acknowledge the UK's Medical Research Council (MRC) for funding.

References

1. Bastiaens PIH, Pepperkok R (2000) Observing proteins in their natural habitat: the living cell. Trends Biochem Sci 25:631–637
2. Bastiaens PIH, Squire A (1999) Fluorescence lifetime imaging microscopy: spatial resolution of biochemical processes in the cell. Trends Cell Biol 9:48–52
3. Verveer PJ, Bastiaens PI (2008) Quantitative microscopy and systems biology: seeing the whole picture. Histochem Cell Biol 130(5): 833–843
4. Wouters FS, Verveer PJ, Bastiaens PI (2001) Imaging biochemistry inside cells. Trends Cell Biol 11(5):203–211
5. Levitt JA, Matthews DR, Ameer-Beg SM, Suhling K (2009) Fluorescence lifetime and polarization-resolved imaging in cell biology. Curr Opin Biotechnol 20:28–36
6. Förster T (1946) Energiewanderung und Fluoreszenz. Naturwissenschaften 33 (6):166–175, translated into English by Klaus Suhling, J Biomed Opt 17(1), 011002, 2012
7. Birch DJS (2011) Fluorescence detections and directions. Meas Sci Technol 22(5):052002
8. Piston DW (2010) Fluorescence anisotropy of protein complexes in living cells. Biophys J 99 (6):1685–1686
9. Axelrod D (1979) Carbocyanine dye orientation in red cell membrane studied by microscopic fluorescence polarization. Biophys J 26 (3):557–573
10. Ha T, Laurence TA, Chemla DS, Weiss S (1999) Polarization spectroscopy of single fluorescent molecules. J Phys Chem B 103 (33):6839–6850
11. Yan YL, Marriott G (2003) Fluorescence resonance energy transfer imaging microscopy and fluorescence polarization imaging microscopy. Methods Enzymol 360:561–580
12. Fisz JJ (2009) Another treatment of fluorescence polarization microspectroscopy and imaging. J Phys Chem A 113(15):3505–3516
13. Fisz JJ (2007) Another look at magic-angle-detected fluorescence and emission anisotropy decays in fluorescence microscopy. J Phys Chem A 111(50):12867–12870
14. Fisz JJ (2007) Fluorescence polarization spectroscopy at combined high-aperture excitation and detection: application to one-photon-excitation fluorescence microscopy. J Phys Chem A 111(35):8606–8621
15. Koshioka M, Sasaki K, Masuhara H (1995) Time-dependent fluorescence depolarization analysis in 3-dimensional microspectroscopy. Appl Spectrosc 49(2):224–228
16. Fixler D, Namer Y, Yishay Y, Deutsch M (2006) Influence of fluorescence anisotropy on fluorescence intensity and lifetime measurement: theory, simulations and experiments. IEEE Trans Biomed Eng 53(6):1141–1152
17. Axelrod D (1989) Fluorescence polarization microscopy. Methods Cell Biol 30:333–352
18. Becker W (2005) Advanced time-correlated single photon counting techniques. In: Castleman AWJ et al (eds) Springer series in chemical physics, vol 81. Springer, NY, p 401
19. Suhling K, Siegel J, Lanigan PMP et al (2004) Time-resolved fluorescence anisotropy imaging applied to live cells. Opt Lett 29(6):584–586
20. Birch DJS (2001) Multiphoton excited fluorescence spectroscopy of biomolecular systems. Spectrochim Acta A Mol Biomol Spectrosc 57(11):2313–2336
21. Lidke DS, Nagy P, Barisas BG et al (2003) Imaging molecular interactions in cells by dynamic and static fluorescence anisotropy (rFLIM and emFRET). Biochem Soc Trans 31(5):1020–1027
22. Dix JA, Verkman AS (1990) Mapping of fluorescence anisotropy in living cells by ratio imaging. Applications to cytoplasmic viscosity. Biophys J 57:231–240
23. Swaminathan R, Hoang CP, Verkman AS (1997) Photobleaching recovery and anisotropy decay of green fluorescent protein GFP-S65T in solution and cells: cytoplasmic viscosity probed by green fluorescent protein translational and rotational diffusion. Biophys J 72:1900–1907
24. Gough AH, Taylor DL (1993) Fluorescence anisotropy imaging microscopy maps calmodulin-binding during cellular contraction and locomotion. J Cell Biol 121(5):1095–1107

25. Levitt JA, Chung PH, Kuimova MK et al (2011) Fluorescence anisotropy of molecular rotors. Chemphyschem 12(3):662–672
26. Bigelow CE, Conover DL, Foster TH (2003) Confocal fluorescence spectroscopy and anisotropy imaging system. Opt Lett 28(9):695–697
27. Foster TH, Pearson BD, Mitra S, Bigelow CE (2005) Fluorescence anisotropy imaging reveals localization of meso-tetrahydroxyphenyl chlorin in the nuclear envelope. Photochem Photobiol 81(6):1544–1547
28. Bigelow CE, Vishwasrao HD, Frelinger JG, Foster TH (2004) Imaging enzyme activity with polarization-sensitive confocal fluorescence microscopy. J Microsc (Oxf) 215:24–33
29. Li W, Wang Y, Shao HR et al (2007) Probing rotation dynamics of biomolecules using polarization based fluorescence microscopy. Microsc Res Tech 70(4):390–395
30. Cao Z, Huang CC, Tan W (2006) Nuclease resistance of telomere-like oligonucleotides monitored in live cells by fluorescence anisotropy imaging. Anal Chem 78:1478–1484
31. Mattheyses AL, Hoppe AD, Axelrod D (2004) Polarized fluorescence resonance energy transfer microscopy. Biophys J 87(4):2787–2797
32. Rizzo MA, Piston DW (2005) High-contrast imaging of fluorescent protein FRET by fluorescence polarization microscopy. Biophys J 88(2):L14–L16
33. Matthews DR, Carlin LM, Ofo E et al (2010) Time-lapse FRET microscopy using fluorescence anisotropy. J Microsc (Oxf) 237(1): 51–62
34. Piston DW, Rizzo MA (2008) FRET by fluorescence polarization microscopy. Methods Cell Biol 85:415–30
35. Squire A, Verveer PJ, Rocks O, Bastiaens PI (2004) Red-edge anisotropy microscopy enables dynamic imaging of homo-FRET between green fluorescent proteins in cells. J Struct Biol 147(1):62–69
36. Keating SM, Wensel TG (1991) Nanosecond fluorescence microscopy. Emission kinetics of fura-2 in single cells. Biophys J 59(1):186–202
37. Spitz JA, Polard V, Maksimenko A et al (2007) Assessment of cellular actin dynamics by measurement of fluorescence anisotropy. Anal Biochem 367(1):95–103
38. Tramier M, Kemnitz K, Durieux C et al (2000) Restrained torsional dynamics of nuclear DNA in living proliferative mammalian cells. Biophys J 78(5):2614–2627
39. Benninger RKP, Hofmann O, McGinty J et al (2005) Time-resolved fluorescence imaging of solvent interactions in microfluidic devices. Opt Express 13(16):6275–6285
40. Clayton AHA, Hanley QS, Arndt-Jovin DJ et al (2002) Dynamic fluorescence anisotropy imaging microscopy in the frequency domain (rFLIM). Biophys J 83:1631–1649
41. Botchway SW, Lewis AM, Stubbs CD (2011) Development of fluorophore dynamics imaging as a probe for lipid domains in model vesicles and cell membranes. Eur Biophys J Biophys Lett 40(2):131–141
42. Benninger RKP, Önfelt B, Neil MAA et al (2005) Fluorescence imaging of two-photon linear dichroism: cholesterol depletion disrupts molecular orientation in cell membranes. Biophys J 88:609–622
43. Reeve JE, Corbett AD, Boczarow I et al (2012) Probing the orientational distribution of dyes in membranes through multiphoton microscopy. Biophys J 103(5):907–917
44. Brack AS, Brandmeier BD, Ferguson RE et al (2004) Bifunctional rhodamine probes of myosin regulatory light chain orientation in relaxed skeletal muscle fibers. Biophys J 86(4): 2329–2341
45. Mojzisova H, Olesiak J, Zielinski M et al (2009) Polarization-sensitive two-photon microscopy study of the organization of liquid-crystalline DNA. Biophys J 97(8):2348–2357
46. Dale RE, Eisinger J, Blumberg WE (1979) The orientational freedom of molecular probes. Biophys J 26:161–194
47. Dos Remedios CG, Moens PDJ (1995) Fluorescence resonance energy transfer spectroscopy is a reliable "ruler" for measuring structural changes in proteins. Dispelling the problem of the unknown orientation factor. J Struct Biol 115(2):175–185
48. van der Meer BW (2002) Kappa-squared: from nuisance to new sense. J Biotechnol 82(3): 181–96
49. Jares-Erijman EA, Jovin TM (2003) FRET imaging. Nat Biotechnol 21(11):1387–1396
50. Varma R, Mayor S (1998) GPI-anchored proteins are organized in submicron domains at the cell surface. Nature 394:798–801
51. Sharma P, Varma R, Sarasij RC et al (2004) Nanoscale organization of multiple GPI-anchored proteins in living cell membranes. Cell 116(4):577–589
52. Vishwasrao HD, Trifilieff P, Kandel ER (2012) In vivo imaging of the actin polymerization state with two-photon fluorescence anisotropy. Biophys J 102(5):1204–1214
53. van Ham TJ, Esposito A, Kumita JR et al (2010) Towards multiparametric fluorescent imaging of amyloid formation: studies of a YFP model of α-synuclein aggregation. J Mol Biol 395(3):627–642

54. Steinmeyer R, Harms GS (2009) Fluorescence resonance energy transfer and anisotropy reveals both hetero- and homo-energy transfer in the pleckstrin homology-domain and the parathyroid hormone-receptor. Microsc Res Tech 72(1):12–21
55. Gautier I, Tramier M, Durieux C et al (2001) Homo-FRET microscopy in living cells to measure monomer-dimer transition of GFP-tagged proteins. Biophys J 80(6):3000–3008
56. Bader AN, Hofman EG, Voortman J et al (2009) Homo-FRET imaging enables quantification of protein cluster sizes with subcellular resolution. Biophys J 97(9):2613–2622
57. Bader AN, Hofman EG, van Bergen en Henegouwen PM, Gerritsen HC (2007) Imaging of protein cluster sizes by means of confocal time-gated fluorescence anisotropy microscopy. Opt Express 15(11): 6934–6945
58. Bader AN, Hoetzl S, Hofman EG et al (2011) Homo-FRET imaging as a tool to quantify protein and lipid clustering. Chemphyschem 12(3):475–483
59. Yeow EKL, Clayton AHA (2007) Enumeration of oligomerization states of membrane proteins in living cells by homo-FRET spectroscopy and microscopy: theory and application. Biophys J 92(9):3098–3104
60. Thaler C, Koushik SV, Puhl HL et al (2009) Structural rearrangement of CaMKII alpha catalytic domains encodes activation. Proc Natl Acad Sci U S A 106(15):6369–6374
61. Nguyen TA, Sarkar P, Veetil JV et al (2012) Fluorescence polarization and fluctuation analysis monitors subunit proximity, stoichiometry, and protein complex hydrodynamics. PLoS One 7(5):e38209
62. Devauges V, Marquer C, Lecart S et al (2012) Homodimerization of amyloid precursor protein at the plasma membrane: a homoFRET study by time-resolved fluorescence anisotropy imaging. PLoS One 7(9):e44434
63. Vogel SS, Thaler C, Blank PS, Koushik SV (2010) Time-resolved fluorescence anisotropy. In: Ammasi Periasamy RMC (ed.) Microscopy in biology and medicine. Chapman & Hall: Boca Raton, FL. pp. 245–288
64. Woolley P, Steinhäuser KG, Epe B (1987) Forster-type energy-transfer—simultaneous forward and reverse transfer between unlike fluorophores. Biophys Chem 26(2–3):367–374
65. Porter GB (1972) Reversible energy-transfer. Theor Chim Acta 24(2–3):265–270
66. Willaert K, Loewenthal R, Sancho J et al (1992) Determination of the excited-state lifetimes of the tryptophan residues in barnase, via multifrequency phase fluorometry of tryptophan mutants. Biochemistry 31(3):711–716
67. Jung G, Ma YZ, Prall BS, Fleming GR (2005) Ultrafast fluorescence depolarisation in the yellow fluorescent protein due to its dimerization. Chemphyschem 6(8):1628–1632
68. Gratton E, Breusegem S, Sutin J et al (2003) Fluorescence lifetime imaging for the two-photon microscope: time-domain and frequency-domain methods. J Biomed Opt 8(3):381–390
69. Suhling K (2006) Fluorescence lifetime imaging. In: Stephens D (ed) Cell imaging. Scion, Bloxham, pp 219–245
70. Suhling K, French PMW, Phillips D (2005) Time-resolved fluorescence microscopy. Photochem Photobiol Sci 4:13–22
71. Esposito A, Bader AN, Schlachter SC et al (2011) Design and application of a confocal microscope for spectrally resolved anisotropy imaging. Opt Express 19(3):2546–2555
72. Philip J, Carlsson K (2003) Theoretical investigation of the signal-to-noise ratio in fluorescence lifetime imaging. J Opt Soc Am A 20(2):368–379
73. Esposito A, Gerritsen HC, Wouters FS (2007) Optimizing frequency-domain fluorescence lifetime sensing for high-throughput applications: photon economy and acquisition speed. J Opt Soc Am A 24:3261
74. Pelet S, Previte MJR, So PTC (2006) Comparing the quantification of Förster resonance energy transfer measurement accuracies based on intensity, spectral, and lifetime imaging. J Biomed Opt 11(3):34017
75. Gee ML, Lensun L, Smith TA, Scholes CA (2004) Time-resolved evanescent wave-induced fluorescence anisotropy for the determination of molecular conformational changes of proteins at an interface. Eur Biophys J Biophys Lett 33(2):130–139
76. Smith TA, Gee ML, Scholes CA (2005) Time-resolved evanescent wave-induced fluorescence anisotropy measurements. In: Lakowicz CDGJR (ed) Reviews in fluorescence. Springer, New York, pp 245–271
77. Bruns T, Strauss WSL, Schneckenburger H (2008) Total internal reflection fluorescence lifetime and anisotropy screening of cell membrane dynamics. J Biomed Opt 13(4):041317
78. Jameson DM, Ross JA (2010) Fluorescence polarization/anisotropy in diagnostics and imaging. Chem Rev 110(5):2685–2708
79. Gradinaru CC, Marushchak DO, Samim M, Krull UJ (2010) Fluorescence anisotropy: from single molecules to live cells. Analyst 135(3):452–459
80. Yengo CM, Berger CL (2010) Fluorescence anisotropy and resonance energy transfer: powerful tools for measuring real time protein

dynamics in a physiological environment. Curr Opin Pharmacol 10(6):731–737

81. Chan FTS, Kaminski CF, Schierle GSK (2011) HomoFRET fluorescence anisotropy imaging as a tool to study molecular self-assembly in live cells. Chemphyschem 12(3):500–509
82. Tramier M, Coppey-Moisan M (2008) Fluorescence anisotropy imaging microscopy for Homo-FRET in living cells. In: Kevin FS (ed) Methods in cell biology. Academic, New York, pp 395–414
83. Barber PR, Ameer-Beg SM, Gilbey J et al (2009) Multiphoton time-domain fluorescence lifetime imaging microscopy: practical application to protein–protein interactions using global analysis. J R Soc Interface 6: S93–S105
84. Vishwasrao HD, Heikal AA, Kasischke KA, Webb WW (2005) Conformational dependence of intracellular NADH on metabolic state revealed by associated fluorescence anisotropy. J Biol Chem 280(26):25119–25126
85. Harris CM, Selinger BK (1979) Single-photon decay spectroscopy. 2. Pile-up problem. Aust J Chem 32(10):2111–2129
86. Suhling K, McLoskey D, Birch DJS (1996) Multiplexed single-photon counting. II. The statistical theory of time-correlated measurements. Rev Sci Instrum 67(6):2238–2246
87. vandeVen M, Ameloot M, Valeur B, Boens NI (2005) Pitfalls and their remedies in time-resolved fluorescence spectroscopy and microscopy. J Fluoresc 15(3):377–413
88. Cross AJ, Fleming GR (1984) Analysis of time-resolved fluorescence anisotropy decays. Biophys J 46:45–56
89. Birch DJS, Imhof RE, Dutch A (1984) Differential pulse fluorometry using matched photomultipliers—a new method of measuring fluorescence lifetimes. J Phys E Sci Instrum 17(5):417–418
90. Birch DJS, Imhof RE, Dutch A (1984) Pulse fluorometry using simultaneous acquisition of fluorescence and excitation. Rev Sci Instrum 55(8):1255–1264
91. Lidke KA, Rieger B, Lidke DS, Jovin TM (2005) The role of photon statistics in fluorescence anisotropy imaging. IEEE Trans Image Process 14(9):1237–1245
92. Koushik SV, Vogel SS (2008) Energy migration alters the fluorescence lifetime of Cerulean: implications for fluorescence lifetime imaging Forster resonance energy transfer measurements. J Biomed Opt 13(3):031204

Chapter 23

Multimodal Fluorescence Imaging Spectroscopy

Martijn H.W. Stopel, Christian Blum, and Vinod Subramaniam

Abstract

Multimodal fluorescence imaging is a versatile method that has a wide application range from biological studies to materials science. Typical observables in multimodal fluorescence imaging are intensity, lifetime, excitation, and emission spectra which are recorded at chosen locations at the sample. This chapter describes how to build instrumentation that allows for multimodal fluorescence imaging and explains data analysis procedures for the observables.

Key words Fluorescence, Lifetime, Excitation, Emission, Spectroscopy, Multimodal imaging, Supercontinuum light source, Time-correlated single photon counting

1 Introduction

Fluorescence is characterized by the absorption of light, followed by spontaneous emission of light that is typically "red-shifted." The red-shifted light can be easily separated from the illumination light by choosing a suitable filter set. In fluorescence microscopy, this spectral separation of absorbed and emitted light is exploited to highlight and study specific domains spatially, rendered visible against a dark background. Nevertheless, these methods are of limited use in their ability to analyze the dynamics, interactions, and physical environment of molecules when unassisted by spectroscopy. Imaging spectroscopy methods enable the extension of simple spatial analyses to demonstrate function, co-localization, and molecular interaction at the nanoscale [1].

Multimodal imaging spectroscopy combines imaging with a range of spectroscopic observables to image macro- and microscopic objects, while at the same time obtaining detailed physicochemical information at the micro- and nanoscale. The parameters that are typically measured are fluorescence intensity, lifetime [2], polarization [3], and emission spectra [4]. At the moment, there is

Yves Engelborghs and Antonie J.W.G. Visser (eds.), *Fluorescence Spectroscopy and Microscopy: Methods and Protocols*, Methods in Molecular Biology, vol. 1076, DOI 10.1007/978-1-62703-649-8_23,

an increasing interest in supercontinuum (SC) white light sources [5–7]. These light sources provide the flexibility to pick any desired excitation wavelength between 400 nm and >2,000 nm and allow for time-correlated single photon counting (TCSPC) experiments to accurately measure fluorescence lifetime. Even more important is that SC light sources give the possibility to perform excitation spectroscopy [5, 6, 8], in addition to the commonly measured observables in multimodal microscopy mentioned above. In this chapter, we describe how to assemble and operate a versatile hyperspectral fluorescence lifetime imaging microscopy (FLIM) setup and discuss important practical bottlenecks that are crucial for optimum functionality of the setup.

2 Assembly Procedure

There are many ways to implement multimodal imaging spectroscopy. This chapter describes how to build instrumentation that is capable of both wide-field fluorescence imaging and single-emitter confocal imaging spectroscopy, although the latter method will be the main focus of this chapter. The instrumentation incorporates a supercontinuum white light laser in combination with an acousto-optical tunable filter (AOTF) to provide any desired excitation wavelength in the range between 400 nm and 2,000 nm and allows for fast switching between different wavelengths within the full visible range. The confocal detectors have a sensitivity that is capable of detecting single photons, which not only provides a high signal-to-background ratio but also allows one to study single emitting fluorophores (quantum emitters). The complete setup scheme for the multimodal fluorescence lifetime imaging setup is shown in Fig. 1. This section describes how to assemble each segment and also contains a list of the key components that are needed during each assembly step. Trivial components such as optical mounts, screws, and other minor accessories are not mentioned in these lists but should be included accordingly for the listed modules. The majority of the components used during the assembly of this setup were ordered at Thorlabs Inc. (Dachau/München, Germany). The remaining parts are from different manufacturers and are *italicized* in the equipment lists.

2.1 Microscope Body

The microscope body acts as a solid platform that holds the sample and controls the pathways of light, including the movement of the imaging microscope objective to focus light onto the sample. A convenient configuration is achieved by illuminating the sample through the back-port of the microscope and detecting through one of the other ports. For optimal stability and flexibility, *see* **Note 1**.

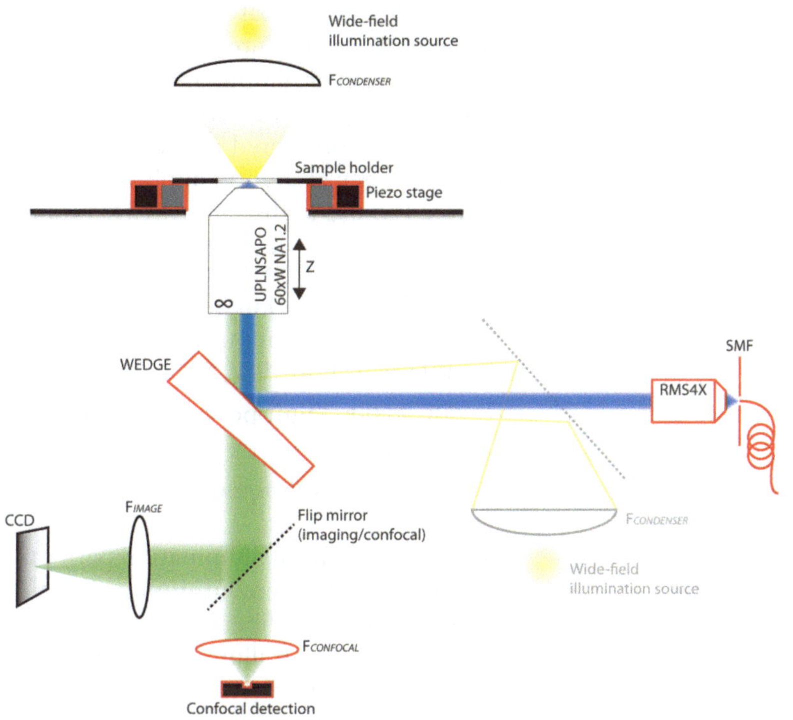

Fig. 1 Schematic overview of the multimodal fluorescence imaging setup. The parts that are highlighted in *red* are parts that are essential for the confocal illumination and detection configuration. Without these highlighted components, the setup would operate in conventional fluorescence wide-field imaging configuration. Adding the highlighted components will not degrade the performance in the wide-field imaging configuration. A wide-field illumination source is depicted in this figure for the conventional transmission illumination, but the same light source can be used as well to provide epi-illumination (*shaded*) through the same light path as the confocal illumination, when focused on the back focal plane of the imaging microscope objective. Filters should be added to this scheme to spectrally separate excitation and emission light. Excitation filters should be placed after the excitation source (wide-field illumination or confocal illumination), and the emission filters should be placed after the wedge

- *Olympus IX71 body.*
- *Base stage plate (IX2-SP), to mount the sample holder and piezo stage.*
- *Binoculars (U-BI90).*

2.2 Excitation Sources

2.2.1 Wide-Field Illumination

Wide-field illumination can be achieved either by standard transmission illumination by a light source mounted on the illumination pillar above the sample stage or in an epi-illumination configuration through the back-port of the microscope. This combination enables work with transparent as well as nontransparent substrates or samples. The goal of the alignment procedure is to create a homogeneous and intense illumination pattern across the full field of view. Köhler illumination is a commonly used illumination scheme in light microscopy to achieve such illumination and is usually not considered as a critical alignment step. For both high light intensity and good applicability for fluorescence, it is recommended to use high-power LEDs or a Mercury lamp (*see* **Note 2**).

- *Illumination pillar, condenser.*
- M470L2-C1: High-power blue LED with collimator and adapter for Olympus BX/IX (additional power LEDs are available as well for UV, VIS, and NIR spectral ranges).
- DC2100: High-power LED driver—1–100 kHz modulation option.
- *(Optional) Olympus 100 W Mercury Lamp.*

2.2.2 Confocal Illumination

Coupling Light into the Single-Mode Fiber

Before the confocal illumination path can be aligned, the light from the SC400-PP supercontinuum excitation source must be coupled into a single-mode fiber to spatially clean up the excitation light and provide a Gaussian illumination profile from a point source at the other end of the fiber. Use a kinematic mirror mount to fix the collimator of the supercontinuum laser and couple the collimated white light beam through the acousto-optical tunable filter (AOTF) with a precision of $\pm 0.5°$ with respect to the normal of the AOTF crystal (caution, high radiation, do not use full power for alignment). This AOTF crystal diffracts only a small spectral band by applying a fixed radio-frequency (RF) signal and blocks remaining unwanted wavelengths. Use the manufacturer's software to open a channel of the AOTF to transmit a narrow band of visible light at the back-end of the AOTF. Three different beams can be seen at the back: the zeroth order (white light), which should be blocked with a proper beam block, and two first-order diffracted beams (well-defined bands). Only one of the first-order beams is collinear with the crystal and does not displace when the AOTF is tuned to a different wavelength. This collinear beam needs to be coupled into a single-mode fiber (P1-460A) using two mirrors, incoupling optics (A240TM-A), and a precision XYZ stage (MBT613D/M).

- *Supercontinuum white light source (SC400-PP) from Fianium Ltd. (Southampton, UK).*
- 1 mount for the output fiber of the SC400-PP laser.
- *Crystal Technologies broadband acousto-optical tunable filter (AOTF) + RF controller.*
- 1 rotation stage for the AOTF.
- 1 beam block to block stray light and unwanted orders from the AOTF.

- 2 broadband visible light mirrors + 2 kinematic mirror mounts.
- A240TM-A aspheric lens, $f = 8$ mm + E12RMS adapter.
- MBT613D/M 3-axis precision fiber incoupling platform.
- P1-460A single-mode optical fiber.

Acousto-optical Tunable Filter Calibration

The calibration of the AOTF allows one to choose any desired excitation wavelength and opens up possibilities for excitation spectroscopy and hyperspectral imaging. Calibration of the AOTF is done by applying different RF frequencies to the crystal and measuring the corresponding transmission wavelengths using a calibrated spectrograph (CCS100/M). The peak transmission wavelength as a function of RF frequency is the required calibration curve and should be fitted with a cubic polynomial fit. The corresponding polynomial coefficients are used in custom written software to convert the desired wavelength to the RF frequency that has to be applied to the crystal.

- *CCS100 USB spectrograph*

Collimation of the Excitation Beam

The best confocal illumination spot is achieved by projecting a small point source (single-mode fiber exit) into a small diffraction-limited illumination focus spot at the sample plane. Before the light enters the infinity-corrected imaging microscope objective, the light should be collimated, and its beam diameter should roughly match the back aperture of the imaging objective. The infinity-corrected RMS4X collimation objective can be used to collimate the light coming out of the single-mode fiber (NA 0.1) and slightly overfill the back aperture, which produces the sharpest confocal illumination spot (for more details, *see* **Note 3**).

- MBT616D/M 3-axis precision stage.
- 1 kinematic mirror mount for XY tilt angles + FC connector adapter to mount the fiber.
- RMS4X objective to collimate the output light.
- 1 SM1-RMS external-internal ring adapter and 1 thick 30 mm cage plate to mount the RMS4X objective.
- *500 mm telescope to accurately collimate laser light (Möller-Wedel FR500/65/14.7).*

Aligning the Excitation Beam to the Central Optical Axis

After collimating the excitation beam, the excitation beam needs to be aligned to the central optical axis of the imaging objective. First, place the mirrors on a breadboard at the same height as the back-entrance of the microscope. Position two mirrors which are mounted in robust gimbal mounts such that the excitation beam is directed in a Z-angled configuration and reflected by the wedge along the central optical axis of the imaging objective (*see* **Note 4**).

- 2 gimbal mirror mounts + 2 mirrors.
- Beam splitter: BSF10-A—beam sampler, which deflects the ghost beam.

2.3 Detection Path

2.3.1 Imaging CCD

The imaging CCD is used to obtain a picture of a full area at one instance. A good test sample to align the CCD would be a standard microscopy test sample of the plant *Convallaria majalis* (Lily of the valley, Fig. 2), because it is highly autofluorescent and shows a large spectral diversity. Illuminate the Convallaria sample using the wide-field illumination source and sharply focus the sample using the microscope's binoculars. Direct the fluorescence emission light towards the CCD and focus the fluorescence emission light onto the CCD. Be sure to use the proper filters to spectrally separate the excitation light (typically, a band-pass filter) and emission light (typically, a long-pass filter).

- *Test sample: Convallaria majalis (Lily of the valley).*
- Imaging lens (180 mm ø2 inch achromat, or tube lens of the microscope).
- *CMOS or CCD camera (Zeiss Axiocam or Andor iXon3).*
- *Filters, for example: FF02-447/60 band-pass for the excitation light and BLP488R long-pass for the emission light (same filters can be used for the confocal fluorescence microscopy configuration).*

2.3.2 Confocal Detectors

The confocal detector acts as a point detector, which measures fluorescence emission from a well-defined volume. To constrain the detection to this volume, it is necessary to remove as much stray light as possible (*see* **Note 5**). After proper shielding, the dark noise level should match the specified level given by the

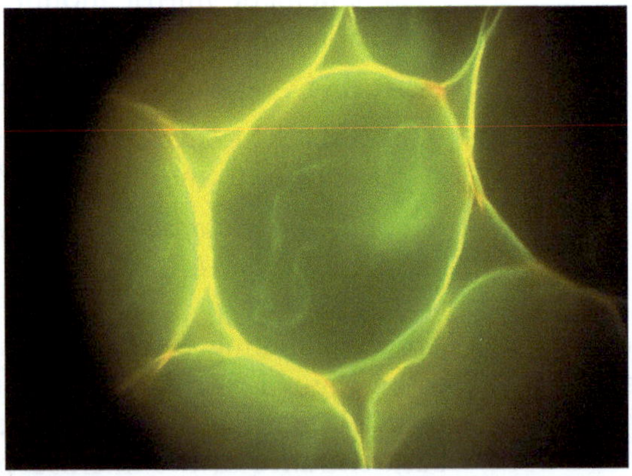

Fig. 2 Wide-field fluorescence image of *Convallaria majalis* (Lily of the valley). Lifetime and spectral data can be recorded once the confocal detectors have been aligned (Fig. 4)

manufacturer of the detector. Since the confocal detection volume is static, the sample has to be raster-scanned through this detection volume to record intensity and lifetime (avalanche photodiode) and emission spectra (spectrograph) at each point. The incorporation of the scanning stage into the confocal detection scheme is explained at the end of this section.

Avalanche Photodiode

The single photon counting avalanche photodiode (APD from Micro-Photon-Devices, Bolzano, Italy) is capable of counting photons one by one and allows for accurate time-resolved lifetime and intensity analyses of the fluorophores (time-correlated single photon counting, TCSPC, *see* **Note 6**). Focus the laser light onto the upper side (glass-air interface) of the thin cover glass. Mount the APD on a XYZ stage and direct the reflected laser beam towards the center of the detector. Use a 50 mm achromat lens to roughly focus the reflected laser beam onto the APD's chip without adjusting the XYZ stage. Connect the electronic synchronization output of the Fianium SC400-PP to the A-PPI pulse inverter and connect it to the "sync" input of the Becker & Hickl TCSPC card. Directly connect the output signal of the APD to the "stop" input of the TCSPC card. Use an oscilloscope to find the minimum trigger level of both trigger lines and configure the software triggering accordingly. Read out the intensity at the APD and align the APD using the XYZ stage to obtain the maximum intensity as possible (attenuate the light if necessary). Next, spin-coat a thin film of an efficient fluorophore (e.g., Rhodamine 6G) and mount this sample at the sample plane. Insert the appropriate filters (same filter set as used for the wide-field imaging procedure) and illuminate this sample with the excitation beam. Record fluorescence emission intensity of the sample and focus the fluorescent light onto the APD using the XYZ stage, until the maximum intensity has been reached. Now, the APD has been aligned to operate in the confocal detection configuration. Depending on the quality of the alignment and stray light shielding, this configuration is capable of detecting emission of single fluorophores.

- Achromat lens, $f = 50$ mm, ø1 inch.
- *PicoQuant MPD-APD SPAD, single photon avalanche photo diode from Micro-Photon-Devices.*
- XYZ stage, PT3/M or PT3A/M.
- *Becker & Hickl TCSPC card (SPC-830) for single photon counting + A-PPI pulse inverter.*

Spectrograph

The spectrograph (*see* **Note 7**) records the emission spectrum of the light coming from the confocal detection volume. The emission light that originates from the confocal detection volume is coupled into a multimode fiber that leads to the spectrograph. The light at the exit of the multimode fiber acts as a point source, which is angularly dispersed via a dispersive element (prism, grating, or acousto-optical tunable filter) and projected onto a camera (typically rectangular EMCCD cameras) to record the emission spectra with a high spectral resolution of approximately 1 nm. The image at the camera is built up from multiple images of the point source at different wavelengths and typically shows a rainbow-colored line for a white light point source. A schematic diagram of a prism spectrograph is shown in Fig. 3. For easy alignment, it is highly recommended to mount a tilt-adjustable ø2″ mirror onto a linear stage in between the imaging lens and the camera. This allows for precise focus adjustments of the spectral image at the CCD plane, while the camera is fixed.

- Multimode fiber incoupling
  - Achromat lens, $f = 30$ mm, ø1 inch.
  - One 30 mm thick cage plate + four 6 mm rods.
  - One 30 mm cage Z-translator to hold the multimode fiber, using a SM1-multimode fiber connector adapter.
  - One 50 μm aperture multimode fiber (0.22 NA).
- Visible light spectrograph
  - SM1-multimode fiber connector adapter.
  - Achromat collimation lens, $f = 200$ mm, ø2 inch.
  - Achromat imaging lens, $f = 300$ mm, ø2 inch.
  - Equilateral prism, 50 mm.
  - *Andor Newton DU971 EMCCD camera.*

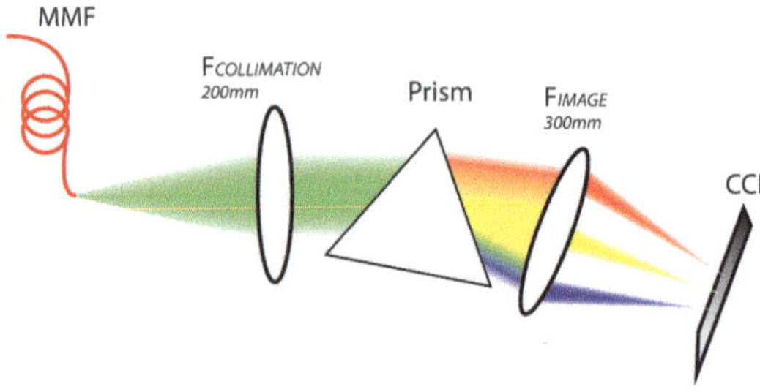

Fig. 3 Schematic diagram of a prism spectrograph. Emission from the confocal detection volume is coupled into the multimode fiber (MMF), and the fiber exit acts as a point source. The light of this point source is collimated by the first lens. The prism angularly disperses the collimated beam, and the imaging lens images the point source onto the CCD. The angular dispersion by the prism causes a lateral displacement of the point source image at the CCD, which is wavelength dependent

Spectrograph Calibration

When the spectrograph alignment has been accomplished, the CCD needs a calibration to relate the pixel axis of the CCD to wavelength. A calibration source (HL-2000-CAL) provides distinct spectral lines that are easily identified by comparing the recorded spectrum at the CCD with the emission spectrum of the calibration source. A cubic polynomial fit is used to calibrate the spectrograph CCD.

- *Ocean Optics HL-2000-CAL, Wavelength calibration source.*

Scanning, Synchronization Scheme, and Cabling

In the confocal configuration, it is possible to monitor emission from a well-defined fixed volume with high-accuracy and high-detection sensitivity. However, it is not possible to laterally displace this detection volume. Therefore, a piezo scanning stage is needed to the scan the sample through the detection volume, with a precision of 2 nm and a full-range repeatability of 10 nm, which is especially important for studying single emitters, since these have to be revisited after a quick localization scan. For each step, fluorescence emission is recorded using one of the confocal detectors. Accurate timing of this detection and synchronization with scanning is therefore a key aspect for confocal scanning microscopes. The best way to synchronize scanning and detection is communicating through hardware triggering. The scanning stage controller (PI E-710) serves as the master controller and triggers the detectors to record data, using TTL pulse trigger lines. Four scanning triggers can be accessed at the "digital-IO" port at the back of the scanning stage controller. The first three trigger lines are the "pixel," "line," and "frame" triggers, and the fourth trigger line indicates when the "illumination" should be active and can be used to control the activity of the excitation source. The recording time of the detectors is set beforehand.

The four scanning triggers are connected directly to the TCSPC card, and the "pixel" trigger line is split to trigger the EMCCD camera that is used for the spectrograph. The "illumination" line is also split and is used to switch the laser source on and off while scanning (*see* **Note 8**).

- *Piezo scanning stage (PI P-527).*
- *Scanning stage controller (PI E-710 digital controller).*
- *Connectors and cables for the trigger lines.*

3 Sample Preparation

There are many different ways to prepare a sample depending on the size, transparency, brightness of the fluorophores, etc. One much used sample preparation method to study the photo-physical properties of single emitters is to embed the emitters in a thin polymer film to spatially separate and localize the emitters. To prepare such a sample very clean coverslips are needed to avoid contaminating impurities. A good way to clean coverslips is by placing them in an ozone cleaner for at least one hour.

Dilute fluorophores in 2 %wt. polyvinyl alcohol (PVA) to a concentration below 1 nM and spin cast this dilute solution onto the clean coverslip at 6,000 RPM for 30 s.

- *#1.5 coverslips (EMS diasum).*
- *Ozone cleaner (UV/Ozone ProCleaner Plus, Bioforce, San Diego, CA).*
- *Eppendorf tubes, pipets, and pipet tips.*
- *Fluorophores in spectroscopy grade water (Invitrogen's eFluor650 quantum dots are good fluorophores to test the performance of the setup).*
- *2 %wt. filtered polyvinyl alcohol (PVA) in spectroscopy grade water.*
- *Spin-coater.*

4 Measurement Procedure

Put a drop of immersion liquid on top of the immersion objective and mount a sample at the sample holder which is fixed to the piezo scanning stage. Switch on the laser and focus the excitation light beam on the upper interface of the coverslip (caution, high-power radiation). Direct the emission light towards the confocal detector (s). Insert appropriate spectral filters at the excitation and emission path. Define the scan settings of the piezo scanning stage and apply the same settings to TCSPC hardware or Andor Newton spectrograph camera. The TCSPC hardware can be either configured to store recorded data in its RAM memory (memory configuration required) or configured to record first-in-first-out (FIFO) data, which is read out for each pixel. For spectral imaging, the Andor Newton camera is set to "kinetic" readout mode with "fast external" triggering. The kinetic series length equals the total number of pixels that will be scanned during the measurement. After arming the confocal detector(s), the detectors will wait for scan triggers from the scanning stage controller and will start recording when the scanning stage controller starts moving the stage from pixel to pixel.

The multimodal microscope can record intensity, lifetime, excitation spectra, and emission spectra in the configuration described in this chapter. From a practical perspective, it is recommended to fix the excitation wavelength and first record an intensity map to find the right orientation of the sample (confocal microscopy)

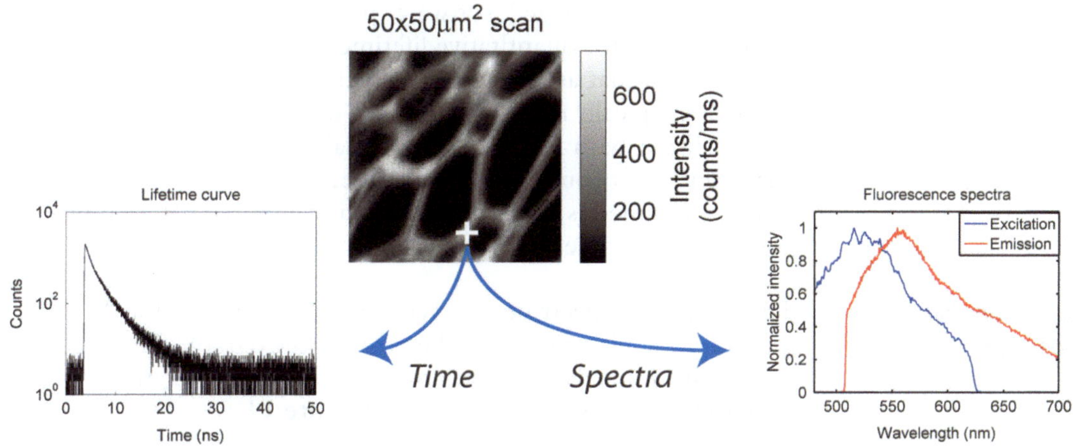

Fig. 4 Record lifetime and spectral data of *Convallaria majalis* at specific locations of the intensity map. First, the intensity map is measured by raster scanning an area to find the right orientation of the sample. Lifetime curves and excitation and emission spectra are recorded at a specified location, indicated by the *cross*. The sharp drops in the spectra indicate the cutoff edges of the filters

and secondly record a lifetime map and spectral maps (Fig. 4, *see* **Note 9**). Here, it is very important to choose appropriate short-pass filters and long-pass filters as described in the section "confocal detectors." For both excitation and emission spectra, it is very important to measure background spectra as well, since these are required for proper data analysis.

5 Analysis

The "measurement procedure" describes how to record intensity, lifetime, excitation, and emission spectra for each pixel. Once this data has been acquired, there are a few post-processing steps required to correct this data, depending on the data type. These corrections will be described for each observable in separate paragraphs below. After applying the appropriate corrections, the temporal and spectral properties of each pixel are known.

5.1 Intensity and Lifetime

The intensity is calculated from the number of photons of the recorded FIFO photon stream (*see* **Note 10**). In case that the background (number of uncorrelated photons) is high compared to the fluorescence signal, one can use the decay curve to filter out uncorrelated background photons. A threshold can be used that defines occurrences in the decay curve that correspond to fluorescence photons, which are then separated from uncorrelated photons. The number of fluorescence photons is then used to calculate the intensity map.

The "SPC Image" software package of Becker and Hickl GmbH allows for quantitative lifetime analyses of the fluorescence decay curves that are measured at each pixel. The measured decay curves are typically fitted with a mono-exponential decay model but can be fitted with a multi-exponential decay model if the decay characteristic is complex. The lifetime, which is the time-constant of the spontaneous emission process, is calculated for each pixel to construct a lifetime map.

5.2 Emission Spectrum

The emission spectrum is recorded with the Andor EMCCD camera, which is attached to the prism spectrograph. This recorded spectrum (S) needs to be background (B) subtracted and corrected for spectral response (R) (Eq. 1) (*see* **Note 11**):

$$S_{\mathrm{em}} = \frac{S - B}{R(\lambda)} \tag{1}$$

5.3 Excitation Spectrum

The excitation spectrum resembles the absorbance spectrum in case the quantum efficiency remains constant for all excitation wavelengths. Proper corrections should involve background subtraction and wavelength dependent photon flux corrections, as indicated by Eq. 2:

$$S_{\mathrm{exc}} = \frac{S - B}{\Phi_{\mathrm{d}}} \tag{2}$$

where S_{exc} is the corrected excitation spectrum that is calculated from the measured fluorescence intensity at different wavelengths S, background spectrum B, and photon flux density Φ_{d}. The photon flux density is the number of photons per second (Φ) per unit area, where the diffraction-limited area A scales with λ^2 (Eq. 3). The photon flux (Φ) is determined from the illumination power P divided by the energy of a single photon E_{ph}, which inversely scales with λ:

$$\Phi_{\mathrm{d}}(\lambda) = \frac{\Phi(\lambda)}{A(\lambda)} = \frac{P(\lambda)}{E_{\mathrm{ph}} \cdot A(\lambda)} \sim \frac{P(\lambda)}{\lambda} \tag{3}$$

6 Notes

1. *Optimizing stability and flexibility of the microscope body.* Before starting the assembly procedure, it is highly recommended to selectively remove or add optical elements inside the microscope body to improve stability and flexibility. These optical elements cause aberrations and losses and would additionally require even more optical elements to collimate light and direct this light towards the detectors. First, the tube lens

needs to be removed, which can be accessed below the plate (IX2-CA2X-2) that resides directly under the filter-cube turret. There also is a lens system present behind the observation tube (attachment point for the binoculars) which needs to be removed. This lens system compensates the tube lens and optically shortens the light path distance but is not useful anymore in the design of Fig. 1. One more superfluous glass element can be found at the left side-port slider. This glass cube formerly shortened the light path but only causes losses in the configuration of Fig. 1. Once these elements have been removed, the bottom port mirror needs to be replaced with a flipable bottom port mirror, so emission light can exit through the bottom of the microscope without losses. The flipable bottom port mirror needs to be adjusted by the screws at its bottom to realign the light path from the imaging objective towards the binoculars. When all these adjustments have been made, lift the microscope at least 7 cm and fix it on robust stilts (ø1.5″ posts) to allow for a 45° angled mirror below the bottom port.

– *Bottom port flip mirror (IX2-TVR)*

2. *Wide-field light sources.*
   The sources that are used for wide-field illumination are preferably light sources with low coherence, because coherent sources cause interference at the illumination pattern. For coherent light sources, one should implement additional diffusive optical elements to scramble the phase of the coherent light. The best choices for wide-field illumination sources in the design of Fig. 1 are a tungsten halogen lamp and high-power light-emitting diodes (LEDs) for transmission microscopy and high-power LEDs and a mercury lamp for epi-illumination.
3. *Collimating the excitation beam.*
   It is recommended to use infinity-corrected optics to collimate the excitation light beam by changing the distance between the fiber exit and collimation objective. When the excitation light is collimated, one should see a symmetric and crisp focus through the infinity-corrected optics. One very critical aspect of the collimation procedure is the alignment of the angle and position of the fiber exit with respect to the central optical axis of the collimation objective. The best way to optimize angle and position is by judging the quality of the beam profile, which should be Gaussian shaped. A controlled procedure to align the angle and position of the fiber exit is (1) to place a mirror perpendicular to the central optical axis and align the fiber exit in lateral dimensions (XY) until the back reflected light roughly hits the core of the fiber. (2) Then, the mirror needs to be removed again, and the XY tilt angles are adjusted until

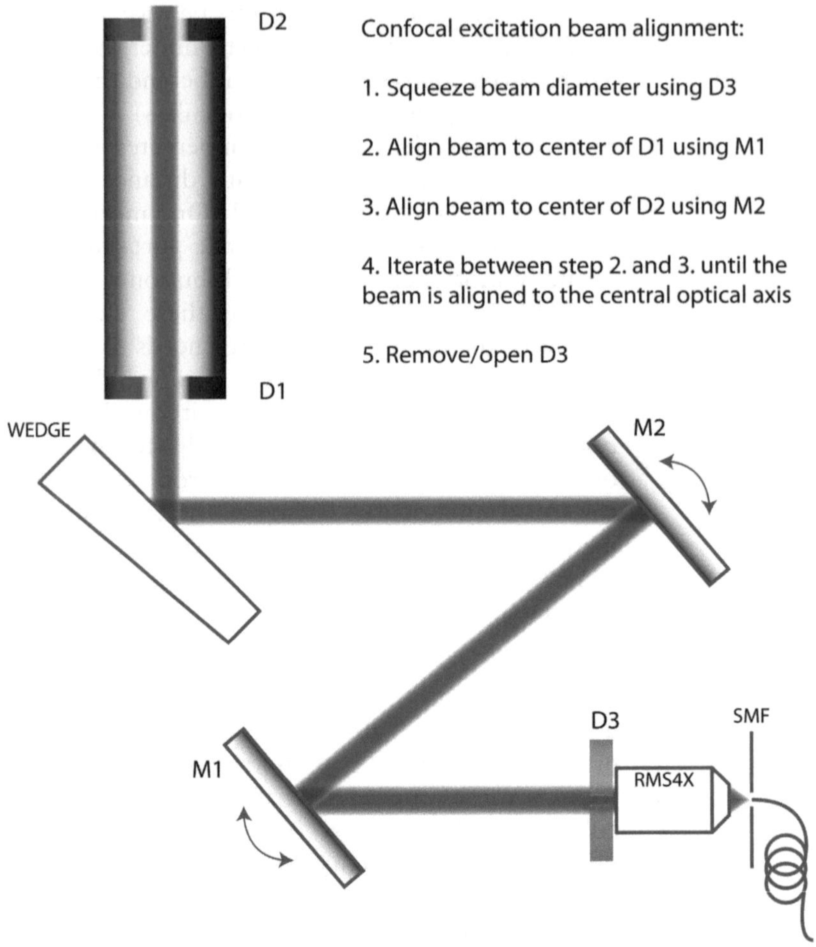

Fig. 5 Alignment procedure of the confocal illumination path. Use a diaphragm (D3) to shrink the beam diameter and then use the first mirror (M1) to align the light to the center of the first diaphragm (D1) and the second mirror (M2) to align the light to the center of the second diaphragm (D2). Iterate between these steps to align the beam to the central optical axis

the beam profile is Gaussian shaped and does not show shades around the edges of the beam profile. Iterate between these two steps to obtain a beam with good quality and collimate the beam using the depth translation (Z-axis).

4. *Aligning the excitation beam to the central optical axis.*
The easiest way to align the excitation beam is using two mirrors and a long alignment tube which has two diaphragms; one at the beginning and one at the end (Fig. 5). Use the alignment tube as a replacement for the imaging objective to precisely align the excitation beam to the central optical axis of the imaging objective. Since the beam diameter is broad, mount and close a ring diaphragm at the back of the collimation objective to shrink the beam diameter, which makes it a lot

easier to align the excitation beam using the two diaphragms on the alignment tube.

- Alignment tube: SM1 tubes and 2 diaphragms for proper alignment.
- RMS-SM1 external-internal adapter ring to align the laser light to the center optical axis of the imaging objective.
- 1 ring diaphragm.

5. *Suppressing unwanted light.*
   The main sources of stray light are leakage of excitation light through emission filters or simply stray light from reflections in the environment. Leaking excitation light can be suppressed by choosing the proper filter set. These filters should have an optical density of at least 6 at the operable spectral ranges. Further, proper boxes are needed to shield the detectors and emission beam path from environmental stray light.
6. *Avalanche photodiode (APD) and TCSPC.*
   The avalanche photodiode (APD) gives a signal when a photon is registered but does not provide photon timing data. The Becker & Hickl TCSPC card is required to record this signal and resolve absolute photon arrival times ("macrotimes") and arrival times relative to the excitation pulse ("microtimes") from the APD's signal.
7. *Spectrograph type.*
   Either a grating or a prism spectrograph can be used for multimodal imaging spectroscopy, although it is recommended to use a prism spectrograph, because its losses are less than 20 %.
8. *Electronic "on/off" switching of the SC400-PP laser.*
   To switch the Fianium SC400-PP "on" or "off" via a TTL line at approximately 1 kHz speed, one needs to solder a 15 way D connector, where pin 4 is attached to "Puma enable" and pin 8 to "ground."
9. *Recording data.*
   Excitation spectra, intensity, and lifetime maps can be recorded all at once if the TCSPC card records FIFO images at different excitation wavelengths. Emission spectra are recorded subsequently using the same scan settings but at a fixed excitation wavelength.
10. *First-in-first-out (FIFO) photon streams.*
    The photon streams are recorded with the APD and TCSPC card for each pixel. For each photon, both "macrotime" and "microtime" information was recorded. The "macrotime" is the time relative to the start of the recording, and "microtime" is the time delay between the detected photon and the excitation pulse. Macrotimes are used to construct an intensity trace, and microtimes are used for decay dynamics, which is

generally calculated by making a histogram of the "microtimes" of all of the recorded photons. As with the intensity trace, "microtimes" can also be used to construct decay traces, by calculating the average photon arrival time per unit time.

11. *Spectral response correction of the emission spectrum.*
The spectral response correction is not required if the spectral response is approximately constant along the spectral range of the recorded emission spectrum.

References

1. Blum C, Cesa Y, Escalante M, Subramaniam V (2009) Multimode microscopy: spectral and lifetime imaging. J R Soc Interface 6 (Suppl 1):S35–S43. doi:10.1098/rsif.2008.0356.focus
2. Schierle GSK, Bertoncini CW, Chan FTS et al (2011) A FRET Sensor for Non-Invasive Imaging of Amyloid Formation in Vivo. Chemphyschem 12(3):673–680. doi:10.1002/cphc.201000996
3. Chan FTS, Kaminski CF, Schierle GSK (2011) HomoFRET Fluorescence Anisotropy Imaging as a Tool to Study Molecular Self-Assembly in Live Cells. Chemphyschem 12(3):500–509. doi:10.1002/cphc.201000833
4. Fereidouni F, Bader AN, Gerritsen HC (2012) Spectral phasor analysis allows rapid and reliable unmixing of fluorescence microscopy spectral images. Opt Express 20 (12):12729–12741
5. Frank JH, Elder AD, Swartling J et al (2007) A white light confocal microscope for spectrally resolved multidimensional imaging. J Microsc 227 (3):203–215. doi:10.1111/j.1365-2818.2007.01803.x
6. Owen DM, Auksorius E, Manning HB et al (2007) Excitation-resolved hyperspectral fluorescence lifetime imaging using a UV-extended supercontinuum source. Opt Lett 32(23):3408–3410
7. Betz T, Teipel J, Koch D et al (2005) Excitation beyond the monochromatic laser limit: simultaneous 3-D confocal and multiphoton microscopy with a tapered fiber as white-light laser source. J Biomed Opt 10(5):054009–054009. doi:10.1117/1.2114788
8. Blum C, Schleifenbaum F, Stopel M et al (2011) Room temperature excitation spectroscopy of single quantum dots. Beilstein J Nanotechnol 2:516–524. doi:10.3762/bjnano.2.56

Part VII

Fluorescence Microscopy: Fluorescence Fluctuation Spectroscopy (FFS)

Chapter 24

Application of Fluorescence Correlation Spectroscopy (FCS) to Measure the Dynamics of Fluorescent Proteins in Living Cells

Thomas Weidemann

Abstract

Fluorescence correlation spectroscopy (FCS) can add dynamic molecular information to images of live cells. For example, a confocal laser scanning microscope (CLSM) equipped with an accessory FCS unit provides the possibility to first image the spatial distribution of a fluorescent protein before probing its mobility within defined regions of interest. Whereas specific protein–protein interactions are preferably assayed with a dual-color approach, single-color FCS can still provide valuable information about the size of the diffusing entities and potential interactions with other, nonfluorescent, proteins or subcellular structures. Because number fluctuations are measured, the concentrations of freely diffusing complexes and their state of oligomerization are accessible.

Key words Fluorescence correlation spectroscopy, Green fluorescent protein, Heterologous expression, Confocal laser scanning microscopy, Diffusion coefficient, Obstructed diffusion, Protein–protein interactions, Molecular brightness, Intracellular concentrations

1 Introduction

With FCS the diffusion properties and concentrations of fluorescent particles are measured in free solution by recording and correlating a fluorescence signal originating from an illuminated tiny, subfemtoliter-sized spot. The microscopic observation volume is usually implemented by a confocal optical alignment, for example, as provided by a CLSM. FCS is measurable in solutions from higher pico- to lower micromolar concentrations (~100 pM to 1 μM). For too high concentrations, the relative fluctuations are too small to be measurable. For too low concentrations, very long measurement times are needed to collect a sufficient number of molecular events. As a rule of thumb, solutions at 1 nanomolar (1 nM) are well suited for measuring FCS in free solution as well as in living cells with recordings in the time scale of several minutes ($1\ \text{nM} \times 1\ \text{fl} \times N_A \approx 1$

Yves Engelborghs and Antonie J.W.G. Visser (eds.), *Fluorescence Spectroscopy and Microscopy: Methods and Protocols*, Methods in Molecular Biology, vol. 1076, DOI 10.1007/978-1-62703-649-8_24,

particle in the observation volume). The output of a single-color FCS measurement is the time-correlated fluorescence signal displayed as an autocorrelation curve (ACF, e.g., *see* Figs. 2–5). Pairs of parameters for both correlation times of the decay (on the abscissa) and corresponding contributions to the fluctuation amplitude (on the ordinate) are extracted by fitting appropriate model functions to the ACF. Fitting provides a set of times constants which have to be assigned to underlying physical processes like fluorophore blinking and the different mobility states of the particles.

An overview about FCS theory, application, and instrumentation can be found in numerous papers [1–7]. The autocorrelation function (ACF; $G(\tau)$) is generated from the fluorescence signal $F(t)$

$$G(\tau) = \frac{\langle F(t) \cdot F(t+\tau) \rangle}{\langle F \rangle^2} - 1 \tag{1}$$

and evaluated by the experimenter with a set of assumptions. For example, a widely used model function for 3D-diffusion with triplet-blinking is of the form [8]

$$G(\tau) = \frac{1}{cV_{\text{eff}}} \left(\frac{1 - \theta_{\text{nf}} + \theta_{\text{nf}} \cdot \mathrm{e}^{-\tau/\tau_{\text{nf}}}}{1 - \theta_{\text{nf}}} \right) \sum_{i=1}^{3} \theta_i \left(1 + \frac{\tau}{\tau_{\text{d},i}}\right)^{-1} \left(1 + \frac{\tau}{S^2 \cdot \tau_{\text{d},i}}\right)^{-1/2}, \tag{2}$$

assuming a three-dimensional (3D) Gaussian shape of the observation volume $V_{\text{eff}} = \pi^{3/2} S w_0^3$ and a single nonfluorescent state (*see* **Note 1**). In this model, the observation volume is assumed to be stretched along the optical axis z by a factor $S = z_0/w_0$. The beam waist w_0 denotes the radius along which the intensity profile decays from the center to $1/e^2$. The time constant for populating nonfluorescent states τ_{nf} and up to 3 diffusion times $\tau_{\text{d},i}$ as well as their fractional contributions to the fluctuation amplitude, θ_{nf} and θ_i, can be extracted as parameters. If the fluorescent ensemble is homogeneous in molecular brightness, the inverse, dark state-corrected intercept of the ACF equals the average number of fluorescent molecules in the detection volume V_{eff}

$$\frac{1}{cV_{\text{eff}}} = \frac{1}{N_{\text{ob}}} = (1 - \theta_{\text{nf}}) G(0), \tag{3}$$

In the following, I describe a tutorial of how to perform a single-point, single-color FCS experiment with continuous wave laser excitation inside living cells. The goal is to illustrate how a cytoplasmic protein fused to enhanced green fluorescent protein (EGFP) can be characterized under native conditions. The approach comprises cloning of the EGFP-fusion construct, characterization of the microscope with an organic dye (Alexa488), and probing of the EGFP sensor with increasing environmental complexity. A vector system is introduced for expression of the fluorescent protein in concentrations suitable for FCS. Finally, application of the method is illustrated with data obtained for EGFP-tagged human Janus kinase 3 (JAK3) in the cytoplasm of an epithelial cancer cell line.

2 Materials

1. Mammalian expression vector (pEGFP-N1, Clontech).
2. Inducible mammalian vector expression system (Complete Control; Stratagene).
3. Inducible human adherent cell line (ER293; Stratagene).
4. Culture media (Dulbecco's Modified Eagle Medium (DMEM) with 10 % FCS; Gibco).
5. Transfection reagent (Attractene; Quiagen).
6. Air buffer (150 mM NaCl, 20 mM Hepes pH 7.4, 15 mM Glucose, 46 mM Trehalose, 5.4 mM KCl, 0.85 mM $MgSO_4$, 1.7 mM $CaCl_2$, 0.15 mg/ml BSA).
7. Human fibronectin (lyophilized, Roche).
8. Chambered cover glass #1 (8-well Cat. No.:#155411, Nunc).
9. Nuclear stain (DRAQ5, Biostatus Limited, UK).
10. AlexaFluor 488-TFP (CatNo.:#30005, Invitrogen/Molecular Probes, USA).
11. Recombinant EGFP (CatNo.:#4999-100, BioVision, USA).
12. Confocal microscope equipped with FCS unit (e.g., Confocor3, Zeiss, Germany).

3 Methods

The protocol includes (1) the preparation of the samples and cells, (2) adjustment of the microscope, (3) the required control measurements, and (4) intracellular measurements and interpretation of the data.

3.1 Expression of Fluorescent Protein in Culture Cells

Heterologous expression of EGFP-tagged fusion constructs is now standard in cell biology. However, many vector systems contain powerful viral promoters and produce much higher expression levels as they are needed for FCS. For example, cytomegalovirus (CMV)-driven transcription of EGFP with the widely used vector *pEGFP-N1* (Clontech) will accumulate μM concentration of fluorescent protein inside the cells. In addition, due to stochastic

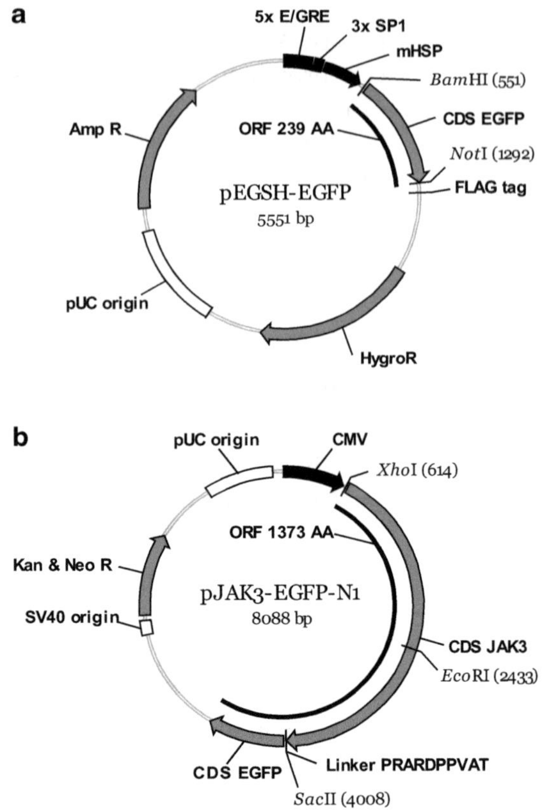

Fig. 1 Vector maps. (**a**) Schematic drawing of pEGSH-EGFP expression plasmid for cytoplasmic EGFP. (**b**) Schematic drawing of expression plasmid pJAK3-EGFP-N1 for the expression of C terminally EGFP-tagged JAK3

incorporation of plasmids, the individual cellular expression levels vary over three orders of magnitude and thus charge the experimenter with extra time to search and select suitable cells for FCS analysis. Therefore, it is convenient to lower expression by using either very weak promoters or inducible expression systems (*see* **Note 2**).

3.1.1 Cloning

DNA manipulation was performed by standard recombinant DNA methods. The vector *pEGSH-EGFP* was generated by subcloning the coding sequence of EGFP from *pEGFPP-N1* (Clontech) into *pEGSH* (Stratagene) via BamH1/NotI restriction sites (Fig. 1a). JAK3 was inserted into *pEGFP-N1* via XhoI/SacII as described [9]. *pJAK3-EGFP* was kindly donated by Sigrun Hofmann (Fig. 1b).

3.1.2 Transient Transfection of ER293 Culture Cells

ER293 cells were passaged in DMEM supplemented with 10 % fetal bovine serum and 300 μg/ml G418. Subconfluent cells were harvested by trypsination and diluted to 8 × 10^4 cells/ml.

The chambered 8-well cover slips were coated by covering the glass surface with 10 μg/ml human fibronectin in PBS followed by incubation of 30 min at 37 °C. Fibronectin was aspirated and 200 μl cell suspension seeded immediately into the well. After 24 h, the medium was replaced prior to adding 50 μl of transfection mix (100 ng DNA and 0.38 μl Attractene in OptiMEM). Three hours after transfection, the supernatant was replaced with fresh medium. The cells can be observed after 12–24 h for up to 2 days, depending on confluence.

3.1.3 Prepare for Imaging

From a biological perspective, observing cells under physiological conditions at 37 °C and CO_2 containing, humidified atmosphere may seem ideal; however, there are drawbacks for FCS analysis. Cell motility is much faster; therefore, the internal membrane systems are in constant motion. Under such conditions, it is difficult to find homogeneous regions within the cytoplasm which are not constantly crossed by mitochondria and other organelles. Organelles may exclude the fluorescent protein. Dark (or bright) organelles which traverse the detection volume lead to dramatic intensity changes which render the correlation data worthless.

The measurements described here have been performed at room temperature (22 °C). Cells were mounted on cover slides on top of a microscope stage. Using culture media as a supernatant for intracellular FCS measurements features several disadvantages. Due to high concentrations of supplemented proteins and pH indicators, it provides high levels of background signal. In addition, when removing the cells from the incubator, the CO_2 release leads to a rapid increase of pH, which may induce unwanted physiological stress. Therefore, cells were observed in a sterile translucent air buffer (for the composition, *see* Subheading 2) [10]. The solution contained a Hepes-based pH buffering system, physiological salts, and trehalose to alleviate osmotic stress. Measurements in air buffer can be performed for several hours.

Although the intracellular EGFP distribution provides some contrast, for proper positioning inside the cell, it may be beneficial to stain the nucleus. Here, I used DRAQ5, a 633 nm excitable intercalating dye in the far red, which does not interfere with the EGFP emission window. For staining, a 5 mM stock was freshly diluted 1:1,000 in air buffer, incubated 5–10 min at RT, and washed once in air buffer.

3.2 The Microscopic Settings

Irregardless of differences in technical implementation, the microscopic system has to be characterized with a common set of parameters prior to investigative FCS measurements.

3.2.1 Confocal Alignment

By using a solution of organic dye as a fluorescent standard, like Alexa488, the system has to be optimized for a good confocal alignment. While positioning the focus in the solution, confocality is improved by iteratively moving the pinhole and the aperture of the photon counting detector to maximum signal position. With a ConfoCor3, this has been automated, and the user just activates a software routine. When the alignment is far off, it is useful to start with higher, e.g., μM concentrations of dye. At maximum signal, the correction collar of the objective has to be adjusted for maximum counts per particle (CPP) or, if the product of the intercept of the ACF and the mean intensity is not monitored by the control software, the amplitude of the ACF (*see* **Note 3**). For the ConfoCor3 microscope, it was sufficient to optimize simply the count rate since the maxima for intensity and amplitude did coincide; however, this may not always be the case. As a result, one should be able to record correlation data with at least 10 kHz counts per particle.

Generally, the quality of the alignment is validated by fitting a 1-component model function (Eq. 2 for $i = 1$) to the measured ACF for Alexa488 and to check, whether the residuals omit a systematic deviation (*see* Fig. 2c, d). Systematically biased residuals may indicate that the point spread function deviates from a 3D Gaussian distribution. In such a case, the application of analytical model functions like Equation (2) would not be justified.

3.2.2 Pinhole

The optimal pinhole size is determined by the maximum counts per particle detectable for a given excitation intensity. Diameters between 50 and 90 μm are commonly used. With a simple home-made confocal optical setup using a pinhole between 50 and 70 μm in combination with an Olympus 63× 1.2 W objective, dwell times for Alexa488 between 35 and 50 μs can be measured [11, 12]. With a ConfoCor3, the beam additionally passes a system of projecting lenses between the pinhole and the back aperture of the objective resulting in dwell times of 25 μs, even when using a 90 μm pinhole in combination with a lower magnification objective Zeiss 40× 1.2 W (Fig. 1a, d).

3.2.3 Laser

Diffusion times can be measured provided that the average signal remains constant within the measurement time of several minutes. To compare the counts per particle values between different sets of measurements, one has to assure long-term stability of the laser. E.g. using the 488 nm line of an Argon ion laser may require pre-warming of several hours. However, the absolute laser power is usually not accessible during the experiment. Recording 10 kHz per particle from a nanomolar solution of Alexa488 in water corresponds to an excitation intensity of about 2 kW/cm^2 as measured with a 40× 1.2 W objective [13]. Thus, for FCS measurements in a diffraction-limited confocal spot ($w_0 \cong 200$ nm), several μW are sufficient for excitation.

3.3 Measuring FCS

The following section lists the individual steps of how to perform an FCS measurement and, more importantly, shows with examples how the user can interpret the measured parameters.

3.3.1 Procedure for Intracellular FCS

1. Switch on the Argon ion laser at least 1 h before measuring (488 nm line).
2. Take the cells out of the incubator, eventually perform nuclear stain, wash, and transfer into air buffer. Give the cover glass 15 min to cool down (*see* **Note 4**).
3. Clean the bottom of the chambered cover glass with 100 % pure ethanol (pro analysis), dry with tissue, and avoid finger tips.
4. Place a 50–100 μl droplet of immersion water on the objective and mount the sample on the microscope stage.
5. Calibrate the *z*-position by focusing an object or looking at the upper back reflection of laser light at the glass/water interface.
6. Move 50–100 μm into a well containing a 5–25 nM solution of Alexa488 in 10 mM Tris-HCl pH 8.
7. Check the confocal alignment.
8. Adjust the correction collar at the objective for maximum intensity or, even more accurately, for maximum counts per particle (CPP).
9. Adjust excitation power for the Alexa488 standard solution between 5 and 10 kHz per particle.
10. Measure a triplicate of correlation curves 12 × 10 s at three positions → determine the axis ratio S and calculate V_{eff} to calibrate of the focal volume (*see* **Note 5**).
11. Move into a well with the cells at $z = 3$–5 μm.
12. Search for suitable fluorescent cells by visual inspection or confocal scanning at low magnification.
13. Position the cell in the center and zoom in (field of view ~50 × 50 μm; *see* **Note 6**).
14. Position the focus in the subcellular region of interest and readjust the excitation intensity.
15. Measure multiple runs of 10–30 s depending on diffusion times (*see* **Note 7**).
16. Discard runs for which the intensity visibly changes on the time scale of seconds; very slow fluctuations or bleaching of more than 10 % along the trace (*see* **Note 8**).
17. Fit the remaining correlation curves with Eq. 2 (*see* **Note 9**).
18. Measure the background fluorescence in non-transfected cells (*see* **Note 10**).

3.3.2 Control Measurements with Alexa488

Figure 2 summarizes an exercise in which a dilute nanomolar aqueous solution of Alexa488 was measured at different *z*-positions along the optical axis. The diffusion times and the average numbers of observed particles in the detection volume were derived from a 1-component fit and plotted against the focal depth (Fig. 2a, b).

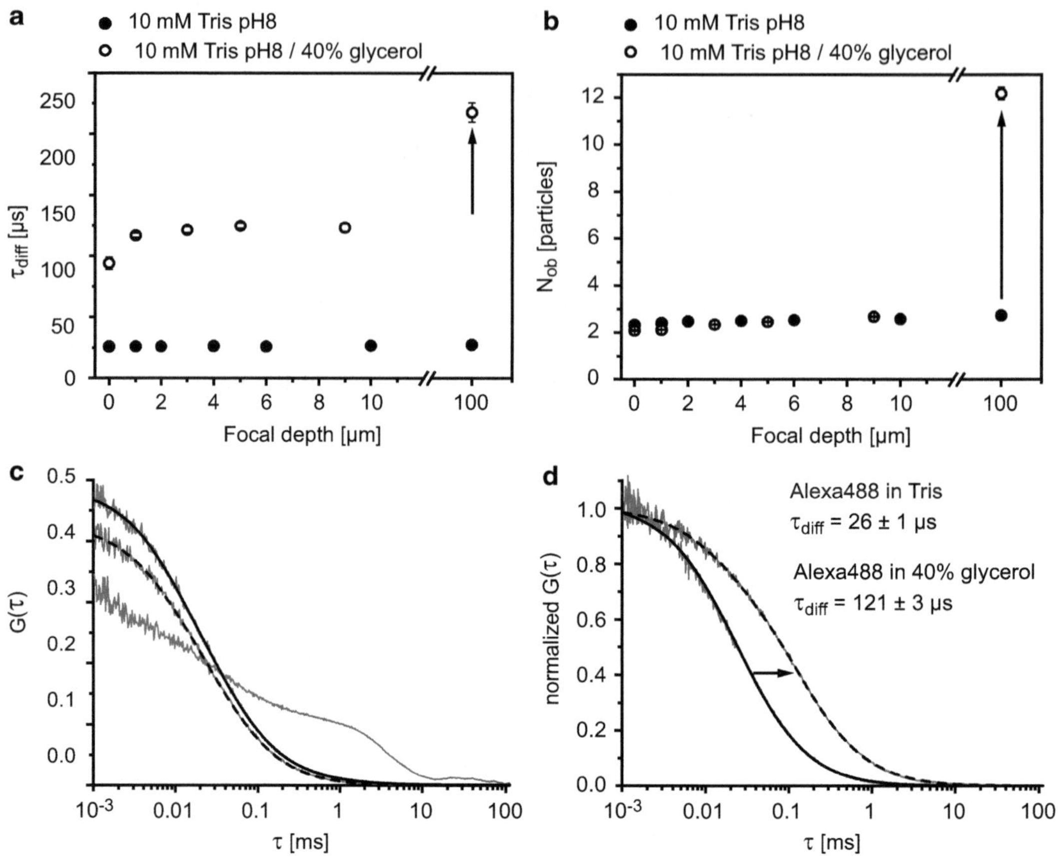

Fig. 2 Control measurements using an organic dye. Diffusion times (**a**) and particle numbers N_{ob} (**b**) of Alexa488 in Tris buffer containing 0 % (*closed circles*) or 40 % glycerol (*open circles*). (**c**) ACFs at different distances from the glass surface $z = -1$ μm, $z = 1$ μm (*dashed fit*), and $z = 10$ μm (*straight fit*). (**d**) Normalized ACF for Alexa488 in Tris buffer (*straight fit*) or Tris buffer containing 40 % glycerol (*dashed fit*)

In Tris buffer, Alexa488 showed a diffusion time of ~25 μs. Both the diffusion time and the inverse amplitudes, indicating in average two particles in the focal volume, were quite stable over the whole range of 1–100 μm penetration depth. Figure 2c displays the corresponding correlation curves for $z = -1$, 0, and 10 μm. While the ACF measured at 1 and 10 μm appear similar, a step of 1 μm below glass surface totally distorts the shape of the ACF. Due to uncertainty of imaging-based positioning (±1 μm), placement close to the glass surface deserves some caution.

The focal depth suddenly matters in solutions with higher refractive index than water ($n = 1.333$) for which the water immersion objectives are optimized. To illustrate this effect, I used PBS supplemented with 40 % glycerol. 40 % glycerol is about 3.6-fold more viscous than water and has a significantly higher refractive index ($n = 1.384$) and thus mimics properties as typically encountered in the cytoplasm of cells ($n = 1.36$–1.39) [14, 15]. As can be seen in Fig. 2d, the diffusion time of Alexa488 in glycerol measured at $z = 4$ μm was four- to fivefold larger than determined in aqueous buffer, and this merely reflects changes in viscosity. However, the viscosity-dependent changes were additionally increased at $z = 100$ μm due to the blurred and enlarged detection volume (Fig. 2a, b) [16]. In summary, even within 40 % glycerol, positioning of the laser focus within the z-range of 2–10 μm yields stable results for both diffusion times and particle numbers, and this range matches quite nicely the dimensions of a cellular monolayer.

3.3.3 Control Measurements with Purified EGFP

Adjustment of excitation power is crucial for intracellular measurements. In this section this issue is illustrated with solution measurements. Let us first compare purified recombinant EGFP with Alexa488 in aqueous buffers for different excitation powers. Figure 3a shows the diffusion time for EGFP is overall increased (~100 μs) in comparison to Alexa488 (~25 μs), which can be

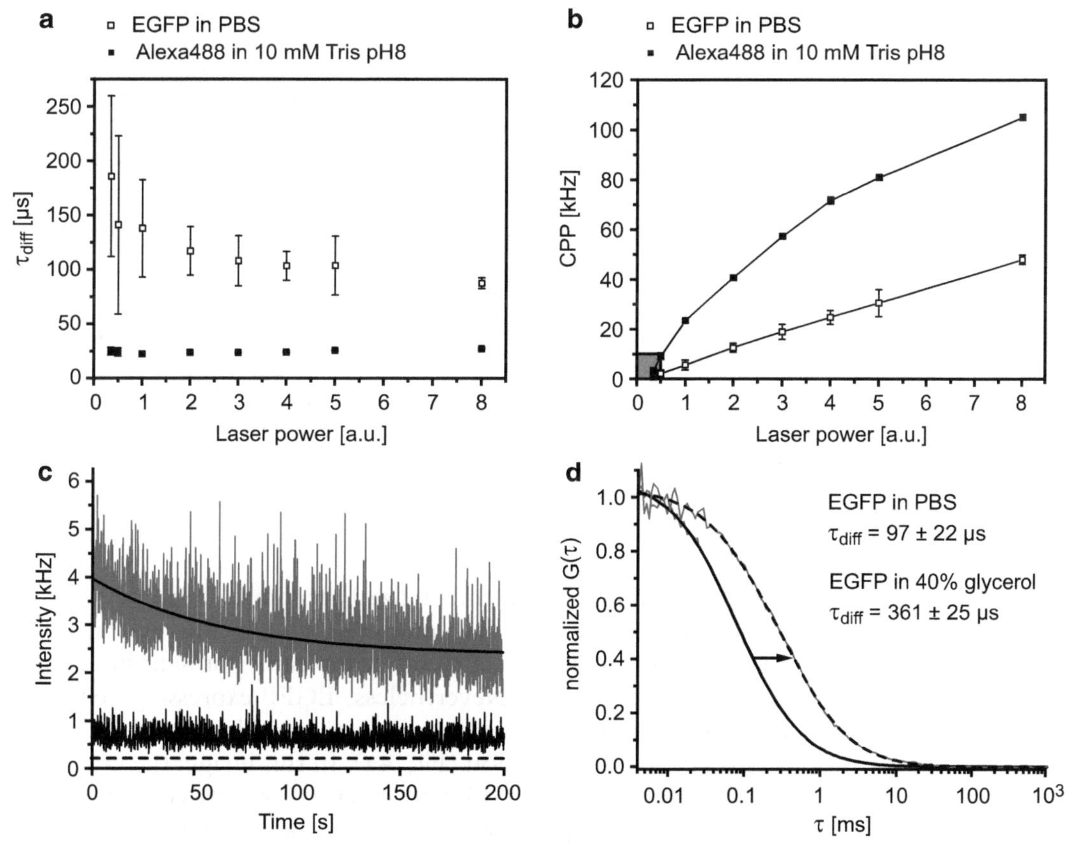

Fig. 3 Control measurements of recombinant EGFP. Diffusion times (**a**) and CPP (**b**) of Alexa488 and recombinant EGFP in aqueous buffers for increasing excitation power. (**c**) Intensity trace for EGFP in PBS/ 40 % glycerol measured at two excitation powers as used for generating the ACFs in (**d**): EGFP in PBS (*straight fit*) and in 40 % glycerol (*dashed fit*)

attributed to their difference in molecular size (*see* **Note 11**). Interestingly, Alexa488 and EGFP behave differently with respect to higher excitation powers. We see that the diffusion times for EGFP systematically decrease due to pronounced bleaching of molecules in the illuminated area. In contrast, the diffusion times of Alexa488 seem stable for up to 40 kHz per particle, yet even slightly increase from 25 to 27 μs (Fig. 3a). Evaluating the molecular brightness reveals a perfect linear rise of the CPP for EGFP, whereas the molecular brightness of Alexa488 increases nonlinearly (Fig. 3b). Both effects, the increased diffusion time and the nonlinearity of the CPP of Alexa488, were attributed to saturation effects of the fluorophores in the excited state [17, 18].

In intracellular applications, two aspects further slow down the diffusion times: the cytoplasm is 4–5 times more viscous than water, and the genetic fusion of a protein of interest increases the molecular size. FCS analysis of slow particles at low concentrations may require to readjust the excitation power. Figure 3c shows intensity traces of the recombinant EGFP solution ($N_{ob} = 0.6$ particles) in PBS containing 40 % glycerol. As apparent from the decay, it was not possible to measure EGFP in 40 % glycerol at the same excitation power as in pure PBS (not shown). In glycerol the longer residence time of EGFP in the observation volume (*see* also ACFs in Fig. 3d) leads to pronounced bleaching (Fig. 3c, gray trace fitted with a single exponential). Therefore, to obtain a stable intensity trace, the excitation power had to be further reduced down to a factor of 3 above background (Fig. 3c, black trace above the dashed straight line). Note that a bleaching-induced intensity decay at higher concentrations may be covered by a rapid influx of fluorescent molecules from the bulk. Hence, even stable intensity traces just reflect a steady state and bleaching can never be fully eliminated [11]. The problem of bleaching is even aggravated in cellular experiments because of potential interactions with other, nonfluorescent structures and the comparably small finite volume of the intracellular reservoir.

3.3.4 *Intracellular FCS*

Cells are highly compartmentalized structures. The wide use of EGFP in cell biology has mounted convincing evidence that EGFP is inert with respect to other molecules present in mammalian culture cells [19]. Nevertheless, EGFP expressing cells show contrast because not all volume elements inside the cell are equally accessible by pure diffusion (Fig. 4a). In this image, the gray values nonlinearly represent EGFP fluorescence for three cells: two highly expressing (dark) and one with very weak expression (light gray). If we look at the dark cells, we can easily distinguish the nucleus from the cytoplasm. Within the nucleus, the nucleolus and heterochromatin associated to the nuclear membrane can be distinguished. In the cytoplasm it is apparent that multiple vesicles exclude EGFP fluorescence and these endogenous membranes are

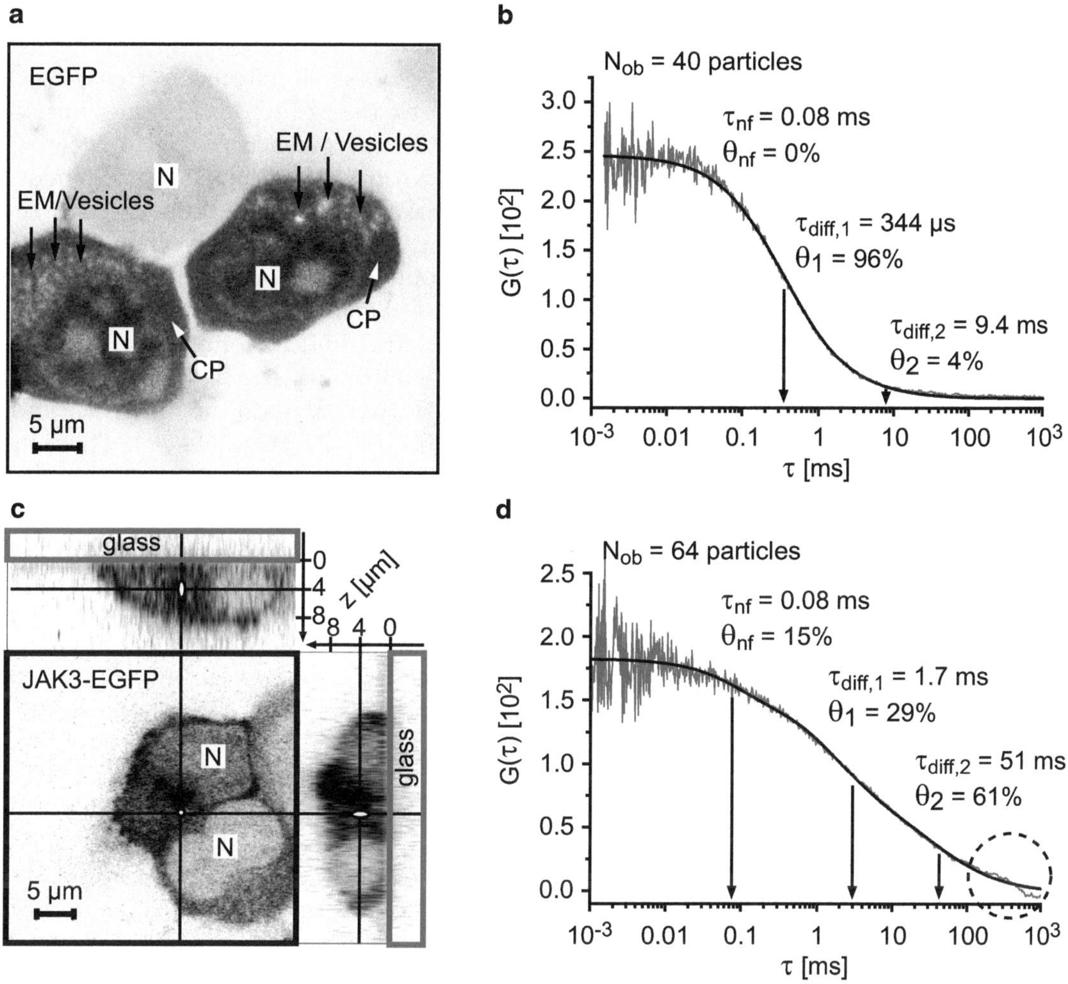

Fig. 4 Intracellular FCS. (**a**) Confocal cross section of ER293 cells expressing cytoplasmic EGFP. (**b**) ACF measured for cytoplasmic EGFP. *Subcellular structures: nucleus (N), homogeneous area of cytoplasm (CP), endogenous membranes like ER and Golgi (EM). (**c**) *z*-stack of confocal cross sections of a HEK293T cell expressing JAK3-EGFP. Position and dimension of the focal spot as positioned in the cytoplasm is indicated. (**d**) ACF measured at the spot with 10 s runs. *Dashed circle* indicates uncertainty at longer correlation times

likely associated with the endoplasmic reticulum, the Golgi apparatus, or mitochondria. These vesicular structures are preferentially located at one side of the nucleus. Cytoplasmic regions at the opposite side (indicated by arrows, CP) are more homogeneous and, when positioning the laser focus for FCS, provide a better stability. An example ACF is shown in Fig. 4b. As expected, diffusion time for free EGFP in the cytoplasm (344 μs) is of the same order as measured in 40 % glycerol (361 μs, Fig. 3d). Due to the low excitation power, the fraction of molecules in a nonfluorescent dark state was negligible. However, for fitting, a small fraction of an additional slow diffusion time (9.4 ms) had to be covered by a 2-component fit model (Eq. 2 for $i = 2$).

Such slow contributions likely represent slow instabilities inside the cellular environment.

Figure 4c shows a confocal cross section from a *z*-stack of a cell expressing JAK3-EGFP. Due to its size and a missing nuclear localization signal, JAK3-EGFP fluorescence is largely excluded from the nucleus. While the cytoplasm appears as homogeneous as for free EGFP, a short measurement (12 × 10 s) reveals the fundamental difference. Fitting of the ACF clearly requires a 2-component model and the second, slower correlation time (51 ms) contributes with significant 61 %. Even the shorter correlation time cannot be assigned to disperse JAK3-EGFP monomers (153 kDa). The diffusion time (1.7 ms) is significantly larger than what can be estimated for 3D diffusion (~613 μs) based on the molecular mass and the diffusion time of free EGFP (27 kDa, 344 μs). Thus, the diffusion data indicate that JAK3 forms some kind of complex. However, the second, diffusion time (51 ms) seems too slow as to represent complexes and rather reflects properties of the cellular environment. Crowding effects, obstructed diffusion due to obstacles, or transient binding have been proposed to produce these long-tail correlations [20, 21].

The ACF in Fig. 4d was chosen as a typical but nonoptimal example. Although positioning at $z = 4$ μm nicely locates inside the cell, the lateral position is a bit too close to the endogenous membrane systems (Fig. 4c). Moreover, the length of a single run (10 s) was set the same as for free EGFP (Fig. 4b). Comparison of the two correlation curves shows that the ACF of JAK3-EGFP is worse defined and does not fully approach to zero (dashed circle). As a result, the fitting suffers from a weaker significance. As mentioned, the duration of the runs have to be increased for slower particles.

Figure 5 shows a final data set measured in different positions of a small patch of transfected cells showing varying expression levels of JAK3-EGFP. The corresponding intensity traces (8 × 30 s) and correlation curves are shown (Fig. 5b, c). The reader may notice the reciprocal dependence of the correlation amplitude and the darkness of the measured area reflecting the protein concentrations at each spot. The intensity traces show that only a subset of the runs could be used for fitting (gray shaded). The data are summarized in the table. The intensities varied between 23 and 304 kHz emitted from average numbers of 8.1–63 particles in the observation volume. Comparing the columns "N_{ob}" with "CPP" shows that the molecular brightness correlates with intracellular concentration of JAK3-EGFP. Because fluorescent particles measured at the same laser power exhibit the same CPP, even for different concentrations, one has to conclude that the JAK3-EGFP molecules oligomerize. This observation seems to be corroborated by increasing diffusion times τ_1, which we associated with molecular size (*see* **Note 12**).

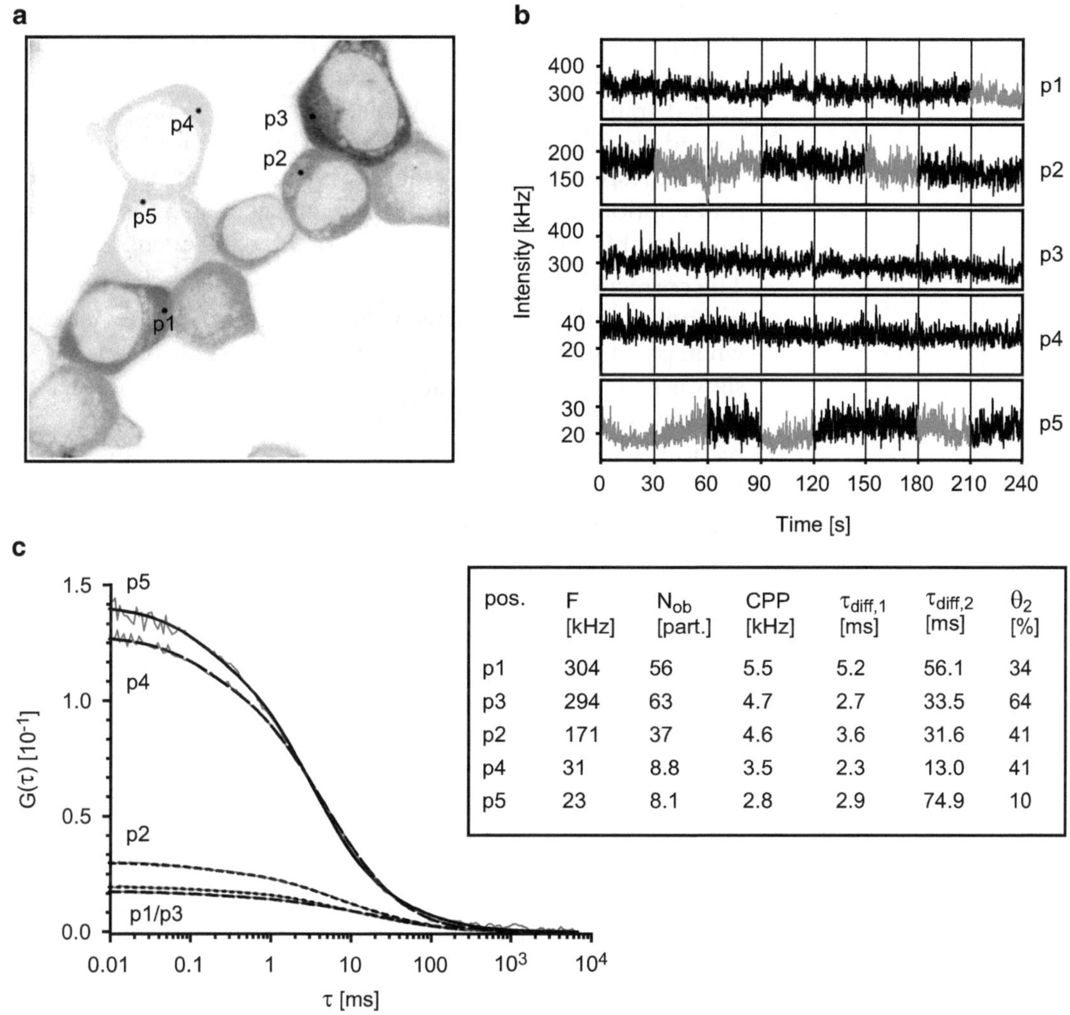

pos.	F [kHz]	N_{ob} [part.]	CPP [kHz]	$\tau_{diff,1}$ [ms]	$\tau_{diff,2}$ [ms]	θ_2 [%]
p1	304	56	5.5	5.2	56.1	34
p3	294	63	4.7	2.7	33.5	64
p2	171	37	4.6	3.6	31.6	41
p4	31	8.8	3.5	2.3	13.0	41
p5	23	8.1	2.8	2.9	74.9	10

Fig. 5 FCS analysis of JAK3-EGFP in the cytoplasm. (**a**) Confocal cross section of a cluster of HEK293T cells expressing JAK3-EGFP. (**b**) Time traces of measurements at indicated positions p1–p5 and the corresponding ACF in (**c**). Fit results are summarized in a table (*inset*)

4 Notes

1. Using a single triplet term is a simplification. The rate constants for intersystem crossing, i.e., population of the triplet state, depend on the chemical environment of the fluorophores. Moreover, for autofluorescent proteins, two dark states with different rate constants have been assigned reflecting the two protonation sites of the chromophore [22, 23]. From a practical point of view these differences are difficult to resolve, especially for intracellular FCS measurements at low excitation power. Thus, it is common to apply the model containing a single triplet term for each color species.

2. For expressing free EGFP the Complete Control® system from Stratagene was used (www.stratagene.com). In brief, human embryonic kidney cells (ER293) stably express the modular engineered transcription factor VgEcR and serve as recipients for a second plasmid containing the gene of interest under control of a VgEcR-responsive DNA sequence. It turned out that the basic transcription level is sufficient for confocal imaging in combination with FCS, even in the absence of inducing agent.
3. The correction collar at the objective corrects for the difference in optical path length, which the excitation beam experiences when passing cover slips of varying thickness. Using #1.5 cover slips is suitable for most of the high aperture water immersion objectives. If optimization of the count rate enforces the correction collar at an extreme position, this may indicate a wrong type of cover glass.
4. For cell seeding it is useful to combine nonfluorescent control cells, the transfected cells, and a free chamber for the standard Alexa488 on the same chambered cover glass. By this, the glass properties are the same for all measurements.
5. For single-point FCS, in contrast to scanning FCS [24], one needs to calibrate the observation volume in order to transform diffusion times into diffusion coefficients. This is done by calculating $D_{\text{meas}} = D_{\text{st}} \cdot \tau_{\text{st}}/\tau_{\text{meas}}$ from the measured diffusion time τ_{meas} and the standard τ_{st} for which the diffusion coefficient D_{st} is known. The diffusion coefficient for Alexa488 in 10 mM Tris pH 8 is $(4.14 \pm 0.1) \times 10^{-10}$ m^2/s [25]. Since the mobility of large proteins in the cytoplasm is highly anomalous, it may be better to compare the correlation times directly for different proteins or mutants. As stated in the introduction, particle numbers can be converted into concentrations by applying $V_{\text{eff}} = \pi^{3/2} S w_0^3$. The axis ratio S and the diffusion time trade during the fit and thus the axis ratio is usually a poorly determined parameter. To achieve consistency, it is quite common to determine S from standard ($3 < S < 7$) and to fix this value for all fitting procedures of cellular measurements.
6. The shape of the observation volume changes when the laser beam is tilted away from the optical axis. Measuring at positions close to the border of a low magnification image gives different values as compared to the centered position. For our setup (ConfoCor3), we determined a deviation of ~3 % in diffusion time and of ~4 % in particle numbers when moving 50 μm in *x* or *y* from the optical axis.
7. The run length should exceed the slowest correlation time at least by a factor of 1,000. The rational of how many runs should be taken is based on the heterogeneity of the signal and the overall duration of illuminating the cell. If many aggregates or vesicles are expected, it is better to minimize the length per run and to

take a larger series. By this, the "bad" runs can be sorted out. If more than one position in the same cell is to be measured, one should not illuminate longer than 1–2 min per spot. Cells respond to the irradiation. After several minutes, the cytoskeleton tends to change its state and the cells may move.

8. This procedure is justified when these slow fluctuations can be clearly assigned to movements of large cellular organelles or the cell as a whole. FCS relies on quasi-equilibrium conditions which in a complex system like a cell can only be approximated.
9. We used Matlab scripts to fit the data, and their description is beyond the scope of this article. However, it is straightforward to use data evaluation programs like Origin; import the ACF versus lag time τ (control software for the hardware correlator usually export the ACF as an ASCII file) and fit with the analytical function of Eq. 2 with τ as an independent variable. Channels below 1 μs should be omitted if in this region the curve shows a steep artificial increase due to afterpulsing of the photon counting avalanche detector (APD). For fitting, the number of correlation times should be increased only stepwise, in order to define the minimal set of free parameters. It is also important to determine reasonable starting values from which the fit-routine can evolve the function into a minimum for χ^2.
10. Uncorrelated background intensity (BG) changes both the intensity and the particle numbers from which the CPP value is derived. While the correction of the signal (F_{tot}) is just a subtraction $F_{\text{part}} = F_{\text{tot}} - \text{BG}$, the dependence of the amplitude is nonlinear $N_{\text{cor}} = (1 + \text{BG}/F)^{-2} G^{-1}(0)$. For example, 5 % of uncorrelated background drops the amplitude by almost 10 %, and hence, the apparent particle numbers have to be corrected by the factor 0.907. Because the particle derived signal itself is only 0.95 times the measured intensity, this cancels in part, and the CPP value will be a factor of 1.05 off. Because the background cannot be determined for each position individually, it is a workaround to measure with a signal-to-noise ratio above 20 and to account for background with an average value as determined in nonfluorescent cells.
11. To judge such differences, a quick calculation is useful: assuming a globular shape and identical density for both particles, the diffusion time is proportional to the cubic root of the ratio in molecular masses (EGFP ~27 kDa, Alexa488 ~0.8 kDa → 3.2). The values for EGFP in Fig. 3a are somewhat higher than expected, which may account for less dense packed masses within the protein.
12. Pure dimerization cannot be resolved by changes in diffusion time. It was determined that a factor of 1.6 in diffusion

coefficient can be discriminated under optimal conditions and this corresponds to a fourfold difference in molecular size if a globular shape is assumed [26].

Acknowledgments

I thank Wolfgang Staroske for providing his Matlab fitting script, Karin Crell, and Ellen Sieber for their assistance in cell culture, and Paul Müller for proof reading.

References

1. Bacia K, Kim SA, Schwille P (2006) Fluorescence cross-correlation spectroscopy in living cells. Nat Methods 3:83–89
2. Bacia K, Schwille P (2007) Practical guidelines for dual-colour fluorescence cross-correlation spectroscopy. Nat Protoc 2:2842–2856
3. Haustein E, Schwille P (2003) Ultrasensitive investigations of biological systems by fluorescence correlation spectroscopy. Methods 29:153–166
4. Hess ST, Huang S, Heikal AA, Webb WW (2002) Biological and chemical applications of fluorescence correlation spectroscopy: a review. Biochemistry 41:697–705
5. Kim SA, Heinze KG, Schwille P (2007) Fluorescence correlation spectroscopy in living cells. Nat Methods 4:963–973
6. Weidemann T, Wachsmuth M, Tewes M et al (2002) Analysis of ligand binding by two-colour fluorescence cross-correlation spectroscopy. Single Mol 3:49–61
7. Widengren J, Rigler R (1998) Fluorescence correlation spectroscopy as a tool to investigate chemical reactions in solutions and on cell surfaces. Cell Mol Biol (Noisy-le-Grand) 44:857–879
8. Widengren J, Mets Ü, Rigler R (1995) Fluorescence correlation spectroscopy of triplet states in solution: a theoretical and experimental study. J Phys Chem 99:13368–13379
9. Hofmann SR, Lam AQ, Frank S et al (2004) Jak3-independent trafficking of the common gamma chain receptor subunit: chaperone function of Jaks revisited. Mol Cell Biol 24:5039–5049
10. Ohrt T, Merkle D, Birkenfeld K et al (2006) In situ fluorescence analysis demonstrates active siRNA exclusion from the nucleus by Exportin 5. Nucleic Acids Res 34:1369–1380
11. Wachsmuth M, Weidemann T, Muller G et al (2003) Analyzing intracellular binding and diffusion with continuous fluorescence photobleaching. Biophys J 84:3353–3363
12. Weidemann T, Wachsmuth M, Knoch TA et al (2003) Counting nucleosomes in living cells with a combination of fluorescence correlation spectroscopy and confocal imaging. J Mol Biol 334:229–240
13. Petrov E, Schwille P (2008) State of the art and novel trends in fluorescence correlation spectroscopy. In: Resch-Genger U (ed) Standardization and quality assurance in fluorescence measurements ii: bioanalytical and biomedical applications. Springer, Berlin. ISBN 978-3-540-70570-3
14. Choi W, Fang-Yen C, Badizadegan K et al (2007) Tomographic phase microscopy. Nat Methods 4:717–719
15. Lide DR (2004) Handbook of chemistry and physics. CRC, Bota Raton, FL. pp 8–66
16. Enderlein J, Gregor I, Patra D et al (2005) Performance of fluorescence correlation spectroscopy for measuring diffusion and concentration. Chemphyschem 6:2324–2336
17. Gregor I, Patra D, Enderlein J (2005) Optical saturation in fluorescence correlation spectroscopy under continuous-wave and pulsed excitation. Chemphyschem 6:164–170
18. Nishimura G, Kinjo M (2004) Systematic error in fluorescence correlation measurements identified by a simple saturation model of fluorescence. Anal Chem 76:1963–1970
19. Zhang J, Campbell RE, Ting AY, Tsien RY (2002) Creating new fluorescent probes for cell biology. Nat Rev Mol Cell Biol 3:906–918
20. Wachsmuth M, Waldeck W, Langowski J (2000) Anomalous diffusion of fluorescent probes inside living cell nuclei investigated by spatially-resolved fluorescence correlation spectroscopy. J Mol Biol 298:677–689
21. Weiss M, Elsner M, Kartberg F, Nilsson T (2004) Anomalous subdiffusion is a measure

for cytoplasmic crowding in living cells. Biophys J 87:3518–3524

22. Haupts U, Maiti S, Schwille P, Webb WW (1998) Dynamics of fluorescence fluctuations in green fluorescent protein observed by fluorescence correlation spectroscopy. Proc Natl Acad Sci USA 95:13573–13578
23. Weber W, Helms V, McCammon JA, Langhoff PW (1999) Shedding light on the dark and weakly fluorescent states of green fluorescent proteins. Proc Natl Acad Sci USA 96:6177–6182
24. Petrášek Z, Schwille P (2008) Precise measurement of diffusion coefficients using scanning fluorescence correlation spectroscopy. Biophys J 94:1437–1448
25. Petrov EP, Ohrt T, Winkler RG, Schwille P (2006) Diffusion and segmental dynamics of double-stranded DNA. Phys Rev Lett 97:258101
26. Meseth U, Wohland T, Rigler R, Vogel H (1999) Resolution of fluorescence correlation measurements. Biophys J 76:1619–1631

Chapter 25

Fluorescence Cross-Correlation Spectroscopy (FCCS) in Living Cells

Xiaoxiao Ma, Yong Hwee Foo, and Thorsten Wohland

Abstract

Fluorescence cross-correlation spectroscopy (FCCS) is a single-molecule sensitive technique to quantitatively study interactions among fluorescently tagged biomolecules. Besides the initial implementation as dual-color FCCS (DC-FCCS), FCCS has several powerful derivatives, including single-wavelength FCCS (SW-FCCS), two-photon FCCS (TP-FCCS), and pulsed interleaved excitation FCCS (PIE-FCCS). However, to apply FCCS successfully, one needs to be familiar with procedures ranging from fluorescent labeling, instrumentation setup and alignment, sample preparation, and data analysis. Here, we describe the procedures to apply FCCS in various biological samples ranging from live cells to in vivo measurements, with the focus on DC-FCCS and SW-FCCS.

Key words Fluorescence cross-correlation spectroscopy (FCCS), DC-FCCS, SW-FCCS, TP-FCCS, Molecular interactions, Single molecule spectroscopy, Quantitative biology, Zebrafish imaging

1 Introduction

Fluorescence cross-correlation spectroscopy (FCCS) is an extension of fluorescence correlation spectroscopy (FCS). FCS analyzes the signals from particles labeled with a single fluorescent species and extracts information about concentrations and molecular dynamics. On the other hand, FCCS analyzes the fluorescence signal from two or more particle species that are each tagged with a fluorophore of a distinct emission spectrum. The signals from the distinct fluorophores are recorded in separate wavelength-dependent detection channels, and the signals are cross-correlated to retrieve information about the interaction between the different fluorophore species. Because FCCS relies on the cross-correlation of different signals, it reports directly on molecular interactions and is not dependent on any changes in diffusion coefficients or mass as is FCS [1] or on the interactions between the labels as is the case in Förster resonance energy transfer (FRET) [2]. Therefore, FCCS is particularly

Yves Engelborghs and Antonie J.W.G. Visser (eds.), *Fluorescence Spectroscopy and Microscopy: Methods and Protocols*, Methods in Molecular Biology, vol. 1076, DOI 10.1007/978-1-62703-649-8_25,

useful in *quantitating* molecular interactions that happen between molecules with similar molecular weight [3] and is complementary to imaging techniques, providing essentially the same spatial resolution but adding information on molecular interactions [4].

FCCS was first demonstrated by J. Rička and Th. Binkert in 1989 using scattered light from nonfluorescent latex particles in the blue channel and fluorescence from labeled latex particles in the green channel [5]. This technique found widespread application when Schwille and colleagues successfully measured DNA-DNA renaturation through detecting and cross-correlating the fluorescently tagged DNA in 1997 using a DC-FCCS setup [6], in which the different dyes are excited with separate lasers. Since then the application of FCCS ranges from the quantitation of biomolecular interactions, protein oligomerizations, to enzyme activity [7–11].

DC-FCCS, while a very powerful technique, suffers also from problems. Initially, the alignment of two lasers to precisely the same spot was difficult; cross talk between the two detection channels and moving structures, e.g., the plasma membrane of living cells, gave rise to a false-positive cross-correlation background. Therefore new FCCS modalities were suggested. Single-wavelength FCCS (SW-FCCS) has the advantage of easier calibration, simpler setup, and elimination of nonideal overlap of excitation volumes by exciting several fluorophores simultaneously using one single laser line [3, 10–12]. Similarly, two-photon excitation FCCS (TP-FCCS) uses a single laser line for the two-photon excitation of two separate fluorophores [13]. Although the counts per particle per second (cps) achieved by TP-FCCS is generally lower than that of DC-FCCS due to photobleaching [14] and saturation [15], TP-FCCS provides potentially a wider selection of fluorophore pairs compared to SW-FCCS [16]. Another derivative, pulsed interleaved excitation FCCS (PIE-FCCS) uses two lasers that are pulsed alternatively to overcome the spectral cross talk among fluorophores. The laser pulses are alternated in the range of nanoseconds such that the emission and excitation cycles of one type of fluorophore are completed before the next fluorophore is excited [17]. PIE-FCCS thus generates crosstalk-free cross-correlations. In addition, it can determine contributions from FRET, which can be used to correct FCCS data [18, 19]. Cell and membrane movement can be corrected by using scanning FCCS [20] (*see* **Note 3**).

Here, we describe a DC-FCCS/SW-FCCS procedure starting from sample preparation and emphasizing calibration and measurements which can be used for the study of biomolecular interactions in cells and zebrafish embryos [3, 10, 11, 21, 22]. PIE-FCCS is discussed in another protocol in the same publication.

2 Materials

2.1 Fluorophores for FCCS Experiments

The most important characteristic of fluorophore pairs required for FCCS measurements is the separation of spectra. For DC-FCCS, separated excitation and emission spectra are used for selective excitation and the reduction of cross talk, respectively. For SW-FCCS and TP-FCCS, the fluorophores need to have similar excitation spectra for one- or two-photon excitation, respectively. Besides the excitation and emission spectra, the aspects to be considered include molecular brightness (dependent on extinction coefficient and quantum yield), photostability, and available laser lines [4]. The fluorophores used include organic dyes, quantum dots and nanoparticles, and fluorescent proteins [23] (*see* **Note 4**).

2.1.1 Organic Dyes

There is a wide range of commercially available organic dyes with different excitation and emission spectra serving all possible laser lines, labeling a wide range of biological structures, and with very good photophysical properties. The obstacle for the use in live cell and in vivo measurements is the problem of introducing these extrinsic labels specifically and with defined labeling ratios into the sample. Several approaches have been introduced and allow linking the dye to molecules of interests, including synthesized DNA [6], genetically labeled recombinant proteins [24, 25], labeled antibodies [25–28], and extrinsic labeling of proteins through specific reactions with synthetic probes [29, 30]. Besides, pure dye solutions are widely used for system calibration and provide the basis for comparing the performance and measurements of different systems.

2.1.2 Quantum Dots (QDs) and Nanoparticles

QDs are high-quantum-yield fluorescent semiconductor nanoparticles, which can be excited with a single wavelength in the blue region and have size- or composition-dependent emission spectra [31–33]. They are therefore potential probes for SW-FCCS [12], but they can suffer from aggregation making cross-correlation measurements difficult. For FCCS measurements, commercial QD streptavidin conjugates are reported for studying binding between G-protein-coupled receptors and its ligands [13, 34, 35]. Other nanoparticles with potential use in FCCS are fluorescent silica particles [36] and nanodiamonds with color centers [37, 38].

2.1.3 Fluorescent Proteins (FPs)

FPs are arguably the most widely used fluorophores for FCCS. They can be genetically encoded and thus guarantee specific labeling of the protein of interest with a defined stoichiometry. Although their photostability and brightness is inferior to organic dyes and QDs, and their complicated photophysics can make the interpretation of FCCS curves difficult (*see* **Note 1**), the ease of their usage and the fact that many protein-FP constructs are widely available, make them a good choice, as long as the relative concentration are considered (*see* **Note 2**). In addition, new FP variants with improved photophysical properties are continuously produced. For example, mKeima, a far-red fluorescent protein endowed with a large Stokes shift, is reported to be able to combine with both ECFP and EGFP for SW-FCCS [39]. Another FP with a large Stokes shift, LSSmOrange, is used to develop four-color single-laser FCCS solely based on FPs [40]. The most frequently used FP pair for DC-FCCS and SW-FCCS is mainly green and red pairs especially EGFP vs. mRFP/mCherry [10, 11, 21, 41], while TP-FCCS usually uses FPs with a large Stokes shift such as Cerulean and mCherry [42] or dyes with high two-photon absorption cross section such as Cyanine-5 (Cy5) and Rhodamine Green (RhGr) [43].

2.2 Instrumentation

FCS and FCCS instrumentation is commercially available from several manufacturers. Alternatively, commercial confocal microscopes can also be customized for the particular need of the user. The general FCCS setup and the correlation functions recorded are shown in Fig. 1, and more information about constructing FCCS systems can be found in literature [4, 6, 44]. The basic parameters for the parts of a standard system are as follows:

1. Laser: To avoid photobleaching and saturation, which will alter auto- and cross-correlation functions (ACF/CCF), a laser power under 30 μW before entering the objective (in the case of an NA of 1.2) is recommended in most cases, and 5–30 μW are widely used [43, 45–48]. Use laser power as low as possible to reduce photobleaching but high enough to obtain good signal-to-noise ratio. The laser power should be constant for all experiments. A cps as low as 500 can be sufficient to generate useful CCFs [49], although typically cps of 1,000–5,000 is reached.
2. Excitation/emission filters and the dichroic mirror: Filters should be chosen as a balance of maximizing the photons collected and minimizing cross talk. Filters with steep cut-ons/cutoffs are produced and help maximize light efficiency. The dichroic mirror should be selected to reduce cross talk as much as possible (*see* **Note 3**).
3. Microscope objective: High-numerical aperture objectives with chromatic correction are recommended. For live biological samples, which are usually in an aqueous environment and in which deep penetration into tissue is required, water immersion objectives are more suitable than oil immersion objectives to avoid aberrations due to the refractive index mismatch between immersion and mounting media [50]. The objective correction collar, present on most microscope objectives used

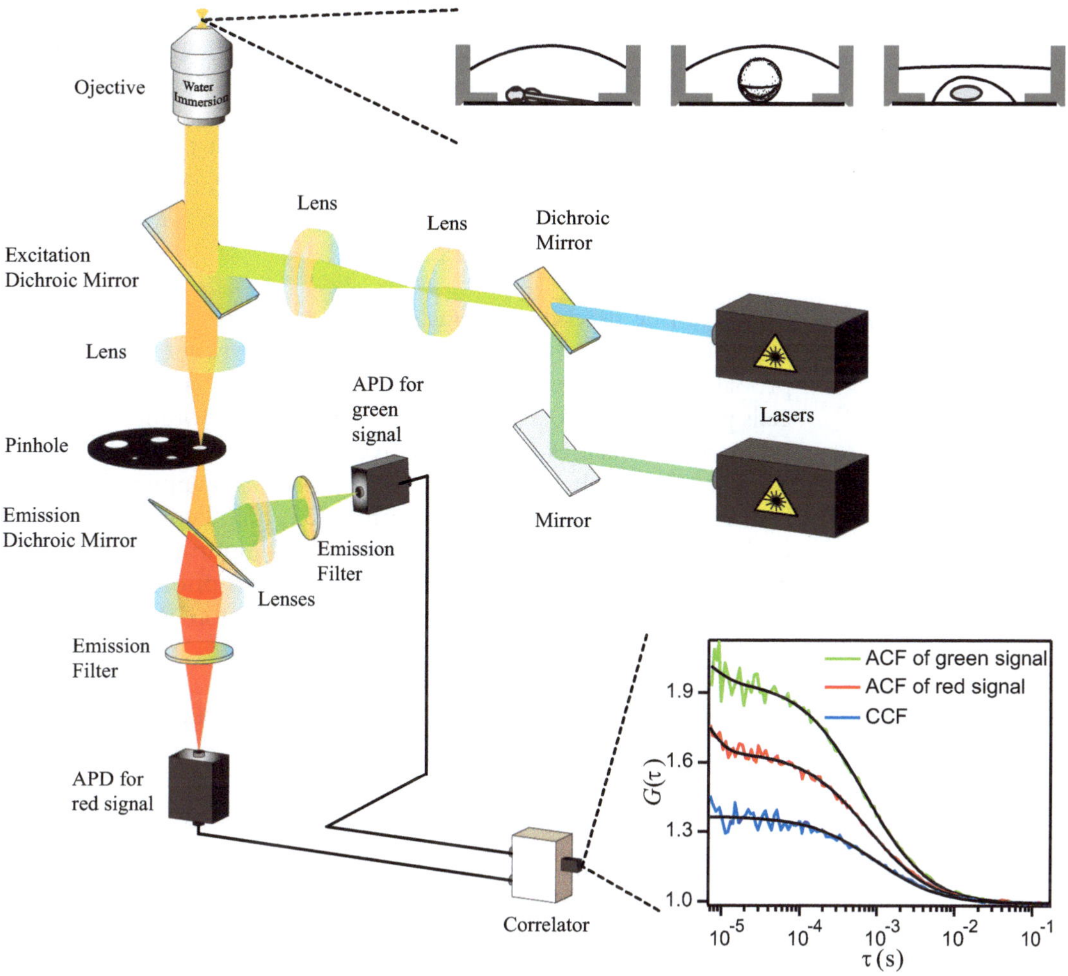

Fig. 1 General setup of an FCCS instrument and the illustration of ACF/CCF recorded

in FCCS, is designed to correct for the different thickness from different cover slides/glass bottom dishes. It must be adjusted every time before changing to another type of cover slips or chambers (*see* Subheading 3.2, **step 4**).

4. Pinhole: The pinhole of a confocal microscope is used to constrain the detected light to the observation volume. A 60×, NA 1.2 water immersion objective with a pinhole size of 50 μm is practical to most biological samples and provides a focal volume of ~0.5–1 fL in size [11, 41] (*see* more in Subheading 3.3).

5. Detector: Avalanche photodiodes (APDs) are highly sensitive photon counting devices and are widely used as the detectors for FCCS measurements. Photomultiplier tubes (PMTs), although they typically posses a lower quantum efficiency, can also be used.

6. Correlator: In order to synchronize the signal from different channels and generate cross-correlations, both hardware and software correlators are widely used [17, 51–54]. While hardware correlators used to be preferred due to their better time resolution, the enhanced flexibility and low cost of software correlators and the continuing increase in speed and computational power of desktop computers make them a good alternative.

2.3 Choice of Sample

FCCS can be used for various samples, ranging from solutions to cell cultures and organisms. At the time of writing, live cells are probably the most widely used system for FCCS application, since cells provide a platform to mimic the natural environment for most biological molecules and are relatively easy to handle compared to organisms. That being said, FCCS has also been applied in organisms such as Drosophila and zebrafish [22, 55]. Penetration depth in living organisms can go as deep as 50–80 μm in FCS using one-photon excitation, or at least 200 μm with two-photon excitation, without adversely affecting the measured diffusion coefficient [22] (*see* Subheading 3.1).

3 Methods

3.1 Sample Preparation

3.1.1 Preparation of Dye Solutions for Calibration

1. Stock solution: A defined amount of a solid organic dye is dissolved in DMSO or methanol to make a highly concentrated stock, typically in the mM range. It is stored at −20 °C or according to the instructions from the manufacturers. The dyes should not be exposed to light.
2. Concentration determination: The stock can be diluted in deionized water or buffer solutions, depending on the dye and molar extinction coefficient, to the μM range so that the concentration can be confirmed using the molar extinction coefficient of the dye and the absorbance measurement in a spectrophotometer.
3. Working solution: For FCCS measurement, the optimal concentration is in the 1–10 nM range [21, 56], corresponding to 0.3–6 molecules per focal volume (~0.5–1 fL). Working solutions can be kept at room temperature. Some organic dyes can easily attach to the plastic walls of their container. Hence shaking vigorously (or vortexing) before use is necessary. This can in some cases lead to deviations in measured concentrations when small volumes of dye solution are stored.

3.1.2 Cell Preparation

Recombinant fusion proteins for experiments with FPs can be transfected into cells by a variety of methods [57, 58], and protocols for cell transfections are provided by the relevant companies.

In our experience, the different cell lines show different preferences to different methods and parameters being used, so the method and parameters have to be optimized for each case (*see* **Note 2** for expression level control).

3.1.3 Embryo Specimen Preparation (Using Zebrafish as an Example)

Microinjection of RNA can be used to study proteins in early stage embryos, within a few hours postfertilization (hpf), while DNA microinjection is suitable for studies at a later stage, typically a few days postfertilization (dpf) [22].

The procedure is as follows: 2–10 pg of RNA/DNA is injected into a single cell or the yolk at an early stage, typically 1–32 cell stage. The embryos are incubated in egg water at 28.5 °C before being mounted in 0.5 % low-melting-temperature agar. The mounting procedure is performed under a stereoscope. To prevent pigmentation, PTU (0.003 % 1-phenyl-2-thiourea in 10 % Hank's saline) is added at 18 hpf if a longer incubation period is needed. At stages after development of the central nervous system, onset of heartbeat and muscle contraction, the zebrafish might have to be anesthetized. This is accomplished by using, e.g., 0.01 % or ~500 μM tricaine [59, 60] (*see* **Note 2** for expression level control).

3.2 Measurement Procedure

1. System setup: Select the filters and dichroic mirrors according to the laser line(s) and fluorophores used. As overexposure of the APD to ambient room light can damage the APD, the room should be kept dark or make sure no ambient light can reach the APDs. In addition, ambient light can increase the background counts and influence the amplitudes of correlation functions. Count rates above 800 kHz should be avoided during measurements to prevent detector nonlinearities [4].
2. Laser power determination: In typical confocal systems, the actual laser power is indicated as the percentage of the current. The dependence of the laser power on this percentage value is not linear, and the actual laser power has to be determined using a power meter. The laser power can be measured by removing the objective and placing the power meter in the laser's path. Repeat the laser power measurement on a regular basis to confirm the stability of the lasers. Typically the laser power used is between 5 and 30 μW (*see* also Subheading 2.2). The cps values can vary for different systems and depend on the quality of the alignment, the filters used, and the overall transmission efficiency of the system. So they can vary even during a single measurement day, especially when the instrument room undergoes large temperature changes, and thus should be monitored. Atto 488 excited by a 5 μW 488 nm laser yields ~5–15 kHz with a 510 AF23 band-pass filter and 60×, NA 1.2 water immersion objective. Under the same experimental

conditions, a cps of 3 kHz can be obtained from enhanced green fluorescent protein (EGFP) in live cells.

3. Placement of the sample container: Type 1.5 (thickness 0.16–0.19 mm) and type 1.0 (thickness 0.13–0.16 mm) cover slides or glass bottom dishes are commonly used for the samples since they are usually within the range of the objective's correction collar, 0.13–0.21 mm. Location of the sample by imaging is easy for a biological sample. However, for a solution sample or in cases where the position of the focal volume relative to the cover glass is important, the back reflection of the laser can be used to locate the position of the cover glass. This can be achieved by moving the laser focus from outside the cover glass into the sample while acquiring an image. The back reflection of the laser can be seen whenever the focal volume reaches the lower and upper side of the glass. Continue moving the objective at least 5 μm upward into the solution after seeing the upper side of the glass slide to avoid any distortion of the focal volume due to the glass slide.
4. Calibration/pinhole adjustment: Before any measurement, calibration of the FCCS system is required for the optimal size of the focal volume as well as the optical efficiency of the system. The pinhole will reduce the signal-to-noise ratio if it is smaller than the optimal size, while the upper limits of a pinhole size can also be determined [61, 62] (*see* Subheading 3.3).
5. Background determination: The background counts stem from four sources. The first source is ambient light from the room. This can be avoided by switching off all lights in the room. The second source is fluorescent impurities either from the buffer or from molecules in live cells (autofluorescence). They contribute about 300–500 Hz for buffer and 500–1,500 Hz for Chinese hamster ovary (CHO) cells for a laser power of 30 μW at 488 nm and fluorescence collected from 510 to 540 nm. Note that these values change with excitation wavelength and detection range but should be on the same order of magnitude. The third source is the detector, which possesses a certain amount of dark counts (20–500 counts per second, depending on the model). The dark counts, which are the signal the detector reports even in the total absence of light [63], are constant. In addition the detector produces additional counts due to afterpulsing, i.e., the recording of additional counts directly after actual photon detection events due to the detection method [63]. Afterpulsing has a finite probability to occur after each photon detection event, and thus this background is dependent on the actual counts detected. The last source, which is perhaps the least common, is due to the poor or wrong selection of dichroics and filters. This unintentionally results in back scattering of the laser, which is not

filtered properly. This is easily detected by turning on the laser without any sample to check if there is any increase in the count rate. Background count rate should be determined for both channels before any experiments for data correction.

6. Measurements: To obtain a precise result, concentrations of the fluorophores lower than 1 pM or higher than 100 nM are not recommended [64]. The count rate should range between 1 and 800 kHz. The measurement times depend on the signal-to-noise ratio of the particular measurement but typically range between 5 and 60 s. The measurement times in live cells should be kept as low as possible to avoid artifacts from cell movement and photobleaching but should be long enough to allow proper data analysis. Repeat the measurements 3–10 times for each position. Although there is no strict consensus on how long one can measure on a cell without inducing cell damage, to stay conservative we recommend measuring not more than three points in a cell where on each point one collects three consecutive measurements of 10–30 s with the settings as described here. For membrane measurements, ensuring that the membrane is in the center of the observation volume is important [65, 66]. There are two common ways to focus correctly at the upper membrane of a living cell. One is by observing the intensity trace: focus at the center of the nucleus first, and then while observing the intensity trace move the focus up until the strongest fluctuations, which stem from molecules in the plasma membrane, are observed. Another possibility is by using live imaging: focus at the center of the nucleus first, and then move the focus up until only the highest part of the upper membrane is in focus.
7. Data analysis: Data analysis in FCS/FCCS is arguably the most difficult part, and new methods for the analysis of correlation data are being investigated to avoid data overinterpretation and helping in deciding between the numerous possible fitting models [67]. During fitting of the ACF and CCF, a few things have to be considered. First, the fitting range of the correlation functions has to be decided. The correlation function at times shorter than 1 μs is usually attributable to either the photodynamics of the fluorophore or the detector's afterpulsing and is not fitted along. On the other end of the time scale of the correlation function, the function should have already decayed to 0 or 1 (depending on the correlation scheme), which is generally 0.1–1 s. There is no fixed rule for the exact range to fit the correlation function as samples, and fluorophore differs in each experiment. However, the same set of experiments should have a consistent fitting range. The second issue to note is the model used to fit the correlation function. In general, the ACFs are fitted using the simplest model consistent

with the data, while the CCF is fitted with a single-diffusion component model without photodynamic contribution. Except in the case of strong cross talk, the photodynamics in the two channels are not correlated. Yet, in FCCS measurements often more than a single-diffusion time is found for a particular molecule. This might be because the molecules under investigation can be found in different states. For instance, a cytosolic fast diffusing molecule can bind to a membrane protein, or molecules in a lipid membrane might partition into different domains. However, if true interactions exist there must be at least one characteristic time, which is the same in all three correlation functions (two ACFs and the CCF). So the binding can only be confirmed by the increase in $G_{\mathrm{CCF}}/G_{\mathrm{ACF}}$ (see next paragraph) compared to the negative control and similar diffusion times in the correlation functions [11].

The most common parameter obtained from FCCS in regard to the binding strength of a pair of molecules is the ratio between the amplitudes $G(0)$ of the CCF and ACF ($G_{\mathrm{CCF}}(0)/G_{\mathrm{ACF}}(0)$) [41]. However, this ratio is not corrected for background and cross talk. Hence the concentrations of the non-bound molecules and complex are calculated after correcting for background and cross talk [6, 10, 11]. Furthermore, with the calculated concentrations, the dissociation constant K_{d} can be estimated by different schemes as reported in literature [21, 41, 68]. A long description of this is out of the scope of this chapter, but details about data fitting can be found in many articles regarding the fitting of FCS data [1, 56, 69] and the evaluation of FCCS amplitudes for the quantification of molecular interactions [6, 10, 11, 21, 41, 68, 70] and Bayesian methods [71, 72].

3.3 FCCS System Calibration

The calibration dye used for the system should have a similar spectrum as the fluorophores used in the experiments and have a known diffusion coefficient.

After placing a droplet of the calibration dye (for SW-FCCS or TP-FCCS) or dye-mixture (for DC-FCCS) on the glass cover slide and placing the focus into the solution as described above, observe the intensity trace and adjust the following:

1. Pinhole adjustment and detection path adjustment: In a single pinhole system (most confocal microscopes), the pinhole adjustment is relatively easy. Most of the time, the pinhole size and position on a modern confocal microscope are automatically adjusted based on the laser line used. If adjustments are required, refer to the manufacturer's instructions on how to adjust the position of the pinhole. Adjustment should be done in the x–y or z direction to optimize the count rate of the corresponding calibration dye with the shorter wavelength and

the corresponding laser. The pinhole size has to be adapted to the microscope objective. For the 60×, NA 1.2 water immersion objective discussed here, a typical pinhole size is 50 μm (fixed pinholes) or 1 Airy unit (adjustable pinholes). Once a measurement is recorded, fit the curves to obtain the structure parameter SP (ratio of the axial to radial size of the focal volume, i.e. z_0/ω_0) and cps (count rate divided by number of particles N). Repeat the measurement-and-data-fitting step until a SP value of ~4–6 and maximum cps are reached. In DC-FCCS, check that SP and cps are also optimal for the second laser. If different pinhole alignments are needed for the two lasers to optimize the alignment, the lasers might not overlap optimally, and a realignment of the lasers might be required (refer to the instrument manual or contact the manufacturer). In SW-FCCS and TP-FCCS, the alignment is needed only once since only one laser is used. Alternatively, a dye with a large Stokes shift can be used [4]. Another procedure is to use a dye with a broad emission spectrum such as Rhodamine 110. The emission signal is equally divided into the two detection channels to obtain the two ACFs and the CCF. In a well-adjusted system, all three curves should perfectly overlap [27].

2. Objective correction collar adjustment: Switch on the laser and acquire a correlation function. Then adjust the collar and redo the measurement. Repeat the adjustment-and-measurement step for a few times until the highest cps is obtained. Likewise, the amplitude of the correlation function can be used instead of the cps (since cps = count rate × amplitude).
3. Determination of the effective volumes of both detection channels: Once SP and the diffusion times of the calibration dyes are determined, the effective volume of the respective detection channels can be calculated using $V_i = \pi^{3/2}\omega_{0,i}{}^2 z_{0,i}$ (where i is the detection channel).
4. Calibration of the CCF: Calibration of the CCF can be performed by using tandem dyes such as tandem FPs [41] or doubly labeled DNA oligos [6]. The maximum CCF to ACF amplitude ratio is usually about 0.5 if the system is properly aligned [41].

4 Notes

1. *Corrections of factors affecting FCCS calibration and quantitation* The ratio of the CCF/ACF amplitude, $G_{CCF}(0)/G_{ACF}(0)$, varies usually from 0 to 0.5 (assuming a 1:1 stoichiometry) and is never reported nearing the theoretical value of 1. The factors affecting the ratio are discussed in the following paragraphs, and more details can be found in reference [41]:

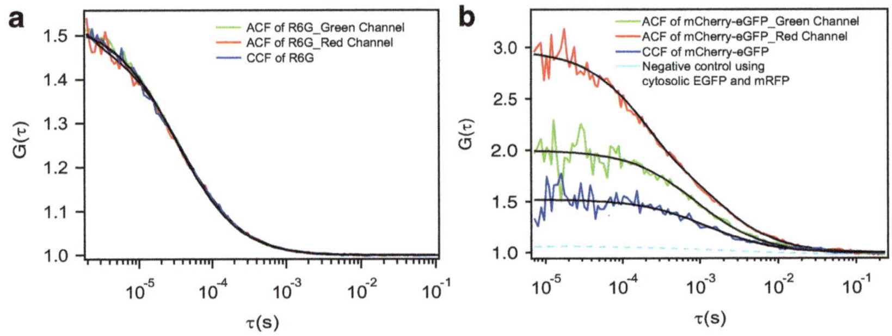

Fig. 2 Illustration of calibration curves of Rhodamine 6G (R6G) and tandem FPs (mCherry-EGFP). To do calibration in living cells, EYFP can be used with similar results [41]

Effective focal volume: In SW-FCCS, due to the different emission spectra, the effective volume of the long-wavelength fluorophores is larger than that of short-wavelength ones, resulting in non-perfect overlap of the observation volumes. The size ratio of the EGFP to mRFP/mCherry effective volumes was estimate to be about 1.4 [41].

FRET: In cases where FRET is high, the CCF will be reduced. This can be corrected [21, 73].

Nonfluorescent FPs: Nonfluorescent FPs can occur due to photobleaching (during the measurement or due to the imaging of the cells before the measurement), fluorophore dark states [74], and slow fluorophore maturation [68]. Generally speaking, the lower the amount of the nonfluorescent fraction, the higher the amplitude ratio. For example, in Ref [41, 74], it was found that under normal conditions, only 40–50 % of mCherry/mRFP are actually fluorescent. The non-matured mCherry portion can be accounted for in order to include the non-detected complex [41]. If the photobleaching is minimized by the reduction of laser power, $G_{\mathrm{CCF}}(0)/G_{\mathrm{ACFr}}(0)$ can increase significantly.

Endogenous proteins: Endogenous proteins reduce the cross-correlation amplitude. As long as the amount of endogenous proteins can be determined, one can correct for the effect of endogenous proteins [41] (*see* Fig. 2).

2. *Expression level control*

One important advantage of FCCS, compared to many other techniques, is that it can measure at low concentrations, which are on the same order as the physiological expression levels of proteins of interests. The $G_{\mathrm{CCF}}(0)/G_{\mathrm{ACF}}(0)$ ratio gives either the amount of green or red molecules in complexes. Thus its value is dependent on the ratio of the red- to green-labeled molecules. For example, given the same pair of interacting molecules, one with overabundance of red-labeled molecules relative to the green-labeled molecules will have a higher $G_{\text{CCF}}(0)/G_{\text{ACFr}}(0)$ compared to the same pair of interacting molecules with similar concentration of red- and green-labeled molecules. In theory, controls and all experiments should be performed under similar "red to green" concentration ratios. However, samples such as transfected cells tend to have a range of concentration ratios. In this case, "red to green" concentration ratios of about 0.5–2 typically provide good results and minimize false positives.

Stable cell lines often overexpress the protein of interest and thus are difficult to measure in FCCS. This can be overcome by screening for a stable cell line with an appropriate expression level if several stable lines are available. Better control can be exerted by transient transfection in which the expression level can be adjusted by the amount of DNA used in the transfection [75]. Usually in a transiently transfected sample, a relatively wide range of expression levels exist in a single dish, and one can search for appropriate cells. For FCCS when measuring molecular interactions, the concentration of the two different proteins should be within a factor of 2 (at maximum) of their concentrations, to avoid artifacts of cross talk and false-positive cross-correlations [21].

Similarly in zebrafish studies, transgenic animals provide a single expression level that is often optimized for imaging but is too high for FCCS measurements. Here the expression level is controllable in microinjection in early stage embryos since the expression level of the protein of interest is generally positively related to the amount of DNA/RNA used for microinjection. To exclude the effects of endogenous proteins, morpholinos can be used [76, 77].

In Drosophila embryos, lower temperature can be used to decrease the expression levels [22].

3. *False-positive and false-negative cross-correlations*

False-positive cross-correlations can result from cross talk, aggregation, or cell movement in which all fluorescently tagged molecules move in concert.

Cross talk is the leakage of photons from one channel to another (typically from a shorter wavelength (green) into the detector of a longer wavelength (red)), due to the overlap of emission spectra, the usual long tails of the emission spectra, in particular for FPs, and sometimes a choice of emission filters with wide bandwidth. False-positive results from cross talk can be avoided by reducing the amount of cps from the green fluorophore leaking into the red channel relative to the cps from the red molecule in the red channel. This can be achieved by lowering the green laser power, increasing the red laser power, using a brighter red fluorophore, or avoiding a green fluorophore that is much brighter than the red fluorophore. However, one should be careful with the problems of lowering or increasing the laser power. For details refer to [78].

Aggregation is more difficult to avoid, but in some cases aggregates can be identified by clear peaks in the intensity traces, which can then be eliminated before the calculation of the correlation functions [70].

Cell movement sometimes introduces artificial components into the ACF and false-positive results in CCF. The artifacts of cell and in particular membrane movement can be overcome by scanning FCCS [20].

False-negative cross-correlations can be caused by the nonoptimal detection volume overlap [4], photobleaching, nonfluorescent proteins, or competition from endogenous proteins [41].

4. *The influence of fluorescent tags on interactions*

   The fluorescent tags could influence the behaviors of the molecule of interest, such as protein localization, aggregation, and hydrophobic and hydrophilic affinities. Some FPs are reported to tend to form weak dimers or tetramers [79, 80]. So a basic functional study should be done after attaching fluorescent tags to the proteins of interest.

References

1. Meseth U, Wohland T, Rigler R et al (1999) Resolution of fluorescence correlation measurements. Biophys J 76(3):1619–1631
2. Erickson M, Moon D, Yue D (2003) DsRed as a potential FRET partner with CFP and GFP. Biophys J 85(1):599–611
3. Hwang LC, Wohland T (2004) Dual-color fluorescence cross-correlation spectroscopy using single laser wavelength excitation. Chemphyschem 5(4):549–551
4. Bacia K, Schwille P (2007) Practical guidelines for dual-color fluorescence cross-correlation spectroscopy. Nat Protoc 2(11):2842–2856
5. Rička J, Binkert T (1989) Direct measurement of a distinct correlation function by fluorescence cross correlation. Phys Rev A 39(5):2646–2652
6. Schwille P, Meyer-Almes FJ, Rigler R (1997) Dual-color fluorescence cross-correlation spectroscopy for multicomponent diffusional analysis in solution. Biophys J 72(4):1878–1886
7. Tudor C, Feige JN, Pingali H et al (2007) Association with coregulators is the major determinant governing peroxisome proliferator-activated receptor mobility in living cells. J Biol Chem 282(7):4417–4426
8. Weidemann T, Worch R, Kurgonaite K et al (2011) Single cell analysis of ligand binding and complex formation of interleukin-4 receptor subunits. Biophys J 101(10):2360–2369
9. Worch R, Bokel C, Hofinger S et al (2010) Focus on composition and interaction potential of single-pass transmembrane domains. Proteomics 10(23):4196–4208
10. Liu P, Sudhaharan T, Koh RM et al (2007) Investigation of the dimerization of proteins from the epidermal growth factor receptor family by single wavelength fluorescence cross-correlation spectroscopy. Biophys J 93(2):684–698
11. Ma X, Ahmed S, Wohland T (2011) EGFR activation monitored by SW-FCCS in live cells. Front Biosci (Elite Ed) 3:22–32
12. Hwang LC, Wohland T (2005) Single wavelength excitation fluorescence cross-correlation spectroscopy with spectrally similar fluorophores: resolution for binding studies. J Chem Phys 122(11):114708
13. Swift JL, Burger MC, Massotte D et al (2007) Two-photon excitation fluorescence cross-correlation assay for ligand-receptor binding:

cell membrane nanopatches containing the human micro-opioid receptor. Anal Chem 79(17):6783–6791

14. Dittrich PS, Schwille P (2001) Photobleaching and stabilization of fluorophores used for single-molecule analysis with one- and two-photon excitation. Appl Phys B-Lasers O 73(8):829–837
15. Berland K, Shen GQ (2003) Excitation saturation in two-photon fluorescence correlation spectroscopy. Appl Optics 42(27):5566–5576
16. Hwang LC, Wohland T (2007) Recent advances in fluorescence cross-correlation spectroscopy. Cell Biochem Biophys 49(1):1–13
17. Thews E, Gerken M, Eckert R et al (2005) Cross talk free fluorescence cross correlation spectroscopy in live cells. Biophys J 89(3):2069–2076
18. Lamb DC, Muller BK, Brauchle C (2005) Enhancing the sensitivity of fluorescence correlation spectroscopy by using time-correlated single photon counting. Curr Pharm Biotechnol 6(5):405–414
19. Müller BK, Zaychikov E, Brauchle C et al (2005) Pulsed interleaved excitation. Biophys J 89(5):3508–3522
20. Betaneli V, Petrov EP, Schwille P (2012) The role of lipids in VDAC oligomerization. Biophys J 102(3):523–531
21. Shi XK, Foo YH, Sudhaharan T et al (2009) Determination of dissociation constants in living zebrafish embryos with single wavelength fluorescence cross-correlation spectroscopy. Biophys J 97(2):678–686
22. Shi XK, Teo LS, Pan XT et al (2009) Probing events with single molecule sensitivity in Zebrafish and Drosophila embryos by fluorescence correlation spectroscopy. Dev Dynam 238 (12):3156–3167
23. Hotzer B, Medintz IL, Hildebrandt N (2012) Fluorescence in nanobiotechnology: sophisticated fluorophores for novel applications. Small 8(15):2297–2326
24. Doi N, Takashima H, Kinjo M et al (2002) Novel fluorescence labeling and high-throughput assay technologies for in vitro analysis of protein interactions. Genome Res 12(3): 487–492
25. Wu CY, Huang CK, Chung CY et al (2011) Probing the binding kinetics of proinflammatory cytokine-antibody interactions using dual color fluorescence cross correlation spectroscopy. Analyst 136(10):2111–2118
26. Hansen K, Ruttekolk IR, Glauner H et al (2009) The in vitro biological activity of the HLA-DR-binding clinical IgG4 antibody 1D09C3 is a consequence of the disruption of cell aggregates and can be abrogated by Fab arm exchange. Mol Immunol 46(16): 3269–3277
27. Ruan Q, Tetin SY (2008) Applications of dual-color fluorescence cross-correlation spectroscopy in antibody binding studies. Anal Biochem 374(1):182–195
28. Stromqvist J, Johansson S, Xu L et al (2011) A modified FCCS procedure applied to Ly49A-MHC class I cis-interaction studies in cell membranes. Biophys J 101(5):1257–1269
29. Oyama R, Takashima H, Yonezawa M et al (2006) Protein-protein interaction analysis by C-terminally specific fluorescence labeling and fluorescence cross-correlation spectroscopy. Nucleic Acids Res 34(14):e102
30. Zhou X, Tang Y, Xing D (2011) One-step homogeneous protein detection based on aptamer probe and fluorescence cross-correlation spectroscopy. Anal Chem 83(8):2906–2912
31. Alivisatos AP, Gu W, Larabell C (2005) Quantum dots as cellular probes. Annual Review of Biomedical Engineering 7:55–76
32. Pinaud F, Michalet X, Bentolila LA et al (2006) Advances in fluorescence imaging with quantum dot bio-probes. Biomaterials 27(9): 1679–1687
33. Chan WC, Maxwell DJ, Gao X et al (2002) Luminescent quantum dots for multiplexed biological detection and imaging. Curr Opin Biotechnol 13(1):40–46
34. Swift JL, Burger MC, Cramb DT (2009) A quantum dot-labeled ligand-receptor binding assay for G protein-coupled receptors contained in minimally purified membrane nanopatches. Methods Mol Biol 552:329–341
35. Xie C, Dong CQ, Ren JC (2008) Fluorescence cross-correlation spectroscopy using single wavelength laser excitation. Chem J Chin U 29(5):897–901
36. Ow H, Larson DR, Srivastava M et al (2005) Bright and stable core-shell fluorescent silica nanoparticles. Nano Lett 5(1):113–117
37. Hui YY, Zhang B, Chang YC et al (2010) Two-photon fluorescence correlation spectroscopy of lipid-encapsulated fluorescent nanodiamonds in living cells. Opt Express 18 (6):5896–5905
38. Neugart F, Zappe A, Jelezko F et al (2007) Dynamics of diamond nanoparticles in solution and cells. Nano Lett 7(12):3588–3591
39. Kogure T, Kawano H, Abe Y et al (2008) Fluorescence imaging using a fluorescent protein with a large Stokes shift. Methods 45(3): 223–226
40. Shcherbakova DM, Hink MA, Joosen L et al (2012) An orange fluorescent protein with a

large Stokes shift for single-excitation multicolor FCCS and FRET imaging. J Am Chem Soc 134(18):7913–7923

41. Foo YH, Naredi-Rainer N, Lamb DC et al (2012) Factors affecting the quantification of biomolecular interactions by fluorescence cross-correlation spectroscopy. Biophys J 102 (5):1174–1183
42. Savatier J, Jalaguier S, Ferguson ML et al (2010) Estrogen receptor interactions and dynamics monitored in live cells by fluorescence cross-correlation spectroscopy. Biochemistry-US 49(4):772–781
43. Wenger J, Gerard D, Lenne PF et al (2006) Dual-color fluorescence cross-correlation spectroscopy in a single nanoaperture : towards rapid multicomponent screening at high concentrations. Opt Express 14(25):12206–12216
44. Pan X, Foo W, Lim W et al (2007) Multifunctional fluorescence correlation microscope for intracellular and microfluidic measurements. Rev Sci Instrum 78(5):053711
45. Gregor I, Patra D, Enderlein J (2005) Optical saturation in fluorescence correlation spectroscopy under continuous-wave and pulsed excitation. Chemphyschem 6(1):164–170
46. Glauner H, Ruttekolk IR, Hansen K et al (2010) Simultaneous detection of intracellular target and off-target binding of small molecule cancer drugs at nanomolar concentrations. Brit J Pharmacol 160(4):958–970
47. Chiantia S, Ries J, Schwille P (2009) Fluorescence correlation spectroscopy in membrane structure elucidation. Biochim Biophys Acta 1788(1):225–233
48. Ivashyna O, Garcia-Saez AJ, Ries J et al (2009) Detergent-activated BAX protein is a monomer. J Biol Chem 284(36):23935–23946
49. Mütze J, Ohrt T, Schwille P (2009) Fluorescence correlation spectroscopy in vivo. Laser Photon Rev 5(1):52–67
50. Hell S, Reiner G, Cremer C et al (1993) Aberrations in confocal fluorescence microscopy induced by mismatches in refractive index. J Microsc 169(3):391–405
51. Benda A, Fagul'ova V, Deyneka A et al (2006) Fluorescence lifetime correlation spectroscopy combined with lifetime tuning: new perspectives in supported phospholipid bilayer research. Langmuir ACS J Surf Colloids 22(23):9580–9585
52. Magatti D, Ferri F (2003) 25 ns software correlator for photon and fluorescence correlation spectroscopy. Rev Sci Instrum 74(2): 1135–1144
53. Mocsár G, Kreith B, Buchholz J et al (2012) Multiplexed multiple-τ auto- and cross-correlators on a single FPGA. Rev Sci Instrum 83(4):046101–046103
54. Wahl M, Gregor I, Patting M et al (2003) Fast calculation of fluorescence correlation data with asynchronous time-correlated single-photon counting. Opt Express 11(26): 3583–3591
55. Ries J, Yu SR, Burkhardt M et al (2009) Modular scanning FCS quantifies receptor-ligand interactions in living multicellular organisms. Nat Methods 6(9):643–645
56. Wohland T, Friedrich-Bénet K, Pick H et al (2001) The characterization of a transmembrane receptor by fluorescence correlation spectroscopy. In: Single Molecular Spectroscopy, Nobel Conference Lectures, Springer, Heidelberg, p. 195–210.
57. Chu G, Hayakawa H, Berg P (1987) Electroporation for the efficient transfection of mammalian cells with DNA. Nucleic Acids Res 15(3):1311–1326
58. Zaragosi LE, Billon N, Ailhaud G et al (2007) Nucleofection is a valuable transfection method for transient and stable transgene expression in adipose tissue-derived stem cells. Stem Cells 25(3):790–797
59. Malone MH, Sciaky N, Stalheim L et al (2007) Laser-scanning velocimetry: a confocal microscopy method for quantitative measurement of cardiovascular performance in zebrafish embryos and larvae. BMC Biotechnol 7:40
60. Wohland T, Shi X, Sankaran J et al (2010) Single plane illumination fluorescence correlation spectroscopy (SPIM-FCS) probes inhomogeneous three-dimensional environments. Opt Express 18(10):10627–10641
61. Hess ST, Webb WW (2002) Focal volume optics and experimental artifacts in confocal fluorescence correlation spectroscopy. Biophys J 83(4):2300–2317
62. Rigler R, Mets Ü, Widengren J et al (1993) Fluorescence correlation spectroscopy with high count rate and low background: analysis of translational diffusion. Eur Biophs J 22(3):169–175
63. Cova S, Ghioni M, Lacaita A et al (1996) Avalanche photodiodes and quenching circuits for single-photon detection. Appl Opt 35(12): 1956–1976
64. Sezgin E, Schwille P (2011) Fluorescence techniques to study lipid dynamics. Cold Spring Harb Perspect Biol 3(11):a009803
65. Milon S, Hovius R, Vogel H et al (2003) Factors influencing fluorescence correlation spectroscopy measurements on membranes: simulations and experiments. Chem Phys 288(2–3):171–186

66. Humpolickova J, Gielen E, Benda A et al (2006) Probing diffusion laws within cellular membranes by Z-scan fluorescence correlation spectroscopy. Biophys J 91(3):L23–L25
67. Shi X, Wohland T (2010) Fluorescence correlation spectroscopy. In: Diaspro A (ed) Nanoscopy and multidimensional optical fluorescence microscopy. Taylor and Francis, Boca Raton, FL
68. Maeder CI, Hink MA, Kinkhabwala A et al (2007) Spatial regulation of Fus3 MAP kinase activity through a reaction-diffusion mechanism in yeast pheromone signalling. Nat Cell Biol 9(11):1319–1326
69. Saffarian S, Elson EL (2003) Statistical analysis of fluorescence correlation spectroscopy: the standard deviation and bias. Biophys J 84(3):2030–2042
70. Ries J, Petrášek Z, García-Sáez AJ et al (2010) A comprehensive framework for fluorescence cross-correlation spectroscopy. New J Phys 12(11):113009 (113032pp)
71. Guo SM, He J, Monnier N et al (2012) Bayesian approach to the analysis of fluorescence correlation spectroscopy data II: application to simulated and in vitro data. Anal Chem 84(9):3880–3888
72. He J, Guo SM, Bathe M (2012) Bayesian approach to the analysis of fluorescence correlation spectroscopy data I: theory. Anal Chem 84(9):3871–3879
73. Kohl T, Heinze KG, Kuhlemann R et al (2002) A protease assay for two-photon crosscorrelation and FRET analysis based solely on fluorescent proteins. Proc Natl Acad Sci USA 99(19):12161–12166
74. Hillesheim LN, Chen Y, Muller JD (2006) Dual-color photon counting histogram analysis of mRFP1 and EGFP in living cells. Biophys J 91(11):4273–4284
75. Smith EM, Mueller JD (2012) The statistics of protein expression ratios for cellular fluorescence studies. Eur Biophys J EBJ 41(3):341–352
76. Draper BW, Morcos PA, Kimmel CB (2001) Inhibition of zebrafish fgf8 pre-mRNA splicing with morpholino oligos: a quantifiable method for gene knockdown. Genesis 30(3):154–156
77. Nasevicius A, Ekker SC (2000) Effective targeted gene 'knockdown' in zebrafish. Nat Genet 26(2):216–220
78. Bacia K, Petrasek Z, Schwille P (2012) Correcting for spectral crosstalk in dual-color fluorescence cross-correlation spectroscopy. Chemphyschem 13(5):1221–1231
79. Shaner NC, Steinbach PA, Tsien RY (2005) A guide to choosing fluorescent proteins. Nat Methods 2(12):905–909
80. Muller-Taubenberger A, Anderson KI (2007) Recent advances using green and red fluorescent protein variants. Appl Microbiol Biot 77(1):1–12

Quantifying Lipid-Protein Interaction by Fluorescence Correlation Spectroscopy (FCS)

Ana M. Melo, Manuel Prieto, and Ana Coutinho

Abstract

Fluorescence correlation spectroscopy (FCS) is a powerful method to investigate molecular interactions based on the variation of diffusion properties at the single-molecule level. This technique allows studying quantitatively the interaction of fluorescently labeled proteins/peptides with lipid vesicles. Here, we describe how to acquire and analyze FCS partition data in order to accurately determine the protein/peptide partition coefficients between the aqueous and lipid phases. It is shown that the recovery of unbiased partition coefficients from FCS partition curves (fractional amplitude of the bound species versus lipid concentration) requires considering explicitly the Poissonian loading of the lipid vesicles with the fluorescently labeled protein in order to account for the variable liposome brightness in each sample. Additionally, the impact of a trace amount of a fluorescent non-binding component on the partition curves determined by FCS is also discussed.

Key words Protein/peptide partition coefficients, Poisson distribution, Molecular brightness, Electrostatic interactions

1 Introduction

The single-molecule-based fluorescence correlation spectroscopy (FCS) is a suitable and powerful technique to investigate a large variety of dynamic processes and to quantify molecular interactions both in vitro and in vivo [1–3]. This technique measures fluorescence intensity fluctuations of a small number of fluorescently labeled molecules diffusing into and out of a very small focal volume (Fig. 1a, b). Therefore, any process that affects these fluctuations, e.g., ligand-macromolecule binding, internal macromolecule dynamics, intersystem crossing, translational and rotational diffusion, and excited state reactions, can be detected from FCS data [1–3].

Yves Engelborghs and Antonie J.W.G. Visser (eds.), *Fluorescence Spectroscopy and Microscopy: Methods and Protocols*, Methods in Molecular Biology, vol. 1076, DOI 10.1007/978-1-62703-649-8_26,

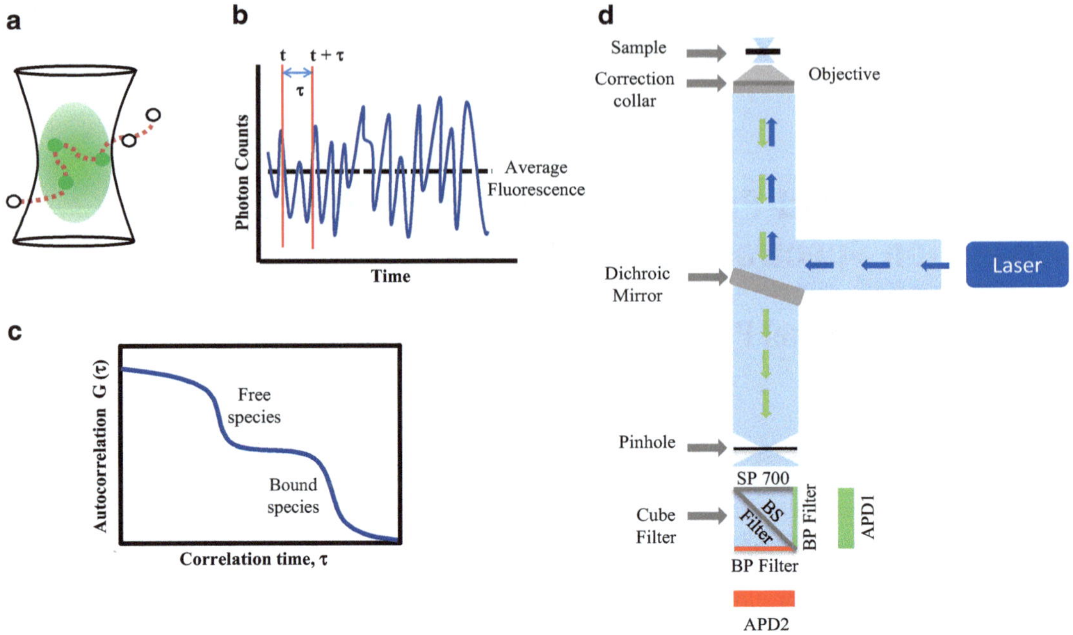

Fig. 1 Fluorescence correlation spectroscopy. (**a**) Fluorescently labeled molecules diffuse into and out of the ellipsoidal shape observation volume, (**b**) causing fluctuations in the fluorescence intensity signal over time. This signal is then processed using an autocorrelation algorithm to obtain the AC curve of the diffusing species. (**c**) Plot of an idealized two-component AC curve related to protein/peptide partition studies. Using single-color FCS, the free fluorescently labeled molecules can be easily distinguished from the slower-diffusing species (liposomes containing one or more bound fluorescently labeled proteins/peptides). (**d**) FCS setup. The laser beam is focused by a high-numerical objective on the sample. The fluorescence light is then collected by the same objective, passes the dichroic mirror and is focused onto a pinhole to reduce the excitation volume. Finally, the fluorescence signal is detected by an APD after passing through an emission band-pass filter cube. This filter cube has a 700 nm short band-pass filter (SP 700) and a beam splitter (BS Filter), which separate the emission light into two channels containing band-pass filters (BP Filter)

The normalized autocorrelation (AC) function, $G(\tau)$, of fluorescence fluctuations at a given time t, $\delta F(t)$, and at a later time $t+\tau$, $\delta F(t+\tau)$, is given by

$$G(\tau) = \frac{\langle \delta F(t) \cdot \delta F(t+\tau) \rangle}{\langle F(t) \rangle^2} \quad (1)$$

where $F(t)$ is the fluorescence intensity as a function of time t and $\delta F(t) = F(t) - \langle F(t) \rangle$. The angular brackets refer to time averaging and τ is the correlation time. Therefore, the AC function is a measure of the self-similarity of the fluorescence signal at time t and after a given delay time τ.

To obtain the physical parameters of interest, the experimental AC curves are fitted by the appropriate mathematical model describing the characteristics of the system under study. For a single diffusing species undergoing Brownian motion in a three-dimensional

(3D) Gaussian focus volume element, the AC function that includes the triplet-state formation is described by [3]:

$$G(\tau) = \frac{1}{N}\left(1 + \frac{T}{1-T}e^{-\tau/\tau_T}\right)\left(1 + \frac{\tau}{\tau_D}\right)^{-1}\left(1 + \frac{\tau}{S^2\tau_D}\right)^{-\frac{1}{2}} \quad (2)$$

where N is the average number of fluorescent molecules in the laser focus, and T and τ_T are the fraction of fluorophores in the triplet state and triplet lifetime, respectively. The parameter τ_D is the correlation time of the particle, which corresponds to the characteristic average diffusion time during which a molecule resides in the observation confocal volume with axial (ω_z) to lateral (ω_{xy}) dimensions S ($S = \omega_z/\omega_{xy}$). τ_D is related to the diffusion coefficient of the molecule, D, through the Einstein relation, $\tau_D = \omega_{xy}^2/4D$.

For multiple independent fluorescent species, the AC function is a linear combination of the AC curve for each species i, $G_i(\tau)$, weighted by the square of its molecular brightness, B_i [4]:

$$G(\tau) = \sum_{i=1}^{n} B_i^2 N_i^2 G_i(\tau) \Bigg/ \left[\sum_{i=1}^{n} B_i N_i\right]^2 \quad (3)$$

where N_i is the mean particle number of the ith species. The molecular brightness is an important parameter in FCS measurements as it controls the signal-to-noise ratio of the measurements. This parameter depends on several molecular intrinsic properties of the conjugated fluorophore under study (absorption cross-section and fluorescence quantum yield) as well as on the detection efficiency of the confocal microscope.

Considering that $G_i(\tau) = 1/N_i \cdot g_i(\tau)$, Eq. 3 can be rewritten as

$$G(\tau) = \frac{1}{N_{av}} \cdot \sum_{i=1}^{n} A_i \cdot g_i(\tau) \quad (4)$$

with $N_{av} = \sum_{i=1}^{n} q_i N_i$ and $A_i = q_i^2 N_i \Big/ \left(\sum_{i=1}^{n} q_i N_i\right)$. q_i is the ratio between the brightness of the ith component relatively to a given component chosen as a reference.

1.1 Determination of Partition Coefficients by FCS (Two-Component Model M2)

The determination of partition coefficients of proteins/peptides to lipid vesicles is usually the first step in any lipid-protein/peptide interaction study. For the determination of this parameter, several spectroscopic techniques, including different fluorescence methods [5], have often been employed. Since the pioneer work of Rusu et al. in 2004 [6], FCS has emerged as an important alternative option. FCS produces reliable partition measurements using nanomolar peptide/protein concentrations; these are between 10 and 100 times lower than those required for other techniques, such as

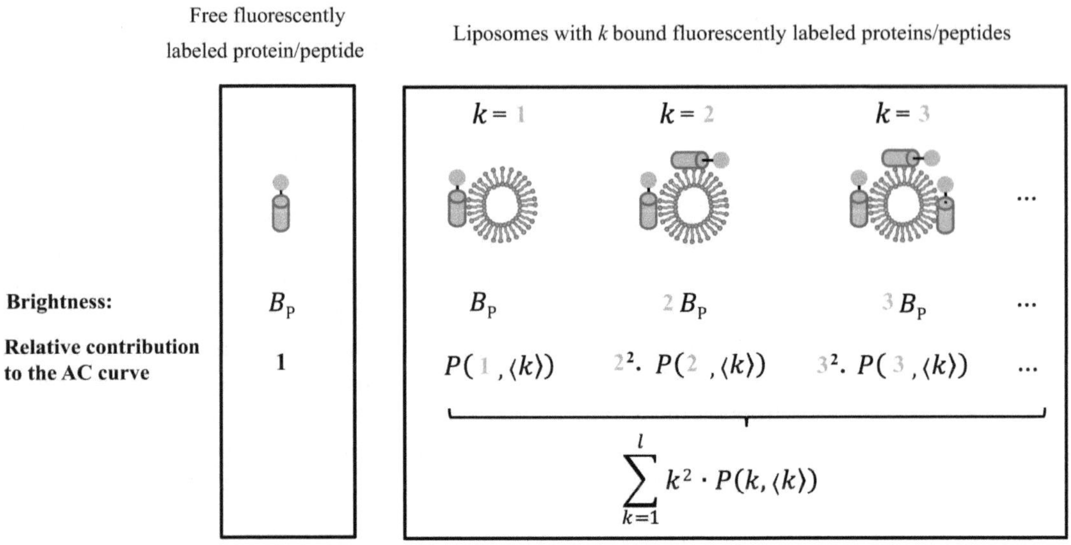

Fig. 2 To accurately determine partition coefficients by FCS, the statistical distribution of fluorescently labeled proteins/peptides per lipid vesicle must be considered in order to take into account the effect of variable liposome brightness at low lipid concentrations. $P(k, \langle k \rangle)$ is the Poisson probability of finding a liposome with k bound proteins/peptides

circular dichroism (CD) and fluorescence spectroscopy. In addition, FCS measurements can be carried out rapidly and therefore a series of AC curves that allows for statistical analysis of the partition behavior can be collected within a short period after first mixing the conjugated protein/peptide and the vesicles together. In order to carry out FCS measurements, however, it is necessary to label the protein/peptide with an adequate fluorescent dye that has low triplet-state excitation, a high quantum yield, and high photostability.

Fluorescence correlation spectroscopy depends on molecular mobilities, enabling to identify molecular interactions based on a change of diffusion times. In a single-color FCS partition study, the only species in solution that can be distinguished according to this parameter are the fast-diffusing free fluorescently labeled protein/peptide (*index P*) and the much slower-diffusing lipid vesicles containing one or more bound fluorescently labeled proteins/peptides (*index ves*) (Fig. 1c). The contribution of each of these fluorescent species to the AC function is strongly biased by its brightness as it is shown by Eq. 3 (i.e., a liposome with two fluorescently labeled proteins/peptides is exactly twice as bright as a lipid vesicle with a single protein/peptide, but it will contribute four times as much to the two-component AC curve) (Fig. 2). The way this problem affects the analysis of FCS partition data was addressed in detail in a previous study by considering the statistical (Poissonian) distribution of the fluorescently labeled protein/peptide among the lipid vesicles [7]. The model derived in that work was based on the following assumptions:

1. Protein/peptide partition to lipid vesicles is a noncooperative process. The mole fraction of the fluorescently labeled bound species to the liposomes, x_m, can be calculated from

$$x_m = \frac{K_P \cdot [L]_{ac}}{[W] + K_P \cdot [L]_{ac}} \tag{5}$$

   where K_P is the mole fraction partitioning coefficient [5], $[W]$ is the water concentration, and $[L]_{ac}$ is the accessible lipid concentration. This last parameter is half of total lipid concentration, when the peptide/protein binds only to the outer leaflet of the membrane.

2. The number of fluorescently labeled protein/peptide molecules bound per vesicle follows a Poisson distribution [8]. Therefore the probability of finding a liposome with k bound proteins/peptides (Fig. 2), is described by

$$P(k, \langle k \rangle) = \frac{\langle k \rangle^{k} e^{-\langle k \rangle}}{k!} \tag{6}$$

   where $\langle k \rangle$ is the average number of membrane-bound proteins/peptides per liposome. In addition, the number of lipid vesicles with one or more fluorescently conjugated protein/peptide bound, N_{ves}^{P}, can be calculated:

$$N_{ves}^{P} = N_{ves}^{t}(1 - P(0, \langle k \rangle)) \cong N_{ves}^{t} \sum_{k=1}^{l} \frac{\langle k \rangle^{k} e^{-\langle k \rangle}}{k!} \tag{7}$$

   where N_{ves}^{t} and $P(0, \langle k \rangle)$ are the total number of lipid vesicles and the probability of finding liposomes without membrane-bound protein/peptide, respectively.

3. The liposome population is monodisperse and its morphology is not affected by protein/peptide partition, so $\langle k \rangle$ will be

$$\langle k \rangle = \frac{x_m [P]_t \mu}{[L]_t} \tag{8}$$

   where $[P]_t$ and $[L]_t$ are the total protein/peptide and lipid concentrations, respectively, and μ is the average number of lipid molecules in a liposome.

4. The bound fluorescently labeled protein/peptide does not dissociate from a liposome during its diffusion through the confocal volume (frozen exchange regime) [9].

5. The brightness of the labeled proteins/peptides is not affected by its partition to the liposomes.

According to the previous assumptions, and using Eq. 4, the *absolute amplitudes* of the AC associated to each diffusing species present in a FCS partition experiment are given by

$$A_{\mathrm{P}}^{\mathrm{M2}} = \frac{N_{\mathrm{P}}^{\mathrm{f}}}{N_{\mathrm{P}}^{\mathrm{f}} + \left(\sum_{k=1}^{l} k \cdot P(k, \langle k \rangle)\right) N_{\mathrm{ves}}^{\mathrm{t}}} \tag{9}$$

$$A_{\mathrm{ves}}^{\mathrm{M2}} = \frac{\left(\sum_{k=1}^{l} k^2 \cdot P(k, \langle k \rangle)\right) N_{\mathrm{ves}}^{\mathrm{t}}}{N_{\mathrm{P}}^{\mathrm{f}} + \left(\sum_{k=1}^{l} k \cdot P(k, \langle k \rangle)\right) N_{\mathrm{ves}}^{\mathrm{t}}} \tag{10}$$

where $N_{\mathrm{P}}^{\mathrm{f}}$ is the mean number of free conjugated proteins/peptides in solution.

The membrane-bound fraction of the fluorescently labeled protein/peptide, x_{m}, can then be directly calculated using the absolute amplitude of the AC curve associated with the fast-diffusing species, $A_{\mathrm{P}}^{\mathrm{M2}}$. In fact, Eq. 9 shows that $A_{\mathrm{P}}^{\mathrm{M2}}$ is equal to the mole fraction of free protein/peptide in solution, x_{f}, as its denominator is equal to $N_{\mathrm{P}}^{\mathrm{t}}$, the total number of protein/peptide molecules present in each sample. Consequently, $x_{\mathrm{m}} = 1 - A_{\mathrm{P}}^{\mathrm{M2}}$. It should also be noted that $x_{\mathrm{m}} \neq A_{\mathrm{ves}}^{\mathrm{M2}}$ because $A_{\mathrm{ves}}^{\mathrm{M2}}$ is weighted by the factor k^2 on its numerator.

However, the fitting equations most commonly available in commercial software programs to analyze the experimental AC curves make use of *fractional*, f_i, instead of *absolute amplitudes*, A_i, as weighting factors:

$$G(\tau) = \frac{1}{\langle N \rangle} \sum_{i=1}^{n} f_i \cdot g_i(\tau) \tag{11}$$

with $f_i = A_i \Big/ \sum_{i=1}^{n} A_i$ and $\sum_{i=1}^{n} f_i = 1$.

For the system under study, $1/\langle N \rangle$ is described by the more complex Eq. 12:

$$\frac{1}{\langle N \rangle} = \frac{N_{\mathrm{P}}^{\mathrm{f}} + \left(\sum_{k=1}^{l} k^2 \cdot P(k, \langle k \rangle)\right) N_{\mathrm{ves}}^{\mathrm{t}}}{\left(N_{\mathrm{P}}^{\mathrm{t}}\right)^2} \tag{12}$$

In this case, and according to Eqs. 9, 10, and 11, the *fractional amplitudes* of the AC curve associated to each species can be calculated from

$$f_{\mathrm{P}}^{\mathrm{M2}} = \frac{N_{\mathrm{P}}^{\mathrm{f}}}{N_{\mathrm{P}}^{\mathrm{f}} + \left(\sum_{k=1}^{l} k^2 \cdot P(k, \langle k \rangle)\right) N_{\mathrm{ves}}^{\mathrm{t}}} \tag{13}$$

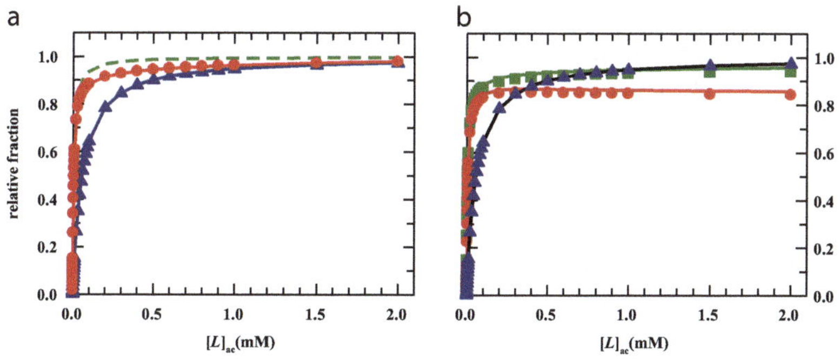

Fig. 3 (**a**) Two-component model, M2: the fractional amplitude calculated for the slow-diffusing species using Eq. 14 ($f_{\mathrm{ves}}^{\mathrm{M2}}$, *filled circle*) converges to the bound mole fraction of fluorescently labeled proteins/peptides [x_{m}, *filled triangle* (Eq. 5 with $K_{\mathrm{P}} = 1 \times 10^6$)] only at saturating lipid concentrations. The K_{p} obtained from fitting directly Eq. 5 to the data of $f_{\mathrm{ves}}^{\mathrm{M2}}$ versus $[L]_{\mathrm{ac}}$ (*dashed green curve*, $K_{\mathrm{p}}^{\mathrm{fitted}} = 8.5 \times 10^6$) was almost tenfold higher than the input value used in the simulation. (**b**) Three-component model, M3: the presence of a vestigial amount of a very bright free dye in solution causes an apparent maximal protein bound fraction lower than 1.0. The variation of (**a**) $f_{\mathrm{ves}}^{\mathrm{M2}}$ (*filled circle*) and (**b**) $f_{\mathrm{ves}}^{\mathrm{M3}}$ (*filled square* and *filled circle*) as a function of accessible lipid concentration was calculated considering the Poisson distribution of 10 nM fluorescently labeled proteins/peptides among lipid vesicles using $\mu = 100{,}000$ and $K_{\mathrm{P}} = 1 \times 10^6$. (**b**) The additional presence of 5 % of free dye (non-binding fluorescent component) with $q = 1$ (*filled square*) and $q = 2$ (*filled circle*) was also included in the simulation (**b**) (Reprinted from Melo et al., 2011 with permission. Copyright from Elsevier)

$$f_{\mathrm{ves}}^{\mathrm{M2}} = \frac{\left(\sum_{k=1}^{l} k^2 \cdot P(k, \langle k \rangle)\right) N_{\mathrm{ves}}^{\mathrm{t}}}{N_{\mathrm{P}}^{\mathrm{f}} + \left(\sum_{k=1}^{l} k^2 \cdot P(k, \langle k \rangle)\right) N_{\mathrm{ves}}^{\mathrm{t}}} \tag{14}$$

Equations 13 and 14 show that both fractional amplitudes, $f_{\mathrm{P}}^{\mathrm{M2}}$ and $f_{\mathrm{ves}}^{\mathrm{M2}}$, are now different from the free- and membrane-bound molar fractions of the conjugated protein/peptide, respectively, again due to the presence of the weighting factor k^2 on their numerator and/or denominator. This is an important observation frequently overlooked in the literature that can lead to an overestimation of the partition coefficients determined by FCS [7]. Figure 3a shows that $f_{\mathrm{ves}}^{\mathrm{M2}}$ converges to x_{m} only when very high lipid concentrations are used in the partition experiment (infinite dilution regime, i.e., when each lipid vesicle contains no more than one membrane-bound protein on average). These two parameters are substantially different at lower lipid concentrations due to the high average number of membrane-bound proteins/peptides per liposome obtained under these experimental conditions ($\langle k \rangle \gg 1$). This effect causes the fitted K_{p} to the data of $f_{\mathrm{ves}}^{\mathrm{M2}}$ versus accessible lipid concentration to be much larger than the input value used in the simulation (Fig. 3a).

Accordingly, a correct quantitative analysis of FCS partition data obtained as fractional amplitude of the bound species versus accessible lipid concentration requires using a nonlinear iterative least-squares procedure that takes into account the statistical (Poissonian) distribution of the fluorescently labeled protein among the ensemble of liposomes to not overestimate the recovered K_p values (*see* Subheading 3.5).

1.2 Determination of Partition Coefficients by FCS in the Presence of a Non-binding Fluorescent Component (Free Dye) (Three-Component Model M3)

The vestigial presence of free unreacted dye in solution is often detected in FCS binding studies [10–13]. This situation can be particularly problematic in FCS partition experiments carried out with negatively charged liposomes because some fluorophores are anionic (e.g., Alexa 488). Therefore, strong electrostatic repulsive interactions between these two components produce a non-binding fluorescent species in solution. In order to evaluate how this non-binding component can affect the FCS partition curves, the model derived in the previous section was extended by including the presence of a third species in solution, namely, unconjugated free dye (three-component model M3) [7].

Using the same set of assumptions as described in Subheading 1.1, it can be easily derived that the *fractional amplitude* associated with the slowest-diffusing species in this three-component system (liposomes with at least one fluorescently labeled protein bound), $f_{\mathrm{ves}}^{\mathrm{M3}}$, is now given by [7]

$$f_{\mathrm{ves}}^{\mathrm{M3}} = \frac{\left(\sum_{k=1}^{l} k^2 \cdot P(k, \langle k \rangle)\right) N_{\mathrm{ves}}^{\mathrm{t}}}{q^2 \cdot N_{\mathrm{dye}} + N_{\mathrm{P}}^{\mathrm{f}} + \left(\sum_{k=1}^{l} k^2 \cdot P(k, \langle k \rangle)\right) N_{\mathrm{ves}}^{\mathrm{t}}} \tag{15}$$

In this equation, N_{dye} is the mean number of free dye molecules in solution and q is the ratio between the brightness of the free dye relatively to the fluorescently labeled protein ($q = B_{\mathrm{dye}}/B_{\mathrm{P}}$). Equation 15 shows that in this case the fractional amplitude associated with the slowest-diffusing species converges to a limiting value lower than 1 at lipid saturating concentrations due to the presence of the factor $q^2 \cdot N_{\mathrm{dye}}$ in its denominator. The higher the q value is (i.e., the brighter the free fluorophore is relatively to the conjugated protein), the lower the plateau value presented by the FCS partition curve of $f_{\mathrm{ves}}^{\mathrm{M3}}$ versus accessible lipid concentration, as illustrated in Fig. 3b. Therefore, the presence of only a trace amount of a very bright free dye in solution can have a strong impact on the experimental FCS partition curves.

1.3 Demonstration of the Application of the FCS Method in Measuring Membrane–Water Partition Coefficients

The partition studies of Alexa488 fluorescently labeled lysozyme (Lz-A488) towards large unilamellar vesicles (LUVs) containing variable mol% of phosphatidylserine will be used here to illustrate how to measure K_p values using FCS [7].

2 Materials

2.1 Labeling of Lysozyme with Alexa 488

1. Lysozyme from chicken egg white (EC 3.2.1.17) (Fluka Biochemika, Buchs, Switzerland).
2. Alexa Fluor 488 SE (carboxylic acid, succinimidyl ester, mixed isomers) (Alexa 488) (Molecular Probes, Eugene, OR).
3. Dimethyl sulfoxide (DMSO) of spectroscopic grade (Merck, Darmstadt, Germany).
4. Hydroxylamine hydrochloride (Sigma Chemical Co., St. Louis, MO).
5. Protein labeling buffer: 0.1 M sodium bicarbonate buffer, pH 8.3 (*see* **Note 1**).
6. Elution buffer: 20 mM HEPES, 0.1 mM EDTA, pH 7.4 (*see* **Note 1**).
7. Chromatography column XK 16/40 connected to a peristaltic pump P-1 to control the eluent flow (GE Healthcare, Uppsala, Sweden).
8. Sephadex G-25 (Sigma Chemical Co., St. Louis, MO).

2.2 Liposome Preparation

1. Lipid mixture: stock solutions of 1-palmitoyl-2-oleoyl-*sn*-glycero-3-phosphocholine (POPC) and 1-palmitoyl-2-oleoyl-*sn*-glycero-3-phosphoserine (POPS) were prepared by dissolving the phospholipids (Avanti Polar Lipids, Alabaster, AL) in chloroform (Merck, Darmstadt, Germany) (*see* **Note 2**).
2. 2-(4,4-Difluoro-5methyl-4-bora-3a,4a-diaza-*s*-indacene-3-dodecanoyl)-1-hexadecanoyl-*sn*-glycero-3-phosphocholine (BODIPY-PC) (Molecular Probes, Eugene, OR) was dissolved in chloroform (Merck, Darmstadt, Germany).
3. Buffer: 20 mM HEPES, 0.1 mM EDTA, pH 7.4 (*see* **Note 1**).
4. Polycarbonate membranes of 100-nm pore diameter (Whatman, Kent, United Kingdom).
5. Mini-extruder (Avanti Polar Lipids, Alabaster, AL) (*see* **Note 3**).

2.3 FCS Instrumentation

FCS measurements were performed using a Leica SP5 TCS confocal inverted microscope (Leica Microsystems, Wetzlar, Germany) equipped with (Fig. 1d):

1. Argon-ion laser. The 488 nm line was used to excite the Alexa 488 fluorophore.
2. A 63×, NA 1.2 water immersion objective (Zeiss, Jena, Germany) with a correction ring (*see* **Note 4**).
3. Emission band-pass filter cube (Leica Microsystems, Wetzlar, Germany). The Alexa 488/Alexa 633 filter cube (band-pass filters 500–550 and 607–683 nm; beam splitter at 560 nm) was used.

4. A dual channel ISS Alba fluorescence correlation detector with APDs was connected to the confocal inverted microscope.

Most of the outlined methods below can be applied to other commercial FCS setups.

2.4 FCS Measurements

1. 8-Well chamber slides (Ibidi, Martinsried, Germany) pre-coated with 10 % bovine serum albumin (BSA) solution for 2 h (Sigma Chemical Co., St. Louis, MO) (*see* **Notes 5** and **6**).
2. Rhodamine 110 (Molecular Probes, Eugene, OR) dissolved into spectroscopic grade methanol (*see* **Note 7**).
3. Buffer: 20 mM HEPES, 0.1 mM EDTA, pH 7.4 (*see* **Note 8**).

2.5 Software for Analyzing FCS Data

1. Analysis of experimental AC curves was performed with ISS Vista software (Champaign, IL), which allows choosing the number of fluorescent species under consideration and fixing and/or linking several parameters of the fitting equations between a set of several AC curves. The fitting equations available in the software to describe two- and three-component systems employ fractional amplitudes as weighting factors. This program uses a Marquardt–Levenberg nonlinear least-squares fitting routine, and the goodness of the fits can be judged by the recovered χ^2 and visual inspection of a random distribution of the weighted residuals.
2. Analysis of experimental FCS partition curves (fractional amplitudes as a function of accessible lipid concentrations) (*see* Subheadings 1.1, 1.2 and 3.5) was carried out using the Solver Add-in of the Microsoft Excel spreadsheet program (Microsoft, Redmond, WA). The precision of the fitted parameters was estimated using the macros Solver Statistics.

3 Methods

3.1 Labeling of Lysozyme with Alexa 488

A general experimental procedure suitable for protein labeling with amine-reactive probes provided by Molecular Probes [14] was used:

1. Prepare 1 mL of lysozyme solution at 0.75 mM in 0.1 M sodium bicarbonate buffer, pH 8.3 (*see* **Note 9**).
2. Immediately before use, dissolve 1 mg of Alexa 488 in 80 μL DMSO.
3. Add the reactive dye dropwise to the lysozyme solution under stirring.
4. Incubate the reaction mixture for 2 h at room temperature protected from light with continuous gentle agitation.
5. Add 100 μL of 1.2 M hydroxylamine hydrochloride, pH 8–8.5, to stop the labeling reaction. This reagent should be freshly prepared.

6. Incubate the reaction mixture again under the same conditions for 1 h.
7. Centrifuge the sample at 10,000 × *g* for 10 min to remove all precipitate material from the sample. Carefully remove the supernatant that will be applied to the chromatographic column.
8. Separate Lz-A488 from unreacted free dye by gel filtration through a Sephadex G-25 column equilibrated in 20 mM HEPES, 0.1 mM EDTA, pH 7.4. Use a flow rate of 1.5 mL/min and collect 1 mL fractions.
9. Determine spectrophotometrically the final protein concentration and labeling stoichiometry, *D/P*, for all the fractions corresponding to the first chromatographic peak (elution of Lz-A488). For this purpose, measure the absorbance of each collected fraction at 280 nm and at the maximum absorption wavelength of Alexa 488, λ_{max} (*see* **Note 10**).
10. Pool the more concentrated protein fractions that present a uniform labeling stoichiometry and keep the samples at 4 °C. These fractions will be used in the FCS partition studies (*see* **Notes 11** and **12**).

3.2 Liposome Preparation

FCS partition studies were performed with large unilamellar vesicles (LUVs) containing POPC mixtures with variable mol% of POPS prepared by the standard extrusion technique [15, 16]. When the FCS measurements were used to characterize the diffusing properties of each LUV population prepared by measuring its diffusion correlation time, BODIPY-PC was co-solubilized in the lipid mixture at 1:10,000 molar ratio (Fig. 4a).

1. Prepare the desired lipid mixture to a final lipid concentration of 5–6 mM by mixing the adequate volumes of each phospholipid stock solution in a round-bottom flask.
2. Dry the solvent out using a gentle nitrogen flow and leave the samples under an oil pump vacuum overnight to remove any residual chloroform.
3. Hydrate the phospholipid mixtures with 20 mM HEPES, 0.1 mM EDTA, pH 7.4 for 30 min and resuspend the lipid films repeatedly by vortexing until all lipid is removed from the flask wall.
4. Perform ten freeze-thaw cycles using liquid nitrogen and a 50 °C water bath to equilibrate and homogenize the sample.
5. Extrude the lipid suspension at least 21 times using 100-nm pore size polycarbonate membranes. The mini-extruder should be assembled according to the instructions provided by the manufacturer.

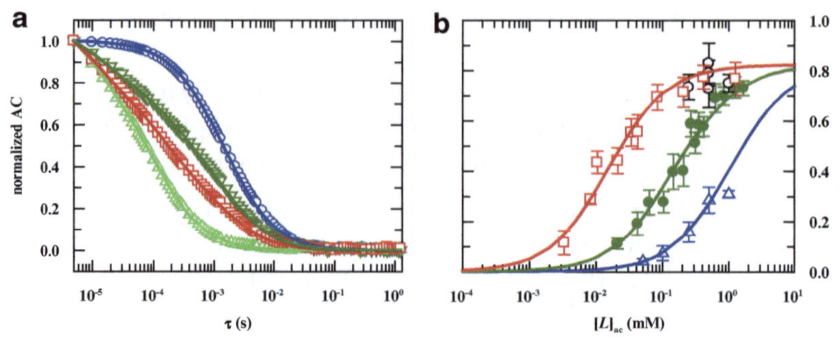

Fig. 4 Lysozyme binding to POPS-containing liposomes is dominantly driven by electrostatic interactions. (**a**) Normalized AC curves of Lz-A488 (10 nM; $D/P = 0.67$) in the absence (*open triangle*) and in the presence of 0.42 mM (*open square*) and 1.25 mM (*inverted triangle*) POPC:POPS 80:20 LUVs and of BODIPY-PC-labeled liposomes prepared with the same lipidic composition (*open circle*). The lines are the best fit of Eqs. 2 and 22 to one- and two-component systems, respectively. (**b**) The experimental fractional amplitudes, f_2, obtained for POPC liposomes containing 10 mol% (*open triangle*), 20 mol% (*filled circle*), 30 mol% (*open square*), and 50 mol% (*open circle*) of POPS are plotted as a function of the accessible lipid concentration in the sample. The *solid curves* are the best fit obtained in a global analysis of the binding data according to a three-component model ($\mu = 100{,}000$ and $q = 1.9$ were fixed in the analysis): the fitting parameters obtained were 0.4 nM for the concentration of free Alexa 488 dye and $K_p = (4.5 \pm 0.7) \times 10^4$, $K_p = (1.9 \pm 0.1) \times 10^5$, and $K_p = (7.0 \pm 0.4) \times 10^5$ for 10 mol%, 20 mol%, and 30 mol% POPS-containing liposomes, respectively (Reprinted from Melo et al., 2011 with permission. Copyright from Elsevier)

6. The resulting stock solution is stored at 4 °C and the lipid vesicles are usually used within 2 days of preparation.

3.3 FCS Sample Preparation

Confocal volume calibration: A ~20 μM stock solution of Rhodamine 110 in methanol is first prepared and kept stored at 4 °C in the dark. Before the FCS measurements, dilute this solution to 5 and 10 nM final concentrations in the adequate buffer/water. These dilutions should not be reused.

Characterization of individual species: Dilute the solutions of free dye (Alexa 488), LUVs containing a small amount of BODIPY-PC (1:10,000 labeled phospholipid/total lipid molar ratio) and fluorescently labeled protein (Lz-A488) to nM range (*see* **Note 13**).

FCS partition studies: Prepare several independent solutions with a constant concentration of fluorescently conjugated protein at a nM concentration and with increasing total lipid concentration.

3.4 FCS Measurements

1. Switch on the excitation laser around 30 min prior to the measurements to allow for stabilization.
2. Select the FCS configuration of the microscope that best suits the spectral properties of the fluorophore under study (laser line and channel of the filter cube).

3. Carefully, add a droplet of water to the objective which is suitable to perform FCS measurements.
4. Place a pre-coated chamber slide on the microscope stage and add 400 μL of sample to each well.
5. Set the pinhole to 1 Airy unit to guarantee confocality. In the Leica SP5 TCS confocal inverted microscope, 1 Airy corresponds to 111.44 μm for the objective used in this study.
6. To correct for the variable thickness of each observation chamber, adjust the correction collar of the objective by focusing the reflection image generated at the transition of the coverslip/ buffer interface.
7. Place the laser focus at ~200 μm above the top surface of the cover slide [6, 7, 17].
8. Adjust the laser power that will be used during the FCS measurements by obtaining a reasonable count rate, preferentially at least 10× the background count rate (*see* **Notes 14** and **15**).
9. Perform a calibration FCS measurement by recording ten AC curves of 20 s each for both diluted dye solutions prepared. By following the steps described below, fit the AC curves globally and estimate the diffusion time of Rhodamine 110 and the structure parameter, *S*. If the setup is properly aligned, these results should be reproducible on a daily basis (*see* **Note 16**).
10. For each assay mixture used in the partition study, perform an FCS measurement for 10 × 20 s. An increase in diffusion time should be detected upon the addition of increasing amounts of lipid vesicles to each protein sample (Fig. 4a) (*see* **Note 17**).
11. Before starting the analysis of the data, all AC curves obtained should be scrutinized. AC curves whose count rate is not constant over time should be discarded. These fluctuations in intensity can be indicative of adsorption to the chamber walls, photobleaching or aggregation of the sample (*see* **Notes 6, 15, 18,** and **19**).

3.5 FCS Data Analysis

Calibration of confocal volume:

1. Begin by globally fitting the experimental AC curves obtained for 5 and 10 nM rhodamine 110 with a 3D Gaussian model that assumes a triplet term and a single diffusing species (Eq. 2). In these analysis, S, T, τ_T and τ_D should be linked between all AC curves. When a good optical alignment is reached, the structure parameter, S, normally ranges from 4 to 6 (*see* **Notes 16**, **20** and **21**).

Characterization of individual species:

Before analyzing FCS data from heterogeneous samples, it is essential to characterize the individual species by FCS in order to have independent knowledge of the different parameters, namely, triplet lifetime, percentage of triplet state and diffusion correlation time of each species. These can later be fixed when studying more complex samples.

1. Perform a fit of the experimental AC curves of each individual species with Eq. 2, but now fix the value of S, which was previously determined in the calibration of the FCS setup.
2. According to fitting results, calculate the diffusion coefficient of protein (*see* **Note 22**) and the hydrodynamic radius of liposomes (*see* **Note 23**). Then, compare the values obtained with those reported in the literature or expected.

Determination of protein/peptide–lipid partition coefficients. Two different situations can be anticipated.

A. The analysis software uses ***absolute amplitudes*** as weighting factors in the fitting equation.
   1. Fix S in the fitting equation to a two-component system used in the analysis of the AC curves (Eq. 4 with $n = 2$). To reduce the strong statistical correlation between the fit parameters, the diffusion time associated with the slowest component must be linked between all the AC curves under analysis. Additionally, the triplet lifetime and diffusion time related to the free fluorescently labeled protein are fixed at the previously determined values.
   2. Calculate the membrane-bound fraction of the fluorescently labeled protein/peptide, x_m, in each sample using the absolute amplitude of the AC curve associated with the fast-diffusing species, A_1 ($x_m = 1 - A_1$).
   3. Determine the partition coefficient of the protein by fitting Eq. 5 to the experimental data of $x_m = 1 - A_1$ as a function of accessible lipid concentration.

B. The analysis software uses ***fractional amplitudes*** as the weighting factors in the fitting equation.
   1. First, the ***fractional amplitude*** related to the slow-diffusing lipid vesicles with one or more bound fluorescently labeled proteins/peptides, f_2, must be determined by fitting (*see* **Note 24**). Implement the same analysis strategy of the AC curves as described above in order to minimize the error associated with f_2.
   2. In a spreadsheet, guess an initial value for the partition coefficient of the conjugated protein/peptide and calculate the average number of membrane-bound proteins per lipid vesicle, $\langle k \rangle$, for each lipid concentration studied using Eq. 8. A reasonable estimate for the average number of lipid molecules in a liposome for 100-nm LUVs is

100,000. Compute the probabilities of finding a lipid vesicle with k bound proteins/peptides ($k = 1, 2, \ldots$) using Eq. 6 and, finally, $f_{\mathrm{ves}}^{\mathrm{M2}}$, using Eq. 14. In our study, the Poisson distribution was approximated by a finite sum with a cutoff l set to 20, to guarantee that under the experimental conditions used $\sum_{k=1}^{l} P(k, \langle k \rangle) \cong 1$.

3. Compare these calculated fractional amplitudes, $f_{\mathrm{ves}}^{\mathrm{M2}}$, to the experimental values, f_2, and minimize the sum of their square deviations by using Solver in an iterative procedure by changing the value initially guessed for K_{p}.
4. If the *fractional amplitude* associated with the slow-diffusing liposomes asymptotically reaches a plateau value lower than 1 as the accessible lipid concentration of the samples is increased, a non-binding fluorescent component (usually free dye) is present in the system (*see* **Note 25**).
5. Repeat the iterative fitting strategy described above to determine K_{p} and the concentration of free dye present in each sample by comparing the calculated $f_{\mathrm{ves}}^{\mathrm{M3}}$ (Eq. 15) to the experimental values of the *fractional amplitude* associated with the slow-diffusing liposomes, f_2 (*see* **Note 26**). The q value must be determined independently by performing control steady-state fluorescence measurements and its value should be kept fix in the analyses.

The partition curves relative to Lz-A488 binding to liposomes prepared with variable anionic phospholipid content are shown in Fig. 4b. The data was globally analyzed as described above by linking the value of [Dye] between all the partition curves used in the fitting procedure. It should be noted that although a plateau value of $f_2^{\max}$~0.80 was reached in these experiments, the true molar fraction of the freely diffusing dye was much lower than 20 % (~5.6 %) because its molecular brightness was approximately two-fold higher than that from Lz-A488. This study confirmed that lysozyme binding to negatively charged liposomes is dominantly driven by electrostatic interactions [7].

4 Notes

1. All buffer solutions must be prepared using ultrapure water, e.g., produced by a Milli-Q system (specific resistance >18 MΩ cm).
2. Concentrations of lipid stock solutions in organic solvents can change during storage due to solvent evaporation. The

concentration should be frequently confirmed using phosphate analysis [18].

3. It is essential to prepare monodisperse lipid vesicles in order to be able to analyze the AC curves obtained during the partition study using a two-component fitting function. Also the liposomes must be unilamellar in order that the available lipid for interaction is precisely defined in each experiment.
4. Note that these objectives require a specific thickness of coverglass (usually no. 1 coverslips).
5. As an alternative, Labtek chambered coverglass (Nunc, Wiesbaden, Germany) can be used.
6. Surface adsorption of the sample to the chamber walls can cause problems in FCS since very low sample concentrations are used. This effect can be minimized by pre-coating the surface with a noninteracting agent. In FCS partition studies, SUVs (prepared with a noninteracting lipid) or BSA are usually used [6, 7]. The samples should be loaded after extensive rinsing of the wells with buffer.
7. Choose a dye with a known diffusion coefficient [19] and also that allow measurements to be done with the same laser/filter configuration as the sample. Acquire AC curves using the same conditions as the sample.
8. Always filter the buffers prior to the FCS measurements using 0.22 μm cellulose acetate filters.
9. The amine terminus of proteins can sometimes be selectively labeled by using a buffer close to neutral pH. The sample labeling ratio, D/P, can be changed by exploring different dye-to-protein molar ratios in the reaction mixture.
10. The labeling stoichiometry, D/P, of a sample can be calculated using,

$$D/P = [D]/[P] \tag{16}$$

where $[D]$ and $[P]$ are the dye and protein concentrations in the conjugate, respectively. They are determined from

$$[D] = \frac{\mathrm{Abs}_{\mathrm{D}}^{\lambda\,\mathrm{max}}}{\varepsilon_{\mathrm{D}}^{\lambda\,\mathrm{max}}} \tag{17}$$

and

$$[P] = \frac{\mathrm{Abs}_{\mathrm{P}}^{280} - \mathrm{Abs}_{\mathrm{D}}^{\lambda\,\mathrm{max}} \cdot \mathrm{CF}}{\varepsilon_{\mathrm{P}}^{280}} \tag{18}$$

where $\mathrm{Abs}_{\mathrm{D}}^{\lambda\,\mathrm{max}}$ and $\varepsilon_{\mathrm{D}}^{\lambda\,\mathrm{max}}$ are the absorbance and the extinction coefficient of free dye at λ_{max}, respectively. $\mathrm{Abs}_{\mathrm{P}}^{280}$ and $\varepsilon_{\mathrm{P}}^{280}$ are the absorbance and extinction coefficient of the protein in conjugate at 280 nm, and CF is the correction factor due to the fluorophore's absorbance at 280 nm:

$$CF = \frac{\varepsilon_{D}^{280}}{\varepsilon_{D}^{\lambda\,max}} \tag{19}$$

where ε_{D}^{280} is the extinction coefficient of free dye at 280 nm. For the system under study, D/P was calculated using the extinction coefficient of the dye provided by the supplier [14] and that of lysozyme in the UV [20].

11. It should always be checked that the protein structure was not significantly altered by the labeling procedure by performing control CD measurements and activity assays when applicable.
12. Additional purification steps of the fluorescently labeled protein after its conjugation reaction can be performed to guarantee that all free unreacted dye is removed from the sample. Particularly, in case that the unreacted free dye also binds nonspecifically to the protein under study by predominantly electrostatic interactions, the gel filtration chromatography can be carried out using a high ionic strength buffer to minimize this effect. If the protein stability/purification yields are compromised by the inclusion of these extra purification steps in the protocol, FCS partition data can still be analyzed by performing a nonlinear iterative least-squares fit of the three-component model to the experimental data of fractional amplitude of the bound species, f_2 versus accessible lipid concentration as it is presented in detail in Subheading 3.5.
13. Always centrifuge and/or filter the protein sample with a low-protein binding filter before FCS measurements.
14. The laser power should be adjusted in order that the triplet fraction is not higher than 20 %.
15. Photobleaching causes a reduction of the sample fluorescent count rate and an artificial decrease in the diffusion time of the particle at high laser powers. During optimization of FCS data acquisition, it is important to check whether the diffusion times change with the laser power.
16. It is essential to pay attention not only to the structure parameter S but also to other FCS parameters of the calibration dye, namely, its triplet lifetime, percentage of triplet state and diffusion correlation time. Any change in these parameters can report misalignments in the system. Note that FCS is highly sensitive to moderate misalignment that is not perceived when confocal imaging is performed. For FCS measurements, a perfect centering of the excitation laser on the back aperture of the microscope objective and a perfect conjugation of the

pinhole with the excitation spot are required. In the case of commercial setups, some of the alignment settings may not be accessible to the user and company maintenance may be required. A too large structure parameter and diffusion coefficient values are indicative of a misalignment of the system.

17. Before starting the FCS measurements, confirm that the conjugated protein does not change its brightness upon binding to the lipid vesicles using steady-state fluorescence measurements. Alternatively, this can be done by comparing the counts per second (CPS) measured by FCS for the fluorescently labeled protein alone and after the addition of a saturating concentration of lipid vesicles. The number of bound conjugated proteins/peptides should be sufficiently low that self-quenching does not occur at low liposome concentrations.

18. Large spikes in the fluorescence signal and rapid changes in the amplitude or shape of AC curves are indicative of the presence of aggregates. Aggregates have a high impact in AC curves, since each particle contributes with the square of its brightness to the AC function. Although they can be removed from the raw data, this is not advisable in partition studies because they cannot be distinguished from liposomes with a low number of conjugated proteins bound also present in more complex mixtures. So, it is essential to always centrifuge and filter the protein sample with a low-protein binding filter before FCS measurements.

19. Due to the low-protein concentrations used in FCS measurements, the binding of the fluorescently labeled protein/peptide does not usually change significantly the surface charge density of the lipid vesicles. This minimizes the occurrence of electrostatically driven, protein-mediated liposome aggregation. However, when this phenomenon takes place, it can be easily detected in the FCS measurements by a dramatic change in the shape of the AC curves [7, 21]. Therefore, FCS partition measurements are easily uncoupled from this adverse effect.

20. Global analysis of the data can help to constrain fitting parameters, improving the accuracy of fitting results. The number of free fitting parameters in FCS data analysis should always be minimized or linked between several AC curves obtained under the same experimental conditions.

21. The term relative to the triplet-state formation should be included in the fitting equation to the AC curves depending on the dyes and excitation intensity used. In specific situations, this term can be neglected.

22. The protein/peptide diffusion coefficient, D_P, can be calculated from

$$D_P = \frac{\tau_{D_{Rhod110}}}{\tau_{D_P}} \cdot D_{Rhod110} \tag{20}$$

where $\tau_{D_{Rhod110}}$ and τ_{D_P} are the diffusion times of Rhodamine 110 and of the fluorescently labeled protein/peptide, respectively. $D_{rhod110}$ is the known diffusion coefficient of Rhodamine 110 [19].

23. The diffusion coefficient is related to the hydrodynamic radius, r_H, of the particle by the Stokes–Einstein equation:

$$D = \frac{k_B T}{6\pi\eta r_H} \tag{21}$$

where η is the viscosity of the medium, T is the temperature, and k_B is the Boltzmann constant.

24. The AC function for 3D Gaussian diffusion of a two-component system is given by

$$G_D(\tau) = \frac{1}{\langle N \rangle}\left(1 + \frac{T_1}{1-T_1} \cdot \exp\left(-\frac{\tau}{\tau_{T_1}}\right)\right)\left(1 + \frac{T_2}{1-T_2} \cdot \exp\left(-\frac{\tau}{\tau_{T_2}}\right)\right) \times \sum_{i=1}^{2} f_i \left(1 + \frac{\tau}{\tau_{D_i}}\right)^{-1} \cdot \left(1 + \frac{\tau}{S^2 \tau_{D_i}}\right)^{-\frac{1}{2}} \tag{22}$$

where T_i is the fraction of fluorescent component i in the triplet state, τ_{T_i} and τ_{D_i} are their triplet and diffusion lifetimes. Finally, f_i is the fractional contribution of each component to the AC function, with $f_2 = 1 - f_1$ (model M2: free labeled protein ($i = 1$) and liposomes containing at least one conjugated protein bound ($i = 2$)).

25. It should be noted that the fluorescent non-binding component detected can be due to the presence of a non-membrane-competent form of the protein in solution [17].

26. Although the simultaneous presence in solution of three fluorescent species with different diffusion coefficients (free dye, free fluorescently labeled protein and lipid vesicles with at least one conjugated protein bound, respectively) implies that the analysis of the experimental AC curves should be done using a three-component model, we found that reliable partitioning data was still obtained when the experimental AC curves were fitted using a simplifying two-component equation (Eq. 22). Probably, the large difference between the diffusion times of the free dye and Alexa488-conjugated lysozyme, on one hand, and the liposomes, on the other hand, allowed to temporally "separate" the fluorescence fluctuations from each set of components. In fact, it is known that a difference of a factor of about 7–8 in molecular weight is needed to unambiguously distinguish two species by their diffusion times [22].

Acknowledgments

This work was supported by Fundação para a Ciência e Tecnologia (Project PTDC/QUI-BIQ/099947/2008, Project PTDC/QUI-BIQ/112067/2009 and Ph.D. grant SFRH/BD/61723/2009 to A.M.).

References

1. Rigler R, Elson ES (eds) (2001) Fluorescence correlation spectroscopy: theory and applications. Springer, Berlin
2. Hess ST, Huang SH, Heikal AA, Webb WW (2002) Biological and chemical applications of fluorescence correlation spectroscopy: a review. Biochemistry 41(3):697–705
3. Haustein E, Schwille P (2007) Fluorescence correlation spectroscopy: novel variations of an established technique. Annu Rev Biophys 36:151–169
4. Thompson NL (1991) Fluorescence correlation spectroscopy. In: Lakowicz J (ed) Topics in fluorescence spectroscopy techniques, vol 1. Plenum Press, New York, NY, pp 337–378
5. Santos NC, Prieto M, Castanho M (2003) Quantifying molecular partition into model systems of biomembranes: an emphasis on optical spectroscopic methods. Biochim Biophys Acta 1612(2):123–135
6. Rusu L, Gambhir A, McLaughlin S, Rädler J (2004) Fluorescence correlation spectroscopy studies of peptide and protein binding to phospholipid vesicles. Biophys J 87(2):1044–1053
7. Melo AM, Prieto M, Coutinho A (2011) The effect of variable liposome brightness on quantifying lipid-protein interactions using fluorescence correlation spectroscopy. Biochim Biophys Acta-Biomembr 1808 (10):2559–2568
8. Yu LL, Tan MY, Ho B et al (2006) Determination of critical micelle concentrations and aggregation numbers by fluorescence correlation spectroscopy: aggregation of a lipopolysaccharide. Anal Chim Acta 556(1):216–225
9. Al-Soufi W, Reija B, Novo M et al (2005) Fluorescence correlation spectroscopy, a tool to investigate supramolecular dynamics: inclusion complexes of pyronines with cyclodextrin. J Am Chem Soc 127 (24):8775–8784
10. Xu H, Frank J, Trier U et al (2001) Interaction of fluorescence labeled single-stranded DNA with hexameric DNA-helicase RepA: a photon and fluorescence correlation spectroscopy study. Biochemistry 40(24):7211–7218
11. Krouglova T, Vercammen J, Engelborghs Y (2004) Correct diffusion coefficients of proteins in fluorescence correlation spectroscopy. Application to tubulin oligomers induced by Mg^{2+} and paclitaxel. Biophys J 87 (4):2635–2646
12. Nath S, Meuvis J, Hendrix J et al (2010) Early aggregation steps in alpha-synuclein as measured by FCS and FRET: evidence for a contagious conformational change. Biophys J 98(7):1302–1311
13. Reija B, Monterroso B, Jimenez M et al (2011) Development of a homogeneous fluorescence anisotropy assay to monitor and measure FtsZ assembly in solution. Anal Biochem 418 (1):89–96
14. Haugland RP (ed) (2005) Handbook of fluorescent probes and research chemicals, 10th edn. Molecular Probes, Eugene, OR
15. Hope MJ, Bally MB, Mayer LD et al (1986) Generation of multilamellar and unilamellar phospholipid vesicles. Chem Phys Lipids 40 (2–4):89–107
16. Mayer LD, Hope MJ, Cullis PR (1986) Vesicles of variable sizes produced by a rapid extrusion procedure. Biochim Biophys Acta 858 (1):161–168
17. Posokhov YO, Rodnin MV, Lu L, Ladokhin AS (2008) Membrane insertion pathway of annexin B12: thermodynamic and kinetic characterization by fluorescence correlation spectroscopy and fluorescence quenching. Biochemistry 47(18):5078–5087
18. McClare CWF (1971) An accurate and convenient organic phosphorus assay. Anal Biochem 39(2):527–530

19. Gendron PO, Avaltroni F, Wilkinson KJ (2008) Diffusion coefficients of several rhodamine derivatives as determined by pulsed field gradient-nuclear magnetic resonance and fluorescence correlation spectroscopy. J Fluoresc 18(6):1093–1101
20. Pace CN, Vajdos F, Fee L et al (1995) How to measure and predict the molar absorption-coefficient of a protein. Protein Sci 4 (11):2411–2423
21. Elbaum-Garfinkle S, Ramlall T, Rhoades E (2010) The role of the lipid bilayer in tau aggregation. Biophys J 98(11):2722–2730
22. Meseth U, Wohland T, Rigler R, Vogel H (1999) Resolution of fluorescence correlation measurements. Biophys J 76(3):1619–1631

Chapter 27

PET-FCS: Probing Rapid Structural Fluctuations of Proteins and Nucleic Acids by Single-Molecule Fluorescence Quenching

Markus Sauer and Hannes Neuweiler

Abstract

Quenching of organic fluorophores by aromatic amino acids and DNA nucleotides with expelled electron donating properties allows the study of conformational dynamics of biomolecules. Efficient fluorescence quenching via photoinduced electron transfer (PET) requires van der Waals contact and can be used as reporter for structural fluctuations at the 1-nm scale in proteins, peptides, and nucleic acids. The combination of PET with fluorescence correlation spectroscopy (FCS) establishes a powerful method (PET-FCS) to study equilibrium dynamics at the single-molecule level on time scales from nano- to milliseconds. We delineate the fundamentals of PET-based fluorescence quenching, reporter engineering, instrumental and experimental design, and provide examples.

Key words Fluorescence quenching, Photoinduced electron transfer, Fluorescence correlation spectroscopy, Protein folding and dynamics, Intrinsically disordered proteins

1 Introduction

1.1 General Introduction

The ability of polypeptides and nucleic acids to spontaneously adopt unique three-dimensional folds in aqueous solution is one of the most fundamental examples of biological self-assembly. Following the traditional structure-function relationship, function is determined by the exact three-dimensional arrangement of building blocks, i.e., amino acids and nucleotides. The discovery of proteins that natively lack regular structure, so-called intrinsically disordered proteins, challenges this traditional view and makes a mechanism of function difficult to comprehend [1, 2]. An estimated 40 % of proteins in the human proteome contain intrinsic disorder, highlighting their importance in higher organisms [3]. Disordered proteins are inherently dynamic, and the role of

Yves Engelborghs and Antonie J.W.G. Visser (eds.), *Fluorescence Spectroscopy and Microscopy: Methods and Protocols*, Methods in Molecular Biology, vol. 1076, DOI 10.1007/978-1-62703-649-8_27,

dynamics for function has been recognized in recent years [4]. Protein conformational motions occur on broad length and time scales and are often modulated by binding.

In order to monitor motions within biopolymers, changes in distance between specific sites need to be measured as function of time. From the spectroscopic point of view, any energy transfer method such as resonant or nonresonant energy transfer, proton transfer, or electron transfer can be used to reveal conformational changes, as long as the change of spectroscopic signal results from alterations in the interaction distance or geometry of extrinsic or intrinsic probes. In this context, fluorescence spectroscopy is a method of choice because emission properties of chromophores are sensitive to their short-range and long-range environment. The exquisite sensitivity of modern fluorescence spectroscopy allows detection down to the ultimate limit of single molecules. In recent years, single-molecule (sm) fluorescence has arrived as integral part of the spectroscopic tool box for molecular biology [5, 6]. Sm methods detect heterogeneity that is hidden in the ensemble average signal of conventional spectroscopy. The approach is therefore well suited to study structural changes of proteins and nucleic acids, molecules that can be considered dynamically heterogeneous in structural terms. The perhaps most popular and applied fluorescence reporter system for distance changes at a molecular scale is fluorescence resonance energy transfer (FRET) [5, 7–9]. FRET involves the radiationless transfer of excited-state energy from a donor (D) to an acceptor (A) fluorophore, a photophysical process that occurs through space via dipole-dipole coupling. The FRET signal scales with $1/(1 + (R/R_0)^6)$, where R_0 is the so-called Förster radius of 50 % transfer efficiency and ranges typically between 2 and 8 nm, depending on the photophysical properties of D and A as well as environmental parameters. In the case of smFRET, however, R_0 is typically on the order of 5 nm, due to the stringent demands on brightness restricting the selection of fluorophores. Thus, smFRET monitors distance changes typically around 5 nm.

Photoinduced electron transfer (PET), on the other hand, is a photochemical process that quenches fluorescence emission through charge transfer between fluorophore (F) and quencher (Q) at van der Waals contact [6, 10, 11]. Contact typically occurs at <1 nm distance separation between the geometric centers of F and Q [12]. But PET fluorescence quenching has also energetic constraints. Only if the reduction and oxidation potential of F and Q match, the fluorescence-quenching electron transfer can occur spontaneously. Among all natural occurring building blocks of proteins and nucleic acids, it is only the amino acid tryptophan (Trp) and the nucleobase guanosine (G) that have oxidation potentials low enough to serve as efficient electron donors in PET reactions with extrinsic labels. This unique redox property makes them an exquisite tool to site-specifically transform protein and nucleic acid conformational changes into on/off fluorescence fluctuations of labels placed at specific sites within a structure. When observed in highly dilute solution on confocal fluorescence microscopes, structural fluctuations emerge as fluorescence fluctuations from individual molecules and can be analyzed by correlation spectroscopy. In recent years, PET in combination with fluorescence correlation spectroscopy (PET-FCS) has arrived as valuable tool to study fast conformational dynamics of peptides [13–15], proteins [16–20], and nucleic acids [21, 22].

In the following two subsections, we introduce the basics of PET and its combination with FCS.

1.2 Photoinduced Electron Transfer (PET) Reactions

Fluorophores are capable of reporting changes in their local environment. Dependent on the photophysical properties of the fluorophore, changes in the microenvironment are accompanied by alterations in fluorescence characteristics like spectrum, quantum yield, or lifetime. If excitation occurs through absorption of light in a fluorophore (usually an organic molecule with a delocalized aromatic π-electron system), its reactivity generally increases enabling redox reactions that are energetically forbidden in the singlet ground state. In PET experiments, a single fluorophore and a quencher are incorporated into the biopolymer chain. Depending on oxidation and reduction potentials, electron transfer from the fluorophore, being the donor D, to its interaction partner, the acceptor A, or vice versa can occur, resulting in non-radiative dissipation of excited-state energy. In order to analyze and interpret experimental results from fluorescence quenching caused by PET, the transfer mechanism and its distance dependence has to be well understood. In general, quenching of fluorophores in the first excited singlet state by electron donors or acceptors results in charge separation and the formation of a radical ion pair $A^{-\bullet}D^{+\bullet}$, which returns to the ground state via charge recombination (Fig. 1). The efficiency of charge separation is controlled by the relation between the free energy of the reaction, the reorganization energy, and the distance between donor and acceptor.

To estimate the efficiency of photoinduced charge separation, the associated change in free energy, ΔG_{cs}, can be estimated using the Rehm-Weller equation [23]

$$\Delta G_{cs} = E_{ox} - E_{red} - E_{0,0} + C, \qquad (1)$$

where E_{ox} and E_{red} are the first one-electron oxidation potential of the donor and the first one-electron reduction potential of the acceptor, respectively, in solvent. $E_{0,0}$ is the energy of the zero-zero transition to the lowest excited singlet state, and C is the solvent-dependent Coulombic interaction energy (which can be neglected in moderately polar environment). The reaction rate follows an exponential D-A distance dependence with a characteristic length scale on the order of few Angström [24]. Evidence for

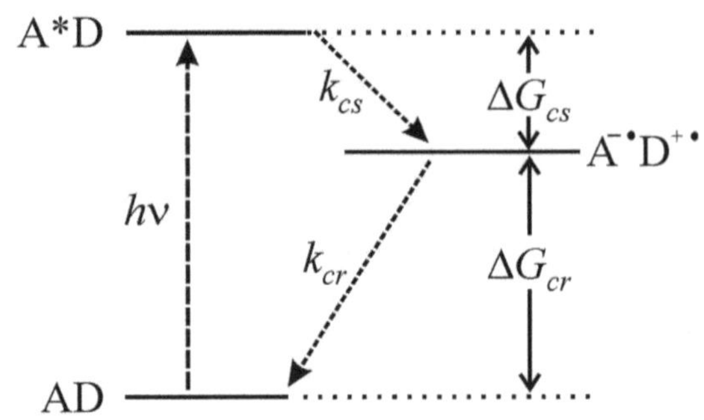

Fig. 1 Kinetic and thermodynamic scheme of PET charge separation and charge recombination. *A* represents the acceptor fluorophore, and *D* is the electron donor, e.g., Trp. (k_{cs}: rate constant for charge separation; k_{cr}: rate constant for charge recombination; ΔG_{cs}: free energy change for charge separation; ΔG_{cr}: free energy change for charge recombination)

PET reactions can be derived from experiments showing the transient existence of charged-separated species. Measured redox potentials allow the estimation of PET efficiencies based on the Rehm-Weller formalism (Eq. 1) and strongly support the photochemical process.

The kinetics of PET reactions between electron donors like Trp or G and fluorophores are ultrafast, i.e., in the femtosecond (fs) to picosecond (ps) time regime [25, 26]. Examples for PET between fluorophores and planar aromatic ring systems, such as the indole moiety of Trp or the fused pyrimidine-imidazole ring system present in G, including redox potentials can be found in the literature [11, 27–29]. The oxidation potentials of Trp (E_{ox} = 0.60–1.00 V versus SCE) and guanine (E_{ox} = ~1.24 V versus SCE) are unique among all naturally occurring amino acids and nucleobases. In combination with suitable fluorophores, Trp and G in polypeptides and oligonucleotides serve as selective electron donors in PET reactions. The oxazine derivative MR121, for instance, has a reduction potential of E_{red} = −0.42 V vs. SCE (measured in acetonitrile) and a transition energy of ~1.9 eV, such that PET from Trp or G to MR121 is exergonic (ΔG_{cs} between −0.7 and −0.2 eV) according to Eq. 1. Note that commercial oxazine fluorophores, like ATTO 655 or ATTO Oxa11, have the chromophore of MR121. Besides oxazine derivatives, rhodamine and BODIPY fluorophores are also prone to being efficiently quenched by electron donors [11].

Interactions between fluorophore and quencher can either be "dynamic" or "static." Whereas collisional quenching of fluorophores by suitable compounds (electron donors or acceptors) is reflected in a decrease in fluorescence lifetime and intensity ("dynamic quenching"), complex formation between the fluorophore and quencher leads to efficient "static quenching" without observed changes in fluorescence lifetime. These molecular complexes feature stacked interaction geometries [12] that are essentially nonfluorescent and dissociate much slower than the fluorescence decay time. Thus, association and dissociation of fluorophore and quencher leads to digital off/on fluorescence switching, the kinetics of which depend on the microscopic rate constants of complex formation and dissociation.

The formation of efficiently quenched fluorophore/quencher complexes in aqueous solution is the basis for using them as reporter for rapid conformational fluctuations. Through site-specific attachment of fluorophore and quencher, distance variations between chosen attachment points can be observed spectroscopically and analyzed temporally.

1.3 Combining PET with Fluorescence Correlation Spectroscopy

Fluorescence correlation spectroscopy (FCS) analyzes the kinetics of fluorescence fluctuations at equilibrium through statistical analysis of intensity time traces. Virtually any process that gives rise to a fluctuation of fluorescence emission can be analyzed via FCS. With the advent of modern fluorescence microscopy techniques and equipment within the past decade, FCS has been integrated as single-molecule technique for molecular biology. One main application of classical FCS is to measure the time constants of molecular diffusion, a quantity that is related to the size and shape of a molecule as well as interactions within its environment. The technique can thus provide information on biomolecular structure and function. For details on FCS and its applications, the reader is referred to reviews [30–32].

In FCS fluorescence intensity time traces are recorded from molecules diffusing through a spatially well-defined excitation/detection volume by Brownian motion. Typically, micro-ensembles of molecules are probed from pM to nM samples, yielding about 1–100 molecules at a time in the detection focus depending on microscope setup geometry. Fluorescence fluctuations are analyzed by calculating the autocorrelation function (ACF), $G(\tau)$, of the recorded fluorescence intensity time trace:

$$G(\tau) = \frac{< I(t) \cdot I(t+\tau) >}{I(t)^2}, \tag{2}$$

where $I(t)$ is the fluorescence intensity at time t, $I(t + \tau)$ is the fluorescence intensity after a time lag τ, and $< >$ denotes the time average over the total observation time. Under appropriate experimental conditions, constantly emitting fluorescent labels give rise to a stochastic fluctuation of the detector signal as molecules diffuse in and out of the focal volume (Fig. 2). This process results in a decay of the ACF with a characteristic time constant τ_D representing the experimental diffusion time constant. τ_D is typically on the order of one ms and depends, besides on the size of the observation

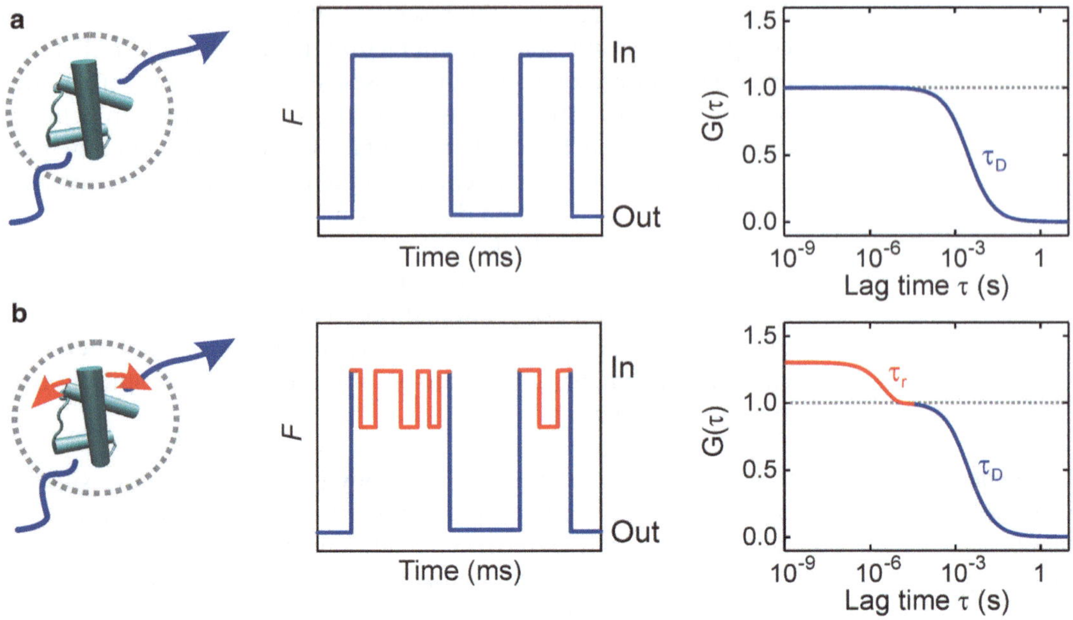

Fig. 2 Probing protein dynamics by PET-FCS. Fluorescently modified proteins diffuse through a confined detection volume by Brownian motion, causing a stochastic fluctuation of the fluorescence signal (*F*) at the detector. (**a**) A protein that shows constant fluorescence emission gives rise to fluctuations that result in a decay of the ACF ($G(\tau)$) with a characteristic time constant of translational diffusion τ_D (*blue*). (**b**) Kinetics of conformational motions that are faster than τ_D and modulate fluorescence emission of the extrinsic label emerges as additional decay in the ACF with an observed relaxation time constant τ_r (*red*) (Adapted with permission from ref. 20. Copyright 2011, Elsevier)

volume, on the size and shape of the molecule, temperature, solution viscosity, and specific interactions within the diffusion matrix.

Through incorporation of a suitable fluorophore and Trp or G as quenchers at specific sites within a biomolecule, conformational changes that induce fluorophore/quencher association and dissociation are transformed into changes of fluorescence emission intensity. If equilibrium conformational fluctuations occur on time scales faster than τ_D, corresponding kinetics emerge as additional decay in the ACF with an observed relaxation time constant τ_r (Fig. 2). This represents the basic idea of a PET-FCS experiment.

Assuming two-state equilibrium between fluorescent and fluorescence-quenched conformational states A and B, the observed relaxation time constant τ_r contains the microscopic rate constants k_1 and k_{-1} (Scheme 1): $\tau_r = 1/(k_1 + k_{-1})$. Deconvolution requires knowledge of the equilibrium constant $K_{eq} = k_1/k_{-1}$, which may be obtained from independent experiments. The amplitude a_r of the ACF decay contains K_{eq} but is convoluted with the change of fluorescence brightness associated with the transition from state A to B. The quenched state B may, e.g., be an ensemble of states showing residual fluorescence that results from rapid

$$A \underset{k_{-1}}{\overset{k_1}{\rightleftharpoons}} B$$

Scheme 1 Two-state equilibrium of conformational transition between states A and B

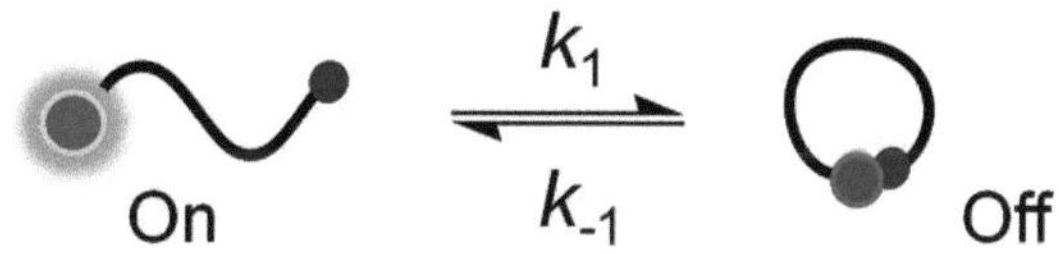

Scheme 2 Two-state equilibrium of loop closure

conformational dynamics of the chain segment separating fluorophore and quencher. B may therefore be a dim state, in which case K_{eq} does not equal a_r.

PET-FCS is particularly useful for the measurement of kinetics of loop closure within unfolded polypeptides or single-stranded nucleic acids. Loop closure is an early event in folding, and the underlying kinetics is a useful parameter to investigate structure and dynamics of denatured states or intrinsically disordered proteins. Intra-chain complex formation of fluorophore and Trp or G within an unfolded polypeptide or nucleic acid closes a loop and results in off-switching of fluorescence emission. The underlying kinetics can be described as in Scheme 2.

For loop closure, the microscopic rate constants can be calculated from τ_r and a_r because the closed fluorescence-quenched configuration is assumed to be a true off-state that results from a stacked complex between fluorophore and quencher. In this case, a_r equals the equilibrium constant K_{eq} of loop closure.

2 Materials

2.1 The PET-FCS Microscope

A typical microscope setup for PET-FCS experiments is illustrated in Fig. 3. The basic components are an inverted fluorescence microscope body equipped with an objective lens, a continuous-wave laser as excitation source, two avalanche photodiodes (APDs) as detection devices, a digital hardware correlator or PC plug-in card for data acquisition plus correlation software, and optical elements. The wavelength of the laser light is selected to match the electronic absorption band of the fluorophore (*see* **Note 1**). The laser should feature a well-defined (ideally spherical) beam profile, as well as temporally stable emission intensity. The laser light beam is guided into the back aperture of the microscope body via optical

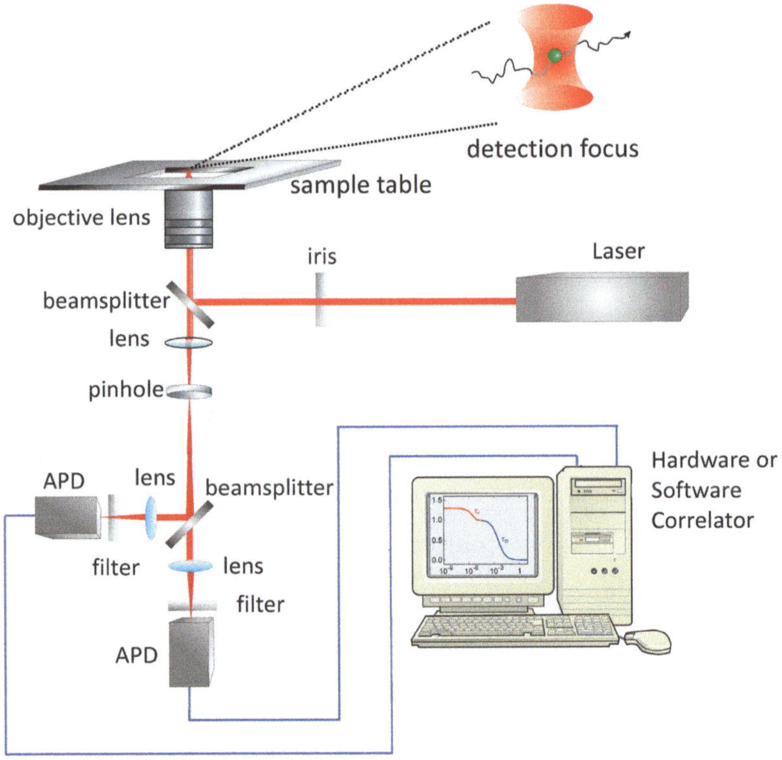

Fig. 3 PET-FCS microscope setup. The figure illustrates the basic components of a setup for PET-FCS experiments, as detailed in the main text

mirrors. The laser light power at the back aperture of the microscope can be conveniently adjusted using an optical density filter positioned between laser and microscope. The laser light beam profile can be shaped and size-adjusted using an iris mounted to the back aperture of the microscope body (*see* **Note 2**). Suitable objective lenses have oil immersion medium, a 60× magnification, and high numerical aperture (e.g., NA 1.4 or higher). Fluorescence light from the sample is collected by the same objective lens and passed through a dichroic beam splitter that separates fluorescence from excitation light. The fluorescence light leaves the microscope body and is guided by mirrors onto a 50 % nonpolarizing cubic beam splitter that shares the light equally between two single-photon sensitive APDs. Band-pass or long-pass optical filters are placed between the exit of the microscope body and the cubic beam splitter to eliminate residual background from excitation light. A lens positioned in the fluorescence light path is used to focus the fluorescence light onto the active zone of APDs. In practice, fiber-coupled APDs equipped with 100-μm optical fibers are recommended as detector devices because they are

(1) convenient to adjust in the fluorescence light path using x, y, z optical stages and (2) serve as spatial filter for the fluorescence light before it passes onto the photodiode, thereby replacing the pinhole in conventional confocal setups. Alternatively, the setup may be equipped with a 100-μm pinhole as spatial filter in the detection path. Two APDs are required in order to bypass their dead time and after-pulsing effects through cross-correlation of detector signals yielding a pseudo-ACF (*see* **Note 3**). Cross-correlation can either be performed by recording the detector signals individually using a PC plug-in card, followed by software correlation, or directly by using a digital hardware correlator device in cross-correlation mode. The advantage of hardware correlation is that the evolution of the ACF can be directly observed in real time, alleviating troubleshooting during data acquisition. Further, the size of an ACF output file from a hardware device is small (typically <1 MB), while the time trace recorded over a macroscopic time window for subsequent software correlation can easily exceed 1 GB. The time resolution of modern hardware correlators is on the order of 1 ns, i.e., fast enough to capture most biologically relevant time scales. The disadvantage of hardware correlation, however, is that the raw fluorescence intensity time traces are lost since they are analyzed on the fly and that the lag-time window to calculate the ACF cannot be varied.

Finally, it is important to control the temperature of the sample accurately because both diffusion and relaxation kinetics are temperature sensitive. Temperature control can be achieved by, e.g., constructing an objective-type heater from a hollow metal ring (ideally made from copper) that fits onto the objective lens. Coolant from a water bath is circulated through the metal ring and transfers heat via the objective lens and the immersion medium rapidly into the sample (*see* **Note 4**).

2.2 Solution Requirements and Sample Preparation

2.2.1 Solvent Requirements and Additives

PET-FCS measurements are generally performed in aqueous solution. The formation of stacked, fluorescence-quenched complexes between fluorophore and quencher are the prerequisite to obtain measureable ACF decay amplitudes containing the kinetics of conformational fluctuations. Since the formation of these complexes is driven by hydrophobic forces, organic solvents or chaotropes at high concentrations, such as guanidinium chloride or urea, should be avoided (*see* **Note 5**). At the low sample concentrations applied, surface adsorption of the sample to glass slides is a common problem. The observation of reduced or slowly decaying detector count rates is indicative. To minimize glass adsorption, the application of a mixture of 0.05 % Tween-20 and 0.3 mg/ml bovine serum albumin (BSA) as solution additive is always recommended (*see* **Note 6**). Additionally, the surface may be coated with suitable polymers (*see* Subheading 2.2.3).

2.2.2 Sample Concentration

Data are recorded from micro-ensembles of typically 10–50 sample molecules at a time in the detection focus. This corresponds to sample concentrations on the order of 1 nM depending on setup geometry. However, the dynamic range of concentrations that can be applied is rather large (upper pM to lower nM range). The observed time constants of ACF decays are independent on sample concentration if no intermolecular interactions are present. Also, the signal to noise of an ACF does not vary much with concentration. An increase of sample concentration increases the photon flux at the detectors, but the effect tends to cancel with the reduction of fluctuation contrast as signals from individual molecules are correlated over a larger fluorescence background.

2.2.3 Sample Preparation

All samples should be filtered through a 0.2-μm syringe filter before measurement (*see* **Note 7**). The sample is then transferred onto a cavity glass slide and covered with a quartz glass slip for microscopy. Alternatively, a sample chamber with lid for microscopy can be used. In order to suppress sample-glass interactions, the glass surface may additionally be passivized using suitable polymers. For instance, poly-L-lysine inverts the negative surface charge caused by the siloxy groups of quartz (*see* **Note 8**). After the sample is placed on the microscope table, the excitation/detection focus is moved into the sample. 2–3 Min temperature equilibration should be allowed before measurement.

3 Methods

3.1 Reporter Design

In PET-FCS an extrinsic fluorophore and a Trp or G are placed at specific sites within a polypeptide or nucleic acid such that the conformational change of interest is accompanied by the formation or disruption of fluorophore and Trp or G interactions. The polymer motion is then signaled by a change of fluorescence emission.

3.1.1 Selection of the Fluorophore

For PET-FCS fluorophore selection is vital. As for any single-molecule experiment, high brightness and photostability are required. Today, a fair selection of commercial organic fluorophores that fulfill these requirements is available (e.g., Alexa Fluor and ATTO dyes). The selected fluorophore needs to exhibit suitable photochemical properties (redox potential) to be efficiently quenched by Trp or G via PET. Oxazine, rhodamine, and fluorescein derivatives fulfill this requirement [28]. In general, oxazine fluorophores are superior because they lack fast photophysical fluorescence fluctuations at moderate excitation energies, which emerge as additional decay in the sub-ms time domain of the ACF. In contrast to oxazines, such fluctuations, originating from, e.g., triplet states, are omnipresent in most dyes. Moreover, oxazine dyes are most efficiently quenched by electron donors via PET as compared with other fluorophores [27, 28]. Finally, the selected fluorophore needs to be available as reactive dye to be covalently linked to a protein or nucleic acid using standard chemistry protocols.

3.1.2 Incorporation of the Fluorophore

In order to probe a defined conformational coordinate, the fluorophore needs to be labeled site specifically to the macromolecule. Most commercial reactive fluorophores target selectively amino or thiol groups to form a covalent linkage. Corresponding labeling protocols are standard and well established (*see* **Note 9**). The polypeptide amino terminus contains a primary amine that can be modified. But, the ε-amino group of the side chain of lysine is an even more reactive target due to its higher pK_a value. In natural polypeptide chains, lysines are omnipresent and give rise to non-specific labeling using amino-reactive dyes (*see* **Note 10**). However, the amino terminus of a polypeptide can be selectively targeted in the presence of lysine at tailored solution pH [33]. The most common way to site-specifically incorporate a dye into a polypeptide is via thiol modification of a single cysteine (Cys) side chain. In proteins or peptides, Cys can easily be incorporated site specifically using conventional mutagenesis protocols or during solid-phase synthesis. In mutagenesis experiments, however, great care has to be taken to avoid conserved side chains that are involved in structure or function. These can be identified through visual inspection of structures or through sequence alignment. Many natural proteins and peptides contain Cys side chains that are often involved in structure-stabilizing covalent disulfide linkages. Natural Cys can lead to loss of specificity during fluorescence modification carried out under reductive conditions. But, site-specific modification may still be achieved [20]. Alternatively, advanced labeling protocols involving enzymatic transfer of the fluorophore onto a short recognition sequence facilitate specific labeling in the presence of native Cys [34].

Peptides and oligonucleotides are efficiently produced using commercial or in-house solid-phase synthesis. Here, the site-specific incorporation of fluorophores after synthesis is comparatively straightforward. Nucleic acids do naturally not contain reactive amines. But nucleobases synthetically modified with primary amines can be incorporated site-specifically during synthesis and subsequently modified using amino-reactive dyes. Alkyne- and azido-modified building blocks are also available for solid-phase synthesis of both oligonucleotides and peptides. After synthesis, the incorporated building block can be covalently labeled using a modified fluorophore and copper-catalyzed click chemistry [35].

3.1.3 Incorporation of the Quencher

Trp and G serve as selective electron donors in PET-FCS experiments. They are natural building blocks of polypeptides and nucleic acids. Trp side chains natively present within the target sequence may be used advantageously for PET-FCS [13, 19]. If more than one natural Trp is present, excess Trp may be mutated to, e.g., phenylalanine eliminating interference [19]. Alternatively, Trp can be incorporated into proteins at selected sites via site-directed mutagenesis. Solvent-exposed sites should be chosen, and mutation of conserved, structurally, or functionally important side chains should be avoided. The structural and functional integrity of the modified construct should always be assessed carefully. In case of solid-phase synthesis, Trp or G can conveniently be incorporated as synthetic building blocks.

3.2 Data Acquisition and Analysis

3.2.1 Data Acquisition

In PET-FCS, the excitation power of the illuminating light beam is an important experimental parameter. The signal to noise is proportional to $\nu\sqrt{T}$, where ν is the average number of detected photons per molecule, and T is the total measurement time [30]. From this relation it becomes clear that maximizing the excitation power is much more efficient than increasing the measurement time in order to optimize the signal-to-noise. ν scales with excitation power below saturation. However, many fluorophores exhibit photophysical fluctuations caused by, e.g., triplet states that emerge as sub-ms decay in the ACF. Such fluctuations can be suppressed by reduction of excitation power. Finding the right excitation power for PET-FCS is therefore an optimization problem. In this regard, oxazine fluorophores are superior because they are essentially void of fast photophysical fluctuations even at fairly high excitation energies of several hundreds of μW. On the other hand, at such elevated levels of excitation power, photobleaching can occur during passage of molecules through the detection focus. The observation of reduced τ_{D} values at increasing levels of excitation power is indicative. However, observed relaxation kinetics of conformational fluctuations are independent on excitation power. For very fast processes on ns time scales, extended measurement times should be applied, since the photon flux at short time increments τ is considerably reduced. The total measurement time for a PET-FCS experiment at an excitation power of typically 400 μW, using oxazine fluorescence, can vary between 10 min and 1 h.

3.2.2 Data Analysis

The recorded ACFs are fitted to analytical models containing the kinetic parameters of interest. For a fluorescent molecule diffusing through a detection focus, there is an analytical solution to the ACF using an equation for diffusion in two dimensions[30]:

$$G(\tau) = \frac{1}{N}\left(1 + \frac{\tau}{\tau_D}\right)^{-1}, \qquad (3)$$

where N is the average number of molecules in the detection focus. A model for diffusion in two dimensions can be applied if the lateral dimension (z) of the excitation/detection focus is much larger than the horizontal dimensions (x, y). In practice, this is the case for most microscope setups.

In PET-FCS conformational fluctuations faster than τ_D emerge as decay of the ACF in addition to the one resulting from molecular diffusion through the detection focus. Assuming an equilibrium between two conformational states A and B (Scheme 1), equation (3) can be extended to [30]:

$$G(\tau) = \frac{1}{N}\left(1 + \frac{\tau}{\tau_D}\right)^{-1}\left(1 + a_r \exp\left(-\frac{\tau}{\tau_r}\right)\right) \quad (4)$$

where a_r is the amplitude and τ_r the time constant of the observed relaxation. If the molecule undergoes more than one conformational transition, which are independent from each other, Eq. 4 can be extended to a sum of exponential relaxations:

$$G(\tau) = \frac{1}{N}\left(1 + \frac{\tau}{\tau_D}\right)^{-1}\left(1 + \sum_n a_n \exp\left(-\frac{\tau}{\tau_n}\right)\right) \quad (5)$$

The observed time constants τ_n may be analyzed microscopically if the corresponding equilibrium constants are known, as outlined above in Subheading 1.3.

3.3 Examples

In the following, we review two recent studies from protein science showing how kinetics of structural fluctuations measured by PET-FCS can provide new insights into fundamental aspects of molecular biology.

3.3.1 Segmental Chain Motions of an Intrinsically Disordered Transactivation Domain

The tumor suppressor protein p53 is a transcription factor that is critically involved in the protection of higher organisms against cancer [36]. p53 coordinates responses to external cellular stress factors and initiates cell cycle arrest and DNA repair or apoptosis to permanently remove damage. Mutations in the p53 sequence contribute to 50 % of human cancer cases [37]. Activity of this transcription factor is controlled through its N-terminal transactivation domain (p53 TAD) that is intrinsically disordered. Activity is modulated through interactions with regulatory proteins and extensive posttranslational modifications. p53 TAD resembles properties of a fully unfolded protein domain with some nascent secondary structural elements (Fig. 4a), and the mechanisms of regulation are difficult to understand from a structural perspective.

To investigate how posttranslational phosphorylation and binding of the negative regulator MDM2 affect interactions and dynamics within p53 TAD, chain motions of four consecutive segments were measured using the PET-FCS approach [19]. Two of a total of three native Trp residues present in the 93-residue sequence were used as individual and selective quenchers of an

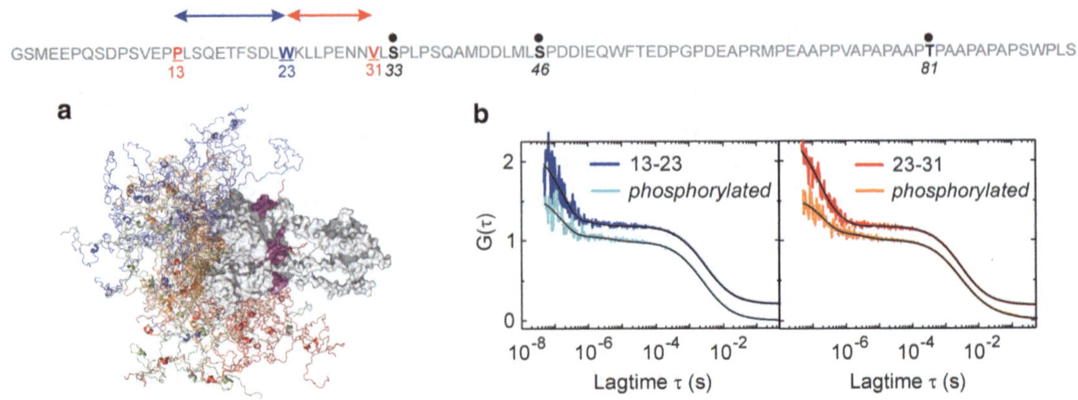

Fig. 4 Segmental chain motions of the intrinsically disordered p53 TAD. (**a**) Structural model of the tetrameric p53 protein. The *gray space-filling* representation shows the tetramerization domain. The DNA-binding domain is shown in *magenta*. The intrinsically disordered ensemble of conformations of the TADs of the four protomers is shown in different colors and *backbone* representation. (**b**) PET-FCS data of loop closure kinetics from segments 13–23 and 23–31 of p53 TAD. Recorded ACFs before and after phosphorylation at sequence positions 33, 46, and 81 are shown. ACFs are normalized to *N* and offset along the *y*-axis for reasons of clarity. *Black lines* are data fits to a model for a single diffusion and a sum of two single-exponential relaxations (Eq. 5). The sequence of p53 TAD in the one-letter amino acid code is shown on *top* of the panel. Positions of fluorescence modification via Cys mutations are shown in *red*. The native Trp serving as quencher is highlighted *blue* (W23). The probed chain segments are indicated by *arrows*. Positions of phosphorylation are indicated by *black dots* (Panels **a** and **b** adapted with permission from refs. 40 and 19. Copyrights 2008 and 2012, National Academy of Science USA and American Chemical Society, respectively)

extrinsic label. The remaining two Trps were silenced by mutating them to phenylalanine. The oxazine derivative ATTO Oxa11 was labeled site specifically to single Cys side chains that were incorporated at different positions along the sequence via mutagenesis.

PET-FCS revealed multi-exponential and site-dependent loop closure kinetics on the ns-to-μs time scale, which deviated from those expected for a randomly coiled polymer (Fig. 4b). Enzymatic phosphorylation changed the pattern of PET-FCS decays and kinetics of loop closure significantly, as did the binding of MDM2. Surprisingly, changes in kinetics were observed not only in the targeted sequence segments but also in remote regions. Results showed that local structural modifications within the intrinsically disordered protein, induced by phosphorylation or binding, are transmitted to remote sites through transient interaction networks. The study revealed allosteric capacity in the mechanism of regulation of p53 activity through its N-terminal domain.

3.3.2 Helix Motions of a Partially Folded Protein

Protein folding continues to be a central topic of molecular biology [3]. In this fundamental example of biological self-assembly, a nascent, linear polypeptide chain spontaneously adopts a 3D structure in aqueous solution. To date, the basic mechanism of this process is only understood in parts. The engrailed homeodomain (EnHD) belongs to the class of ultrafast folding proteins with a well-characterized pathway of folding. EnHD is a 61-residue three-helix bundle containing a helix2-turn-helix3 motif and folds via a populated intermediate within 20 μs [38]. It represents a rare example of a protein where the solution structure of the folding intermediate was solved at atomic detail using NMR spectroscopy [39]. The folding intermediate folds in 2 μs and is a well-structured helix2-turn-helix3 motif with the extended N-terminal helix1 being loose without tertiary interactions (Fig. 5a). In order to form the native fold, helix1 has to dock onto the helix2-turn-helix3 motif. It was a puzzle why the helix2-turn-helix3 motif folds in 2 μs, while the final docking of helix1 takes ten times longer.

To address this question a PET reporter (ATTO Oxa11) was moved along the sequence of an EnHD mutant that forms the folding intermediate [17]. ATTO Oxa11 was introduced at eight individual positions via Cys mutations spanning the entire sequence area of EnHD. A Trp side chain in the core of the native protein is solvent-exposed in the intermediate and served as fluorescence quencher (Fig. 5a). Recorded ACFs showed complex and site-dependent, multi-exponential sub-ms decays reporting on folding kinetics (Fig. 5b). Based on the wealth of previous kinetic and

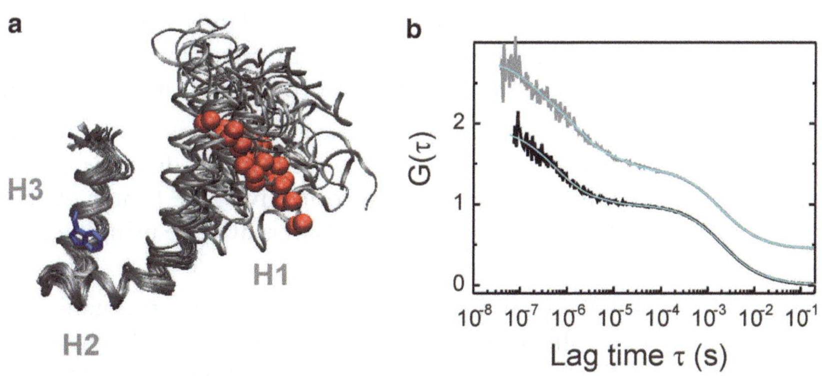

Fig. 5 Helix motions of the folding intermediate of EnHD. (**a**) *Backbone* representation of the ensemble of 25 lowest-energy conformers of the folding intermediate of EnHD determined by NMR (pdb id 1ZTR). Helices 1, 2, and 3 are indicated. The side chain of the native Trp on helix3 serving as PET fluorescence quencher is shown in *blue stick* representation. The ATTO Oxa11 label at sequence position 14 on helix1 is illustrated as *red sphere*. (**b**) ACFs recorded at high (*gray*) and low (*black*) solution ionic strength where the EnHD conformational ensemble is biased towards folded and intermediate state, respectively. The *cyan lines* represent data fits to a model for a single diffusion and three independent single-exponential relaxations (Eq. 5). Data are normalized to *N* and offset along the *y*-axis for reasons of clarity (Adapted with permission from ref. 17. Copyright 2010, National Academy of Science USA)

structural data available for this protein, all kinetic phases were assigned to specific events. PET-FCS data showed that helix1 in the folding intermediate was moving comparatively slowly at 10 μs, motions that were constraint by nonnative interactions [17]. The 10-μs time constant of helix motion was similar to that of overall folding and provided the missing kinetic link to explain why the final folding event of EnHD was slow.

4 Notes

1. The absorption and emission bands of organic fluorophores are comparatively broad. It is of advantage to excite the fluorophore at wavelengths slightly shorter than the absorption maximum in order to facilitate the collection of fluorescence photons from the entire emission band, using suitable bandpass or long-pass optical filters.
2. The laser beam diameter may be slightly smaller than the diameter of the back aperture of the lens, resulting in a larger excitation/detection focus and consequently a larger observed value of τ_D. A large τ_D is desirable because it increases the upper time limit of observable PET-FCS relaxations (Fig. 2) and maximizes the number of photons that can be collected from molecules during prolonged residence time in the detection focus.
3. APDs show characteristic after-pulsing effects leading to ACF decay artifacts on the 100-ns time scale. This limits the time resolution to ~1 μs in single-detector setup geometries. The application of two APDs in cross-correlation mode bypasses after-pulsing effects and provides ns time resolution.
4. Most commercial objective lenses tolerate temperatures between 4 and 50 °C, setting the dynamic range of sample temperatures using an objective-type heater. A thermoelectric Peltier element is a good alternative to using a circulating water bath.
5. Chaotropes at high concentration or organic solvents, like alcohols or acetonitrile, abolish the formation of stacked fluorescence-quenching complexes of fluorophore and quencher, which are required to obtain PET-FCS decays of significant amplitude. However, the effects are concentration dependent, and measureable amplitudes may still be obtained in aqueous solutions containing low concentrations of chaotrope or organic solvents. Note also that guanidinium chloride at high concentration induces microsecond photophysical fluctuations of minor amplitude in oxazine fluorophores. Photophysical fluctuations can be identified by excitation light

power-dependent measurements and eliminated by recording PET-FCS data at reduced excitation energies.

6. If measurements are carried out in buffered solutions at physiological pH, the use of phosphate as buffer salt is recommended. Phosphate ions are quite effective in shielding molecular glass-surface adsorption of biological samples compared with amine buffers such as MOPS or HEPES. The applied chemicals for spectroscopy should all be high grade (e.g., biochemical or molecular biology grade).
7. The presence of higher-order oligomeric species, aggregates, or particles in sample solutions is a common source of artifacts in FCS experiments. The observation of significantly increased values of τ_D and/or erroneous data fitting is indicative. The presence of aggregates may also be observed as spikes identified in the macroscopic count-rate time traces of the detector signal. Filtration of the sample prior measurement, using a 0.2 μm syringe filter, eliminates these artifacts.
8. Poly-L-lysine is an amino acid-based polycation that will minimize the adsorptivity of net positively charged molecules and proteins on glass surfaces. The net charge of a protein at a given pH can be estimated from its isoelectric point. For polylysine coating of glass, the surface of the sample chamber is incubated with ~5 mg/ml poly-L-lysine hydrobromide in water for 5 min or longer, followed by three washing steps, before the sample is applied.
9. Most reactive oxazine derivatives are readily soluble in acetonitrile, a solvent that can be removed quickly in vacuum. Rapid lyophilization alleviates the preparation of fluorophore aliquots and their storage as dry solid at −20 °C where they can be kept for up to 3 years. Reactive fluorophores in solution can degrade quickly. Labeling should be performed following the protocol of the manufacturer. For thiol modification, in practice, 2 h incubation at 25 °C or an overnight step at 4 °C using five- to tenfold molar excess of fluorophore over protein in aqueous buffered solution at pH 7, in the presence of a tenfold molar excess of TCEP as reductive agent, is recommended.
10. For amino modification of proteins with oxazine fluorophores, the application of only a twofold molar excess of amino-reactive dye (NHS derivative) over protein is recommended. Incubation can be done in, e.g., 50 mM sodium hydrogen carbonate solution (pH 8) for 1 h at 25 °C. This protocol yields a degree of labeling around one, i.e., an average stoichiometry of one conjugated dye per protein molecule. However, labeling efficiencies may vary between proteins, depending on sequence and surface properties.

References

1. Dyson HJ, Wright PE (2005) Intrinsically unstructured proteins and their functions. Nat Rev Mol Cell Biol 6(3):197–208
2. Dunker AK, Silman I, Uversky VN, Sussman JL (2008) Function and structure of inherently disordered proteins. Curr Opin Struct Biol 18(6):756–764
3. Fersht AR (2008) From the first protein structures to our current knowledge of protein folding: delights and scepticisms. Nat Rev Mol Cell Biol 9(8):650–654
4. Henzler-Wildman K, Kern D (2007) Dynamic personalities of proteins. Nature 450(7172): 964–972
5. Kapanidis AN, Strick T (2009) Biology, one molecule at a time. Trends Biochem Sci 34 (5):234–243
6. Sauer M, Hofkens J, Enderlein J (2011) Handbook of fluorescence spectroscopy and imaging: from single molecules to ensembles. Wiley-VCH, Weinheim, Germany
7. Weiss S (1999) Fluorescence spectroscopy of single biomolecules. Science 283(5408): 1676–1683
8. Selvin PR (2000) The renaissance of fluorescence resonance energy transfer. Nat Struct Biol 7(9):730–734
9. Schuler B, Eaton WA (2008) Protein folding studied by single-molecule FRET. Curr Opin Struct Biol 18(1):16–26
10. Neuweiler H, Sauer M (2004) Using photoinduced charge transfer reactions to study conformational dynamics of biopolymers at the single-molecule level. Curr Pharm Biotechnol 5(3):285–298
11. Doose S, Neuweiler H, Sauer M (2009) Fluorescence quenching by photoinduced electron transfer: a reporter for conformational dynamics of macromolecules. Chemphyschem 10(9–10):1389–1398
12. Vaiana AC et al (2003) Fluorescence quenching of dyes by tryptophan: interactions at atomic detail from combination of experiment and computer simulation. J Am Chem Soc 125(47):14564–14572
13. Neuweiler H, Doose S, Sauer M (2005) A microscopic view of miniprotein folding: enhanced folding efficiency through formation of an intermediate. Proc Natl Acad Sci USA 102(46):16650–16655
14. Neuweiler H, Lollmann M, Doose S, Sauer M (2007) Dynamics of unfolded polypeptide chains in crowded environment studied by fluorescence correlation spectroscopy. J Mol Biol 365(3):856–869
15. Bollmann S et al (2011) Conformational flexibility of glycosylated peptides. Chemphyschem 12(16):2907–2911
16. Neuweiler H, Johnson CM, Fersht AR (2009) Direct observation of ultrafast folding and denatured state dynamics in single protein molecules. Proc Natl Acad Sci USA 106(44):18569–18574
17. Neuweiler H, Banachewicz W, Fersht AR (2010) Kinetics of chain motions within a protein-folding intermediate. Proc Natl Acad Sci USA 107(51):22106–22110
18. Teufel DP, Johnson CM, Lum JK, Neuweiler H (2011) Backbone-driven collapse in unfolded protein chains. J Mol Biol 409(2):250–262
19. Lum JK, Neuweiler H, Fersht AR (2012) Long-range modulation of chain motions within the intrinsically disordered transactivation domain of tumor suppressor p53. J Am Chem Soc 134(3):1617–1622
20. Jensen MH, Sukumaran M, Johnson CM et al (2011) Intrinsic motions in the N-terminal domain of an ionotropic glutamate receptor detected by fluorescence correlation spectroscopy. J Mol Biol 414(1):96–105
21. Kim J, Doose S, Neuweiler H, Sauer M (2006) The initial step of DNA hairpin folding: a kinetic analysis using fluorescence correlation spectroscopy. Nucleic Acids Res 34(9): 2516–2527
22. Schüttpelz M et al (2008) Changes in conformational dynamics of mRNA upon AtGRP7 binding studied by fluorescence correlation spectroscopy. J Am Chem Soc 130(29): 9507–9513
23. Rehm D, Weller A (1969) Kinetik und Mechanismus der Elektronenübertragung bei der Fluoreszenzlöschung in Acetonitril. Ber-Bunsenges Phys Chem 73:834–839
24. Adams DM et al (2003) Charge transfer on the nanoscale: current status. J Phys Chem B 107(28):6668–6697
25. Zhong D, Zewail AH (2001) Femtosecond dynamics of flavoproteins: charge separation and recombination in riboflavin (vitamin B2)-binding protein and in glucose oxidase enzyme. Proc Natl Acad Sci USA 98(21): 11867–11872
26. Li X, Zhu R, Yu A, Zhao XS (2011) Ultrafast photoinduced electron transfer between tetramethylrhodamine and guanosine in aqueous solution. J Phys Chem B 115(19):6265–6271
27. Doose S, Neuweiler H, Sauer M (2005) A close look at fluorescence quenching of organic dyes by tryptophan. Chemphyschem 6(11):2277–2285

28. Marme N, Knemeyer JP, Sauer M, Wolfrum J (2003) Inter- and intramolecular fluorescence quenching of organic dyes by tryptophan. Bioconjug Chem 14(6):1133–1139
29. Heinlein T, Knemeyer JP, Piestert O, Sauer M (2003) Photoinduced electron transfer between fluorescent dyes and guanosine residues in DNA-hairpins. J Phys Chem B 107(31):7957–7964
30. Krichevsky O, Bonnet G (2002) Fluorescence correlation spectroscopy: the technique and its applications. Rep Prog Phys 65(2):251–297
31. Hess ST, Huang S, Heikal AA, Webb WW (2002) Biological and chemical applications of fluorescence correlation spectroscopy: a review. Biochemistry 41(3):697–705
32. Haustein E, Schwille P (2007) Fluorescence correlation spectroscopy: novel variations of an established technique. Annu Rev Biophys Biomol Struct 36:151–169
33. Selo I, Negroni L, Creminon C et al (1996) Preferential labeling of alpha-amino N-terminal groups in peptides by biotin: application to the detection of specific anti-peptide antibodies by enzyme immunoassays. J Immunol Methods 199(2):127–138
34. Yin J, Lin AJ, Golan DE, Walsh CT (2006) Site-specific protein labeling by Sfp phosphopantetheinyl transferase. Nat Protoc 1(1): 280–285
35. Lallana E, Riguera R, Fernandez-Megia E (2011) Reliable and efficient procedures for the conjugation of biomolecules through Huisgen azide-alkyne cycloadditions. Angew Chem Int Ed Engl 50(38):8794–8804
36. Vousden KH, Lane DP (2007) p53 in health and disease. Nat Rev Mol Cell Biol 8(4): 275–283
37. Joerger AC, Fersht AR (2008) Structural biology of the tumor suppressor p53. Annu Rev Biochem 77:557–582
38. Mayor U et al (2003) The complete folding pathway of a protein from nanoseconds to microseconds. Nature 421(6925):863–867
39. Religa TL, Markson JS, Mayor U et al (2005) Solution structure of a protein denatured state and folding intermediate. Nature 437(7061): 1053–1056
40. Wells M et al (2008) Structure of tumor suppressor p53 and its intrinsically disordered N-terminal transactivation domain. Proc Natl Acad Sci USA 105(15):5762–5767

Chapter 28

Z-Scan Fluorescence Correlation Spectroscopy as a Tool for Diffusion Measurements in Planar Lipid Membranes

Tomáš Steinberger, Radek Macháň, and Martin Hof

Abstract

Studies of lateral diffusion are used for the characterization of the dynamics of biological membranes. One of the techniques that can be used for this purpose is fluorescence correlation spectroscopy (FCS), which belongs to the single-molecule techniques. Unfortunately, FCS measurements, when performed in planar lipid systems, are associated with a few sources of inaccuracy in the determination of the lateral diffusion coefficient. The main problems are related to the imperfect positioning of the laser focus relative to the plane of the sample. Another source of inaccuracy is the requirement for external calibration of the detection volume size. This protocol introduces a calibration-free method called *Z*-scan fluorescence correlation spectroscopy (*Z*-scan FCS), which is based on the determination of the diffusion time and particle number in steps along the optical (*z*-) axis by sequential FCS measurements. *Z*-scan FCS could be employed for diffusion measurements in planar membrane model systems—supported phospholipid bilayers (SPBs) and giant unilamellar vesicles (GUVs) and also in biological membranes. A result from measurements in SPBs is also presented in the protocol as a principle example of the *Z*-scan technique.

Key words *Z*-scan fluorescence correlation spectroscopy, Supported phospholipid bilayers, Lateral diffusion, FCS diffusion law, Humpolíčková plot, Microdomains

1 Introduction

This chapter is dedicated to *Z*-scan fluorescence correlation spectroscopy (*Z*-scan FCS) which is based on fluorescence correlation spectroscopy (FCS); therefore we mention the basic principles of FCS at first. FCS is an experimental technique which was introduced in the early 1970s [1–4], but has become widespread only much later when confocal microscopes equipped with single-photon detectors with low noise, lasers with high temporal stability, and high-quality objectives became available [5–7]. Such instrumentation based on confocal microscopes is the most common in FCS nowadays.

Thanks to the confocal setup, where a pinhole is positioned in the image plane and is adjusted to overlap with the image of the

Yves Engelborghs and Antonie J.W.G. Visser (eds.), *Fluorescence Spectroscopy and Microscopy: Methods and Protocols*, Methods in Molecular Biology, vol. 1076, DOI 10.1007/978-1-62703-649-8_28,

focal spot, the signal from out-of-focus fluorescence is suppressed. Fluorescence emission from a dye is effectively collected by the detector only from a small volume element of the sample (detection volume), which is typically about 1 μm^3 (1 fl) or smaller [5, 6, 8–10]. The size and shape of this effective detection volume is given by the so-called point-spread function, which is the product of the excitation light intensity distribution within the sample and of the light-collection efficiency, which is determined by the detection optics (especially by the confocal aperture) [5, 6, 10, 11]. The radius of the detection volume within the focal plane (the *xy* plane) is diffraction limited and ranges from 200 to 300 nm. The characteristic dimension along the optical axis of the microscope (the *z*-axis) is a few micrometers [5, 6, 10–16]. The profile of the laser beam within the sample is Gaussian in the *x* and the *y* directions and Lorentzian in the *z*-direction, but the shape of the effective detection volume is often approximated by a three-dimensional Gaussian [5, 9, 10, 17].

A typical source of the signal in FCS measurements is the diffusion of a few molecules through the detection volume, resulting in fluctuations in the fluorescence signal above the level of the background noise. The basic scheme of the FCS procedure starts by recording the intensity fluctuations with sufficient temporal resolution (at least an order of magnitude better than the timescale of the studied process), followed by the analysis of this recorded signal via autocorrelation functions $G(\tau)$. The physically relevant parameters, i.e., diffusion time τ_D and particle number N, are then extracted from the obtained $G(\tau)$ by nonlinear fitting with an appropriate theoretical model taking into account the processes which may occur in the sample. The standard deviation of the fit of the autocorrelation curve can be obtained using the method described by Wohland et al. [18].

The normalized autocorrelation function $G(\tau)$ (*see* Eq. 1) is calculated from recorded time traces of fluorescence intensity $I(t)$. $G(\tau)$ indicates the probability of finding a molecule in the detection volume at time $t + \tau$, given a molecule was present there at time t. Pointed brackets in Eq. 1 represent averaging over all values of t:

$$G(\tau) = \frac{\langle I(t)\, I(t+\tau)\rangle}{\langle I(t)\rangle^2}. \tag{1}$$

The shape of the autocorrelation function is given by the timescale of the fluctuations of the fluorescence intensity $I(t)$ originating from the detection volume. The sources of intensity fluctuations are translational diffusion into and out of the detection volume, ligand– macromolecule binding, rotational diffusion, internal macromolecule dynamics, intersystem crossing to nonfluorescent triplet state, and excited-state reactions [19]. Fluctuations due to diffusion occur on a timescale of hundreds of microseconds to seconds, while fluctuations due to photochemical processes are much faster. This enables separation each of their contributions to the recorded time traces [5, 6, 20–22].

The detection volume in the case of confocal FCS has been described above. The theoretical model of autocorrelation function $G(\tau)$ for three-dimensional Brownian diffusion in a Gaussian volume element is given by [7]

$$G(\tau) = 1 + \frac{\gamma_{3D}}{N} \frac{1}{1 + (\tau/\tau_D)} \left(\frac{1}{1 + (\tau/\tau_D)\ (1/S^2)} \right)^{1/2}, \quad (2)$$

where the structure parameter S is the ratio ω_z/ω_0 of the characteristic dimensions of the detection volume along the optical z-axis and in the xy plane, respectively. ω_z and ω_0 are defined as distances at which the intensity drops by factor $1/e^2$. τ_D is the diffusion time, and N is the particle number (number of fluorescent particles within the detection volume). Correction factor γ_{3D} is defined by Thompson [7] to take into account the intensity profile in the focus, and it depends on experimental geometry. For a three-dimensional Gaussian profile γ_{3D} equals $2^{-3/2}$ [7]. The dimensions of the detection volume can be determined by a calibration measurement of the diffusion of a molecule with known diffusion coefficient D, fitting the data to Eq. 2 and using Eq. 3 [17].

Diffusion coefficient D is an important parameter for mobility description of molecules in biological membranes. Coefficient D can be extracted directly from the experimentally obtained τ_D by using the following expression:

$$D = \frac{\omega_0^2}{4\ \tau_D}, \quad (3)$$

where ω_0 is the radius of the detection volume and is defined as the radial distance from the optical axis, at which intensity drops by $1/e^2$ [17].

If the transition of fluorophores to nonfluorescent triplet state due to intersystem crossing cannot be neglected, Eq. 2 has to be extended by an average fraction of dye molecules in the triplet state T and an intersystem crossing relaxation time τ_T into following fitting function [23, 24]:

$$G_T(\tau) = 1 + [1 - T + T\ \exp(-\tau/\tau_T)] \frac{\gamma_{3D}}{N(1-T)} \frac{1}{1 + (\tau/\tau_D)} \left(\frac{1}{1 + (\tau/\tau_D)\ (1/S)^2} \right)^{1/2}. \quad (4)$$

The expression for the autocorrelation function is simplified when the diffusion is restricted to two dimensions, for example, in phospholipid bilayers or other planar systems. Unfortunately, when FCS is measured in these structures, it is accompanied by a few artifacts (see below). In case of two-dimensional diffusion, Eq. 4 can be simplified into [7, 25]:

$$G_{2D}(\tau) = 1 + \frac{\gamma_{2D}}{N} \frac{1}{1 + (\tau/\tau_D)}, \tag{5}$$

where the correction factor γ_{2D} is 0.5 in the case of the two-dimensional Gaussian profile [7].

If introduction of intersystem crossing into two-dimensional Eq. 5 is necessary, the parameters T and τ_T extend the equation to [17, 23–25]:

$$G_{2DT}(\tau) = 1 + [1 - T + T \exp(-\tau/\tau_T)] \frac{\gamma_{2D}}{N(1 - T)} \frac{1}{1 + (\tau/\tau_D)}. \tag{6}$$

In case that M fluorophores with different diffusion times and detection efficiency contribute to the recorded signal, the following general equation should be used [7, 22, 25]:

$$G_{2DTM}(\tau) = 1 + \frac{\gamma_{2D} \sum_{i=1}^{M} \left(\frac{Q_i}{Q_1}\right)^2 F_i\, g_i(\tau)}{N \left[\sum_{i=1}^{M} \frac{Q_i}{Q_1} F_i\right]^2}, \tag{7}$$

where F_i is fraction of fluorophores diffusing with diffusion time τ_{Di}, and Q_i represents their detection efficiency which depends on quantum yields, extinction coefficients, and overlap of emission spectra with transmission region of the optical filters in the detection pathway. Term $g_i(\tau)$ incorporates triplet parameters and diffusion time for i-fluorophore into Eq. 7 and is defined as [7, 22, 25]:

$$g_i(\tau) = \frac{1 - T_i + T_i \exp(-\tau/\tau_{Ti})}{1 - T_i} \frac{1}{1 + (\tau/\tau_{Di})}. \tag{8}$$

The probability of errors introduced to the fits of the autocorrelation function using Eq. 7 increases with increasing number (M) of components due to the higher number of free parameters [26, 27].

Equation 7 is reduced when only one type of fluorophore is present within the sample, but it could have two different diffusion times (for instance when a fraction of fluorophores F_1 is bound to larger molecules and the rest is free). If it can be assumed that Q and triplet properties are not affected by binding, Eq. 7 is simplified in [7, 23]:

$$G_{2DT}(\tau) = 1 + [1 - T + T \exp(-\tau/\tau_T)] \frac{\gamma_{2D}}{N(1 - T)} \left(\frac{F_1}{1 + (\tau/\tau_{D1})} + \frac{1 - F_1}{1 + (\tau/\tau_{D2})}\right), \tag{9}$$

where τ_{D1} is the diffusion time of bound fluorophore, and τ_{D2} is analogously for free fluorophore.

The first source of artifacts in FCS is related to high excitation intensities which are not necessary for performing measurement as proved in [11, 28, 29]. High excitation intensities can even lead to artifacts due to photobleaching, when a time trace of fluorescence intensity during the measurement is no longer stationary (which is one of the assumptions behind the above mentioned theoretical models of autocorrelation functions). The probability, that fluorescent species undergo photobleaching, is related to the time spent in the illuminated volume (which is longer for slowly diffusing molecules) and grows with increasing laser power. Optical saturation is the other effect related to the excitation intensity. This saturation occurs when an increased number of molecules within the detection volume is not in the ground state but in an excited or triplet state, leading to the loss of proportionality between excitation and fluorescence intensities [5, 6, 10, 11, 28].

Other artifacts that can influence the results of FCS are caused by deviations of the real shape of the detection volume from its form assumed in the derivation of the above mentioned theoretical models for the diffusion coefficient. A detailed discussion of the sources of those artifacts (like laser beam astigmatism, refractive index mismatch between the sample and objective immersion medium, or imperfect pinhole adjustment) and numerical simulations showing how they influence the physically relevant parameters obtained by FCS (τ_D and N) is presented in studies by Jörg Enderlein and his group.[5, 6, 11, 28]

The most important source of artifacts in FCS measurements in phospholipid bilayers and other planar system is positioning of the sample with respect to the focus along the z-axis. Those artifacts are specific for planar samples and are not encountered in FCS investigations of three-dimensional diffusion in bulk. The thickness of a membrane bilayer is about 5 nm and therefore several orders of magnitude smaller than the characteristic dimension ω_z of the detection volume. Different positions of bilayer with respect to the focus result in different diameters of illuminated bilayer area (thanks to divergence of laser beam) and thus in different measured values of τ_D and N (Fig. 1). There are different ways to overcome this problem, for example, the recently developed two-focus FCS [9, 30, 31], which requires a nonstandard setup using two alternatively pulsed polarized lasers. The Z-scan method [17, 29, 32], on the other hand, can be easily performed with a standard FCS setup allowing scanning along the z-axis.

Z-scan fluorescence correlation spectroscopy (Z-scan FCS) overcomes the positioning problem and is a calibration-free method; this means that there is no requirement for the calibration measurement of the detection volume dimensions. Z-scan FCS can

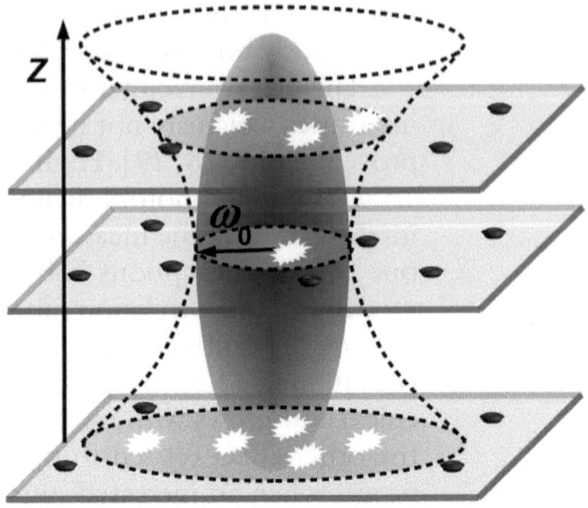

Fig. 1 Scheme of *Z*-scan positioning problem of planar samples. *Z*-positioning of planar sample out of the focus plane results in a larger detection area due to the divergence of the laser beam and, thus, higher *N* and τ_D. The interior of the ellipsoid marks the volume where photon detection efficiency is higher than $1/e^2$ of its maximum

be employed for diffusion monitoring of labelled membrane lipids or membrane associated molecules such as proteins and also for characterizing the binding kinetics of these entities to planar membrane model systems—supported phospholipid bilayers (SPBs) and giant unilamellar vesicles (GUVs)—as well as to suitable biological membranes. When comparing GUVs and SPBs, Przybylo et al. [29] found the diffusion in GUVs to be more than 2 times higher than in SPBs, which suggests coupling of the bilayer with the support; on the other hand, GUVs membranes suffer from thermal undulations [33]. Another typical field of biophysics investigations is the study of the dependence of the *D* of proteins or lipids, on their content in membrane composition.

The *Z*-scan method is based on the determination of τ_D and *N* in steps along the optical axis (*z*-axis) by sequential FCS measurements (*see* Fig. 2). Assuming a Lorentzian profile of the effective detection volume along the *z*-axis, dependencies of τ_D and *N* on the *z*-coordinate (Fig. 3) are the following [34, 35]:

$$N(z) = \pi C \omega_0^2 \left(1 + \frac{\lambda_0^2 z^2}{\pi^2 n^2 \omega_0^4}\right), \tag{10}$$

$$\tau_D(z) = \frac{\omega_0^2}{4D}\left(1 + \frac{\lambda_0^2 z^2}{\pi^2 n^2 \omega_0^4}\right), \tag{11}$$

where C is the average concentration of diffusing fluorescence molecules in the illuminated area, ω_0 is the $1/e^2$ radius of the beam in the focal plane (the waist of the beam; the term

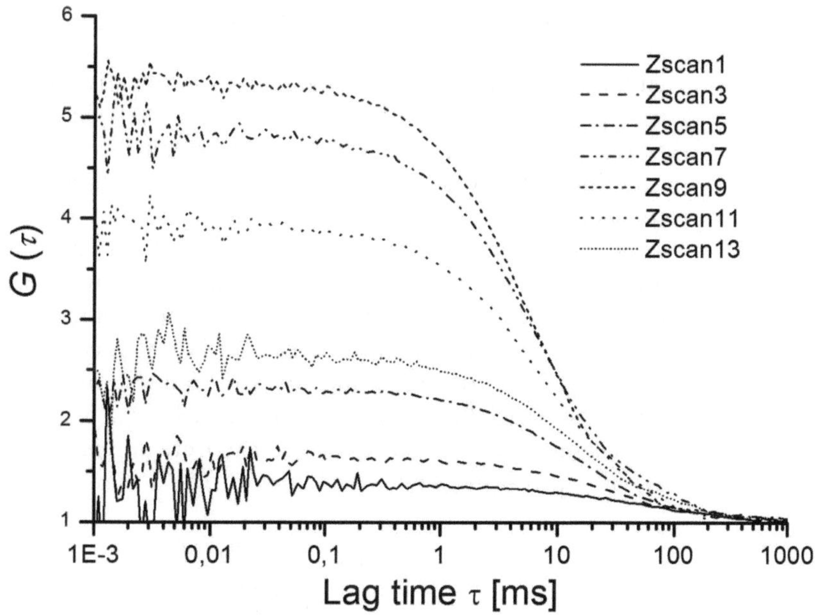

Fig. 2 Resulting *Z*-scan autocorrelations. Measurements with 0.2 μm steps were performed with starting position (*Z*-scan 1) 1 μm below estimated *z*-position of the focal plane. It can be concluded that *Z*-scan 8 and *Z*-scan 9 were performed when the focus plane and the sample were in the nearest vicinity (where autocorrelation amplitude *G*(0) is the highest, thanks to its indirect proportionality to *N*)

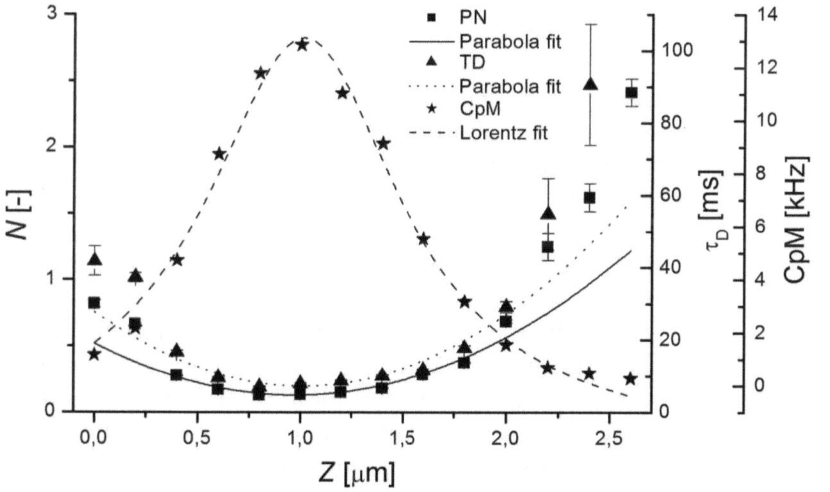

Fig. 3 Dependencies of N(z)—*triangle*, τ_D(z)—*squares* and the photon count per molecule—*stars* (CpM—the total fluorescence count rate divided by *N*) on sample positioning are depicted. Values $N(z)$ and $\tau_D(z)$ are fitted with parabolas described by Eqs. 10 and 11 while CpM(z) is fitted with a Lorentzian function describing the axial intensity profile in the focus

$\pi C\omega_0^2 = N_0$ is the particle number in the beam waist), λ_0 is the wavelength of the excitation light in vacuum, z is the axial distance between the sample position and the focal plane, and n is the refractive index of the medium [17, 22, 25, 32, 36].

A theoretical model, called the FCS diffusion law, was formulated by Wawrezinieck et al. [37] for characterization of hindered lateral diffusion in membranes. This model is valid only when membranes contain diffusion-preventing obstacles in low concentrations, and their size is sufficiently small in comparison to the ω_0 value. Otherwise, if this assumption is not fulfilled, it is necessary to take into account a theoretical fitting model for description of anomalous diffusion. Humpolíčková et al. utilized diffusion law analysis (based on determination of diffusion times in dependence on alterations in size of detection area) and character of the *Z*-scan FCS (changes in dimensions of the illuminated membrane area while performing scanning in the *z*-direction) and elegantly combined the diffusion law equation [37] with Eqs. 10 and 11. Resulting equation expresses the theoretical linear dependence of τ_D on *z*-position [32]:

$$\tau_D(z) = t_0 + \frac{\omega_0^2}{4D_{\text{eff}}} \frac{N(z)}{N_0}, \tag{12}$$

where D_{eff} is the effective diffusion coefficient, $N(z)$ number of particles in the detection area at a given *z*-coordinate, and t_0 is intersection point with the diffusion time axis, and its value gives information about the presence of obstacles in the membrane [32, 37].

The basic issue of this combination method is plotting the dependency of the measured $\tau_D(z)$ on $N(z)/N_0$ (sometimes called Humpolíčková plots) and fitting these data by Eq. 12. Figure 4 is an example of such a Humpolíčková plot of pure SPBs and SBPs after treatment by halictine-1 peptide [38, 39]. The obtained value t_0 indirectly reveals information about the presence and dimensions of membrane structures. Monte Carlo simulations made by Wawrezinieck et al. [37] showed that if fluorophores move by free Brownian diffusion in membranes without any inhomogeneities, then $t_0 = 0$; the same holds if only fully impermeable obstacles are in the membrane (they only lead to a decrease of *D*). Secondly they assumed isolated lipid rafts—microdomains in membrane; those act as a trap for diffusing fluorescent species, which can enter them, and their movement is hindered inside the microdomains. In such cases t_0 value is positive. Finally simulations were done for the case that membranes are divided into corrals (within which fluorophores diffuse freely) separated by partially permeable barriers. $t_0 < 0$ was found in that case. The absolute value of t_0 yields information on dimensions of the obstacles [22, 36, 37].

More detailed information about FCS and *Z*-scan FCS diffusion measurements is provided in reviews by Macháň et al. [25, 36] and by Štefl et al. [22]. Other techniques, which are widely used for determining lateral diffusion coefficient *D*, are fluorescence recovery after photobleaching (FRAP) and single particle tracking (SPT) or single dye tracing (SDT). These three approaches (FRAP, SPT, and several FCS techniques) for characterization of lateral diffusion

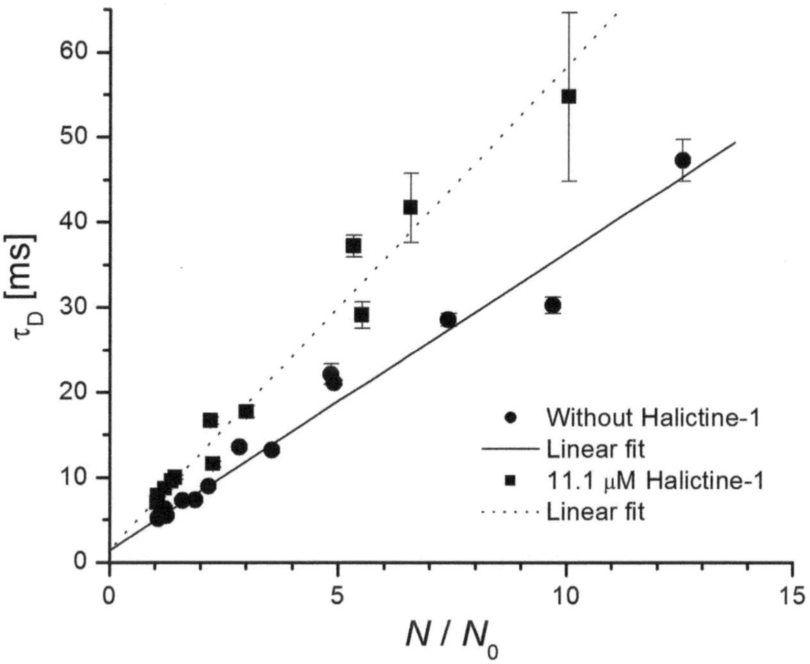

Fig. 4 Humpolíčková plot of diffusion times for a partially charged SPB membrane composed of POPC (70 %), POPG (30 %) and ATTO 644-DOPE (10^{-3} %) in dependence on relative particle number N/N_0. *Circles* are for pure SPB, and *squares* are for SPB with presence of 11.1 μM halictine-1 peptide [38, 39]

in planar lipid bilayers were systematically compared by the group of Wohland et al. [40]. Other possibility how to determine lateral D is, for example, the scanning FCS method and a broad range of image correlation spectroscopy methods [22, 25].

2 Materials

The *Z*-scan technique is presented in the protocol on a principle example: *Z*-scan measurement of lateral diffusion in partially charged SPBs. General requirements on *Z*-scan instrument are discussed in Subheading 2.1, **items 1–5** at first. In Subheading 2.2, **items 1–6,** materials and laboratory instruments are described that are needed for repeating the principle example.

2.1 Z-Scan Instrument Equipment

1. FCS Instrument: Confocal microscope with high numerical aperture objective and with single-photon sensitivity detector, typically single-photon avalanche diode (SPAD) detector or photomultiplier tubes PMT.
2. Laser with light wavelength ideally at the absorption peak of dye fluorescence as an excitation source.
3. *Z*-axis scanner unit which moves either with objective or with sample stage.

4. Dichroic mirror and excitation and emission filters selected according to laser excitation wavelength and dye emission properties.
5. Hardware correlator or software capable to correlate intensity time traces recorded by the instrument.

2.2 Example Experiment

1. Microscope Sample Chamber (*see* **Note 1**): Lab-Tek borosilicate chamber (Nalge Nunc International, Rochester, New York, USA)
2. Buffer A without Ca: 2 mM EDTA, 10 mM HEPES, 150 mM NaCl, pH 7.4. Weigh all substances and put them into a glass beaker. Add appropriate volume of Milli-Q water and start stirring. To speed up the dissolution of the reagents, it is possible to slightly increase the temperature (*see* **Note 2**). When all reagents are dissolved, adjust pH with concentrated NaOH, while stirring. Finally pass the buffer through a 0.2 μm syringe filter.
3. Buffer B with Ca (*see* **Note 3**): 2 mM $CaCl_2$, 10 mM HEPES, 150 mM NaCl, pH 7.4. Prepare the buffer as described in the previous step.
4. Membrane Lipid Compounds (*see* **Note 4**): 1-palmitoyl-2-oleoyl-*sn*-glycero-3-phosphocholine (POPC) as a neutral lipid and 1-palmitoyl-2-oleoyl-*sn*-glycero-3-phospho-(1′-rac-glycerol) (POPG) as a negatively charged lipid. Both were obtained from Avanti Polar Lipids (Alabaster, AL, USA) and stored in chloroform at −20 °C.
5. Membrane Labelling: Use ATTO 655-NHS-ester (ATTO-TEC, Siegen, Germany) labelled 1,2-dioleoyl-*sn*-glycero-3-phosphoethanolamine (DOPE). Lipid was covalently labelled as described previously in [4]. Store in chloroform at −20 °C.
6. SUVs Preparation: Nitrogen and vacuum evaporator, vortex mixer, sonicator, and centrifuge.

3 Methods

The method section is divided into four parts; the first two are dedicated to the preparatory procedures, followed by the measurement itself and the diffusion law analysis of the obtained data in the last part.

3.1 Preparation of SPBs

When lipids with high transition temperature are used, *see* **Note 4**.

1. Prepare mixture of labelled and unlabelled lipids (*see* **Note 5**) by mixing appropriate amounts (*see* **Note 6**) of desired lipids dissolved in organic solvent (e.g., chloroform).

2. Evaporate organic solvent with gentle streaming of dry nitrogen until there are no visible droplets (*see* **Note** 7).
3. Place the sample under vacuum for 1 h at least, to remove remaining solvent (*see* **Note 8**).
4. Add appropriate volume of buffer A without Ca and resuspend lipid mixture by high-speed vortexing for 2–4 min (*see* **Note 9**).
5. Create small unilamellar vesicles (SUVs) by the sonication technique (*see* **Note 10**) for a time until the sample is transparent or only very slightly opalescent (*see* **Note 11**).
6. When a probe sonicator is used, next step is elimination of impurities released from the tip by 10 min centrifugation at 16,000 × *g*.
7. SUV samples can be stored for a few days at 4 °C.
8. Prepare SPBs in microscope sample chamber (*see* **Note 1**) by mixing the SUVs solution and buffer B with Ca (final lipid concentration 100 μM, *see* **Note 12**) and letting the mixture incubate at a dark place for 15–30 min.
9. Remove redundant vesicles by flushing out with ten times volume excess of buffer A without Ca; the bottom of the cuvette should be covered by buffer all the time.

3.2 Setup of FCS Instrument

The procedure given below should be adjusted to the operational instructions of the used instrument. Maximal laser intensity achieved in **step 1** is used as an indicator of well alignment of the excitation laser pathway; for next steps, it is necessary to adjust the laser power to an acceptable level for the purpose of damage prevention of sensitive apparatus parts such as detectors. Check if the amount of the immersion liquid is sufficient, every time, when a sample is exchanged.

1. Maximize excitation laser power into the microscope by alignment of input light pathway (*see* **Note 13**).
2. Using reflection of excitation laser beam on glass-air interface (*see* **Note 14**), finely tune the excitation laser pathway to obtain a symmetric shape of the reflected beam positioned in the center of the field of view.
3. Using reflection of excitation laser beam on glass-air interface (*see* **Note 14**), align emission light pathway directing the emission beam to the center of detector. Do not use emission filters for this step.
4. Tune *x*, *y*, *z* pinhole positions (*see* **Note 15**) with usage of free dye solution (*see* **Note 16**) and find positions where the measured fluorescence intensity is the highest (*see* **Note 13**).

5. Adjust the correction ring on the objective by searching for position corresponding to highest fluorescence intensity (*see* **Note 13**).
6. For verification of setup, tune the laser power for FCS measurement in solution (*see* **Note 17**), measure FCS in testing a solution of free dye (nM) with known dye diffusion coefficient, fit autocorrelation data with appropriate model for 3D diffusion—Eq. 2 or 4 (if the dye has a significant triplet fraction) and consider results.
7. Measure average background signal and keep the value, in order to correct *N* for the influence of uncorrelated background (*see* **Note 18**).
8. Apply droplet of immersion liquid on the objective lens and put SPB-sample inside microscope. Find *z*-position where the signal from the membrane dye is the highest (*see* **Note 20**).
9. Tune laser power on suitable level for FCS measurement on membrane-type sample (*see* **Note 17**). You can also check intensity trace, if there are noticeable fluctuations and are appropriately higher than background signal.

3.3 Z-Scan Measurement

Mechanical stability of microscope focusing is a crucial factor influencing the quality of data in *Z*-scan FCS. Relaxation of mechanical parts can lead to significant deformation in the parabolic dependencies of $\tau_D(z)$ and $N(z)$.

It is reasonable to perform minimally six *Z*-scan measurements (Subheading 3.3, **steps 1–4**) in different *xy* points in the membrane to increase the relevance of the results before continuing with Subheading 3.3, **step 5**. An *xy* image of the membrane is useful for selection of measurement points.

1. Find the membrane surface by tuning the *z*-position to achieve the highest fluorescence intensity of the lipid dye (*see* **Note 20**).
2. Set the upper and lower limits in *Z*-scanner driver and measure set of FCS curves (Fig. 2). The typical range is 1 μm above and below *z*-position, where the highest fluorescence intensity is found; take 10–20 measurements in equidistant positions within that range with acquisition time 1–2 min for each step (*see* **Note 21**).
3. Fit individual FCS curves by adequate theoretical model (Eqs. 5; 6; 7; 9) and extract $N(z)$ and $\tau_D(z)$ for each *z*-position. It is also advisable to preserve the errors for the weighting of the acquired values.
4. Fit (*see* **Note 22**) received $N(z)$ by Eq. 10 with weighting by instrumental errors, analogously fit $\tau_D(z)$ by Eq. 11 (Fig. 3).
5. Add tested substance and let it incubate for adequate time at a dark place.

6. Check the presence of immersion liquid, insert sample, and repeat **steps 1–6**.
7. Finally export results from each scan for final comparison.

3.4 Advanced Evaluation of Z-Scan Data

1. For each Z-scan, count value $N(z)/N_0$ for each z-position, where $N(z)$ is particle number received from fit in (Subheading 3.3, **step 3**), and N_0 is the particle number in SPB obtained in (Subheading 3.3, **step 4**).
2. Create a Humpolíčková plot for given Z-scan; it means plot of dependencies of $\tau_D(z)$ on $N(z)/N_0$ together with $\tau_D(z)$-errors (Fig. 4).
3. Fit this plot with linear fit Eq. 12 and set $\tau_D(z)$-errors as weights.
4. The value of the intercept t_0 can be directly determined from the linear regression results (*see* **Note 23**).

4 Notes

1. Surfaces at which SPBs are formed have to be hydrophilic. This is achieved by removing any organic residues from the surface by detergents, extensive rinsing with water, and drying in a stream of nitrogen just prior the use. Exposing the surface to ozone or using plasma cleaning can be also applied. The Lab-Tek cuvette used in our example experiment has a bottom made of borosilicate glass. SPBs can be also prepared on other surfaces for example on mica or silicon (a special cuvette for measuring on arbitrary surface is presented in [17]). Borosilicate glass cover slides are a more affordable alternative to Lab-Tek cuvettes.
2. EDTA is a considerably acidic agent and poorly soluble under acidic conditions. pH of the solution should therefore be controlled during dissolving of EDTA, for example, by adding NaOH.
3. If you plan to use phosphate buffer due to required experiment conditions, do not prepare phosphate buffer with Ca content, because Ca precipitates as calcium phosphate in phosphate buffers. In this case, buffer without Ca content could be used for SPBs formation, but incubation time has to be prolonged. Calcium is used for facilitation of bilayers formation, this effect is the strongest for mica surface. For silica, presence of calcium has a slightly accelerating effect and for glass, calcium has no essential influence on vesicle adsorption and formation of bilayers. Other effects influencing SBPs stability were studied by Beneš et al. [41].
4. Key SUVs preparation steps have to be performed at temperatures above phase transition to liquid phase of all lipids in the

mixture. Thus, if lipids with phase transition temperature above room temperature are chosen, preheated buffer (Subheading 3.1, **step 4**) and heating bath (Subheading 3.1, **step 5**) (Subheading 3.1, **step 8**) should be used.

5. Membrane Labelling: The signal-to-noise ratio is highest when there is on average about one fluorescent molecule within the detection volume [17, 42]. This means low concentrations (nM) of fluorescent molecules (fluorescently labelled lipid analogs, peptides, or proteins, whose mobility is studied) are sufficient for performing FCS. There is a range of values of the optimal ratio of labelled to unlabelled phospholipid molecules for planar membrane models, a range could be, for example, from $1{:}5 \times 10^5$ (Benda et al. [17]) to $1{:}2 \times 10^4$ (Forstner et al. [43]). The optimal ratio could be estimated by calculation with taking into account size of detection area and area per one lipid; after all, the ratio should be adjusted when checked for given experimental conditions.
6. Use Hamilton glass syringe for manipulation with lipids dissolved in organic solvents. Firstly, there is drop risk, because of solvent gas partial pressure compensation in free volume above liquid in the micropipette tip. Secondly organic solvent can extract molecules from plastic tips.
7. Keep bottom of the test tube in water bath to protect the sample against evaporation freezing effect.
8. The best way is to prepare samples the day before and keep them in evaporator overnight.
9. This is a key step, when turbid suspension of micelles is formed thanks to the hydrophobic effect. Suggested time is only tentative, vortex sample until entire lipid film is hydrated.
10. Keep bottom of the test tube in water bath for overheating protection.
11. We sonicate 1–2 ml sample by probe sonicator for 20 min at 7 W power.
12. Minimal lipid concentration can be checked by calculation of required lipid amount for covering all hydrophilic surfaces, which are in contact with sample, with double layer of lipids with known area per one lipid molecule.
13. It is useful to record maximal value for comparison with next alignment result.
14. The strongest signal is obtained when reflection on mirror is used.
15. Start tuning with *x*-, *y*-axis, then tune *z*-axis and repeat procedure in this order until the maximal intensity is achieved.

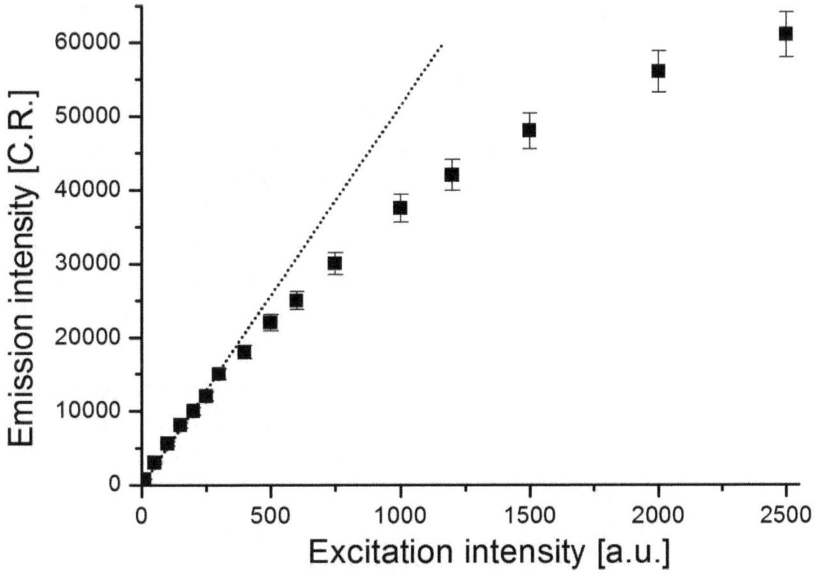

Fig. 5 Saturation characteristic of 5.6 nM carboxyfluorescein (Sigma-Aldrich, St. Louis, USA), solution in buffer A indicates suitable setup of laser power for FCS measurements. Emission intensity is linearly proportional to excitation intensity for laser powers below a certain threshold (approximately 300 a.u. in this case) above which saturation effects are manifested

16. Used dye should have similar emission characteristics. Suitable concentration is 10–100 nM depending on dye brightness. Be careful with proper laser power so that detector will not be saturated. For example, it is recommended that count rate on SPAD detectors does not exceed 1×10^6 cps.
17. Suitable laser power range can be estimated experimentally by measurement dependency of fluorescence count rate on increasing laser power. This dependence should be linear until appearance of saturation effect (*see* Fig. 5).
18. Background (caused by scattered and stray ambient light as well as detector noise) may lead to apparent higher values of N. This artifact is especially significant when fluorescence signal is comparable to (or lower than) background signal, that means for example when low concentration of fluorophore or a fluorophore with low brightness is used. Background correction is therefore needed in such cases to obtain correct values of N. First possible strategy is based on calculation of background correction factor χ^2 from measured background level as presented in Rüttinger et al. [44] (*see* **Note 19**).

$$\chi^2 = 1 + \langle b \rangle \Big/ \langle f \rangle^2, \tag{13}$$

where $\langle b \rangle$ is average background intensity, and $\langle f \rangle$ is the pure fluorescence contribution to the total averaged measured signal $\langle I \rangle$, $\langle f \rangle = \langle I \rangle - \langle b \rangle$. Frequently, detector noise is the most important source of background; $\langle b \rangle$ can be experimentally determined, for example, by a measurement with excitation light turned off. If contribution of scattered light to the background cannot be neglected, $\langle b \rangle$ can be measured using an unlabelled sample (an identical sample without fluorescence dye). Corrected N is obtained when the apparent (measured) N is divided by achieved χ^2. An alternative strategy differentiates fluorescence from background using the differences in photon arrival histograms. Pulsed excitation and time-correlated single-photon counting unit are necessary to employ that strategy, see more in [45, 46].

19. Procedures described in [44] are connected with estimation of the exact size and shape of the confocal volume, which are not necessary when a calibration-free technique such as *Z*-scan is performed.
20. You can also use reflection on glass-buffer interface to locate position of the solid support surface.
21. Proper estimation of those experimental conditions is crucial for quality of *Z*-scan results. It is necessary to find a compromise between the requirement of long acquisition time needed for achieving high statistical relevance of correlation functions and the requirement of stable sample positioning in time. Longer acquisition times increase the probability that sample positioning will be unfavorably affected by mechanical drifts in the experimental setup. Resulting inaccuracy in *z*-position of the sample distorts *Z*-scan parabolas and introduces systematic errors into results.
22. FCS curves measured far from the focal plane suffer from low signal-to-noise ratio. Thus values of N and τ_D are obtained from those curves with high error; it may be advisable to exclude them from parabolic fits in cases when they are outlying and would unfavorably affected resulting parabolas.
23. The value of t_0 does not necessarily prove presence of particular type of inhomogeneities in membrane. It only proves certain deviation from free diffusion which can support conclusions concerning type of membrane inhomogeneities based on findings of other methods or on theoretical assumptions.

Acknowledgements

The authors thank Martin Štefl for his comments and suggestions and collaborators from Institute of Organic Chemistry and Biochemistry of ASCR for providing halictine peptide. This work was supported by Czech Science Foundation through Grant No. P208/10/0376 and by Ministry of Education, Youth and Sports of the Czech Republic via BIO-OPT-XUV Research Team Advancement at the Faculty of Biomedical Engineering, Czech Technical University in Prague, registration nr.: MEYS ESF Project CZ.1.07/2.3.00/20.0092. M.H. acknowledges financial support by AS CR via Praemium Academiae award.

References

1. Elson E, Magde D (1974) Fluorescence correlation spectroscopy. I. Conceptual basis and theory. Biopolymers 13:1–27
2. Magde D, Webb WW, Elson E (1972) Thermodynamic fluctuations in a reacting system--measurement by fluorescence correlation spectroscopy. Phys Rev Lett 29:705–708
3. Magde D, Elson EL, Webb WW (1974) Fluorescence correlation spectroscopy. II. An experimental realization. Biopolymers 13:29–61
4. Štefl M, Kułakowska A, Hof M (2009) Simultaneous characterization of lateral lipid and prothrombin diffusion coefficients by z-scan fluorescence correlation spectroscopy. Biophys J 97:L01–L03
5. Enderlein J, Gregor I, Patra D et al (2005) Performance of fluorescence correlation spectroscopy for measuring diffusion and concentration. Chemphyschem 6:2324–2336
6. Enderlein J, Gregor I, Patra D et al (2004) Art and artefacts of fluorescence correlation spectroscopy. Curr Pharm Biotechnol 5:155–161
7. Thompson NL (1991) Fluorescence correlation spectroscopy. In: Lakowicz JR (ed) Topics in fluorescence spectroscopy. Vol 1 Plenum, New York, pp 337–378
8. Beneš M, Billy D, Hermens WT et al (2002) Muscovite (mica) allows the characterisation of supported bilayers by ellipsometry and confocal fluorescence correlation spectroscopy. Biol Chem 383:337–341
9. Dertinger T, Pacheco V, von der Hocht I et al (2007) Two-focus fluorescence correlation spectroscopy: a new tool for accurate and absolute diffusion measurements. Chemphyschem 8:433–443
10. Miszta A, Macháň R, Hermens WT et al (2010) Peptide-membrane interactions studied by ellipsometry, laser scanning microscopy, and z-scan fluorescence correlation spectroscopy. In: Castanho MA (ed) Membrane-active peptides: methods and results on structure and function. International University Line, California, pp 217–245
11. Gregor I, Patra D, Enderlein J (2005) Optical saturation in fluorescence correlation spectroscopy under continuous-wave and pulsed excitation. Chemphyschem 6:164–170
12. Enderlein J, Gregor I, Patra D et al (2005) Statistical analysis of diffusion coefficient determination by fluorescence correlation spectroscopy. J Fluoresc 15:415–422
13. Benda A, Fagul'ová V, Deyneka A et al (2006) Fluorescence lifetime correlation spectroscopy combined with lifetime tuning: new perspectives in supported phospholipid bilayer research. Langmuir 22:9580–9585
14. Enderlein J (1991) Autowave propagation in a piecewise linearized schlögl model with inhomogeneously distributed catalyst. Phys Lett A 156:429–435
15. Enderlein J (2000) Theoretical study of detection of a dipole emitter through an objective with high numerical aperture. Opt Lett 25:634–636
16. Enderlein J, Ambrose WP (1997) Optical collection efficiency function in single-molecule detection experiments. Appl Opt 36:5298–5302
17. Benda A, Beneš M, Mareček V et al (2003) How to determine diffusion coefficients in planar phospholipid systems by confocal fluorescence correlation spectroscopy. Langmuir 19:4120–4126
18. Wohland T, Rigler R, Vogel H (2001) The standard deviation in fluorescence correlation spectroscopy. Biophys J 80:2987–2999
19. Lakowicz JR (2006) Principles of fluorescence spectroscopy. Springer, New York
20. Ries J, Schwille P (2008) New concepts for fluorescence correlation spectroscopy on membranes. Phys Chem Chem Phys 10:3487–3497
21. Hess ST, Huang S, Heikal AA et al (2002) Biological and chemical applications of fluorescence correlation spectroscopy: a review. Biochemistry 41:697–705
22. Štefl M, Macháň R, Hof M (2011) Z-scan fluorescence correlation spectroscopy: a powerful tool for determination of lateral diffusion in biological systems. In: Geddes CD (ed) Reviews in fluorescence 2009. Reviews in fluorescence. Springer, New York, pp 321–344

23. Schwille P, Oehlenschlager F, Walter N (1996) Quantitative hybridization kinetics of dna probes to rna in solution followed by diffusional fluorescence correlation analysis. Biochemistry 35:10182–10193
24. Widengren J, Mets U, Rigler R (1995) Fluorescence correlation spectroscopy of triplet states in solution: a theoretical and experimental study. J Phys Chem 99:13368–13379
25. Macháň R, Hof M (2010) Lipid diffusion in planar membranes investigated by fluorescence correlation spectroscopy. BBA-Biomembranes 1798:1377–1391
26. Donsmark J, Rischel C (2007) Fluorescence correlation spectroscopy at the oil-water interface: hard disk diffusion behavior in dilute beta-lactoglobulin layers precedes monolayer formation. Langmuir 23:6614–6623
27. Provencher SW (1982) A constrained regularization method for inverting data represented by linear algebraic or integral equations. Comput Phys Commun 27:213–227
28. Enderlein J (2005) Dependence of the optical saturation of fluorescence on rotational diffusion. Chem Phys Lett 410:452–456
29. Przybylo M, Sýkora J, Humpolíčková J et al (2006) Lipid diffusion in giant unilamellar vesicles is more than 2 times faster than in supported phospholipid bilayers under identical conditions. Langmuir 22:9096–9099
30. Ries J, Schwille P (2006) Studying slow membrane dynamics with continuous wave scanning fluorescence correlation spectroscopy. Biophys J 91:1915–1924
31. Chiantia S, Ries J, Kahya N et al (2006) Combined afm and two-focus sfcs study of raft-exhibiting model membranes. Chemphyschem 7:2409–2418
32. Humpolíčková J, Gielen E, Benda A et al (2006) Probing diffusion laws within cellular membranes by z-scan fluorescence correlation spectroscopy. Biophys J 91:L23–L25
33. Milon S, Hovius R, Vogel H et al (2003) Factors influencing fluorescence correlation spectroscopy measurements on membranes: simulations and experiments. Chem Phys 288:171–186
34. Palmer AG, Thompson NL (1989) Optical spatial intensity profiles for high order autocorrelation in fluorescence spectroscopy. Appl Opt 28:1214–1220
35. Sorscher SM, Klein MP (1980) Profile of a focused collimated laser beam near the focal minimum characterized by fluorescence correlation spectroscopy. Rev Sci Instrum 51:98–102
36. Macháň R, Hof M (2010) Recent developments in fluorescence correlation spectroscopy for diffusion measurements in planar lipid membranes. Int J Mol Sci 11:427–457
37. Wawrezinieck L, Rigneault H, Marguet D et al (2005) Fluorescence correlation spectroscopy diffusion laws to probe the submicron cell membrane organization. Biophys J 89:4029–4042
38. Monincová L, Buděšínský M, Slaninová J et al (2010) Novel antimicrobial peptides from the venom of the eusocial bee halictus sexcinctus (hymenoptera: halictidae) and their analogs. Amino Acids 39:763–775
39. Macháň R (2012) Supported lipid bilayers and antimicrobial peptides: characterized by ellipsometry and fluorescence correlation spectroscopy. LAP LAMBERT Academic, Germany
40. Guo L, Har JY, Sankaran J et al (2008) Molecular diffusion measurement in lipid bilayers over wide concentration ranges: a comparative study. Chemphyschem 9:721–728
41. Beneš M, Billy D, Benda A et al (2004) Surface-dependent transitions during self-assembly of phospholipid membranes on mica, silica, and glass. Langmuir 20:10129–10137
42. Kask P, Gunther R, Axhausen P (1997) Statistical accuracy in fluorescence fluctuation experiments. Eur Biophys J Biophys Lett 25:163–169
43. Forstner MB, Yee CK, Parikh AN et al (2006) Lipid lateral mobility and membrane phase structure modulation by protein binding. J Am Chem Soc 128:15221–15227
44. Rüttinger S, Buschmann V, Krämer B et al (2008) Comparison and accuracy of methods to determine the confocal volume for quantitative fluorescence correlation spectroscopy. J Microsc 232:343–352
45. Böhmer M, Wahl M, Rahn HJ et al (2002) Time-resolved fluorescence correlation spectroscopy. Chem Phys Lett 353:439–445
46. Kapusta P, Wahl M, Benda A et al (2007) Fluorescence lifetime correlation spectroscopy. J Fluoresc 17:43–48

Chapter 29

Scanning Fluorescence Correlation Spectroscopy (SFCS) with a Scan Path Perpendicular to the Membrane Plane

Paul Müller, Petra Schwille, and Thomas Weidemann

Abstract

Scanning fluorescence correlation spectroscopy (SFCS) with a scan path perpendicular to the membrane plane was introduced to measure diffusion and interactions of fluorescent components in free-standing biomembranes. Using a confocal laser scanning microscope (CLSM), the open detection volume is repeatedly scanned through the membrane at a kHz frequency. The fluorescence photons emitted from the detection volume are continuously recorded and stored in a file. While the accessory hardware requirements for a conventional CLSM are minimal, data evaluation can pose a bottleneck. The photon events must be assigned to each scan, in which the maximum signal intensities have to be detected, binned, and aligned between the scans, in order to derive the membrane-related intensity fluctuations of one spot. Finally, this time-dependent signal must be correlated and evaluated by well-known FCS model functions. Here we provide two platform-independent, open source software tools (*PyScanFCS* and *PyCorrFit*) that allow to perform all of these steps and to establish perpendicular SFCS in its one- or two-focus as well as its single- or dual-color modality.

Key words Scanning fluorescence correlation spectroscopy (SFCS), Fluorescence correlation spectroscopy (FCS), Fluorescence cross-correlation spectroscopy (FCCS), Membrane diffusion, Giant unilamellar vesicles (GUV), Diffusion, Protein–protein interaction, Ligand binding

1 Introduction

In recent years, fluorescence correlation spectroscopy (FCS) has become a well-established method in biomedical research [1, 2]. In addition to confocal laser scanning microscopy (CLSM), where the spatial fluorescence intensity distribution is sampled, FCS records the time-dependent intensity fluctuations originating from a microscopic spot. With sufficient sensitivity and time resolution of the detectors, the signal fluctuations will reflect not only

Yves Engelborghs and Antonie J.W.G. Visser (eds.), *Fluorescence Spectroscopy and Microscopy: Methods and Protocols*, Methods in Molecular Biology, vol. 1076, DOI 10.1007/978-1-62703-649-8_29,

instrumental noise but also the stochastic Brownian motion of the fluorescent molecules performing random walks through the illuminated, open detection volume. Thus, the time-dependent signal contains information about molecular properties which can be extracted and analyzed (*see* Chapter 24).

The average time required by molecules to "walk" through the detection volume (diffusion time) depends on its shape and size and is therefore characteristic for the optical setup. The mobility of the molecules depends on their shape and size, as well as the viscosity of the surrounding medium (*see* Chapter 24). Because the viscosity of a lipid bilayer or a cellular plasma membrane is about 100 times larger than that of the aqueous media, the diffusion times are slowed down to the same degree. In addition, the diffusion is confined to a very thin membrane plane of 5–10 nm thickness. Using diffraction limited illumination, lateral, two-dimensional diffusion times of lipids are typically in the millisecond (ms) regime and single-pass transmembrane receptors are even 20–50 times slower.

Perpendicular scanning (SFCS) was introduced to measure mobility and interactions in the lipid bilayer of giant unilamellar vesicles (GUVs) [3, 4]. Slower diffusion of fluorescent molecules in combination with small concentrations produces rare events with longer lifetime, and therefore, it is sufficient to collect data with an appropriate sampling frequency. In perpendicular line SFCS [3, 5, 6], this is achieved by guiding the focused laser beam repeatedly along a straight line with kHz frequency perpendicular through the membrane plane (Fig. 1). Scanning is performed by means of the conventional control software of a CLSM. At all times, the emitted photons are collected with single-photon counting detectors, e.g., avalanche photo diodes (APDs). The signal is digitalized by a separate device and stored in a file. Here we used a USB connected hardware correlator; however, other time-correlated single-photon counting (TCSPC) cards may also be suitable. Since the file contains all photon incidents along the scan path, the signal originating from the spot where the scan path crosses the membrane plane must be extracted post-measurement.

For this purpose, we developed a program (*PyScanFCS*). The correct periodicity (scan cycle time) corresponding to the repetition rate of each scan is determined by means of Fourier analysis of the photon arrival times. For visual inspection, the aligned scan intensities are displayed in a kymograph containing the number of scans on the *x*-axis and the intensity along the scan path on the *y*-axis. The time range displaying homogeneous intensity can be manually selected. The number of bins along the scan path containing membrane-related intensity can be defined and corrected for systematic bleaching during the course of the measurement (typically several minutes). Finally, the resulting intensity trace is time-correlated by a multiple τ algorithm. To evaluate the correlation

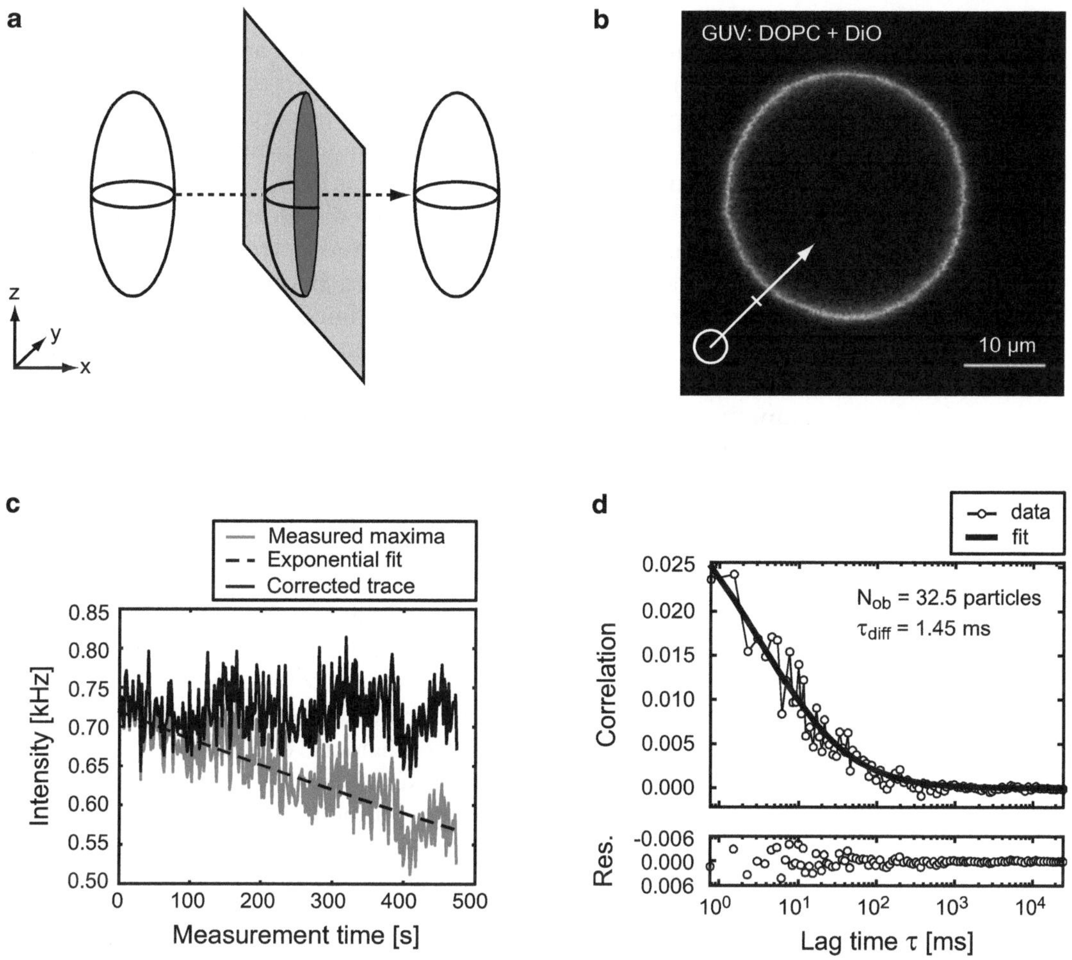

Fig. 1 Single-focus perpendicular SFCS. (**a**) Graphical representation of the open detection volume as an ellipsoid, stretched along the optical axis *z*. Lateral movement on a perpendicular scan path (*dashed arrow*) through a two-dimensional membrane plane (*light gray*) creates an elliptical, two-dimensional detection area (*dark gray*). (**b**) Equatorial confocal image of a GUV composed of DOPC containing 0.007 % of the fluorescent lipid DiO. The scan path (*white arrow*) was positioned perpendicular to the membrane plane at a 45° angle. (**c**) Intensity trace of the maxima (*gray*) along each scan is plotted against the measurement time (500 s $= 6.5 \times 10^5$ scans with a cycle time 0.768 ms). During the measurement, the intensity dropped exponentially (*dashed*) by about 25 % due to bleaching, and was back-corrected (*black*) [6]. The magnitude of the intensity values (here about 0.75 kHz) depends on the exact procedure to assign membrane-related fluorescence, in particular the number of bins used to sample each line scan (here 70) and the number of bins assigned to the membrane area (here 7: 1 maximum plus 6 adjacent bins). (**d**) Autocorrelation curve of the corrected intensity trace from (**c**) was fitted with a model function for two-dimensional diffusion through an elliptical Gaussian shaped detection volume (Eq. 2)

data, the curves can be analyzed with a second program (*PyCorrFit*) written for fitting different model functions containing molecular parameters like diffusion times, diffusion coefficients, and particle numbers.

Compared to single-point confocal FCS, perpendicular SFCS has several advantages when applied to membranes [5, 7]. Free-standing membranes are usually large structures showing slow thermal motion like undulations. During the measurement, the vesicle or the cell may move or change shape. This is a problem for conventional single-point confocal FCS, because the detection volume is fixed in space; even minor displacements of the fluorescent membrane with respect to a fixed detection volume cause a detrimental bias in correlation curves. Because in SFCS the maximum intensity on each path is determined post-measurement, such positional instabilities can be corrected. Additionally, the laser focus will be outside the membrane most of the time and therefore the amount of bleaching is reduced. However, a potential disadvantage of the SFCS approach is a low average signal-to-noise ratio and accordingly prolonged measurement times.

We demonstrate how to obtain and evaluate perpendicular SFCS data by measuring the diffusion of the 488 nm excitable, fluorescent lipid 3,3′-dioctadecyloxacarbocyanine (DiO) in GUVs composed of 1,2-dioleoyl-*sn*-glycero-3-phosphocholine (DOPC). Scanning is either performed on a single line (single-focus SFCS) or on a frame composed of two lines (two-focus SFCS). Two-focus SFCS takes advantage of the fixed distance between the scan paths to obtain calibrated diffusion coefficients [8]. Additional scan modalities like dual-color scanning fluorescence cross-correlation spectroscopy (SFCCS) are briefly outlined. Dual-color SFCCS is a powerful approach to address interactions between differently labelled chemical species in biomembranes [9, 10].

2 Materials

2.1 Preparation of GUVs

1. 10 ml 200 mM sucrose in H_2O.
2. 10 ml 200 mM glucose in H_2O.
3. Preparation of lipid mixture: Take DOPC stock solution (20 mg/ml in chloroform) and DiO (5 μM in chloroform) from −20 °C and dilute with chloroform to final concentration of 2 mg/ml DOPC and a relative concentration of DiO in DOPC of 0.007 mol% (Avanti Polar Lipids, Alabaster, Alabama, USA). 5–10 μl is used to prepare GUVs in the Teflon chamber.
4. Bovine serum albumin (BSA), 2 mg/ml in H_2O (Sigma-Aldrich, Munich, Germany).
5. Vacuum chamber for drying samples (Vacuubrand, Wertheim, Germany).
6. Homemade Teflon chamber containing two platinum electrodes; capacity 300 μl (*see* **Note 1**).
7. Electrical sine wave generator (TTi TG315 function generator, Huntingdon, Cambridgeshire, UK).

2.2 Microscopy

1. 8-well chambered cover slides (LabTek II, #1.5, Nunc, Thermo Scientific, USA).
2. Diffusion standard to calibrate the detection volume: hydrolyzed AlexaFluor 488 succinimidylester (Life technologies/ Molecular Probes, USA). Prepare a 25 nM solution in 10 mM Tris-HCl, pH 8.
3. Purified enhanced green fluorescent protein (EGFP) (CatNo.: #4999-100, BioVision, USA).

2.3 Hardware for SFCS Data Acquisition

CLSM setups equipped with photon counting detectors (e.g., a LSM510 with accessory Confocor3 unit, Zeiss, Germany) offer direct access to the signal by means of BNC-connected coaxial cables attached to the APDs. If the CLSM does not possess photon counting detectors, one can use the fiber-out exit and feed the signal into external APDs [3]. For this, one needs:

1. APDs (Perkin-Elmer Optoelectronics, USA).
2. Multimode fibers with MM connectors, ASF105/125Y (Thorlabs, Newton, New Jersey, USA).
3. Positioning system (miniature *xyz* stage) with FC adapter to align the optical fiber (Owis, Staufen, Germany).
4. USB connectable hardware correlator to record the photon history (Model Flex02-12D/B, www.correlator.com, USA/ China) (*see* **Note 2**).

2.4 Software for SFCS Data Acquisition

The data processing software (*PyScanFCS* and *PyCorrFit*) has been tested on Windows XP/7 and Ubuntu Linux (*see* **Note 3**).

1. Software for photon history recording, *Photon.exe* (shipped with the Flex correlator Flex02-12D/B, www.correlator.com).
2. Program to generate correlation curves from the photon history file *PyScanFCS* (free download at http://pyscanfcs.craban.de). Due to constraints in memory of 32-bit systems, we recommend to use a 64-bit computer with either Windows 7 or an Ubuntu AMD64 operating system.
3. Fitting program to extract relevant parameters from correlation curves, *PyCorrFit* (free download at http://pycorrfit.craban.de).

3 Methods

3.1 Preparation of GUVs by Electroformation

1. Clean the cap and reservoir of the Teflon chamber two times with 80 % ethanol and with hot tap water.
2. Rinse the Teflon container and the cap with distilled water and dry both with pressurized air.

3. Using a pipette, distribute a total 5 μl of the 2 mg/ml lipid–dye mixture evenly on both platinum electrodes pointing towards the inside of the chamber.
4. To remove the chloroform, let the lipid mixture dry in air for 2 min and further in a vacuum chamber for 10 min.
5. Fill the Teflon container with 300 μl of 200 mM sucrose solution and close the lid. The platinum electrodes should be completely covered by sucrose solution.
6. Apply an alternating electric field of 2.15 V at 10 Hz for 1.5 h. During this phase, the GUVs form at the electrodes.
7. Decrease the frequency to 2 Hz for another 20 min to allow the GUVs to detach from the electrodes.
8. Incubate the Labtek chamber with 2 mg/ml bovine serum albumin (BSA) for 20 min.
9. Wash the Labtek chamber five times with 200 mM glucose solution.
10. Fill the Labtek chamber with 400 μl of 200 mM glucose solution.
11. Add 50 μl of the GUV-sucrose solution. Cut off the tip when pipetting GUVs to prevent them from bursting due to shear forces (*see* **Note 4**).

3.2 Define the Microscope Settings for SFCS

3.2.1 Calibration with Alexa488 Solution

1. Prepare the 8-well Labtek containing a 1–50 nM Alexa488 solution and the GUV solution in adjacent wells. Clean the glass bottom with ultra pure ethanol and mount on the stage.
2. Define settings for excitation and detection (e.g., a dichroic 488/633 followed by BG35 to eliminate scattered UV light and a cleaning band pass (505–560) for Alexa488 and DiO emission).
3. Set the CLSM to measure in a single spot (with LSM510 and external APDs this is possible by selecting scan mode “Spot” and performing a time series).
4. Position the focus in solution at 50 μm above the glass bottom (*see* **Note 5**).
5. Adjust laser power to 50–100 kHz and maximize the count rate at the APD by iterative alignment of pinhole position (by means of the LSM software, Confocor3) and the external fiber coupling (by moving the external miniature *xyz* stage).
6. Adjust the correction collar of the objective to the thickness of the Labtek cover glass by maximizing the count rate.
7. Record three autocorrelation curves (1–2 min) at three different lateral positions and save as a ~.sin file (Flex correlator software) or ~.fcs file (Confocor3).

8. Evaluate the data with *PyCorrFit* using a confocal 3D-diffusion model with triplet contribution to obtain the average diffusion time and the axis ratio ("structure parameter") of the detection volume. The residuals of the fit should not show a systematic bias, therefore indicating proper alignment and nearly Gaussian geometry of the point spread function. Calculate the effective focal volume (*see* **Note 6**).

3.2.2 Calibration for Two-Focus SFCS

To take advantage of the calibration-free determination of diffusion coefficients with two-focus SFCS, the signal from two parallel scan paths is cross-correlated [8] (Fig. 2). However, to evaluate the cross-correlation curves, one needs to know their physical distance *d*. This can be determined by bleaching a layer of EGFP adsorbed

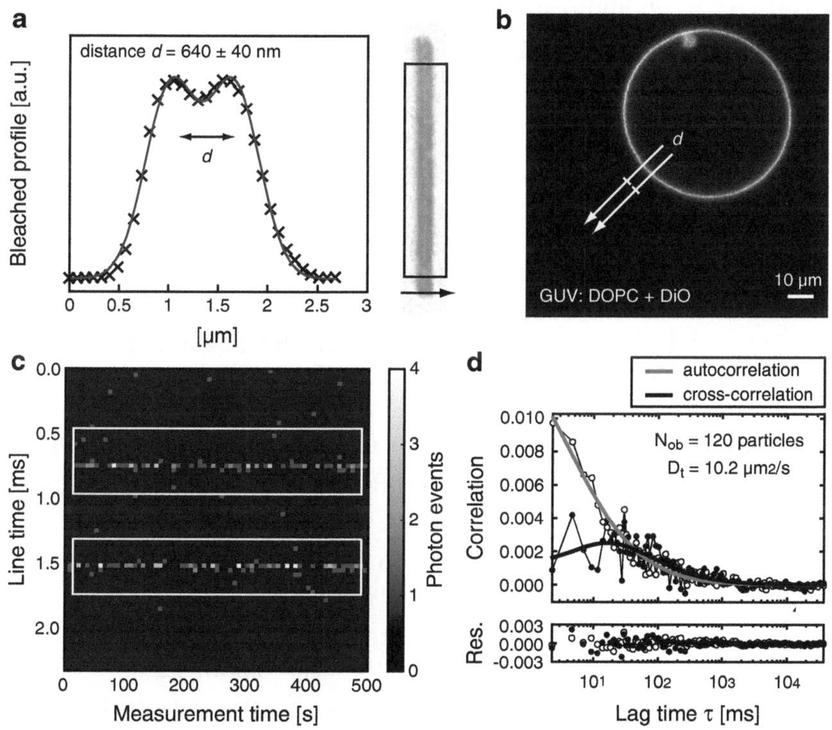

Fig. 2 Two-focus perpendicular SFCS. (**a**) Calibration procedure. The image shows EGFP protein adsorbed to the glass surface after continuous scanning for about 1 min (a reference pre-bleach image was subtracted). The *two dark lines* representing the parallel scan path were integrated (*rectangular area*, *ImageJ*) and the profile fitted with Eq. 1 to derive the mean distance *d*. (**b**) Equatorial confocal image of a GUV composed of DOPC containing 0.007 % of the fluorescent lipid DiO. The scan path (*white arrow*) was positioned perpendicular to the membrane plane at a 45° angle. (**c**) Kymograph containing the binned (70) photon counts along the scan path (*y*-axis, in total 2.3 ms) during the measurement time (*x*-axis, in total 500 s). Since the focal volume passes the membrane twice during each scan cycle, two horizontal lines appear in the plot. (**d**) Averaged autocorrelation and cross-correlation curves were fitted with Eqs. 2 and 3, respectively. The diffusion coefficient was calculated by applying Eq. 4

to the glass bottom of the well and fitting a double Gaussian to the intensity profile (Fig. 2a):

$$P(d) = \frac{A}{\sqrt{2\pi}\sigma}\left[e^{-\frac{1}{2}\left(\frac{x-\mu}{\sigma}\right)^2} + e^{-\frac{1}{2}\left(\frac{x-\mu-d}{\sigma}\right)^2}\right] + B, \quad (1)$$

where x is a spatial coordinate normal to the scan path, μ is the location of the first maximum, σ the standard deviation, and d is the distance between the maxima. The parameters A and B describe the amplitude and offset of the cross section. Because of hysteresis effects, not every scanning procedure reproduces two parallel beam paths (*see* **Note 7**). A sufficiently accurate two-focus scan could be achieved by the following sequence (LSM software, Zeiss):

1. Scan mode "Multitrack" with the first track laser off and the second track laser on; change tracks between each scanned line.
2. Bidirectional scanning.
3. Zoom 12, angle: 45°.
4. Acquisition: Frame mode 32 × 2 pixels.

By this, the first line is scanned forward without laser. The second line is scanned backward with the laser on and fluorescence is detected (first detected scan). The second line is again scanned, but forward, without laser. The scan cycle is completed with a scan of the first line backward while the laser is on (second detected scan). As seen in the bleached image (Fig. 2a, right panel), there is only minimal hysteresis at the beginning and at the end of the scan path.

3.3 Data Acquisition: One-Focus SFCS

1. Perform Alexa488 calibration according to Subheading 3.2.1.
2. Use the CLSM software to image the equatorial plane of a GUV.
3. Switch to line scanning and "crop" mode and define a scan path perpendicular to the membrane (in the LSM) (*see* **Note 8**).
4. Set the scan mode to "unidirectional" at maximum scan rate (e.g., LSM510, 0.786 ms at "zoom 12").
5. Direct the emission light to the APD (Confocor) or via the fiber-out exit to the external detector.
6. Execute the software *Photon.exe* to record the photon stream. Choose the file format 16 bit. Set the acquisition time between 200 and 500 s. Set the channels to "A only" since this protocol uses only one APD. Make sure that the APD is connected to the "input A" of the Flex correlator.
7. To measure, first start a continuous line scan in the LSM software and then start file recording in *Photon.exe*.

The program *Photon.exe* creates a ~.dat file (*see* **Note 9**) that contains the entire photon history of the measurement. The file has to be processed using *PyScanFCS*, described in Subheading 3.6.

3.4 Data Acquisition: Two-Focus SFCS

1. Perform Alexa488 calibration according to Subheading 3.2.1.
2. Use the CLSM software to image the equatorial plane of a GUV.
3. Apply the exact measurement settings (zoom, scanning speed, scanning angle) that were recorded during the calibration of the distance *d* between the scanned lines in Subheading 3.2.2. Position the scan path in the image (*see* **Note 8**).
4. Direct the emission light to the APD (Confocor) or via the fiber-out exit to the external detector.
5. Execute the software *Photon.exe* to record the photon stream. Choose the file format 16 bit. Set the acquisition time between 200 and 500 s. Set the channels to "A only" since this protocol uses only one APD. Make sure that the APD is connected to the "input A" of the Flex correlator.
6. To measure, first start continuous frame scanning in the LSM software and then start file recording in *Photon.exe*.

The program *Photon.exe* creates a ~.dat file (*see* **Note 9**) that contains the entire photon history of the measurement. The file has to be processed using *PyScanFCS*, described in Subheading 3.6.

3.5 Data Acquisition: Dual-Color Applications

In dual-color SFCCS, the emission pathway is separated by an appropriate beam splitter and the signal is detected by two APDs. The scanning settings to measure dual-color SFCS are the same as for "one-focus SFCS" as described in Subheading 3.3, with the extension that both laser lines and their corresponding detection pathways are simultaneously used. For "alternating excitation" the color channels are scanned sequentially (LSM510, "Multitrack mode"). Since two APDs are now connected to the Flex correlator in the program *Photon.exe*, channels A and B have to be selected for data recording. SFCCS with alternating excitation is a convenient way to circumvent cross talk between the color channels, for example, when observing green and red fluorescent proteins [9, 10].

3.6 Data Processing with PyScanFCS

PyScanFCS can extract membrane-related intensity fluctuations from the photon history of the scan and calculate experimental correlation functions (Fig. 3).

1. Loading the data. Open an experimental data file (File | Open photon events .dat file). In the "Pre-binning" box, set the bins per scan cycle to 70 and the number of used photon events to 1,000 (*see* **Note 10**). Set the bin width to 5 μs and click "Calculate and plot."

a
b
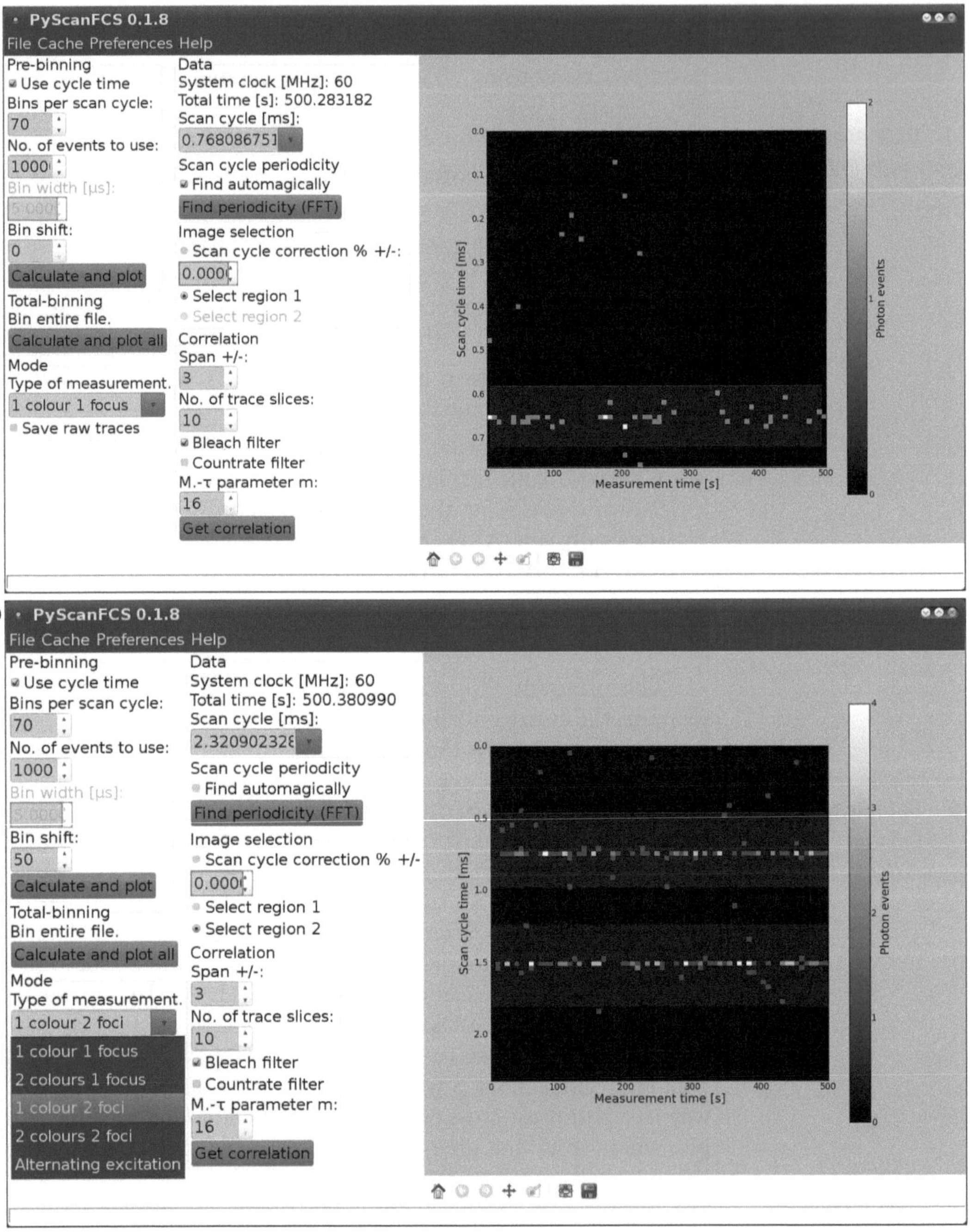

Fig. 3 *PyScanFCS* is a tool to generate correlation curves from a stored photon stream. (**a**) Screenshot showing the "1-colour-1-focus" application when evaluating a single-focus FCS measurement. The complete file (500 s) was loaded and binned (70). By means of FFT, the program determined a cycle time of 0.768 ms. Prior to calculating the correlation functions, the horizontal intensity trace was selected by dragging a rectangle (gray shaded). (**b**) Example for "2-colour-1-focus" application (see pull-down menu for other measurement modes). "Bin shift" (50) was used to place the horizontal intensity traces in the center of the kymograph window

2. The scan cycle time needs to be found. In the “Scan cycle periodicity” box, check “Find automatically” and click “Find periodicity (FFT).” The program performs a Fourier transform of the photon arrival times on the first 1,000 events and searches for a maximum amplitude. Check the value of the cycle time found by *PyScanFCS*. If this value does not match the magnitude as expected from the microscope settings, uncheck “Find automatically” and click “Find periodicity (FFT)” again. A new interactive window pops up displaying all amplitudes of the FFT and their corresponding cycle times. Manually select the expected cycle time by clicking on both sides of the peak (*see* **Note 11**).
3. Binning of the entire file. In the “Pre-binning” box, check the “Use cycle time” option. The displayed cycle time will be used to calculate the kymograph. The plot on the right hand side should show a line of bins which represents the intensity trace measured in the membrane. If this line is tilted, the cycle time has to be corrected: Select “Scan cycle correction” in the “Image selection” box and manually drag a line along the intensity trace inside the kymograph window. Clicking again “Calculate and plot” will reproduce a corrected, horizontal line. Increase the “No. of events to use” stepwise by a factor of 10 and subsequently correct the cycle time until the entire ~. dat file has been processed, i.e., the total measurement time has been plotted.
4. Trace selection in the kymograph. If one-focus SFCS was performed, a single trace should be visible in the kymograph. If two-focus SFCS was performed, two traces representing the two parallel scan paths should be visible. In the “Mode” box, select “1-colour-1-focus” or “1-colour-2-foci” according to your measurement. In the “Image selection” box, choose “Select region 1” and draw a rectangle comprising a region of homogeneous intensity. For two-focus SFCS, “Select region 2” is used to manually select the second trace.
5. Calculation of the intensity trace and the correlation curve. In the “Correlation” box, define the number of neighboring bins along the scan path that will be added up to the final number of counts for this time point. With a resolution of 70 bins per scan cycle, a “span” of ± 3 is a good measure. The trace can be split in a number of slices each of which will be correlated separately. This is useful when inhomogeneities like temporary signal loss or bright aggregates appear during the measurement. In *PyCorrFit*, the correlation curves containing the corresponding bias can be sorted out (*see* **Note 12**). Check “Bleach filter” if you want to correct for bleaching during the measurement [6]. *PyScanFCS* is able to perform an approximation of the actual count rate using the “Countrate filter” (not done here).

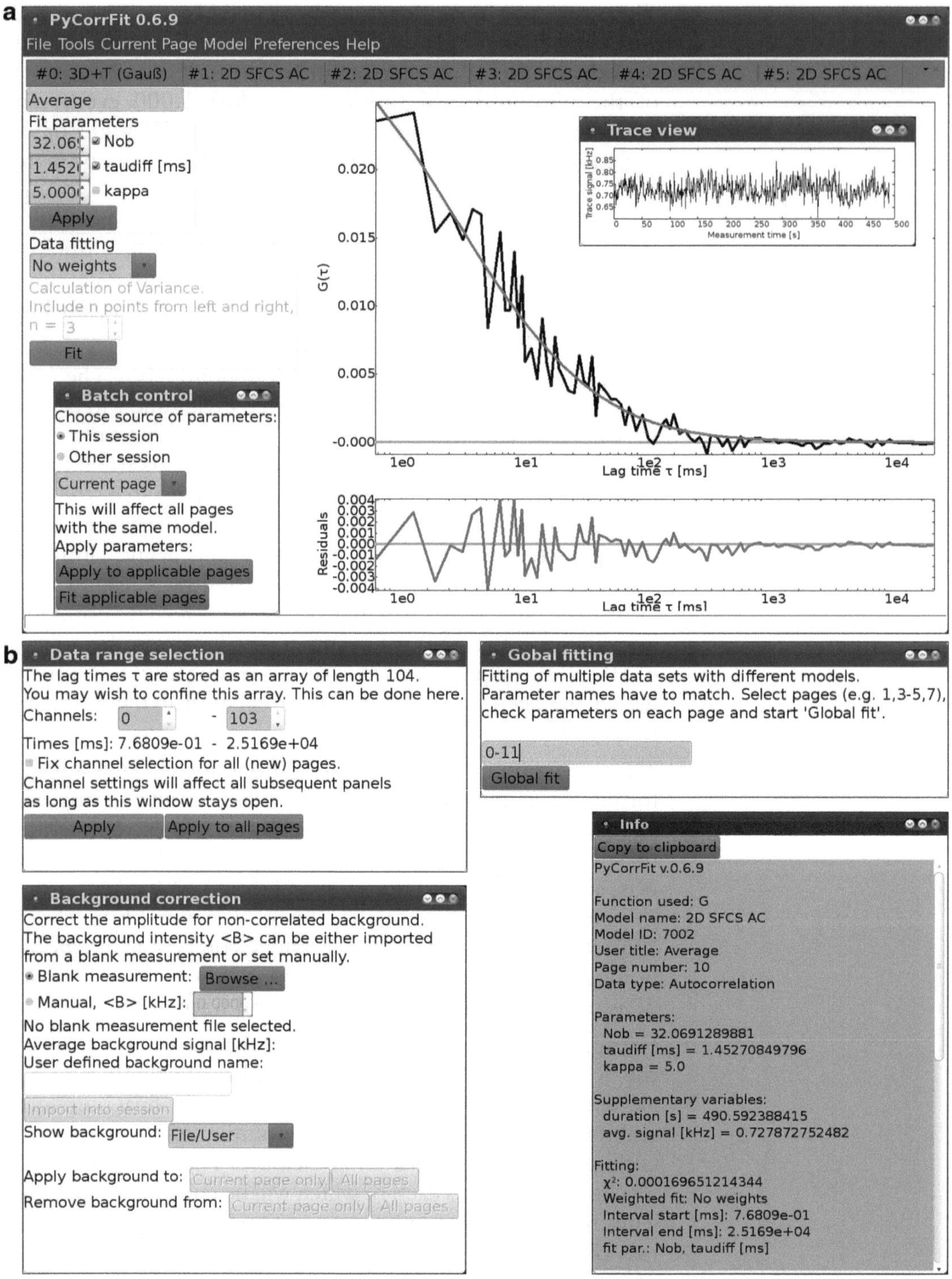

Fig. 4 *PyCorrFit* is a tool to evaluate correlation functions. (**a**) Main window showing the menu bar, tabs containing experimental or averaged correlation curves and the fitting parameters. The set of fitting parameters is defined by the model function which must be loaded beforehand. The experimental correlation data (*black*) and the actual fit function (*gray*) are shown in a semi-log plot together with the residuals

The real count rate at the membrane is about 2 orders of magnitudes greater than shown in Fig. 1c. The magnitude of the intensity does not affect the calculation of the correlation curves. The theoretical background of the count rate filter is described in the *PyScanFCS* documentation. The parameter *m* of the multiple-τ algorithm should be set to 16. Finally, click on "Get correlation." This will calculate the correlation curves and save them as a ~.zip file inside the measurement folder.

3.7 Data Processing with PyCorrFit

PyCorrFit can fit analytical model functions to experimental correlation curves (Fig. 4). For measurements in GUVs, a two-dimensional diffusion model was applied:

$$G(\tau) = \frac{1}{N_{\mathrm{ob}}}\left(1 + \frac{\tau}{\tau_{\mathrm{diff,2D}}}\right)^{-1/2}\left(1 + \frac{\tau}{S^2 \cdot \tau_{\mathrm{diff,2D}}}\right)^{-1/2} \tag{2}$$

containing the diffusion time $\tau_{\mathrm{diff,2D}}$, the particle number N_{ob}, and the structure parameter $S = z_0/w_0$, describing the stretched detection volume, as parameters. For two-focus SFCS, a cross-correlation function can be fitted to determine the radius of the beam waist w_0 using the previously calibrated distance d between the parallel scan paths.

$$G_{\times}(\tau) = G(\tau) \cdot \exp\left[\frac{-d^2}{w_0^2\left(1 + \tau/\tau_{\mathrm{diff,2D}}\right)}\right]. \tag{3}$$

The beam waist w_0 is needed to convert diffusion times into diffusion coefficients

$$D_{\mathrm{2D}} = \frac{w_0^2}{4\tau_{\mathrm{diff,2D}}}. \tag{4}$$

Evaluation of the correlation data comprises the following steps:

1. Import the appropriate model function via "File | Import model function" (*see* **Note 13**). Import "ExampleFunc_SFCS_1C_2-D_Autocorrelation.txt" for one-focus SFCS (Eq. 2) and additionally import "ExampleFunc_SFCS_1C_2D_Cross-correlation.txt"

Fig. 4 (continued) displaying the difference between the data and the fit. Note that "Trace view" displays the corresponding intensities and allows for removing outliers. With "Batch control" the user can act on multiple panels and therefore allow controlled fitting of large data sets. (**b**) Examples for additional functions: "Data range selection" restricts the number of time channels of the correlation data as used for fitting, "Background correction" accounts for non-correlated intensity to determine absolute concentrations, and "Global fit" allows for simultaneous fitting of data sets containing shared parameters in their model functions. At any time, the user can call information panels specifying the current fitting parameters and models for two-focus SFCS (Eq. 3). The imported autocorrelation and cross-correlation model functions are now available in the menu "Model | User", denoted as "2D SFCS AC" and "2D SFCS CC," respectively.

2. Import the ~.zip file containing the correlation data created by *PyScanFCS* via "File | Load single data file." As a model select the previously imported model function "User: 2D SFCS AC." For two-focus SFCS, you will be asked which correlation curves to import. First choose the autocorrelation (AC) curves and select the model "User: 2D SFCS AC." The cross-correlation (CC) curves need to be imported separately with the model "User: 2D SFCS CC." Note that the data import function associates each correlation curve with a particular model.
3. Now several tabs are shown, each containing a correlation curve corresponding to the number of intensity traces (slices of the measurement time). Open "Tools | Data range selection" to eventually narrow the window of lag times as used for fitting. Eventually apply a "Background correction" for non-correlated intensity. Check parameters (e.g., τ_{diff} and N_{ob}) that should be varied during the fitting procedure and start by clicking "Fit." Use "Tools | Batch control" to apply the same settings when fitting multiple curves (tabs) assigned to the same model. Browse through the results while checking the intensity traces with "Tools | Trace view" in a separate window. Correlation curves with unstable intensities can be removed by simply closing the tab.
4. Using "Tools | Average data," multiple correlation curves can be averaged; the new curve will be shown in a new tab and can be fitted separately. To evaluate two-focus SFCS data, it is necessary to create an average for both AC and CC functions and save each in a separate ~.csv file by "Current Page | Save data (*.csv)." Clear the session with "File | Clear session" and reimport the saved ~.csv files with their corresponding model functions "2D SFCS AC" and "2D SFCS CC" via "File | Load single data file." Since autocorrelation and cross-correlation curves share parameters, a global fit can be applied. Activate the "Tools | Global fitting" menu. In each tab, check the parameters that should be varied during the fitting procedure (τ_{diff}, N_{ob}, w_0) and start by clicking on "Gobal fit." The fit results can be used to calculate the diffusion coefficient according to Eq. 4.
5. After fitting, the panel displays the data, the parameter values and the corresponding model function in the main window. In addition, you can use "Tools | Page info," which contains a comprehensive description of the current status for each tab. Finally, you can save the entire session by "File | Save session"

or each individual correlation curve via "Current Page | Save data (*.csv)." In both cases, the fit parameters will be saved as well. More information and a full documentation of *PyCorrFit* are available online at http://pycorrfit.craban.de.

4 Notes

1. The cylindrical chamber (1 cm diameter, 2 cm height) is made out of a Teflon cap which can be screwed onto a reservoir. The cap is pierced by a pair of platinum electrodes protruding 1 cm on each side. The distance between the wires is 4 mm. The platinum wires can be connected to a power supply and are used to generate an electric field inside the chamber.
2. Unfortunately, the photon history recorder mode does not work with the newer correlator version Flex02-01D/C. Please consult www.correlator.com or use other time-correlated single-photon counting (TCSPC) cards and convert the data accordingly (*see* **Note 9**).
3. The two software packages, *PyCorrFit* and *PyScanFCS*, enable the user to derive molecular parameters from an SFCS experiment. The program *PyScanFCS* is used to extract experimental correlation curves from the photon stream recorded during scanning. The correlation curves are stored on the hard disk and can be evaluated by *PyCorrFit*, a generic front-end tool for fitting of model functions. *PyCorrFit* contains additional features exceeding SFCS applications that are described in detail in the documentation. Both programs are available from our download page (http://fcstools.craban.de). The source code (*Python*) and the executables are free software published under the GNU general public license. There are also pre-built binaries/executables for Windows and Ubuntu Linux.
4. When the osmolarity outside and inside of the GUVs differ, the GUVs are instable and may burst. Therefore, the molarity on both sides of the vesicle membrane should be the same. Due to the larger molecular weight of sucrose (inside) than glucose as contained in the glucose/sucrose mixture (outside), the negative buoyancy of the GUVs enforces their sedimentation to the glass bottom where they can be conveniently observed with an inverted microscope.
5. Turn the course and fine-focus to move the detection volume from the glass bottom into solution and verify the sudden increase in signal intensity. At the glass bottom reset the position offset to zero, then move 50 μm upward.
6. To transform diffusion times into diffusion coefficients one has to calculate $D = D_{st} \cdot \tau_{diff,st}/\tau_{diff}$ from the measured diffusion

time τ_{diff} and the standard $\tau_{\text{diff,st}}$ for which the diffusion coefficient D_{st} is known. The diffusion coefficient for Alexa488 in 10 mM Tris-HCl pH 8 measured at 22 °C is $(4.35 \pm 0.1) \times 10^{-10}$ m^2/s [11]. The focal volume can be derived by applying $V_{eff} = \pi^{3/2} S w_0^3$. Note that the axis ratio S and the diffusion time trade during the fit and thus the axis ratio is usually a poorly determined parameter. To achieve consistency, it is quite common to determine S in solution for the standard like Alexa488 ($3 < S < 8$) and to fix this value for all measurements.

7. The scan path should be checked for several angles. For example, the parallel arrangement was not maintained when scanning horizontally at 0°. Furthermore, our chosen settings in "multitrack mode" are obviously not optimal in terms of time resolution. Using other instruments, one may achieve hysteresis-free bidirectional scanning with correspondingly shorter scan cycle times.
8. The center of the scanning line should be positioned close to the membrane but outside of the GUV. Since the GUV membrane bends inward, excitation and bleaching above and below the equatorial plane is therefore minimized.
9. *PyScanFCS* can only import ~.dat files or pre-binned ~.fits files (see online documentation at http://pyscanfcs.craban.de). ~. dat files are created by the software *Photon.exe* that ships with the correlators Flex02-12D/B. If you are using a different photon history recording device, you will have to transform your data files to the ~.dat file format. The format stores the time span between two photon events in units of the system clock. The data format is as follows:
   - The value of the first byte (uint8) identifies the format of the file: 8, 16, or 32 bit.
   - The second byte (uint8) identifies the system clock, usually 60 MHz.
   - The rest of the file contains the time span between two photon events. The time unit is 1/60 MHz = 16.67 ns.
   - 8 bit format is not supported.
   - 16 bit format. Each word (2 bytes, uint16) represents a photon event. For low intensities, values of 0xFFFF may occur. In this case, the following 4 bytes (uint32) represent a photon event.
   - 32 bit format. Each dword (4 bytes, uint32) represents a photon event. Can be created from 16 bit files using *Photon.exe*. *PyScanFCS* saves ~.dat files in this format.

10. The binning resolution is an empirical value balancing computation time and spatial scanning resolution. Under conditions as used here, membrane-related photon events usually appear in 1–3 adjacent bins. Initially limiting the number of photon events to 1,000 speeds up the determination of the cycle time by the FFT.
11. In the FFT spectrum the maximum amplitude does not always reflect the correct scan cycle time. Moreover the scan periodicity as displayed by the LSM software may differ from the real physical value.
12. The remaining time span of the split traces should still exceed the diffusion times by at least a factor of 1,000.
13. *PyCorrFit* can import external, user defined model functions. An external model function must be defined and stored in a ~.txt file with a simple syntax. Examples and instructions on how to create such a model functions can be found on the download page (http://pycorrfit.craban.de). As in standard confocal SFCS, here we utilize well-known autocorrelation function derived for Gaussian detection profiles.

Acknowledgments

We thank Fabian Heinemann and Eugene Petrov for helpful discussions.

References

1. Kim SA, Heinze KG, Schwille P (2007) Fluorescence correlation spectroscopy in living cells. Nat Methods 4:963–973
2. Liu P, Ahmed S, Wohland T (2008) The F-techniques: advances in receptor protein studies. Trends Endocrinol Metab 19:181–190
3. Ries J, Schwille P (2006) Studying slow membrane dynamics with continuous wave scanning fluorescence correlation spectroscopy. Biophys J 91:1915–1924
4. Ruan Q, Cheng MA, Levi M, Gratton E, Mantulin WW (2004) Spatial-temporal studies of membrane dynamics: scanning fluorescence correlation spectroscopy (SFCS). Biophys J 87:1260–1267
5. Heinemann F, Betaneli V, Thomas FA, Schwille P (2012) Quantifying lipid diffusion by fluorescence correlation spectroscopy: a critical treatise. Langmuir 28(37):13395–133404
6. Ries J, Chiantia S, Schwille P (2009) Accurate determination of membrane dynamics with line-scan FCS. Biophys J 96:1999–2008
7. Ries J, Schwille P (2008) New concepts for fluorescence correlation spectroscopy on membranes. Phys Chem Chem Phys 10:3487–3497
8. Dertinger T, Pacheco V, von der Hocht I, Hartmann R, Gregor I, Enderlein J (2007) Two-focus fluorescence correlation spectroscopy: a new tool for accurate and absolute diffusion measurements. Chemphyschem 8:433–443
9. Ries J, Yu S, Burkhardt M, Brand M, Schwille P (2009) Modular scanning FCS quantifies receptor-ligand interactions in living multicellular organisms. Nat Methods 6:643–645
10. Worch R, Bökel C, Höfinger S, Schwille P, Weidemann T (2010) Focus on composition and interaction potential of single-pass transmembrane domains. Proteomics 10:4196–4208
11. Petrášek Z, Schwille P (2008) Precise measurement of diffusion coefficients using scanning fluorescence correlation spectroscopy. Biophys J 94:1437–1448

Chapter 30

Implementation and Application of Pulsed Interleaved Excitation for Dual-Color FCS and RICS

Jelle Hendrix and Don C. Lamb

Abstract

Pulsed interleaved excitation (PIE) employs pulsed laser sources that are interleaved such that differentially colored fluorophores can be measured or imaged quasi-simultaneously in the absence of spectral crosstalk. PIE improves the robustness and reduces data analysis complexity of many fluorescence techniques, such as fluorescence cross-correlation spectroscopy (FCCS) and raster image cross-correlation spectroscopy (ccRICS), two methods used for quantitative investigation of molecular interactions in vitro and in living cells. However, as PIE is most often used for fluorescence fluctuation spectroscopy and burst analysis experiments and utilizes time-correlated single-photon counting detection and advanced optoelectronics, it has remained a technique that is mostly used by specialized single-molecule research groups. This protocols chapter provides an accessible overview of PIE for anyone considering implementing the method on a homebuilt or commercial microscope. We give details on the instrumentation, data collection and analysis software, on how to properly set up and align a PIE microscope, and finally, on how to perform proper dual-color FCS and RICS experiments.

Key words Pulsed interleaved excitation, Raster image correlation spectroscopy, Time-correlated single-photon counting, Fluorescence cross-correlation spectroscopy, Fluorescence correlation spectroscopy, Confocal laser scanning microscopy, FCS, FCCS, RICS, ccRICS, CLSM

1 Introduction

Fluorescence cross-correlation spectroscopy (FCCS) and raster image cross-correlation spectroscopy (ccRICS) are very powerful fluorescence fluctuation spectroscopy (FFS) techniques for investigating dynamics, concentrations, and molecular interactions of fluorescently labeled biomolecules in vitro and in living cells [1, 2]. FCCS is the dual-color extension of fluorescence correlation spectroscopy (FCS), the technique with which dynamics and absolute concentrations of diffusing molecules are determined from point fluorescence fluctuation spectroscopy measurements by temporal autocorrelation analysis [3]. ccRICS is the analogous extension of the raster image correlation spectroscopy (RICS) method,

Yves Engelborghs and Antonie J.W.G. Visser (eds.), *Fluorescence Spectroscopy and Microscopy: Methods and Protocols*, Methods in Molecular Biology, vol. 1076, DOI 10.1007/978-1-62703-649-8_30,

the technique with which dynamics and absolute concentrations of diffusing molecules are determined from fluorescence fluctuation imaging experiments by spatial autocorrelation analysis [4, 5]. However, due to spectral crosstalk of commonly used fluorophores for FFS, FCCS and ccRICS suffer from a relatively low sensitivity in terms of measuring interactions. Under some circumstances, it is possible to correct experimental data a posteriori for crosstalk [6, 7]. Quantitative analysis of molecular interactions inside living cells with FCCS in the presence of considerable crosstalk has been shown to be possible, but requires complex and careful data analysis [8, 9]. Therefore, it is desirable to avoid crosstalk altogether to reduce complexity and increase robustness of data analysis.

Pulsed interleaved excitation (PIE) was based on the earlier work of Kapanidis and co-workers [10] and was originally developed to avoid problems of spectral crosstalk during FCCS experiments and to provide stoichiometry information from burst analysis experiments [11]. In addition, PIE can be used to increase the robustness of many quantitative fluorescence techniques. For example, it allows the Förster resonance energy transfer (FRET) efficiency to be calculated directly from FCCS experiments without a priori knowledge of the labeling degree [11], it increases the accuracy of single-pair FRET experiments with multiparameter fluorescence detection [12], it allows for quantitative fluorescence fluctuation and lifetime imaging [13], and it is commonly used in dual-focus FCS for accurate measurements of the diffusion coefficient [14]. At present, many biological systems have been studied using PIE; for an overview, *see* ref. 15.

PIE is typically performed in combination with time-correlated single-photon counting (TCSPC) detection, a technique in which the time delay of a detected fluorescence photon is measured relative to the laser excitation pulse (of a duration typically $\lesssim$100 ps) with extremely high temporal resolution [16]. When using TCSPC, fluorescence is typically measured on two timescales: the macroscopic timescale, or "macrotime", that is, the arrival time of the photons with respect to the start of the experiment with a time resolution of typically 30–100 ns, and the microscopic timescale, or "microtime," which is the time between the excitation pulse and emission of the detected photons with a time resolution of a few ps (*see* Fig. 1a). A histogram of the photon microtimes in a TCSPC experiment can be used for analyzing the fluorescence lifetime (*see* Fig. 1b) while the macrotime traces can be used for, inter alia, autocorrelation analysis (*see* Fig. 1c). When TCSPC is combined with confocal laser scanning microscopy (CLSM), confocal images can be constructed from the macrotimes (*see* Fig. 1d). The "trick" used in PIE lies in the fact that, in addition to just employing two (or more) pulsed lasers and two (or more) detectors, the lasers are perfectly synchronized with a time delay between them that is smaller than the repetition period of

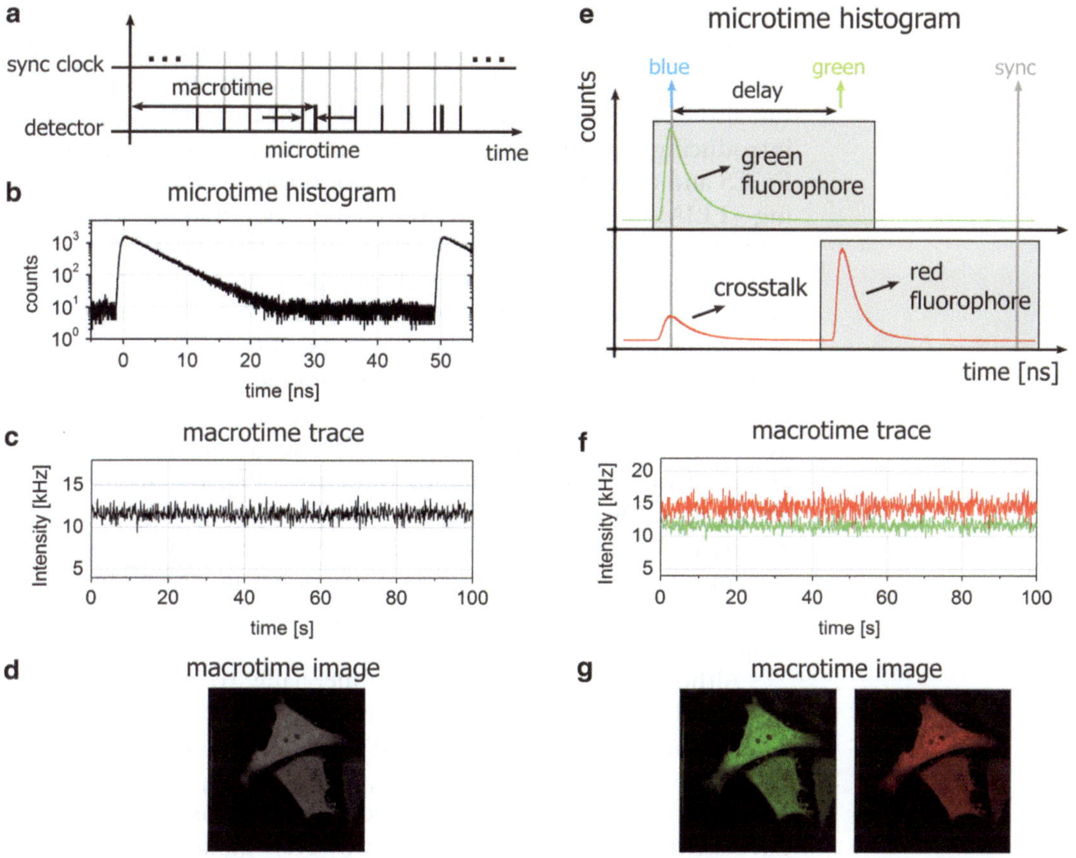

Fig. 1 Illustration of TCSPC and PIE. (**a**) The basic principle of TCSPC is shown where both the macro- and microtime of each detected photon are recorded in the experiment. (**b**) By histogramming the microtimes of all detected photons, the fluorescence decay of the fluorophore is observable. (**c**) By consecutive binning of the macrotimes of the detected photons, an intensity or macrotime trace is obtained. (**d**) When recording data with a CLSM, the macrotimes of the TCSPC data can be binned into image pixels and in this way the confocal image can be displayed. (**e**) Illustration of a two-color excitation two-color detection PIE experiment with TCSPC detection. The selected time gates for further analysis are hatched in *gray*. Species-specific macrotime traces (**f**) and images (**g**) can be calculated from the macrotimes of the photons within the selected time gates. As there is no crosstalk in the macrotime images, the clear signal in both channels indicates that both fluorophores are present in the imaged cells

the lasers, i.e., they are interleaved. This creates a new, temporal separation of emitted photons in addition to their common spectral and spatial separation. Indeed, as illustrated in Fig. 1e for a two-color PIE microscope, crosstalk of the green fluorophore into the red detection channel is temporally well separated from direct emission of the red fluorophore when the delay between the two lasers is set appropriately. By setting defined time gates [17] in the microtime histograms, species-specific macrotime traces and/or images can be constructed (*see* Fig. 1f, g).

Nowadays, the PIE technique is accessible to a large scientific community. Commercial ready-to-use point-measurement PIE microscopes exist that allow i.a. FCCS experiments with PIE. PIE can also be integrated into commercial CLSM systems (*see* **Note 1**), introducing quantitative fluorescence fluctuation analysis (i.a. FCCS and cross-RICS) and multicolor fluorescence lifetime imaging (FLIM) capabilities into the system. Upgrading an existing CLSM can be advantageous because it can be both time and cost saving when the microscope is already present. Moreover, most, if not all, functionalities of the microscope can subsequently be used in combination with PIE. Different TCSPC hardware manufacturers provide ready-to-use hardware upgrade "kits" for commercial CLSMs and additionally offer excellent technical support, avoiding the need for experience with assembling advanced optics and electronics. Finally, building ones' own PIE system also has its advantages: the microscope can be specifically designed to the researchers' needs, usually costs significantly less than a commercial system, and can be modified and upgraded in a straightforward fashion. It does, however, require expertise in assembling microscopes.

Although PIE is available commercially, to date it has been employed mostly by specialized single-molecule research groups. This is due, in part, to the required expertise in fluorescence correlation spectroscopy (FCS), TCSPC, and microscopy, if a PIE microscope is to be built ab initio. To make PIE more easily accessible, we have written this chapter as a step-by-step guide for the correct implementation and application of PIE, be it on a completely homebuilt setup or on an upgraded commercial CLSM. We give a comprehensible overview of the instrumentation, data collection, and analysis software that is suitable for PIE, discuss how to properly set up the excitation and detection equipment, give advice how to align a PIE microscope for optimal performance, and finally, give a detailed protocol for doing point-based and image-based measurements and subsequent PIE-FCCS and PIE-RICS analyses. Our hope is that by reading this chapter, nonexpert research groups interested in PIE will finally take the step to implement it, thereby profit from the benefits, possibilities, and robustness that PIE offers.

2 Materials

2.1 Instrumentation

1. Lasers: For pulsed interleaved excitation, pulsed laser sources are required. In principle, picosecond (ps)-pulsed lasers with high output stability and preferentially collimated, TEM_{00} output are well suited for PIE (*see* **Note 2**). Most importantly, all lasers that are used simultaneously need to be synchronized (*see* Subheading 3.1). It is possible to use continuous-wave

lasers and modulate the intensity using an acousto-optically modulated (AOM) as was performed in the original report of PIE, but at the cost of losing the fluorescence lifetime information [11]. Nowadays many suitable lasers are available, such as (but not limited to):

- Externally triggered ps-pulsed quasi-monochromatic diode lasers covering a wide range of the visible spectrum.
- Fixed-frequency or externally triggered ps-pulsed white-light supercontinuum lasers providing ultimate flexibility in the choice of excitation wavelengths.
- Fixed-frequency fs–ps-pulsed tunable fiber lasers covering the visible spectrum and providing extremely high spectral purity.
- An AOM-modulated CW laser [11].

As the lasers need to be synchronized, only one fixed-frequency laser can be used in a PIE microscope at the most, while all other lasers should be synchronized by external triggering. The laser power requirements for PIE are similar as for normal FCS experiments, i.e., roughly 1–100 μW per laser, as measured at the sample, should be sufficient for most applications. Because of losses in combining the lasers, coupling them into the single-mode fiber and then into the sample, the power output of the laser should be significantly more than 100 μW. For most applications, an average laser output power of 1 mW is more than sufficient.

2. Electronic delay box: The electronic delay box introduces a variable delay in an electrical signal. We use electronic delay boxes to both shift the laser pulses with respect to each other and adjust the arrival time at the TCSPC hardware of the signal from the detectors. Delay boxes are not necessary for PIE, but are extremely convenient. In our case, the delay boxes are homebuilt and can delay signal in steps of 1 ns from 1 to 32 ns.
3. Attenuators: Lasers can be attenuated manually by partially blocking the beam path (*see* **Note 3**) or with a neutral density filter or electronically with an acousto-optical tunable filter (AOTF). An AOTF has the advantage that different lasers can be attenuated independently and in an automated fashion.
4. Single-mode (polarization maintaining, *see* **Note 4**) optical fibers: Different lasers should be combined into a single fiber (*see* **Note 5**) and a broadband achromatic collimator should be used to collimate the light into a free beam again before directing the light to the microscope (*see* **Note 6**). Optical fibers can also be used to delay pulses optically (*see* **Note 7**).
5. Optics: In general, highest-quality, antireflection-coated optics should to be used. More specifically, silver mirrors provide high reflection in the visible spectrum but are difficult to clean.

Achromatic doublet lenses minimize optical aberrations. Di-/polychroics, excitation and emission filters should have sharp cut-in/cut-out and a high degree of reflection and/or transmission (*see* Fig. 2b). Thicker substrates for the polychroics are desirable to avoid astigmatism in the beam (*see* **Note 8**). Excitation filters are only needed if the laser is not monochromatic. Emission filters additionally should block all excitation wavelengths by at least OD6 and higher if measuring on surfaces where a higher fraction of laser light is reflected into the detection path.

6. Fast mirror scanner: When scanning is desired, scanning mirrors will need to be incorporated into the system. Different types exist, but for most imaging applications, two closely spaced galvanometric mirrors, or "galvo," are sufficient. Depending on the manufacturer and controller electronics used, the position of the galvo can be controlled via an analog signal or digitally. A feedback control can be used for calibration purposes, but is generally not needed during imaging experiments. Of importance, a resonant mirror allows for higher imaging speeds, but does not allow specific positioning of the point spread function (PSF) without physically moving the sample.
7. Kepler-type telescope: To maximize the scanning speed, the size of the galvo mirrors needs to be kept to a minimum. Hence, it is often desirable to expand the beam after the scanner to fill the back aperture of the objective (*see* **Note 9**). In addition, the focal plane of the mirrors needs to be imaged on the back aperture of the objective to ensure the excitation beam is constant during scanning. To do this, a telescope is placed between the scanner and the objective. We use two achromatic planoconvex lenses mounted in a cage for stability (*see* **Note 10**).
8. Microscope base: For a homebuilt PIE microscope, any microscope base can be used. It is important that 100 % of the light can be directed to the port where the detection optics are mounted. Microscopes with more available input ports also allow more functionality to be incorporated into the system.
9. Objective lens: The objective is the heart of the confocal microscope and only the best quality objectives should be used. High-NA water (NA $\geq$ 1.2) or oil (NA $\geq$ 1.3) immersion objectives with corrections for chromatic and/or spherical aberration (*see* **Note 11**) and with a correction collar for coverslip thickness and/or refractive index (n-) mismatches are desirable (*see* **Note 12**).
10. Immersion solution: For long measurements using a water-immersion objective, nonvolatile immersion liquids with $n = 1.33$ should be used.

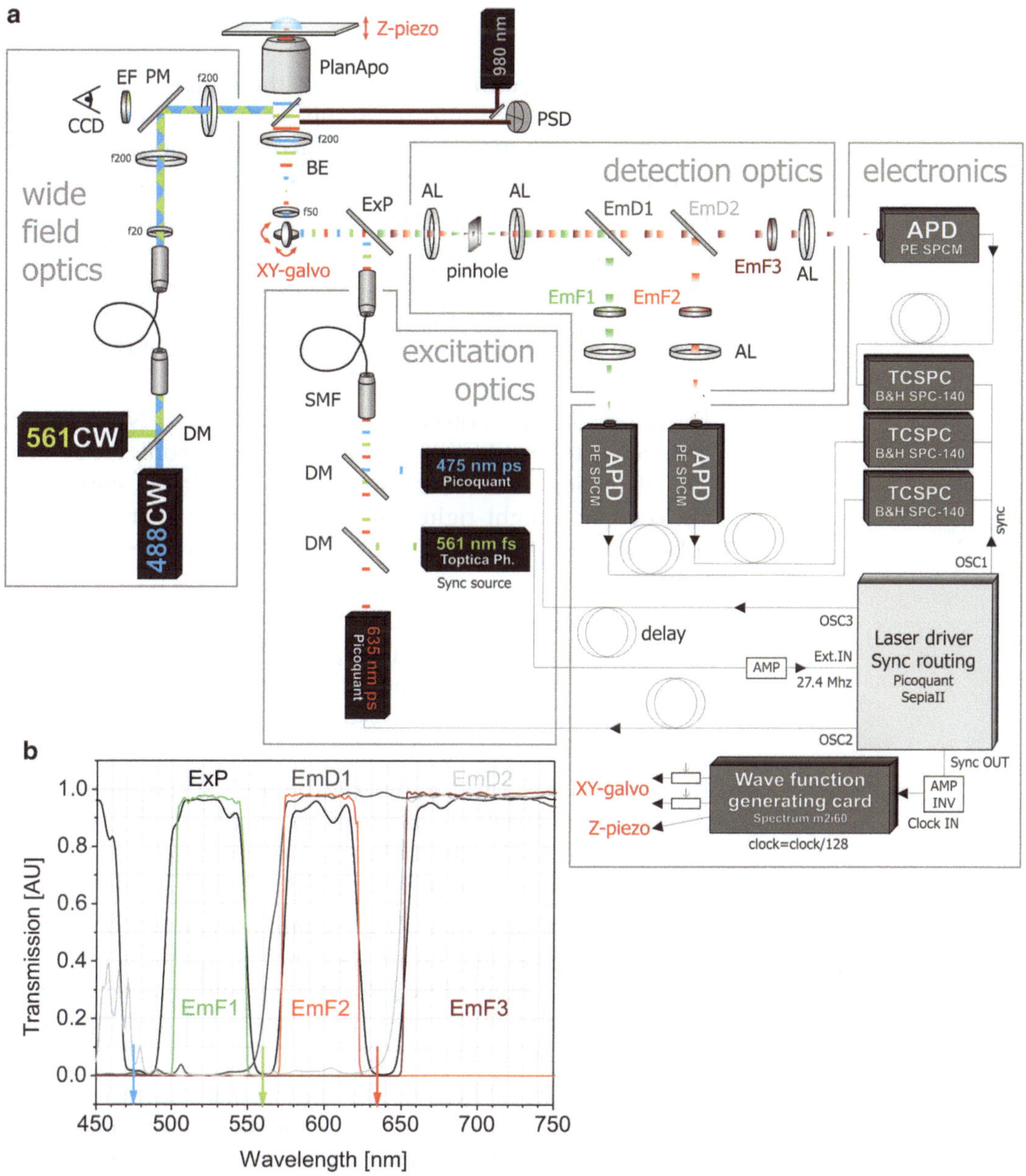

Fig. 2 Schematic illustration of our 3-color PIE microscope. (**a**) A frequency-doubled Erbium-doped fiber laser pulsing at 27.4 MHz determines the clock of the whole system. A fraction of its 561-nm output is converted to an electrical signal and fed to the laser driver of the 475-nm and 635-nm lasers that are consequently synchronized to the same frequency. A variable electronic delay box delays the latter two lasers with respect to the 561-nm laser for proper 3-color PIE. Three independent TCSPC devices, each connected to an avalanche photodiode (APD) detector, and an arbitrary wave-function generating PCI card controlling the galvo and a Z-actuator, are synchronized with the master clock through the laser driver. Abbreviations: ExP = excitation polychroic mirror, EmF1–3 = emission filters, AL = achromatic lens, BE = beam expander, DM = dichroic mirror, EmD1-2 = emission dichroics mirror, SMF = single-mode fiber. (**b**) Spectrum of the optical elements for filtering on the PIE microscope. The excitation polychroic efficiently reflects the three laser lines depicted with an *arrow*, while maximally transmitting the emission light. After the pinhole, the emission dichroics furthermore separate the emission into three large spectral windows, and the emission filters define the specific spectral detection windows

11. Z-actuator: To control the focus of the sample, a Z-actuator can be incorporated into the setup. Either the sample or the objective can be mounted on a Z-actuator. The Z-actuator is usually a piezoelectric element with subnanometer resolution.
12. Pinhole: A 50–100-μm diameter pinhole works best for most applications. For multicolor detection, a single pinhole is sufficient for defining the confocal volume and the different colors can be separated after the pinhole. In general, a larger pinhole results in a higher apparent molecular brightness at the cost of a slightly larger PSF (*see* **Note 13** and ref. 18).
13. Detectors: Single-photon counting detectors (*see* **Note 14**) are connected to the TCSPC device via an electronic delay box. The detection path after the pinhole should be completely enclosed in a light-tight container of some sort (a box or tubes) to shield the sensitive area of the detectors from stray light.
14. TCSPC device: To optimally use the lifetime information available from pulsed excitation, TCSPC detection is desired. Preferably, each detector should be connected to an independent TCSPC device to avoid dead time related artifacts (*see* **Note 15**).
15. Router: If necessary, a router can be used at low count rates. A router allows multiple detectors to be connected to the same TCSPC device. This device sends digital detector routing bits along with the photons to the TCSPC device (*see* **Note 15**).
16. Trigger: To ensure that all electronics are synchronized, an external trigger is needed to start the measurement. This can be done with a simple DAQ device that sends out a TTL pulse when the experiments begins, which ensures all TCSPC devices start measuring simultaneously.
17. Widefield optics: When performing cellular experiments, it is useful to have a widefield module on the microscope. In this way, cellular fluorescence can be observed directly through the eyepiece or on a camera, making it quick and easy to select cells with appropriate expression levels for the confocal measurements.

2.2 Sample Preparation

1. Sample holder: To avoid evaporation of the sample at small volumes and to protect the sample from contaminations, a closed-lid sample holder is preferred. Be sure the coverslip thickness matches the specifications of the objective (*see* **Notes 16** and **17**).
2. Fluorophores for PIE: Suitable properties of fluorophores used in PIE experiments are good photostability, no significant spontaneous degradation, little or no fast photophysics, negligible nonspecific absorption and good solubility. Many organic

dyes exist that possess all these properties. If photostability or fast photophysics are a concern, such as when using fluorescent proteins, combining PIE with fast CLSM imaging provides an excellent alternative to point measurements [13].

3. Reference dyes: To calibrate the system and check the sensitivity, reference dyes with known diffusion coefficients (*D*) should be used. Be aware that the *D* of dyes depends significantly on their functionalization [19]. In our experience, ATTO488 carboxylic acid (ATTO-TEC GmbH, Siegen, Deutschland), Rhodamine B and ATTO655 carboxylic acid (ATTO-TEC) perform very well as reference dyes for 475-nm, 561-nm and 635-nm excitation respectively (*see* **Note 18**).
4. Buffers: Care needs to be taken that the buffers used exhibit as little autofluorescence as possible. This can be particularly challenging when the buffer contains a number of salts. In vitro experiments can be performed with sample volumes as small as 10–20 μL (*see* **Note 19**).

2.3 Acquisitioning and Analysis Programs

1. Acquisitioning software: For a homebuilt PIE microscope, no special acquisitioning software is needed beyond the manufacturer's software to control the TCSPC hardware when only point-based measurements are performed. When imaging with PIE is to be performed on a commercial CLSM, usually the manufacturer's software, together with the software controlling the TCSPC hardware are used to acquire images. For our homebuilt CLSM, we use an acquisitioning program written in the C# programming language to control and synchronize the galvo and TCSPC cards and to provide a user interface for displaying PIE images in real time (*see* **Note 20**).
2. Point-analysis software: Packages for analysis of experiments with PIE are available from different manufacturers. We use an *in-house* developed program, PIE Analysis with Matlab or simply "PAM." With PAM, the following analysis methods can be applied to PIE data: individual and global FCS/FCCS analysis (*see* Subheading 3.4), fluorescence lifetime correlation spectroscopy [20], brightness analysis by photon counting histogram [21] or cumulant analysis [22], pair correlation analysis [23], line/circle scanning FCS [24, 25], single-pair FRET by burst analysis and multiparameter fluorescence detection (MFD) [26, 27], photon distribution analysis [28], fluorescence lifetime analysis and time-resolved anisotropy analysis. In addition, various Monte-Carlo simulation algorithms are available.
3. Image-analysis software: We also have an *in-house* developed program for CLSM image analysis, called Microtime Image Analyzer or "MIA." With MIA, the following analysis methods

can be applied to PIE imaging data: individual, global and lifetime-weighted raster image correlation spectroscopy [29] (*see* Subheading 3.4), number and brightness analysis [30], phasor analysis [31] and fluorescence lifetime imaging.

3 Methods

3.1 Setting Up PIE Excitation Equipment

The first step in using a PIE microscope is setting up the excitation equipment according to the type and number of fluorophores under investigation as well as their fluorescence lifetime. Be sure to be well acquainted with laser safety before proceeding with any steps involving lasers.

1. Master clock: A master clock is needed to synchronize the instrumentation. This is typically taken as the frequency at which one of the pulsed lasers is run. We illustrate in Fig. 2 how the frequency of a fixed-repetition rate laser is used as the master clock for all other lasers, as synchronization source for the TCSPC hardware, as well as for the wave-function generating card controlling the galvo and Z-actuator. The system shown in Fig. 2 is a schematic of our actual three-color PIE microscope.
2. Frequency: The number of excitation lines determines the repetition period (and thus frequency) of the lasers (*see* Fig. 3a and **Note 21**), and conversely, a fixed-frequency laser determines the maximum number of fluorophores that can be imaged optimally and simultaneously with PIE. Keep in mind that the molecular brightness (in Hz) will decrease at lower repetition frequencies [11].
3. Delay: The delay between laser pulses should be set according to the fluorescence lifetime of the previously excited fluorophore, as illustrated in Fig. 3b for two-color excitation, two-color detection. For externally triggered lasers, the delay can be set with an electronic delay box between the laser driver and laser head. For lasers that cannot be triggered electronically, a delay can be set with an optical fiber (*see* **Note 7**) or a free-beam delay line.
4. Pre-/post-fiber separation: After adjusting the delay time, combine all lasers into a single fiber (*see* **Notes 5** and **22**). A factory prealigned beam combiner can also be used to combine fiber-coupled lasers into one single-mode fiber when the excitation path is not changed regularly. This approach has the advantage that no pre-fiber alignment is necessary once the excitation path is set up.

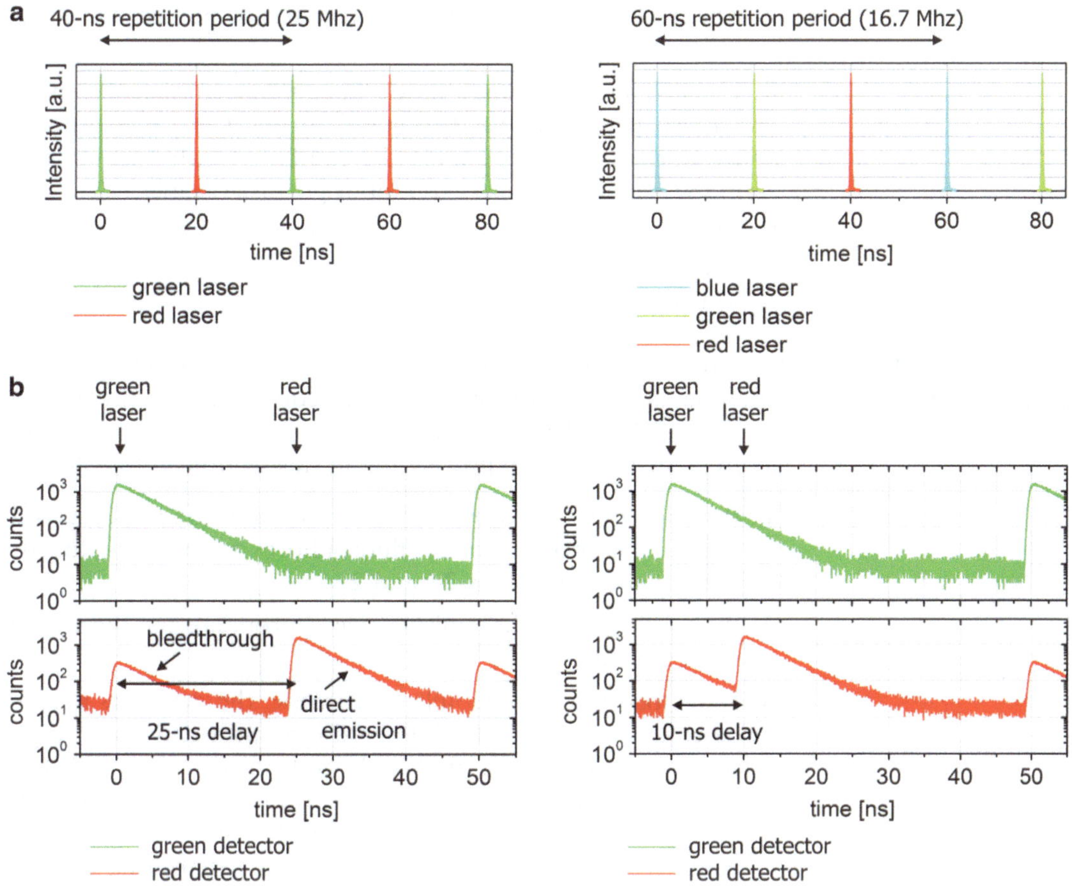

Fig. 3 Laser settings for PIE. (**a**) Schematics showing the synchronized, interleaved laser pulses on the nanosecond timescale. The more excitation lines in a single PIE experiment, the lower the repetition frequency needs to be for each, as illustrated for a two-color (*left*) and three-color (*right*) PIE setup. (**b**) Since TCSPC detection is used, the fluorescence lifetime of each fluorophore is available in a PIE experiment, as illustrated for a two-color excitation, two-color detection PIE microscope. The green fluorophore, excited with the green laser, will emit primarily in the green detection channel with a residual, though non-negligible spectral bleedthrough into the red detection channel. The red fluorophore, excited with the red laser, will emit only in the red detection channel. The delay of different lasers determines the performance of the PIE setup. By setting the delay of the red laser correctly (25 ns, *left panels*), bleedthrough and direct emission of the red fluorophore can be perfectly separated temporally, while, for a too short delay of the red laser (10 ns, *right*), temporal overlap of the green signal after green excitation into the red detection channel with red excitation will occur

5. Power: Depending on the experiment, a laser power between 0.1 and 100 μW in the sample is needed. Use attenuators to achieve the proper laser power (*see* **Note 23**).

There are some important considerations related to the excitation module when upgrading a commercial CLSM to PIE.

6. For convenience and safety, lasers for PIE should be combined in a single, closed laser combining unit.

7. The necessary excitation polychroic has to be installed in the microscope.
8. Usually, an extra single-mode fiber input is available on the CLSM that can be used to couple in the PIE lasers. Alternatively, the PIE lasers can also be combined with existing lasers in the same fiber, if no extra input port is available.
9. The timing (master clock, frequency, delay) and power of the PIE lasers is usually controlled independently of the CLSM software.

3.2 Setting Up PIE Detection Equipment

Next, the detectors and TCSPC hardware need to be set up for PIE experiments. It is important to be well acquainted with the dos and don'ts of TCSPC hardware, so we suggest referring to the manufacturers' manual before proceeding in this section.

1. Emission optics: Choose and install the emission optics according to the spectra of the fluorophores to be used for the experiments (*see* **Notes 24** and **25**).
2. Pinhole: Choose the pinhole size appropriate for the experiment (*see* **Note 13**).
3. Detectors: On a homebuilt microscope, the detectors need to be connected to the TCSPC hardware, provided their output signal has the correct amplitude and polarity. Also the delay between photon detection and recording of the TTL pulse by the TCSPC card should be similar for all detectors. On a commercial CLSM with photon counting detectors, the output of the detector can be split with one input directly connected to the external TCSPC hardware and the other to the detector input of the CLSM. When appropriate detectors are not available within the CLSM, a fiber-based output of the LSM unit can be used to guide the emission light to a light-tight single-photon counting detection box that carries the emission optics, detectors and TCSPC hardware.
4. Galvo synchronization: On a homebuilt microscope, the TCSPC and galvo hardware can usually be controlled *in sync* by letting the master clock control the galvo, as is the case on our PIE setup (Fig. 2). On a commercial CLSM, the galvo operates independently from the PIE hardware, creating the need for another means of synchronizing the detected signal and location of the excitation spot. Most often, the galvo hardware transmits a digital signal to the TCSPC hardware containing pixel, line and/or frame bits during image scanning. These bits can be written directly into the recorded photon files as pseudo-photons, which allows a posteriori histogramming of photons into the individual pixels in an image. For more information regarding synchronization of the galvos, it is advisable to contact the manufacturer of the microscope (*see* **Note 27**).

5. Count rate: Fluctuation techniques analyze fluctuations in fluorescence intensity and, as such, are performed at very low concentrations. Under normal conditions, the single-to-noise ratio of a FCS experiment depends on the molecular brightness of the fluorophore but is independent of concentration within certain limits [11]. Hence, it is not necessary to measure at high count rates (>500 kHz). For some applications, such as accurate fluorescence lifetime or brightness measurements, it is even recommended to measure at low (1–100 kHz) count rates (*see* **Notes 15** and **27**).
6. PIE channels: Because of the microtime information available with TCSPC detection, analysis of PIE data can be performed with subsets of photons in a single detection channel. This is done by setting a defined microtime gate or "PIE channel" [17]. PIE channels add an extra dimension for data analysis (*see* Subheading 3.4). At high count rates, the PIE channel can usually be set to encompass the complete fluorescence decay of a fluorophore (Fig. 4a). With two-color PIE, PIE channels allow green fluorophore crosstalk to be separated from red fluorophore direct emission. At lower count rates, scattering will likely contribute significantly to the signal; a narrower time gate can then be used to increase the SNR of the measurement by omitting the scattering photons from the PIE channel (Fig. 4b). Finally, since different detectors are usually used for PIE, it can be beneficial to simply combine the photons from different PIE channels (Fig. 4c). For example, the crosstalk of the green fluorophore on the red detector can be combined with its direct emission on the green detector to increase the SNR when direct excitation of the red fluorophore with green light is insignificant.

3.3 Aligning the Microscope

Any good PIE experiment starts with a well-aligned microscope. Here we list the important steps that need to be completed before experiments can be performed accurately and with the highest sensitivity and signal-to-noise.

1. Choose which objective will be used for the experiments and from now on, use only this objective (*see* **Notes 11**, **12**, and **16**).
2. Pre-fiber excitation alignment: Measure the amplitude and variance of the fiber output intensity for each laser independently with a sensitive power meter. When the beam is not well aligned into the fiber, the variance of the laser intensity will be high. If necessary, use two independent mirrors per laser to guide the respective laser beam into the fiber (*see* **Note 28**).
3. Post-fiber excitation alignment: Load an empty sample holder, focus at the top of the coverslip and observe the pattern of the

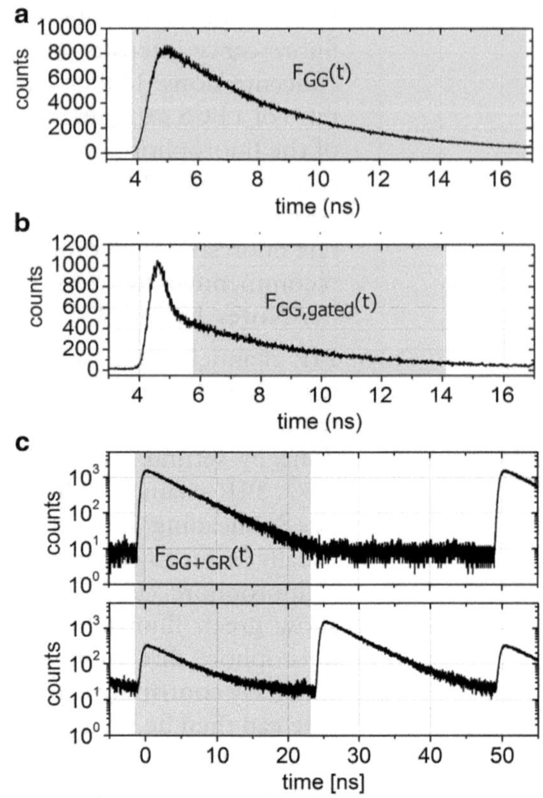

Fig. 4 The use of microtime PIE channels. (**a**) In samples with high count rates, the microtime gate can be set to encompass the complete fluorescence decay. (**b**) At lower count rates, a quantitative correlation analysis can be performed by omitted photons from scattering using a narrower time gate. (**c**) Photons from two different detectors can be combined into the same time gate, for example, when they are known to originate from the same fluorophore (adapted from ([13])

reflected excitation light. When the reflection does not remain perfectly centrosymmetrical as the objective is moved up or down, realign the post-fiber excitation part with two mirrors.

4. Post-fiber detection alignment: Load a nanomolar solution of a reference dye (*see* **Notes 12**, **18**, and **29**) and analyze the temporal autocorrelation function (ACF). Because PIE is used, different reference dyes can even be mixed and measured simultaneously. Characterize the alignment of the detection path by calculating the lateral and axial radius of the PSF, as well as the apparent molecular brightness of the fluorophore. For the same sample, sample holder, objective lens and post-fiber laser power, these parameters should remain constant over time. If necessary, realign the pinhole and/or detectors (*see* **Note 30**). If detector alignment does not work, perform a rough detector alignment (*see* **Note 31**).

5. Fine detector alignment: Because fluorescence lifetimes are commonly measured with PIE, an important alignment step is optimizing the instrument response function (IRF) of the system. For many detectors, the width of the IRF depends on where the photons impinge on the detector. To measure the IRF, strongly attenuate the laser (~OD6 or more), load a water sample and focus in solution. Block the laser completely, remove the emission filter in front of the detector, and start a continuous 1-s TCSPC measurement. Slowly unblock the laser while monitoring the signal. Adjust the laser power to be able to accurately judge the width of the IRF and adjust the detector position to obtain the sharpest IRF.
6. Telescope alignment: The telescope should be aligned so that the excitation beam is centered in the field of view, goes through the center of the optics, and is not deviated by the telescope. In addition, the focal plane in between the galvo mirrors should be imaged via the telescope onto the back focal plane of the objective (which does not necessarily coincide with the physical aperture at the back of the objective). After the telescope is aligned, two tests should be performed: (a) the size and shape of the PSF should be measured as a function of the *x*- and *y*-mirror position (*see* **Note 32**) and (b) the scan range of the individual mirrors should be calibrated to correspond to the actual movement of the beam (*see* **Note 33**).

We have included parameters obtained with reference dyes on two of our microscopes (the small PSF CLSM from Fig. 2 and a large PSF MFD microscope). These values are setup-specific and should therefore not be taken as absolute references, but can be used as a guideline when building a PIE setup (*see* Table 1).

3.4 Dual-Color Point Measurements and Imaging with PIE

When the excitation and detection modules are set up and the microscope is aligned, experiments can be performed.

1. Determine the optimal laser power. The laser power should be set to obtain a reasonable fluorophore brightness (*see* **Notes 3, 13,** and **34**). For most experiments, a laser power of 1–10 μW per laser in the sample yields satisfactory results. However, studying the laser power dependence of any fluorophore a priori to determine its photophysical behavior on the PIE microscope is strongly recommended and is common practice for any fluctuation technique.
2. Load the sample (*see* **Notes 16**, **17**, **19**, and **29**) and shield the area around the objective front aperture from potential surrounding background light.
3. Focus appropriately; depending on the measurement time and a possible *n*-mismatch (*see* **Notes 12** and **35**).

Table 1
Typical PSF size values and dye brightness, as measured with point-FCS using PIE

PIE-CLSM					PIE-MFD			
Dye	**λ_{exc} (nm)**		**Water**	**Oil**	**Dye**	**λ_{exc} (nm)**		**Water**
ATTO488	475	ω_r (μm)	0.20	0.14	ATTO488	480	ω_r (μm)	0.50
		ω_z (μm)	1.00	0.85			ω_z (μm)	2.50
		ε (kHz/μW)	10.0	17.5			ε (kHz/μW)	1.0
ATTO565	561	ω_r (μm)	0.24	0.17	ATTO532	532	ω_r (μm)	0.50
		ω_z (μm)	1.20	1.10			ω_z (μm)	2.50
		ε (kHz/μW)	18.1	18.0			ε (kHz/μW)	2.9
ATTO655	635	ω_r (μm)	0.25	0.18	ATTO655	635	ω_r (μm)	0.50
		ω_z (μm)	1.20	1.10			ω_z (μm)	2.50
		ε (kHz/μW)	5.6	6.0			ε (kHz/μW)	0.5

The PIE-CLSM (Fig. 2) uses a small PSF, and has a repetition rate of 27.4 MHz. The PIE-MFD setup uses a large PSF, and has a repetition rate of 26.7 MHz. The water objective is a Nikon CFI Plan Apo VC 60XWI (NA 1.2) and the oil objective is a Nikon CFI Apo TIRF 100X Oil (NA 1.49). The average peak molecular brightness ε [39] was measured under non-saturating conditions [41]

4. For a point measurement, select the position of the galvo (if present) and the measurement time.
5. For an imaging experiment, choose the offset and range of the galvo, the frame time (or scan speed) and the number of frames.
6. Start the measurement. The TCSPC hardware will record the microtime and macrotime at the arrival of each photon and possibly also a detector routing bit. If imaging was performed on a commercial CLSM, the TCSPC hardware may also save the pixel-, line- and/or frame bits into the photon files, which can be used later on to calculate the images.

3.5 Dual-Color FCS and RICS Analysis

In this protocol, we will focus on FFS-type analyses on PIE data, in particular dual-color FCS (or FCCS) and dual-color RICS (or ccRICS). Many different types of analyses can be performed, but this is beyond the scope of this protocol (for an overview of what our software is capable of, *see* Subheading 2.3).

1. The difference between FFS analysis with and without PIE is, to put it simply, the availability of the PIE channels. As we will illustrate with some concrete experiments, in addition to allowing crosstalk-free FFS experiments, PIE channels allow the experimenter to question obtained FFS data in very useful ways, for example: What is the concentration of free and interacting molecules? What is the binding affinity? Is the green fluorophore exhibiting FRET towards the red fluorophore? Can the FRET efficiency be calculated from FFS data?

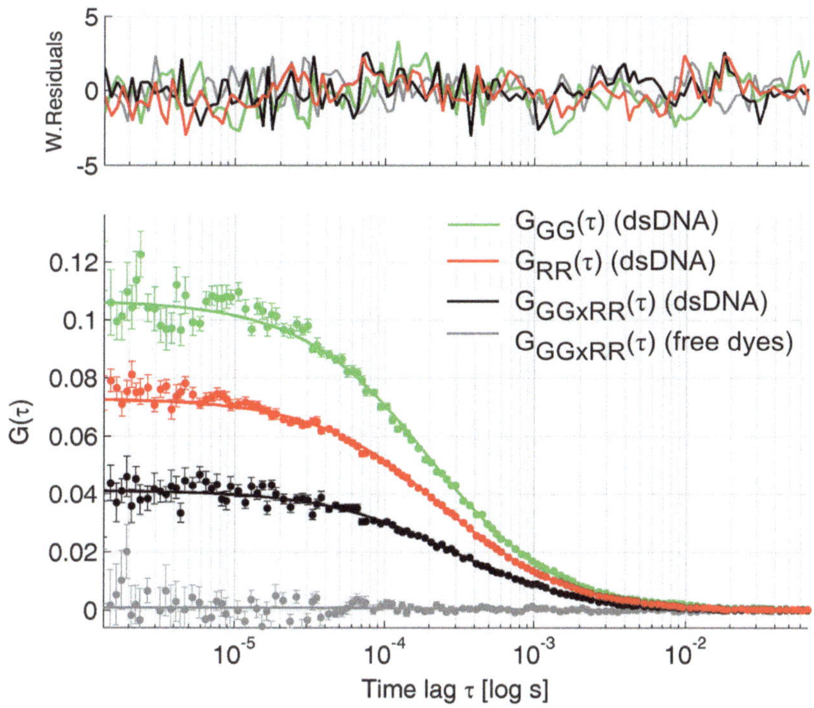

Fig. 5 PIE-FCCS analysis of a dual-labeled double-stranded DNA molecule. The auto- and cross-correlation function of a double-stranded DNA molecule is shown where one strand is labeled with Cy2 and the other with ATTO647N. As a control, the cross-correlation function of two dyes with a similar fluorescence spectrum, ATTO488 and ATTO655, is shown (*gray*). Fits to the auto- and cross-correlation functions are shown as *solid lines*. The weighted residuals are shown in the *upper panel*

2. Point FFS analysis starts with reading the recorded photon files per detector, bit by bit, and creating tables with the photon microtimes and macrotimes per detector to allow selection of the different PIE channels to be analyzed. Dual-color microtime histograms such as the one shown in Fig. 1e allow species-specific PIE channels to be defined and dual-color macrotime traces as shown in Fig. 1f give an initial indication of the concentration of fluorophores.
3. Different PIE channels can be selected for performing temporal auto- and cross-correlation analyses of the respective macrotime traces (*see* **Note 36**). An example is shown in Fig. 5 where a Cy2/ATTO647N dual-labeled double-stranded DNA molecule freely diffusing in solution was measured. As the PIE channels are specifically used to select photons for correlation analysis, the correlation functions can be analyzed with remarkably simple fit models, outlined in Table 2, from which the absolute concentration and the degree of double-labeling can be directly calculated (*see* **Note 37**). To show that PIE-FCCS

Table 2
Parameters, PIE channels and fit models for dual-color FCS and RICS analysis

Parameters	
$\gamma = 2^{-3/2}$	Geometrical factor for a 3D Gaussian PSF.
F_i, D_i and n	Fractional contribution and diffusion coefficient for species i. n is the total number of species present in the sample.
ω_r, ω_z and $V = \left(\frac{\pi}{2}\right)^{3/2}\omega_r^2\omega_z$	Lateral and axial PSF waist (radius) respectively, of volume V (*see* **Notes 39** and **40**).
E	FRET efficiency (*see* **Note 41**).
N_G, N_R, and N_{GR}	Average number of green-only labeled, red-only labeled and double-labeled particles respectively (*see* **Note 42**).
$N_{G_T} = N_G + N_{GR}$ and $N_{R_T} = N_R + N_{GR}$	Average number of particles carrying a green label and a red label respectively (*see* **Note 42**).
$f_{GR,G_T} = N_{GR}/N_{G_T}$; $f_{GR,R_T} = N_{GR}/N_{R_T}$ and $f_{G_T,R_T} = N_{G_T}/N_{R_T}$	Fractions of various labeled species.
F_j, τ_j and m	Fractional contribution and relaxation time of blinking process j. m is the total number of blinking processes observed in the sample.
Fit models	
$G_{FCS,D}(\tau) = \boldsymbol{G} \cdot \sum_{i=1}^{n}\left[\left(1+\frac{4D_i\tau}{\omega_r^2}\right)^{-1}\left(1+\frac{4D_i\tau}{\omega_z^2}\right)^{-1/2}\right]$	FCS correlation function for a 3D Gaussian probe volume of molecules undergoing free diffusion[a]. The amplitude factor, $\boldsymbol{G}$, is given below.
$G_{FCS}(\tau) = \prod_{j=1}^{m}\left(1+\frac{F_j}{1-F_j}\exp\left(-\frac{\tau}{\tau_j}\right)\right) \cdot G_{FCS,D}(\tau)$	Generalized model for FCS and FCCS analysis including a blinking term, with τ the correlation lag time.

$G_{\mathrm{RICS,D}}(\xi,\psi) = \boldsymbol{G}\cdot\sum_{i=1}^{n}\left[\boldsymbol{F_i}\left(1+\frac{4\boldsymbol{D_i}(\tau_{\mathrm{p}}\xi+\tau_{\mathrm{l}}\psi)}{\omega_r^2}\right)^{-1} \times \left(1+\frac{4\boldsymbol{D_i}(\tau_{\mathrm{p}}\xi+\tau_{\mathrm{l}}\psi)}{\omega_r^2}\right)^{-1/2} \times \exp\left(-\frac{\delta r^2(\xi^2+\psi^2)}{\omega_r^2+4\boldsymbol{D_i}(\tau_{\mathrm{p}}\xi+\tau_{\mathrm{l}}\psi)}\right)\right]$	RICS correlation function for freely diffusing molecules[a], where τ_{p} is the pixel dwell time, τ_{l} is the line time and δr is the pixel size (*see* **Note 40**). The amplitude factor, ***G***, is given below.
$G_{\mathrm{RICS}}(\xi,\psi) = \prod_{j=1}^{m}\left(1+\frac{F_j}{1-F_j}\exp\left(-\frac{\tau_{\mathrm{p}}\xi+\tau_{\mathrm{l}}\psi}{\tau_j}\right)\right)\cdot G_{\mathrm{RICS,D}}(\xi,\psi)$	Generalized model for RICS and ccRICS analysis including a blinking term, where ξ and ψ are the spatial correlation lag in the fast and slow scanning axes respectively.
PIE channels	
GG	Photons after green excitation in the green emission channel.
GR	Photons after green excitation in the red emission channel.
RR	Photons after red excitation in the red emission channel.
GG + GR	All photons detected after green excitation.
GR + RR	All photons detected in the red emission channel.
$\varepsilon_{\mathrm{GG}}$	Brightness of the fluorophore in PIE channel GG, calculated as the ratio between the count rate in that PIE channel and N_{GG} (*see* **Note 34**).
Values for G	
$\boldsymbol{G}_{\mathrm{GG}} = \frac{\gamma}{N_{\mathrm{G_T}}}$	Green fluorophore ACF, no FRET.
$\boldsymbol{G}_{\mathrm{GG}} = \left(\frac{1+(E^2+2E)f_{\mathrm{GR,G_T}}}{(1-Ef_{\mathrm{GR,G_T}})^2}\right)\frac{\gamma}{N_{\mathrm{G_T}}}$	Green fluorophore ACF in the presence of FRET.
$\boldsymbol{G}_{\mathrm{GG+GR}} = \frac{\gamma}{N_{\mathrm{G_T}}}$	Green fluorophore ACF, irrespective of FRET[b,c].
$\boldsymbol{G}_{\mathrm{GR+RR}} = \left(\frac{\beta^2 f_{\mathrm{G_T,R_T}}+2\beta_{\mathrm{GR,R_T}}+1}{(\beta f_{\mathrm{G_T,R_T}}+1)^2}\right)\frac{\gamma}{N_{\mathrm{R_T}}}$	Red fluorophore ACF, with crosstalk but no FRET.

(continued)

Table 2
(continued)

$G_{\mathrm{RR}} = \frac{\gamma}{N_{\mathrm{R_T}}}$	Red fluorophore ACF, irrespective of FRET or crosstalk.
$G_{\mathrm{GG\times(GR+RR)}} = \frac{\gamma\beta N_{\mathrm{G}} + \gamma(\beta+1)N_{\mathrm{GR}}}{N_{\mathrm{GT}}(N_{\mathrm{R_T}} + \beta N_{\mathrm{G_T}})}$	CCF, no FRET but with crosstalk.
$G_{\mathrm{GG\times RR}} = \frac{\gamma N_{\mathrm{GR}}}{N_{\mathrm{G_T}} N_{\mathrm{R_T}}}$	CCF, no FRET, irrespective of crosstalk.
$G_{\mathrm{GG\times RR}} = \left(\frac{1-E}{1-Ef_{\mathrm{GR,G_T}}}\right)\frac{\gamma N_{\mathrm{GR}}}{N_{\mathrm{G_T}} N_{\mathrm{R_T}}}$	CCF in the presence of FRET, irrespective of crosstalk.
$G_{\mathrm{(GG+GR)\times RR}} = \frac{\gamma N_{\mathrm{GR}}}{N_{\mathrm{G_T}} N_{\mathrm{R_T}}}$	CCF, irrespective of FRET or crosstalk[b,c].

[a]Independent from the value of ***G***, it is assumed that the brightness of fluorophores is homogenous for different diffusion components
[b]Negligible misguided excitation (excitation with the laser used for exciting the green fluorophore) of the red fluorophore is assumed
[c]The product of quantum yield and detection efficiency in the respective PIE channel should be similar for the donor and acceptor fluorophore

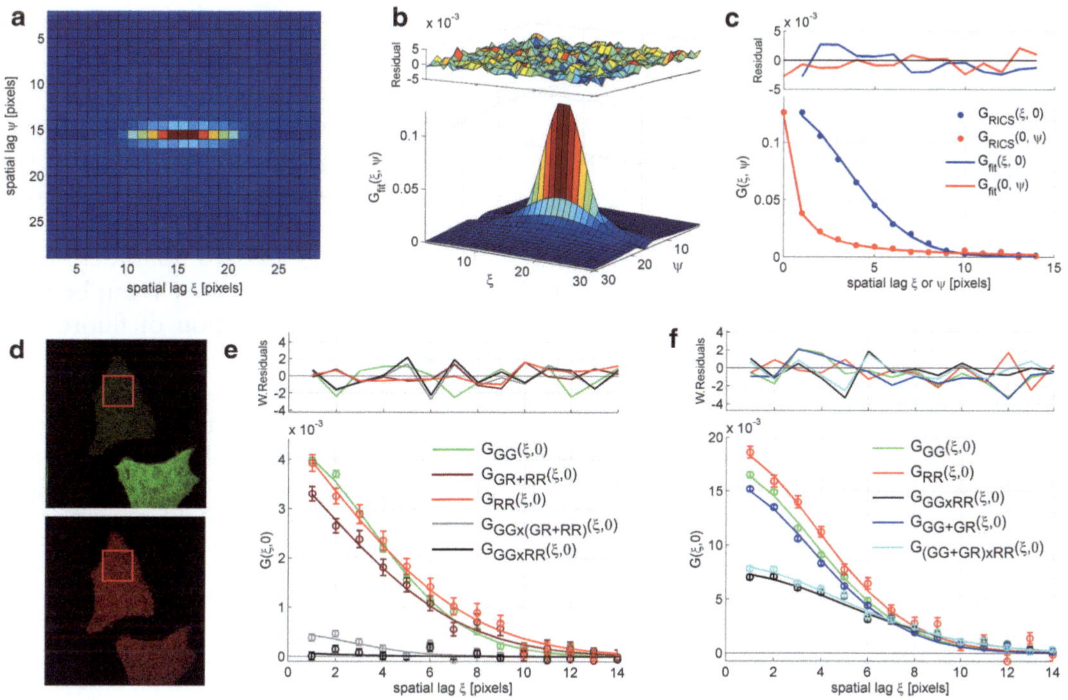

Fig. 6 PIE-RICS analysis of fluorescent proteins in cells. (**a–c**) Single-color PIE-RICS analysis of Venus fluorescent protein in HeLa cells. The spatial correlation function is shown in (**a**) 2D, (**b**) 3D and (**c**) in *x* and *y* cross-sections. (**d–f**) Dual-color PIE-RICS analysis of eGFP and mCherry in HeLa cells. (**d**) Macrotime images of cells expressing eGFP and mCherry. The region of interest selected for the RICS analysis is shown in *red*. (**e**, **f**) ccRICS analysis of cells expressing (**e**) eGFP and mCherry and (**f**) an eGFP-mCherry fusion protein. The different correlation functions are explained in Subheading 3.5 (adapted from [13])

really has a zero-amplitude cross-correlation function for non-interacting dyes, we performed the same experiment on two freely diffusing dyes, ATTO488 and ATTO655 (*see* Fig. 5).

4. Image FFS analysis starts with reading the recorded photon files per detector and per image frame, creating tables with the photon microtimes and macrotimes per detector, with all image frames pooled, to allow selection of the different PIE channels to be analyzed. Subsequently, the macrotimes in the different PIE channels are histogrammed into the individual images using the available scanning time information (pixel, line and frame time). The resulting images can be spatially auto- and/or cross-correlated [4].

5. An example of a single-color RICS experiment is shown in Fig. 6a–c, where Venus fluorescent protein was measured in solution. Knowing the image timing information (pixel dwell time τ_p and line time τ_l), the RICS image can be analyzed with a general analysis model for RICS data, outlined in Table 2.

A well-aligned scanner with excellent linearity (i.e., uniform pixel sizes and pixel times) is needed for performing RICS.

6. The advantage of PIE becomes clear when performing dual-color RICS. In Fig. 6d,e, we analyzed two fluorescent proteins, eGFP and mCherry, independently and freely diffusing in the same HeLa cell. In contrast to the crosstalk-contaminated $G_{GG\times(GR+RR)}(\xi,\ 0)$ cross-correlation function (CCF), the crosstalk-free $G_{GG\times RR}(\xi,\ 0)$ CCF has a zero amplitude. Moreover, the amplitude of the crosstalk-free red ACF can be used directly to calculate the correct concentration of fluorescent mCherry. In Fig. 6f, we expressed a fusion protein of eGFP and mCherry. Clearly, the crosstalk-free $G_{GG\times RR}(\xi,\ 0)$ CCF shows the co-diffusion of the two proteins in the tandem (*see* **Note 37**). Moreover, the nonidentical amplitudes of the two CCFs and the two green fluorophore ACFs clearly provides evidence that the two fluorescent proteins exhibit FRET in the fusion construct, as follows directly from the fit models in Table 2 (*see* **Note 38**).

4 Notes

1. The only requirement for combining PIE (or TCSPC) with a commercial CLSM is that the image scanning hardware is able to send out position signals that the TCSPC hardware can interpret. One can also obtain upgrade packages commercially that upgrade any major confocal microscope system into a PIE confocal microscope.
2. Two-photon excitation is, in principle, also compatible with PIE, as long as the fluorophores can be specifically excited at different wavelengths. However, two-photon PIE has not been performed to date. This is due to the fact that two-photon absorption spectra are typically broader than one-photon absorption spectra, making it challenging to find the specific excitation wavelengths for fluorophores. In addition, it would require the use of two synchronized two-photon laser sources, which is expensive and technically challenging. It is also possible to perform FCCS experiments with single-wavelength excitation using either one-photon [32] or two-photon excitation [33]. Single-wavelength FCCS has the advantage that the two detection channels have the identical excitation volume, but does not benefit from many of the advantages that PIE offers.
3. A useful trick for attenuating lasers when they are coupled into a single-mode fiber is to manually block a portion of the excitation beam, for example, by using an adjustable mirror holder that is just lying around. Place the mirror holder partly in the pre-fiber beam path of the laser and use the screws that are

normally used for adjusting the mirror, to move the holder accurately into or out of the beam path. Although this procedure distorts the pre-fiber beam shape, the post-fiber beam shape at the output of the single-mode fiber is still perfectly Gaussian.

4. For some applications, such as multiparameter fluorescence detection (MFD), in which i.a. the fluorescence anisotropy of fluorophores is measured [34], or dual-focus FCS, where diffusion coefficients can be measured accurately [14], the polarization of the excitation beam is extremely important. In these cases, the use of a polarization maintaining fiber is recommended.
5. Initially, we used different fibers for different excitation wavelengths to offset the wavelength dependence of the diffraction-limited spot size to obtain similar excitation volumes for the different wavelengths. However, in practice, using a single single-mode fiber for coupling all the lasers into the microscope simplifies the alignment immensely and ensures a stable overlap of excitation spots of different wavelength, irrespective of the pre- or post-fiber alignment. Hence, we have obtained better results when coupling all wavelengths into a single fiber.
6. The quality of the combined free laser beam is preserved by minimizing the optics in the free-beam part and by avoiding tension on thin optical elements.
7. When using an optical fiber to delay a laser pulse, be aware that for most fibers, the index of refraction is approximately 1.5 meaning that about ~20 cm fiber corresponds to a delay of 1 ns.
8. Reflecting optics such as polychroics and mirrors should be thick enough ($\geq$3 mm) to avoid introducing astigmatism in the optical path, which is detrimental to the PSF shape [35]. Rather than clamping optics, which might bend them and thereby create astigmatism, transparent nail polish can be used to fix optics in the beam path.
9. For cellular PIE fluctuation imaging experiments, a small PSF is necessary, so the beam is expanded ~5-fold to ensure a slight overfilling of the back aperture of the objective. For PIE-MFD experiments, a large PSF is necessary, so the beam outputted by the collimator (~3 mm diameter) is typically left unexpanded.
10. When working with planoconvex lenses, remember that both sides of the lens should equally contribute to the focusing/collimation of the beam path. Hence, the flat surface of the lens should always be towards to the focus.
11. Especially when PIE is used in combination with laser scanning, an objective that is well corrected for spherical aberrations is necessary.

12. If a significant n-mismatch does exist, for example, when measuring with an oil-immersion objective in aqueous solution or when an aqueous solution with $n \neq 1.33$ is measured with a water-immersion objective, the PSF will deform the further it is positioned from the coverslip [36]. Under these conditions, measurements should be performed at a constant distance close to ($\lesssim$10 μm) the coverslip.
13. If fluorophores are not particularly photostable, a larger pinhole allows measurements with reasonable brightness at lower laser power. With a larger pinhole, the background and detected Raman scattering is also increased. However, with fluorescently pure buffers and good emission filters that block a significant amount of the detected Raman scattering, this is not a problem. On the other hand, a small pinhole should be used when a small PSF size is important, for example, when performing 3D-resolved measurements, when measuring at high concentrations or for measurements in live cells. Occasionally, one should verify that the pinhole is clean and in good shape. This can be done with a low-magnification microscope.
14. Points to consider when selecting a detector for PIE experiments are: quantum efficiency in the detected spectral range (higher is better), afterpulsing probability (lower is better), width of the instrument response function (IRF) (larger transit time spread decreases the accuracy of lifetime measurements; smaller is better) and stability (minimal jitter), count rate dependence of the IRF position (for accurate lifetime imaging, the IRF should not shift with the count rate), dead time (although the detector dead time is usually lower than that of the TCSPC system; lower is better) and dark counts (when measurements at low signal are to be performed; lower is better).
15. Dead time of the electronics can be a limiting factor in PIE experiments. When a FCCS experiment is performed with two detectors connected to a single TCSPC device, dead time of the TCSPC electronics results in an anti-correlation between the two detectors [11]. Even when using a single detector, dead time can be a serious issue when the statistics of photon emission are used for analysis [37]. If artifacts related to the dead time are small, they can usually be corrected for a posteriori, but we suggest limiting dead time problems as much as possible by using an independent TCSPC device for each detector, by measuring at low count rates (~100 kHz) and by using TCSPC hardware with a low dead time (<100 ns if possible).
16. Often, different objective lenses perform better with different coverslip thicknesses (#1 vs. #1.5), so we suggest trying out different types and brands.

17. To limit nonspecific absorption of protein samples to the sample holder, you can coat the coverslip prior to loading the fluorophore solution. Different coating agents exist, including proteins, carbohydrates, peptide polymers and lipids. We often use 1 mg/mL of BSA protein, which we allow to incubate for 10 min prior to washing with measurement buffer and loading the fluorophore solution.
18. Many dyes, such as Rhodamine B, absorb to coverslips. This causes a high background from immobilized fluorophores when FCS measurements are performed close to the surface.
19. Always load the same volume of the fluorophore solution or less as the volume of the solution you have used to coat the coverslip.
20. Calculating fluorescence images in quasi-real time with photon counting detection is a demanding task. Therefore, we recommend using a fast and dedicated computer for controlling the microscope as well as a low-level programming language such as C++/C# for data acquisition. Also, we recommend saving data from different detectors and/or image frames in different small files instead of saving all the data from an experiment in one large file. This has the advantage that only parts of the data can be loaded and analyzed without the need for loading the complete experiment into the analysis software. This procedure also saves computer memory.
21. As a rule of thumb, 80/(1 + the number of excitation lines) MHz is the optimal frequency of the PIE system. For dual-color PIE, for example, this is 26.67 MHz.
22. Fine alignment of the PSFs from the multiple lasers is not possible when all lasers are coupled into the same fiber. Thus, the wavelength dependence of refraction in the objective lens will cause the PSFs to be slightly bigger for higher wavelengths. Moreover, lenses in the optical path introduce, to certain degree, axial and lateral chromatic aberration. Lateral chromatic aberrations, in particular, can be easily quantified by dual-color image correlation analysis of a sample containing immobilized multicolor bead with diameters smaller than the PSF [38].
23. We suggest frequently measuring the laser power during experiments by installing a power meter head on a flip mount in the post-fiber beam path. To preserve the shape of the laser IRF, we suggest always running lasers at the same hardware setting and using attenuators to perform the final adjust power adjustment. Lastly, we suggest running the lasers under conditions where they are stable. When low excitation intensity is needed, it is better to run the lasers at higher power and

attenuate the laser intensity than to run the lasers close to their lasing threshold.

24. Physically mount the emission filter on or close to the detector to ensure that only filtered light can reach the detector. This limits background counts. Also, all laser lines used for excitation should be suppressed by all emission filters to at least 6 ODs.
25. If the anisotropy-free lifetime of a fluorophore is to be measured using a single detector, a polarizer at 54.7° with respect to the orientation of the excitation polarization needs to be installed in the detection path, at the cost of lower signal-to-noise.
26. For more information on combining commercial CLSM galvos with TCSPC hardware, we refer the reader to literature from specific manufacturers (http://www.picoquant.com/products/lsm_kit/lsm_synchronisation.htm and the TCSPC handbook of Becker&Hickl).
27. The count rate at which correct PIE experiments can be performed depends on the application and typically ranges anywhere between 0 and 1 MHz. For correct estimations of the fluorescence lifetime with TCSPC, for example, the maximum count rate is advised to be below 1/100th of the laser frequency, to avoid photon pile-up. For correct brightness measurements, the dead time of the TCSPC hardware has to be accounted for, which also sets a limit on the maximum count rate [13]. Finally, at high concentrations, (a) lower laser powers are needed to avoid detector saturation, which result in lower molecular brightness and hence a lower signal-to-noise and (b) due to the lower amplitude of the correlation function coming from molecular fluctuations, other sources of correlated noise can contribute significantly to the overall correlation signal and lower the experimental accuracy. In any case, the typical count rate of a PIE/TCSPC measurement with single-photon counting detectors is much lower than the maximum allowed count rate for safe operation.
28. For a well-aligned laser, the maximum post-fiber laser intensity is always similar, and its variance below 1 %. When using a broadband fiber coupler for combining all lasers into a single fiber is used, a compromise in the coupling efficiency has to be made. Generally however, enough laser light can still be coupled in per laser and the coupling stability does not suffer from the lower (~30 %) coupling efficiency.
29. Variations in the coverslip thickness do occur, even for coverslips of the same type. We suggest when working with multiwell chambers to always leave one chamber free for characterization of the PSF with reference fluorophores.

30. For detector alignment, remove the pinhole (or fully open a variable pinhole) and align the detectors by maximizing the signal in all three dimensions (horizontal, vertical and along the optical axis). Then, install and align the pinhole in three dimensions. The axial position of the pinhole does not need frequent realigning since it is determined only by the collimation of the excitation beam.
31. For a rough detector alignment (e.g., during construction of the microscope), leave the detector turned off, remove the pinhole, load a concentrated fluorophore sample and use a relatively high laser power. In the detection path, the fluorescence emission beam from the fluorophore can be easily seen in a dark room and used for positioning the detector visually.
32. The alignment of the galvo and/or telescope can be tested by comparing the ACF of reference dyes at different positions of the galvo mirrors. For a well-built system, variations in the detected molecular brightness and the shape of the PSF should be minimal as a function of galvo position.
33. When controlling the galvo with an analog signal, an easy and accurate calibration of the scanner can be performed by placing potentiometers between the wave-function generating card and the galvo. A microscopic ruler, such as the 1951-USAF pattern, can be imaged on the microscope and the x- and y-potentiometers adjusted until the measured range corresponds to the set range.
34. An average peak molecular brightness [39] of at least 1 kHz/molecule usually generates adequate data for fluorescence fluctuation experiments, but to limit measurement times and increase signal-to-noise, we recommend a brightness of at least 3 kHz/molecule. At all times, avoid artifacts by determining the range of laser powers where the molecular brightness scales linearly with excitation intensity [35].
35. When experiments are performed on timescales longer than a couple of minutes or when surface-immobilized molecules are studied, a drift control device should be installed to keep a constant distance between the coverslip and PSF.
36. To calculate temporal correlations in real time, we use a specially developed algorithm [40].
37. The existence of bimolecular interactions can always be proven in a straightforward manner simply by calculating the $G_{\mathrm{GG\times RR}}(\tau)$ (FCS) or $G_{\mathrm{GG\times RR}}(\xi, \psi)$ (RICS) CCF. Because of the complete absence of crosstalk in this CCF, even low-affinity interactions can be shown with the highest sensitivity and certainty.

38. When part of the green fluorophores exhibits FRET, the green fluorophore population exhibits a nonhomogeneously distributed brightness, rendering FFS analysis without PIE complicated. In dual-color PIE experiments however, by adding the photons in the *GR* PIE channel back to the *GG* channel, FRET induced quenching of the green fluorophore is effectively undone, and FFS analysis can be performed with straightforward fit models (*see* Table 2).
39. For a cross-correlation between two concentric PSFs, the lateral and axial waists of the cross-correlation volume are the root mean square of the individual PSF waists [1].
40. Because the fast-scanning *x*-mirror translates the PSF while photons are being registered during the pixel dwell time, the effective size of the PSF along this scanning axis is somewhat elongated. This effect becomes apparent especially at the relatively fast scan speeds needed for RICS. To take this into account in the RICS analysis and because ω_r is an independent fitting parameter, we suggest performing a calibration RICS measurement to determine the effective ω_r for RICS, instead of using the value obtained from an FCS calibration measurement.
41. With PIE, the FRET efficiency can be conveniently determined from FCS/RICS data by fitting the fluorescence decay histograms [13]. With the appropriate assumptions, *E* can even be calculated directly from FCCS data [11].
42. To convert particle numbers into concentrations, divide *N* by the effective V_{PSF} of the respective PIE channel and by Avogadro's number.

Acknowledgements

J.H. wishes to acknowledge the Research Foundation Flanders for a postdoctoral fellowship and travel grant. D.C.L. gratefully acknowledges the financial support of the Deutsche Forschungsgemeinschaft through the SFB1032 and the Nanoinitiative Munich (NIM) and of the LMU via the Center for NanoScience (CeNS) and the LMUinnovativ BioImaging Network. We thank Dr. Matthias Höller, Dr. Volodymyr Kudryavtsev, Dipl. Chem. Waldemar Schrimpf and Dr. Martin Sikor for helping develop the data acquisition and analysis software. Dr. Nick Smisdom and Prof. Marcel Ameloot (Hasselt University, Belgium) are thanked for fruitful discussions.

References

1. Schwille P, Meyer-Almes FJ, Rigler R (1997) Dual-color fluorescence cross-correlation spectroscopy for multicomponent diffusional analysis in solution. Biophys J 72:1878–1886
2. Digman MA, Wiseman PW, Horwitz AR, Gratton E (2009) Detecting protein complexes in living cells from laser scanning confocal image sequences by the cross correlation raster image spectroscopy method. Biophys J 96: 707–716
3. Elson EL, Magde D (1974) Fluorescence correlation spectroscopy. 1. Conceptual basis and theory. Biopolymers 13:1–27
4. Petersen NO, Hoddelius PL, Wiseman PW, Seger O, Magnusson KE (1993) Quantitation of membrane receptor distributions by image correlation spectroscopy: concept and application. Biophys J 65:1135–1146
5. Digman MA, Sengupta P, Wiseman PW, Brown CM, Horwitz AR, Gratton E (2005) Fluctuation correlation spectroscopy with a laser-scanning microscope: exploiting the hidden time structure. Biophys J 88:L33–L36
6. Bacia K, Petrasek Z, Schwille P (2012) Correcting for spectral cross-talk in dual-color fluorescence cross-correlation spectroscopy. Chemphyschem 13:1221–1231
7. Rigler R, Foldes-Papp Z, Meyer-Almes FJ, Sammet C, Volcker M, Schnetz A (1998) Fluorescence cross-correlation: a new concept for polymerase chain reaction. J Biotechnol 63:97–109
8. Oyama R, Takashima H, Yonezawa M, Doi N, Miyamoto-Sato E, Kinjo M, Yanagawa H (2006) Protein-protein interaction analysis by C-terminally specific fluorescence labeling and fluorescence cross-correlation spectroscopy. Nucleic Acids Res 34:e102
9. Shi X, Foo YH, Sudhaharan T, Chong S-W, Korzh V, Ahmed S, Wohland T (2009) Determination of dissociation constants in living zebrafish embryos with single wavelength fluorescence cross-correlation spectroscopy. Biophys J 97:678–686
10. Kapanidis AN, Lee NK, Laurence TA, Doose S, Margeat E, Weiss S (2004) Fluorescence-aided molecule sorting: analysis of structure and interactions by alternating-laser excitation of single molecules. Proc Natl Acad Sci U S A 101:8936–8941
11. Müller BK, Zaychikov E, Bräuchle C, Lamb DC (2005) Pulsed interleaved excitation. Biophys J 89:3508–3522
12. Kudryavtsev V, Sikor M, Kalinin S, Mokranjac D, Seidel CA, Lamb DC (2012) Combining MFD and PIE for accurate single-pair Förster resonance energy transfer measurements. Chemphyschem 13:1060–1078
13. Hendrix J, Schrimpf W, Höller M, Lamb DC (2013) Pulsed interleaved excitation fluctuation imaging. Biophys J 105:848–861
14. Dertinger T, Pacheco V, von der Hocht I, Hartmann R, Gregor I, Enderlein J (2007) Two-focus fluorescence correlation spectroscopy: a new tool for accurate and absolute diffusion measurements. Chemphyschem 8: 433–443
15. Hendrix J, Lamb DC (2013) Pulsed interleaved excitation: principles and applications. Methods Enzymol 518:205–243
16. Becker W, Hickl H, Zander C, Drexhage KH, Sauer M, Siebert S, Wolfrum J (1999) Time-resolved detection and identification of single analyte molecules in microcapillaries by time-correlated single-photon counting (TCSPC). Rev Sci Instrum 70:1835
17. Lamb DC, Schenk A, Rocker C, Scalfi-Happ C, Nienhaus GU (2000) Sensitivity enhancement in fluorescence correlation spectroscopy of multiple species using time-gated detection. Biophys J 79:1129–1138
18. Naredi-Rainer N, Prescher J, Hartschuh A, Lamb DC (2013) Confocal microscopy. In: Kubitscheck U (ed) Fluorescence microscopy—from principles to biological applications, vol. 1. Wiley-VCH, Weinheim, p 527
19. Kapusta P (2010) Absolute diffusion coefficients: compilation of reference data for FCS calibration. Picoquant Application Note. http://www.picoquant.com/technotes/appnote_diffusion_coefficients.pdf.
20. Böhmer M, Wahl M, Rahn HJ, Erdmann R, Enderlein J (2002) Time-resolved fluorescence correlation spectroscopy. Chem Phys Lett 353:439–445
21. Chen Y, Muller JD, So PT, Gratton E (1999) The photon counting histogram in fluorescence fluctuation spectroscopy. Biophys J 77:553–567
22. Müller JD (2004) Cumulant analysis in fluorescence fluctuation spectroscopy. Biophys J 86:3981–3992
23. Digman MA, Gratton E (2009) Imaging barriers to diffusion by pair correlation functions. Biophys J 97:665–673
24. Ries J, Schwille P (2006) Studying slow membrane dynamics with continuous wave scanning fluorescence correlation spectroscopy. Biophys J 91:1915–1924

25. Petrasek Z, Schwille P (2008) Precise measurement of diffusion coefficients using scanning fluorescence correlation spectroscopy. Biophys J 94:1437–1448
26. Deniz AA, Dahan M, Grunwell JR, Ha TJ, Faulhaber AE, Chemla DS, Weiss S, Schultz PG (1999) Single-pair fluorescence resonance energy transfer on freely diffusing molecules: observation of Forster distance dependence and subpopulations. Proc Natl Acad Sci U S A 96:3670–3675
27. Eggeling C, Berger S, Brand L, Fries JR, Schaffer J, Volkmer A, Seidel CAM (2001) Data registration and selective single-molecule analysis using multi-parameter fluorescence detection. J Biotechnol 86:163–180
28. Antonik M, Felekyan S, Gaiduk A, Seidel CA (2006) Separating structural heterogeneities from stochastic variations in fluorescence resonance energy transfer distributions via photon distribution analysis. J Phys Chem B 110: 6970–6978
29. Digman MA, Brown CM, Sengupta P, Wiseman PW, Horwitz AR, Gratton E (2005) Measuring fast dynamics in solutions and cells with a laser scanning microscope. Biophys J 89: 1317–1327
30. Digman MA, Dalal R, Horwitz AF, Gratton E (2008) Mapping the number of molecules and brightness in the laser scanning microscope. Biophys J 94:2320–2332
31. Digman MA, Caiolfa VR, Zamai M, Gratton E (2008) The phasor approach to fluorescence lifetime imaging analysis. Biophys J 94: L14–L16
32. Hwang LC, Wohland T (2004) Dual-color fluorescence cross-correlation spectroscopy using single laser wavelength excitation. Chemphyschem 5:549–551
33. Savatier J, Jalaguier S, Ferguson ML, Cavailles V, Royer CA (2010) Estrogen receptor interactions and dynamics monitored in live cells by fluorescence cross-correlation spectroscopy. Biochemistry 49:772–781
34. Widengren J, Kudryavtsev V, Antonik M, Berger S, Gerken M, Seidel CA (2006) Single-molecule detection and identification of multiple species by multiparameter fluorescence detection. Anal Chem 78:2039–2050
35. Dertinger T, Loman A, Ewers B, Muller CB, Kramer B, Enderlein J (2008) The optics and performance of dual-focus fluorescence correlation spectroscopy. Opt Express 16:14353–14368
36. Enderlein J, Gregor I, Patra D, Dertinger T, Kaupp UB (2005) Performance of fluorescence correlation spectroscopy for measuring diffusion and concentration. Chemphyschem 6:2324–2336
37. Hillesheim LN, Muller JD (2003) The photon counting histogram in fluorescence fluctuation spectroscopy with non-ideal photodetectors. Biophys J 85:1948–1958
38. Wiseman PW, Squier JA, Ellisman MH, Wilson KR (2000) Two-photon image correlation spectroscopy and image cross-correlation spectroscopy. J Microsc 200:14–25
39. Ivanchenko S, Lamb DC (2011) Fluorescence correlation spectroscopy: principles and developments. In: Brnjas-Kraljevic J, Pifat-Mrzljak G (eds) Supramolecular Structure and Function, vol. 10. Springer, Berlin, pp 1–30
40. Wahl M, Gregor I, Patting M, Enderlein J (2003) Fast calculation of fluorescence correlation data with asynchronous time-correlated single-photon counting. Opt Express 11:3583–3591
41. Gregor I, Patra D, Enderlein J (2005) Optical saturation in fluorescence correlation spectroscopy under continuous-wave and pulsed excitation. Chemphyschem 6:164–170

Chapter 31

Quantitative Study of Protein–Protein Interactions in Live Cell by Dual-Color Fluorescence Correlation Spectroscopy

Sergi Padilla-Parra, Nicolas Audugé, Maïté Coppey-Moisan, and Marc Tramier

Abstract

Dual-color FCS is a powerful method to monitor protein–protein interactions in living cells. The main idea is based on the cross-correlation analysis of temporal fluorescence intensity fluctuations of two fluorescent proteins to obtain their co-diffusion and relative concentration. But, when performing these experiments, the spectral overlap in the emission of the two colors produces an artifact that corrupts the cross-correlation data: spectral bleed-through. We have shown that problems with cross talk are overcome with Fluorescence Lifetime Correlation Spectroscopy (FLCS). FLCS applied to dual-color cross-correlation, utilizing for example eGFP and mCherry fluorescent proteins, allows the determination of protein–protein interactions in living cells without the need of spectral bleed-through calibration. Here, we present in detail how this methodology can be implemented using a commercial setup (Microtime from PicoQuant, SP8 SMD from Leica or any conventional confocal with PicoQuant TCSPC module, and also with a Becker and Hickl TCSPC module). The dual-color FLCS experimental procedure where the different laser intensities do not have to be controlled during the experiment constitutes a very powerful technique to quantitatively study protein interactions in live samples.

Key words FCS, FCCS, FLCS, Quantitative fluorescence microscopy, TCSPC

1 Introduction

The study of the dynamics of macromolecular complexes becomes essential to investigate molecular mechanisms in live cells. Different approaches are used to investigate protein–protein interactions in live cells. To this end, we are here interested in dual-color fluorescence correlation spectroscopy (FCS) [1]. The technique consists on the cross-correlation analysis of temporal fluorescence intensity fluctuations of two color fluorophores to follow the co-diffusion of two proteins labeled with these fluorescent tags. Dual-color FCS is performed at a single point and is capable of detecting interactions

Yves Engelborghs and Antonie J.W.G. Visser (eds.), *Fluorescence Spectroscopy and Microscopy: Methods and Protocols*, Methods in Molecular Biology, vol. 1076, DOI 10.1007/978-1-62703-649-8_31,

between proteins in which FRET does not necessarily occur, therefore provides complementary information relative to FRET in imaging mode. This approach overcomes the main limitation of FRET that requires the close proximity of the two fluorescent tags. The only requirement in order to detect interactions using dual-color FLCS is that both fluorescent proteins should diffuse and/or transiently interact with high molecular weight complexes or mobile subcellular structures. Recently, this approach was used in live cells giving novel insights on the study of macromolecular complexes, such as the MAP kinase complexes in yeast pheromone signaling [2], the dimerization of EGF receptors [3], the T cell receptor complexes [4], and the virus RNA polymerase assembly [5]. But couples of protein tags commonly used in live cells (ECFP/EYFP and EGFP/mCherry) present spectral overlap in their emission giving rise to spectral bleed-through that corrupts the cross-correlation data [5]. The cross-talk contribution from the shorter wavelengths emission species in the longer wavelengths channel is usually minimized by choosing red-shifted band-pass filter for the longer wavelengths channel and/or by decreasing the intensity of the laser exciting the shorter wavelengths emission species. But these tricks decrease the brightness of both fluorescent species, a parameter usually critical in dual-color FCS in live cells. The cross-talk can also be calibrated and its amount retrieved during data analysis [6]. But nonetheless it impedes the study of very low amounts of protein–protein interactions since at least 10 % of cross talk is always observed when working with green and red fluorescent proteins. The use of novel fluorescent proteins exhibiting a large Stokes shift in dual-color FCS experiments decreases the contribution of the spectral cross talk to the cross-correlation [7, 8], but the weak photo stability of some of them introduces other difficulties.

To eliminate spectral bleed-through in dual-color FCS, the switching method or pulsed interleaved excitation was introduced [9–11]. In this method, two alternating laser beams are controlled by an acousto-optic modulator and the synchronized correlation signals in both detectors are calculated as cross-correlation functions with no cross-talk. These approaches necessitate homemade developed confocal setups and cannot be easily implemented to commercial microscopes. A similar method was recently implemented in dual-color scanning FCS with a conventional confocal microscope to quantify ligand-receptor interactions in cell membranes of living zebrafish embryos [12]. In membranes, the diffusion of protein complexes is slow, which can be monitored by using the sequential dual-color scanning method. However, the speed of acquisition of line-scanning FCS in sequential dual color is too slow to investigate intracellular co-diffusion.

Here, we propose a FCCS-based method to overcome the EGFP and mCherry spectral cross talk in quantitative studies of protein–protein interactions in live cells by using a commercial confocal system (Microtime from PicoQuant, SP8 SMD from Leica or any conventional confocal TCSPC module) (e.g., PicoQuant or Becker & Hickl). Our approach is based on Fluorescence Lifetime Correlation Spectroscopy (FLCS). The use of a pulsed laser to perform FCS takes advantage of Time Correlated Single Photon Counting (TCSPC) to differentiate the contribution of two different fluorescence lifetime species to FCS file (FLCS, [13]). Indeed, each single photon is recorded with its global arrival time and the delay time with respect to the corresponding laser pulse. The FLCS method relies on the calculation of lifetime filters for each species having different fluorescence decay signatures in the TCSPC histogram. Each fluorescent component, in the fluorescent time trace, having different lifetime properties can be resolved on the basis of its fluorescence decay signature. FLCS was already used in spectroscopy to suppress scattered light and detector afterpulses [14], to study dynamic transition between two fluorescence lifetime states [15], to resolve different lifetime species [16], and to measure precise concentration values as low as picomolar of fluorophore [17]. FLCS was also employed in live cells to quantify the interaction of EGFR-neutralizing antibody labeled with Alexa488 and EGFR-GFP by fluorescence lifetime cross-correlation measurements [18]. Recently, we have shown that FLCS applied to dual-color EGFP and mCherry cross-correlation allows the determination of protein–protein interactions in living cells without any spectral bleed-through artifact [19].

By using two lasers to excite simultaneously two spectrally different fluorophores, a pulsed laser (470 nm which can excite a green fluorescent species, i.e., EGFP) and a continuous laser (561 nm to excite a red fluorescent protein, i.e., mCherry), dual-color FLCS provides a way to remove the spectral bleed-through of the green species into the red detection channel. The photons coming from the green species detected in the red channel are time correlated to the pulsed laser (*see* the fluorescence decay histogram in Fig. 1a, *top*) in contrast to the photons coming from the red species which do not harbor any fluorescence decay signature (*see* the constant emission histogram in Fig. 1a, *middle*). When both fluorophores are excited together, the fluorescence histogram contains the two signals (fluorescence decay for the spectral bleed-through of the green species and constant histogram for the red species, Fig. 1a, *bottom*). Lifetime filter functions are calculated from the two-component fluorescence decay analysis acquired in the red channel. For each time-resolved channel, the lifetime filter value gives the statistical weight of the counted photon for its corresponding species. In Fig. 1b are shown the filter corresponding to the fluorescence decay species (in blue) and this corresponding to the steady-state fluorescence (in pink). These different filters can be selected during auto- and cross-correlation process of the absolute time traces to take into account the different

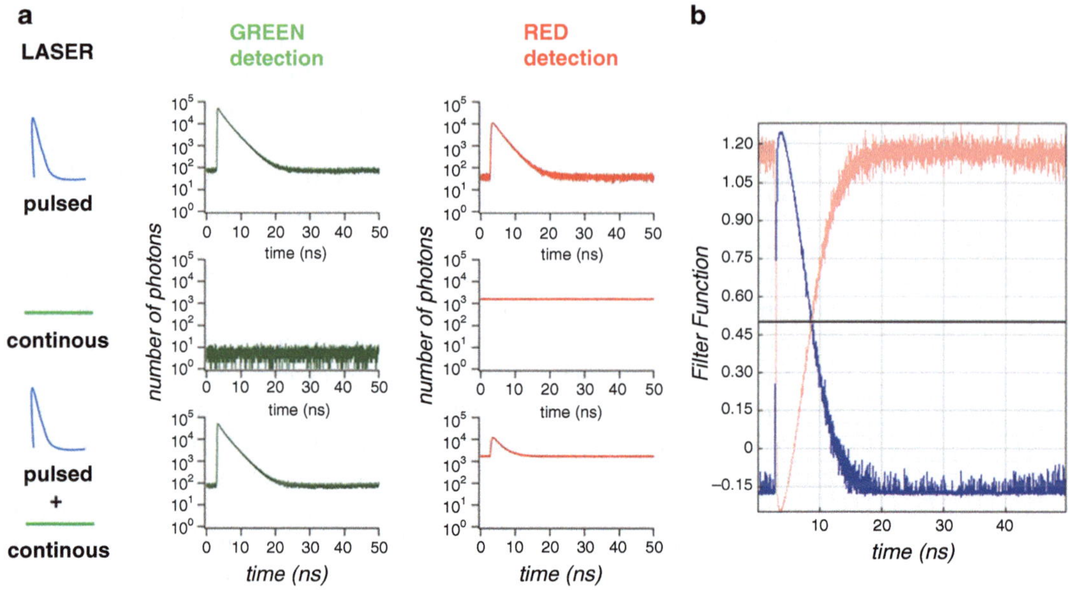

Fig. 1 Dual-color FLCS approach and fluorescence lifetime filters. (**a**) Fluorescence time-correlation histograms of GFP and mCherry fluorescence detected in the *green* (*left*) and *red* (*right*) channels by using 470 nm pulsed laser alone (*upper*), 561 nm CW laser (*middle*), and both (*lower*). (**b**) Fluorescence lifetime filters of the fluorescence decay (*blue*) and the steady-state fluorescence (*pink*) calculated by Symphotime software (PicoQuant) for the fluorescence decay histogram of both GFP and mCherry in the red channel detection by using 470 nm pulsed and 561 nm CW laser excitation (*red signal* in **a**)

weighting of each individual photon. By using the lifetime filter of the fluorescence decay of EGFP in the green channel and the filter of the steady-state fluorescence of mCherry in the red channel during cross-correlation calculation of the green and red time trace, the spectral bleed-through artifact is completely removed. We stress the potentiality of the technique when applied to quantify protein–protein interactions in vivo where protein A is fused to EGFP and protein B is fused to mCherry. The cross-correlation amplitude is directly dependent on the quantity of protein complexes and is not related to both the relative concentration of green and red species and the relative laser intensities used during the experiment. The dual-color FLCS experimental procedure where the different laser intensities do not have to be controlled during the experiment is extremely powerful to quantitatively study protein interactions in live samples.

2 Materials

2.1 Live Cells Sample Preparation

Dual-color FLCS has been optimized for green and red fluorescent proteins. In this protocol, we present the technique to be employed to detect very low amounts of interacting proteins in He-La cells (i.e., A-GFP + B-mCherry where A and B are the proteins of interest). The method can be adapted to any other cell line or other live sample expressing the two proteins, with the limitation of having sufficiently low amount of fluorescent proteins for a FCS experiment (*see* **Note 1**).

1. Culture He-La cells in Dulbecco's modified eagle medium containing 10 % fetal bovine serum (PAA Laboratories GmbH, Pasching, Austria) at 37 °C in a humidified atmosphere of 5 % CO_2.
2. Seed the cells on 32-mm round glass coverslip at a density of 2×10^5 cells. When cells are around 50–70 % of confluence, perform a transient transfection using the proteins A-GFP and B-mCherry. Also, perform a second transfection using EGFP and mCherry expressing plasmids and mCherry-EGFP tandem expressing plasmid with a total amount of 1 μg of expression vectors (*see* **Note 2**) using Nanofectin I (PAA).
3. 24 h after transfection, mount the cells in an open observation chamber with special medium without phenol red (we use special DMEM-F12 without phenol red, B12 vitamin, riboflavin and supplemented with 20 mM Hepes, and L-glutamine from PAA) and with low amount of serum (with He-La cells no serum is required) to prevent autofluorescence from the sample. Add 1 ml of parafilm oil (Merck) to avoid evaporation at 37 °C.
4. Use a microscope incubation chamber (for example PeCon GmbH or LIS) to perform live cell measurements at 37 °C, if possible with 5 % CO_2 flux.

2.2 Microscopy Setup

A confocal light scanning microscope equipped with pulsed laser excitation, electronics to carry out Time Correlated Single Photon Counting (TCSPC), and single photon counting detectors (i.e., SPADS and/or APD) is needed to perform FLCS measurements. We carried out all measurements presented here employing the PicoQuant Microtime 200 [16, 17, 19] but other solutions are available in the market, for instance, the SP8/SP5 combined with the SMD module by Leica (Manheim, Germany) or the combination of any confocal with a PicoQuant TCSPC module. In the rest of the protocol, we describe how to carry dual-color FLCS with a Microtime 200, but the reader should be able to translate this description to his/her own system. Figure 2 shows the optical pathway. Two laser excitation sources (if we are to use GFP and mCherry, a pulsed blue laser operating at 470 nm and a continuous green laser operating at 561 nm) are directed into a water objective with a high numerical aperture objective (60×, NA 1.3) mounted in an inverted microscope and is directly focused onto co-transfected live cells. The fluorescence emission is focused through a pinhole

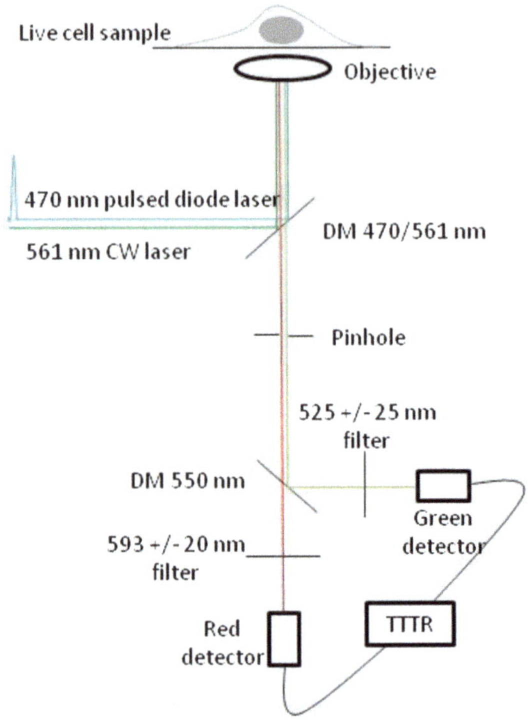

Fig. 2 Dual-color FLCS microscope. Two lasers (470 nm pulsed and 561 nm CW) are conducted into epi-fluorescence microscope for live sample measurement. The confocal fluorescence selected by a pinhole is split into green and red channels. Single photon counting detectors (SPADs) are used in Time-Tagged Time-Resolved (TTTR) mode

and the corresponding emission filters (510DF40, 580DF20) and is collected by two avalanche photodiode detectors (SPAD 1 and SPAD 2). The PicoQuant TimeHarp 300 module is employed in a Time-Tagged Time-Resolved (TTTR) mode (*see* **Note 3**) and is coupled to a PC that pilots the data acquisition and analysis using the PicoQuant Symphotime software:

1. Select the pulsed diode tuned at 470 nm (PicoQuant) to excite EGFP and a continuous DPSS laser at 561 nm (Coherent) to excite mCherry.
2. Utilize a water immersion objective with a high numerical aperture (for instance, we recommend a 60×, NA 1.3) mounted on an inverted microscope.
3. Select a double dichroic mirror (DM 470/571 nm) and adjust the pinhole aperture for the two colors (*see* **Note 4**).
4. Select the right emission filters for GFP and mCherry (525 ± 25 for the green channel and 593 ± 20 for the red channel) (*see* **Note 5**).

5. Utilize as detection two avalanche photodiodes (for instance, SPAD from Perkin Elmer, Tau-SPAD or MPD SPAD from PicoQuant, *see* **Note 6**).

3 Methods

We present here the protocol for dual-color FLCS in the context of A-GFP and B-mCherry co-diffusion study to reveal protein–protein interactions. We suppose that the system is already calibrated in terms of conventional FCS (calibration of the confocal volume, *see* **Note 7**). We also encourage readers using GFP and mCherry alone and in tandem expressed in cells to validate the technique with their own system ([19], *see* **Note 8**).

3.1 Dual-Color FLCS in Live Cells

1. Choose, using the mercury lamp set at low intensities, the right cell in which to park the confocal beam. Picking up the right cells (low expressing levels of fluorophore) is essential in order to get the right data (*see* **Note 9**).
2. Park the beam in different areas of the cell and check the intensity trace (it should not be higher than 200 kHz) by adjusting the different laser intensities. The correction of the spectral bleed-through by FLCS is not related to the laser intensities. For example, try diminishing the power in the green laser (561 nm excitation) only, this will have a drastic effect in false cross-correlation (bleed-through coming from green photons detected in the red channel). This is only true with conventional dual-color FCS approaches, but not in dual-color FLCS. This is crucial to investigate very low amounts of proteins in interaction and makes FLCS a unique technique in terms of sensitivity. To increase the laser power in this scenario for both channels, but especially for the mCherry channel, is crucial, since the relatively low sensitivity of the SPADS in the red part of the spectrum and the lower photostability of mCherry compared to GFP require particular intensity conditions that may change from cell to cell. This is why the independence of FLCS to intensity (as is the case in FLIM) constitutes a great improvement compared to FCCS.
3. Check that the shape of the intensity trace is steady over time (indicating that your measurement is not affected by photobleaching) in both channels. Decrease the laser intensities if photobleaching occurs.
4. Acquire the whole data for about 30 s to 1 min in a TTTR file which contains all the single photon events with their respective absolute time, correlated time, and detection channels. Using Symphotime software, it is easy to play TTTR file to recover the different histograms: fluorescence decays of both channels and time trace of both channels. Symphotime is also designed to

calculate auto- and cross-correlations of the different time traces, particularly using calculated lifetime filter from the fluorescence decays (see below).

3.2 Fitting the Data Using Lifetime Filters

In dual-color FLCS, the lifetime filter functions (f_a and f_b) are directly calculated from the fluorescence decay coming from both green and red channels. Indeed, in the red channel, this fluorescence decay histogram ($Decay^{\mathrm{r}}(i)$) contains "green" photons (excited with the blue pulsed laser) and "red" photons (excited with the *CW* green laser). In the following equation, i corresponds to the time-resolved channel (in ns) and r indicates the red channel. The histogram ($Decay^{\mathrm{r}}(i)$) can be expressed then as a linear combination of two normalized histograms:

$$Decay^{\mathrm{r}}(i) = w_a^{\mathrm{r}} \cdot a(i) + w_b^{\mathrm{r}} \cdot b(i) \tag{1}$$

where w_a^{r} and w_b^{r} are the amplitudes of the photon count contribution of GFP and mCherry, respectively; and $a(i)$ and $b(i)$ are fractions defined as follows:

$$a(i) = \frac{A(i)}{\sum A(i)} \quad \text{and} \quad b(i) = \frac{B(i)}{\sum B(i)} \tag{2}$$

with $A(i)$ and $B(i)$ corresponding to the histograms of the EGFP fluorescence decay species (Fig. 1a, green curve) and of the steady-state fluorescence of mCherry (Fig. 1a, constant histogram in pink), respectively.

Now, we can define the corresponding lifetime filters:

$$< \sum f_a^{\mathrm{r}}(i) \cdot Decay^{\mathrm{r}}(i) = w_a^{\mathrm{r}} > \quad \text{and} \quad < \sum f_b^{\mathrm{r}}(i) \cdot Decay^{\mathrm{r}}(i) = w_b^{\mathrm{r}} > \tag{3}$$

$f_a^{\mathrm{r}}(i)$ and $f_b^{\mathrm{r}}(i)$, are the "lifetime filter" functions. And the sum is over the time interval of the bin size used to calculate w.

Next, it is straightforward to show how the autocorrelation functions for the two different channels ($G_{\mathrm{g}}(\tau)$, $G_{\mathrm{r}}(\tau)$) and the cross-correlation ($G_{\mathrm{cross}}(\tau)$) are defined as follows:

$$G_{\mathrm{g}}(\tau) = < \left[\sum f_a^{\mathrm{g}}(i) \cdot I_i^{\mathrm{g}}(t)\right]\left[\sum f_a^{\mathrm{g}}(i) \cdot I_i^{\mathrm{g}}(t+\tau)\right] > \tag{4}$$

$$G_{\mathrm{r}}(\tau) = < \left[\sum f_a^{\mathrm{r}}(i) \cdot I_i^{\mathrm{r}}(t)\right]\left[\sum f_a^{\mathrm{r}}(i) \cdot I_i^{\mathrm{r}}(t+\tau)\right] > \tag{5}$$

$$G_{\mathrm{gr}}(\tau) = < \left[\sum f_a^{\mathrm{g}}(i) \cdot I_i^{\mathrm{g}}(t)\right]\left[\sum f_a^{\mathrm{r}}(i) \cdot I_i^{\mathrm{r}}(t+\tau)\right] > \tag{6}$$

where $f_a^{\mathrm{g}}, f_a^{\mathrm{r}}$ are the lifetime filters for the fluorescence decay of the GFP in the green channel and the filter for the steady-state fluorescence in the red channel, respectively, and $I_t^{\mathrm{g}}(t)$, $I_i^{\mathrm{r}}(t)$ are the fluorescence intensities in the green and red channels, respectively.

It is possible to develop a homemade program for the different calculations presented above. Using Symphotime, it is straightforward to obtain the filtered auto- and cross-correlation functions:

1. Play the TTTR file in Symphotime to recover the time trace of the experiment. At this step, it is possible to select only part of the time trace, for example, if bleaching occurs only at the end of the experiment or if a burst appears to remove it.
2. Click on "lifetime filter" icon. A window appears showing fluorescence decay histogram on the left and lifetime filter on the right (as shown in Fig. 1b). Select the green detector to bring up the green fluorescence decay and calculate background and fluorescence decay filters by adjusting the background and the decay linear combination (this last adjustment is done automatically but sometimes it is necessary to adjust it by hand). Validate and close the window.
3. Click on "add filter" icon, and another identical window appears. Select the red detector to bring up the red fluorescence decay and calculate background and fluorescence decay filters. Validate and close the window. Now you have four filters: background and fluorescence decay for both green and red detectors.
4. Before calculating auto- and cross-correlation functions in the Symphotime software, you have the option to select the desired filter for the different time traces. It is then straightforward to choose fluorescence decay filter for the green detector (photons coming from GFP after pulsed excitation) and background filter for the red detector (photons coming from mCherry after CW excitation).
5. Inspect by eye the two autocorrelation functions to validate that you are able to measure the diffusion of the two color species. At this step, if the autocorrelation does not show an amplitude (too high protein concentrations or immobile proteins) or if the curve is too noisy (too low fluorescence brightness, or too much autofluorescence), one has to start again the acquisition. If both green and red autocorrelations have the desired good shape, one can determine the diffusion of both green and red proteins by fitting the autocorrelation curves and one can inspect the cross-correlation.
6. Inspect by eye the dual-color cross-correlation obtained with lifetime filters. If it does not exhibit any amplitude, no codiffusion was measured (*see* Fig. 3a, blue curve as example). On the contrary, the amplitude is directly attributed to codiffusion (*see* Fig. 3c, blue curve as example) since spectral bleed-through is corrected by lifetime filters (*see* **Note 8**).

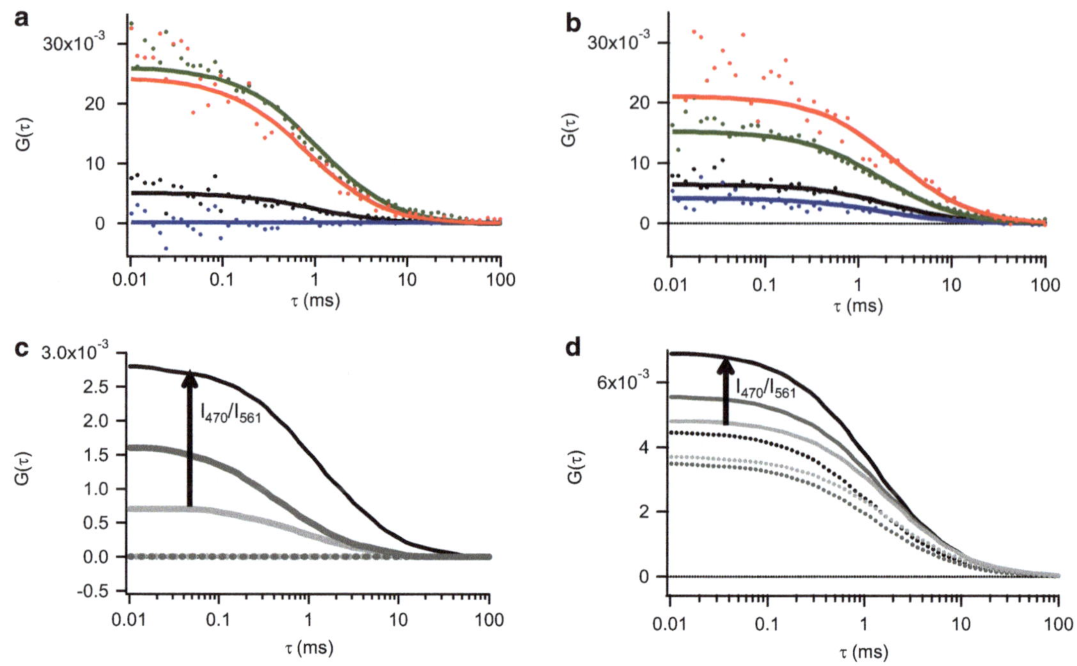

Fig. 3 Dual-color FLCS in live cell expressing EGFP and mCherry alone and in tandem. (**a–b**) Experimental auto- and cross-correlation curves (*dots*) and fitting curves with a single species Brownian diffusion model (*line*) are presented for green autocorrelation (*green*), red autocorrelation (*red*), conventional cross-correlation (*black*), and lifetime filter cross-correlation (*blue*) for a representative cell expressing EGFP and mCherry in separate diffusion (**a**) and expressing EGFP and mCherry in co-diffusion (**b**). (**c–d**) Conventional cross-correlation (*solid line*) and the lifetime filter cross-correlation (*dotted line*) for three consecutive measurements on the same focal volume by decreasing 561 nm laser power without changing the power of the 470 nm laser (from *grey* to *black*) for a cell expressing EGFP and mCherry in separate diffusion (**c**) and expressing EGFP and mCherry in co-diffusion (**d**)

7. Auto- and cross-correlations without lifetime filters can be also calculated. In this case, the dual-color cross-correlation curve always exhibits some amplitude (*see* Fig. 3a, black curve as example) since the spectral bleed-through artifact is always present.

8. To quantify the correlation functions, a fit can be done using, for example, the 3D diffusion with one diffusing species equation:

$$G(\tau) = 1 + G^0 \times \left(1 + \frac{\tau}{\tau_D}\right)^{-1} \times \left(1 + S^2 \cdot \frac{\tau}{\tau_D}\right)^{-1/2} \quad (7)$$

where G^0 is the amplitude, τ_D is the residence time, and S the structural parameter (*see* **Note 7** for the calibration of these parameters). Other models can be chosen (*see* **Note 10**). These fits can be achieved easily through Symphotime. One can use the residence time (or the diffusion coefficient) to interpret the diffusion of the protein complexes. The auto- and cross-correlation amplitudes (G^0_G, G^0_R, and G^0_x for green, red, and cross-correlation) are very helpful to quantify protein interactions: Protein A fused to GFP and protein B fused to mCherry since $G^0{}_x/G^0{}_G = [AB]/[B]$ and $G^0{}_x/G^0{}_R = [AB]/[A]$. With this quantification and sufficient data with different quantity of protein A and protein B, one can obtain the dissociation constant of the complex and relative concentration in live cells.

4 Notes

1. Other cells (e.g., CHOK1, 3T3, etc.) could also be employed, but each cell line should be optimized independently for transient transfection. As pointed out in the protocol, it is crucial to have low expressing cells to avoid confocal volume saturation.
2. Depending on the transfection agent, it is wise to test different DNA plasmid concentrations and different volumes of agent; be careful with cell mortality when increasing DNA concentrations and the volume of agent! In that regard, stable cell lines co-expressing A-GFP and B-mCherry would be desirable; this would avoid the calibration step and optimization for FP expression. These experiments can be carried out in double transgenic animals.
3. In the TTTR file, each single photon is recorded with its global arrival time (Time-Tagged) and the delay time with respect to the corresponding laser pulse (Time-Resolved). Histograms with the delay time give fluorescence decays, and histograms with the global arrival time can be used to form a fluorescence intensity time trace, making all related analysis possible (TCSPC fitting, FCS calculation, time trace TCSPC, time-gated FCS, FLCS, etc.).
4. In the Microtime 200, one has to align the pinhole manually. Also the dichroic and emission filters have to be selected manually and installed in the right place. This is only true in the case of using the Microtime 200 or a homemade system capable of FLCS measurements; in the case of using the Leica LSM SP5 (or SP8) SMD, the optical path can be directly set with the LAS software. This is also valid for another confocal using a PicoQuant attachment like Zeiss 780, for example. Here the pinhole adjustment can be carried out easily with the Zen acquisition software; in other confocal microscopes the pinhole orientation is fixed (i.e., SP5 and SP8 form Leica).
5. Since green spectral bleed-through in the red channel is not problematic in the dual-color FLCS method, red filters for maximum mCherry detection to optimize the brightness can be used.

In conventional dual-color FCS, red-shifted band-pass filters are used to minimize the spectral bleed-through with the consequence of analyzing mCherry with lower brightness.

6. For dual-color FLCS methodology, the temporal response of the detector is not crucial and conventional SPAD or tau-SPAD are sufficient and convenient in the red since their sensitivity is better in the red than MPD SPAD. In contrast, to use the system for fluorescence lifetime measurements using TCSPC fitting MPD, SPAD have about 100 ps temporal response while other SPAD have 600 ps (also depending on photon counting rate). A good compromise would be to combine MPD SPAD in the green channel (which will be usable for EGFP fluorescence lifetime measurements) and tau-SPAD in the red channel for the best mCherry sensitivity in dual-color FLCS measurements. Finally, new hybrid detectors with their high sensitivity, high dynamic (count rate), and high temporal response would be the new standard for all these approaches.

7. In the calibration step, FLCS is not different from any calibration carried out with an FCS apparatus.

   The first approach is to measure the correlation amplitude of samples with known concentration. In order to find the confocal volume, one could prepare a set of solutions (for instance, Alexa 488 and Alexa 568 (Invitrogen) MiliQ water with the next concentration approx: 0.3, 0.75, 1, 1.5, and 3 nM). The solutions employed in the calibration are in the nM regime, and in this case an intensity of around 15–50 kHz is more than enough to obtain a reliable signal. This approach gives a better accuracy and reveals the suitable concentration range to determine the confocal volume. It is always desirable to work with relatively low levels of intensity, although unwanted photophysical effects could occur with high laser power (i.e., the passage to the triplet state (dark species) and/or photobleaching). If the measured signal fluctuations are due to diffusion, the amplitude of the autocorrelation (G_0) equals the inverse number of molecules present in the effective detection volume (V_{eff}). This method allows calculating this volume according to

$$V_{\mathrm{eff}} = 1/G^0 \cdot N_{\mathrm{eff}} \tag{8}$$

where N_{eff} is the number of particles in the effective detection volume.

   The second approach is to approximate the effective volume by a 3-dimensional Gaussian shape. The autocorrelation function coming from the different chromophore solutions can be fitted to a single-component model equation:

$$G(\tau) = 1 + N_{\mathrm{eff}}^{-1} \times \left(1 + \frac{\tau}{\tau_{\mathrm{D}}}\right)^{-1} \times \left(1 + S^2 \cdot \frac{\tau}{\tau_{\mathrm{D}}}\right)^{-1/2} \tag{9}$$

where τ_D is the residence time (it can be interpreted as the average time a molecule needs to transverse the detection volume by diffusion) and S (the structural parameter, $S = waist\,(w_0)/axial\ radius\,(z)$). If the diffusion coefficient (D) and the concentration are known, the size of the effective volume can be extracted from a global fit taking into account that the residence time is a function of the waist (w_0) and the diffusion coefficient (D):

$$\tau_D = \frac{w_0^2}{4D} \tag{10}$$

and also that the number of particles (N_{eff}) in the effective detection volume (V_{eff}) is a function of the axial radius (z) and the waist (w_0):

$$V_{eff} = \pi^3/2 \cdot \omega_0^2 \cdot z \tag{11}$$

Results for the waist and axial focal radius corresponding to the values obtained from the autocorrelation curve fits applied on each one of the curves can give for the blue laser $w_0 = 0.25$ μm and $z = 1.5$ μm and for the green laser $w_0 = 0.32$ μm and $z = 1.9$ μm.

The third approach to determine the effective volume is to image subresolution fluorescent beads (100 nm). Subresolution beads can be considered like a point and be used to scan the confocal volume. The sections in *xy*, *yz*, and *xz* directions were fitted with a 2-dimensional Gaussian distribution in order to determine w_0 and z. Moreover, this approach is interesting to test the detection volume overlap. In cross-correlation analysis, a displacement between the two detection volumes reduces the cross-correlation amplitude. This issue is avoided when only one laser line is used for excitation as in two-photon excitation. Note that the thickness of the coverslip for the calibration should be the same one for live cell measurements. For example, an approximate adjustment for the coverslip thickness #1.5 would correspond to 160–190 mm in the collar objective. One can play between 160 and 190 mm to find the right adjustment for the collar to compensate for nonoptimal axial detection.

8. The efficient removal of spectral bleed-through from EGFP emission into the red detection channel by dual-color FLCS was validated on co-transfected cells expressing free EGFP and mCherry [19]. The cross-correlation curves calculated without (black) and with the lifetime filters (blue), displayed in Fig. 3a, show the efficiency of the use of FLCS to get rid of the green fluorescence detected into the red channel. The positive control was carried out by measuring the amplitude of the cross-correlation when EGFP and mCherry are linked together in the tandem molecule. Applying the lifetime filters to the calculation of the cross-correlation leads to a significant amount of

cross-correlation amplitude for the tandem (Fig. 3b, black curve: without FLCS; blue curve: with FLCS). Moreover, the correction of the spectral bleed-through by FLCS is independent of the respective green and red fluorescence intensities and thus to the percentage of spectral bleed-through in the red detector channel. Indeed, we show that decreasing the 561 nm laser power without changing the power of the 470 nm pulsed laser leads to an increase of the green bleed-through in the red channel that is completely eliminated by using FLCS in cells expressing free EGFP and mCherry (Fig. 3c). In contrast, for cell expressing EGFP and mCherry in tandem, the residual cross-correlation amplitude after lifetime filters application does not change significantly (Fig. 3d). This result is important for the quantification of the amount of interacting proteins in living cells. Indeed, the quantity of spectral bleed-through (and its repercussion to the cross-correlation amplitude) depends on both the relative concentration of green and red species (which is generally not controlled in live cells) and on the relative laser intensities used during the experiment. To use a conventional methodology to get rid of spectral bleed-through by subtracting its quantity to the measured cross-correlation amplitude [6], the two laser powers need to be defined before the experiment. Here, it is convenient to adapt the green and the red laser powers during the experiment, for example, to increase the signal-to-noise ratio or to take care of photobleaching.

9. Have ready a coverslip with non-transfected cells under the same conditions to test for autofluorescence employing the same settings you found for low expressing cells; this measurement will tell you how much distortion you get from autofluorescence.

10. The general equation describing protein motions can be expressed as follows:

$$G(\tau) = \frac{1}{V_{\text{eff}}} \cdot \frac{\sum_i \eta_i < C_i > M_i(\tau)}{(\sum_i \eta_i < C_i >)^2} \tag{12}$$

where η_i is the brightness and C_i is the concentration of the species i. $M_i(\tau)$ is the motion described by the species i and can be, for example,

free 3D diffusion:

$$M_i(\tau) = \frac{1}{\left(1 + \frac{\tau}{\tau_{di}}\right) \cdot \sqrt{1 + S^2 \cdot \frac{\tau}{\tau_{di}}}} \tag{13}$$

Free 2D diffusion:

$$M_i(\tau) = \frac{1}{\left(1 + \frac{\tau}{\tau_{di}}\right)} \tag{14}$$

Active transport with velocity v_i:

$$M_i(\tau) = \mathrm{e}^{-\left(\frac{\tau_i v_i}{\omega_0}\right)^2} \tag{15}$$

where τ_i is the residence time of the species i, z is the axial radius, and w_0 is the waist of the effective volume.

To choose the right model for the fitting procedure, assumptions about the behavior of the biological system have to be made (i.e., number of species, motion of each species, compartmentalization). It is clear that the shape of the auto- or cross-correlation curve cannot give clear evidence to choose the right model. The 3-dimensional diffusion model can be used in a general way, in particular for cytosolic proteins autocorrelation and cross-correlation, and robustness of the fitting procedure can be tested by adjusting the number of populations. Moreover, unwanted photophysical effects, as the passage to the triplet state, could occur and interfere with the fitting procedure. Models can be adjusted to take care of this kind of phenomenon, for example, with 3-dimensional model:

$$G(\tau) = \left(1 - T + T \cdot \mathrm{e}^{-(\tau/\tau_{\mathrm{triplet}})}\right) \cdot \sum \frac{1}{\left(1 + \frac{\tau}{\tau_{\mathrm{d}i}}\right) \cdot \sqrt{1 + S^2 \cdot \frac{\tau}{\tau_{\mathrm{d}i}}}} \tag{16}$$

where T is the triplet fraction of molecules and τ_{triplet} is the lifetime of the triplet state.

Cross-correlation can clearly detect interaction between proteins in the same diffusing complex, but autocorrelation analysis can also detect protein interactions if the labeled molecule binds to a much larger molecule that diffuses significantly more slowly than does the free labeled molecule. When this occurs, the autocorrelation data will be well fit by a model with two diffusing components. This approach has been used in live cells but unfortunately it does not provide estimates of association and dissociation rates and furthermore is not strictly applicable to immobile binding sites. More recently, models taking into account these biological parameters have been developed [20].

References

1. Bacia K, Majoul IV, Schwille P (2002) Probing the endocytic pathway in live cells using dual-color fluorescence cross-correlation analysis. Biophys J 83:1184–1193
2. Maeder CI, Hink MA, Kinkhabwala A et al (2007) Spatial regulation of Fus3 MAP kinase activity through a reaction-diffusion mechanism in yeast pheromone signalling. Nat Cell Biol 9:1319–1326
3. Liu P, Sudhaharan T, Koh RM et al (2007) Investigation of the dimerization of proteins from the epidermal growth factor receptor family by single wavelength fluorescence cross-correlation spectroscopy. Biophys J 93:684–698
4. Lillemeier BF, Mörtelmaier MA, Forstner MB et al (2010) TCR and Lat are expressed on separate protein islands on T cell membranes

and concatenate during activation. Nat Immunol 11:90–96
5. Huet S, Avilov SV, Ferbitz L et al (2010) Nuclear import and assembly of influenza A virus RNA polymerase studied in live cells by fluorescence cross-correlation spectroscopy. J Virol 84:1254–1264
6. Bacia K, Schwille P (2007) Practical guidelines for dual-color fluorescence cross-correlation spectroscopy. Nat Protoc 2:2842–2856
7. Kogure T, Karasawa S, Araki T et al (2006) A fluorescent variant of a protein from the stony coral Montipora facilitates dual-color single-laser fluorescence cross-correlation spectroscopy. Nat Biotechnol 24:577–581
8. Shcherbakova DM, Hink MA, Joosen L et al (2012) An orange fluorescent protein with a large Stokes shift for single-excitation multicolor FCCS and FRET imaging. J Am Chem Soc 134:7913–7923
9. Müller BK, Zaychikov E, Bräuchle C, Lamb DC (2005) Pulsed interleaved excitation. Biophys J 89:3508–3522
10. Thews E, Gerken M, Eckert R et al (2005) Cross talk free fluorescence cross correlation spectroscopy in live cells. Biophys J 89:2069–2076
11. Takahashi Y, Nishimura J, Suzuki A et al (2008) Cross-talk-free fluorescence cross-correlation spectroscopy by the switching method. Cell Struct Funct 33:143–150
12. Ries J, Yu SR, Burkhardt M et al (2009) Modular scanning FCS quantifies receptor-ligand interactions in living multicellular organisms. Nat Methods 6:643–645
13. Böhmer M, Enderlein J (2003) Fluorescence spectroscopy of single molecules under ambient conditions: methodology and technology. Chemphyschem 4:793–808
14. Kapusta P, Wahl M, Benda A et al (2007) Fluorescence lifetime correlation spectroscopy. J Fluoresc 17:43–48
15. Gregor I, Enderlein J (2007) Time-resolved methods in biophysics. 3. Fluorescence lifetime correlation spectroscopy. Photochem Photobiol Sci 6:13–18
16. Ray K, Zhang J, Lakowicz JR (2008) Fluorescence lifetime correlation spectroscopic study of fluorophore-labeled silver nanoparticles. Anal Chem 80:7313–7318
17. Rüttinger S, Kapusta P, Patting M et al (2010) On the resolution capabilities and limits of fluorescence lifetime correlation spectroscopy (FLCS) measurements. J Fluoresc 20: 105–114
18. Chen J, Irudayaraj J (2010) Fluorescence lifetime cross correlation spectroscopy resolves EGFR and antagonist interaction in live cells. Anal Chem 82:6415–6421
19. Padilla-Parra S, Audugé N, Coppey-Moisan M, Tramier M (2011) Dual-color fluorescence lifetime correlation spectroscopy to quantify protein-protein interactions in live cell. Microsc Res Tech 74:788–793
20. Michelman-Ribeiro A, Mazza D, Rosales T et al (2009) Direct measurement of association and dissociation rates of DNA binding in live cells by fluorescence correlation spectroscopy. Biophys J 97(1):337–346

Chapter 32

Brightness Experiments

Patrick J. Macdonald, Jolene Johnson, Yan Chen, and Joachim D. Mueller

Abstract

This chapter presents an overview of quantitative fluorescence brightness experiments with special emphasis on single-color measurements of protein homo-interactions inside living cells. We discuss practical considerations in the choice of the fluorescent labels and the calibration measurements necessary for quantitative interpretation of brightness experiments. Methods to identify and avoid common pitfalls, such as bleaching and saturation, are addressed. We examine the interpretation of brightness data with moment analysis. In particular, we focus on how to avoid or correct for undersampling, as well as how to characterize and adjust for photon detector effects. We conclude by describing brightness titration experiments which determine the binding curve and stoichiometry of a protein from apparent brightness data.

Key words Stoichiometry, Fluctuation, Titration, Two-photon, Photon count moments, Autocorrelation, Dead time, Afterpulsing, Photobleaching

1 Introduction

Fluorescently tagged biomolecules passing through a small optical observation volume give rise to fluorescence bursts. The frequency, duration, and amplitude of these bursts are the main sources of information for most fluorescence fluctuation spectroscopy (FFS) experiments. Here we focus on brightness, which is closely connected to the amplitude of the fluorescence fluctuations [1]. Brightness represents the average photon count rate emitted by a fluorescent bioparticle and is typically expressed in units of counts per second (cps). The main application discussed in this chapter is the calculation of the average number of fluorescently tagged biomolecules per molecular complex. Thus, brightness analysis provides information about the average stoichiometry of a sample. While brightness experiments can be performed both in cells and solution, we emphasize cellular applications, which are the most challenging and most interesting applications of brightness analysis.

Yves Engelborghs and Antonie J.W.G. Visser (eds.), *Fluorescence Spectroscopy and Microscopy: Methods and Protocols*, Methods in Molecular Biology, vol. 1076, DOI 10.1007/978-1-62703-649-8_32,

Our group focuses on two-photon excitation measurements, which are reflected in this chapter. Nevertheless, the selection of fluorophores, measurement of cells, and analysis of data present challenges that are largely independent of the excitation method. We discuss common pitfalls and suggest strategies to avoid them. The potential topics relevant to brightness experiments are vast and occasionally intricate. To keep the subject focused and the material manageable, we restrict our discussion to single-color FFS experiments, analysis of photon count moments, and brightness titration. In our experience, single-color brightness titration provides the most straightforward brightness method to quantify protein homo-interactions in cells.

2 Methods

2.1 The Fluorescent Label

The fluorescent label is the fundamental tool of brightness experiments, and great care should be taken in its selection and characterization. In this chapter we limit ourselves to genetic tagging, i.e., the use of fluorescent proteins as labels. The number of fluorescent proteins available is now quite large and continues to grow, but not all are suitable for brightness measurements. Brightness is important because it encodes stoichiometry. To emphasize this point and to simplify discussion, we introduce the concept of dimensionless brightness (also referred to as normalized brightness),

$$b = \frac{\lambda_{\text{sample}}}{\lambda_{\text{label}}} \tag{1}$$

where λ_{sample} is the brightness of the sample and λ_{label} is the brightness of the label alone (Fig. 1a). By this definition, $b = 1$ represents a monomeric sample (Fig. 1b), and $b = 2$ is a dimeric sample (Fig. 1c). In later sections, we will critically evaluate this straightforward relationship between brightness and protein stoichiometry and generalize it.

2.1.1 Fluorophore Brightness Properties

Because the determination of stoichiometry is predicated on one label per protein, the selected fluorophore label must exist as a monomer in its natural state. Fluorescence proteins that self-associate to form dimers or tetramers ought not to be used for brightness experiments because the measured interactions stem, at least partially, from the labels rather than from the proteins of interest. Therefore these labels are not faithful reporters of the interactions of the protein of interest.

One also needs to be aware of additional properties of the fluorescent label. Ideally, a fluorescent protein exhibits a single brightness state. In this case, the normalized brightness directly reflects the oligomeric state of the protein, i.e., an n-mer labeled protein complex has a brightness $b = n$. Enhanced green

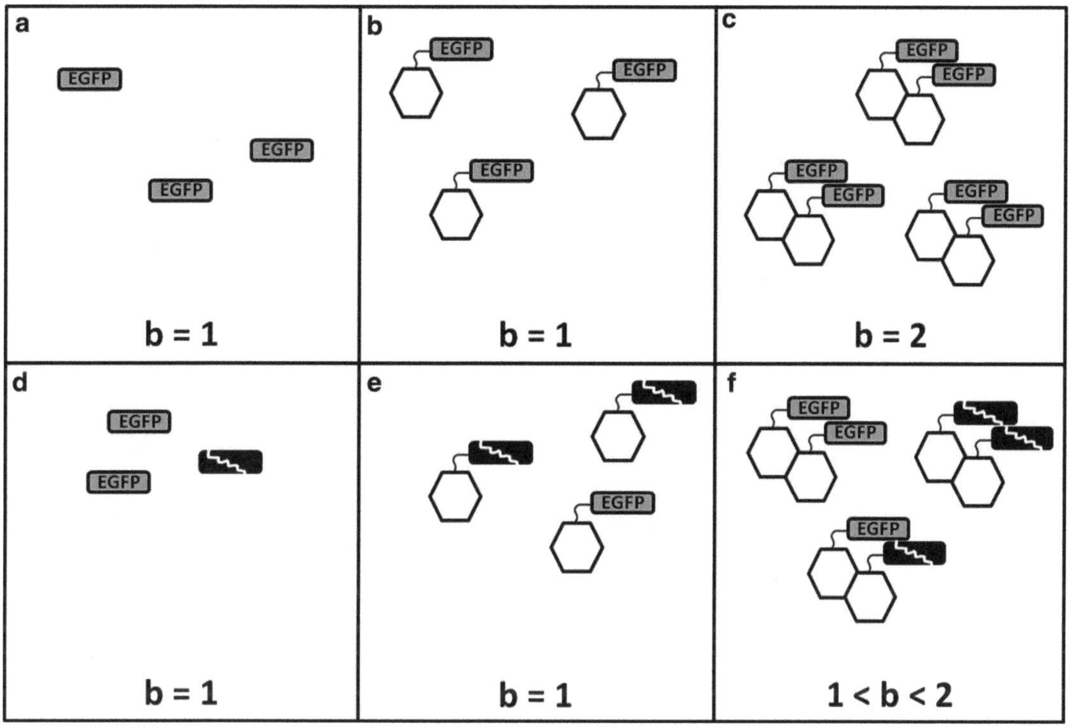

Fig. 1 Brightness, labels, and stoichiometry. The top rank indicates a conceptual picture for the fluorophore EGFP with a single brightness state. Once the inherent brightness of the fluorescent label (**a**) is identified, the information can be used to identify the oligomeric state of a protein of interest (depicted as a *hexagon*). If the protein is monomeric (**b**), the brightness returns a normalized value of 1, while if the protein is dimeric (**c**), the normalized brightness equals 2. However, if a fraction of the label is in a nonfluorescent state, neither the label calibration (**d**) nor a monomeric protein (**e**) will detect the presence of the dark label. In contrast, for a dimeric protein (**f**), the presence of dark label induces a mixture of brightness species, which returns an apparent brightness that falls between the monomer and the dimer

fluorescent protein (EGFP) and enhanced yellow fluorescent protein (EYFP) each have, for all practical purposes, only one brightness state. Unfortunately, some fluorescent proteins exist in more than a single brightness state. We first discuss the special case where a fraction of the labels is nonfluorescent. The nonfluorescent state of a label is also referred to as a dark state. The presence of labels in the dark state destroys straightforward interpretation of the normalized brightness and can lead to misinterpretation of the data. For example, consider a dimeric sample with fluorophore labels, some fraction of which is in a dark state (Fig. 1d). In a monomer sample the brightness will be unaffected, as only the fluorophores which are "on" will be detected (Fig. 1e). However, in a dimer sample, a mixture of bright and dark labels will lead to three populations: bright-bright, dark-dark, and dark-bright (Fig. 1f). The first species returns the expected normalized brightness of $b = 2$, the second species is nonfluorescent and thus not detected, and the third species returns $b = 1$. An FFS experiment of such a mixture usually identifies only a single brightness (*see* **Note 1**), which we refer to as apparent brightness λ. Its value represents the ensemble average over all brightness species (*see* **Note 2**). The specific value depends on the fractional populations of the three species depicted in Fig. 1f, but it is clear that the apparent brightness for this case will fall between 1 and 2 [2].

Because dark states complicate the interpretation of brightness experiments, it is best to avoid the various conditions that introduce dark states. As an example, photobleaching of the sample introduces a population of labels in the dark state. Similarly, the slow maturation kinetics of some fluorescent proteins causes a significant delay between the folding of the protein and the onset of fluorescence, which generates a population of proteins with nonfluorescent labels. Thus, it is prudent to avoid slowly maturing fluorescent proteins, experimental conditions that lead to photobleaching, and fluorescent proteins with an intrinsic population of dark states (*see* **Note 3**).

The presence of a bright and a dark state is just a special case of brightness heterogeneity of the label (*see* **Note 4**). The red fluorescent protein mCherry is another example of brightness heterogeneity and has been successfully modeled as a mixture of a long-lived bright and dim state [3]. Consequently, an mCherry-labeled dimer consists of a mixture of three brightness species (bright-bright, dim-dim, and dim-bright), which leads to a normalized brightness of less than 2. In general, brightness heterogeneity should be avoided, because it also destroys the simple relationship between stoichiometry and normalized brightness. However, brightness heterogeneity of the label does not necessarily prohibit its use in brightness experiments as long as the brightness states of the label are well characterized [3]. We provide a concrete example in Subheading 2.2.

Finally, some fluorophores are brighter than others at an equivalent excitation power. It is difficult to get good signal statistics from dim proteins especially in cells which have non-negligible autofluorescence. Enhancing brightness by increasing the excitation power is limited by the onset of photobleaching. Thus, the selection of the label should also take the intrinsic brightness of the fluorescent protein into account.

2.1.2 Photobleaching and Saturation

Photobleaching is the permanent loss of fluorescence in a fluorophore caused by photon-induced chemical damage. The literature [4–6] rates fluorescent proteins based on photostability and one must choose the most stable fluorophore that is suitable for the proposed experiment. It is also necessary to consider the saturation level of the fluorophore. A fluorophore in the excited state has a finite lifetime before emitting a photon and relaxing back to the ground state. Because of this, there is a maximum rate beyond which the fluorophore will not respond to changes in the excitation intensity as expected. Brightness experiments should stay well below this regime.

To characterize a fluorophore with regard to these phenomena, we begin by performing a power study, measuring the fluorophore at a series of excitation powers and recording the results. For two-photon excitation, both the fluorescence intensity and brightness should scale with the square of the excitation power, but brightness is the more sensitive parameter (*see* **Note 5**). A plot of the measured brightness as a function of power squared (for 2-photon excitation) visually identifies the power at which the brightness is starting to be sublinear. We choose a point approximately 10 % below that divergence as the maximum allowable excitation power. Photobleaching may be detected through a decrease in the fluorescence intensity during the course of the measurement because the cumulative damage leads to a reduction of the fluorescent label population within the finite volume of the cell. The autocorrelation function can also detect photobleaching through a reduction in diffusion time, but this only becomes apparent for severe photobleaching. We prefer the much more sensitive approach of monitoring the fluorescence intensity over a long data acquisition. Regardless, detectable levels of photobleaching are clearly detrimental to brightness measurements as they create a population of nonfluorescent labels and reduce the normalized brightness in the same fashion as was discussed for long-lived dark states. Therefore, one should select a photostable fluorophore and choose conditions that ensure negligible photobleaching.

On the whole, the fluorophore is to a fluorescence experimentalist as is the voice to an opera singer. If one is not aware of and comfortable with all its nuances, strengths, and flaws, the result will not be worth the price of the ticket.

2.2 Calibrations

Another critical feature of a quantitative brightness measurement is a careful calibration. By its nature, the fluctuation intensity signal is noisy and provides little in the way of observable phenomena or immediate evidence of underlying problems. Furthermore, the power of brightness analysis to determine stoichiometry is directly dependent on the brightness of the protein complex relative to the brightness of the chosen label. Therefore, both the successful application of the technique and the careful interpretation of inherently noisy data are founded upon control experiments. Calibrations are required on two levels. The first is a daily measurement of a well-known dye sample. This serves as a standard for the long-term performance of the instrument and an alert to any problems with the laser, the instrument, or the alignment thereof. One common instrument issue is a straightforward—but undocumented—alteration by the previous user. In addition, small particles of dust or finger prints on the optics, slight relaxations of a mirror or dichroic filter, and—in one notable instance—a spider web in the laser cavity have each caused a deterioration in instrument performance that we might not have easily or quickly detected without the record of a daily instrument calibration. Daily brightness (*see* **Note 5**) calibrations prevent such issues from spoiling an experiment and provide a baseline for further optimization. In our calibration procedure, we prepare a quantity of a bright dye solution and then aliquot and freeze it, so that each day's experiment begins with a control measurement performed in the same way, on the same sample, and at the same distance focused into the sample (*see* **Note 6**). In this way, we track the operation of the instrument not only from day to day but from year to year.

The second level of calibration is the sample calibration that establishes the baseline brightness for a given sample preparation. The basic sample calibrations for any brightness experiment in cells involve measuring both the monomeric label and a dimeric construct of said label (generated by cloning a second EGFP sequence into a vector alongside the original EGFP sequence) [2]. These measurements provide controls to relate brightness to stoichiometry. At the simplest level, the monomer sample provides a simple reference point for normalized brightness, and the dimer sample indicates that the brightness measurement and analysis are proceeding as expected, namely,

$$b_n = n \tag{2}$$

where b_n is the normalized brightness of an n-mer sample. Under further consideration, there is more to be learned. Both calibrations should cover a broad range of concentrations by selecting cells that have high, medium, and low expression levels of the protein. A good label will have a constant brightness independent of concentration, demonstrating that brightness is a robust value and that the label will not induce interactions on its own. A successful dimer calibration should return a normalized brightness very close to 2. This indicates the absence of significant brightness heterogeneity of the label, such as introduced by photobleaching, unmatured fluorophores, or dark states. Table 1 shows monomer/dimer label calibrations in different cell lines for the three fluorophores most commonly used in our lab. EGFP and EYFP show good doubling within measurement error, indicating the presence of a single brightness state. However, mCherry has a value less than 2, revealing the presence of brightness heterogeneity of the label. In this case, the expected brightness of an n-mer sample can still be determined from a good monomer/dimer label calibration:

$$b_n = (n-1)b_2 - (n-2)b_1 \tag{3}$$

where b_2 is the experimentally determined brightness from the dimeric label calibration and $b_1 \equiv 1$. (Note that Eq. 3 reduces to

Table 1
Fluorescent protein dimer calibrations

Normalized brightness, *b*	$EYFP_2$	$EGFP_2$	$mCherry_2$
293T cells	1.95 ± 0.06	1.90 ± 0.07	1.62 ± 0.12
COS-1 cells	1.92 ± 0.07	1.95 ± 0.05	1.65 ± 0.10

The normalized brightness results of a monomer/dimer label calibration of three fluorescent proteins in two different cell lines. Each result is based on two experiments in which approximately 15 cells expressing monomers and 15 cells expressing dimers were measured. EGFP and EYFP display good doubling within error. The lower values of mCherry indicate the presence of brightness inhomogeneity of the label

Eq. 2 when $b_2 = 2$.) It is always best to choose a fluorophore with a single, uniform brightness state and thus have the simple brightness-stoichiometry relationship. However, when circumstances (like color choice) drive the fluorophore selection, proteins with brightness heterogeneity, like mCherry, can still be made to work in quantitative brightness experiments, provided that those fluorophores are well characterized. Brightness experiments can quickly and simply provide stoichiometric information, but the experiments must be well designed and carefully interpreted. Good calibrations are at the core of this methodology (*see* **Note 7**).

2.3 Cell Measurements

The first practical consideration in cell measurements is the choice of cell line. In most cases, this will be driven by the relevant biology, but it is worth mentioning that choosing cells that are easy to transfect, adhere well to surfaces, and are large in size simplifies brightness measurements. Having chosen a cell line, it is important to become familiar with it. Cells expressing the labeled protein of interest should look just the same as untransfected cells. Otherwise, a cell under stress may exhibit abnormal biology, adding scatter or a bias to the brightness data. We consider it good practice to make a brief inspection of every cell before measurement to ensure that the morphology appears healthy.

The next goal is to avoid photobleaching of the labeled proteins. Not all of the prepared cells express the labeled protein, so efficient identification and selection of measureable cells requires viewing the cells under epifluorescent light. This also makes it possible to more evenly sample across a broad protein concentration range. However, it is very important to limit such wide-field excitation of the cells to preclude photobleaching, so we endeavor to make our cell selections "briskly," only turning on the lamp when actively viewing the cells. Before beginning the experiment, select a cell and illuminate it under epifluorescence for a long period while monitoring the fluorescence intensity. Having identified the characteristic time for decay of the fluorescence signal, one can select a shorter period during which epifluorescent photobleaching

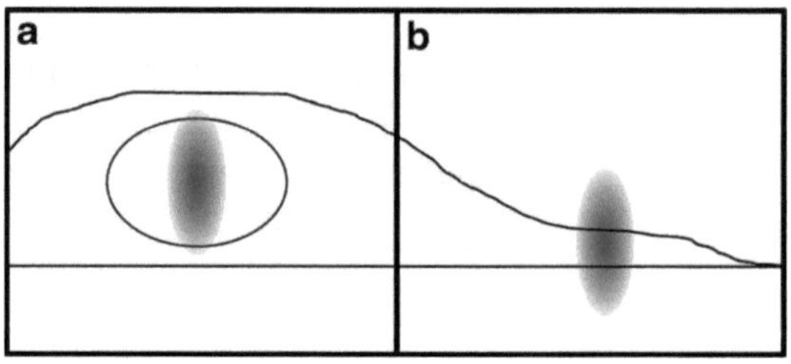

Fig. 2 Cell geometry. The relationship between the PSF of the laser excitation and cell geometry plays an important role in the interpretation of quantitative brightness experiments. When measuring in the nuclei of most cell lines, the PSF will be completely contained (**a**) and the sample can be considered effectively infinite in thickness. (**b**) In thin samples, like cell cytoplasm, the PSF will likely extend beyond the sample, limiting the region of excitation accessible to the fluorophores. In such a case, or when membrane-bound protein fractions create inhomogeneous layers, a z-scan approach and theory must be used to recover accurate data

is negligible. Similarly, it is vital to limit the photobleaching caused by the laser excitation while measuring. Continuous decay in the fluorescence intensity during the measurement points toward a reduction of the fluorescent population due to photobleaching. We recommend taking some measurements 5× longer than the typical acquisition. If there is no decay in the fluorescence trace, photobleaching effects may be considered negligible at that power for that fluorophore (*see* **Note 8**). If photobleaching is detected, reduce the excitation power. However, reducing the excitation power also produces a weaker fluctuation signal which imposes a lower limit on this approach. In the event that one cannot identify an excitation power that has a sufficiently strong signal and avoids photobleaching, it is necessary to redesign the experiment.

Lastly, when planning a cell measurement, it is important to consider the geometry of the cell and protein localization within it. For a nuclear protein, we position the excitation volume in the nucleus, which, aside from the easily seen nucleoli, is an open space free from organelles and internal membranes. Importantly, the nucleus is typically large enough to completely contain the excitation volume as depicted in Fig. 2a. In this scenario, there is usually little cause for concern (*see* **Note 9**). However, if the protein must be measured in the cytoplasm, the height of the cell will most likely be shorter than the excitation volume. In this case (Fig. 2b), the proteins are restricted to the central, high-intensity region of the excitation volume which will bias the calculated brightness upwards. Using a z-scan approach can identify and correct for such thickness-dependent artifacts [7].

We now turn to the execution of cell brightness measurements. We present here the protocol for performing a basic monomer/dimer calibration of the fluorescent label in cells:

1. Focus on the cell layer under bright field illumination. Being close to the correct focus will reduce the cells' exposure time under epifluorescent light.
2. Using epifluorescent light (of the correct wavelength for the selected fluorophore), quickly find a cell that expresses the fluorescent protein (*see* **Note 10**). We often search the cell population in a raster-like pattern to avoid unknowingly remeasuring cells.
3. Having found a cell that expresses the fluorophore, return to bright field illumination to focus and to ensure that the cell appears healthy and exhibits the proper morphology. Position the cell for measurement, avoiding large organelles and membranes.
4. Open the laser excitation shutter and focus to maximize fluorescence intensity on the photon detector. In a uniform sample, this will position the excitation light at the vertical center of the sample (*see* **Note 11**).
5. Take 30–60 s of data and calculate the brightness.
6. Repeat for 10–15 cells, using epifluorescence to select cells with low, medium, and high expression levels. This establishes that the label brightness is independent of concentration and determines the standard deviation of the brightness (~10 %).
7. Repeat for 10–15 cells expressing the dimeric fluorophore construct.

The brightness experiments on the labeled protein of interest are conducted once the calibration of the fluorescent label is established.

It is important to be aware that the internal regions of cells are not necessarily homogeneous. Nonfluorescent aggregates, vesicles, or other cell components can drift through the excitation volume and temporarily reduce the signal. Similarly, fluorescent aggregates or vesicles containing fluorescent material can cause bright spikes in the intensity trace. Brightness analysis depends on a stationary signal, i.e., a stable, average fluorescence intensity, around which the information-carrying fluctuations occur. Because even a single large spike or drop in intensity can skew the average intensity, we do not treat the measurement as a single data set. Whenever the strength of the signal statistics permits, we chop the data up into segments, analyze and display each segment individually, and then average the results. This serves two purposes. First, drops or spikes in the intensity trace are transient and will bias the average only in one of the divided segments. Thus, the averaged brightness per segment provides a much more robust value than analysis of the complete data set [8]. Second, the individual segment analysis provides a kind of miniature time series, and observing the brightness variation across sequential segments can generate a much clearer picture of the homogeneity (or inhomogeneity) of the sample than could be determined from a single brightness value or from the intensity trace alone. We emphasize again that the strength of brightness analysis and FFS as a quantitative technique relies on accurate interpretation of the data.

2.4 Analysis of Brightness Data

This section focuses on the calculation of brightness from experimental data. Most experiments are conducted on samples representing a mixture of brightness species. The signal statistics from a single FFS measurement are frequently insufficient to directly resolve a mixture of two or more brightness species, particularly in cellular FFS experiments (*see* **Note 1**). Given these constraints the best approach is to forgo direct resolution of brightness species in favor of calculating the apparent brightness $\tilde{\lambda}$ of the sample. Recall that the apparent brightness reflects the average brightness of all the fluorescent species present in the sample (*see* **Note 2**) and is the most straightforward and robust brightness analysis of FFS data.

2.4.1 Brightness and Autocorrelation Function

Ideally, an FFS instrument provides access to the raw photon counts, which provides maximum flexibility in conducting brightness analysis. However, even if the instrument only grants access to the autocorrelation function (ACF), the apparent brightness can be determined. The first step requires a fit of the autocorrelation function to find the fluctuation amplitude $g(0)$. In the second step, the apparent brightness is calculated from the fluctuation amplitude, the average fluorescence intensity $\langle F \rangle$, and the shape factor γ_2,

$$\tilde{\lambda} = \frac{g(0)\langle F \rangle}{\gamma_2}. \quad (4)$$

Although this analysis is straightforward, there are a few challenges. First and foremost, correct identification of the fluctuation amplitude requires proper treatment of afterpulsing and kinetic processes in fitting the autocorrelation function. Second, the cellular environment can give rise to autocorrelation functions with amplitude offsets and decay shapes that are challenging to fit. Third, the noise in the calculated autocorrelation function can be sufficiently large that a determination of the fluctuation amplitude is not feasible.

2.4.2 Brightness and Photon Count Moments

Access to the raw photon counts of an FFS experiments allows a direct calculation of the photon count moments. Each data point represents the number of photons k detected during a time interval T. The first two moments of the photon counts determine Mandel's Q-parameter

$$Q_T = \frac{\langle \Delta k^2 \rangle_T - \langle k \rangle_T}{\langle k \rangle_T} = \gamma_2 \tilde{\lambda} \frac{B_2(T, \tau_D)}{T}, \quad (5)$$

which in turn can be related to the apparent brightness. The subscript T is included to highlight that moments were calculated from photon counts sampled with a time interval T. Note that unlike Mandel's Q-parameter, the apparent brightness is model dependent, because the value of the shape factor γ_2 is determined by the point spread function (PSF) model (*see* **Note 12**) as is the case in ACF analysis (Eq. 4).

The second important factor is the binning function $B_2(T, \tau_D)$ (*see* **Note 13**). This function takes into account the diffusion of fluorophores during the sampling time T. The diffusion time τ_D represents the average time for a fluorophore to pass through the FFS observation volume. If we sample the data much slower than the diffusion time, we average away most of the fluctuations, which we refer to as undersampling. The binning function corrects for the undersampling effect. If we sample much faster than the diffusion time, no undersampling occurs. In this limit the binning function is approximated by the sampling time squared, $B_2(T, \tau_D) \approx T^2$, and Mandel's Q-parameter is

$$Q_T = \gamma_2 \tilde{\lambda} T, \tag{6}$$

which provides a straightforward calculation of the apparent brightness. As a rule of thumb, undersampling effects are negligible when the sampling time is factor of 10 faster than the diffusion time (*see* **Note 14**).

2.4.3 Sampling Time

While it is generally a good idea to avoid undersampling and thus determine brightness using Eq. 6, signal-to-noise considerations might force one to rethink fast sampling. Mandel's Q-parameter is a crucial measure of the signal to noise of an FFS experiment (*see* **Note 15**). Faster sampling leads to a reduction of Q_T and thus to a reduction in the signal to noise of the FFS experiment. These ideas are fully developed in the literature [9]. Because of the complexity and highly mathematical nature of this study, we will limit ourselves to the most important conclusions which may be understood intuitively. Both too fast and too slow sampling result in a reduction of the signal-to-noise ratio. Fast sampling results in very few photon counts from the fluorophore per sample time T, while long sampling times result in averaging over most fluctuations. Between these two extremes is a sweet spot that maximizes the number of collected photons, while minimizing the detrimental effects of undersampling. Quantitative analysis shows that the optimal sampling time is on the order of the diffusion time, $T \approx \tau_D$.

We rarely use such a slow sampling time, because the necessary corrections due to undersampling require knowledge of the diffusion time. The uncertainty of measuring the diffusion time from the ACF in cells leads to uncertainty in the correction factor. Thus, we typically start out with a sampling time that is more than one order of magnitude faster than the estimated diffusion time to avoid this complication. Only if the signal to noise of the measurement is insufficient do we increase T stepwise (*see* **Note 16**) to find conditions that allow for brightness analysis with the smallest undersampling correction possible (*see* **Note 17**).

2.4.4 Detector Artifacts

All brightness experiments that employ photodetectors, such as avalanche photodiodes (APDs), must contend with two primary artifacts, dead time and afterpulsing. Dead time is a fixed period after the collection of a photon during which the detector cannot register any other events. This temporary "blindness" leads to a decrease in the number of photons detected and is of particular concern at high fluorescence intensities. At high count rates more events are lost since there is a greater probability of additional photons arriving while the detector is blind to new events. This leads to an artificial decrease of the brightness. Afterpulses are spurious events that occur with a small probability after the detection of an actual photon and lead to an increase in the number of apparent photons detected, which artificially increases the brightness. However, for some APDs the afterpulse probability is sufficiently low ($p_A < 0.002$) that its effect on FFS experiments is usually negligible.

FFS experiments measure nonideal photon count moments, which are denoted by $\langle k \rangle'_T$ and $\langle \Delta k^2 \rangle'_T$. Thus, the measured Q-parameter is given by

$$Q'_T = \frac{\langle \Delta k^2 \rangle'_T - \langle k \rangle'_T}{\langle k \rangle'_T} == \gamma_2 \tilde{\lambda}' \frac{B_2(T, \tau_D)}{T}, \tag{7}$$

which leads to a dead time- and afterpulsing-biased brightness $\tilde{\lambda}'$. Methods to correct for this bias have been developed and tested [10, 11]. Two detector parameters, the dead time $\tau_\dagger$ and the afterpulsing probability p_A, are needed to account for the detector bias. The next section discusses how these parameters are determined experimentally. Because the treatment of nonideal photon count moments leads to intricate formulas, we discuss here the case of first-order correction in dead time and afterpulsing.

The following two equations relate the ideal Q_T and $\langle k \rangle_T$ to the nonideal Q'_T and $\langle k \rangle'_T$,

$$Q_T + p_A(2 + Q_T) - \tau_\dagger T \left(2\langle k \rangle_T + 2Q_T + 3\langle k \rangle_T Q_T - Q_T^2 + 3Q_T^2 \frac{\gamma_3}{\gamma_2^2} \frac{B_3 T}{B_2^2} \right) - Q'_T = 0 \tag{8}$$

$$\langle k \rangle_T \left(1 + p_A - \tau_\dagger T Q_T \right) - \langle k \rangle_T^2 \tau_\dagger T - \langle k \rangle'_T = 0 \tag{9}$$

Because we measure Q'_T and $\langle k \rangle'_T$ and not Q_T and $\langle k \rangle_T$, we need to solve the two coupled equations in an iterative process. Since the average counts change only slightly in the presence of dead time and afterpulsing, we first use Eq. 9 to find an initial Q_T value by making the approximation $\langle k\rangle'_T \approx \langle k\rangle_T$. Using this Q_T value, we then find a new approximation for $\langle k\rangle_T$ using Eq. 8. This new value of $\langle k\rangle_T$ is then reinserted into Eq. 9 to find an improved approximation for Q_T. This procedure is repeated iteratively until the series of Q_T values converge, which takes only a couple of iterations. Once Q_T is found, the apparent brightness is calculated with Eq. 5 as previously discussed.

Note that Eq. 8 contains the factor $3Q_T^2 \frac{\gamma_3}{\gamma_2^2} \frac{B_3 T}{B_2^2}$, which requires the third-order shape factor γ_3 and the third-order binning function $B_3(T, \tau_D)$. The value of γ_3 depends on the PSF model and is easy to determine (*see* **Note 18**). However, because there are no analytical formulas describing B_3, it has to be determined numerically [9]. This complication can be bypassed by choosing a short enough sampling time T. In the absence of undersampling, the above factor reduces to $3Q_T^2 \frac{\gamma_3}{\gamma_2^2}$, which is straightforward to evaluate.

It is essential to perform a control experiment that demonstrates the absence of detector artifacts in brightness experiments. This is best accomplished by measuring a bright dye over a wide concentration range. We frequently use a solution of Alexa 488 in 50 % glycerol/water. The concentration of the initial solution is chosen so that the fluorescence intensity is around 10^6 cps and an FFS measurement is recorded. Next, the sample is diluted successively followed by an FFS measurement after each dilution step. Because dilution only changes the concentration but not the brightness of the sample, we expect to observe a constant brightness as a function of concentration. However, if we directly calculate the brightness λ' from the uncorrected moments (Eq. 7), a strong decrease of brightness with fluorescence intensity is observed (Fig. 3). To avoid this bias we account for the nonideal detector by using Eqs. 8 and 9. The detector used for the control experiment has a dead time of 26 ns and afterpulsing probability of 0.0026. The corrected brightness λ is, within experimental uncertainty, independent of the fluorescence intensity (Fig. 3). This result demonstrates that analysis of brightness data will be free of detector artifacts provided the fluorescence intensity is $\leq 10^6$ cps.

2.4.5 Detector Characterization

While detectors often include factory specifications for dead time and afterpulsing, these numbers need to be independently verified to perform accurate experiments. A very highly concentrated fluorescent dye solution ($c < 10$ μM) is used to imitate a constant fluorescence intensity light source. Mandel's Q-parameter characterizes the width of the photon count distribution. For a constant intensity light source, the photon count distribution is Poissonian, which results in $Q_T = 0$. However, the nonideal detector effects modify the Q_T value of a constant intensity light source as follows:

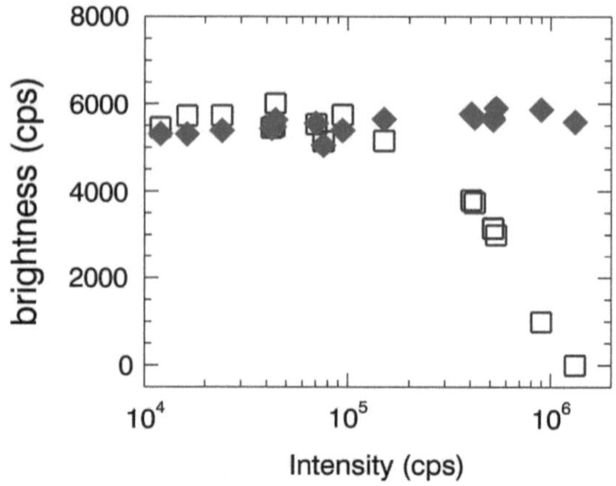

Fig. 3 Brightness as a function of dye concentration. The fluorescent dye Alexa 488 dissolved in a buffer solution of 50 % glycerol and 50 % HEPES buffer was measured by FFS. The dye solution was successively diluted with buffer and an FFS measurement was conducted after each dilution step. The brightness was calculated and graphed versus the fluorescence intensity of the sample for all dilutions. The reduction in brightness (*hollow squares*) with increasing intensity is due to detector dead time. The *diamonds* represent the brightness calculated with nonideal detector effects taken into account (Eqs. 8 and 9) and result in a constant brightness independent of intensity. A dead time of 26 ns and an afterpulsing probability of 0.0026 are used to model the detector

$$Q_T = -2\tau_{\dagger}\langle F\rangle + 2p_A. \tag{10}$$

Q_T is calculated as a function of fluorescence intensity $\langle F\rangle$ of the sample, which is varied by changing the power of the excitation light using neutral density filters. We graph Q_T versus $\langle F\rangle$ and fit the data to a straight line (*see* **Note 19**). The slope and intercept determine the dead time and the afterpulsing probability.

2.5 Brightness Titration

Apparent brightness represents the brightness averaged over all the species present in a sample. Because the complexity of the sample has been reduced to a single numerical value, it is prudent to couple it with another observable that leads to predictable changes in apparent brightness. We are specifically interested in protein association. Because protein association is concentration dependent, a binding curve of the association process can be constructed by systematic variation of the protein concentration and simultaneous brightness measurements of the sample. As cells vary in the level of protein expressed, we have access to a wide concentration range by selecting cells from low to high expression levels. For each cell, the normalized apparent brightness b is graphed as a function of the fluorescently labeled protein concentration. Such a plot characterizes the binding curve of a cellular protein in its native environment and is referred to as brightness titration. The principle of the brightness titration plot is

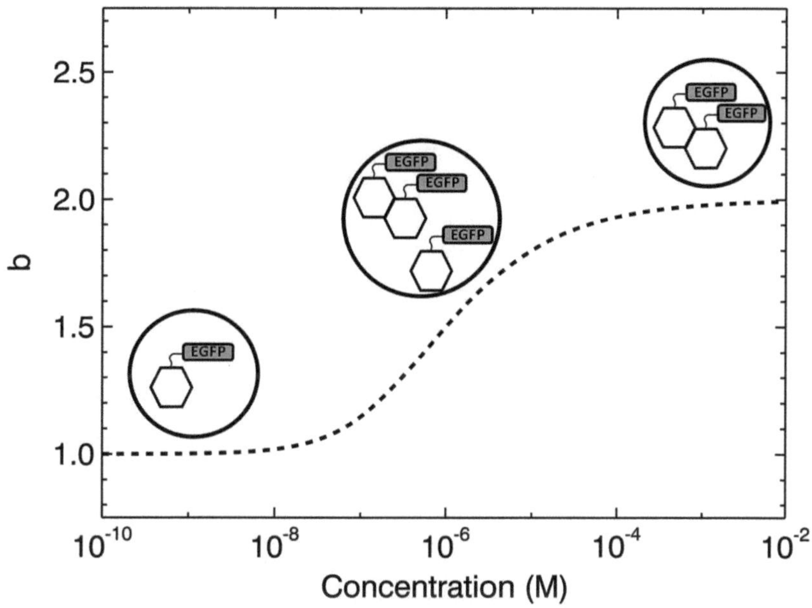

Fig. 4 Brightness titration. We display the curve for a hypothetical monomer-dimer equilibrium of a protein with dissociation constant $K_d = 1$ μM. By selecting different cells that express varying levels of the protein, we can determine normalized, apparent brightness as a function of protein concentration. At low concentration, the protein is primarily monomeric and has $b = 1$, while at high concentrations the protein is primarily dimeric and has $b = 2$. In between these extremes, there is a mixed population and the apparent brightness returns an average weighted by the fractional intensity of each species

illustrated in Fig. 4, which shows the curve of a hypothetical monomer-dimer protein equilibrium. The normalized brightness is close to 1 at low concentrations, which indicates a monomeric protein state. Because higher concentrations favor dimerization, the normalized brightness climbs with increasing concentration and saturates at $b = 2$, indicating the presence of a dimeric protein population. The brightness values at intermediate concentrations are between 1 and 2, which indicates a sample containing a mixture of monomeric and dimeric proteins. Thus the brightness titration curve provides information about the average degree of association of a protein sample and the midpoint of the curve indicates the strength of interaction. The scatter in the data points also provides useful feedback about the quality of the experiment (*see* **Note 20**).

3 Notes

1. The conditions of single-color brightness experiments in the cell typically prohibit the resolution of individual brightness species from a mixed population. Photobleaching-safe

excitation powers and dim fluorescent proteins (compared to organic dyes) reduce the signal strength. At the same time, transfected proteins often express at high concentration which, while increasing the average signal, causes the fluctuations (with their valuable information) to represent a smaller fraction of the total signal. Therefore, in the cellular environment we focus on the average or apparent brightness.

2. The apparent brightness for a mixture of m species with brightness λ_i and number of molecules N_i for the ith species is usually defined by

$$\tilde{\lambda} = \frac{\sum_{i=1}^{m} \lambda_i^2 N_i}{\sum_{i=1}^{m} \lambda_i N_i} \tag{11}$$

Because this equation is quadratic in the brightness λ_i, it is useful to rewrite it using the fractional intensity f_i of species i,

$$f_i = \frac{\langle F_i \rangle}{\langle F \rangle} = \frac{\lambda_i N_i}{\sum_{i=1}^{N} \lambda_i N_i}, \tag{12}$$

where $\langle F_i \rangle$ is the fluorescence intensity of species i and $\langle F \rangle$ is the total fluorescence of the sample. Combining these two equations demonstrates that the apparent brightness reflects a fractional intensity weighted brightness of the sample,

$$\tilde{\lambda} = \sum_{i=1}^{N} f_i \lambda_i. \tag{13}$$

3. Some fluorescent proteins display a fast blinking process which complicates brightness analysis. For example, fast switching between a dark and bright fluorophore state results in blinking. A cross-correlation measurement with a 50/50 beam splitter provides a straightforward method to identify the presence of blinking. We confirmed that neither EGFP, nor EYFP, nor mCherry display a fast blinking process with two-photon excitation at the powers used in our experiments.

4. Some fluorescent proteins undergo photoconversion at certain excitation wavelengths, which also introduces brightness heterogeneity of the label.

5. Fluorescence intensity is equal to the product of the brightness and the number of molecules present in the observation volume. A decrease in brightness may be compensated by an increase in the number of particles, caused, for instance, by a broadened excitation volume due to saturation or a change in instrument alignment. For this reason brightness provides a better diagnostic measure than fluorescence intensity.

6. Fluorescence fluctuation experiments are typically performed with oil or water immersion objectives. Oil immersion objectives unlike water immersion objectives do not suffer evaporation of the immersion liquid, which is convenient for long measurements and may be necessary for use with blower-style stage heaters. However, oil immersion objectives are subject to spherical aberrations, stemming from an index of refraction mismatch between the oil and the sample medium, that lead to significant intensity and brightness variations when the distance of focus into the sample changes by more than a few micrometers. For this reason, calibration measurements with oil immersion objectives should always be conducted at the same height. Water immersion objectives do not suffer from the same artifact provided the correct coverslip thickness is used.
7. When designing a brightness experiment, it is important to think beyond the basic brightness controls and consider what other conditions may affect the quality of the data, e.g., background effects and spectral crosstalk. For instance, it is very important to know the fluorescence intensity and brightness of the sample medium in which the protein is suspended. Thus, particularly when beginning a new series of experiments, FFS measurements should be taken of the buffer alone or of an untransfected cell. On a known system these measurements are less crucial unless the experiments are conducted at a low concentration that is close to the background level, in which case is again very important to have parallel and freshly prepared "blank" controls. When the fluorescence signal exceeds the background by a factor of 10, background effects will be less than 10 % and may be considered negligible as they fall within the expected noise of brightness experiments.
8. When measuring the excitation power of the laser after it passes through the objective, we consider 0.2–0.3 mW to be a relatively safe power for brightness measurements on EGFP.
9. If proteins have different stoichiometric behaviors in different cell compartments (nucleus, nuclear membrane, cytoplasm, plasma membrane) or appear at different concentrations, the coexcitation of multiple compartments makes the data difficult to interpret. For example, since the excitation light does not have a uniform intensity, a particle passing through the central high-intensity region appears brighter than one which passes through the dimmer tail region. Conventional FFS assumes that a uniform population of diffusing particles explores the whole excitation volume, and so a spatiotemporal average provides a good characterization of the sample. If, however, a low concentration of protein in the nucleus fills the central region, while a high concentration of protein bound to the nuclear membrane is excited only by the much dimmer tail, the

calculated brightness is artificially reduced. In this scenario, one must employ a z-scan approach to identify the layers and develop a model to fit and correctly interpret the data [7].

10. When employing epifluorescent light, set the light source to the lowest power at which cells in the desired concentration range are visible. We use an iris in the light path so that only the eyepiece/camera field of view is illuminated in an additional effort to avoid widespread photobleaching.
11. Beware of immobilized or very slowly diffusing proteins such as those bound to a membrane. Even under conditions where proteins in solution experience negligible photobleaching, immobilized proteins endure constant exposure and so can bleach away completely during the time it takes to focus the laser into the sample. To test for this, we first use an extremely low power (just achieves a visible fluorescence signal above background) to maximize the intensity and so focus in the sample. We may also move laterally a small distance to a new location before opening the shutter at the usual laser power. A rapid intensity decrease in the first split-second points to a protein fraction that is immobilized.
12. Three different PSF models are widely used. The three-dimensional Gaussian PSF with $\gamma_2 = 1/(2\sqrt{2})$, the two-dimensional Gaussian PSF with $\gamma_2 = 1/2$, and the squared Gaussian-Lorentzian PSF with $\gamma_2 = 3/16$. Additional modifications to these PSF models or any other PSF model require an explicit calculation of the shape factor γ_2 for consistent interpretation of brightness values. Because reported brightness values depend on the PSF, it is imperative that the assumed PSF model is explicitly mentioned in any brightness study.
13. The binning function B_2 depends on the PSF model. Analytical forms of the binning function are available for the two-dimensional Gaussian PSF,

$$B_{2,2\mathrm{DG}}(T, \tau_\mathrm{D}) = -2\tau_\mathrm{D}\left(T + (T + \tau_\mathrm{D})\, Log \frac{\tau_\mathrm{D}}{T + \tau_\mathrm{D}}\right) \tag{14}$$

and three-dimensional Gaussian PSF,

$$B_{2,3\mathrm{DG}}(T, \tau_\mathrm{D}) = \frac{4r\tau_\mathrm{D}^2}{s}\left(rs - s\sqrt{r^2 + x} - (1 + x)\, Log\left[\frac{(r - s)(s + \sqrt{r^2 + x})}{\sqrt{1 + x}}\right]\right), \tag{15}$$

which uses the dimensionless sample time, $x = T/\tau_\mathrm{D}$, the parameter $s = \sqrt{r^2 - 1}$, and r which represents the axial to radial beam waist ratio. The binning function $B_{2,\mathrm{sGL}}$ for the squared Gaussian-Lorentzian PSF is well approximated by the binning function $B_{2,3\mathrm{GL}}$ for the three-dimensional Gaussian PSF with a beam waist ratio of ~5.

14. For a dimensionless sampling time of $x = T/\tau_D = 0.1$, the brightness correction due to the binning function is less than 10 %. Because brightness measurements in cells typically have an intrinsic uncertainty of 10 %, the effect of undersampling on brightness is negligible.
15. In earlier publications [1], we used an alternative brightness definition, $\varepsilon = Q_T/\gamma_2$, which, in the absence of undersampling, directly reflects the number of photon counts per sampling time T, $\varepsilon = \tilde{\lambda} T$. This parameter has been crucial in photon counting histogram (PCH) analysis, because it is a measure of the deviation of the PCH from a Poissonian distribution, which contains all the information about the sample. Measuring brightness values and resolving species becomes easier the larger the parameter ε. Because γ_2 is just a numerical factor, Q_T and ε share the same significance in determining the quality of an FFS experiment.
16. Of course it is not necessary to retake data if the original sampling time T is too fast for brightness analysis. The sequence of photon counts can be rebinned to a new sampling time $2T$ in software by adding up neighboring photon counts to create a new sequence. The same principle allows us to create any integer multiple of the original sampling time T.
17. Analyzing the photon count moments simultaneously at many different sampling times instead of at just a single sampling time is a more powerful approach and is known as TIFCA (time-integrated fluorescence cumulant analysis), where cumulants are essentially a special type of moments with properties especially suited for FFS experiments. TIFCA is more robust than simple photon count moment analysis, determines the diffusion time of the sample, and by including higher-order moments provides information about a mixture of brightness species.
18. For convenience, we report the third-order shape factor for the two-dimensional Gaussian PSF ($\gamma_3 = 1/3$), the three-dimensional Gaussian PSF ($\gamma_3 = 1/(3\sqrt{3})$), and the squared Gaussian-Lorentzian PSF ($\gamma_3 = 35/384$).
19. Note that Eq. 10 is an approximation. The linear relationship between Q_T and $\langle F \rangle$ breaks down at sufficiently high intensities. Additionally, this equation measures the total dead time of the system. If the dead time is significantly higher than the manufacturer's specifications, this may be due to dead time of the data acquisition card or other hardware issues.
20. We expect, for a well-behaved system, that the brightness titration curve varies smoothly with concentration. It has been our experience that decent FFS experiments in cells lead to a scatter in brightness with a standard deviation of ~10 %. Thus, brightness titration data that vary smoothly with concentration on top of the expected scatter indicate a successful experiment.

Acknowledgments

This work was supported by grants from the National Institutes of Health (GM64589 & GM091743) and the National Science Foundation (PHY-0346782).

References

1. Chen Y, Mueller JD, So PT et al (1999) The photon counting histogram in fluorescence fluctuation spectroscopy. Biophys J 77:553–567
2. Chen Y, Wei L-N, Mueller JD (2003) Probing protein oligomerization in living cells with fluorescence fluctuation spectroscopy. Proc Natl Acad Sci USA 100:15492–15497
3. Wu B, Chen Y, Mueller JD (2009) Fluorescence fluctuation spectroscopy of mCherry in living cells. Biophys J 96:2391–2404
4. Day RN, Davidson MW (2009) The fluorescent protein palette: tools for cellular imaging. Chem Soc Rev 38:2887
5. Day RN, Davidson MW (2009) *Aequorea victoria* fluorescent proteins. Zeiss Online Campus. http://zeiss-campus.magnet.fsu.edu/articles/probes/jellyfishfps.html. Accessed 01 September 2012
6. Piston DW, Robertson RE, Day R.N., et al. (2009) Anthozoa fluorescent proteins. Zeiss Online Campus. http://zeiss-campus.magnet.fsu.edu/articles/probes/anthozoafps.html. Accessed 01 September 2012
7. Macdonald PJ, Chen Y, Wang X et al (2010) Brightness analysis by z-scan fluorescence fluctuation spectroscopy for the study of protein interactions within living cells. Biophys J 99:979–988
8. Chen Y, Mueller JD, Ruan Q et al (2002) Molecular brightness characterization of EGFP in vivo by fluorescence fluctuation spectroscopy. Biophys J 82:133–144
9. Wu B, Mueller JD (2005) Time-integrated fluorescence cumulant analysis in fluorescence fluctuation spectroscopy. Biophys J 89:2721–2735
10. Hillesheim LN, Mueller JD (2003) The photon counting histogram in fluorescence fluctuation spectroscopy with non-ideal photodetectors. Biophys J 85:1948–1958
11. Sanchez-Andres A, Chen Y, Mueller JD (2005) Molecular brightness determined from a generalized form of Mandel's Q-parameter. Biophys J 89:3531–3547

Chapter 33

Global Analysis of Autocorrelation Functions and Photon Counting Distributions in Fluorescence Fluctuation Spectroscopy

Victor V. Skakun, Anatoli V. Digris, and Vladimir V. Apanasovich

Abstract

In fluorescence correlation spectroscopy (FCS) and photon counting histogram (PCH) analysis, the same experimental fluorescence intensity fluctuations are used, but each analytical method focuses on a different property of the signal. The time-dependent decay of the correlation of fluorescence fluctuations is measured in FCS yielding molecular diffusion coefficients and triplet-state parameters such as fraction and decay time. The amplitude distribution of these fluctuations is calculated by PCH analysis yielding the molecular brightness. Both FCS and PCH give information about the molecular concentration. Here we describe a global analysis protocol that simultaneously recovers relevant and common parameters in model functions of FCS and PCH from a single fluorescence fluctuation trace. Application of a global analysis approach allows increasing the information content available from a single measurement that results in more accurate values of molecular diffusion coefficients and triplet-state parameters and also in robust, time-independent estimates of molecular brightness and number of molecules.

Key words Fluorescence fluctuation spectroscopy, Fluorescence correlation spectroscopy, Photon counting histogram, Fluorescence intensity distribution analysis, Global analysis

1 Introduction

Fluorescence correlation spectroscopy (FCS), originally introduced by Elson et al. [1–3] in the early 1970s, has become a widely used technique for studying various dynamic molecular processes. It has found applications in measuring local concentrations, mobility coefficients, reaction rates, and detection of intermolecular interactions in vitro and in vivo [4–7]. The sensitivity and noninvasive nature of FCS has made it an important technique for studying molecular processes in cells and thus a useful tool for biochemists, biophysicists, and biologists. After developing new data analysis methods, this technique was renamed to a more general term: fluorescence fluctuation spectroscopy (FFS).

Yves Engelborghs and Antonie J.W.G. Visser (eds.), *Fluorescence Spectroscopy and Microscopy: Methods and Protocols*, Methods in Molecular Biology, vol. 1076, DOI 10.1007/978-1-62703-649-8_33,

Fluorescence fluctuation methods are based on the detection of tiny, spontaneous fluctuations in fluorescence intensity caused by deviations from thermal equilibrium in an open system. These fluctuations can arise, e.g., because of diffusion of fluorescent molecules in and out of a well-defined observation volume generated by a focused laser beam. The intensity fluctuations can be monitored and autocorrelated over time as in fluorescence correlation spectroscopy [6, 8] or a distribution of photon counts can be analyzed as in photon counting histogram (PCH) analysis [9, 10] or in fluorescence intensity distribution analysis (FIDA) [11, 12].

Performing fluorescence fluctuation spectroscopy experiments is relatively straightforward. However, the theory for FFS is based on several assumptions, which do not always hold in a real experimental situation. For example, FFS analysis assumes that the experimental observation volume has a Gaussian shape, which is not always true especially for in vivo measurements [13]. Dead time and afterpulsing of the photodetectors may influence FIDA and PCH analysis [14]. There are a number of optical factors that influence FFS measurements such as cover-slide thickness, refractive index of the sample and optical saturation that should be taken into account for proper analysis [15, 16]. In a confocal microscope configuration employing one-photon excitation, a PCH model with a three-dimensional Gaussian observation volume profile does not adequately fit the experimental fluorescence fluctuation data [17, 18]. In addition, there is a clear effect of bin time on FIDA and PCH analysis, which influences the estimation of molecular concentration and brightness [19, 20]. While it is crucial that the experimental setup must be optimized to be close to ideal conditions, one should also be aware of all these factors in the data analysis procedure to avoid erroneous fits and misinterpretation of the experimental data.

Dependence of model parameters in PCH and FIDA on the bin time makes it difficult to recover the true time-independent brightness and number of molecules from just a single measured photon counting distribution (PCD). PCD refers here to the measured and theoretical distribution of photon counts, while PCH is a commonly used term to denote the method of analysis. An attempt to fit PCD calculated at a very small bin time to avoid any influence of diffusion and triplet kinetics usually fails if the detected photon count rate is not high enough. The PCD calculated at such conditions has only a few data points, which makes it difficult to fit the data well. To get time-independent brightness and number of molecules from a fit of PCD calculated at higher bin time, knowledge of diffusion and triplet parameters is required [19, 20].

A number of extensions of PCH and FIDA methods enabling simultaneous determination of diffusion coefficients and brightness of molecules were proposed up to this time: fluorescence intensity multiple distribution analysis (FIMDA) [19], photon counting multiple histogram (PCMH) analysis [20], master equations fluorescence intensity distribution analysis (ME-FIDA) [21], and the method, originated from the numerical solution of the diffusion equation [22].

The fit of a set of PCDs calculated at different bin times in FIMDA [19] or PCMH analysis [20] enables recovering of all parameters (including diffusion times, triplet fraction and decay time) from a single analysis. However, both FIMDA and PCMH have a drawback. These methods use the correction in the form of an integral of the time-dependent part of the FCS model. It means that the same correction value can be achieved for a different functional form of the integrand; for example, one can set the diffusion time to a very small value and compensate this by appropriate value of triplet parameters and vice versa. Therefore, after global analysis of several PCDs, the fit can be satisfactory, but results in erroneous values for diffusion, triplet, and (also) other parameters.

Although it is well possible to analyze FFS data separately by original FCS and PCH/FIDA methods or their extensions and to estimate all parameters of interest, it is advantageous to combine the complementary FFS methods in a global analysis and to increase the information content available from a single measurement. FCS has been established as a robust method for estimation of triplet parameters (fraction and decay time) and diffusion times (or coefficients) [8]. Vice versa, after correcting for the bin-time effect (using obtained estimates of triplet and diffusion parameters), PCH/FIDA analysis yields correct time-independent estimates for the brightness that finally helps to obtain correct concentrations in FCS, if the sample contains two (or more) components having different brightness values. Global analysis of ACF and PCD increases sensitivity and accuracy of the analysis while keeping the total number of fit parameters almost unchanged. Therefore, it results in more accurate values of triplet-state and diffusion parameters and also in robust, time-independent estimates of molecular brightness and number of molecules for a relatively broad range of measurement conditions.

The growing number of applications of FFS methods demands for new approaches in data processing, aiming at increased speed and robustness. Iterative algorithms of parameter estimation, although proven to be universal and accurate, require some initial guesses (IG) for the unknown parameters. If the IG are in close proximity to the optimal values of unknown parameters, they can significantly increase the efficiency and accuracy of the fit. If the target criterion surface has a complex shape with many local minima, the possibility to reach the global minimum directly depends on the quality of IG. Being an essential component of any data processing technology, IG become especially important in case of PCH/FIDA, since even with apparently reasonable, and physically admissible but randomly chosen IG, the iterative procedure may converge to situations where for a certain combination of parameters, the PCH/FIDA model cannot be numerically evaluated [23]. Iterative fitting with generated IG proved to be more robust and at least five times faster than with an arbitrarily chosen IG [23]. It is also important that a reliable algorithm of IG generation reduces user participation and thereby leads to a more standardized and automated procedure.

In this chapter we describe the global analysis protocol of ACF and PCD. The use of the method is demonstrated through fitting of experimental fluorescence fluctuation data obtained from monomeric and dimeric forms of enhanced green-fluorescent protein (eGFP) in aqueous buffer.

2 Materials

2.1 Samples

Monomeric and dimeric eGFP were purified according to a recently described procedure [24]. The two eGFP molecules in the dimer were linked by six amino acids (GSGSGS). Purified monomeric and dimeric eGFP solutions were in 50 mM TRIS buffer (pH 8.0). For the measurements, 200-μl solutions were added to an 8-chambered cover glass (Lab-Tek, Nalge Nunc International Corp., USA). Rhodamine 110 (R110) (Invitrogen) in water was used for calibration measurements.

2.2 Instrumentation

The measurements were performed on the ConfoCor 2-LSM 510 combination setup (Carl Zeiss, Jena, Germany) detailed in [25–27]. eGFP was excited with the 488 nm line from an argon-ion laser (excitation intensity in the range 10–40 μW) focused into the sample with a water immersion C-Apochromat 40× objective lens N.A. 1.2 (Zeiss). After passing through the main beam splitters HFT 488/633, the fluorescence was filtered with a band pass 505–550 and detected with an avalanche photodiode (APD). The pinhole for confocal detection was set at 70 μm. The microscope was controlled by Zeiss AIM 3.2 software. Raw intensity fluctuation data consisting of up to 4×10^6 photons were collected from single measurements. The data collection time ranged between 30 and 120 s.

3 Methods

Global analysis of ACF and PCD is performed by the iterative least-squares method with the Marquardt-Levenberg optimization [28] (*see* details of its realization in Chapter 10 by Digris et al. in the same volume). To successfully apply it in the analysis of ACF and PCD, one has to know:

1. How to calculate the FCS model?
2. How to calculate the PCH model?
3. How to generate initial guesses for fit parameters?
4. How to perform the linking of fit parameters?
5. Which criterion to use to stop the fit?

Answers to these questions are given in the subsequent subsections.

3.1 Calculation of the FCS Model

The FCS model is calculated using the general form

$$G(\tau) = G_{\text{inf}} + X_{\text{kinetics}}(\tau)\sum_i \frac{q_{0\text{eff } i}^2 N_{0\text{eff } i}}{\left(\sum_j q_{0\text{eff } j} N_{0\text{eff } j}\right)^2} G_{\text{motion } i}(\tau), \quad (1)$$

where $G_{\text{inf}} = G(\infty)$, $N_{0\text{eff}}$ is the number of molecules in the effective volume $V_{\text{eff}} = \chi_1^2/\chi_2$, $\chi_k = \int_V [B(\mathbf{r})/B_0]^k \, dV$, $q_{0\text{eff}}$ is the specific brightness of molecules (in photon counts per second per molecule), $X_{\text{kinetics}}(\tau)$ denotes a kinetic process, and $G_{\text{motion}}(\tau)$ describes the type of motion of the particles. Subscript 0 denotes that the given parameter does not depend on time. $B(\mathbf{r})$ is the brightness profile function, which is the product of excitation intensity and detection efficiency. The kinetic process and type of motion depend on the experimental conditions.

To demonstrate the application of the global analysis method, we used the model describing L independent molecular species diffusing freely in a 3D-Gaussian-shaped observation volume and undergoing singlet-triplet transitions [29]:

$$G(t) = G_{\text{inf}} + \frac{1 - F_{\text{trip}} + F_{\text{trip}} e^{-t/\tau_{\text{trip}}}}{(1 - F_{\text{trip}})\left(\sum_j q_{0\text{eff } j} N_{0\text{eff } j}\right)^2} \times \sum_{i=1}^{L} \frac{q_{0\text{eff } i}^2 N_{0\text{eff } i}}{(1 + t/\tau_{\text{diff } i})\sqrt{(1 + t/a^2\tau_{\text{diff } i})}}, \quad (2)$$

where F_{trip} and τ_{trip} are, respectively, the fraction and the relaxation time of molecules in the triplet state, $a = z_0/\omega_0$, ω_0 and z_0 are, respectively, the lateral and axial radii of the confocal detection volume, and τ_{diff} is the lateral diffusion time, which is related to the diffusion coefficient D_{tran} via $\tau_{\text{diff}} = {\omega_0}^2/(4D_{\text{tran}})$. All molecules are supposed to have the same triplet-state parameters.

3.2 Calculation of the PCH Model

In contrast to the FCS model, which is represented by a relatively simple equation, the PCH model does not have a closed-form solution. It is calculated by a numerical algorithm that includes several steps. Here we describe the algorithm that calculates the PCD $P(k)$ via successive convolutions of single-molecular PCD

$p^{(1)}(k)$ [9]. This algorithm allows accurately calculating the PCD using a wide range of model parameters, whereas other algorithms [11, 30] fail to calculate the model due to various numerical problems. The algorithm steps are described below (*see* **Note 1** *for details*):

1. Calculate the PCD $\tilde{P}(k) = \tilde{H}(k)/M$, where $\tilde{H}(k)$ is the measured photon counting histogram and M is the total number of bins that can be readily calculated from the histogram itself: $M = \sum_k \tilde{H}(k)$, if $\tilde{H}(k)$ decays to zero.
2. Calculate the binning correction factor $B_2(T)$ for each diffusion component as

$$B_2(T) = \frac{2}{T^2} \int_0^T (T - t) g(t)\, dt, \tag{3}$$

where T is the counting time interval (bin time), $g(t)$ is a time-dependent term of the autocorrelation function in FCS

$$g(t) = \frac{1 - F_{\text{trip}} + F_{\text{trip}} e^{-t/\tau_{\text{trip}}}}{1 - F_{\text{trip}}} \frac{1}{(1 + t/\tau_{\text{diff}}) \sqrt{(1 + t/a^2 \tau_{\text{diff}})}} \tag{4}$$

and a, F_{trip}, τ_{trip}, τ_{diff} are fit parameters. The integral in Eq. 3 is calculated numerically as a sum of two integrals (from 0 to t_{br} and from t_{br} to T) by the function *qadrat* [31]. The value of t_{br} is calculated as a minimum of function: $H(t) = -(s(0) - s(t))(T - t)$, where $s(t) = (T - t)g(t)$. The minimum of function can be calculated, e.g., by the function *mimin* [31].
3. Calculate $q_{\text{eff}} = q_{0\text{eff}} B_2(T)$, $N_{\text{eff}} = N_{0\text{eff}}/B_2(T)$, where $q_{0\text{eff}}$ and $N_{0\text{eff}}$ are fit parameters, for each brightness component (for the sake of simplicity we omitted subscripts, which denote the component index). For a multicomponent system where each brightness component has a different molecular weight, the correction should be calculated using the proper diffusion component.
4. Calculate the single-molecular PCD $p^{(1)}(k)$, $k = 1, 2, \ldots$ for each brightness component:

$$p^{(1)}(k) = \frac{1 + F_2}{(1 + F_1)^2} \left[p_G^{(1)}(k) + \frac{1}{k!\Theta} \sum_{n=k}^{\infty} \frac{(-1)^{n-k} (q_{\text{eff}} T)^n F_n}{(n - k)!(2n)^{3/2}} \right], \tag{5}$$

where F_n, $n = 1, 2$ are instrumental out-of-focus correction parameters (F_n are fit parameters) and

$$p_G^{(1)}(k) = \frac{1}{\Theta \sqrt{\pi} k} \int_0^{\infty} \gamma(k, q_{\text{eff}} T e^{-x^2})\, dx. \tag{6}$$

The incomplete gamma function $\gamma(a, x)$ is calculated as described in [32] (*gammp*) and the infinite integral is calculated as a sum of two definite integrals (from 0 to 2.4 and from 2.4 to 20). The integration limits were defined empirically by investigating the behavior of the integrand for many combinations of its parameters. The additional integral was added to ensure the acceptable accuracy of the integration if the integrand does not decay to zero at 2.4. Integration is performed numerically by the function *qadrat* [31]. The parameter Θ is varied depending on the product of q_{eff} and T. We define empirically: if($q_{\text{eff}}T < 10$) $\Theta = 1$; else if($q_{\text{eff}}T < 50$) $\Theta = 6$; else if($q_{\text{eff}}T < 100$) $\Theta = 12$; else $\Theta = 20$;

5. Calculate $p^{(1)}(0) = 1 - \sum_k p^{(1)}(k), \quad k = 1, 2, \ldots.$
6. Calculate $P(k)$ for each brightness component:

$$P(k) = \sum_{M=0}^{\infty} p^{(M)}(k)\, Poi(M, \Theta N_{\text{eff}}), \tag{7}$$

where $p^{(M)}(k) = \underbrace{p^{(1)} \otimes \cdots \otimes p^{(1)}}_{M \text{ times}}(k)$ is M-times convolution of the single-molecule PCD and $Poi(k,\eta)$ denotes the Poisson distribution with the mean value η. Do it in a recurrent way: set all elements in the array $P(k)$ to zero; at $M = 0$ calculate $Poi(0, \Theta N) = e^{-Q\,N}$ and $p^{(0)}(k) = \begin{cases} 1, & k = 0, \\ 0, & k \neq 0, \end{cases}$ multiply results, and add the product to the sum; at $M = 1$ calculate $Poi(1, \Theta N) = Poi(0, \Theta N) \times \Theta N$, multiply the result to $p^{(1)}(k)$, and add the product to the sum; at $M = 2, 3, \ldots$ calculate $Poi(M, \Theta N) = Poi(M - 1, \Theta N) \times \Theta N / M$, perform a convolution $p^{(M)}(k) = p^{(M-1)}(k) \otimes p^{(1)}(k)$, multiply the results, and add the product to the sum. Stop the calculation of the sum when $Poi(M, \Theta N) < 10^{-8}$ after reaching its maximum. Such recurrent calculation of both Poisson distribution and M-times convolution increases the computation efficiency of the algorithm.

7. Calculate the total PCD. The PCD of a number of independent species is given by a convolution of PCD of each species

$$P(k) = P(k, N_{\text{eff}\,1}, q_1) \otimes \cdots \otimes P(k, N_{\text{eff}\,n}, q_n). \tag{8}$$

8. Perform the final convolution of the obtained $P(k)$ with the background term

$$P(k, \lambda) = P(k) \otimes Poi(k, \lambda T), \tag{9}$$

where λ is the background count rate (fit parameter), if necessary.

9. Perform the correction on afterpulses and dead time if necessary. Correction on afterpulses and dead time can be made

according to the algorithm described in [21]. Correction for afterpulses is done by the following formula:

$$P_{\mathrm{AP\ corr}}(k) = \sum_{j=0}^{m} P_0(k-j)\, P_{\mathrm{binomial}}(j; k-j, p_{\mathrm{ap}}), \tag{10}$$

where p_{ap} is the afterpulsing probability, m is a number of points in PCD, $P_0(k)$ is the ideal PCD (i.e., without the correction), and P_{binomial} is the binomial distribution

$$P_{\mathrm{binimial}}(j; M, p) = \frac{M!}{j!(M-j)!} p^j (1-p)^{M-j}, \quad j = 0, 1, \ldots, M. \tag{11}$$

This algorithm works well, if afterpulses are always counted in the same bin together with the main events. This assumption is satisfied for avalanche photodiodes where afterpulses are delayed on a relatively short time (25–40 ns). Correction for dead time is done by the following formula:

$$P_{\mathrm{DT\ corr}}(k) = \sum_{j=0}^{\infty} P_0(k+j) P_{\mathrm{binomial}}\left(j; k+j, \frac{(k+j)\tau_{dt}}{T+(k+j)\tau_{dt}}\right), \tag{12}$$

where τ_{dt} is the detector dead time. Parameters p_{ap}, τ_{dt} are fit parameters.

3.3 Generation of Initial Guesses in Photon Counting Histogram Analysis

Here we describe the efficient algorithm of generation of initial guesses for "main" parameters ($q_{0\mathrm{eff}\ i}$ and $N_{0\mathrm{eff}\ i}$, $i = 1{,}2$ and F_1, F_2) of one- and two-component PCH models:

1. Calculate the first four cumulants $\tilde{K}_n$ of the measured PCD. The experimental fluorescence factorial cumulants are calculated through fluorescence factorial moments $\tilde{F}_n$

$$\tilde{K}_n = \tilde{F}_n - \sum_{i=1}^{n-1} \frac{(n-1)!}{i!(n-i-1)!} \tilde{K}_{n-i} \tilde{F}_i, \tag{13}$$

$$\tilde{F}_n = \langle (k-1)\ldots(k-n+1) \rangle = \sum_{k=n}^{m} k(k-1)\ldots(k-n+1)\tilde{P}(k), \tag{14}$$

where $\tilde{P}(k)$ is a measured PCD and the angular brackets indicate averaging with the set of probabilities $\tilde{P}(k)$. If a photon counting histogram $\tilde{H}(k)$ is measured instead of PCD $\tilde{P}(k)$, one must perform the normalization: $\tilde{P}(k) = \tilde{H}(k)/M$.

2. For the case of a one-component model, calculate IG using equations summarized in Table 1 (*see* **Note 2** *for details*).
3. Calculate $B_2(T)$ as described in Subheading 3.2, **step 2**.

Table 1
Initial guesses for PCH analysis for a one-component system

Parameter fixing	IG
λ, F_1, F_2 are fixed	$N_{\mathrm{eff}} = \frac{(\tilde{K}_1 - \lambda T)^2}{\tilde{K}_2}$, $q_{\mathrm{eff}} = \frac{(1+F_1)\tilde{K}_2}{\gamma_2(1+F_2)(\tilde{K}_1 - \lambda T)T}$
F_1, F_2 are fixed	$N_{\mathrm{eff}} = \frac{\gamma_3^2(F_1+1)^2\tilde{K}_2^3}{\gamma_2^4(F_2+1)^4\tilde{K}_3^2}$, $q_{\mathrm{eff}} = \frac{\gamma_2(1+F_2)\tilde{K}_3}{\gamma_3\tilde{K}_2 T}$, $\lambda = \frac{\tilde{K}_1}{T} - \frac{\gamma_3(F_1+1)\tilde{K}_2^2}{\gamma_2^2(F_2+1)^2\tilde{K}_3 T}$
λ, F_1 are fixed	$N_{\mathrm{eff}} = \frac{(\tilde{K}_1 - \lambda T)^2}{\tilde{K}_2}$, $q_{\mathrm{eff}} = \sqrt{\frac{(1+F_1)\tilde{K}_3}{\gamma_3(\tilde{K}_1 - \lambda T)}}$, $F_2 = \frac{\tilde{K}_2\sqrt{\gamma_3(1+F_1)}}{\gamma_2\sqrt{\tilde{K}_3(K_1 - \lambda)}} - 1$
λ, F_2 are fixed	$N_{\mathrm{eff}} = \frac{(\tilde{K}_1 - \lambda T)^2}{\tilde{K}_2}$, $q_{\mathrm{eff}} = \frac{\gamma_2(1+F_2)\tilde{K}_3}{\gamma_3\tilde{K}_2 T}$, $F_1 = \frac{\gamma_2^2(F_2+1)^2\tilde{K}_3(K_1 - \lambda T)}{\gamma_3\tilde{K}_2^2} - 1$
λ is fixed	$N_{\mathrm{eff}} = \frac{(\tilde{K}_1 - \lambda T)^2}{\tilde{K}_2}$, $q_{\mathrm{eff}} = \frac{\gamma_3\tilde{K}_4}{\gamma_4\tilde{K}_3 T}$, $F_1 = \frac{\gamma_3^3\tilde{K}_4^2(\tilde{K}_1 - \lambda T)}{\gamma_4^2\tilde{K}_3^3} - 1$, $F_2 = \frac{\tilde{K}_2\tilde{K}_4\gamma_3^2}{\tilde{K}_3^2\gamma_2\gamma_4} - 1$

4. Calculate the time-independent estimates of brightness and number of molecules: $q_{0\,\mathrm{eff}\,i} = q_{\mathrm{eff}\,i}/B_2(T)$, $N_{0\,\mathrm{eff}\,i} = N_{\mathrm{eff}\,i}B_2(T)$. Do it independently for each brightness component using diffusion parameters from the proper diffusion component.
5. For a two-component model with known background λ and known instrumental parameters F_1, F_2, the system of nonlinear equations (Eq. 37) (*see* **Note 2** *for details*) can be reduced to a third-order polynomial with respect to $q_{\mathrm{eff}\,2}$

$$\begin{aligned}(A_2^3A_3 - A_3^2A_1)q_{\mathrm{eff}\,2}^3 + (2A_1A_3A_4 - A_2^2A_4 - A_2A_3^2)q_{\mathrm{eff}\,2}^2 \\ + (A_3^3 - A_1A_4^2)q_{\mathrm{eff}\,2} + (A_4^2A_2 - A_3^2A_4) = 0,\end{aligned} \quad (15)$$

where

$$\begin{aligned}A_1 &= (1+F_1)(\tilde{K}_1 - \lambda T)/\gamma_2(1+F_2)T, \\ A_2 &= (1+F_1)^2\tilde{K}_2\big/\gamma_2^2(1+F_2)^2T^2, \\ A_3 &= (1+F_1)^2\tilde{K}_3/\gamma_2\gamma_3(1+F_2)T^3, \\ A_4 &= (1+F_1)^2\tilde{K}_4/\gamma_2\gamma_4(1+F_2)T^4.\end{aligned} \quad (16)$$

The gamma factors γ_n in Eq. 16 are calculated as follows (for a 3D Gaussian profile):

$$\gamma_n = \chi_n/\chi_1 = 1/n^{3/2}. \quad (17)$$

Seeking roots of the polynomial can be done numerically as described in [32]. The estimates of other parameters are calculated as

$$N_{\mathrm{eff}\,2} = \frac{A_2A_4 - A_3^2}{A_2q_{\mathrm{eff}\,2}^4 - 2A_3q_{\mathrm{eff}\,2}^3 + A_4q_{\mathrm{eff}\,2}^2}, \quad (18)$$

$$q_{\text{eff }1} = \frac{A_2 - N_{\text{eff }2}q_{\text{eff }2}^2}{A_1 - N_{\text{eff }2}q_{\text{eff }2}}, \qquad N_{\text{eff }1} = \frac{A_1 - N_{\text{eff }2}q_{\text{eff }2}}{q_{\text{eff }1}}. \tag{19}$$

Since in this case a number of IG is available (*see* **Note 2**), we developed a procedure of their selection. To select the best set of IG, do the following:

(a) Reject the IG that are physically unacceptable or residing outside predefined margins.

(b) Calculate the χ^2 criterion for remaining sets of IG. Set χ^2 to the largest floating value if calculation of the theoretical $P(k)$ has failed due to any reason or no roots have been found.

(c) Search the obtained matrix for IG with minimal χ^2.

Finally, apply bin-time correction as described in **steps 3** and **4**.

6. For a two-component model with known background and unknown instrumental parameters, perform the following steps:

(a) Subdivide an acceptable range of instrumental parameters F_1 and F_2 (e.g., estimated from a calibration measurement) into a number of sections.

(b) For each pair of F_1 and F_2, find the IG as described in **step 5** (without **step c**).

(c) Search the obtained matrix for the IG with minimal χ^2.

3.4 The Protocol of Global Analysis of ACF and PCD

Global analysis of ACF and PCD is demonstrated through fitting of experimental fluorescence fluctuation data obtained from monomeric and dimeric forms of enhanced green-fluorescent protein (eGFP) in aqueous buffer. Measured data used in the analysis were acquired as described in Subheading 2 and were stored in raw binary data files. Analysis is performed by the FFS Data Processor 2.3 package [33]:

1. Calculate the ACF and several PCDs. The global analysis of ACF and PCD allows extraction of time-independent values for brightness and concentration for molecular samples using just a single PCD and a single ACF. However, taking into account that PCD usually has a low number of data points, it is advisable to calculate several PCDs to increase the brightness information content of experimental data. We calculated the ACF and two PCDs from raw data with bin times of 0.2 μs, 40 μs, and 120 μs, respectively. The choice of bin times for calculation of PCDs depends on the count rate. The bin time should not be too short where the PCD has just a few data points and too long where the PCD tends to obtain a Gaussian shape.
2. Calculate weighting factors [34] of ACF and PCDs. Weighting factors of the ACF are calculated by the algorithm proposed by Wohland et al. [35]. The intensity trace is subdivided into

a number of non-overlapping sub-traces and the local autocorrelation function is calculated from each sub-trace. Then the standard deviations for each point of the ACF are obtained from these local autocorrelation functions. Finally, weighting factors of the ACF are calculated by dividing the obtained deviations by the square root of the number of sub-traces. This algorithm does not depend on the type of the ACF time scale (quasi-logarithmic or linear) and FCS model parameters (i.e., it does not require any prior knowledge about the explored system) and usually results in good estimates of standard deviations of the ACF. Weighting factors of the PCD are calculated as standard deviations of a binomial distribution given by the following: $\sigma_k = \sqrt{mp_k(1 - p_k)}$, where m is the total number of bins and $p_k = \tilde{P}(k)$.

3. Fit the ACF to the FCS model (Eq. 2) in order to get proper initial values for the parameters F_{trip}, τ_{trip}, $\tau_{\mathrm{diff}\ i}$, a, and G_{inf}. Performing the preliminary analysis of ACF is advisable, because IG for the bin-time correcting parameters F_{trip}, τ_{trip}, $\tau_{\mathrm{diff}\ i}$, and a of the PCH model (*see* Eqs. 3 and 4) are not calculated. Initial guesses for these parameters cannot be obtained from a separate PCD. Although they can be estimated from a set of PCDs (with a number greater than or equal to the number of estimated parameters), their quality will be low if the number of PCDs included in the analysis is not high enough (as compared to the number of points in ACF). It was found that the global analysis of some PCDs (as well as some PCDs and one or several ACF together) started from arbitrary IG are often trapped in local minima due to the integral form of the applied correction [33]. Therefore, starting the global fit from nearly optimal values of F_{trip}, τ_{trip}, $\tau_{\mathrm{diff}\ i}$, a, and G_{inf} prevents the analysis from trapping in local minima and increases its stability.

4. Fix F_{trip}, τ_{trip}, $\tau_{\mathrm{diff}\ i}$, and a to obtained values. This step is mandatory for the FFS Data Processor because the data processor averages fit parameter values during the parameter linkage, unless at least one parameter included in the group is not fixed to some value.

5. Add PCDs to the analysis, configure the PCH model by setting on the diffusion and triplet correction, and form parameter groups. The parameters $N_{0\mathrm{eff}\ i}$, $q_{0\mathrm{eff}\ i}$, F_{trip}, τ_{trip}, $\tau_{\mathrm{diff}\ i}$, and a should be linked across all models (FCS and PCH) and F_1, F_2 across the PCH models. If one is going to correct the PCH model for dead time and afterpulses, two additional parameter groups can be formed.

6. Generate IG for $N_{0\mathrm{eff}\ i}$, $q_{0\mathrm{eff}\ i}$, F_1, F_2 as described in Subheading 3.3.

Table 2
Initial guesses and best fit results of global analysis of autocorrelation function (ACF) and two-photon counting distributions (PCDs) of monomeric and dimeric eGFP

Parameter	eGFP monomer IG	eGFP monomer best fit	eGFP dimer IG	eGFP dimer best fit
F_1	0.77	0.60 ± 0.03	0.79	0.65 ± 0.01
F_2	0.079	0.037 ± 0.007	0.070	0.037 ± 0.003
a	5.0 (fixed)	5.0 (fixed)	5.0 (fixed)	5.0 (fixed)
F_{trip}	0.16	0.16 ± 0.01	0.12	0.12 ± 0.02
τ_{trip} (μs)	3.84	3.80 ± 0.56	2.23	2.29 ± 0.37
τ_{diff} (μs)	62.9	62.7 ± 0.52	96.94	97.0 ± 0.45
N_{0eff}	1.604	1.609 ± 0.008	1.70	1.70 ± 0.004
q_{0eff} (cpsm)	38,410	54,140 ± 670	66,990	95,380 ± 580
q (cpsm)		34,990		59,810
χ^2_{gl}		0.99		1.31
$\chi^2_{gl\ mod}$		1.05		1.68

ACF was calculated with bin time 0.2 μs. PCD's were calculated with bin times of 40 and 120 μs. The data consisted of 1.2×10^6 photons for eGFP monomer and 2.2×10^6 photons for eGFP dimer. Initial guesses and best fit results and both χ^2 global criteria, standard and modified, are presented for comparison. Confidence intervals at the 67 % confidence level were calculated by the asymptotic standard-errors method as described in [40]. The true value of the brightness q was calculated via Eq. 41 (*see* **Note 3**)

7. Perform the iterative fit. Calculate the FCS model as described in Subheading 3.1. Calculate the PCH model as described in Subheading 3.2. Calculate the global χ^2 criterion as described in **Note 4**.
8. Unfix parameters F_{trip}, τ_{trip}, $\tau_{diff\ i}$, and a.
9. Perform the iterative fit again. **Step 8** can be done directly after **step 6** if you are sure that IG for all parameters were generated accurately. **Step 9** can be omitted then.

Analysis results of monomeric and dimeric eGFP are summarized in Table 2. Graphical results are shown in Fig. 1 (monomeric eGFP) and Fig. 2 (dimeric eGFP). The PCH model was calculated using Eqs. 5–7. The FCS model was calculated by Eq. 2. Parameter G_{inf} was fixed to unity, parameter λ was fixed to zero, and parameter a was fixed to the value obtained from the calibration measurement (Rho 110 in water). Although parameters F_1, F_2 are descriptive parameters of the used setup, we did not fix them to the value obtained from the calibration measurement, as even small deviations of these parameters from the best-fit value will affect the resulted PCD substantially. It is a consequence of the high sensitivity of the PCH model to the value of correction

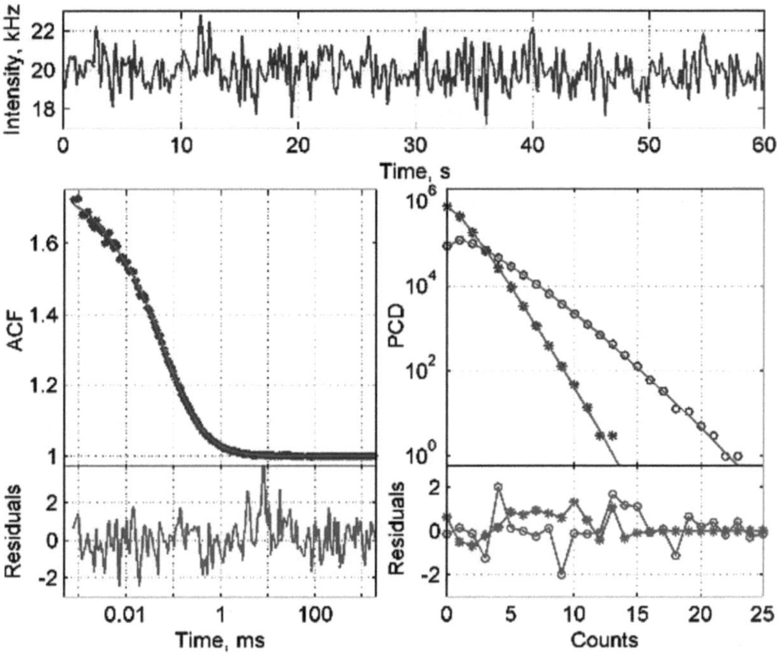

Fig. 1 Results of global analysis of autocorrelation function (ACF) and photon counting distributions (PCDs) of monomeric eGFP. PCD's were calculated with bin times of 40 μs (*closed circles*) and 120 μs (*open circles*). The experimental intensity fluctuations are shown on *top*. Residuals are plotted below each curve. Recovered parameters and criterion values are presented in Table 2

parameters F_1 and F_2. Therefore, it is advisable to link these correction parameters in the global analysis of several measurements only if they do not differ much. Initial guesses, generated by the algorithm described in Subheading 3, are presented for comparison. Analysis, being performed on a typical personal computer (Intel Core 2 Duo 2 GHz, 2 Gb RAM), lasted not more than a few seconds. After inspection of the analysis results, we can conclude that the global analysis of ACF and PCD fits the experimental data well. Brightness values and diffusion times were recovered with a good precision as indicated by calculated confidential intervals.

We obtained the ratio of the dimer and monomer brightness equal to 1.71, which is less than the expected value of 2. This can be explained by the presence of resonance energy transfer in dimeric eGFP [33, 36]. From the obtained ratio of the diffusion times (τ_{diff}) for monomeric and dimeric eGFP (0.646, expected value is 0.707), we concluded that eGFP move like rodlike molecules, which is in excellent agreement with the results of recent FCS experiments on eGFP oligomers [37].

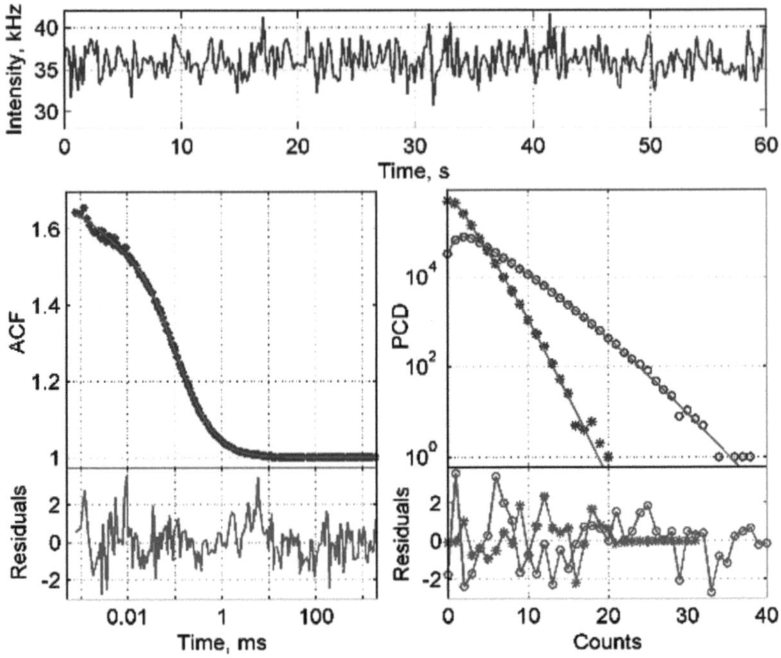

Fig. 2 Results of global analysis of autocorrelation function (ACF) and photon counting distributions (PCDs) of dimeric eGFP. PCD's were calculated with bin times of 40 μs (*closed circles*) and 120 μs (*open circles*). The experimental intensity fluctuations are shown on *top*. Residuals are plotted below each curve. Recovered parameters and criterion values are presented in Table 2

4 Notes

1. *Theory of photon counting histogram analysis*

   In PCH the total PCD from a number of molecules located in the closed volume V_0 is calculated by successive convolutions of single-molecule PCDs [9]

$$p_{V_0}^{(1)}(k) = \frac{1}{V_0} \int_{V_0} Poi(k, qTB(\mathbf{r}))d\mathbf{r}, \tag{20}$$

   where $p^{(1)}(k)$ denotes a single-molecule PCD and Poi(k,η) denotes the Poisson distribution with the mean value η. The brightness q is defined as $q = I_0\sigma_a Q\kappa$, where I_0 is the excitation intensity in the focus σ_a and Q are molecular absorption cross section and fluorescence quantum yield, respectively, and κ is the detection efficiency of the confocal setup. It is assumed that the contribution of each single molecule to the recorded photon trace is independent and the emission intensity is constant during the counting time interval T. $B(\mathbf{r})$ is assumed to be a 3D Gaussian

$$B(\mathbf{r}) = B_0 \exp[-2(x^2 + y^2)/\omega_0^2 - 2z^2/z_0^2] \tag{21}$$

with $B_0 = B(0)$. To reduce the three-dimensional integral in Eq. 20 to a one-dimensional integral, let us apply the following transformation of the coordinate system:

$$\begin{cases} x = r\omega_0 \cos\varphi \sin\alpha \\ y = r\omega_0 \sin\varphi \sin\alpha \\ z = z_0 r \cos\alpha \end{cases} \tag{22}$$

where r is the length of the radius vector and φ and α are angles. The Jacobian of this transformation is

$$\partial(x, y, z)/\partial(r, \varphi, \alpha) = \omega_0^2 z_0 r^2 \sin\alpha. \tag{23}$$

Equation 21 simplifies to

$$B(r) = B_0 e^{-2r^2}, \quad r = \sqrt{x^2/\omega_0^2 + y^2/\omega_0^2 + z^2/z_0^2}. \tag{24}$$

The size of the volume V_0 with given radius r_0 depends on w_0, z_0

$$V_0 = \int_0^{r_0}\int_0^{2\pi}\int_0^{\pi} r^2 \omega_0^2 z_0 \sin\alpha \, \mathrm{d}r \, \mathrm{d}\varphi \, \mathrm{d}\alpha = \frac{4\pi}{3} r_0^3 \omega_0^2 z_0. \tag{25}$$

After integration over φ and α, Eq. 20 takes the form

$$\begin{aligned} p_{V_0}^{(1)}(k) &= \frac{4\pi\omega_0^2 z_0}{V_0 k!} \int_0^{r_0} \left(q_{\text{eff}} T \, e^{-2r^2}\right)^k e^{-q_{\text{eff}} T e^{-2r^2}} r^2 \, \mathrm{d}r \\ &= \frac{3}{r_0^3 k!} \int_0^{r_0} \left(q_{\text{eff}} T e^{-2r^2}\right)^k e^{-q_{\text{eff}} T \, e^{-2r^2}} r^2 \, \mathrm{d}r \end{aligned} , \tag{26}$$

where we define $q_{\text{eff}} = B_0 q$. Note that the obtained probability distribution is normalized: $\sum_k p_{V_0}^{(1)}(k) = 1$ for any r_0. In FFS the observation volume is open and the effective size is usually estimated as $V_{\text{eff}} = \chi_1^2/\chi_2$. When the observation volume is open, the choice of r_0 is crucial. The size of V_0 grows proportionally to r_0^3. Therefore, the number of molecules in this volume and consequently the number of convolutions needed for calculation of the total PCD grow very rapidly making the algorithm inefficient. The radius length must be taken as less as possible so that it completely covers the region where molecules are excited and emit photons.

The integrand in Eq. 26 decays rapidly with increasing of r_0 for any $k > 0$. It allows to set the upper integration limit in Eq. 26 to infinity and to calculate Eq. 26 in a more efficient way [9] through the incomplete gamma function $\gamma(a, x) = \int_0^x e^{-t} t^{a-1} \, \mathrm{d}t$

$$p^{(1)}(k) = \frac{3}{4r_0^3 k} \int_0^{\infty} \gamma(k, q_{\text{eff}} T e^{-x^2}) \mathrm{d}x, \quad k = 1, 2, \ldots, \tag{27}$$

$$p^{(1)}(0) = 1 - \sum\nolimits_k p^{(1)}(k), \quad k = 1, 2, \ldots. \tag{28}$$

The photon counting distribution $P(k)$ of a number of molecules of concentration C in an open observation volume of size V_0 is the weighted average of $p^{(1)}(k)$ [9]

$$P(k) = \sum_{M=0}^{\infty} p^{(M)}(k)\, Poi(M, V_0 C), \tag{29}$$

where

$$p^{(M)}(k) = \underbrace{p^{(1)} \otimes \cdots \otimes p^{(1)}}_{M \text{ times}}(k), \quad p^{(0)}(k) = \begin{cases} 1, & k = 0, \\ 0, & k \neq 0. \end{cases} \tag{30}$$

Single-molecular PCD (*see* Eq. 25) depends on unknown parameters ω_0 and z_0. To eliminate unknown parameters, let us replace the product $V_0 C$ by $V_0 N_{\text{eff}}/V_{\text{eff}}$ in the equations above and introduce a ratio $\Theta = V_0/V_{\text{eff}} = 4r_0^3/3\sqrt{\pi}$. It results in Eqs. 6 and 7.

It was found that the value of Θ needed for accurate calculation of $P(k)$ depends on the product of $q_{\text{eff}}T$. For small $q_{\text{eff}}T$ (less than 10), it can be safely set to unity, while for greater $q_{\text{eff}}T$ it must be large enough to completely include the observation volume. This is necessary to keep the probability distribution $P(k)$ normalized for any q_{eff} and T. The dependence of Θ on $q_{\text{eff}}T$ can be explained as follows. The effective volume has a fixed size. It depends only on ω_0 and z_0 and does not depend on excitation intensity and molecular brightness. However, the illuminated volume grows with an increase of I_0. As q_{eff} and T are presented in a product in the model, the increase of binning time has the same effect.

The PCD of a number n of independent species is given by a convolution of PCDs of each species (*see* Eq. 8). Correction for the background with mean count rate λ can be done by additional convolution with Poisson distribution with parameters λT; *see* Eq. 9.

Derived expressions often fail to fit experimental data using one-photon excitation due to the large deviation of the actual brightness profile from an assumed 3D Gaussian approximation [11, 18]. To improve the model, Perroud, Huang, and coworkers [17, 18] introduced additional fitting parameters F_n defined as the relative difference between the integral χ_n of the nth power of the normalized-to-unity, actual brightness profile function $B(\mathbf{r})$, and that of its 3D Gaussian approximation $\chi_{G\ n}$.

$$F_n = (\chi_n - \chi_{G\,n})/\chi_{G\,n}. \tag{31}$$

Introducing F_n into the single-molecule PCD leads to Eq. 5 [18] where $p_G^{(1)}(k)$ is calculated by Eq. 6. In most cases only the first-order correction (all F_n equal to zero except F_1) is sufficient to get the best fit to the experimental data.

The PCH model was derived under the assumption that the fluorescence intensity emitted by a molecule during bin time T is constant, which is valid for the limit of short bin times. At arbitrary bin times, both brightness and number of molecules become a function of T. The most practical way to calculate the time-independent brightness and number of molecules is to make the first and second factorial cumulants of PCD exact [19, 20]. According to this theory one has to calculate the so-called binning correction factor [19, 20] (*see* Eqs. 3 and 4) and to correct the brightness and the number of molecules in the following form:

$$\begin{aligned} q_{0\,\mathrm{eff}} &= q_{\mathrm{eff}}/B_2(T), \\ N_{0\,\mathrm{eff}} &= N_{\mathrm{eff}}B_2(T), \end{aligned} \tag{32}$$

where $q_{0\mathrm{eff}}$ and $N_{0\mathrm{eff}}$ are absolute values of brightness and concentration that are independent on T. In general, the binning correction factor can be calculated assuming two or even more diffusing components (such correction can be applied to a mixture of species with approximately equal brightness values but quite different hydrodynamic radii). For a model of multiple brightness components, this correction has to be applied independently to each component. Triplet and diffusion characteristics can be either different or the same for each brightness component.

Although the theories of FCS and PCH/FIDA are well known, several important points have to be taken into consideration when both theories are combined in a global analysis. From the PCH/FIDA side, there are two major points. First, a correction must be applied for out-of-focus emission, which can be done by either a correction of the Gaussian profile via Eq. 31, or using a polynomial profile [11, 19]. Second, PCH/FIDA must be corrected for dynamic processes enabling to obtain time-independent estimates of brightness values and number of molecules. From the FCS side the correction on difference in brightness values must be applied when multiple fluorescent species are present. The last important point is to use the same reference volume in both FCS and PCH/FIDA. The use of the effective volume V_{eff} is preferable, because it allows linking the number of molecules (N) through FCS, PCH, and FIDA models.

2. *Theory of initial guesses for photon counting histogram analysis*
The algorithm of initial guesses is based on the statistical method of moments. The method of moments [38] is widely used in statistics to estimate parameters for any probability distribution, for which the necessary amount of moments can be generated. According to this method the theoretically obtained moments $M_n = M_n(\eta_1, \eta_2, \ldots, \eta_m)$ are equalized to the experimentally obtained ones: $\tilde{M}_n$:

$$M_n(\eta_1, \eta_2, \ldots, \eta_m) = \tilde{M}_n, \quad n = 1, 2, \ldots, m, \tag{33}$$

where $\eta_1, \eta_2, \ldots, \eta_m$ is a set of unknown parameters. The solution of the system of Eq. 33 yields the estimates for $\eta_1, \eta_2, \ldots, \eta_m$. In our case the method of moments can also be reformulated in terms of factorial cumulants K_n that leads to considerable simplification in further mathematical derivations. To calculate theoretical factorial cumulants, it is convenient to introduce the concept of generating function. The generating function is given by [38]

$$G(\xi) = \sum_{k=0}^{\infty} \xi^k P(k), \tag{34}$$

where ξ is a trial variable and the factorial cumulants are calculated as

$$K_n = \left. \frac{d^n \ln G(\xi)}{d\xi^n} \right|_{\xi=1}. \tag{35}$$

Theoretical fluorescence factorial cumulants of PCD with out-of-focus correction can be written as follows:

$$\begin{aligned} K_1 &= \lambda T + \gamma_2 \frac{(1+F_2)}{(1+F_1)} \sum_i N_{\text{eff } i} q_{\text{eff } i} T \\ K_2 &= \gamma_2^2 \frac{(1+F_2)^2}{(1+F_1)^2} \sum_i N_{\text{eff } i} q_{\text{eff } i}^2 T^2 \\ K_n &= \gamma_2 \gamma_n \frac{(1+F_2)}{(1+F_1)^2} \sum_i N_{\text{eff } i} q_{\text{eff } i}^n T^n, \quad n = 3, 4, \ldots, \end{aligned} \tag{36}$$

All unknown parameters can be estimated from a system, where theoretical factorial cumulants (Eq. 36) are made equal to the experimentally obtained ones (Eq. 13). From a practical point of view, however, instability of higher-order factorial cumulants of PCD may result in divergence of the numerical solution. Moreover, systems of nonlinear equations are usually solved by an iterative procedure and thus require initial guesses by its own. Therefore, let us consider more simple cases leading to solutions without iterative calculations.

IG for a one-component system

According to the method of moments one has to define one equation for each estimated parameter N_{eff}, q_{eff}, F_1, F_2 and λ. However, if some parameters are known a priori and fixed to some value or simply ignored (i.e., fixed to zero), there is no need to estimate them. Therefore, one can minimize the number of higher-order cumulants needed for calculation of IG, thus increasing the efficiency and accuracy of the algorithm. The system 36, written for $n = 1,2$, $n = 1,2,3$ and $n = 1,2,3,4$ at $i = 1$ can be solved analytically. The list of obtained expressions for estimating N_{eff}, q_{eff}, F_1, and F_2 at various parameters fixing is summarized in Table 1. To get the time-independent estimates of brightness q_{0eff} and number of molecules N_{0eff} one has to calculate binning correction factor (*see* Eq. 3) and apply finally Eq. 32.

IG for a two-component system

To calculate IG for a two-component system, we have to define seven equations for the factorial cumulants. Although the obtained system can be solved numerically, the solution is not entirely reliable since higher-order factorial cumulants are substantially affected by noise. It is reasonable to reduce the number of equations to four and calculate IG only for brightness and number of molecules assuming that background and instrumental parameters F_1 and F_2 are known. Eq. 36 for $i = 2$ takes the form

$$\begin{aligned}(1+F_1)(K_1-\lambda T)/\gamma_2(1+F_2)T &= N_{eff\,1}q_{eff\,1}+N_{eff\,2}q_{eff\,2}\\ (1+F_1)^2K_2\Big/\gamma_2^2(1+F_2)^2T^2 &= N_{eff\,1}q_{eff\,1}^2+N_{eff\,2}q_{eff\,2}^2\\ (1+F_1)^2K_3/\gamma_2\gamma_3(1+F_2)T^3 &= N_{eff\,1}q_{eff\,1}^3+N_{eff\,2}q_{eff\,2}^3\\ (1+F_1)^2K_4/\gamma_2\gamma_4(1+F_2)T^4 &= N_{eff\,1}q_{eff\,1}^4+N_{eff\,2}q_{eff\,2}^4\,. \end{aligned} \tag{37}$$

The system 37 of nonlinear equations can be solved both analytically and numerically [23]. To rewrite the system 37 in a more compact form, we denote left-hand parts in this system as A_1, A_2, A_3, A_4. From the first equation we get an expression for estimating $N_{eff\,2}$ (*see* Eq. 18) and after substitution of Eq. 18 into the second equation of the system 37, we arrive at the third-order polynomial in respect to $q_{eff\,2}$ (*see* Eq. 15). Finally, $q_{eff\,1}$ and $N_{eff\,1}$ are calculated by Eq. 19. If the discriminant of the polynomial has a positive value, a real-valued analytical solution is available [23]. If the discriminant of the polynomial is negative, two or three real roots are available. They can be found numerically by seeking the roots of the polynomial.

The described method of IG generation can be also used independently for a quick estimation of brightness and concentrations.

3. *Relation between brightness in FCS and PCH*
   Although B_0 can be safely set to unity when no correction is applied, the correction in the form of Eq. 31 makes B_0 and therefore q_{eff} a function of correction parameters F_n. Let us write the theoretical factorial cumulants of a one-component PCD without the out-of-focus correction:

$$\begin{aligned} K_1 &= \lambda T + \gamma_2 \sum_i N_{\text{eff }i} q_i T \\ K_k &= \gamma_2 \gamma_n \sum_i N_{\text{eff }i} q_i^k T^k, \quad k = 2, 3, \ldots \end{aligned} \tag{38}$$

The first two equations of Eqs. 38 and 36 yield the same expression for estimating N_{eff}

$$N_{\text{eff}} = (K_1 - \lambda T)^2 / K_2. \tag{39}$$

However the estimators of brightness are different

$$q = \frac{2\sqrt{2}K_2}{(K_1 - \lambda T)T}, \qquad q_{\text{eff}} = \frac{2\sqrt{2}(1 + F_1)K_2}{(1 + F_2)(K_1 - \lambda T)T} \tag{40}$$

Therefore, when a correction is applied (F_1, $F_2 \neq 0$), the estimated brightness represents the so-called effective or apparent brightness, which must be recalculated into the "true" one

$$q = \frac{(1 + F_2)}{(1 + F_1)} q_{\text{eff}}. \tag{41}$$

While performing FCS and PCH on the same data, one usually likes to relate brightness calculated from the average intensity $<I>$ and estimated number of molecules to the brightness obtained from PCH. From the expression for the first factorial cumulant K_1 (*see* Eq. 36), we derive

$$< I > = \frac{K_1}{T} = \lambda + \gamma_2 \frac{(1 + F_2)}{(1 + F_1)} \sum_i N_{\text{eff }i} q_{\text{eff }i}. \tag{42}$$

The expression for the first factorial cumulant does not change its form, if we replace N_{eff} and q_{eff} by their time-independent variants $N_{0\text{eff}}$ and $q_{0\text{eff}}$ (because $B_2(T) = 1$ for the first cumulant). Therefore, for a one-component model without background, one obtains $< I > = \gamma_2(1 + F_2)\ N_{0\text{ eff}} q_{0\text{ eff}} / (1 + F_1)$. Since the number of molecules in FCS does not depend on time, N_{FCS} is equal to $N_{0\text{eff}}$ and

$$q_{\text{FCS}} = < I > / N_{0\text{ eff}} = \gamma_2(1 + F_2) q_{0\text{ eff}} / (1 + F_1). \tag{43}$$

4. *Global χ^2 criterion*
   Global analysis of experimental data obeying different functional forms may result in overestimation (or underestimation) of some model parameters, if appropriate weighting factors are

not applied to each data point. This is especially important when the number of data points in these functions is quite different, for instance, the ACF usually has 175 experimental data points versus only 10–20 data points in the PCD. Such difference in number of data points leads to a significant difference in the number of degrees of freedom (calculated as number of data points minus number of fit parameters minus one) corresponding to each analyzed curve. Thus, the standard global χ^2 criterion [39] becomes relatively insensitive to small deviations between the measured and model-generated curves that have lower number of data points. To avoid this problem it is necessary to take into account the specific weight of each individual curve that participates in the global analysis. It can be done, if χ^2 is modified, in the following way: [33]

$$\chi^2_{\text{gl mod}} = \frac{N - m - M}{M(N - m + M^{\text{gr}} - 1)} \times \sum_{i=1}^{M} \left(\frac{1}{n_i - m_i - 1} \sum_{j=1}^{n_i} \frac{\left(x_{ij} - F_{ij}^{\text{th}}\right)^2}{\sigma_{ij}^2} \right), \quad (44)$$

where $M^{\text{gr}} = \sum_{i=1}^{M} m_i^{\text{lnk}} - m^{\text{gr}}$, x_{ij}, and σ_{ij} are the measured value and the standard deviation of the *i*th experimental data point in the *j*th measured curve, and F_{ij}^{th} is the model-generated value for this point; M is the number of globally analyzed curves; n_i is the number of points in the *i*th curve; m_i and m_i^{lnk} is the number of free and linked parameters in the *i*th model, respectively; m^{gr} is the number of parameter groups (sets of linked parameters), $N = \sum_{i=1}^{M} n_i$, $m = \sum_{i=1}^{M} m_i$. As follows from Eq. 44, the contribution of each analyzed curve to the value of the global χ^2 is made equal by dividing the sum in brackets by the corresponding number of degrees of freedom (thus obtaining the local χ^2) and finally multiplying the total sum of the local χ^2 by the average number of degrees of freedom $(N - m - M)/M = \sum_{i=1}^{M} (n_i - m_i - 1)/M$. In case of equal number of degrees of freedom in each model ($n_1 = \ldots = n_M$; $m_1 = \ldots = m_M$), Eq. 25 reduces correctly to the standard global χ^2 criterion [39]:

$$\chi^2_{\text{gl}} = \frac{1}{(N - m + M^{\text{gr}} - 1)} \sum_{i=1}^{M} \sum_{j=1}^{n_i} \frac{\left(x_{ij} - F_{ij}^{\text{th}}\right)^2}{\sigma_{ij}^2}. \quad (45)$$

References

1. Magde D, Elson E, Webb WW (1972) Thermodynamic fluctuations in a reaction system: measurement by fluorescence correlation spectroscopy. Phys Rev Lett 29(11):705–708
2. Elson EL, Magde D (1974) Fluorescence correlation spectroscopy. I Conceptual basis and theory. Biopolymers 13(1):29–61
3. Magde D, Elson EL, Webb WW (1974) Fluorescence correlation spectroscopy. II. An experimental realization. Biopolymers 13 (1):29–61. doi:10.1002/bip.1974.360130103
4. Krichevsky O, Bonnet G (2002) Fluorescence correlation spectroscopy: the technique and its applications. Rep Prog Phys 65:251–297
5. Thompson NL, Lieto AM, Allen NW (2002) Recent advances in fluorescence correlation spectroscopy. Curr Opin Struct Biol 12 (5):634–641
6. Haustein E, Schwille P (2007) Fluorescence correlation spectroscopy: novel variations of an established technique. Annu Rev Biophys Biomol Struct 36:151–169. doi:10.1146/annurev.biophys.36.040306.132612
7. Rigler R, Elson E (eds) (2001) Fluorescence correlation spectroscopy. Theory and applications. Springer, Berlin
8. Widengren J, Mets U, Rigler R (1995) Fluorescence correlation spectroscopy of triplet states in solution: a theoretical and experimental study. J Phys Chem 99:13368–13379
9. Chen Y, Müller JD, So PT, Gratton E (1999) The photon counting histogram in fluorescence fluctuation spectroscopy. Biophys J 77 (1):553–567
10. Chen Y, Tekmen M, Hillesheim L, Skinner J, Wu B, Müller JD (2005) Dual-color photon-counting histogram. Biophys J 88(3):2177–2192. doi:10.1529/biophysj.104.048413
11. Kask P, Palo K, Ullmann D, Gall K (1999) Fluorescence-intensity distribution analysis and its application in biomolecular detection technology. Proc Natl Acad Sci U S A 96 (24):13756–13761
12. Kask P, Palo K, Fay N, Brand L, Mets U, Ullmann D, Jungmann J, Pschorr J, Gall K (2000) Two-dimensional fluorescence intensity distribution analysis: theory and applications. Biophys J 78(4):1703–1713
13. Hess ST, Webb WW (2002) Focal volume optics and experimental artifacts in confocal fluorescence correlation spectroscopy. Biophys J 83(4):2300–2317
14. Hillesheim LN, Müller JD (2003) The photon counting histogram in fluorescence fluctuation spectroscopy with non-ideal photodetectors. Biophys J 85(3):1948–1958
15. Enderlein J, Gregor I, Patra D, Dertinger T, Kaupp UB (2005) Performance of fluorescence correlation spectroscopy for measuring diffusion and concentration. Chemphyschem 6 (11):2324–2336
16. Gregor I, Patra D, Enderlein J (2005) Optical saturation in fluorescence correlation spectroscopy under continuous-wave and pulsed excitation. Chemphyschem 6(1):164–170. doi:10.1002/cphc.200400319
17. Perroud TD, Huang B, Wallace MI, Zare RN (2003) Photon counting histogram for one-photon excitation. Chemphyschem 4 (10):1121–1123. doi:10.1002/cphc.200300824
18. Huang B, Perroud TD, Richard N (2004) Photon counting histogram: one-photon excitation. Chemphyschem 5:1523–1531
19. Palo K, Metz U, Jager S, Kask P, Gall K (2000) Fluorescence intensity multiple distributions analysis: concurrent determination of diffusion times and molecular brightness. Biophys J 79 (6):2858–2866
20. Perroud TD, Huang B, Zare RN (2005) Effect of bin time on the photon counting histogram for one-photon excitation. Chemphyschem 6 (5):905–912
21. Palo K, Mets U, Loorits V, Kask P (2006) Calculation of photon count number distributions via master equations. Biophys J 90:2179–2191
22. Gopich IV, Szabo A (2005) Photon counting histograms for diffusive fluorophores. J Phys Chem B 109:17683–17688
23. Skakun VV, Novikov EG, Apanasovich VV, Tanke HJ, Deelder AM, Mayboroda OA (2006) Initial guesses generation for fluorescence intensity distribution analysis. Eur Biophys J 35(5):410–423. doi:10.1007/s00249-006-0048-8
24. Visser AJWG, Laptenok SP, Visser NV, van Hoek A, Birch DJS, Brochon JC, Borst JW (2010) Time-resolved FRET fluorescence spectroscopy of visible fluorescent protein pairs. Eur Biophys J 39:241–253. doi:10.1007/s00249-009-0528-8
25. Hink MA, Shah K, Russinova E, de Vries SC, Visser AJWG (2008) Fluorescence fluctuation analysis of *Arabidopsis thaliana* somatic embryogenesis receptor-like kinase and brassinosteroid insensitive 1 receptor oligomerization. Biophys J 94(3):1052–1062. doi:10.1529/biophysj.107.112003

26. Hink MA, Borst JW, Visser AJWG (2003) Fluorescence correlation spectroscopy of GFP fusion proteins in living plant cells. Methods Enzymol 361:93–112
27. Weisshart K, Jüngel V, Briddon SJ (2004) The LSM 510 META - ConfoCor 2 system: an integrated imaging and spectroscopic platform for single-molecule detection. Curr Pharm Biotechnol 5:135–154
28. Marquardt DW (1963) An algorithm for least-squares estimation of non-linear parameters. J Soc Ind Appl Math 11:431–441
29. Skakun VV, Hink MA, Digris AV, Engel R, Novikov EG, Apanasovich VV, Visser AJ (2005) Global analysis of fluorescence fluctuation data. Eur Biophys J 34(4):323–334
30. Meng F, Ma H (2006) A comparison between photon counting histogram and fluorescence intensity distribution analysis. J Phys Chem B 110(51):25716–25720
31. Lau HT (1995) A numerical library in C for scientists and engineers, Symbolic and numeric computation series. CRC Press, Boca Raton, FL
32. Press WH, Teukolsky SA, Vetterling WT, Flannery BP (1992) Numerical recipes in C: the art of scientific computing, 2nd edn. Cambridge University Press, New York, NY
33. Skakun VV, Engel R, Digris AV, Borst JW, Visser AJ (2011) Global analysis of autocorrelation functions and photon counting distributions. Front Biosci (Elite Ed) 3:489–505, 264 [pii]
34. Bevington PR, Robinson DK (2003) Data reduction and error analysis for the physical sciences, 3rd edn. McGraw-Hill, New York, NY
35. Wohland T, Rigler R, Vogel H (2001) The standard deviation in fluorescence correlation spectroscopy. Biophys J 80(6):2987–2999. doi:10.1016/S0006-3495(01)76264-9, S0006-3495(01)76264-9 [pii]
36. Malikova NP, Visser NV, van Hoek A, Skakun VV, Vysotski ES, Lee J, Visser AJ (2011) Green-fluorescent protein from the bioluminescent jellyfish *Clytia gregaria* is an obligate dimer and does not form a stable complex with the Ca(2+)-discharged photoprotein clytin. Biochemistry 50(20):4232–4241. doi:10.1021/bi101671p
37. Dross N, Spriet C, Zwerger M, Muller G, Waldeck W, Langowski J (2009) Mapping eGFP oligomer mobility in living cell nuclei. PLoS One 4(4):e5041. doi:10.1371/journal.pone.0005041
38. Korn GA, Korn TM (1968) Mathematical handbook for scientists and engineers. Definitions, theorems and formulas for reference and review. Second, enlarged and revised edition. McGraw-Hill Book Company, New York, NY
39. Beechem JM, Gratton E, Ameloot M, Knutson JR, Brand L (1991) Analysis of fluorescence intensity and anisotropy decay data: second generation theory and programs. In: Lakowicz JR (ed) Topics in fluorescence spectroscopy, vol 2. Plenum Press, New York, NY, pp 241–305
40. Johnson ML, Faunt LM (1992) Parameter estimation by least-squares methods. Methods Enzymol 210:1–37

Chapter 34

Simulation of Autocorrelation Function and Photon Counting Distribution in Fluorescence Fluctuation Spectroscopy

Igor P. Shingaryov, Victor V. Skakun, and Vladimir V. Apanasovich

Abstract

In modern fluorescence fluctuation spectroscopy, the autocorrelation function and photon counting distribution are two widely used statistical characteristics of the measured fluctuating fluorescence intensity signal. Applying special analysis methods such as fluorescence correlation spectroscopy (FCS) and photon counting histogram (PCH) to these properties, it is possible to recover values of different parameters of fluorescent molecules such as the concentration, diffusion coefficient, molecular brightness, and kinetic rate constants. The development of new analysis methods is senseless without testing their validity, accuracy, and robustness. The most appropriate check of a method is its application to experimental data. However, sometimes it is more convenient and easier to verify a method on simulated data. Simulation is also useful for better understanding the processes that were modeled during the development of analysis methods. Here, we present two simulation models providing an autocorrelation function and photon counting distribution of a sequence of photon arrival times detected in fluorescence fluctuation spectroscopy.

Key words Fluorescence fluctuation spectroscopy, Fluorescence correlation spectroscopy, Photon counting histogram, Autocorrelation function, Simulation, Stochastic point process

1 Introduction

Fluorescence fluctuation spectroscopy (FFS) methods are widely used in modern biophysical and biochemical research [1]. In FFS the information about dynamics and interactions of fluorescently labeled macromolecules is extracted from the detected fluorescence intensity fluctuations. These fluctuations are caused by various kinetic processes, which alter the number of molecules in a small observation volume and its intrinsic fluorescence properties. A detected fluorescence signal thus contains information about molecular diffusion, photophysical, and chemical dynamics [2–5]. In FFS the commonly used methods for extracting this information are fluorescence correlation spectroscopy (FCS) [6], photon

Yves Engelborghs and Antonie J.W.G. Visser (eds.), *Fluorescence Spectroscopy and Microscopy: Methods and Protocols*, Methods in Molecular Biology, vol. 1076, DOI 10.1007/978-1-62703-649-8_34,

counting histogram (PCH) analysis [7, 9], fluorescence intensity distribution analysis (FIDA) [8] and PCH with out-of-focus correction [9]. FCS extracts the information about the diffusion coefficients, chemical kinetics, and excited-state dynamics of fluorescent molecules at picomolar concentration from the analysis of a temporal autocorrelation function (ACF) of the measured sequence of photon arrival times. Both PCH and FIDA are used to obtain information on the concentration and specific brightness of fluorescent molecules and are based on the least-squares analysis of photon counting distribution (PCD), calculated from the same measured sequence of photon arrival times.

The development of new data analysis methods in FFS requires comprehensive testing of their validity, accuracy, resolvability, and robustness. The testing can be performed by applying a method to data obtained from real experiments. It is also convenient to perform preliminary testing of an analysis method on simulated data, because the use of these data can assist in interpretation and prediction of real experimental results. One typical application of simulations can be the verification of the resolvability of the method under a given set of parameters and the signal-to-noise ratio (S/N).

There are various simulation models of ACF and PCD taking into account translational and anomalous diffusion of fluorescent molecules, photobleaching, triplet-state dynamics, and non-Gaussian brightness profiles [10–12]. All these models are based on sequential simulation of the number of photon counts within a short time interval (or bin time). Therefore, the output of these models is the already binned intensity trace, which is not suitable for calculating ACF and PCD at arbitrary binning times, since binning times must be equal to or a multiple of this interval.

In this chapter we describe two simulation models. The first one simulates translational diffusion of individual fluorescent molecules and their transitions into a non-radiative or dark state (flickering), which are the main processes in fluctuation spectroscopy, and thus enables to obtain a "complete" stream of photon arrival times in FFS. A distinctive feature of this model is that the sequence of photon arrival times is considered as a doubly stochastic Poissonian point process (DSP) [7]. The advantage of such approach is the ability to calculate ACF and PCD and other statistical characteristic of the point process at an arbitrary binning time [13]. The model is well suitable for parallel simulation of photon arrival times from several types of independent molecules, provided that they will be combined into one common point process afterwards. Parallel simulation enables to simulate a point process with long duration and high concentration of molecules. It should be noted that the model is not valid at very high concentrations when the distance between fluorescent molecules becomes shorter and their intermolecular interaction becomes sufficient. The second model was specially developed for the fast simulation of PCD. It enables to simulate the already binned intensity trace from which only the photon counting distribution with given parameters can be calculated. This model does not simulate the diffusion of individual molecules and therefore is simpler and much faster than the first one. The calculation of ACF from the simulated intensity trace is still possible but unreasonable because the information about the diffusion is not involved in the simulation. The model is based on the idea that the number of molecules in the open observation volume and the number of photons emitted by a single molecule follow the Poisson distribution. In addition, it enables to simulate the signal from the "out-of-focus" molecules that opens the possibility to test the analysis methods with the out-of-focus correction [9].

2 Methods

2.1 Simulation of an Autocorrelation Function and Photon Counting Distribution via a Doubly Stochastic Poissonian Point Process

Here we describe the method of simulation of ACF and PCD via DSP. Free translational diffusion of fluorescent molecules through the inhomogeneous Gaussian-shaped observation volume and their transitions into a non-radiative state (flickering) are simulated. Initial model parameters are the dimensions of the modeling area L_x, L_y, L_z in μm ($L_x > \omega_0$, $L_y > \omega_0$, $L_z > z_0$), rate constants k_{AB}, k_{BA} of transitions of molecules into non-radiative state A → B and back B → A in s^{-1}, molecular brightness q in counts per second per molecule (cpsm), diffusion coefficient D in m^2/s, concentration C in nmol/L (nM), and the start and the end modeling time T_0 and T_m in seconds. The constants ω_0 and z_0 (in μm) are parameters of the three-dimensional Gaussian brightness profile [14]

$$B(\mathbf{r}) = B(x, y, z) = B_0 \exp\left(-2(x^2 + y^2)/\omega_0^2 - 2z^2/z_0^2\right). \quad (1)$$

The description of the general scheme of simulation is given below:

1. Set the dimensions of the modeling region and the parameters of the brightness profile $B(\mathbf{r})$.
2. Increase the molecule brightness according to further losses due to transitions of molecules into the non-radiative state $q_s = (1 + P_B)q$, where $P_B = k_{AB}/(k_{AB} + k_{BA})$. Here q is the time-independent "true" brightness ($B_0 = 1$ because no brightness profile correction is applied; *see* Chapter 33 by Skakun et al. in the same volume for details).
3. Get the number of molecules N in the modeling area. The total number of molecules is calculated as the integer closest to the value $N = 8L_xL_yL_z10^{-6}N_A$ if the concentration is given in nmol/L, where $N_A = 6.022 \times 10^{23}$ mol^{-1} is Avogadro's constant.
4. Obtain a sequence of photon arrival times (further simply denoted as events) from a single diffusing molecule switching between radiative and non-radiative states. It is modeled as DSP.

One way to simulate such a process is to reject the events of the stationary Poissonian point process of higher intensity according to the processes that form the second stochastic:

(a) Set the initial coordinates of the molecule according to uniform distribution in the modeling region (*see* **Note 1** for details).

(b) Set the initial state of the molecule (radiative A or non-radiative B) by generation of the random value α uniformly distributed on the interval (0,1) (further simply a basic random value) and checking the condition $\alpha \leq P_A$, where $P_A = k_{BA}/(k_{AB} + k_{BA})$. If the condition is true, then the state is radiative A, else, non-radiative B (*see* **Note 2** for details).

(c) Simulate the event of the stationary Poissonian point process on the time interval $[T_0; T_m]$ with the intensity $\lambda \geq q_s$. The event t_j of the Poissonian point process can be calculated using the recurrent formula $t_j = t_{j-1} - \lambda^{-1} \ln \alpha$, where $t_0 = T_0$ and α is the basic random value.

(d) Simulate molecular diffusion within the time step $\Delta t = t_j - t_{j-1}$. The molecule coordinates x, y, z are shifted with increments Δx, Δy, Δz that are Gaussian-distributed values with mean $\mu = 0$ and variance $\sigma^2 = 2D\Delta t$ [15].

(e) Check the border conditions. One widespread kind of the border conditions is the periodic condition: if a molecule leaves the modeling region, it returns from its opposite side (*see* **Note 3** for details).

(f) Check the condition $\beta_j \leq \Psi(t_j)$, where β_j is a random value, uniformly distributed on $[0; \lambda]$, and $\Psi(t_j) = q_s B(\mathbf{r})$ is the intensity of photon emission at the point with radius vector $\mathbf{r}$ at the time t_j. If this condition is true, go to **step g**; else, ignore the event t_j and go to **step c**.

(g) Check the radiative state of the molecule at time t_j. If the molecule is in the B state, then the event t_j is ignored. This can be done by the generation of an alternate sequence of "on/off" intervals during which the molecule is in either A or B state. This sequence is continuously generated starting from the initial time T_0. The length of the "on" interval is modeled as $\Delta t_A = -\tau_A \ln \alpha$, and the length of the "off" interval is correspondently modeled as $\Delta t_B = -\tau_B \ln \alpha$, where α is the basic random value, $\tau_A = 1/k_{AB}$, and $\tau_B = 1/k_{BA}$ (*see* **Note 2**).

(h) The generation of events (**steps c–g**) is continued while $t_j < T_m$. The **step c** forms the first stochastic of the point process, the so-called shot noise. The next four steps (**d–g**) are intended for the simulation of physical processes (diffusion of molecules through the illuminated volume and their blinking) that forms the second stochastic of the double-stochastic Poissonian point process.

5. Repeat **step 4** for every molecule.
6. Combine all obtained sequences into the resulting sequence with events sorted in ascending order.
7. Calculate ACF, PCD, or other characteristics (e.g., fluorescence factorial cumulants [16]) from the simulated sequence of photon arrival times.

In order to simulate the photon arrival times from a mixture of different molecules, the steps of the algorithm have to be performed independently for each type of molecules. The two last steps have to be done only once in the end of the simulation.

Several simulations have been performed in order to validate the proposed model. The idea was to demonstrate the ability of the model to reproduce real measured data on the basis of the analyzed results. The initial parameters of the model have been set to the values obtained after the global analysis of the solution of monomeric and dimeric forms of eGFP in aqueous buffer solution (*see* details in Chapter 33 by Skakun et al. in the same volume). From the simulated point processes, ACF and two PCDs were calculated and then analyzed globally (separately for the monomer and dimer) accordingly to the protocol described in Chapter 33 by Skakun et al. in the same volume. The analysis was performed in the FFS data processor (http://www.sstcenter.com/software/index.phtml?pageid=idFFSDPFeatures) [17].

The results of analysis of simulated data are shown in Fig. 1 and in Table 1. Good correspondence in the shape of analyzed data and fit parameter values was obtained for the autocorrelation function. However, the experimental PCDs differ from the simulated ones. It could be expected as the model described here does not simulate the out-of-focus emission that was accounted for in the analysis of the measured data. Therefore, we used the "true" brightness in our simulations in order to get rid of the influence of the correction parameter on the brightness value. The correction parameter was set to zero in the analysis of the simulated data.

2.2 Photon Counting Distribution Simulation

Here we describe the algorithm of simulation of PCD for the case of 3D Gaussian approximation of the brightness profile with out-of-focus correction [18]. Note that the algorithm can be easily adopted for other approximations of the brightness profile, e.g., 2D Gaussian or Gaussian-Lorentzian. Initial parameters of the model are the number of molecules $N_{\text{eff } i}$ in the effective volume defined as $V_{\text{eff}} = \chi_1^2/\chi_2, \chi_k = \int_V [B(\mathbf{r})/B_0]^k d\mathbf{r}$ and the effective molecular brightness $q_{\text{eff } i}$ of ith species, $i = 1, 2, \ldots, M$, sampling (binning) time T, the number of sampling intervals m, the number of species M, and the out-of-focus correction parameter F_1 (*see* Chapter 33 by Skakun et al. in the same volume, **Note 1**, for details). The number of sampling intervals m depends on the desired S/N. If S/N is taken

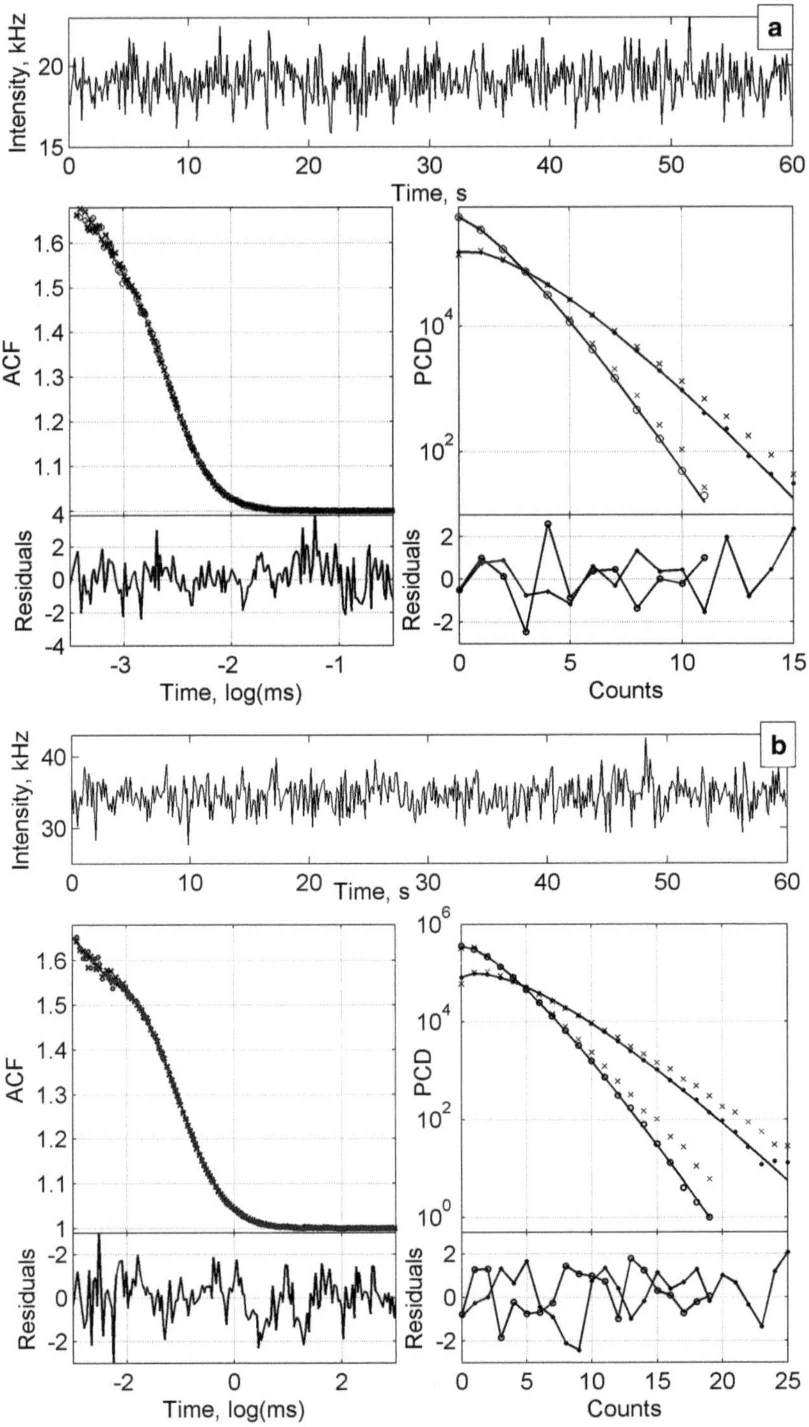

Fig. 1 The results of global analysis of ACF and PCD simulated by the first model. The intensity of the simulated sequence of photon arrival times is plotted on the *top* of each panel. The residuals are plotted in the *bottom* and both simulated and theoretical ACF and PCD are plotted in the *middle*. Simulation was performed with parameters given in Table 1. To reproduce the real experiment, the model parameter values were taken from the analysis of eGFP monomer (**a**) and dimer (**b**) described in Chapter 33 by Skakun et al. in the same volume. One AFC and two PCDs (*circles*) calculated from the simulated point processes with the sampling time 50 μs (*open circles*) and 100 μs (*closed circles*), respectively, were analyzed using the FFS data processor as described in Chapter 33 by Skakun et al. in the same volume. Measured AFC and PCDs (*crosses*) are presented for comparison

Table 1
The results of the global analysis of ACF and PCD simulated by the first method

	eGFP monomer		eGFP dimer	
Parameter	**Initial value**	**Estimation**	**Initial value**	**Estimation**
τ_{trip} (μs)	2.98	2.75 ± 0.38	1.35	1.49 ± 0.13
F_{trip}	0.18	0.21 ± 0.02	0.20	0.20 ± 0.01
τ_D (μs)	62.4	65.0 ± 1.1	95.2	97.69 ± 0.79
N_{eff}	1.611	1.637 ± 0.009	1.70	1.713 ± 0.003
z_0/ω_0	5	4.70 ± 0.26	5	4.79 ± 0.17
$q \times 10^3$ (cpms)	34.49	32.88 ± 0.19	59.32	56.86 ± 0.13
χ^2		1.244		0.979

The initial values of the model parameters were taken so that to reproduce the real measurement of eGFP monomer and dimer described in Chapter 33 by Skakun et al. in the same volume. It results in $k_{AB} = 60{,}402\ s^{-1}$, $k_{BA} = 275{,}167\ s^{-1}$, $C = 3.559$ nM, and $D = 3.606 \times 10^{-10}\ m^2/s$ for the monomer, and $k_{AB} = 140{,}740\ s^{-1}$, $k_{BA} = 600{,}000\ s^{-1}$, $C = 3.757$ nM, and $D = 2.363 \times 10^{-10}\ m^2/s$ for the dimer. These parameters were calculated from N_{eff}, τ_{trip}, F_{trip}, τ_{diff} according to Eqs. 11 (with $F_1 = 0$), 20, and 21 correspondingly (*see* **Note 8**). The profile parameters were $\omega_0 = 0.3$ μm, $z_0 = 1.5$ μm, $L_x = 3$ μm, $L_y = 3$ μm, and $L_z = 15$ μm. Simulation time T_m was set to 60 s. The number of simulated photon arrival times for the monomer was 1.14×10^6 and for the dimer 2.67×10^6, whereas the number of registered photons in the experiment was equal to 1.18×10^6 and 2.17×10^6, respectively. The analysis of simulated data was performed using the FFS data processor as described in Chapter 33 by Skakun et al. in the same volume. Values of χ^2 criterion are also presented

at the maximum $p_{max} = \max_k(P(k))$ of the simulated distribution $P(k)$, the number of intervals m can be calculated as [19]

$$m = (S/N)^2(1 - p_{max})/p_{max}. \tag{2}$$

The steps of the algorithm are presented below:

1. Set the radius r_0 of the observation volume V_0 (*see* **Note 4** for details).
2. Get the number of molecules N_i of the ith species by generating a Poisson random variable with the mean $\bar{N}_i = N_{eff\,i} \frac{4r_0^3}{3\sqrt{\pi}(1+F_1)^2}$ (*see* **Note 5**).
3. Uniformly distribute the molecules over the observation volume V_0 (*see* **Note 6** for details).
4. Get the number of photons emitted by each molecule by generating the Poisson random variable n_{ij} with the mean $\bar{n}_{ij} = q_{eff\,i} T \exp\left(-2r_{ij}^2\right)$, where r_{ij} is the radius-vector length of jth molecule of ith species, $j = 1, 2, \ldots, N_i$, $i = 1, 2, \ldots, M$.
5. Get the number of photons emitted by molecules residing in the out-of-focus region by the generation of a Poisson random variable λ with the mean $\Lambda(T) = \sum_{i=1}^{M} \frac{N_{eff\,i} q_{eff\,i} T F_1}{2\sqrt{2}(1+F_1)^2}$ (*see* **Note 7** for details).

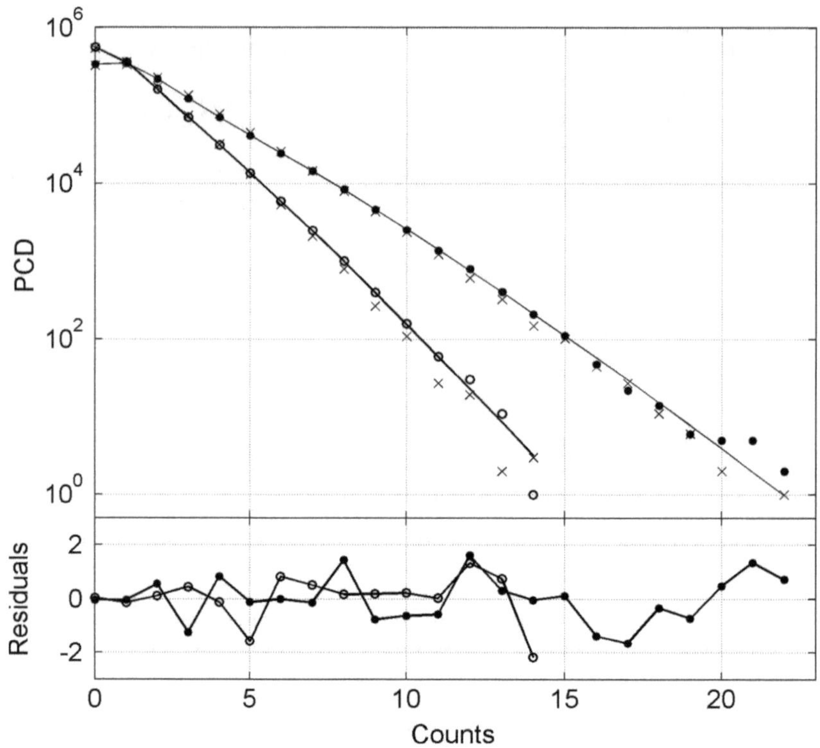

Fig. 2 The results of PCH analysis with the out-of-focus correction of PCDs simulated by the second model. Simulation was performed with parameters given in Table 2. To reproduce the real experiment, the parameter values were taken from the analysis of eGFP monomer (*open circles*) and dimer (*closed circles*) described in Chapter 33 by Skakun et al. in the same volume. The analysis was carried out using the FFS data processor. Measured PCDs (*crosses*) are presented for comparison

6. Summarize all obtained numbers of photons to get the total number of photons $S_T = \sum_{i=1}^{M} \sum_{j=1}^{N_i} n_{ij} + \lambda$ detected in the sampling time interval T.
7. Repeat **steps 2–6** for the next sampling time (m times in total).
8. Calculate the PCD. Any other characteristic properties that are obtained via the PCD (e.g., fluorescence moment and factorial cumulants) can be also calculated. The binning time in the PCD calculation must be equal to or a multiple of the sampling time (e.g., $2T$, $3T$).

In order to validate the proposed model, we performed the simulations using the same initial values that we used for the testing of the first model. In this case we used the effective brightness because of simulation of the out-of-focus signal. PCH analysis with the out-of-focus correction of the simulated PCDs was done in the FFS data processor. The results of analysis are shown in Fig. 2 and in Table 2. Excellent correspondence in the shape of analyzed data and fit parameter values was obtained.

Table 2
The analysis results of PCDs simulated by the second model

	eGFP monomer		eGFP dimer	
Parameter	**Initial value**	**Estimation**	**Initial value**	**Estimation**
N_{eff}	1.611	1.608 ± 0.007	1.70	1.699 ± 0.005
$q_{eff} \times 10^3$ (cpms)	52.33	52.47 ± 0.46	94.33	94.41 ± 0.44
F_1	0.56	0.56 ± 0.01	0.65	0.651 ± 0.005
χ^2	–	1.085	–	0.871

The initial values of the model parameters were taken from the analysis of eGFP monomer and dimer described in Chapter 33 by Skakun et al. in the same volume. The sampling interval $T = 50$ μs and the simulation time $T_m = 60$ s. Analysis was performed using the FFS data processor. Values of χ^2 criterion are also presented

3 Notes

1. If the modeling region is defined in the form of a rectangular parallelepiped, it is convenient to set the origin in the point of intersection of its diagonals. To uniformly distribute molecules in such a parallelepiped, their coordinates can be calculated as follows: $x = (2\alpha_1 - 1)L_x$, $y = (2\alpha_2 - 1)L_y$, $z = (2\alpha_3 - 1)L_z$, where α_i is the basic random value and L_x, L_y, L_z are the halves of the parallelepiped edges.
2. Transitions of fluorescent molecules between states A and B can be described by a system of differential equations of the form [6, 20]

$$\begin{cases} \frac{dS_A(t)}{dt} = k_{BA}S_B(t) - k_{AB}S_A(t), \\ \frac{dS_B(t)}{dt} = k_{AB}S_A(t) - k_{BA}S_B(t), \end{cases} \tag{3}$$

where $S_A(t)$, $S_B(t)$ are the probabilities of observing the molecule in the corresponding state at time t and k_{AB} and k_{BA} are the rate constants of transitions between the corresponding states. Initial conditions are $S_A(0) = 1$, $S_B(0) = 0$. The solution of the system can be represented in the form

$$\begin{aligned} S_A(t) &= \frac{k_{BA}}{k_{AB} + k_{BA}} + \frac{k_{AB}}{k_{AB} + k_{BA}} \exp(-t/\tau), \\ S_B(t) &= \frac{k_{AB}}{k_{AB} + k_{BA}} - \frac{k_{AB}}{k_{AB} + k_{BA}} \exp(-t/\tau), \end{aligned} \tag{4}$$

where we denote $\tau = (k_{AB} + k_{BA})^{-1}$. The probabilities of observing a molecule in state A *or* B (at $t \to \infty$) are

$$P_A = k_{BA}/(k_{AB} + k_{BA}), \qquad P_B = k_{AB}/(k_{AB} + k_{BA}). \tag{5}$$

The first equation of system 3 without the term $k_{BA}S_B(t)$ gives the distribution of the instants of time when the molecule makes a transition from state A to state B: $f_A(t) = \tau_A^{-1}\exp(-t/\tau_A)$, where $\tau_A = 1/k_{AB}$ is the mean residence time in state A (the function $f_B(t) = \tau_B^{-1}\exp(-t/\tau_B)$). Therefore, the probabilities that the molecule changes its state after the time interval Δt_A or Δt_B are

$$p_{\Delta t_A} = 1 - \exp(-\Delta t_A/\tau_A), \qquad p_{\Delta t_B} = 1 - \exp(-\Delta t_B/\tau_B). \quad (6)$$

Finally, the intervals Δt_A or Δt_B can be obtained as

$$\Delta t_A = -\tau_A \ln(\alpha), \qquad \Delta t_B = -\tau_B \ln(\alpha), \quad (7)$$

where α is the basic random value.

3. One way to perform the periodic condition is to change the molecule coordinates according to the rule: $x_{new} = x_{prev} - 2L[x_{prev} + L/2L]_{-\infty}$, where L is the size of the modeling region and $[\ldots]_{-\infty}$ means rounding to $-\infty$.
4. The value of r_0 can be found from the condition

$$\int_0^{r_0}\int_0^{2\pi}\int_0^{\pi} e^{-2r^2} r^2\omega_0^2 z_0 \sin\alpha\, \mathrm{d}r\, \mathrm{d}\varphi\, \mathrm{d}\alpha \Big/ \int_0^{\infty}\int_0^{2\pi}\int_0^{\pi} e^{-2r^2} r^2\omega_0^2 z_0 \sin\alpha\, \mathrm{d}r\, \mathrm{d}\varphi\, \mathrm{d}\alpha = 16\int_0^{r_0} e^{-2r^2} r^2\, \mathrm{d}r \Big/ \sqrt{2\pi}. \to 1. \quad (8)$$

In the equation above we applied the following transformation of the coordinate system:

$$\begin{cases} x = r\omega_0 \cos\varphi \sin\alpha \\ y = r\omega_0 \sin\varphi \sin\alpha \\ z = rz_0 \cos\alpha. \end{cases} \quad (9)$$

where r is the length of the radius vector and φ and α are angles (*see* Chapter 33 by Skakun et al. in the same volume for more details). We used $r_0 = 3$ in our simulations.
5. To determine the average number of molecules $\bar{N}_i$ of the ith species in the volume V_0 defined by the radius vector with the length r_0, let us use the expression related to the average concentration of molecules C_i with the number of molecules in this volume

$$\overline{C}_i = \overline{N}_i/V_0 = N_{\text{eff}\,i}/V_{\text{eff}}. \quad (10)$$

Taking into account that $V_{\text{eff}} = (1 + F_1)^2\pi^{3/2}\omega_0^2 z_0$ and $V_0 = 4\pi r_0^3\omega_0^2 z_0/3$, one obtains

$$\overline{N}_i = \frac{N_{\text{eff}\,i} V_0}{V_{\text{eff}}} = N_{\text{eff}\,i} \frac{4r_0^3}{3\sqrt{\pi}(1+F_1)^2}. \tag{11}$$

6. To uniformly distribute molecules in the volume V_0, one can generate a uniformly distributed in $[-r_0; r_0]$ random value for each molecule coordinate x, y, z and then check the condition $\left(x^2/\omega_0^2 + y^2/\omega_0^2 + z^2/z_0^2\right)^{1/2} \le r_0$. The generation of a triple of random values is repeated until this condition becomes true.
7. To develop the algorithm of simulation of the out-of-focus emission, let us write the generating function (GF) of PCD [8]:

$$G(\xi) = \exp\left\{\sum_{i=1}^{M} \overline{C}_i \int_V (\exp((\xi-1)q_i T W(\mathbf{r})) - 1)\mathrm{d}\mathbf{r}\right\} \tag{12}$$

where ξ is the auxiliary variable, $W(\mathbf{r})$ is the actual brightness profile function, and M is the number of species. After expanding the exponent in Eq. 12 into a Taylor series

$$G(\xi) = \exp\left\{\sum_{i=1}^{M} \bar{C}_i \sum_{k=1}^{\infty} \frac{(\xi-1)^k q_i^k T^k}{k!} \int_V W^k(\mathbf{r})\mathrm{d}\mathbf{r}\right\} \tag{13}$$

and introducing the correction parameters F_k in the form $\int_V W^k(\mathbf{r})\mathrm{d}\mathbf{r} = (1+F_k)\int_V B^k(\mathbf{r})\mathrm{d}\mathbf{r}$ [9], one can see that the GF can be written as a product of two GFs: $G(\xi) = G_G(\xi)G_c(\xi)$. The first GF

$$G_G(\xi) = \exp\left\{\sum_{i=1}^{M} \overline{C}_i \int_V (\exp((\xi-1)q_i T B(\mathbf{r})) - 1)\mathrm{d}\mathbf{r}\right\} \tag{14}$$

accounts for the photons coming from the "main" observation volume described by the Gaussian approximation $B(\mathbf{r})$, and the second one

$$G(\xi) = \exp\left\{\sum_{i=1}^{M} \overline{C}_i \sum_{k=1}^{\infty} \frac{(\xi-1)^k B_0^k q_i^k T^k F_k}{k!} \chi_k\right\} \tag{15}$$

accounts for additional photons detected from the non-Gaussian out-of-focus part of the illuminated volume. In the case of the first-order correction (all $F_k = 0$ except $k = 1$), Eq. 15 can be written as

$$G_c(\xi) = \exp\left\{\sum_{i=1}^{M} \overline{C}_i q_{\text{eff}\,i} T F_1 (\pi/2)^{3/2} \omega_0^2 z_0 (\xi - 1)\right\}, \tag{16}$$

where $q_{\text{eff}\,i} = B_0 q_i = (1 + F_1) q_i$ is the effective brightness that in the case of out-of-focus correction becomes the function of the correction parameter F_1 (*see* [9] and Chapter 33 by Skakun et al. in the same volume for details). Let us rewrite Eq. 16 in the form

$$G_c(\xi) = \exp\{\Lambda(T)(\xi - 1)\} \tag{17}$$

with

$$\Lambda(T) = \sum_{i=1}^{M} \frac{N_{\text{eff}\,i} q_{\text{eff}\,i} T F_1}{2\sqrt{2}(1 + F_1)^2}, \tag{18}$$

where Eqs. 10 and 11 were taken into account. Equation 17 is the GF of the Poisson distribution with the mean $\Lambda(T)$. Therefore, the out-of-focus emission can be simulated by a Poisson distribution with the mean $\Lambda(T)$.

8. In the one-component case with the Gaussian brightness profile, the autocorrelation function $G_{\text{corr}}(t)$ of the fluorescence fluctuations can be expressed as [6, 21]

$$G_{\text{corr}}(t) = 1 + \frac{1}{N_{\text{eff}}} \frac{1 - F_{\text{trip}} + F_{\text{trip}} e^{-t/\tau_{\text{trip}}}}{1 - F_{\text{trip}}} \times \frac{1}{(1 + t/\tau_{\text{diff}})\sqrt{(1 + t/a^2\tau_{\text{diff}})}}, \tag{19}$$

where

$$\tau_{\text{trip}} = (k_{AB} + k_{BA})^{-1}, \quad F_{\text{trip}} = k_{AB}/(k_{AB} + k_{BA}) \tag{20}$$

and

$$\tau_{\text{diff}} = \omega_0^2/4D. \tag{21}$$

References

1. Hess S, Huang S, Heikal A (2002) Biological and chemical applications of fluorescence correlation spectroscopy: a review. Biochemistry 41:697–705
2. Elson E, Qian H (1989) Interpretation of fluorescence correlation spectroscopy and photobleaching recovery in terms of molecular interactions. Methods Cell Biol 30:307–332
3. Maiti S, Haupts U, Webb W (1997) Fluorescence correlation spectroscopy: diagnostics for sparse molecules. Proc Natl Acad Sci USA 94:11753–11757
4. Starr T, Thomson N (2001) Total internal reflection with fluorescence correlation spectroscopy: combined surface reaction and solution diffusion. Biophys J 80:1575–1584
5. Hom E, Verkman A (2002) Analysis of coupled bimolecular reaction kinetics and diffusion by two-color fluorescence correlation spectroscopy: enhanced resolution of kinetics by resonance energy transfer. Biophys J 83:533–546
6. Krichevsky O, Bonnet G (2002) Fluorescence correlation spectroscopy: the technique and its applications. Rep Prog Phys 65:251–297

7. Chen Y, Mueller JD, So P et al (1999) The photon counting histogram in fluorescence fluctuation spectroscopy. Biophys J 77:553–567
8. Kask P, Palo K, Ullmann D et al (1999) Fluorescence-intensity distribution analysis and its application in biomolecular detection technology. Proc Natl Acad Sci USA 96:13756–13761
9. Huang B, Perroud T, Zare R (2004) Photon counting histogram: one-photon excitation. Chemphyschem 5:1523–1531
10. Dix J, Hom E, Verkman A (2006) Fluorescence correlation spectroscopy simulations of photophysical phenomena and molecular interactions: a molecular dynamics/Monte Carlo approach. J Phys Chem B 110:1896–1906
11. Bunfield D, Davis L (1998) Monte Carlo simulation of a single-molecule detection experiment. Appl Optics 37:2315–2326
12. Wohland T, Rigler R, Vogel H (2001) The standard deviation in fluorescence correlation spectroscopy. Biophys J 80:2987–2999
13. Shingaryov IP, Skakun VV, Apanasovich VV (2012) Simulation of a stream of photon counts in single-molecule fluorescence fluctuation spectroscopy. J Appl Spectrosc 78:892–898
14. Rigler R, Mets U, Widengren J et al (1993) Fluorescence correlation spectroscopy with high rate and low background: analysis of translational diffusion. Eur Biophys J 22:169–175
15. MacKeown PK (1997) Stochastic simulation in physics. Springer, Singapore
16. Müller JD (2004) Cumulant Analysis in Fluorescence Fluctuation Spectroscopy. Biophys J 86:3981–3992
17. Skakun VV, Engel R, Digris AV et al (2011) Global analysis of autocorrelation functions and photon counting distributions. Front Biosci (Elite edition) 3:489–505
18. Shingaryov IP, Skakun VV, Apanasovich VV (2009) Photon counts simulation in fluorescence fluctuation spectroscopy. Proc. of Conf. Pattern Recognition and Information Processing (PRIP). Publ. Center of BSU, Minsk, pp 178–182
19. Skakun VV, Novikov EG, Apanasovich VV et al (2006) Initial guesses generation for fluorescence intensity distribution analysis. Eur Biophys J 35:410–423
20. Elson E, Madge D (1974) Fluorescence correlation spectroscopy. Conceptual basis and theory. Biopolymers 13:1–27
21. Malvezzi-Campeggi F, Jahnz M, Heinze K et al (2001) Light-induced flickering of DsRed provides evidence for distinct and interconvertible fluorescent states. Biophys J 81:1776–1785

Part VIII

Fluorescence Microscopy: Single-Molecule Fluorescence Spectroscopy (smFS)

Chapter 35

Single-Molecule Fluorescence of Nucleic Acids

Kaley McCluskey, Euan Shaw, Daniel A. Lafontaine, and J. Carlos Penedo

Abstract

Single-molecule fluorescence studies of nucleic acids are revolutionizing our understanding of fundamental cellular processes related to DNA and RNA processing mechanisms. Detailed molecular insights into DNA repair, replication, transcription, and RNA folding and function are continuously being uncovered by using the full repertoire of single-molecule fluorescence techniques. The fundamental reason behind the stunning growth in the application of single-molecule techniques to study nucleic acid structure and dynamics is the unmatched ability of single-molecule fluorescence, and mostly single-molecule FRET, to resolve heterogeneous static and dynamic populations and identify transient and low-populated states without the need for sample synchronization. New advances in DNA and RNA synthesis, post-synthetic dye-labeling methods, immobilization and passivation strategies, improved dye photophysics, and standardized analysis methods have enabled the implementation of single-molecule techniques beyond specialized laboratories. In this chapter, we introduce the practical aspects of applying single-molecule techniques to investigate nucleic acid structure, dynamics, and function.

Key words Single-molecule detection, DNA, RNA, Förster resonance energy transfer (FRET), Total internal reflection (TIR), Fluorescence labeling

1 Introduction

Over the past decade, single-molecule fluorescence spectroscopy has made new dynamic structural information about biological molecules available with an unprecedented level of detail. Single-molecule microscopy in combination with Förster resonance energy transfer (FRET) enables us to follow the conformational changes of spatially isolated nucleic acids, proteins, and protein-DNA or protein-RNA complexes over time periods ranging from milliseconds to minutes, thus covering the timescale of most biologically relevant processes. By isolating single biomolecules at low concentrations, transient or rare states that would be averaged over in a bulk experiment, altering the result or being lost, can be identified and quantified [1].

Yves Engelborghs and Antonie J.W.G. Visser (eds.), *Fluorescence Spectroscopy and Microscopy: Methods and Protocols*, Methods in Molecular Biology, vol. 1076, DOI 10.1007/978-1-62703-649-8_35,

Single-molecule spectroscopy can be performed on low concentrations of sample diffusing freely in solution, termed fluorescence correlation spectroscopy (FCS), or immobilized on a substrate, generally a glass or quartz slide. Common to all single-molecule experiments is the requirement for a very small excitation volume to reduce background to a low enough level for fluorescence signals to be detectable. In FCS methods, excitation and detection take place at the focal spot of a laser, while with surface-immobilized systems, either a confocal point source or wide-field excitation using total internal reflection fluorescence microscopy (TIRF) can be used [1, 2]. The advantage of surface immobilization over freely diffusing techniques is that individual molecules can be observed for extended periods of time, limited only by the photostability of the reporter dyes, which has greatly improved with the introduction of efficient triplet-state quenchers and oxygen-scavenging systems [3, 4]. In this way, conformational dynamics for a variety of time-scales and environments and entire folding pathways can be determined for each single molecule. This not only identifies transient states as mentioned above but also reveals the natural heterogeneity of folding pathways and lifetimes emerging in recent experiments involving complex nucleic acid structures [5, 6].

In this chapter, we will focus on the measurement of conformational dynamics in single surface-immobilized nucleic acids using FRET and wide-field TIRF techniques. We will describe the most widely used optical setups together with a detailed discussion of the protocols for post-synthetic site-specific labeling of nucleic acids with FRET pairs. Lastly, we will provide a comparative description of the software packages and analysis methods available to date to interpret single-molecule FRET trajectories.

1.1 Single-Molecule Förster Resonance Energy Transfer (smFRET)

Most single-molecule experiments on nucleic acid structure and function involve the measurement of conformational changes via FRET between two fluorophores covalently attached to known positions within the nucleic acid structure (*see* **Note 1**) (Fig. 1a), in which an optically excited donor (D) fluorophore transfers some of its energy to an acceptor (A) with an efficiency E_{app} dependent on the interdye separation [7] (Fig. 1a, b). A trace of fluorescence intensity versus time for one molecule reveals not only structural changes but also their associated lifetimes, thus giving access to a wealth of dynamic information (Fig. 1c). Nucleic acids are particularly suited to surface-immobilized, extended-observation experiments because they do not interact as readily with the glass surface as proteins [8], which require additional surface passivation steps.

The efficiency of FRET between a pair of fluorophores depends on the sixth power of the distance R between them, governed by the equation

$$E = \frac{1}{1 + \left(\frac{R}{R_0}\right)^6} \tag{1}$$

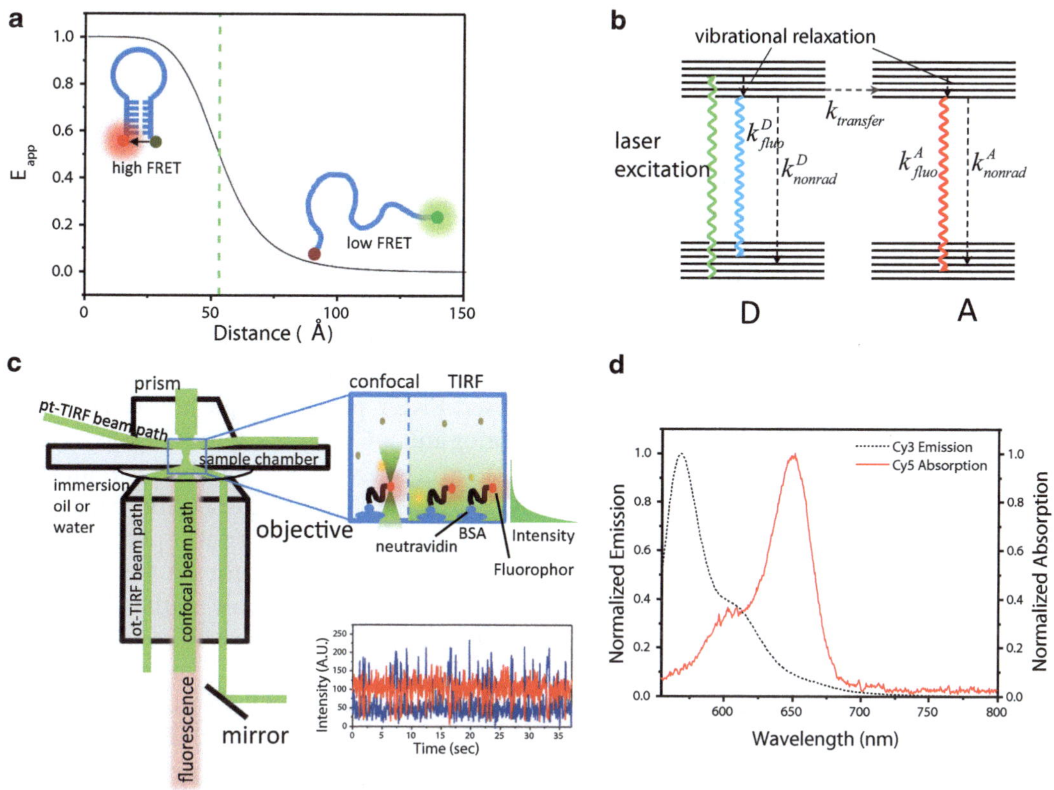

Fig. 1 Principle of FRET microscopy. (**a**) Efficiency of FRET as a function of interdye separation, *R*. E_{FRET} depends on R^{-6}, making changes in FRET-sensitive indicators of dynamics between molecules (intermolecular FRET) or between structural nucleic acid components tagged with fluorescent dyes (intramolecular FRET). (**b**) Jablonski diagram of energy transfer processes in FRET. On the *left*, a donor (*D*) molecule absorbs a photon and enters a higher electronic state. The molecule vibrationally relaxes to the lowest-energy sublevel, from which it de-excites either through fluorescence, nonradiative processes such as collisional, quenching, or through resonant energy transfer to the acceptor (*A*). Once excited, A can also decay through either radiative or nonradiative processes. (**c**) Schematic of the excitation and fluorescence pathways for TIR and single-molecule confocal microscopy. In pTIR, the excitation laser is focused onto the upper face of the sample chamber through a quartz prism (*top left*), while in oTIR, it is coupled onto the lower face through the extreme edge of the microscope objective (*lower left*). In confocal microscopy, the objective focuses the excitation beam as in ordinary confocal microscopy. In all cases, dye fluorescence is collected through the microscope objective, and residual excitation light is filtered or directed away from the detection optics. *Right inset*: excitation volumes for confocal and TIR microscopy. (*Left*) The confocal beam is scanned through the volume, illuminating one molecule at a time at the focal point. (*Right*) The evanescent wave illuminates a small area (~10 μm^2) to a depth of about 150 nm. (**d**) Spectral overlap of Cy3 and Cy5 dyes, a common *D-A* pair. For the energy transfer process in (**b**) to be efficient, the dyes must be in close proximity (3–8 nm) and exhibit significant overlap between the emission spectrum of *D* and the absorption spectrum of *A*, as shown

where R_0, the Förster radius, depends on the spectral properties and orientations of the dyes [7] (*see* **Note 2**). In practice, the apparent FRET efficiency, E_{app}, is calculated from the ratio of acceptor fluorescence to total (donor + acceptor) fluorescence when only the donor is optically excited:

$$E_{\mathrm{app}} = \frac{I_{\mathrm{A}}^{\mathrm{D}}}{\alpha I_{\mathrm{A}}^{\mathrm{D}} + I_{\mathrm{D}}^{\mathrm{D}}} \tag{2}$$

In Eq. 2, $I_A{}^D$ is the intensity of acceptor fluorescence, $I_D{}^D$ is the intensity of donor fluorescence, and α is a correction factor taking account of donor fluorescence leaking into the acceptor detection channel of the dyes. An additional correction factor accounting for differences in quantum yield and detection efficiency is needed to convert E_{app} to the actual FRET efficiency E [7, 9]. The interdye separation can be calculated using Eq. 1, and relative differences in E_{app}, E, or the related quantity ratio A (*see* **Note 3**), can be used to distinguish separation changes as small as 0.34 nm (1 base pair) [10]. The most widely used FRET pair for single-molecule studies contains a Cy3 dye as donor and a Cy5 as acceptor (Fig. 1d). Some reasons for the wide application of Cy3, Cy5, and their variants (i.e., Alexa, ATTO) in smFRET include their photostability (particularly in the presence of oxygen scavengers), appropriate spectral separation between donor and acceptor, similar quantum yield, and the fact that these dyes are commercially available with a variety of functional groups for coupling to both nucleic acids and proteins.

1.2 smFRET Studies of DNA and RNA Dynamics

In recent years, the application of smFRET to nucleic acids has revolutionized our understanding of DNA and RNA dynamics in vitro [11–14]. The variety of nucleic acid structures to which smFRET has been applied includes both single- and double-stranded RNA and DNA, DNA Holliday junctions [13, 15, 16] (Fig. 2a), guanine-quadruplexes (g-quadruplexes) [17–19] (Fig. 2b), and functional RNA motifs involved in catalysis (ribozymes; Fig. 2c–e) [11, 20–24] and gene regulation (riboswitches) (Fig. 2e–j) [25–32].

A g-quadruplex is a three-tiered stack of guanosine tetramers that forms in regions of ssDNA (Fig. 2b) such as the telomere and the promoter regions of some DNAs [17]. Their appearance in these regions has given rise to proposals that they are involved in the inhibition of telomerase activity and in gene transcription, respectively [18]. smFRET has been used to follow the switching of telomeric g-quadruplexes between parallel and antiparallel strand configurations and to measure the lifetimes of the states at varying temperatures and ion concentrations [19]. Studies of promoter-region g-quadruplexes also confirmed that the g-quadruplex can form preferentially over the canonical double helix [17].

Another DNA structure that has received considerable interest because of its biological relevance is the four-way Holliday junction, a four-strand junction formed during DNA replication (Fig. 2a). smFRET studies have addressed both the conformational dynamics of its arms [13], which can stack in several different pairs, and its branch migration rate [15, 16]. Single-molecule confocal FRET was used to measure the distribution of step sizes along a

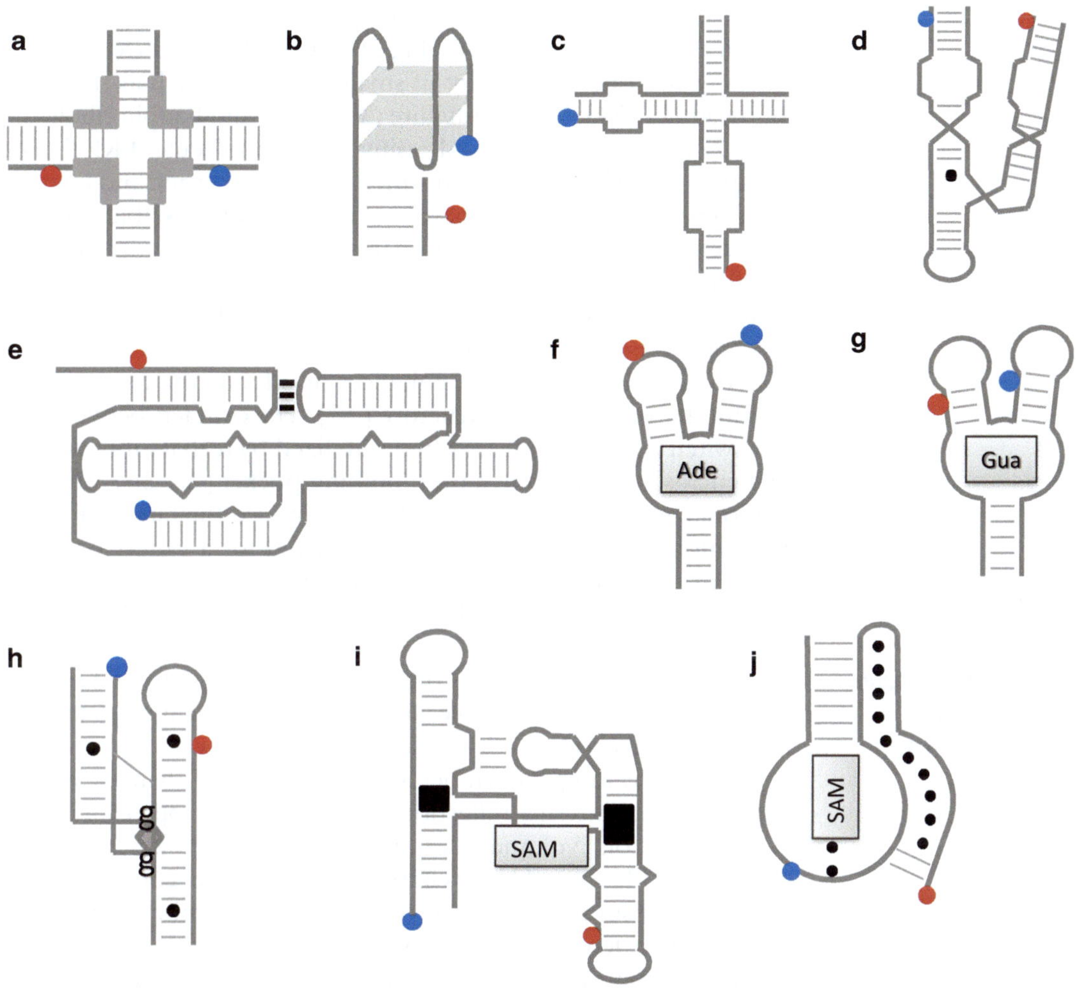

Fig. 2 Schematics of FRET experiments on nucleic acid structures. (**a–j**) illustrate the FRET configurations used to study the structural dynamics of several functional nucleic acid structures. (**a**) The mobile Holliday junction [13, 15, 16]. In this labeling configuration, E_{FRET} changes when the branches migrate across the homologous region (*thick lines*), changing the separation between the dyes. (**b**) The guanine-quadruplex in the antiparallel strand configuration, which would yield a different FRET value from the parallel strand configuration, in which all four "corner" strands have the same directionality [19]. (**c–e**) Small catalytic ribozymes that play a central role in RNA processing by performing self-cleavage reactions. (**c**) The hairpin ribozyme [11, 23], (**d**) the hammerhead ribozyme [21, 24], and (**e**) the Varkud satellite (VS) ribozyme [22]. (**f–j**) Five sample riboswitches, 5′-untranslated mRNA sequences that regulate the expression of downstream genes via conformation changes induced by metabolite binding (metabolites pictured as *shaded boxes*). The riboswitches depicted here specifically bind (**f**) adenine [28, 32], (**g**) guanine [27], (**h**) c-di-GMP [31], (**i**) SAM-I [29], and (**j**) SAM-II [30]

homologous (migration-capable) sequence at one-base pair (bp) resolution and determine its sensitivity to nearby homologous sequences [16].

Three- and four-way helical junctions are also present in complex RNA structures like ribozymes, which are catalytically active RNA sequences that independently process RNA by self-cleavage and/or ligation [11, 20]. The Mg^{2+}-induced folding pathway and catalytic activity of small ribozymes such as the hammerhead [21, 24], Varkud satellite (VS) [22], and hairpin ribozyme [11, 23] have been studied using single-molecule FRET. For instance, smFRET studies confirmed that the cleavage and ligation reactions of the hairpin ribozyme (Fig. 2c) require stable docking between loops on two of the four arms of a four-way junction [23].

Riboswitches are 5′-untranslated mRNA structures involved in gene regulation via ligand-induced structural changes (Fig. 2f–j). Binding of a ligand associated with the upstream gene triggers a structural change that either enhances or inhibits the formation of a promoter or terminator structure [25]. The contacts between the riboswitch and its ligand are often stabilized by tertiary interactions that enclose the ligand in a tight binding pocket [25, 26]. Surface-immobilized FRET experiments have addressed the rates of folding and unfolding in the presence of the ligand for a variety of structures including the riboswitches that sense guanine [27], adenine [28, 32], SAM-I [29], SAM-II [30], and cyclic GMP [31].

1.3 Single-Molecule Total Internal Reflection (sm-TIR)

The system under study and the information to be gained from it determine the most suitable single-molecule technique to use. In TIRF microscopy, a wide-field technique, the excitation light is coupled to the sample chamber at an angle greater than the critical angle for total internal reflection, resulting in an approx. 150 nm deep excitation volume illuminated by the evanescent wave (Fig. 1c) [33, 34]. The excitation light can be coupled either through a prism placed on top of the sample chamber or through the objective lens of the microscope, known as prism type and objective type (pTIR and oTIR), respectively. oTIR requires additional filtering of the excitation light from the fluorescence emission also collected by the objective, but high-numerical aperture (nA > 1.4) oil-immersion objectives reduce the noise cost formerly associated with oTIR [1]. oTIR also leaves the top of the chamber clear, making the simultaneous implementation of other techniques easier. In both forms of TIRF, all the molecules immobilized within a region approx. 10 μm^2, usually 150–200 molecules, are simultaneously imaged on an EMCCD camera operating in the shot-noise limit at maximum frame rate 20–65 frames/s (15–50 ms time resolution) [33].

Scanning confocal microscopy raster-scans a focused excitation beam across the sample, exciting only one molecule at a time. It can scan continually, in which case time resolution is limited by the scanning speed, or stop at each molecule to record a time-dependent trajectory. Since all the detected photons are from the same molecule, a camera is not needed, and a fast APD detector can be used, increasing the time resolution into the sub-millisecond range [1]. In both TIRF and confocal microscopy, a longer

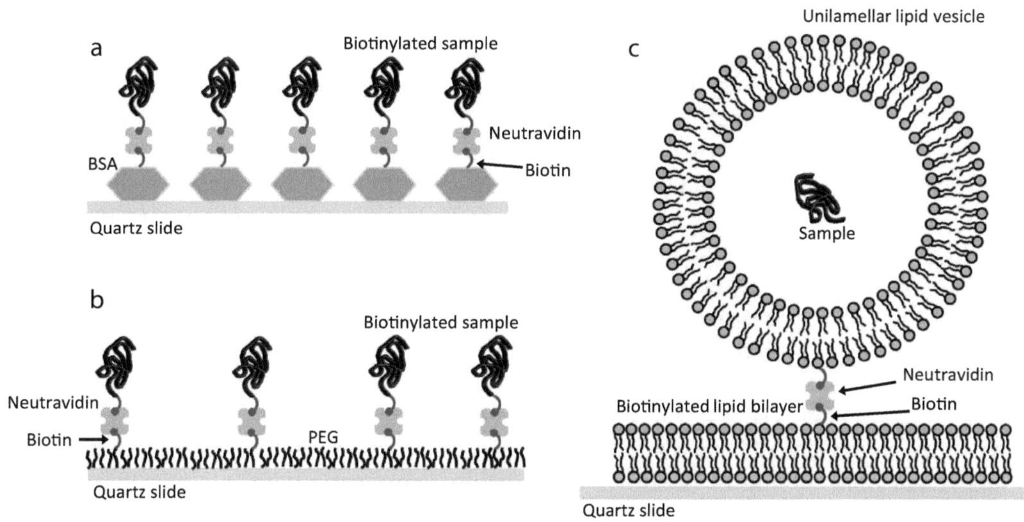

Fig. 3 Methods of sample immobilization. (**a**) Sample immobilization using the non-covalent interaction between biotin-labelled bovine serum albumin (BSA) and NeutrAvidin. The nucleic acid sample is also labelled with biotin, and pM concentrations are injected into the sample chamber and allowed to bind to another of the four NeutrAvidin binding sites. (**b**) To prevent nonspecific interactions between the slide and protein samples, the surface can be passivated with a layer composed of ~1 % biotinylated poly(ethylene glycol) (PEG) and non-biotinylated PEG before adding NeutrAvidin and the sample. (**c**) Samples that cannot be biotinylated can be encapsulated in porous lipid vesicles. The vesicles require a PEG layer or flat lipid bilayer (shown) to prevent them from bursting on contact with the quartz. A small fraction (<0.5 %) of the lipids are biotin labelled, and the "containers" are immobilized via the biotin-neutravidin interaction

integration time (longer time resolution) requires lower laser excitation power, which extends the lifetime of the fluorophores before photobleaching to several minutes or longer under continuous illumination [35].

1.4 Immobilization Methods for sm-TIR

The nucleic acid is usually chemically modified for surface immobilization by attaching a "tether" molecule called biotin (*see* **Note 4**). The sample chamber for TIRF is a microfluidic channel created between the coverslip and a microscope slide separated by two pieces of double-sided sticky tape. The chamber is successively coated with bovine serum albumin (BSA, a highly adherent protein complex) that has been labelled with biotin and then NeutrAvidin, which forms a very strong bond with biotin at one of its four binding sites (Fig. 3a). The biotin-labelled sample can then be flushed into the chamber and captured by the neutravidin [34]. The biotin-neutravidin interaction is resistant to moderate concentrations of denaturant, temperature variations, and illumination in the visible spectrum, making it ideal for most single-molecule fluorescence experiments. However, NeutrAvidin and derivatives are known to undergo chemical degradation under UV illumination.

If the nucleic acid structure does adhere to the glass surface or if the system under investigation involves protein-DNA interactions, it can be passivated using a combination of aminosilanization and polyethylene glycol (PEG) coating with a mixture of biotinylated and non-biotinylated PEG to prevent any nonspecific surface interactions (Fig. 3b). Finally, if tethering the sample molecule with biotin would severely limit its function or if it cannot accommodate enough modifications to include biotin, as is the case with many synthetic RNAs, each dye-labelled molecule can be enclosed in a biotin-labelled lipid vesicle approx. 100–200 nm in diameter, which is then tethered to biotin-neutravidin units on a pre-coated PEG slide or alternatively on a lipid bilayer passivated surface (Fig. 3c) [6, 36, 37].

1.5 Analysis of Time-Dependent Single-Molecule FRET Trajectories

Figure 4 shows typical single-molecule data in a variety of display formats. Figure 4a shows a raw intensity versus time (upper panel) and E_{app} versus time (lower panel) trace for a single molecule, the output of the camera, or APD. (In the case of TIRF, each EMCCD movie must be analyzed by computer to identify the image of each molecule and read out the intensity record of those pixels.) The traces can be analyzed by eye or an automated routine and plotted to produce a histogram of population as a function of E_{app} (Fig. 4b). A static population in a given physical state is expected to have a Gaussian E_{app} distribution with the width determined by shot noise; broader or non-Gaussian peaks indicate the presence of dynamic interconversion between states on a timescale similar to the integration time [38].

Transition rates can be extracted from traces by fitting with a thresholding algorithm or, for complex systems with more than two or three states, hidden Markov modeling, as shown in Fig. 4a (lower panel). The dwell times of the states are then fit with a single or multi-exponential (Fig. 4c), whose time constant(s) corresponds to the rates of the transitions present in the system [39]. A two-dimensional population density plot (Fig. 4d) concisely represents the frequency of transitions between each pair of initial and final FRET states in the system [21, 39].

Surface-immobilized single-molecule techniques offer a valuable, detailed view of nucleic acid structural dynamics at a time when their importance as regulatory and catalytic structures within the cell is just beginning to be realized. In vitro techniques that can isolate important subpopulations and alternate folding pathways and probe environment- and sequence-dependent interactions are shedding light on the complex interactions that make the biological activity of ribozymes and riboswitches in physiological conditions possible. The development of high time resolution, high data throughput techniques is mitigating the perennial impact

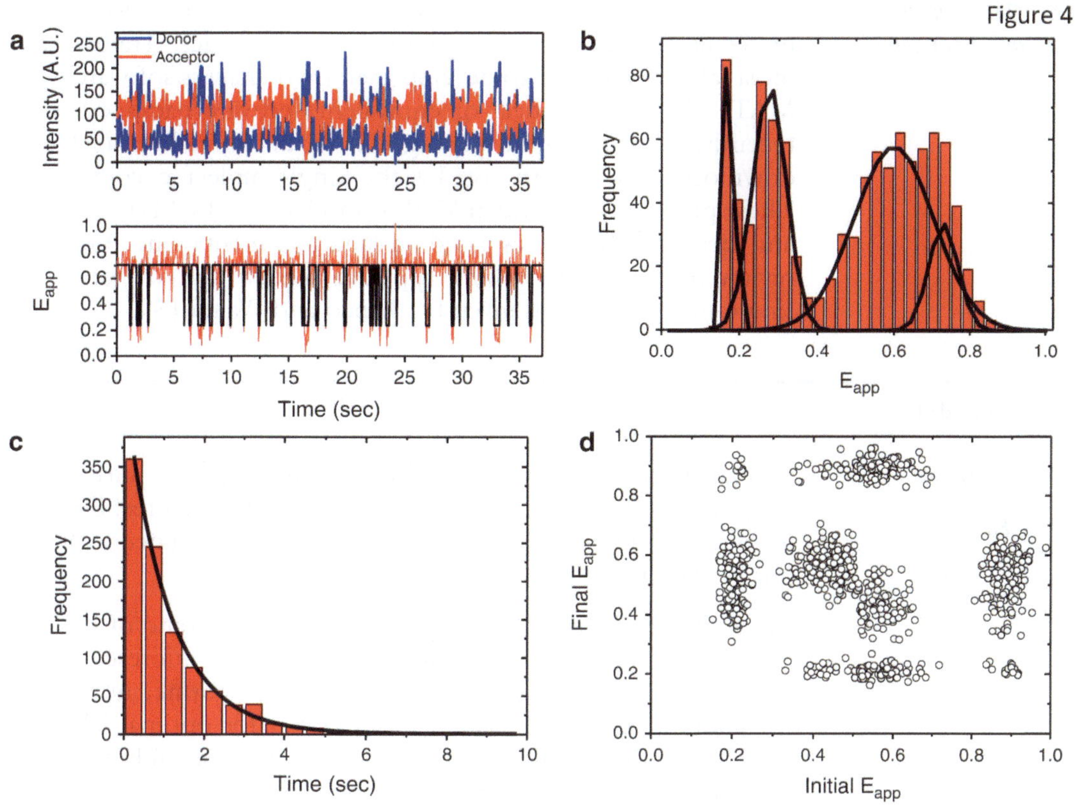

Fig. 4 Analysis of single-molecule data. (**a**, *top panel*) Representative single-molecule trace showing donor and acceptor variations in fluorescence intensity as a function of time for a nucleic acid sample exhibiting two-state dynamics. FRET efficiency can be calculated from the D and A intensities by Eq. 2, resulting in single-molecule trajectory of E_{app} vs. time shown in the *bottom panel*. The *black line* indicates a hidden Markov fit performed with HaMMy, identifying the two FRET states between which the system fluctuates. (**b**) Representative single-molecule histogram of E_{app} values drawn from a simulated four-state distribution. A histogram would be constructed from traces like (**a**) by taking the average FRET value of the first ten frames of each as the representative E_{app} for that trace and constructing the histogram from them. A static, shot-noise-limited distribution should be accurately fit by a sum of Gaussian functions, so a non-Gaussian fit indicates dynamics faster than the time resolution of the experiment, leading to spurious FRET values, or a combination of very similar FRET states. (**c**) Lifetime measurement based on hidden Markov analysis. Following a fit by HaMMy or other Markov modeling software, the distribution of lifetimes of each state can be modeled and fitted by a single or multi-exponential decay curve $e^{-t/\tau}$, where τ is the lifetime of that state. (**d**) A transition density plot, showing all the initial and final states of transitions between the four states of the simulated data set in (**b**). A transition density plot can quickly display the frequency and direction of transitions between any pair of FRET states, and additional information, for example, about lifetime, can be included by color-coding the data points

of dye photophysics on single-molecule experiments, increasing the resolution and reliability of the results. Finally, surface immobilization experimental techniques are easily interconvertible around a basic microscope and detection setup, allowing single-molecule labs great flexibility in the techniques they employ.

1.6 Data Analysis Software Packages

There is no specific commercial software available for single-molecule fluorescence analysis of TIR trajectories. Traditionally, most laboratories had to develop their own analysis software using MATLAB, Mathematica, or equivalent software packages. More recently, some of these packages have been made available to download and free to use by the single-molecule community. These programs use hidden Markov modeling (HMM) to generate and idealized trajectories based on the experimental one. A detailed review of the theory behind applying hidden Markov models to single-molecule data is given by Blanco and Walter [39].

2 Materials

2.1 Chemicals

2.1.1 Buffers

Several specific buffers are required for the labelling, purification, imaging, and long-term storage of nucleic acid structures. Here, we list the most commonly used buffers that can be obtained with high purity from most suppliers:

1. *Labelling buffer*: coupling between a single strand of DNA or RNA carrying a primary amino group and a fluorescence dye containing a succinimidyl ester (NHS) is performed in 100 mM sodium tetraborate (Sigma) at pH 8.5 for optimal labelling. This buffer should be prepared as fresh as possible for maximum labelling efficiency.
2. *Ligation buffer*: in those cases where the RNA or DNA motif is too long for chemical synthesis and it is not possible either to generate the DNA or RNA construct by hybridization of mixtures of single strands, ligation is the method of choice. The ligation of two strands by T4 DNA ligase or T4 RNA ligase requires a specialized buffer containing 5 mM ATP. Fermentas supplies this buffer with a purchase of T4 ligase, but it can also be prepared on site (*see* Subheading 3.3).
3. *Purification buffer*: separation of labelled nucleic acids from free unreacted dye and unlabelled strands is normally performed via denaturing polyacrylamide gel electrophoresis (PAGE). For this 5× TBE is required containing 450 mM Tris-HCl, 450 mM borate, and 10 mM ethylenediaminetetraacetic acid disodium salt (EDTA).
4. *Pegylation buffer for surface functionalization*: 10 mL ultrapure water (ddH_2O) + 84 mg of sodium bicarbonate. This buffer should be freshly prepared every time to achieve maximum pegylation efficiency.
5. *Single-molecule imaging buffer*: 50 mM Tris-HCl (pH 7.8), glucose in a (w/w) from 0.6 to 6 %, 1 % 2-mercaptoethanol,

Table 1
Spectroscopic properties and suppliers of fluorescence dyes suitable for single-molecule fluorescence

Dye	Abs. max (nm)	Em. max (nm)	ε at abs. max (M^{-1} cm^{-1})	Distributors	FRET partners
Cy3	550	565	150,000	GE, IDT	Cy5[a], Cy5.5[a], ATTO647N[a], Fluorescein[b]
Cy5	655	667	250,000	GE, IDT	Cy3[b], Cy3b[b], Cy5.5[a]
Cy5.5	685	706	209,000	GE, IDT	Cy3[b], Cy5[b]
AlexaFluor488	496	519	71,000	Invitrogen	Alexa555[a], Alexa647[a]
AlexaFluor555	555	565	155,000	Invitrogen	Alexa647
AlexaFluor647	650	670	239,000	Invitrogen	Alexa488[b], Alexa555[b]
Fluorescein	496	516	89,000	IDT, Invitrogen	Cy3[a]
TAMRA	559	583	89,000	IDT, Invitrogen	ATTO647N[a], Cy5[a]
Cy3b	559	570	130,000	GE	Cy5[a], ATTO647N[a]
ATTO647N	644	669	150,000	ATTO-TEC	TAMRA[b], ATTO550[b]

[a]Acceptor
[b]Donor

0.1 mg/mL glucose oxidase type II-S from *Aspergillus niger* (Sigma), and 0.02 mg/mL glucose catalase (Sigma). 2-Mercaptoethanol can now be replaced by the more efficient triplet-state quencher Trolox (6-hydroxy-2,5,7,8-tetramethylchromane-2-carboxylic acid, Sigma).

2.1.2 Fluorophores

Small organic dyes are the most commonly used fluorophores for single-molecule experiments on nucleic acids. A good single-molecule fluorescent dye must be bright, photostable, and amenable to chemical modification for attachment to the nucleic acid (*see* **Note 5**). For FRET experiments, the D-A pair must also have a large (>100 nm) spectral separation between emission maxima, sufficient overlap between donor emission and acceptor excitation to promote efficient resonance energy transfer, and, when possible, similar quantum yields to produce unambiguous anticorrelated intensity changes [34]. Cy3 and Cy5 (GE Healthcare) and their corresponding Alexa (Invitrogen) or ATTO (ATTO-TEC) analogues are the most popular dyes for single-molecule studies (Tables 1 and 2 [40, 41]) because of their higher photostability and pH independence compared to fluorescein derivatives. They can be

Table 2
R_0 for some single-molecule FRET pairs used in the cited literature

Donor	Acceptor	R_0 (nm)	Donor	Acceptor	R_0 (nm)
Cy3	Cy5	5.3	Fluorescein	Cy3	5.6
Cy5	Cy5.5	7.3	TAMRA	Cy5	6.5
AlexaFluor555	AlexaFluor647	5.1	Cy3b	ATTO647N	6.1

Table 3
Spectroscopic properties for commonly used Dark Quenchers and corresponding suppliers

Quencher	Abs. max (nm)	ε (M^{-1} cm^{-1})	Effective quencher for dyes	Distributor
BHQ1	534	34,000	Fluorescein, rhodamine derivatives (TAMRA, TMR, Texas Red)	IDT
BHQ2	578	38,000	Cy and Alexa dyes	IDT
QSY7	560	90,000	Green-emitting dyes	Invitrogen
QSY21	660	89,000	Red-emitting dyes	Invitrogen

obtained with either succinimidyl modification (for coupling to a DNA or RNA carrying a primary amino group) or in maleimide format (for coupling to a DNA or RNA carrying a thiol modification).

2.1.3 Dark Quenchers

Recently, non-emitting FRET acceptors known as Dark Quenchers (DQs) have been used instead of emissive acceptors to reduce noise in the data and free regions of the spectrum for additional dyes [42]. The properties of some Dark Quenchers appropriate for the most common FRET dyes are shown in Table 3 [43, 44]. Like fluorescent acceptors, DQs are prone to photoblinking and bleaching and need to be characterized accordingly before use [45]. Furthermore, to reliably distinguish quenching from donor normal intensity fluctuations (blinking, variations in quantum yield), triplet-state quenchers and oxygen-scavenging systems should be used to stabilize the donor emission (*see* **Note 6**), and the remaining baseline fluctuations should be well characterized. The primary advantage of DQs is that they permit an experiment to monitor three vectors simultaneously while using only two regions of the spectrum, simplifying the optical setup and data processing requirements. Also, they enable to investigate weak interactions which otherwise will be difficult to address by TIR due to the high amount of dye-labelled DNA/RNA or protein required.

2.1.4 Synthetic DNA or RNA

Companies such as Integrated DNA Technologies (IDT), Operon, Invitrogen, and Dharmacon (Thermo Scientific) offer custom DNA and RNA synthesis in nano- to micromolar scales of up to 140–150 nt length for DNA and slightly shorter for RNA. The exact 5′, 3′ and internal modifications that may be available will depend on the type of solid-phase synthesis (SPS) that each company provides. For RNA, additional deprotection steps may also be required depending on the company and the specific type of SPS chemistry used to manufacture the strands. For single-molecule immobilization studies, a biotin group is usually incorporated at the 5′ or 3′ end of one strand.

2.1.5 DNA Post-synthesis Fluorescent Labelling

Post-synthetic incorporation of fluorescent labels is normally performed using a succinimidyl ester (NHS) dye derivative and a synthetic nucleic acid strand carrying a primary amino group at a specific position. Although this is the most widely used approach and alternative, using a thiol modification and a maleimide derivative of the dye can also be performed. Incorporation of two specific dyes at known positions of a single strand is also possible using these two orthogonal chemistries. In this case, a single strand carrying a primary amino group and a thiol modification will be sequentially labelled with NHS- and maleimide-dye derivatives, respectively. For these labelling procedures, the following reagents are required:

1. Synthetic single-stranded DNA or RNA carrying a primary amino and/or thiol modification (IDT, Operon, Thermo Scientific).
2. Cy3 NHS or Cy3 maleimide or other modified dye carrying the appropriate coupling group.
3. Dimethyl sulfoxide (e.g., Fisher Scientific).
4. Labeling buffer as described in Subheading 2.1.1 (*see* **Note** 7).
5. Precipitation solutions: 3 M sodium acetate and 100 % ethanol.

2.1.6 Purification of Fluorescently Labelled DNA or RNA by Polyacrylamide Gel Electrophoresis (PAGE)

1. Purification buffer as described in Subheading 2.1.1.
2. Dilution buffer 1×: 7 M urea dissolved in 1× TBE (store at room temperature).
3. 20 % acrylamide/bis solution (19:1) with 7 M urea in 1× TBE. Monomeric acrylamide is neurotoxic, and care should be taken to avoid exposure. Store this solution in the dark at room temperature.
4. 10 % (w/v) ammonium persulfate (APS): solution in water.
5. TEMED (*N*,*N*,*N′*,*N′*-tetramethylethylenediamine).
6. Running buffer (1× TBE).
7. Molecular weight markers: xylene cyanol FF 0.02 % and bromophenol blue 0.02 % dissolved in pure formamide.

2.1.7 Specific Reagents for Single-Molecule TIR Studies

This list consists of the chemical reagents necessary for slide passivation and the immobilization of biological samples. Where no supplier is given, it is assumed that the chemical is sourced from the group's preferred supplier:

1. Spectroscopic grade solvents including acetone, methanol (Sigma), and acetic acid.
2. Imaging buffer as described in Subheading 2.1.1.
3. Quartz microscope slides 1″ × 3″ (QSI, Quartz Scientific, USA) for prism-type TIR.
4. Streptavidin or NeutrAvidin (less nonspecific affinity for glass) can be purchased from a number of suppliers (e.g., Invitrogen, Pierce) and should be dissolved at a concentration of 5 mg/mL in 50 mM Tris-HCl (pH 7.8) and 50 mM NaCl. Store immediately at 4 °C.
5. Carboxylate-modified FluoSpheres, 0.2 μm, Crimson, 2 % Solids (Invitrogen). Store at 4 °C. This solution will be used to prepare a mapping and calibration channel and for initial alignment of the TIR setup.
6. Bovine serum albumin, biotinylated (Sigma). Store at 4 °C.
7. Biotinylated polyethylene glycol (NHS ester), MW 5,000 (Laysan Bio, Nektar Therapeutics). Store at −20 °C.
8. Non-biotinylated polyethylene glycol (NHS ester), MW 5,000 (Laysan Bio, Nektar Therapeutics). Store at −20 °C.
9. 3-(2-Aminoethyl)3-aminopropyltrimethoxysilane (aminosilane) (Fluorochem). Store in the dark.
10. 1,2-Dipalmitoyl-*sn*-glycero-3-phosphoethanolamine-*N*-(cap biotinyl) (Avanti Polar Lipids). Store at −20 °C. Biotinylated DPPE is used to tethering the vesicles to the surface.
11. L-α-Phosphatidylcholine (egg, chicken; 99 %) (Avanti Polar Lipids). Store at −20 °C.
12. 1,2-Dimyristoyl-*sn*-glycero-3-phosphocholine (DMPC) to generate permeable vesicles (Avanti Polar Lipids). Store at −20 °C.
13. α-Hemolysin from *Staphylococcus aureus* (Sigma) to generate permanent pores in the membrane. Once dissolved, store at 4 °C.
14. Mini-extruder (Avanti Polar Lipids).

2.2 Instrumentation for Single-Molecule TIR Studies

Of the two possible single-molecule TIR configurations, prism type and objective type (Fig. 5), only the latter is commercially available (Olympus, PicoQuant). Thus, scientists aiming to implement prism-type TIR because of its higher SNR performance will need to build their own setup. In the next sections, we provide a very

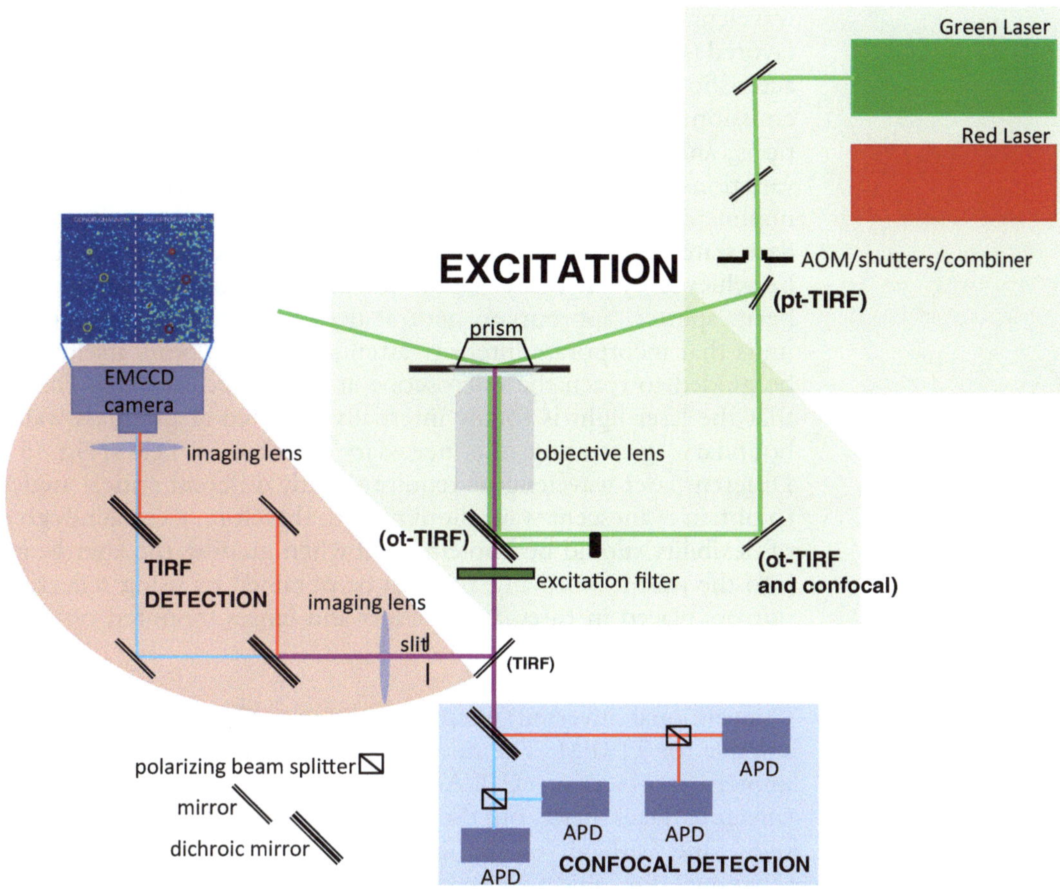

Fig. 5 Excitation and emission optical pathways for prism- and objective-type TIR and confocal FRET experiments. The *upper right corner* shows the excitation optics for a μs alternating-laser excitation (ALEX) experiment, in which an AOM, shutter, or computer-controlled combiner is used to alternate μs pulses of *green* and *red* (donor and acceptor) excitation light. (*Top center*) For pt-TIRF, the beam is deflected to a steering mirror and focused onto the sample chamber by a prism; for ot-TIRF and confocal, the beam is coupled in through the back of the microscope objective. Fluorescence is collected through the objective, and excitation light is filtered out (in the case of pt-TIRF, the excitation beam must be directed away by additional optics). For confocal detection (*bottom center*), the D and A photons are separated by wavelength and/or polarization and detected by an array of APDs. For TIR detection (*left*), the wide-field image is separated into separate D and A images using dichroic mirrors, which are displaced from each other and recombined side by side on an EMCCD camera chip. The D and A spots that correspond to the same molecule are identified in post-processing, and the intensity vs. time traces for those pixels are recorded

brief overview of the main guidelines, and we refer to excellent reviews for a more detailed list of equipment [1, 2].

2.2.1 Excitation Sources

Most of the equipment required for an sm-TIR setup including excitation sources, filters, dichroic mirrors, and detectors will depend on the specific biomarkers incorporated into the nucleic acid of interest. A detailed list of suitable dyes and their excitation wavelengths is shown in Table 1. Continuous laser sources are favored (CrystaLaser, Coherent, LaserQuantum) due to their characteristic properties of coherence, collimation, and high-intensity emission. Laser powers of ~50 mW are enough for most applications, and normally they need to be attenuated to achieve small variations in laser intensity. Attenuation methods are required to minimize power-dependent effects such as blinking (transition to a dark state) and photobleaching of the dyes. Power attenuation may be achieved by the combination of a half-wavelength polarizer and beam splitter, appropriate neutral density filters, or laser power units that incorporate intensity attenuation. The beam itself must be guided to reach the microscope at the correct angle to ensure that the laser light is totally internally reflected at the glass-water boundary, generating the necessary evanescent field (Fig. 5). Different laser wavelengths require slightly different critical angles to obtain evanescent wave illumination; therefore, a certain degree of flexibility should be implemented when guiding the laser beam into the prism. Normally, this can be obtained by using reflecting mirrors placed in three-axis holders and lenses mounted on XYZ stages (Thorlabs, CVI).

2.2.2 Microscope

Conventional inverted microscopes (e.g., Olympus IX-71) are suitable for TIRFM. The microscope must be fitted with a water-immersion objective (UPLAPO60XW, Olympus) for prism type (imaging takes place on the inner surface of the upper quartz slide) and with an oil-immersion objective (UPLSAPO100XO, Olympus) for objective type (imaging takes place in the inner surface of the bottom coverslip). Additional optical components (lenses, mirrors, prisms) and mechanical components (holders, micrometer-driven stages) required for TIRFM are widely available from optomechanical manufacturers (e.g., Thorlabs, CVI). A comprehensive list of optical components can be found in excellent reviews by Walter et al. [1] and Roy et al. [34].

2.2.3 Detectors

The radiation emitted from the sample is collected by the objective and resolved using a system of regular and dichroic mirrors which allowing separation of the donor and acceptor emissions (Figs. 1c and 5). Electron-multiplying charge-coupled devices (EMCCDs) are used for wide-field imaging by TIR in both prism- and objective-type configurations (Andor, Roper Scientific). For single-molecule FRET, resolving the two components of the emission separately onto different halves of the EMCCD chip allows comparison of the anticorrelated intensities of each FRET partner. In point-detection setups based on confocal microscopy, the image is generated using avalanche photodiodes (APDs) as detectors (PerkinElmer, Hamamatsu) in a raster scanning mode. An intrinsic advantage of these detectors compared to EMCCD-based imaging is their increased time resolution, which allows acquisition of single-molecule trajectories with 100–200 μs time resolution compared to the 10–20 ms maximum speed achievable with EMCCDs.

2.2.4 Software Packages for Data Analysis

The most popular packages for analysis of single-molecule fluorescence trajectories to extract structural and dynamic information are listed below together with a brief description of their most important features.

2.2.5 HaMMy

Developed by Prof. Taekjip Ha's group at the University of Illinois [46], HaMMy is a user-friendly HMM-based fitting program, freely available at http://bio.physics.illinois.edu. The only necessary input parameter in the application window is the number of states that are to be fitted, with the option of entering approximate values for each, making it very simple to set up. Input data must be in a *.dat format, consisting of three columns: time, donor intensity, and acceptor intensity. The trace and fit are displayed on the user interface as the program is running, giving an indication of the goodness of fit. There are three files generated per trace: **path.dat* (this file is the same as the input file with two additional columns, the effective FRET and the fitted FRET value), **dwell.dat* (this file contains the length of each step of the fit and may be used in plotting dwell-time and rate information), and **report.dat* (this file is used by the accompanying TDP program, which allows a visual interpretation of the states occupied by the sample under investigation and the frequency of transitions between these states by way of a 2D contour plot).

2.2.6 QuB

QuB was written by Qin and Li of Buffalo University (http://www.qub.buffalo.edu) to be used in analyzing single ion-channel data [47]. However, although not originally intended for use in analyzing smFRET data, it has proven to be a powerful tool in fitting single-molecule traces due to the high degree of control the user has over the program parameters and its ease in analyzing large data sets more quickly than some specifically developed for FRET data [39]. The program also utilizes a model-controlled fitting process which increases the accuracy of the HMM generated by eliminating forbidden transitions from being fitted.

2.2.7 vb-FRET

Chris Wiggins et al. at Columbia University developed vb-FRET (http://vbfret.sourceforge.net), a MATLAB-based program designed to fit FRET traces [48]. In terms of control of fitting parameters, it sits between HaMMy and QuB, but offers the additional advantage of employing Bayesian analysis, namely, applying ideas of maximum likelihood and maximum evidence to the fitting, in order to get the best model possible and minimize overfitting. The fitted traces from these programs may be further analyzed to generate folding and unfolding rates and exported to a graphical package such as Origin. An excellent, detailed review of the mechanisms behind fitting hidden Markov models to FRET data is given by Blanco and Walter [39].

3 Methods

3.1 Post-synthetic Labelling and Purification of Amine-Modified Oligonucleotides

3.1.1 Labelling of Single-Strand Amino-Modified DNA or RNA

Labelling of synthetic nucleic acids modified with primary amine groups requires a slightly basic buffer (sodium tetraborate, pH 8.5). The oligonucleotide solution is mixed with the labelling buffer and NHS- or thiol-modified organic dyes (e.g., cyanine dyes, AlexaFluor) suspended in DMSO and allowed to react. Sodium tetraborate and DMSO are commonly available from chemical supply companies. DMSO is flammable and should be stored in a fire-safe box:

1. Prepare the labelling buffer as described in Subheading 2.1.1.
2. Dissolve 250 μg of the amine-reactive dye in 14 μL DMSO (dimethyl sulfoxide) (*see* **Note 8**).
3. Add to the vial containing the dye:

   7 μL deionized water.

   75 μL labelling buffer.

   100 μg oligonucleotide stock solution (4 μL of 25 μg/μL solution will achieve a total reaction volume of 100 μL).
4. Wrap in aluminum foil and incubate at room temperature for at least 6 h or overnight on a shaker at low speed. If no shaker is available, tap gently to remix every half hour for the first 2 h.

3.1.2 Precipitation of the Reaction Mixture

1. Add to the vial containing the labelled construct:

   0.1 volume of 3 M sodium acetate (e.g., 10 μL into an initial volume of 100 μL).

   2.5 volumes cold 100 % ethanol.
2. Wrapped in foil and store overnight at −20 °C.
3. If purifying the product, prepare a denaturing gel for use the next day (*see* Subheading 3.1.2 below).
4. To retrieve product after incubating, spin at 12,000 × *g* for at least half an hour and then *carefully* extract the solution with a pipette, ensuring that the pellet of precipitated oligonucleotide remains in the vial.
5. Resuspend the pellet in 25 μL of 50 mM Tris-HCl buffer (pH 8).

3.1.3 Single-Strand Purification by Denaturing PAGE

The standard grade acrylamide for polyacrylamide gel electrophoresis (PAGE), in which charged molecules separate by size and global charge as they migrate through a porous gel matrix along a potential gradient, is 19:1 acrylamide/bis-acrylamide, 40 % w/v. Acrylamide is available from most biochemical supply companies, as are the two additives that catalyze its polymerization (ammonium persulfate and tetramethylethylenediamine, TEMED). Although a fully polymerized acrylamide gel is safe, liquid acrylamide is carcinogenic and should never be handled without gloves.

The Day Before Gel Purification

1. To prepare a (20 %, 12 %) denaturing polyacrylamide gel for the purification of oligonucleotides (>40 bp, <40 bp), first thoroughly scrub the glass plates with deionized water and ethanol to remove any dirt that could cause bubbles.
2. To make 50 mL (20 %, 12 %) denaturing gel, combine in a 50 mL plastic tube:

   5 mL 10× concentrated TBE.

   (25, 15) mL 40 %/2 % acrylamide.

   21 g urea.

   Deionized water (top up to 50 mL).
3. Immediately before pouring the gel, add:

   300 μL APS.

   30 μL TEMED (*see* **Note 9**).
4. Mix briefly, then pour between the glass plates, insert the comb, and leave it to polymerize.
5. To store overnight, wrap *completely polymerized* gel in a damp paper towel and cling film, and store at 4 °C.
6. Before use, rinse the wells thoroughly with 1× TBE (Tris-borate-EDTA) to clear any salt accumulated in the wells (*see* **Note 10**).
7. Pre-run the gel for 1 h at 17 W without loading the sample.
8. Add 50 % formamide to the oligonucleotide solution and incubate for 30 s at 90 °C using a heat block to disrupt any secondary structure.
9. Just before loading samples, rewash wells with a syringe. Use xylene cyanol FF and bromophenol blue in 100 % formamide solution as size markers.
10. Run at 17 W (maximum temperature 45 °C or glass plates could crack if too hot) until the labelled and unlabelled bands are clearly separated and the dye markers have migrated 2/3 of the plate length. Remove glass plates and carefully put the acrylamide gel on a Dark Reader transilluminator (DR45M, VWR International) to visualize DNA bands by fluorescence.

11. Carefully cut the desired band from the gel. In a vial, crush it into small pieces and resuspend in 200–400 μL 50 mM Tris-HCl (pH 7.8), depending on the size of the band. Incubate overnight at 4 °C on a shaker.
12. Retrieve, precipitate, and extract the sample as in Subheading 3.1.2 (*see* **Note 11**).

3.1.4 Dye Labelling a Transcribed Oligonucleotide Strand at the 3′ End

A DNA or RNA strand produced by in vitro transcription can also be labelled at the 3′ end by oxidation of the terminal backbone sugar into an aldehyde, forming a Schiff base that will readily form a covalent bond with an amine-modified dye [49]:

1. Incubate the sample for 40 min at 0 °C in the dark [50] with 10 mM sodium periodate ($NaIO_4$) and enough labelling buffer (Subheading 2.1.1) to bring the reaction volume to 100 μL (*see* **Note 12**).
2. Precipitate with sodium acetate and ethanol as in Subheading 3.1.2.
3. For the labelling reaction [49], mix the following in a 400 μL reaction volume:

   200 μg oligonucleotide.

   100 mM sodium acetate, pH 5.5.

   6 mM amine- or azide-modified dye.
4. Incubate 2 h in the dark at room temperature.
5. Precipitate oligonucleotide as in Subheading 3.1.2.

3.2 Hybridization and Purification of Oligonucleotide Strands

3.2.1 Hybridization of Oligonucleotide Strands

1. Combine equimolar quantities of each strand to be hybridized (200–800 pmol are typical) in a vial with 50 mM NaCl and enough Tris-HCl (50 mM, pH 7.8) to bring the total reaction volume to approx. 30–50 μL. Wrap the vial in parafilm and aluminum foil.
2. Preheat a heat bath to 65–70 °C. Turn the bath off and add the vial, leaving it to cool back to room temperature overnight (*see* **Note 13**).

3.2.2 Purification of Hybridized Sample by Gel Electrophoresis

The native gel is prepared in the same way as the denaturing gel in Subheading 3.1.3, with the following exceptions:

1. Do not add urea.
2. The gel does not require pre-running before use.
3. Add 50 % glycerol to the sample in a 1:10 ratio instead of formamide.
4. Run the gel in a cold room at 4 °C and 100–150 V until there is a clear separation between the hybridized construct and single

strands or until the hybridized construct has progressed 2/3 of the way down the gel, generally 6–8 h (*see* **Note 14**).

3.3 Ligation of Oligonucleotide Strands with T4 DNA Ligase or T4 RNA Ligase

Although hybridization of complementary regions will sometimes suffice, ligation of oligonucleotides is often necessary when attaching dye-labelled fragments to DNA/RNA transcripts [49] or joining multiple chemically synthesized strands [51]. T4 DNA ligase or T4 RNA ligase (New England Biolabs, Fermentas) can be used to join two nucleic acid strands with the aid of a "splint" that complements both at the junction point. Ligase enzymes and their specialized ligation buffers require care to prevent them from degradation (*see* **Note 15**):

1. Hybridize the strands as in Subheading 3.2.1, adding the splint strand in approx. 25 % excess over the other strands (*see* **Note 16**).
2. Add to the hybridized mixture:

   4 μL T4 DNA ligase or T4 RNA ligase (determined by the composition of the splint).

   4 μL ligation buffer (400 mM Trizma salt, 100 mM $MgCl_2$, 100 mM DTT (dithiothreitol), 5 mM ATP, pH 7.8).

   2 μL PEG.
3. Incubate in a heat bath for 22 h at 39 °C (*see* **Note 17**).
4. Purify by native gel electrophoresis as in Subheading 3.2.2.

3.4 Slide Passivation and Sample Immobilization

3.4.1 Quartz Slide Cleaning Protocol

1. Carefully remove coverslips from previously used channels by first softening the tape and epoxy, microwaving the slides in tap water for 30 s, and then scraping them off with a razor blade.
2. Scrub both sides with acetone and then with methanol until visibly clean. Transfer to a clean jar for etching.
3. Etch in a 50:50 (v/v) combination of hydrogen peroxide and hydrochloric acid for 2 h (*see* **Note 18**).
4. Pour out the etching solution, and rinse the slides in the jars with tap water.
5. Sonicate following these steps: (a) 10 % solution of Alconox detergent for 15 min, (b) ultrapure water (dd H_2O) for 5 min, (c) acetone for 15 min, (d) dd H_2O for 5 min, (e) 1 M potassium hydroxide (KOH) for 15 min, (f) dd H_2O for 5 min, (g) methanol for 15 min, (h) dd H_2O for 5 min, (i) 1 M KOH for 15 min, and (j) dd H_2O for 5 min.
6. Leave in dd H_2O until needed. Slides will be useable for approximately 1 week after the etching process. If using at a later date, refresh slides using the final three sonication stages described in **step 5** prior to use.

3.4.2 Channel Preparation

1. Dry a pre-cleaned slide with N_2 gas. Burn briefly on both sides and cool with N_2 (*see* **Note 19**).
2. Run two strips of double-sided tape along the slide creating a channel between two predrilled 0.75 mm injection holes. Drilling quartz slides requires diamond drill bits (Eternal Tools, UK) (*see* **Note 20**).
3. Carefully place a glass coverslip on the top of the tape, ensuring that the channel is covered, and press gently using a pipette tip.
4. Seal the ends of the channel with fast-drying transparent epoxy resin (Araldite).

3.4.3 Preparing a Calibration Slide

1. Add 2.5 μL of 1 M HCl to 500 μL of dd H_2O.
2. Add 1 μL of the microsphere commercial stock and sonicate the solution for 10 min (*see* **Note 21**).
3. Prepare a channel as described above and inject the solution into the chamber. Allow this to incubate for 20 min before rinsing with dd H_2O. Repeat this process until the density of fluorescence spots is appropriate (~200–300).
4. Seal around the entire coverslip and over the injection holes with fast-drying epoxy resin. Store the slide in a water box when not in use (*see* **Note 22**).

3.4.4 BSA-NeutrAvidin Immobilization Protocol

This is the simplest means of passivating a slide and sparsely immobilizing single biomolecules. It exploits the interaction between biotinylated bovine serum albumin (BSA) and NeutrAvidin or streptavidin. First, the nonspecific interaction of BSA-bi with the quartz surface is used to coat the slide, and NeutrAvidin is added to form a bridge between the BSA-bi and a biotinylated nucleic acid sample, anchoring it to the surface:

1. Using a slide channel prepared as above, inject buffer solution (50 mM Tris-HCl, pH 8) into the channel and image on the microscope to assess how clean the slide is.
2. If the slide is sufficiently clean, inject 60 μL of a 1 mg/mL biotinylated BSA (bovine serum albumin, Sigma) and incubate for 15 min in a water box.
3. Flush with 100 μL of buffer solution (50 mM Tris-HCl, pH 8) to wash unbound bi-BSA and add 60 μL of 0.2 mg/mL NeutrAvidin (Sigma) (*see* **Note 23**), incubating for a further 15 min in the water box.
4. Flush with 100 μL of buffer solution to wash unbound Neutravidin and add the sample at the desired concentration (normally 100–200 pM of dye-labelled nucleic acid carrying a biotin group) and incubate for 5 min.
5. Prepare 100 μL of imaging buffer as described in Subheading 2.1.1.

6. Wash unbound nucleic acid construct with the imaging buffer and check for sample density in the TIR microscope (200–300 spots are considered an appropriate density).

3.4.5 Pegylation Protocol

While the previous method works well when working exclusively with nucleic acids (e.g., when studying DNA or RNA structure and folding), it is known that BSA coating does not provide enough passivation against nonspecific interactions in studies involving DNA-protein or RNA-protein interactions. For these studies, the surface instead is passivated using a combination of biotinylated and non-biotinylated polyethylene glycol [52] using the protocol outlined below:

1. Clean the quartz slides and coverslips following the protocol outlined previously (*see* **Notes 18–20, 24, 25**).
2. Dry slides individually with N_2 and burn each side under a hot Bunsen flame for about 2 min. Cool with N_2 and transfer to a Falcon tube for storage. Dry coverslips with N_2 and burn both sides (one pass per side) in the Bunsen flame and then transfer to a Falcon tube.
3. First rinse glass jars (VWR International) to be used with methanol, and then refill again with methanol. When the slides and coverslips have been put into the jars, add 2.5 mL of glacial acetic acid and 200 μL of the aminosilane commercial solution (*see* **Note 26**) to each jar and place in a sonicating water bath for approx. 10 min. Empty the solution out and refill with fresh methanol. Remove slides and coverslips; rinse with methanol, dd H_2O, and methanol again before drying with N_2 and storing in a water box.
4. Prepare the pegylation buffer as described in Subheading 2.1.1. Add 1.2 mg of biotinylated PEG and 100 mg of non-biotinylated PEG to 340 μL of pegylation buffer, spinning in a centrifuge to fully dissolve (*see* **Note 27**).
5. Run 60 μL of the PEG solution along the channel between the two injection holes of a previously drilled slide and very gently place a coverslip on top ensuring no bubbles are trapped in the microchannel (*see* **Notes 28** and **29**). Store all prepared channels in a covered water box. Leave all prepared channels in water boxes for 3 h in the dark for maximum pegylation efficiency.
6. Remove coverslip, remembering which side has been pegylated (*see* **Note 30**). Rinse the slide (methanol, dd H_2O, methanol), dry with N_2, and apply double-sided tape as per usual channel construction. Rinse the coverslip (methanol, dd H_2O, methanol), dry in a gentle stream of nitrogen, and place on the channel so as to keep the pegylated surfaces in contact and

press in place. The channel should not be sealed with epoxy at this stage.

7. Wrap the whole construct in aluminum foil and then parafilm, storing it in a Falcon tube in the fridge until use. The channels can be used for 7–10 days after preparation. Be sure to seal the ends of the channels with epoxy prior to use. Alternatively, the assembled microchannel can be sealed under vacuum using a vacuum food sealer. We found that using this approach, preassembled microchannels remain clean for longer periods of time.
8. Add 60 μL of 0.2 mg/mL NeutrAvidin (Sigma), incubating for a further 15 min in the water box. Proceed as described in **steps 4–6** of Subheading 3.4.4.

3.4.6 Vesicle Encapsulation Protocol

Recently, an alternative approach to pegylation-based methods has been developed using vesicle nanocontainers (50–200 nm diameter) to encapsulate the dye-labelled molecule under study [36, 37]. Since the nucleic acid is free to diffuse inside the nanocontainer, this strategy provides a solution-based method that ensures no artifacts due to surface contact are present. The small unilamellar vesicles (SUVs) prepared by extrusion can be made permeable to certain small molecules (i.e., ATP, ADP) or ions (Mg^{2+}) by inserting either transient (using DMPC lipids) or permanent pores (using a bacterial membrane protein, α-hemolysin). This provides a means to easily investigate the effect of environmental changes in real time. Moreover, vesicle encapsulation is also the method of choice when investigating weak protein-nucleic acid complexes by providing an increase in the effective concentration of interacting partners (i.e., encapsulation of a single molecule inside a 200 nm vesicle represents a 50 nM effective concentration). The protocol below is for generating 200 nm non-permeable SUVs using a combination of egg-PC and biotinylated DPPE lipids. For generating permeable SUVs, egg PC should be replaced by DMPC. Alternatively, egg-PC vesicles could be made permanently permeable after immobilization by incubating α-hemolysin directly on the microchannel:

1. Rinse any vials required with chloroform and dry with N_2. Dissolve 20 mg cholesterol in 1,666 μL chloroform in one of these vials (*see* **Note 31**).
2. For each vesicle stock prepared, add the appropriate constituents from:

   250 μL egg-PC stock.

   166 μL cholesterol (*see* **Note 32**).

   1.76 μL biotinylated lipids (DPPE).

3. Under low N_2 pressure, evaporate the chloroform from each vial, taking care not to spread the vesicle solution up the sides of the vial while doing this, and then desiccate under vacuum and in the dark for at least 4–5 h.
4. Rehydrate with 250 μL of buffer solution (50 mM Tris-HCl, 50 mM NaCl, pH 7.8). For an encapsulation experiment, at this point the sample to be contained within the vesicles must also be added. Place the solution into a shaker until completely dissolved and uniformly cloudy.
5. Clean the components of the mini-extruder (Avanti) in the sonic bath, first in a 50:50 mix of detergent and water, and then thoroughly rinse with water. Syringe the lipid solution using a mini-extruder to form the vesicles. The size of the vesicles formed is dictated by the membrane used. For details of the extrusion technique, follow the extruder supplier instructions (*see* **Notes 33** and **34**). When using dye-labelled lipids, immediately rinse with methanol after use to minimize contamination of the extruder.
6. Prepare a channel as described above. At this stage passivation of the quartz surface can be made in different ways depending on the application. PEG passivation followed by NeutrAvidin functionalization as described in Subheading 3.4.5 is one possibility. Alternatively, because vesicles will open and form a lipid bilayer coating on the quartz slide when they touch the surface, it is possible to inject empty vesicles (no molecule encapsulated) prepared as described in the previous steps to create a passivated surface, as described below (*see* **Notes 31** and **32**).
7. Inject 60 μL of the vesicles intended to passivate the surface. Incubate for 1–2 h (4 °C for egg PC and at room temperature for DMPC).
8. Rinse with 60 μL of buffer solution (50 mM Tris-HCl, 50 mM NaCl, pH 7.8). Add 60 μL 0.2 mg/mL of NeutrAvidin and incubate for a further 15 min.
9. Rinse with 60 μL of buffer solution (50 mM Tris-HCl, 50 mM NaCl, pH 7.8). Inject 100–200 pM sample concentration and incubate for 15 min.
10. Flush the channel with 100 μL of buffer solution (50 mM Tris-HCl, 50 mM NaCl, pH 7.8) to remove unbound species and add imagining buffer before viewing on the microscope.

4 Notes

1. When designing an experimental construct, it is vital to place the modifications (dyes, biotin, elongated strands for tethering or added stability, etc.) in locations that will not disrupt the

molecule's native structure. The modification sites must be chosen carefully based on known crystal structures and bioinformatic data [53]. It is important to check that such nucleotides are exposed to solution in the crystal structure such that additional chemical groups will not perturb the native state and that they are non-conserved, implying that they do not play a crucial role in nucleic acid function [54]. A lack of such nucleotides will limit the number of modifications that can be made to the molecule and hence the variety of single-molecule experiments possible. However, 2-aminopurine substitutions can be made even to functionally significant adenines without disrupting the structure. As a check that the modification does not interfere with the molecule's function, assays comparing the labelled molecule's level of activity to the wild type should be performed. These assays are also an important control to check for any effects of surface immobilization [55]. For placement of FRET dyes, the clearest signal will be obtained if the interdye separation changes from $<R_0$, the Förster radius at which E_{app} has fallen to 0.5, to $>R_0$.

2. R_0 can be calculated from the donor quantum yield and fluorescence and emission spectra of the dyes by the following equation:

$$R_0^6 = \frac{9000 \ln 10 k^2 \phi_D J(\lambda)}{128\pi^5 n^4 N_A} \quad (3)$$

where k^2 is the orientation factor, whose average value for a dye free to reorient on timescales faster than its decay lifetime is 2/3, ϕ_D is the donor quantum yield, N_A is Avogadro's number, and n is the index of refraction of the medium [7]. $J(\lambda)$ is the overlap integral of the donor emission and acceptor absorption spectra given by

$$J(\lambda) = \int F_D(\lambda)\varepsilon_A(\lambda)\lambda^4 \, d\lambda \quad (4)$$

where the donor emission spectrum F_D and acceptor absorption spectrum ε_A are normalized to 1 and the acceptor extinction coefficient, respectively [9].

3. The simplest FRET-dependent parameter to calculate from the absorption and emission spectra given by bulk experiments is ratio A, the ratio of acceptor fluorescence when the donor is optically excited, denoted $F_A{}^D$, to the acceptor fluorescence at its own excitation maximum, $F_A{}^A$. Ratio A (rA) is defined by the equation

$$\mathrm{rA} = \frac{F_A^D}{F_A^A} \quad (5)$$

If the absorption spectra of the dyes are also available and the peaks are well separated, rA can be readily converted into FRET efficiency E:

$$E = \frac{\mathrm{rA} - \frac{A_{\mathrm{A}}^{\mathrm{D}}}{F_{\mathrm{A}}^{\mathrm{A}}}}{\frac{A_{\mathrm{A}}^{\mathrm{D}}}{A_{\mathrm{A}}^{\mathrm{A}}}} \qquad (6)$$

In Eq. 6, $A_{\mathrm{A}}{}^{\mathrm{D}}$ denotes the absorption of the acceptor at the donor's peak excitation wavelength, $A_{\mathrm{A}}{}^{\mathrm{A}}$ the acceptor absorption at its own peak wavelength, and $A_{\mathrm{D}}{}^{\mathrm{D}}$ the absorption of the donor at its peak wavelength [9].

4. The biotin tethering group can be directly incorporated at the 5′ or 3′ end of the nucleic acid construct. However, in many situations, it is more convenient to introduce the tether in an extra "handle" strand in the construct. For this, either the 5′ end or 3′ end of the wild-type nucleic acid sequence is elongated by 15–20 extra nucleotides. Then a single-stranded complementary sequence carrying the biotin group is hybridized to this 15–20 nt overhang, providing a handle to attach the nucleic acid construct to the surface.

5. The most common means of conjugating fluorophores to synthetic nucleic acids is by annealing dye with an NHS ester or maleimide group to an amino- or thiol-modified base, which can be site-specifically incorporated at the 3′ or 5′ end or (for amine modifications) any internal site during synthesis [34, 56]. Most synthetic nucleic acids can only accommodate a certain number of modifications. A single DNA strand can be labelled using these two orthogonal chemistries (NHS and thiol modifications) with two FRET dyes and biotin. In contrast, RNA synthesis usually only supports one internal modification and either a 5′ or 3′ modification. In addition to this, chemical RNA synthesis can only produce strands of max. 60–70 nt, so long constructs are often produced in individually labelled pieces and subsequently hybridized or ligated. The choice of where to "cut" the native structure into strands depends on the same factors as the labelling, as discussed in **Note 1**.

6. Intensity fluctuations in the dyes' fluorescence can add noise to single-molecule traces or be mistaken for structural changes, leading to spurious populations and incorrect lifetime calculations. Ultimately, observation time is limited by the photoinduced degradation, termed photobleaching, of the dye [56]. Chemical additives that prolong the dye lifetimes and prevent blinking (entering a temporary dark state) have greatly enhanced the duration and signal-to-noise ratio of single-molecule experiments. Molecular oxygen is an effective

quencher of the non-emitting triplet state, but singlet oxygen also enhances photobleaching. Therefore, oxygen-scavenging systems have been developed and used in combination with alternative triplet-state quenchers. The most common oxygen scavenger is a combination of glucose, glucose oxidase, and catalase (commonly termed "gloxy"), though a system using protocatechuic acid and protocatechuate-3,4-dioxygenase has been shown to be more efficient [3]. As a substitute triplet-state quencher, β-mercaptoethanol (BME) has been used for years. However, although BME does inhibit rapid intensity fluctuations in Cy5, a common acceptor dye, it also induces long (~sec) dark states. A vitamin E analogue, Trolox, does not have this side effect [4] and is frequently used with cyanine dyes. A novel system has recently been proposed to further enhance the stability of non-cyanine dyes such as the Alexa series [57]. Cysteamine (2-mercaptoethylamine) is used to scavenge only radical oxygen, the by-product that oxidizes the dyes, while leaving molecular oxygen in solution to quench the triplet state in combination with Trolox. This combination outperforms the Trolox-gloxy system for dyes whose blinking is dominated by transitions into the triplet state (like the Alexa series, but not the cyanine dyes), making experiments with μs FRET resolution possible with these dyes.

7. For the specific coupling of a dye carrying a maleimide group to a thiol-modified nucleic acid, 10–100 mM phosphate (such as phosphate-buffered saline (PBS)), Tris-HCl, or HEPES buffer with pH between 7.0 and 7.5 should be used.

8. If the amine-reactive dye is not to be used immediately, aliquot and evaporate the DMSO and store at −20 °C until needed. Rehydrate with 14 μL DMSO immediately before use.

9. An APS/TEMED ratio of 10:1 as above will require a few hours to polymerize. Increasing the ratio of TEMED to APS will lead to a faster polymerization time. For example, a 4:1 ratio (e.g., 240 μL APS to 60 μL TEMED) will polymerize in 1 h.

10. TBE is prepared from Tris-HCl, boric acid, and EDTA (ethylenediaminetetraacetic acid). A 1 L 10× stock solution includes 108 g Tris-HCl, 55 g boric acid, and 7.5 g EDTA dissolved in 1 L deionized water and stored at room temperature. A dye such as bromophenol blue is often run in a lane adjacent to the sample to mark its progress down the gel. The marking dye and sample itself both need to be mixed with formamide (for a denaturing gel) or glycerol (for a native gel).

11. The concentration of dye or oligonucleotide in a given sample can be calculated from the absorption spectrum of the construct. The absorption of a sample as a function of concentration is given by the Beer-Lambert law,

$$A = C\varepsilon z \quad (7)$$

where the absorption A is dimensionless, C is the concentration of the sample in mol/L, z is the cuvette optical path length in cm, and ε is the extinction coefficient in units of L/mol cm. A typical absorption cuvette is 1 cm thick, so concentration of a given species, C_i, can be calculated directly from the maximum absorption divided by its extinction coefficient:

$$C_i = \frac{A_{i,\max}}{\varepsilon_i} \quad (8)$$

DNA and RNA exhibit a characteristic absorption peak at 258 nm, so the ratio of the concentration calculated from this peak to the concentration of the dye or dyes will indicate the efficiency of the labelling reaction.

12. Sodium periodate should be prepared shortly before use and stored in the dark. Change from clear or light yellow to dark yellow or brown indicates that it has degraded and should not be used [50].
13. Alternatively, nucleic acid strands can be hybridized by the "snap-cool" method, in which the hybridization mixture is placed in the heat bath at room temperature, heated to 70 °C, then immediately removed and cooled in ice at 0 °C. The snap-cool method works well for the hybridization of comparatively short strands, but constructs that may form significant secondary and tertiary competing structures should be checked for "snap-cool" performance.
14. Since the temperature of a native gel must be kept low to prevent the denaturation of the construct, the gel is run at a low constant voltage, unlike the high power permitted by a denaturing gel. Quantification of hybridization efficiency can best be calculated from the absorption spectrum by determining the molar weight of the hybridized construct and calculating the percentage of the original reactants that were successfully hybridized. The efficiency of incorporation of dye-labelled strands can be calculated as in **Note 11** as a secondary check.
15. To prolong the life of the DNA/RNA ligases, aliquot each vial upon opening, and store in 4–5 μL quantities at −20 °C. Unopened ligase stocks, buffers, and PEG solution should all be stored at −20 °C, and the buffer and PEG should be vortexed for several minutes before use. High NaCl concentrations inhibit ligase activity, so NaCl in excess of 200 mM can be added to terminate ligation.
16. Substrate ligation requires the addition of a phosphate group on the 5′ end of the downstream strand. Synthetic nucleotides

can be ordered with this phosphate added from most nucleic acid suppliers.

17. Incubation times and temperatures may need to be optimized for each experiment. Typical variation in incubation times is from 1 to 24 h, and temperatures in the 35–40 °C range are used. Even under optimal conditions, ligation efficiencies between single RNA strands are typically low (20–40 %) [51].
18. Due to the fumes generated by mixing HCl and hydrogen peroxide, the etching procedure should be performed in the fume hood. Also, to avoid spills, HCl should be added over the hydrogen peroxide. Protective glasses and gloves should be used at all times when handling this mixture.
19. For prism-type imaging and immobilization following the bi-BSA/NeutrAvidin method, only the quartz slides (not the glass coverslips) need burning to remove traces of organic material. 2 s each side of the quartz slide in the blue flame of a butane/propane torch will be enough. Burning for longer periods of time (between a few seconds and <1 min) will enhance the ability of the quartz surface to capture dirt.
20. Diamond-coated drill bits are extremely fragile and should be kept wet while drilling the quartz slide; otherwise, they will break very easily due to the heat generated during drilling. Normally, a diamond drill bit should last enough to drill 5–10 slides.
21. Fluorescently labelled microspheres have a tendency to aggregate, forming big clusters clearly seen when imaging. To avoid this, sonication for at least 10 min is critical.
22. Incubation and storage in water boxes prevent the channels from drying out once liquid has been flushed into the microchannels. A water box can be created from a pipette tip box by filling half of it with water and covering the entire box in black tape or aluminum foil to protect nucleic acid constructs from light.
23. Streptavidin may also be used in this step due to the structural similarities between the two proteins, although it has shown a higher level of nonspecific interactions with the glass surface.
24. Two etching steps may be appropriate before pegylation.
25. The cleaning process makes the coverslips fragile. As a consequence of this, it is prudent to clean twice as many coverslips as you will require compensating for coverslips that may break during the cleaning procedure.
26. The aminosilanization procedure incorporates primary amino groups in the quartz surface for further coupling to biotinylated and non-biotinylated NHS-functionalized PEG polymers.

27. These quantities are for the pegylation of 5 slides. Adjust them as required.
28. The PEG polymer is light sensitive; thus, it is essential to keep it in the dark while it passivates the slide surface.
29. If bubbles are formed, they can be removed by gently tapping with a pipette tip.
30. To help with this, in our lab we wrap electrical tape around one leg of the tweezers and tongs used to move the slides and coverslips. By consistently keeping the taped leg on the same side as the pegylated side of the slide, it is easier to keep track when preparing multiple channels.
31. Chloroform degrades plastic tips when in contact for too long; thus, it is better either to use glass pipette tips or to do the liquid transfer in stages. Furthermore, it is good practice to rinse the tips once with chloroform before transferring the vesicle stock and cholesterol to vials. Due to the small volumes involved, this is not necessary for the biotinylated lipids.
32. Egg PC is in the fluid liquid-crystalline phase at room temperature. At this temperature, lateral diffusion of lipid monomers to which the dye-labelled constructs are immobilized causes the fluorescence spots to diffuse following the lipid monomer. To avoid this, we add normally 30 % cholesterol during the preparation of the vesicle. Cholesterol is known to inhibit lateral diffusion. Alternatively, it is possible to use DPPC lipids which show a gel-to-fluid transition temperature of ~40 °C. At room temperature, these lipids are rigid and lateral diffusion is minimal, providing an adequate immobilization method for vesicles containing the dye-labelled nucleic acid.
33. During extrusion of the lipid contents using the mini-extruder, the number of back and forward passes of the syringes should be an odd number. We normally perform around 21 to ensure homogeneous vesicle size (minimum 11 passes). A transition from a cloudy solution to a clear one indicates successful formation of SUVs.
34. Vesicles may also be formed by sonication of the lipid solutions. In doing this, however, there is no measure of control over the size of the vesicles formed.

References

1. Walter NG, Huang C-Y, Manzo AJ, Sobhy MA (2008) Do-it-yourself guide: how to use the modern single-molecule toolkit. Nat Methods 5:475–489
2. Hohlbein J, Gryte K, Heilemann M, Kapanidis AN (2010) Surfing on a new wave of single-molecule fluorescence methods. Phys Biol 7:031001 (22pp)

3. Aitken CE, Marshall RA, Puglisi JD (2008) An oxygen scavenging system for improved dye stability in single-molecule fluorescence experiments. Biophys J 94:1826–1835
4. Rasnik I, McKinney SA, Ha T (2006) Nonblinking and longlasting single-molecule fluorescence imaging. Nat Methods 3:891–893
5. Ditzler MA, Rueda D, Mo J et al (2008) A rugged free energy landscape separates multiple functional RNA folds throughout denaturation. Nucleic Acids Res 36:7088–7099
6. Okumus B, Wilson TJ, Lilley DMJ, Ha T (2004) Vesicle encapsulation studies reveal that single molecule ribozyme heterogeneities are intrinsic. Biophys J 87:2798–2806
7. Clegg RM (1992) Fluorescence resonance energy transfer and nucleic acids. In: Lilley DMJ, Dahlberg JE (eds) Methods enzymology, vol 211. Academic, San Diego, CA, pp 353–388
8. Santoso Y, Joyce CM, Potapova O et al (2010) Conformational transitions in DNA polymerase I revealed by single-molecule FRET. Proc Natl Acad Sci USA 107:715–720
9. Blouin S, Craggs TD, Lafontaine DA, Penedo JC (2009) Functional studies of DNA-protein interactions using FRET techniques. In: Moss T, Leblanc B (eds) Methods in molecular biology, vol 543. Humana Press, Totowa, NJ, pp 475–502
10. Holden SJ, Uphoff S, Hohlbein J et al (2010) Defining the limits of single-molecule FRET resolution in TIRF microscopy. Biophys J 99:3102–3111
11. Zhuang X, Kim H, Pereira MJB et al (2002) Correlating structural dynamics and function in single ribozyme molecules. Science 296:1473–1476
12. Hohng S, Wilson TJ, Tan E et al (2004) Conformational flexibility of four-way junctions in RNA. J Mol Biol 336:69–79
13. Joo C, McKinney SA, Lilley DMJ, Ha T (2004) Exploring rare conformational species and ionic effects in DNA Holliday Junctions using single-molecule spectroscopy. J Mol Biol 341:739–751
14. Eichhorn CD, Feng J, Suddala KC et al (2011) Unraveling the structural complexity in a single-stranded RNA tail: implications for efficient ligand binding in the prequeuosine riboswitch. Nucleic Acids Res 40:1345–1355
15. Karymov MA, Chinnaraj M, Bogdanov A et al (2008) Structure, dynamics and branch migration of a DNA Holliday Junction: a single-molecule fluorescence and modeling study. Biophys J 95:4372–4383
16. Karymov MA, Bogdanov A, Lyubchenko YL (2008) Single molecule fluorescence analysis of branch migration of Holliday Junctions: effect of DNA sequence. Biophys J 95:1239–1247
17. Shirude PS, Okumus B, Ying L et al (2007) Single-molecule conformational analysis of g-quadruplex formation in the promoter DNA duplex of the proto-oncogene c-kit. J Am Chem Soc 129:7484–7485
18. Jena PV, Shirude PS, Okumus B et al (2009) G-quadruplex DNA bound by a synthetic ligand is highly dynamic. J Am Chem Soc 131:12522–12523
19. Lee JY, Okumus B, Kim DS, Ha T (2005) Extreme conformational diversity in human telomeric DNA. Proc Natl Acad Sci USA 102:18938–18943
20. Steiner M, Karunatilaka KS, Sigel RKO, Rueda D (2008) Single-molecule studies of group II intron ribozymes. Proc Natl Acad Sci USA 105:13853–13858
21. McDowell SE, Jun JM, Walter NG (2010) Long-range tertiary interactions in single hammerhead ribozymes bias motional sampling toward catalytically active conformations. RNA 16:2414–2426
22. Pereira MJB, Nikolova EN, Hiley SL et al (2008) Single VS ribozyme molecules reveal dynamic and hierarchical folding toward catalysis. J Mol Biol 382:496–509
23. Tan E, Wilson TJ, Nahas MK et al (2003) A four-way junction accelerates hairpin ribozyme folding via a discrete intermediate. Proc Natl Acad Sci 100:9308–9313
24. Penedo JC, Wilson TJ, Jayasena SD et al (2004) Folding of the natural hammerhead ribozyme is enhanced by interaction of auxiliary elements. RNA 10:880–888
25. Mandal M, Breaker RR (2004) Gene regulation by riboswitches. Nat Rev Mol Cell Biol 5:451–463
26. Roth A, Winkler W, Regulski E et al (2007) A riboswitch selective for the queuosine precursor $preQ_1$ contains an unusually small aptamer domain. Nat Struct Mol Biol 14:308–317
27. Brenner MD, Scanlan MS, Nahas MK et al (2010) Multivector fluorescence analysis of the *xpt* guanine riboswitch aptamer domain and the conformational role of guanine. Biochemistry 49:1596–1605
28. Lemay J-F, Penedo JC, Tremblay R et al (2006) Folding of the adenine riboswitch. Chem Biol 13:857–868
29. Heppell B, Blouin S, Dussault A-M et al (2011) Molecular insights into the ligand-

controlled organization of the SAM-I riboswitch. Nat Chem Biol 7:384–392

30. Haller A, Rieder U, Aigner M et al (2011) Conformational capture of the SAM-II riboswitch. Nat Chem Biol 7:393–400
31. Wood S, Ferré-D'Amaré AR, Rueda D (2012) Allosteric tertiary interactions preorganize the c-di-GMP riboswitch and accelerate ligand binding. ACS Chem Biol 7:920–927
32. Lemay J-F, Penedo JC, Mulhbacher J, Lafontaine DA (2009) Molecular basis of RNA-mediated gene regulation on the adenine riboswitch by single-molecule approaches. Methods Mol Biol 540:65–76
33. Moerner WE, Fromm DP (2003) Methods of single-molecule fluorescence spectroscopy and microscopy. Rev Sci Instrum 74:3597–3619
34. Roy R, Hohng S, Ha T (2008) A practical guide to single-molecule FRET. Nat Methods 5:507–516
35. Okamoto K, Sannohe Y, Mashimo T et al (2008) G-quadruplex structure of human telomere DNA examined by single molecule FRET and BrG-substitution. Bioorg Med Chem 16:6873–6879
36. Cisse I, Okumus B, Joo C, Ha T (2007) Fueling protein-DNA interactions inside porous nanocontainers. Proc Natl Acad Sci USA 104:12646–12650
37. Ishitsuka Y, Okumus B, Arslan S et al (2010) Temperature-independent porous nanocontainers for single-molecule fluorescence studies. Anal Chem 82:9694–9701
38. Torella JP, Holden SJ, Santoso Y et al (2011) Identifying molecular dynamics in single-molecule FRET experiments with burst variance analysis. Biophys J 100:1568–1577
39. Blanco M, Walter NG (2010) Analysis of complex single-molecule FRET time trajectories. Methods Enyzmol 472:153–178
40. Life Technologies (2012) Fluorophors and their amine-reactive derivatives. In: The Molecular Probes Handbook, 11th Edition. Invitrogen, section 1.3. http://www.invitrogen.com/site/us/en/home/References/Molecular-Probes-The-Handbook.html. Accessed 25 July 2012
41. Integrated DNA Technologies, Inc. (2012) Dyes. IDT. http://eu.idtdna.com/catalog/Modifications/Dyes.aspx. Accessed 25 July 2012
42. Chen J, Tsai A, Petrov A, Puglisi JD (2012) Nonfluorescent quenchers to correlate single-molecule conformational and compositional dynamics. J Am Chem Soc 134:5734–5737
43. Life Technologies (2012) Fluorophors and their amine-reactive derivatives. In: The Molecular Probes Handbook, 11th Edition. Invitrogen, section 1.6. http://www.invitrogen.com/site/us/en/home/References/Molecular-Probes-The-Handbook.html. Accessed 25 July 2012
44. Integrated DNA Technologies, Inc. (2012) Products and Services 2012. IDT. http://eu.idtdna.com/Pages/docs/catalog-product-documentation/idt-product-catalog-2012.pdf?sfvrsn=6. Accessed 25 July 2012
45. Le Reste L, Hohlbein J, Gryte K, Kapanidis AN (2012) Characterization of dark quencher chromophores as nonfluorescent acceptors for single-molecule FRET. Biophys J 102:2658–2668
46. McKinney SA, Joo C, Ha T (2006) Analysis of single-molecule FRET trajectories using hidden Markov modeling. Biophys J 91:1941–1951
47. Qin F, Li L (2004) Model-based fitting of single-channel dwell-time distributions. Biophys J 87:1657–1671
48. Bronson JE, Fei J, Hofman JM et al (2009) Learning rates and states from biophysical time series: a Bayesian approach to model selection and single-molecule FRET data. Biophys J 97:3196–3205
49. Morse DP (2007) Direct selection of RNA beacon aptamers. Biochem Biophys Res Commun 359:94–101
50. Kurata S, Ohtsuki T, Suzuki T, Watanabe K (2003) Quick two-step RNA ligation employing periodate oxidation. Nucleic Acids Res 31: e145
51. Rieder U, Kreutz C, Micura R (2010) Folding of a transcriptionally acting Pre-Q1 riboswitch. Proc Natl Acad Sci USA 107:10804–10809
52. Rasnik I, McKinney SA, Ha T (2005) Surfaces and orientations: much to FRET about? Acc Chem Res 38(7):542–548
53. Walter NG (2001) Structural dynamics of catalytic RNA highlighted by fluorescence resonance energy transfer. Methods 25:19–30
54. Baker JL, Sudarsan N, Weinberg Z et al (2012) Widespread genetic switches and toxicity resistance proteins for fluoride. Science 335:233–235
55. Zhuang X, Bartley LE, Babcock HP et al (2000) A single-molecule study of RNA catalysis and folding. Science 288:2048–2051
56. Ha T, Tinnefeld P (2012) Photophysics of fluorescent probes for single-molecule biophysics and super-resolution imaging. Annu Rev Phys Chem 63:595–617
57. Campos LA, Liu J, Wang X et al (2011) A photoprotection strategy for microsecond-resolution single-molecule fluorescence spectroscopy. Nat Methods 8:143–146

Chapter 36

Photoswitchable Fluorescent Proteins for Superresolution Fluorescence Microscopy Circumventing the Diffraction Limit of Light

Susana Rocha, Herlinde De Keersmaecker, Hiroshi Uji-i, Johan Hofkens, and Hideaki Mizuno

Abstract

In the last two decades, fluorescent proteins became an indispensable tool to noninvasively label a protein in living cells. The discovery of photoswitchable fluorescent proteins expanded the applications of the fluorescent proteins to techniques such as molecular tracking and highlighting on a microscope. Recently, a new microscopic modality to achieve a superresolution circumventing the diffraction limit of light (photoactivated localization microscopy, PALM) has been developed based on the photoswitchable fluorescent proteins. Here we introduce a basic protocol of PALM through the visualization of actin bundles with superresolution.

Key words Actin bundles, Diffraction limit, Fluorescent protein, Photoswitchable fluorescent protein, Superresolution microscopy, PALM

1 Introduction

Fluorescent labeling of molecules is crucial for imaging by fluorescence microscopy. Despite a wide variety of organic fluorophores available to label molecules in vitro, labeling of intracellular molecules under living conditions remains difficult. The gene encoding green fluorescent protein (GFP) was cloned from *Aequorea victoria* in 1992 [1]. GFP and GFP-like fluorescent proteins form a chromophore self-catalytically from its own amino acid residues without any cofactors. Capability of fluorescence emission upon ectopic expression of the gene encoding GFP was shown in 1994 in prokaryotic and eukaryotic cells [2]. These findings provided a way to label proteins in living cells noninvasively by genetic engineering with GFP. A lot of efforts have been made to produce fluorescent proteins with different photophysical properties, such

Yves Engelborghs and Antonie J.W.G. Visser (eds.), *Fluorescence Spectroscopy and Microscopy: Methods and Protocols*, Methods in Molecular Biology, vol. 1076, DOI 10.1007/978-1-62703-649-8_36,

as color variants and brighter fluorescence by mutagenesis and cloning of novel proteins from different sea organisms (for review, *see* [3–7]). On the way, we encountered a series of interesting fluorescent proteins whose fluorescence color/intensity can be regulated upon light illumination, so-called photoswitchable fluorescent proteins [8, 9]. The photoswitchable fluorescent proteins give rise to a new method to track molecules. For example, nucleo-cytoplasmic shuttling of extracellular signal-regulated kinase 1 (ERK1) has been observed by expressing ERK1 fused to a photoswitchable fluorescent protein, Dronpa [9]. This experiment revealed transient bidirectional acceleration of ERK1-Dronpa transport across the plasma membrane upon epidermal growth factor stimulation. Molecular highlighting is another example of an application. By using hippocampal neuron primary culture expressing a photoswitchable fluorescent protein, Kaede, a neuron in the culture can be highlighted with different color upon UV illumination to the specific cell [8, 10]. This method is also applicable to an in vivo system; Sato et al. developed a transgenic zebrafish expressing Kaede in neuronal cells as a tool to label neural morphology [11]. Application of the photoswitchable fluorescent proteins as proteinous media for data storage has also been proposed [12, 13].

So far three classes of photoswitchable fluorescent proteins have been reported: photoactivatable (PAFP), photoconvertible (PCFP), and reversibly switchable (RSFP) (Table 1, Fig. 1). PAFPs are originally nonfluorescent and become fluorescent upon light illumination. The first PAFP, PAGFP, was made by introducing a mutation T203H to wild-type *Aequorea* GFP [14]. PAGFP achieves green fluorescence upon UV illumination. The photoactivation of PAGFP is thought to be the result of decarboxylation from the side chain of E222, which is placed nearby the chromophore, followed by change in the local environment around the chromophore and shift of the chromophore from nonfluorescent protonated to fluorescent anionic form (Fig. 2) [15]. Photoactivatable fluorescent proteins in the red range, PAmRFP1, PAmCherry, and PATagRFP, have also been engineered based on red fluorescent proteins [16–18].

PCFPs change their fluorescent color upon light illumination. The first PCFP reported is Kaede, which was isolated from a stony coral *Trachyphyllia geoffroyi* [8]. Kaede originally emits green fluorescence, and upon UV illumination, the fluorescence is converted to red. The chromophore of Kaede in the green form is 4-(*p*-hydroxybenzylidene)-5-imidazolinone, the same as that of *Aequorea* GFP except the first amino acid residue of the chromophore is histidine (His62) instead of serine. UV illumination evokes a formal β-elimination reaction on His64, results in cleavage of the main chain, and produces a double bond between Cα and Cβ of His62. This double bond expands the π conjugation of the chromophore to the imidazole ring of His62 and shifts the absorption

Table 1
Selected photoswitchable fluorescent proteins

Protein	Quaternary structure[a]	Laser line for excitation[b]	Emission maximum (nm)	Laser line for switching[b]	Available from	Reference
Photoactivatable						
nf → green						
PAGFP	m	488	515	405	Addgene	[14]
nf → red						
PAmRFP1	m	561	605	405	Addgene	[16]
PAmCherry1	m	561	595	405	Clontech	[17]
PATagRFP	m	561	595	405	Evrogen	[18]
Photoconvertible						
Green → red						
Kaede	t	488 (g) 561 (r)	518 (g) 595 (r)	405	Amalgaam	[8]
EosFP	t	488 (g) 561 (r)	516 (g) 581 (r)	405	MoBiTec	[19]
tdEosFP	td	488 (g) 561 (r)	516 (g) 581 (r)	405		[19]
Dendra2	m	488 (g) 561 (r)	507 (g) 573 (r)	405	Clontech	
mEosFP	m	488 (g) 561 (r)	516 (g) 581 (r)	405		[19]
mEos2	m[c]	488 (g) 561 (r)	519 (g) 584 (r)	405	Addgene	[23]
mEos3.1	m	488 (g) 561 (r)	519 (g) 584 (r)	405		[24]
mEos3.2	m	488 (g) 561 (r)	519 (g) 584 (r)	405		[24]
KikGR	t	488 (g) 561 (r)	517 (g) 593 (r)	405	Amalgaam	[20]
mKikGR	m	488 (g) 561 (r)	515 (g) 591 (r)	405	Amalgaam	[22]
Cyan → green						
PSCFP[d]	m	405(c) 488 (g)	468 (c) 511 (g)	405		[25]
PSCFP2[d]	m	405(c) 488 (g)	468 (c) 511 (g)	405	Evrogen	

(continued)

Table 1
(continued)

Protein	Quaternary structure[a]	Laser line for excitation[b]	Emission maximum (nm)	Laser line for switching[b]	Available from	Reference
Reversibly switchable						
Green ↔ nf						
22G[c]	t	488		488 (off) 405 (on)		[9, 39]
Dronpa	m	488	518	488 (off) 405 (on)	Amalgaam	[9]
Dronpa 2	m	488	487	488 (off) 405 (on)		[29]
Dronpa 3	m	488	514	488 (off) 405 (on)	Amalgaam	[29]
bsDronpa	m	488	504	488 (off) 405 (on)		[30]
Padron	m	488	522	488 (on) 405 (on)		[30]
rsEGFP	m	491	510	491 (off) 405 (on)	Abberior	[13]
Dreiklang	m	515	529	405 (off) 369 (on)	Abberior	[33]
Red ↔ nf						
rsCherry	m	561	610	561 (on) 445 (off)		[31]
rsCherryRev	m	561	608	561 (off) 445 (on)		[31]
rsTagRFP	m	561	585	561 (off) 445 (on)		[32]
Photoconvertible/reversibly switchable						
Green ↔ nf/green → red/red ↔ nf						
IrisFP	t	488 (g) 561 (r)	516 (g) 580 (r)	488(g → nf) 405 (nf → g) 405(g → r) 561 (r → nf) 440 (nf → r)		[34]
mIrisFP	m	488 (g) 561 (r)	516 (g) 578 (r)	488(g → nf) 405 (nf → g) 405 (g → r) 561 (r → nf) 440 (nf → r)	MoBiTec	[35]

(continued)

Table 1 (continued)

Protein	Quaternary structure[a]	Laser line for excitation[b]	Emission maximum (nm)	Laser line for switching[b]	Available from	Reference
Niji	m	488 (g) 561 (r)	507 (g) 569 (r)	488(g → nf) 405 (g → r) 405(nf → g) 561 (r → nf) 440 (nf → r)		[36]

[a]Quaternary structure: *t* tetramer; *td* tandem dimer; *m* monomer
[b]Appropriate laser lines for excitation/switching are indicated
[c]Dimer formation at high concentration has been reported
[d]These proteins often categorized into PA due to structural similarity with PAGFP
[e]Photoswitching of 22G was overlooked in the paper [9] due to its very slow switching

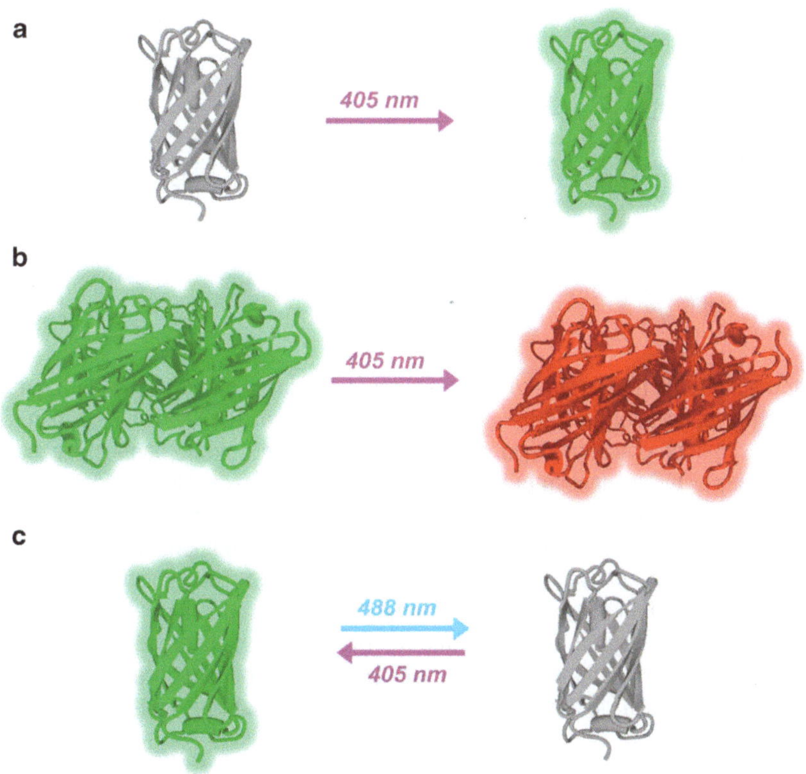

Fig. 1 Different modes of photoswitchable fluorescent proteins. (**a**) PAFP. PAGFP becomes fluorescent upon 405 nm illumination. (**b**) PCFP. Kaede originally emits green fluorescence, and the fluorescence shifts into longer wavelength (*red* region) upon 405 nm illumination. (**c**) RSFP. Dronpa originally emits green fluorescence, loses fluorescence reversibly upon 488 nm illumination, and regains upon 405 nm illumination

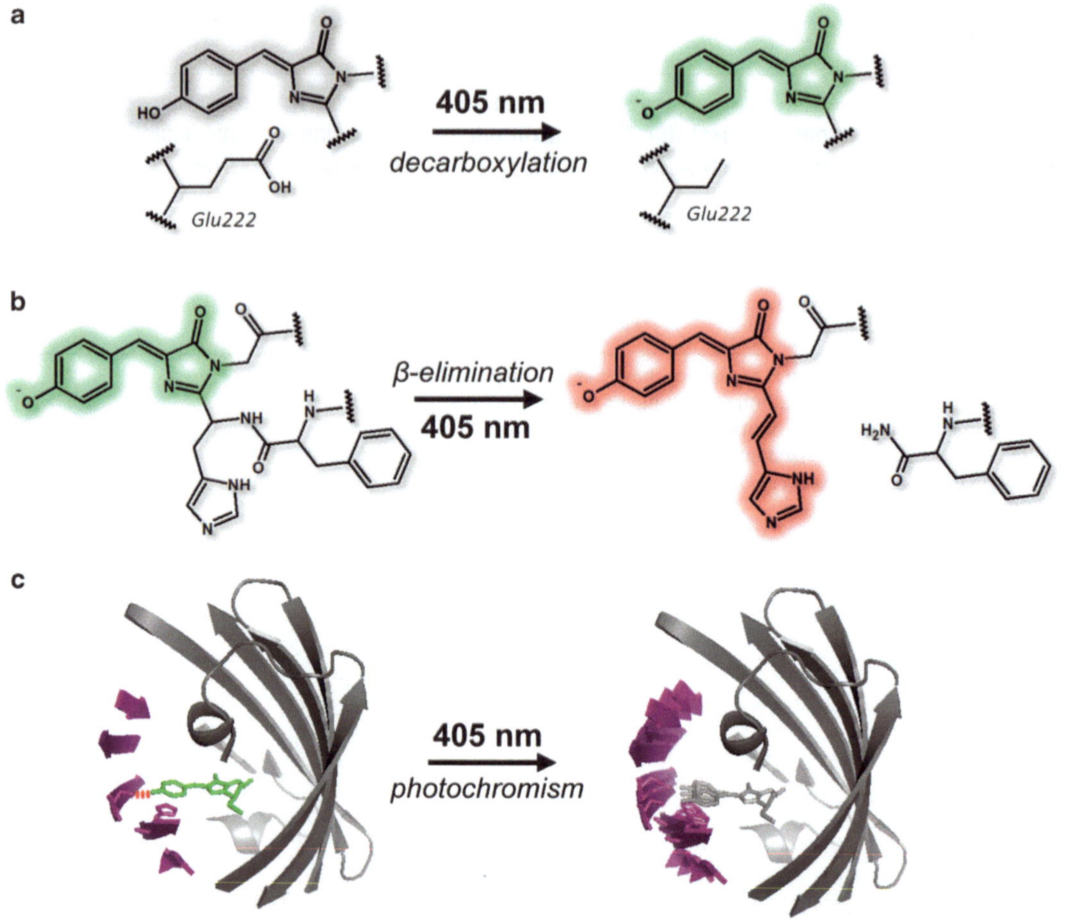

Fig. 2 Molecular mechanisms responsible for the switching behavior of photoswitchable fluorescent proteins. (**a**) Mechanism of photoactivation in PAGFP. The decarboxylation of Glu222 induces changes in the local environment around the chromophore and shifts the chromophore from nonfluorescent protonated to fluorescent anionic form. (**b**) Mechanism of the photoconversion in Kaede. UV illumination induces a formal β-elimination reaction, which results in cleavage of the main chain and produces a double bond between Cα and Cβ of His62. This double bond expands the π conjugation of the chromophore to the His62 imidazole ring and shifts the absorption and emission bands into the *red* region. (**c**) Mechanism of the reversible photoswitching of Dronpa. In the bright state, the anionic chromophore is fixed to the β-barrel via a hydrogen bond (*red dotted line*) and π-π stacking. In the dark state, the hydrogen bond is broken by the protonation of the chromophore, and as a result, part of the β-barrel as well as the chromophore becomes flexible. The structural flexibility in the dark state makes vibrational relaxation the dominant pathway instead of radiational relaxation

and emission bands into the red region (Fig. 2) [10]. The green-to-red conversion is irreversible, since the formal β-elimination is irreversible. Other photoconvertible fluorescent proteins, such as EosFP and KikGR, have been cloned or genetically engineered [19, 20]. A drawback for the use of these proteins is the formation of obligate homotetramer, which can induce artificial molecular interactions of the target molecule and change its behavior. To avoid this issue, monomeric mutants of the photoconvertible fluorescent proteins, such as mEosFP, Dendra2, and mKikGR, have been engineered [19, 21–24]. Most of the PCFPs show green-to-red conversion, but so far only two PCFPs that can be converted from cyan to green fluorescence, i.e., PSCFP and PSCFP2, have been engineered [25]. In this chapter, we categorized these two proteins as PCFP due to their photoswitching property, but they are often categorized into PAFP family due to their structural similarity with PAGFP.

One of the first RSFP reported is Dronpa, which was found on the way to make a monomeric mutant of tetrameric, non-switchable green fluorescent protein, 22G, isolated from Pectiniidae [9]. Dronpa originally emits green fluorescence (bright state), becomes nonfluorescent upon strong blue/green illumination (dark state), and regains fluorescence upon UV illumination. This switching is reversible. Based on X-ray crystallography, *cis/trans* photoisomerization of the chromophore was proposed as the molecular mechanism of the on/off transition of Dronpa [26]. Structural analysis of Dronpa by NMR gave evidence that a part of the β-barrel together with the chromophore becomes flexible only in the dark state, which provides a photochemical explanation for the absence of fluorescence in the dark state; vibrational relaxation becomes dominant in the dark state due to the flexibility (Fig. 2) [10, 27]. Since the switching is based on the photochromism of the protein, this group of proteins is sometimes referred as photochromic fluorescent protein. Several mutants of Dronpa with faster switching have been developed [28–30]. Dronpa and all its mutants show green fluorescence in the bright state. In the same spectrum range, rsEGFP has recently been engineered from *Aequorea* GFP [13]. There are also reversibly photoswitchable fluorescent proteins in the red range, namely, rsCherryRev and rsTagRFP [31, 32]. Most of the RSFP switch to the dark state upon the illumination at the same wavelength as the excitation and switch back to the bright state upon illumination at UV or violet range. Padron and rsCherry are exceptions. There proteins are switched to the bright state upon the illumination at the same wavelength as for excitation and converted to the dark state upon UV-violet illumination [30, 31]. In both cases the wavelength for the excitation is coupled with switching, which limits the application. Recently, an RSFP with fluorescence excitation decoupled from switching, Dreiklang, has been developed. Dreiklang emits yellow fluorescence upon excitation at 515 nm, and it is turned off and on upon illumination at 405 nm and 369 nm, respectively [33].

Some photoswitchable fluorescent proteins have both photoconvertible and reversibly switchable properties. For instance, IrisFP shows a green-to-red photoconversion upon UV illumination, and it can be turned on/off repeatedly in both green and red states [34]. IrisFP was engineered from EosFP and hence, it is tetrameric,

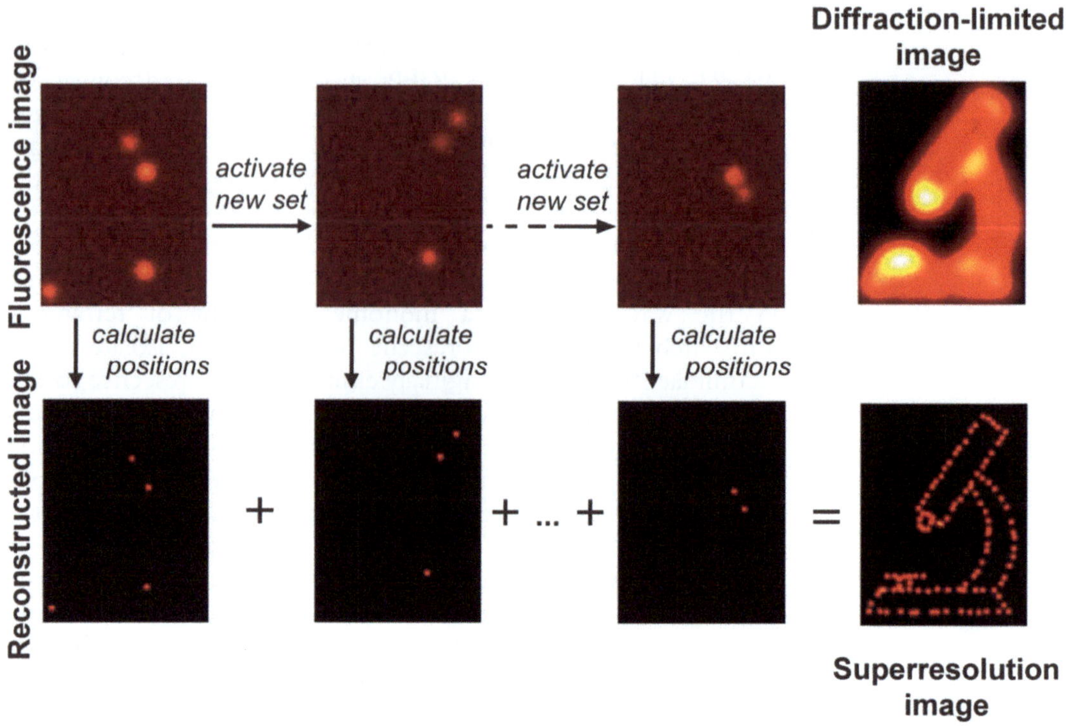

Fig. 3 Principle of PALM. (*Top row*) Diffraction limit fluorescent images. (*Bottom row*) Reconstructed images. In each frame a subset of molecules is converted to an "on-state," and their positions are calculated in nanometer precision through fitting with a 2D Gaussian function. Accumulation of all calculated positions yields a superresolution image (*bottom right*) (images plotted using simulated data)

but the monomeric version of this protein, mIrisFP, and a mutant of Dendra2 with similar properties, Niji, have been developed [35, 36].

Recently, a new superresolution fluorescence microscopic modality, photoactivated localization microscopy (PALM), has been developed based on the photoswitchable fluorescent proteins [37, 38]. In general, the highest achievable resolution of optical microscopy is determined by the diffraction limit of light. In the case of fluorescence microscopy, the best resolution is around 200 nm. PALM uses the stochastic activation of individual photoswitchable fluorophores followed by the precise determination of their position by fitting to render images with a higher resolution (Fig. 3). Weak illumination with a so-called activation laser guarantees that only a very small subset of the fluorophores that label the target molecules are in the emissive state. This subset emits sufficient photons to enable precise localization of the molecule before it enters a dark state or photobleaches. Repeating this process for multiple cycles, each generating a stochastically unique subset of molecules in the emissive state, enables the reconstruction of the image through the calculated position of each molecule.

The resolution of PALM is related to the precision of the fitting, which depends on the number of photon emitted from each molecule and on the level of background. So far the achievable resolution is around 10–20 nm. In this chapter, we will introduce a basic method of PALM by using visualization of actin bundles in cultured cells with KikGR-labeled actin as an example.

2 Materials

2.1 Construction of Gene Encoding a Fluorescent Protein

1. A fluorescent protein expression vector: phKikGR1-MC1 (Amalgaam, Tokyo, Japan) (*see* **Notes 1** and **2**).
2. Pfu DNA polymerase (Thermo Scientific, Rockford, IL, USA).
3. Restriction enzymes: BamHI and NotI (Thermo Scientific).
4. T4 DNA ligase (Thermo Scientific).
5. DH5α chemical competent cells (Life Technologies, Carlsbad, CA, USA) (*see* **Note 3**).
6. PCR Purification Kit (GeneJET; Thermo Scientific).
7. Gel Extraction Kit (GeneJET; Thermo Scientific).
8. Plasmid Miniprep Kit (GeneJET; Thermo Scientific).
9. Luria-Bertani (LB) medium: Dissolve 5 g of yeast extract (Fisher Scientific, Fair Lawn, NJ, USA), 10 g of bactotryptone (Becton Dickinson, Sparks, MD, USA), and 10 g of sodium chloride (Sigma-Aldrich, St. Louis, MO, USA) in 1 l of water. Immediately sterilize by autoclaving.
10. Kanamycin solution: Dissolve kanamycin sulfate (Carl Roth, Karlsruhe, Germany) in water to make 30 mg/ml stock solution.

2.2 Preparation of Cell for PALM

1. Dulbecco's modified Eagle's medium (DMEM) (Life Technologies) (*see* **Note 4**).
2. Fetal bovine serum (FBS) (Life Technologies).
3. Phosphate-buffered saline (PBS) (Life Technologies).
4. Hank's balanced salt solution (HBSS) (Life Technologies).
5. Imaging chamber: 35 mm petri dish, 14 mm glass bottom microwell dish (MatTek, Ashland, MA, USA).
6. Transfection reagent: X-tremeGENE 9 (Roche Applied Science, Basel, Switzerland). Store at 4 °C.
7. 4 % formaldehyde: Dilute 16 % formaldehyde solution (w/v), methanol-free (Thermo Scientific) with PBS. Prepare at the time of use.
8. Fluorescent bead solution: 1:500 dilution of TetraSpeck (Life Technologies) in PBS.

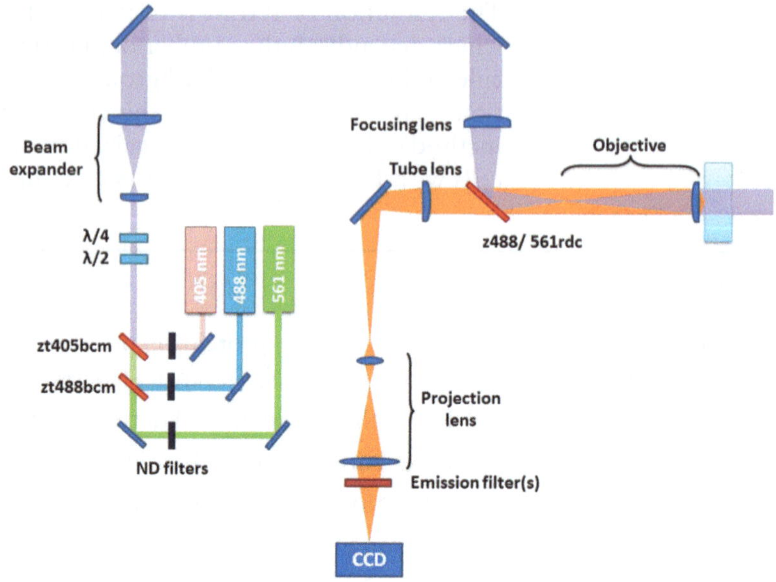

Fig. 4 Schematic drawing of the PALM setup used. The three laser lines are combined with the appropriate dichroic mirrors, expanded 10×, and guided to the objective lens through a 488/561 dichroic mirror. The fluorescence image is further expanded and captured in an EM-CCD camera through the emission filters

2.3 PALM Setup

A commercial PALM setup is available (*see* **Note 5**). The simplest PALM setup can be built in a laboratory with a wide-field fluorescence microscope equipped with two lasers (for switching and excitation of the fluorescent protein), an objective lens with high numerical aperture (NA; typically NA $\geq$ 1.2 for water immersion or NA $\geq$ 1.4 for oil immersion), and an electron-multiplying charge-coupled device (EM-CCD) camera (*see* **Note 6**). Components of our PALM setup for KikGR are as follows. Configuration of the system is shown in Fig. 4:

1. An inverted microscope IX71 (Olympus, Tokyo, Japan).
2. An objective lens APON 60XOTIRF (NA 1.49; Olympus).
3. A laser for switching: 405 nm, 100 mW (Cube, Coherent Inc., Santa Clara, CA, USA).
4. Lasers for excitation: 561 nm, 100 mW (for red channel; Sapphire, Coherent Inc.); 488 nm, 100 mW (for green channel; Sapphire, Coherent).
5. 405 nm and 488 nm dichroic mirrors to combine lasers (zt405bcm and zt488bcm, respectively; Chroma Technology, Bellows Falls, VT, USA).
6. ND filters (Newport Corporation, Irvine, CA, USA) to adjust laser power.

7. $\lambda/4$ and $\lambda/2$ wave plates (zero order for 532 nm, Newport Corporation) to ensure circular polarization of the laser illumination at the sample (*see* **Note 7**).
8. 10× beam expander (Linos, Qioptiq, Luxembourg).
9. Focusing lens (f = 500 mm, BK 7, Newport Corporation) (*see* **Note 8**).
10. A dichroic mirror to separate excitation and emission lights (z488/561rpc, Chroma Technology).
11. Combination of a 572 nm long-pass filter and 595/40 nm band-pass filter (HQ572 LP and HQ595/40 m, respectively; Chroma Technology) to selectively detect red channel emission of KikGR (*see* **Note 9**).
12. A 500 nm long-pass filter (HQ500LP; Chroma Technology) to detect the green channel emission of KikGR.
13. Projection lens 2.5× (Olympus) (*see* **Note 10**).
14. An EM-CCD camera ImagEM Enhanced (Hamamatsu Photonics, Hamamatsu, Japan).

3 Methods

3.1 Plasmid Construction (See Notes 11 and 12)

1. Design forward and reverse primers for human β-actin based on the sequence. Overhang BamHI and NotI site to the forward and reverse primers, respectively.
2. Amplify human β-actin (*see* **Note 13**) by PCR with a high-fidelity polymerase such as Pfu polymerase.
3. Remove primers and the enzyme by using the PCR Purification Kit.
4. Digest the PCR fragment and the phKikGR1-MC1 with BamHI and NotI.
5. Separate the digests by agarose gel electrophoresis, and collect bands corresponding to the gene encoding to β-actin and the vector. Extract the DNA from the agarose gel by using the Gel Extraction Kit.
6. Ligate the fragments with T4 DNA ligase.
7. Transform DH5α competent cells with the ligation product.
8. Inoculate LB medium containing 30 μg/ml of kanamycin.
9. Prepare plasmid by using the Plasmid Miniprep Kit.

3.2 Preparation of Cells

Maintain HeLa cells in DMEM supplemented with 10 % FBS under 5 % CO_2 atmosphere at 37 °C:

1. Seed HeLa cells on the glass bottom dish.
2. Transfect KikGR-β-actin/pMC1 into HeLa cells with X-tremeGENE 9 according to the manufacturer's protocol.

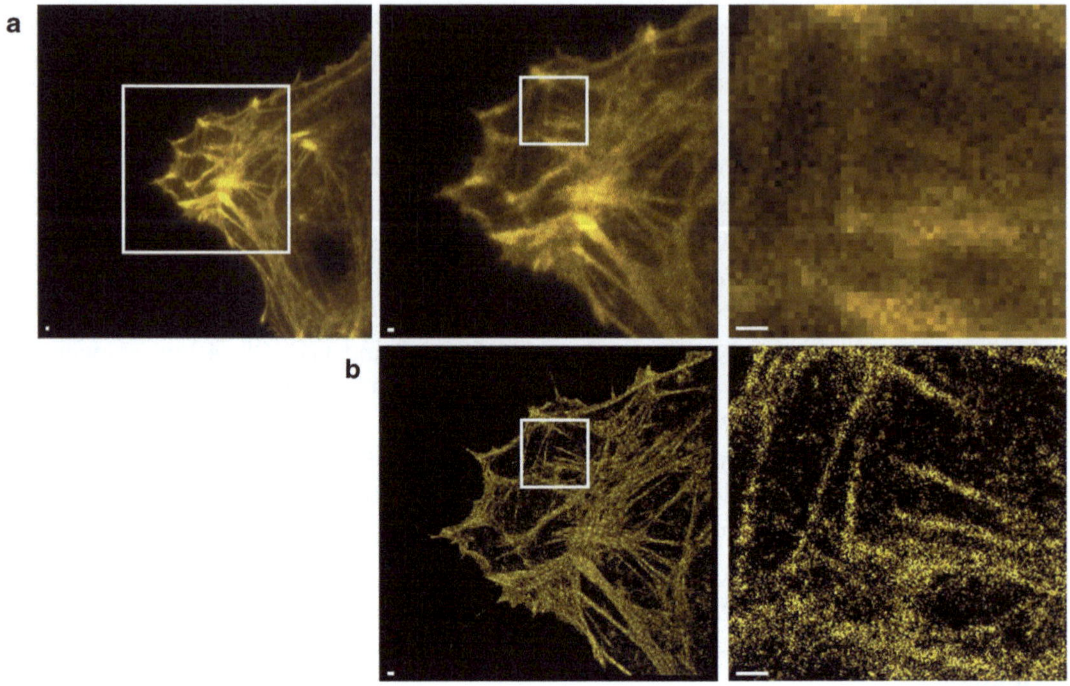

Fig. 5 PALM image of actin bundles labeled with KikGR. (**a**) Diffraction-limited images. (**b**) PALM images at the same regions as (**a**). Expansion of the region indicated by the *box* is shown on the *right side* of the respective panels. Scale bars indicate 500 nm

3. Incubate HeLa cells for 2 days.
4. Fix the cells by replacing the medium with 4 % formaldehyde and place it at 37 °C for 30 min (*see* **Note 14**).
5. Wash cells with 1 ml of PBS 3 times.
6. Remove the entire buffer and add 20 μl of fluorescent bead solution. Wait 10 min and add 1 ml of PBS (*see* **Note 15**).

3.3 Imaging

1. Acquire a diffraction-limited image of the cell with the green fluorescence of KikGR (Fig. 5; *see* **Notes 16** and **17**).
2. Illuminate the specimen with the 561 nm laser line without recording until single molecules are spatially distinct (usually 5–15 min; *see* **Note 18**).
3. Turn on the activation laser (405 nm) (*see* **Note 19**).
4. Optimize the power of the 561 nm laser and the integration time based on the duration of the on-state of the fluorophore (typically 50 ms; *see* **Notes 20** and **21** for more details).
5. Acquire 10,000–20,000 frames (*see* **Note 22**).

3.4 Data Analysis

After acquiring a series of sequential images (a movie), the position of the molecules detected in each frame is calculated from the fitting of the PSF by a 2D Gaussian function. We are using a custom

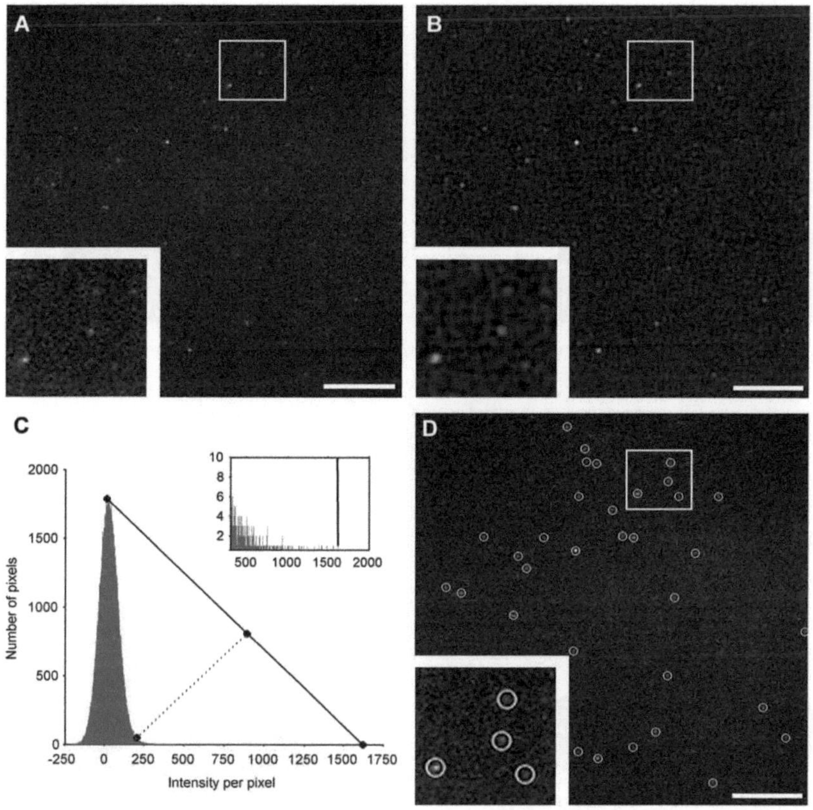

Fig. 6 Algorithmic used for the detection of single molecules. (**a**) Original fluorescence image. (**b**) Image after the filtering process. (**c**) Example of the determination of the image threshold using the triangle method (for more details, *see* **Note 23**). (**d**) Positions corresponding to single molecules as detected by the algorithm. *Left bottom* in each image shows the expansion of the region indicated with the *box*

made algorithm developed in MATLAB® (*see* **Note 23**). The steps used are explained in detail bellow:

1. Background subtraction/filtering.
   Before determining the position of the molecules, the background is subtracted from the image by a combination of smoothing and median filter (Fig. 6).
2. Setting threshold.
   To establish what constitutes a molecule and what is background, a threshold needs to be set. This is done automatically for each frame using the triangle method (*see* **Note 24**). The pixels with intensity higher than the set threshold are identified, and their position saved for further analysis (Fig. 6; *see* **Note 25**).
3. Gaussian fitting.
   For each point detected, an ROI of typically 15 × 15–21 × 21 pixels is selected, and the PSF is fitted with a 2D Gaussian function. The center of the Gaussian function as well as the width and uncertainties of the fit parameters are saved (Fig. 7).

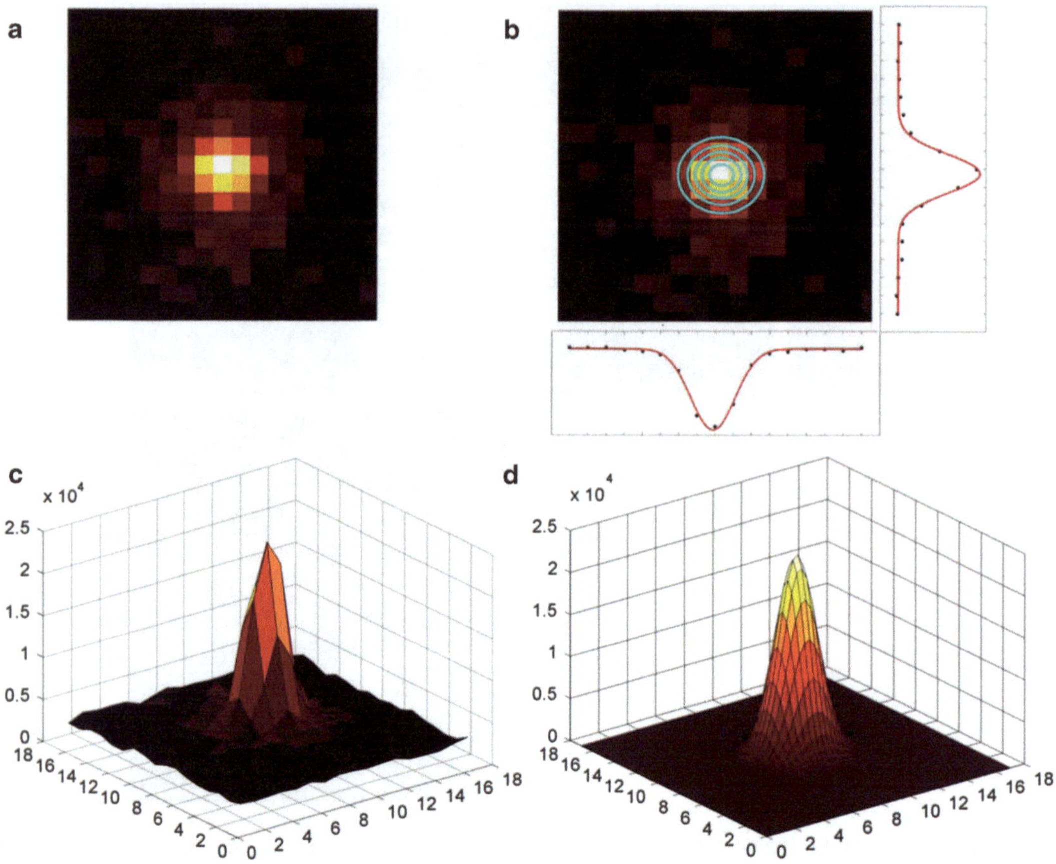

Fig. 7 Fitting of the PSF with a 2D Gaussian function. (**a**, **c**) PSF of a fluorescent bead. (**a**) Intensity plot. (**c**) Surface plot. (**b**, **d**) The fitting result. (**b**) The fitted 2D Gaussian function is shown in *light blue*. The *right* and *bottom* are the *y* and *x* profiles of the fitting (data points are shown in *black*); the fitting is shown in *red*. (**d**) Surface plot of the fitted 2D Gaussian

4. Drift correction.
   The fluorescent beads are used as fiducial markers to compensate movement of the sample during the data acquisition (*see* **Note 26**). Drifting of the positions in each frame is corrected in such a way that the points corresponding to the fiducial markers are within a 10–20 nm precision (Fig. 8).
5. Image rendering.
   Once all the positions have been calculated, the superresolution image can be rendered using three methods:
   (a) Unweighted, point-like plots of the positions of localized molecules (each localized molecule is plotted with the same intensity).
   (b) Weighted plots in which each molecule is plotted with a Gaussian profile (with amplitude proportional to localization precision).

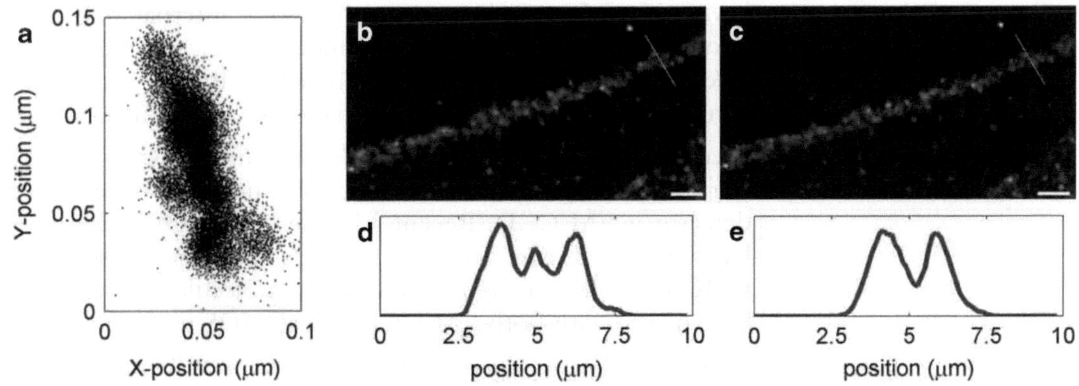

Fig. 8 Compensation of the sample movement during data acquisition. (**a**) Calculated positions of a fluorescent bead during the acquisition. (**b**, **c**) Reconstructed superresolution images before (**b**) and after (**c**) the compensation. (**d**, **e**) Line profile (perpendicular to the length) of the superresolution images before (**d**) and after (**e**) the compensation. Scale bar indicates 500 nm

(c) Density plot calculated by binning all the molecules localized within each pixels (using a bin size of 10–40 nm) (Fig. 5).

4 Notes

1. Most of the genes encoding fluorescent proteins are available from a company or a nonprofit organization. Providers of the genes are listed in Table 1. For a protein not available from those, you can usually get it upon request to the authors of the papers cited. In most of these cases, the usage of the protein is then restricted to research purposes.
2. phKikGR1-MC1 is a vector to express a target protein fused to the C-terminus of KikGR in mammalian cells. It contains cytomegalovirus promoter followed by the gene encoding KikGR and the multiple cloning site [39].
3. Chemical competent cells are commercially available. We can also prepare the competent cells by ourselves. A protocol to prepare high-competency chemical competent cell can be found in [40, 41].
4. To reduce fluorescence background, do not put phenol red in the media.
5. A commercial PALM, Elyra P.1, is available from Carl Zeiss Microscopy (Jena, Germany). This system is the only commercial PALM since Carl Zeiss Microscopy patents for PALM. For noncommercial purpose, researchers can assemble a microscope for PALM by themselves with components from other companies. For example, Olympus and Nikon (Tokyo, Japan)

have wide-field microscopes with a total internal refraction fluorescence (TIRF) illumination module. Leica Microsystems (Wetzlar, Germany) has a system so-called SR GSD, in which a computer-controlled TIRF illumination system is included.

6. Imaging of single molecules in wide-field microscopy requires a sensor with high quantum efficiency. The most widely used detector is EM-CCD. Recent advances in CMOS technology have increased the performance of such sensors (scientific CMOS or sCMOS). Nowadays, single-molecule detection can be achieved with both EM-CCD and sCMOS sensors.
7. Most of the laser emits plane-polarized beam. Placing an appropriate $\lambda/4$ wave plate will turn plane-polarized light into circularly polarized light. The $\lambda/2$ wave plate is placed to compensate the effect of the dichroic mirror in the polarization of the laser beam.
8. In order to achieve wide-field illumination, the laser beam is focused in the back focal plane of the objective lens. TIRF can be obtained by projecting the incident laser beam to the edge of the objective lens. Practically we can mimic it by changing the incident angle, which can be performed by translating the focusing lens perpendicularly to the beam axis.
9. Depending on the photophysical properties of the fluorophore(s), an appropriate filter set combination must be selected. Briefly, the dichroic mirror should reflect both the activation and the excitation laser(s) while maximally transmitting the fluorescence from the sample. Furthermore, the use of a narrow emission bandpass filter reduces the background significantly.
10. The magnification used determines the effective size of a camera pixel within the sample (pxl), and it is one of the most crucial considerations in a PALM setup. If the pixel size is too large, the PSF will be concentrated within too few pixels, creating a large error in the fitting of the 2D Gaussian; if the pixel size is too small, the PSF will be distributed across too many pixels and the S/N per pixel will be too low for accurate determination of the center. According to the Nyquist criterion, the optimal image pixel size is approximately equal to the standard deviation of the PSF (half width of 2–3 pixels).

    The image pixel size is determined by the physical size of the camera pixels (pxl_{camera}) and the total magnification (M) according to $pxl = pxl_{camera}/M$. For the setup used, the pixel is ca. 107 nm (16 μm/(60 × 2.5)).
11. Standard molecular biology is applicable for plasmid construction to express a protein fused to one of the photoswitchable fluorescent proteins. Here we show the protocol to construct human β-actin on phKikGR1-MC1 (KikGR-β-actin/pMC1) vector as an example.

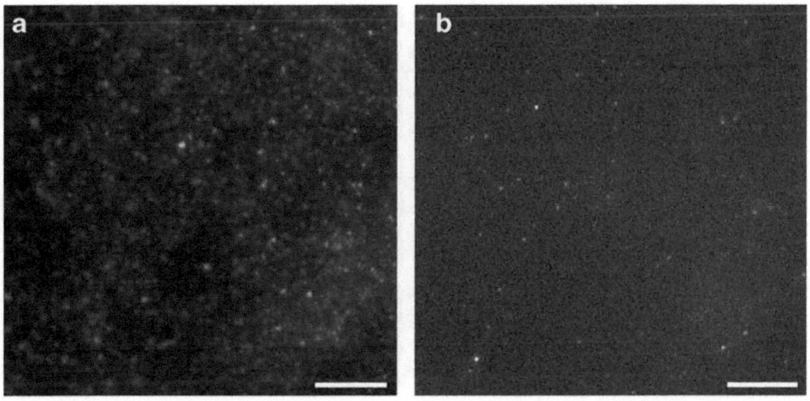

Fig. 9 (**a**) Initial fluorescence image upon excitation of the sample with 561 nm laser light. A lot of preexisting KikGR molecules in the red state overlap each other. (**b**) Fluorescence image after several-minute illumination at 561 nm. During the illumination, most of the preexisting KikGR molecules are bleached out to avoid recording of overlapping molecules, and the specimen is ready for data acquisition. Scale bars indicate 10 μm

12. See manufacturer's protocol for details of each step.
13. The gene encoding β-actin is available from several companies (e.g., Evrogen) or from a nonprofit organization (Addgene).
14. For the majority of structures, fixation with paraformaldehyde/formaldehyde is sufficient. If a better preservation of the structure is necessary, small amounts of glutaraldehyde (0.05–0.2 % (v/v)) can be added, although glutaraldehyde might increase a background fluorescence signal.
15. Fluorescent beads are added as fiducial markers to compensate the movement of the sample during data acquisition.
16. For PCFP and RSFP, a diffraction-limited image can be acquired using the fluorescence of the original on-state form. For PAFP, this is not possible (it is originally in a dark state).
17. For acquisition of the diffraction-limited image, it is advisable to use low laser power and longer excitation time (>100 ms) to avoid bleaching of the FP.
18. During sample manipulation, a small fraction of the molecules is converted to the red form, most probably due to room light. Therefore, before acquiring the data, molecules already in the red form need to be converted to a dark state. The waiting time depends of the type of probe as well as the labeling density (Fig. 9).
19. The resolution of PALM is influenced by the degree of labeling as well as by the switching rates of photoswitchable fluorophore. The photoconversion rate can be optimized by regulating the power of the 405 nm laser. Too strong 405 nm laser illumination might induce the activation of more than one fluorophore per diffraction-limited spot. It gives a fake spot

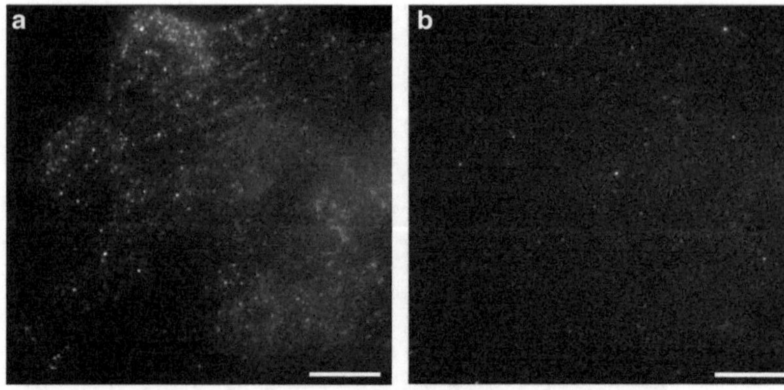

Fig. 10 (**a**) Fluorescence image acquired under too strong illumination of the activation laser. Some photoconverted KikGR molecules overlap each other and cannot be distinguished. (**b**) Fluorescence image acquired under the appropriate intensity of the activation laser. All single molecules can be distinguished. Scale bars indicate 10 μm

between two real molecules and, as a result, reduces the resolution of the final image [42] (Fig. 10).

20. The localization precision (σ_{xy}) of a fluorescent object is given by [43]

$$\sigma_{xy} = \frac{s^2 + q^2/12}{N} + \frac{8\pi s^4 b^2}{q^2 N^2}$$

where s is the standard deviation of the point spread function, N is the total number of photons collected per molecule, q is the size of each pixel in the image, and b is the background noise per pixel.

According to this equation, the localization precision of the molecules (and consequently the resolution of PALM) can be improved by reducing the background and increasing the number of detected photons. While the activation laser can have a power as low as 5 mW, powers >100 mW are recommended for the readout laser as it will directly influence the localization precision.

21. The highest signal-to-noise ratio is obtained by collecting the highest number of photons in each frame. An exposure time longer than the average on time of the fluorophore will result in collecting more background per frame.

22. When selecting the area to be imaged, ensure the presence of at least three fiducial markers in the field of view.

23. The commercial PALM incorporates image-processing algorithms. Open-source software such as QuickPALM (http://code.google.com/p/quickpalm/) and rapidSTORM

(http://www.super-resolution.biozentrum.uni-wuerzburg.de/home/rapidstorm/) are also available.

24. Given a histogram of the distribution of the intensity per pixel, the threshold is calculated by constructing a line between the maximum of frequency (*y*-axis, value *a*) and the highest value of intensity (*x*-axis, value *b*). The distance between the line and the histogram (normal to the line) is computed for all values from *a* to *b*. The level where the distance between the histogram and the line is maximal is the threshold value.
25. Pixels with intensity higher than the threshold are only considered to be a fluorophore if there are more than five connected pixels.
26. The localizations corresponding to the fluorescent beads can be easily distinguished from the fluorescent probes as they are detected consequently in several frames.

Acknowledgements

We thank Dr. A. Miyawaki and Mr. S. Karasawa (Brain Science Institute, RIKEN, Japan) for hKikGR-β-actin/pMC1.

References

1. Prasher DC, Eckenrode VK, Ward WW et al (1992) Primary structure of the Aequorea victoria green-fluorescent protein. Gene 111:229–233
2. Chalfie M, Tu Y, Euskirchen G, Ward W et al (1994) Green fluorescent protein as a marker for gene expression. Science 263:802–805
3. Tsien RY (1998) The green fluorescent protein. Annu Rev Biochem 67:509–544
4. Tsien RY (2005) Building and breeding molecules to spy on cells and tumors. FEBS Lett 579:927–932
5. Miyawaki A (2005) Innovations in the imaging of brain functions using fluorescent proteins. Neuron 48:189–199
6. Newman RH, Fosbrink MD, Zhang J (2011) Genetically encodable fluorescent biosensors for tracking signaling dynamics in living cells. Chem Rev 111:3614–3666
7. Patterson GH (2011) Highlights of the optical highlighter fluorescent proteins. J Microsc 243:1–7
8. Ando R, Hama H, Yamamoto-Hino M et al (2002) An optical marker based on the UV-induced green-to-red photoconversion of a fluorescent protein. Proc Natl Acad Sci 99:12651–12656
9. Ando R, Mizuno H, Miyawaki A (2004) Regulated fast nucleocytoplasmic shuttling observed by reversible protein highlighting. Science 306:1370–1373
10. Mizuno H, Mal TK, Tong KI et al (2003) Photo-induced peptide cleavage in the green-to-red conversion of a fluorescent protein. Mol Cell 12:1051–1058
11. Sato T, Takahoko M, Okamoto H (2006) HuC:Kaede, a useful tool to label neural morphologies in networks in vivo. Genesis 44:136–142
12. Adam V, Mizuno H, Grichine A et al (2010) Data storage based on photochromic and photoconvertible fluorescent proteins. J Biotechnol 149:289–298
13. Grotjohann T, Testa I, Leutenegger M et al (2011) Diffraction-unlimited all-optical imaging and writing with a photochromic GFP. Nature 478:204–208
14. Patterson GH, Lippincott-Schwartz J (2002) A photoactivatable GFP for selective photolabeling of proteins and cells. Science 297:1873–1877
15. Henderson JN, Gepshtein R, Heenan JR et al (2009) Structure and mechanism of the photoactivatable green fluorescent protein. J Am Chem Soc 131:4176–4177
16. Verkhusha VV, Sorkin A (2005) Conversion of the monomeric red fluorescent protein into a photoactivatable probe. Chem Biol 12:279–285
17. Subach FV, Patterson GH, Manley S et al (2009) Photoactivatable mCherry for

high-resolution two-color fluorescence microscopy. Nat Methods 6:153–159

18. Subach FV, Patterson GH, Renz M et al (2010) Bright monomeric photoactivatable red fluorescent protein for two-color super-resolution sptPALM of live cells. J Am Chem Soc 132:6481–6491
19. Wiedenmann J, Ivanchenko S, Oswald F et al (2004) EosFP, a fluorescent marker protein with UV-inducible green-to-red fluorescence conversion. Proc Natl Acad Sci USA 101:15905–15910
20. Tsutsui H, Karasawa S, Shimizu H et al (2005) Semi-rational engineering of a coral fluorescent protein into an efficient highlighter. EMBO Rep 6:233–238
21. Gurskaya NG, Verkhusha VV, Shcheglov AS et al (2006) Engineering of a monomeric green-to-red photoactivatable fluorescent protein induced by blue light. Nat Biotechnol 24:461–465
22. Habuchi S, Tsutsui H, Kochaniak AB et al (2008) mKikGR, a monomeric photoswitchable fluorescent protein. PLoS One 3:e3944
23. McKinney SA, Murphy CS, Hazelwood KL et al (2009) A bright and photostable photoconvertible fluorescent protein. Nat Methods 6:131–133
24. Zhang M, Chang H, Zhang Y et al (2012) Rational design of true monomeric and bright photoactivatable fluorescent proteins. Nat Methods 9:727–729
25. Chudakov DM, Verkhusha VV, Staroverov DB et al (2004) Photoswitchable cyan fluorescent protein for protein tracking. Nat Biotechnol 22:1435–1439
26. Andresen M, Stiel AC, Trowitzsch S et al (2007) Structural basis for reversible photoswitching in Dronpa. Proc Natl Acad Sci USA 104:13005–13009
27. Mizuno H, Mal TK, Wälchli M et al (2010) Molecular basis of photochromism of a fluorescent protein revealed by direct ^{13}C detection under laser illumination. J Biomol NMR 48:237–246
28. Stiel AC, Trowitzsch S, Weber G et al (2007) 1.8 Å bright-state structure of the reversibly switchable fluorescent protein Dronpa guides the generation of fast switching variants. Biochem J 402:35–42
29. Ando R, Flors C, Mizuno H et al (2007) Highlighted generation of fluorescence signals using simultaneous two-color irradiation on Dronpa mutants. Biophys J 92:L97–L99
30. Andresen M, Stiel AC, Fölling J et al (2008) Photoswitchable fluorescent proteins enable monochromatic multilabel imaging and dual color fluorescence nanoscopy. Nat Biotechnol 26:1035–1040
31. Stiel AC, Andresen M, Bock H et al (2008) Generation of monomeric reversibly switchable red fluorescent proteins for far-field fluorescence nanoscopy. Biophys J 95:2989–2997
32. Subach FV, Zhang L, Gadella TW et al (2010) Red fluorescent protein with reversibly photoswitchable absorbance for photochromic FRET. Chem Biol 17:745–755
33. Brakemann T, Stiel AC, Weber G et al (2011) A reversibly photoswitchable GFP-like protein with fluorescence excitation decoupled from switching. Nat Biotechnol 29:942
34. Adam V, Lelimousin M, Boehme S et al (2008) Structural characterization of IrisFP, an optical highlighter undergoing multiple photo-induced transformations. Proc Natl Acad Sci USA 105:18343–18348
35. Fuchs J, Böhme S, Oswald F et al (2010) A photoactivatable marker protein for pulse-chase imaging with superresolution. Nat Methods 7:627–630
36. Adam V, Moeyaert B, David CC et al (2011) Rational design of photoconvertible and biphotochromic fluorescent proteins for advanced microscopy applications. Chem Biol 18:1241–1251
37. Betzig E, Patterson GH, Sougrat R et al (2006) Imaging intracellular fluorescent proteins at nanometer resolution. Science 313:1642–1645
38. Hess ST, Girirajan T, Mason M (2006) Ultra-high resolution imaging by fluorescence photoactivation localization microscopy. Biophys J 91:4258–4272
39. Mizuno H, Dedecker P, Ando R et al (2010) Higher resolution in localization microscopy by slower switching of a photochromic protein. Photochem Photobiol Sci 9:239–248
40. Inoue H, Nojima H, Okayama H (1990) High efficiency transformation of *Escherichia coli* with plasmids. Gene 96:23–28
41. Sambrook J, Russell DW (2001) Molecular cloning: a laboratory manual, 3rd edn. Cold Spring Harbor Laboratory Press, New York, NY
42. van de Linde S, Wolter S, Heilemann M et al (2010) The effect of photoswitching kinetics and labeling densities on super-resolution fluorescence imaging. J Biotechnol 149:260–266
43. Thompson RE, Larson DR, Webb WW (2002) Precise nanometer localization analysis for individual fluorescent probes. Biophys J 82:2775–2783

Index

Yves Engelborghs and Antonie J.W.G. Visser (eds.), *Fluorescence Spectroscopy and Microscopy: Methods and Protocols*, Methods in Molecular Biology, vol. 1076, DOI 10.1007/978-1-62703-649-8,

R

S

T

W

Z

MIX
Papier aus verantwortungsvollen Quellen
Paper from responsible sources
FSC® C105338

If you have any concerns about our products,
you can contact us on
ProductSafety@springernature.com

In case Publisher is established outside the EU,
the EU authorized representative is:
Springer Nature Customer Service Center GmbH
Europaplatz 3, 69115 Heidelberg, Germany

Printed by Libri Plureos GmbH
in Hamburg, Germany